Satellite Altimetry for Earth Sciences

Satellite Altimetry for Earth Sciences

Special Issue Editors

Frédéric Frappart
Ole Andersen
Sergey Lebedev
Guillaume Ramillien

MDPI • Basel • Beijing • Wuhan • Barcelona • Belgrade

MDPI

Special Issue Editors
Frédéric Frappart
Laboratoire d'Etudes en Géophysique et Océanographie Spatiales
France

Ole Andersen
Technical University of Denmark
Denmark

Sergey Lebedev
Geophysical Center of the Russian Academy of Sciences
Russian Federation

Guillaume Ramillien
Géosciences Environnement Toulouse
France

Editorial Office
MDPI
St. Alban-Anlage 66
4052 Basel, Switzerland

This is a reprint of articles from the Special Issue published online in the open access journal *Remote Sensing* (ISSN 2072-4292) from 2017 to 2018 (available at: https://www.mdpi.com/journal/remotesensing/special_issues/SA)

For citation purposes, cite each article independently as indicated on the article page online and as indicated below:

LastName, A.A.; LastName, B.B.; LastName, C.C. Article Title. *Journal Name* **Year**, *Article Number, Page Range.*

ISBN 978-3-03897-680-6 (Pbk)
ISBN 978-3-03897-681-3 (PDF)

Contents

About the Special Issue Editors

Frédéric Frappart received his Ph.D. degree in Geophysics from Université de Toulouse, Toulouse, France, in 2006. He is an engineer from Ecole Nationale Supérieure des Techniques Avancées Bretagne (ENSTA Bretagne, formerly ENSIETA), Brest, France. He has been a researcher at Observatoire Midi-Pyrénées (OMP), Toulouse, France, since 2010, in charge of the scientific applications of radar altimetry over land (hydrology and surface properties) for the Centre de Topographie des Océans et de l'Hydrosphre (CTOH), a French Observation Service dedicated to scientific applications of radar altimetry. He is working on Earth observation techniques devoted to studying the global water cycle and climate.

Ole Andersen is a senior research scientist within the use of satellite altimetry and other remote sensing data for geodetic, geophysical, oceanographic, climatological, and hydrological purposes. He has a Masters degree in Geophysics and a Ph.D. focusing on ocean tide modeling from the University of Copenhagen. He is a member of the science working teams associated with the NASA-CNES TOPEX/POSEIDON, JASON-1, and GRACE satellite missions. Assistant secretary general within the International Association of Geodesy and member of several international study groups within the field of satellite altimetry and board member of the IAS, he has participated in numerous national and international projects. In 1998, he worked as guest scientist at the CSIRO Marine Laboratory and the Antarctic Commonwealth Research centre in Hobart, Australia, and in 2004, at NASA-GSFC. He is currently the project leader on the National Danish research project HYDROGRAV on the use of gravimetric methods for hydrology.

Sergey Lebedev received his PhD degree in Oceanography from the Russian State Oceanographic Institute in 1991. He is a researcher at Laboratory of Geoinformatics and Geomagnetic Studies from the Geophysical Center of the Russian Academy of Sciences. His resaerch interests include the hydro-climatology of the Caspian Sea and the impact of climate change on sea level on Russian coastal areas and inland seas and in the Southern Ocean.

Guillaume Ramillien obtained a PhD in space geophysics from the University Paul Sabatier of Toulouse, France, in January 1998, on the subject of 3-D sea floor topography by inversion of radar satellite altimeter data. He held three successive post-doctoral fellowships (1998–2002), during which his main interest was the exploitation of remote sensing data for characterizing the Earth's surface (hydrology) and sub-surface eophysical processes. Since October 2002, he has been a researcher at the Centre National de la Recherche Scientifique (CNRS), and his current investigations are focused on the time variations of the Earth's gravity field measured by the Gravity Recovery And Climate Experiment (GRACE) satellite mission. He is also interested in the preparation of the future low-altitude gravity missions. He has been the president of the GRGS scientific council since May 2015.

Preface to "Satellite Altimetry for Earth Sciences"

The history of satellite radar altimetry started at the end of the 1960s. The major goal of this new type of sensor was to measure the shape of the Earth. The first altimeter onboard Skylab 3 (1973) had a precision of 0.6 m on one-second-averaged measurements. Its data were mostly used for marine geoid determination. As the technique improved, the accuracy of the one-second-averaged measurement increased to 25 cm on GEOS-3 (1975–1978). This improvement allowed the detection of the major oceanic currents and associated eddy fields. Seasat (1978) reached 0.05 m accuracy on one-second-averaged measurements. In spite of its short 3-month time span, Seasat demonstrated the potential of radar altimetry for the monitoring of the dynamic topography of the sea surface, which varies from 0.1 m to more than 1 m for western boundary currents and eddies. This was followed by the launch of Geosat (1985–1990) and ERS-1 (1991–2000) and then Topex/Poseidon (1992–2005) that started the era of high-precision altimetry. Since then, satellite radar altimetry has commonly been used in a growing number of applications in Earth sciences. For more details about radar altimetry, see References [1,2].

This book contains reviews and recent advances of general interest in the use of remote sensing for hydrology. It is divided by types of surface monitored. The first two contributions present the benefits of the use of the Ka-band for radar altimetry in terms of measurement quality and scientific applications over all types of surfaces. Then, the following contributions are presented over the different types of surface which can be monitored using radar altimetry: Open ocean, coastal areas, land surfaces, and sea ice. Each of these sub-sections is organized as follows: First, the results of calibration/validation are reported, then radar echoes modeling, radar echoes classification, and new algorithms for processing the radar echoes (e.g., retracking algorithms) are presented, and eventually, scientific applications are detailed. Calibration/validation results include evolution of the performances of high-precision radar altimetry missions operating in low resolution mode (LRM), as well as the more recent radar altimetry missions operating in synthetic aperture radar (SAR) mode. Improved corrections for the tropospheric delays allowing an improvement of the error balance over the ocean, including coastal areas, are presented. Improvement of the height estimation using new retracking algorithms is assessed over different surfaces. Several new applications of satellite altimetry are also detailed, such as pycnocline depth, topography of the intertidal zone, or detection of open water in sea-ice-covered regions.

References

1. Frappart, F.; Blumstein, D.; Cazenave, A.; Ramillien, G.; Birol, F.; Morrow, R.; Rémy, F. Satellite Altimetry: Principles and Applications in Earth Sciences. In *Wiley Encyclopedia of Electrical and Electronics Engineering*; John Wiley & Sons, Inc.: Hoboken, NJ, USA, 2017; pp. 1—25, ISBN 047134608X.
2. Stammer, D.; Cazenave, A. *Satellite Altimetry Over Oceans and Land Surfaces*; Taylor & Francis: Boca Raton, FL, USA,2017; ISBN 978-1-4987-4345-7.

Frédéric Frappart, Ole Andersen, Sergey Lebedev, Guillaume Ramillien
Special Issue Editors

remote sensing

MDPI

Article

The Benefits of the Ka-Band as Evidenced from the SARAL/AltiKa Altimetric Mission: Quality Assessment and Unique Characteristics of AltiKa Data

Pascal Bonnefond [1,*], Jacques Verron [2], Jérémie Aublanc [3], K. N. Babu [4], Muriel Bergé-Nguyen [5], Mathilde Cancet [6], Aditya Chaudhary [4], Jean-François Crétaux [5], Frédéric Frappart [5,7], Bruce J. Haines [8], Olivier Laurain [9], Annabelle Ollivier [3], Jean-Christophe Poisson [3], Pierre Prandi [3], Rashmi Sharma [4], Pierre Thibaut [3] and Christopher Watson [10]

[1] SYRTE, Observatoire de Paris, PSL Research University, CNRS, Sorbonne Universités, UPMC Univ. Paris 06, LNE, 75014 Paris, France
[2] Institut des Géosciences de l'Environnement (IGE)/CNRS, 38041 Grenoble, France; Jacques.Verron@univ-grenoble-alpes.fr
[3] Collecte Localisation Satellites (CLS), 31520 Ramonville Saint-Agne, France; jaublanc@cls.fr (J.A.); aollivier@cls.fr (A.O.); jpoisson@cls.fr (J.-C.P.); pprandi@cls.fr (P.P.); pthibaut@cls.fr (P.T.)
[4] Space Applications Centre (ISRO), Ahmedabad 380015, India; kn_babu@sac.isro.gov.in (K.N.B.); aditya.osd@sac.isro.gov.in (A.C.); rashmi@sac.isro.gov.in (R.S.)
[5] Laboratoire d'Etudes en Géophysique et Océanographie Spatiales (LEGOS), 31400 Toulouse, France; Muriel.Berge-Nguyen@cnes.fr (M.B.-N.); Jean-Francois.Cretaux@legos.obs-mip.fr (J.-F.C.); Frederic.Frappart@legos.obs-mip.fr (F.F.)
[6] NOVELTIS, 31670 Labège, France; Mathilde.Cancet@noveltis.fr
[7] Géosciences Environnement Toulouse (GET), 31400 Toulouse, France
[8] Jet Propulsion Laboratory, California Institute of Technology, Pasadena, CA 91109, USA; Bruce.J.Haines@jpl.nasa.gov
[9] Géoazur—Observatoire de la Côte d'Azur, 06905 Sophia-Antipolis, France; Olivier.Laurain@oca.eu
[10] Surveying and Spatial Science Group, School of Geography and Environmental Studies, University of Tasmania, Hobart 7001, Australia; Christopher.Watson@utas.edu.au
*   Correspondence: Pascal.Bonnefond@obspm.fr; Tel.: +33-1-40-51-22-29

Received: 6 December 2017; Accepted: 6 January 2018; Published: 9 January 2018

Abstract: The India-France SARAL/AltiKa mission is the first Ka-band altimetric mission dedicated to oceanography. The mission objectives are primarily the observation of the oceanic mesoscales but also include coastal oceanography, global and regional sea level monitoring, data assimilation, and operational oceanography. The mission ended its nominal phase after 3 years in orbit and began a new phase (drifting orbit) in July 2016. The objective of this paper is to provide a state of the art of the achievements of the SARAL/AltiKa mission in terms of quality assessment and unique characteristics of AltiKa data. It shows that the AltiKa data have similar accuracy at the centimeter level in term of absolute water level whatever the method (from local to global) and the type of water surfaces (ocean and lakes). It shows also that beyond the fact that AltiKa data quality meets the expectations and initial mission requirements, the unique characteristics of the altimeter and the Ka-band offer unique contributions in fields that were previously not fully foreseen.

Keywords: altimetry; Ka-band; data processing; calibration; validation

1. Introduction

Launched in February 2013, the SARAL/AltiKa mission carries the first Ka-band altimeter ever flown [1]. The primary mission objective is the observation of the oceanic mesoscales, as well as coastal oceanography, global and regional sea level monitoring, data assimilation and operational oceanography. Secondary objectives include ice sheet, sea ice and inland waters monitoring. In order to continue the time series and to benefit from the existing mean sea surface, SARAL flew on the same orbit as Envisat. The SARAL/AltiKa mission was considered as a "gap filler" between Envisat (lost in April 2012) and Sentinel-3A (launched in February 2016) and has permitted to continue the sea level measurements on the historical ground track of the ERS-1&2 and Envisat missions. The nominal phase of the mission was planned to end three years after launch with an objective of five years. Since July 2016, SARAL/AltiKa has left its nominal orbit and entered in a Drifting Phase until its end of life. This phase, seamless in term of accuracy, provides a larger data coverage (Figure 1). The new orbit (altitude increased by 1 km) allows to improve the mean sea surface resolution but has been chosen to maintain also a good temporal resolution (15–20 days) for mesoscale studies.

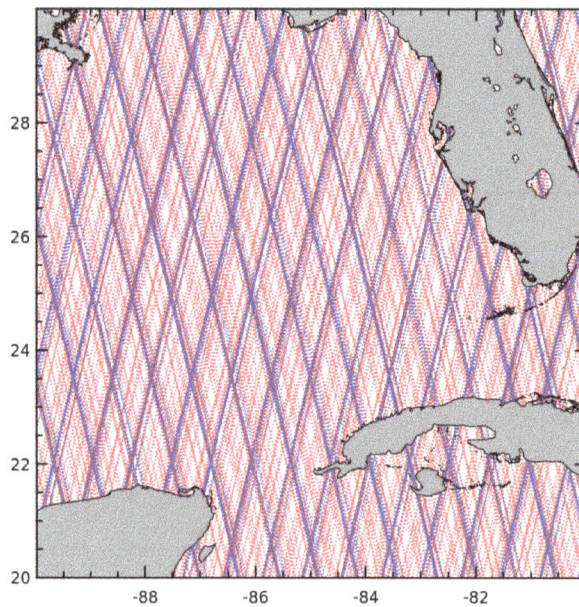

Figure 1. Blue: Repetitive orbit ground track. Red: Drifting orbit ground track.

SARAL/AltiKa's main scientific objective is to provide data products to the oceanographic research user community for studies to improve our knowledge of the ocean mesoscale variability, thanks to the improvement in spatial and vertical resolution brought by SARAL/AltiKa. This main scientific objective is divided into subthemes of mesoscale ocean dynamics: observations, theoretical analyses, modelling, data assimilation, and so forth. This leads to an improvement of our understanding of the climate system through its key ocean component and especially the role and impact of mesoscale features on the climate variability at large spatial and temporal scales. It also contributes to the study of coastal dynamics, which is important for many downstream applications including operational oceanography, which seeks large amounts of in-situ and space-based data.

SARAL/AltiKa's secondary objectives include the monitoring of the main continental water levels (i.e., lakes, rivers, and enclosed seas), the monitoring of large-scale sea level variations, the observation of polar oceans (thanks to the high inclination of its orbit), the analysis and forecast of wave and wind fields, the study of continental ice (thanks to the lower penetration in Ka-band) and sea ice, the access to low rains climatology (enabled in counterpart to the sensitivity of Ka-band to clouds and low rains) and the marine biogeochemistry (notably through the role of the meso- and submeso-scale physics).

The success of SARAL/AltiKa has been made possible by a very fruitful cooperation and efficient joint work between ISRO and CNES and thanks to the highly valuable exchanges between scientists of both countries. The objective of this paper is to highlights salient results in terms of quality assessment and unique characteristics of AltiKa data made since the special issue published in 2015 (Marine Geodesy, 38(S1):1, 2015, [2]). This paper is preceded by a companion paper focusing on the scientific applications where more details about the SARAL/AltiKa mission and its payload are also given [3]. In the first section of this paper, we will focus on the quality assessment of water level through Calibration and Validation and the second section will illustrate some unique contributions of Ka-band information in AltiKa data.

2. Quality Assessment of Water Level through Calibration and Validation

The observation of rising mean sea levels over the Earth's oceans using satellite altimetry has helped inform and shape the debate on global climate change that has emerged over the last two decades [4]. The ongoing monitoring of secular changes in global mean sea level with accuracy tolerances of better than 1 mm/year remains a fundamental science goal within satellite altimetry and represents one of the most challenging objectives in space geodesy. Central to this objective has been the recognition that calibration and validation (cal/val) are vital components of the altimeter measurement system technique. Cal/val defines a multi-disciplinary problem in and of itself, that pushes the limits of available terrestrial, oceanographic and space based observational techniques. Current estimates of regional and global change in mean sea level are only possible with careful and ongoing calibration of altimeter missions. Cross calibration of past, current and future altimeter missions will remain essential for continued sea-level studies.

During the last years, complementary altimetric missions operating concurrently have provided the unprecedented ability to compare measurement systems by undertaking relative calibrations. Studies have demonstrated through global statistics the power of such a technique [5,6]. Given the technical challenges of operating spacecraft for many years in orbit, however, the ability to cross-calibrate multiple missions in this manner cannot be assured. It reinforces the need for a range of complementary calibration methodologies—including a geographically diverse array of in situ absolute calibration sites that can assess changes to instrument behavior in near real time: such facilities, as described in [7], provide in situ sea level measurements in the same reference frame than the satellite altimetry ones and correct them from solid earth vertical motions.

The cal/val activities are focused not only on the important continuity between past, present and future missions but also on the reliability between offshore, coastal and inland altimetric measurement. The ability to sample varied geographically correlated errors and characterize them in an absolute sense are significant benefits of a well-distributed set of calibration sites. There is no doubt, however, that the "calibration task" requires a multifaceted approach, including both in situ calibration sites and regional/global studies.

This section will provide metrics of the SARAL/AltiKa accuracy and precision through absolute calibration sites (Section 2.1), global approach (Section 2.3) and an in between regional approach (Section 2.2). The main achievement of the data quality assessment provided in this section is illustrated in Figure 2 showing that AltiKa data have similar accuracy at the centimeter level in terms of absolute water level whatever the method (from local to global) and the type of water surfaces (ocean and lakes). This illustrates the improvement in altimetry in the last decade thanks to better processing and geophysical corrections but more importantly thanks to new technology (Ka-band, SAR) that paved the way of new frontiers in altimetry. Remaining differences from the various determinations shown in Figure 2 illustrate the characteristics of the studied water level (open-ocean, coastal, inland) but also the remaining errors coming from the methods used and the geodetic datum of each sites.

Another important result is that cal/val activities over lakes are now at the same level of accuracy than the ones performed over decades at historical calibration sites (Bass Strait, Corsica, Crete, and Harvest).

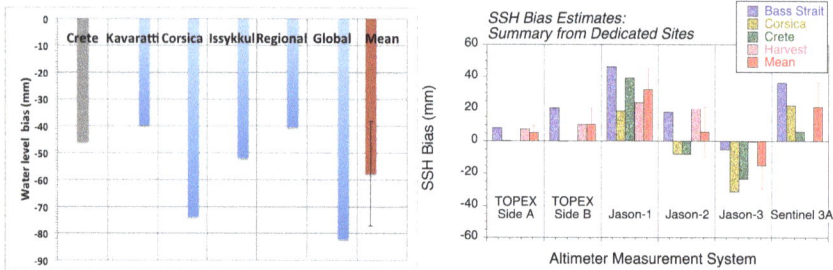

Figure 2. Left (this study): water level bias from Crete [8], Kavaratti (Section 2.1.1), Corsica (Section 2.1.2, Lake Issykul (Section 2.1.3), the regional (Section 2.2) and global (Section 2.3) approaches. The mean value of all the determinations is −54 mm with a standard deviation of 15 mm. **Right** [9]: TOPEX, Jason and Sentinel-3A Sea Surface Height (SSH) biases from in situ calibration sites.

2.1. Water Level Calibration from Open-Ocean, Coastal and Inland Waters In Situ Absolute Calibration Sites

2.1.1. Kavaratti, Altimeters' Calibration Site

The Kavaratti site gives an opportunity to do absolute calibration of altimeters over the tropical region. This calibration site do not show altimeter land contamination since it is in the open ocean, which gives it an added advantage. Kavaratti Island, is located southwest of Indian peninsula, ~450 km from the coast, as a platform of scientific observations, it has historic measurements from many Indian scientific institutions (e.g., Survey of India specific to water level). Figure 3 represents the part of Arabian Sea with Lakshadweep Islands at north and Maldives Islands at south. This permanent facility in Kavaratti is situated near the ground tracks of TOPEX/Jason and SARAL/AltiKa missions. The dedicated calibration facility at Kavaratti includes tide gauges, meteorological and oceanographic instruments, and a bottom pressure recorder [10]. This site has long-term sea level measurement at the tide gauge station TG1, which allows for the precise estimation of drift in sea surface height. The site characterization and calibration experiment and its results can be found in [10].

The absolute calibration of SARAL/AtiKa is carried out for its 23 cycles at this site (Figure 4). The determined absolute bias is −39.8 mm with standard deviation of 23.6 mm. In this AltiKa calibration exercise we have not used any local high resolution tide models, also the effects of wind and atmospheric pressures, and vertical land motion for the precise local geoid estimation are disregarded in determining the absolute Sea Surface Height (SSH) bias. These results and the good agreement with other in situ calibration sites shows that the location is robust and encourages us to continue observations to carry out the absolute calibration of altimetric SSH. The errors in estimation of absolute SARAL/AltiKa bias over this site will be improved through future field experiments.

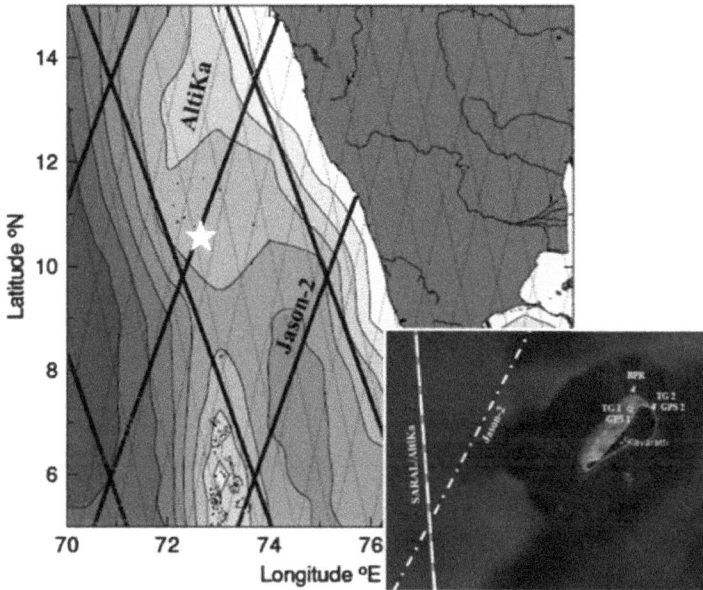

Figure 3. The large map shows the location of Kavaratti island (star) as well as the Jason-2 and SARAL/AltiKa ground tracks. The insert is a zoom of Kavaratti: ISRO's altimeter calibration site, has dual radar tide gauge station and a bottom pressure recorder.

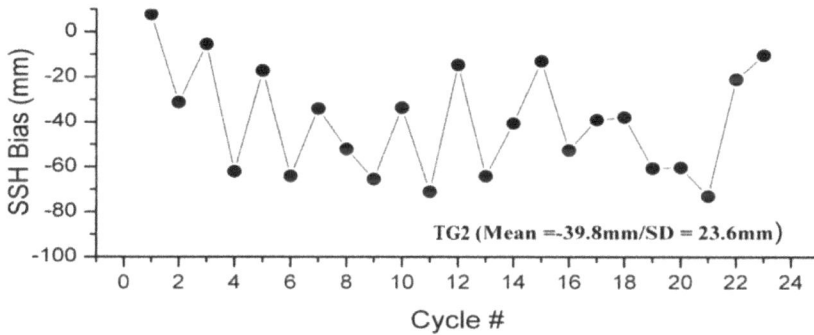

Figure 4. Absolute sea surface height bias of SARAL/AltiKa from cycle 1 to 23 using GDR-T products.

2.1.2. SARAL/AltiKa Absolute Calibration over Corsica during the Nominal and Drifting Phases

The geodetic Corsica site was set up in 1998 in order to perform altimeter calibration of the TOPEX/Poseidon (T/P) mission and subsequently, Jason-1, OSTM/Jason-2 and Jason-3. The scope of the site was widened in 2005 in order to undertake the calibration of the Envisat mission and most recently of SARAL/AltiKa. In [11], we have presented the first results from the latter mission using both indirect and direct calibration/validation approaches. The indirect approach utilizes a coastal tide gauge and, as a consequence, the altimeter derived Sea Surface Height (SSH) needs to be corrected for the geoid slope. The direct approach utilizes a novel GPS-based system deployed offshore under the satellite ground track that permits a direct comparison with the altimeter derived SSH. This latter approach has the advantage being able to be used without any specific infrastructure and have permitted us to make measurements even during the period when the difficulties with the satellite reaction wheels did not permit the ground track to be accurately maintained (from March 2015 until the Drifting Phase in July 2016): this is illustrated in the bottom part of Figure 5 and the SSH biases derived from those configurations (bold red circles) are in agreement with the whole time series. Since the beginning of the Drifting Phase, SARAL/AltiKa have overflown the Corsica calibration sites (Ajaccio and Senetosa) in different configurations illustrated in the right part of Figure 5. The SSH biases have then been computed using Senetosa or Ajaccio tide gauges or even both (cycle 100/pass 735 and cycle 104/pass 677) when the satellite have overflown both geoid areas (shaded in purple on Figure 5 maps). The results for these configurations (bold blue crosses in Figure 5) are also in very good agreement with the whole time series, even in a configuration very close to the coast (∼3–4 km, cycle 100/pass 735). The overflights of both Senetosa and Ajaccio sites have permitted us to compare the SSH bias from Ajaccio and Senetosa tide gauges and we found an offset of 30 mm which is due to an issue in the absolute reference of Ajaccio tide gauge. This problem was identified in [11] and has been solved recently. This is discussed in [12] in which we have performed the Sentinel-3A calibration over both Senetosa and Ajaccio sites in a configuration close to the one encountered with SARAL/AltiKa for cycle 104 and pass 677 (see Figure 5).

In conclusion, in this analysis, after correcting the 30 mm offset in the Ajaccio tide gauge measurements, we demonstrate that using either indirect or direct calibration/validation approach gives similar SSH bias, respectively -74 ± 4 mm and -69 ± 11 mm. Even during the Drifting Phase, we can continue to monitor the SSH bias using both approaches with a good agreement compared to the whole time series and a period close to the initial ones (46 days in average compared to the 35-day repeat period in the nominal phase).

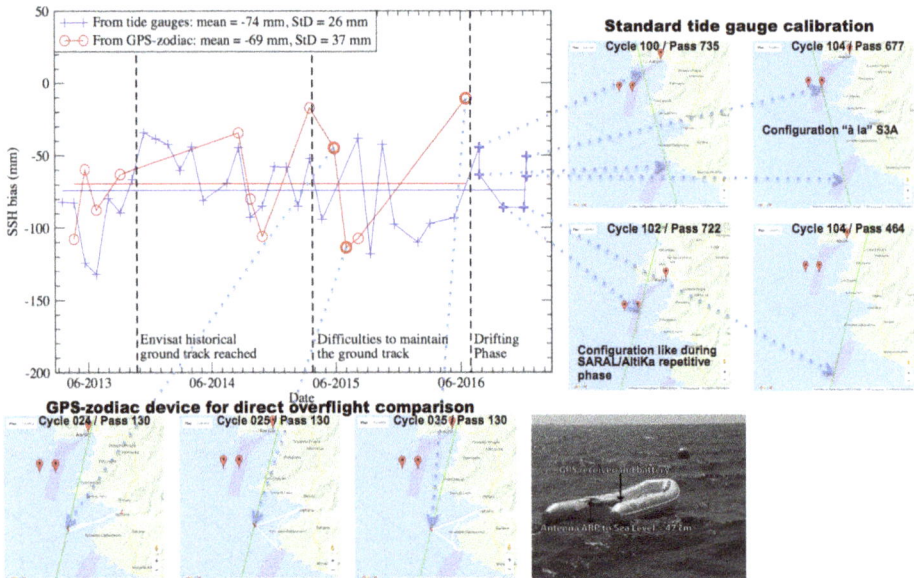

Figure 5. Sea Surface Height (SSH) bias time series for SARAL/AltiKa over Corsica for both direct (GPS-zodiac, red circles) and indirect (tide gauges, blue crosses) approaches. The bold symbols correspond to situations where the satellite was not overflying the calibration sites in a nominal configuration.

2.1.3. Absolute Calibration, Validation of SARAL/AltiKa over the Lake Issykkul

SARAL/AltiKa is used over the continents to measure the water height changes of lakes and rivers [1]. In order to determine the performances of altimeters over inland waters, a dedicated site of calibration/validation has been setup in 2004 over the lake Issykkul. Lake Issykkul (42–43°N, 76–79°E) is a large lake (6000 km^2 of area extent) located in Kyrgyzstan in Central Asia. Initially used as experimental site, thanks to promising first results obtained with TOPEX/Poseidon [13], the Lake Issykkul became permanent cal/val site for several altimeters in orbit. It has served for Jason-1, Jason-2, and Envisat with results coherent with other cal/val sites over the oceans [14,15]. The interest of having cal/val sites over a lake is to benefit from specific conditions of inland water (no tides, no inverse barometer) and to quantify the accuracy of altimeters to measure their water level changes.

Meanwhile, some corrections like the wet or dry tropospheric delay are not measured as accurately as over the ocean. Using ground instrumentation over the lake Issykkul therefore helps quantifying the error budget of altimeters over lakes in general. Since 2004, not less than 15 field campaigns have been done over the Lake Issykkul. The full experimental design is given in [16] and is not described here in details. In few words, it simply consists in measuring the water level changes along the track of the satellite, due to the geoid's gradient (which over the Lake Issykkul reaches several meters over long distance) using a GPS kinematic survey on a boat following the satellite tracks. It allows us to calculate the absolute bias of the altimeter.

First of all, the water level changes of the Lake Issykkul from 2013 to 2016 are calculated every 35 days using the altimetry data from the SARAL/AltiKa instrument. It is compared to the water level changes measured at an historical floating tide gauge located on the north coast of the lake (providing daily water level). It is also compared to hourly water level changes of the Lake measured by a radar installed on the south east lake shoreline. These comparisons allow us to quantify the accuracy of SARAL/AltiKa over the lake Issykkul in particular, which gives an order of magnitude of accuracy of this instrument for large lakes in general. It confirms results obtained in [17] that SARAL/AltiKa allows us to measure water level changes at very high accuracy (3.5 cm for lake Issykkul: Figure 6).

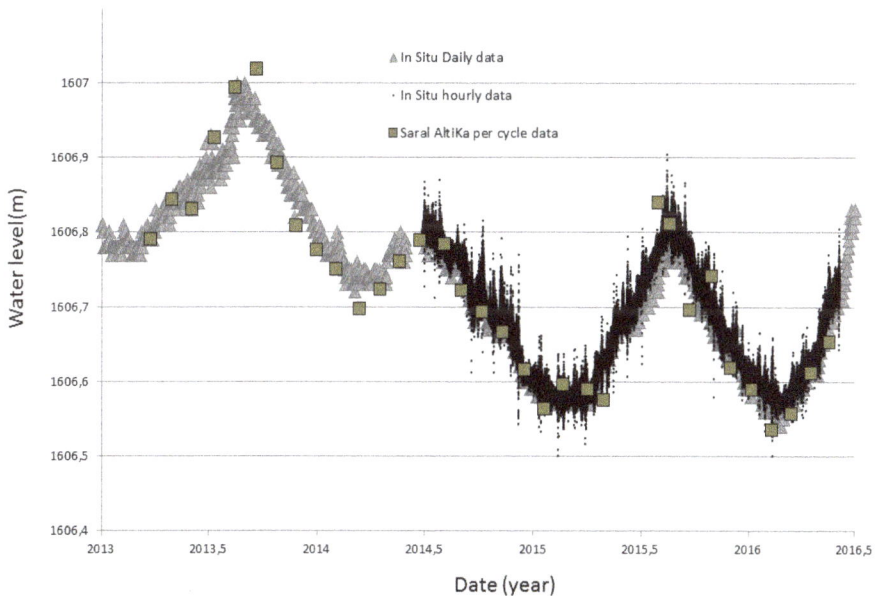

Figure 6. Water level of the lake Issykkul measured by SARAL/AltiKa, a radar and a tide gauge located on respectively east and north coast of the lake, delivering hourly and daily data. RMS of differences between SARAL/AltiKa and the in situ daily data is 3.5 cm.

For specific calculation of altimeter bias of AltiKa, a GPS survey of the track number 554 has been performed during a campaign done on 1 July 2014. The GPS vertical profile along this track is compared to individual measurements of the altimeters over the first 15 cycles using in situ measurements for all corrections:

- Wet and dry tropospheric delay using a permanent GPS and weather stations
- The GPS height antenna using a radar installed just below the GPS on the boat: see [16] for details.
- The in situ daily level changes for the correction of the hydrological signal from cycle 1 to cycle 15. 1 July 2014 is when the GPS survey was done and corresponds to cycle 14.
- The GINS [18] software was used for the calculation of the GPS 3D coordinates in PPP (Precise Point Positioning) mode.

We then compare the GPS vertical coordinates along the track 554 with the individual measurements of the lake altitude above the ellipsoid using AltiKa measurements. We thus perform the calculation of the absolute bias using the ocean and the ice1-OCOG (Offset Centre Of Gravity) retrackers:

- With the ocean retracker, the water level bias, which is averaged along the track 554 and using the first 15 cycles of the satellite is given in Figure 7. The absolute bias obtained is: -52 ± 24 mm.
- With ice1-OCOG retracker, the water level bias, which is averaged along the track 554 and using the first 15 cycles of the satellite, is given in Figure 8. The absolute bias obtained is: $+34 \pm 19$ mm.

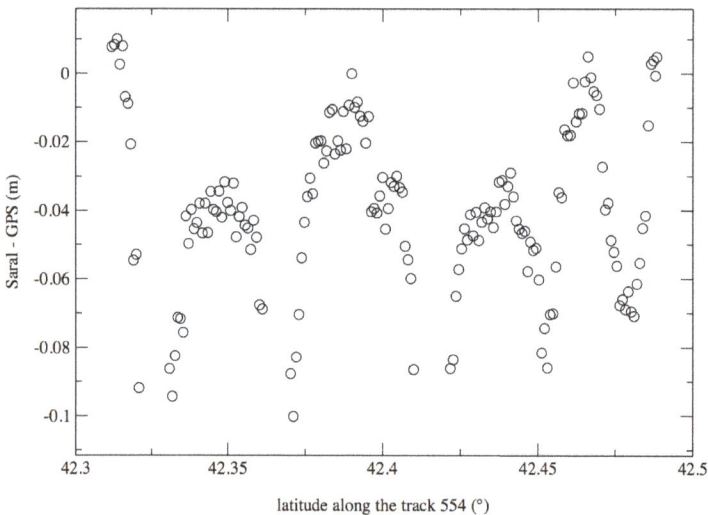

Figure 7. Differences between the GPS profiles and SARAL/AltiKa water altitude above the ellipsoid with ocean retracker at reference points corresponding to the GPS points along the SARAL/AltiKa track 554.

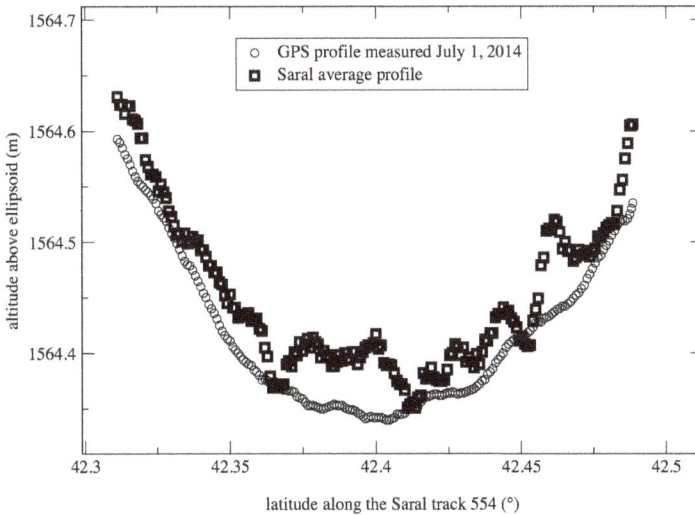

Figure 8. Vertical profiles along the track 554. The altimetry profile is an average over 15 first cycles at reference point corresponding to the GPS points of measurements. Ice1 retracker is represented.

2.2. Regional Calibration of SARAL/AltiKa Sea Surface Height in Corsica, at Harvest and at Bass Strait

In situ calibration of altimeter SSH is usually done at the vertical of a specific cal/val site by direct comparison of the altimeter data with the in situ data. Given that most of the in situ calibration sites (Harvest, Senetosa, Bass Strait and Gavdos) were specifically designed to be located under the TOPEX/Jason orbit, the classical absolute calibration technique can only be used for satellites using this orbit. The regional cal/val technique developed by Noveltis [19,20] aims to assess the altimeter range bias both on satellite passes flying over the calibration site and on satellite passes located several hundreds of kilometres away (Figure 9). In particular, it enables the monitoring of altimetry missions that do not fly directly over the calibration sites, such as SARAL/AltiKa. In principle this technique extends the single site approach to a wider regional scale, thus reinforcing the link between the local and the global cal/val analyses.

The regional method is used to compute the SARAL/AltiKa altimeter biases at three historical TOPEX/Jason-suite calibration sites: the Corsican calibration sites of Senetosa and Ajaccio, the Californian site of Harvest and the Australian site of Bass Strait. These calibration sites are characterized by very different ocean variabilities, which leads to various strategies in terms of corrections for the altimeter SSH.

In Corsica, the ocean variability (from few days to seasons) is rather low (about 50 cm on average), mainly governed by the wind and the atmospheric pressure variations, with low tidal amplitudes (about 20 cm). However, the presence of small islands close to the sites of Ajaccio and Senetosa can have an impact on the quality of the altimeter SSH and the associated corrections, in particular the radiometer wet tropospheric correction. The wet tropospheric correction derived from the ECMWF (European Centre for Medium-Range Weather Forecasts) model is thus used to correct the SARAL/AltiKa SSH in Corsica. In addition, the altimeter and the in situ tide gauge observations are corrected using respectively the COMAPI regional tidal model (Coastal Modeling for Altimetry Product Improvement) [21] for the tides and a global hydrodynamic simulation provided by LEGOS for the DAC (Dynamic Atmospheric Correction) effects [22]. In Harvest and Bass Strait, the ocean variability is much higher than in Corsica (about 2.5 m), with large tidal amplitudes (about 1.5 m). The FES2004 global tidal model [23] and the global DAC simulation is used to correct the altimeter and the in situ observations at both sites. In addition, the radiometer-derived wet tropospheric correction is used to correct the altimeter SSH.

Figure 10 shows the crossover points considered for the computation of the regional SARAL/AltiKa bias estimates at each calibration site. In Harvest and Bass Strait, the number of selected crossover points is rather small in comparison to the number of available points. Large variability is observed in the bias estimates at those surrounding crossover points and they are discarded from the computation (see [19,20] for details). Some previous work on the Envisat mission in Bass Strait [24] showed that using a more recent tidal model with higher resolution (FES2014 global tidal model, [25]) enables us to drastically reduce the variability at some of those points and to reintegrate them in the computation in this specific region. The mean regional bias estimates at each calibration site are given in Table 1. The standard deviation computed over the mean bias estimates at each crossover point is given in the last column. This information shows that the variability of the bias estimates over the considered crossover points is stable from one site to the other, in the order of 1 cm. However, the mean bias estimates and the averaged standard deviation of the bias estimates strongly vary from one site to the other, with mean variability of about 60 mm in Harvest and Bass Strait, twice larger than in Corsica, which is directly linked to local conditions (open ocean with rough seas in Harvest and Bass Strait, sheltered harbours with low ocean variabilities in Corsica). Even after correcting the 30 mm offset in the Ajaccio tide gauge measurements (see Section 2.1.2) the 47 mm difference observed in the bias estimates between Ajaccio and Senetosa, is for the moment unexplained and need further investigation. However, the averaged SSH bias from Ajaccio and Senetosa is 61 mm and is close to the values found from the absolute calibration study (see Section 2.1.2). Further investigation is also needed to understand the ~40 mm difference observed between the bias estimates in Corsica and the estimates in Harvest and Bass Strait. It is more than probable that high precision regional tidal models and dynamical atmospheric correction solutions would improve the computation of accurate altimeter and in situ sea surface heights in these regions.

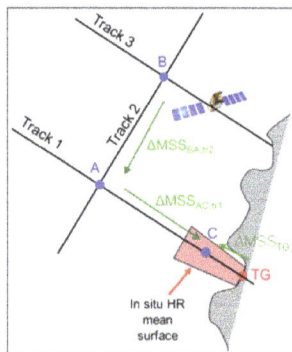

Figure 9. Generic diagram of the regional in situ calibration method. The points A and B represent the crossover points between, respectively, the satellite altimeter tracks 1 and 2, and the tracks 2 and 3. The tracks can belong to different altimeter missions. The in situ high resolution mean surface is used to link the tide gauge (TG) measurements to the altimeter data, and the comparison is done at the point C, located on this surface (adapted from [20]).

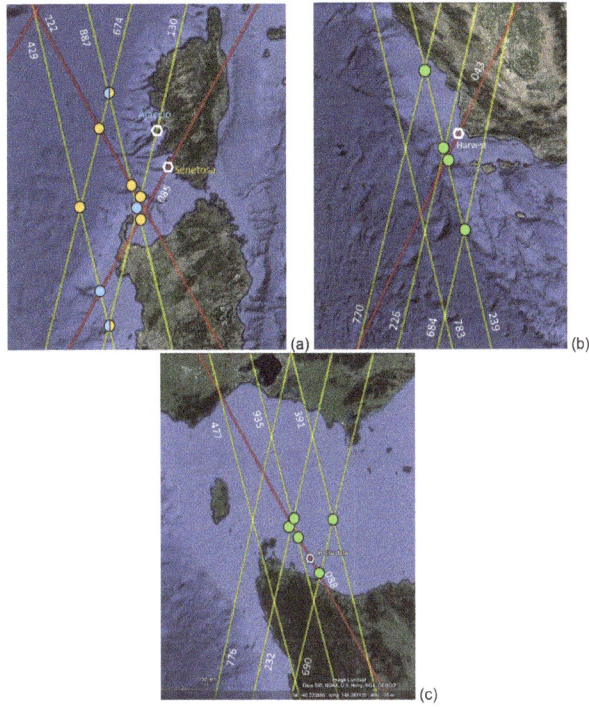

Figure 10. Altimeter crossover points used for the SARAL/AltiKa regional bias estimate in Corsica (a: Ajaccio in blue, Senetosa in orange), Harvest (b: green dots) and Bass Strait (c: green dots). The SARAL/AltiKa and Jason-2 grounds tracks are shown respectively in yellow and red.

Table 1. SARAL/AltiKa regional bias estimates at each site. In Bass Strait, the in situ data available for this study only covers the period until cycle 25.

SARAL/Altika Regional Bias Estimates (mm)	Mean (mm)	Std (mm)	Nb of Cycles	Nb of Xover Points	Std over the Xover Points (mm)
Senetosa (cycle 1–35)	−84 ± 5	32	34	7	5
Ajaccio (cycle 1–35)	−37 ± 6	36	33	6	13
Harvest (cycle 1–35)	−19 ± 12	64	30	4	12
Bass Strait (cycle 1–25)	−22 ± 11	56	24	5	13

2.3. Global Validation of SARAL/AltiKa Sea Surface Height over Ocean

Since the beginning, SARAL/AltiKa data shows a high quality and provides new information with respect to the previous missions [5]. At high latitude, thanks to its orbit reaching 82°, the Arctic Ocean and ice caps are better covered. Near the coast, measurements are more numerous and reliable [26] thanks to its high rate of 40 Hz (one point every 125 m along-track) and to its robust tracking mode (DIODE/Median, except for cycles 10 to 17 in autonomous DIODE mode). Furthermore, high frequency structures are better addressed, thanks to its Ka-band technology [27].

Its data availability is largely over the specification requirements, reaching 99.6% over oceans, Safe Hold Mode periods included, compared to 99.3% on Jason-2. Over other surfaces, the coverage is also very homogeneous as shown on Figure 11. Even the sensitivity of the Ka-band frequency to rain has less impact than expected. Indeed, only 5% of measurements may be not achieved due to rain rates > 1.5 mm/h according to geographic areas [28].

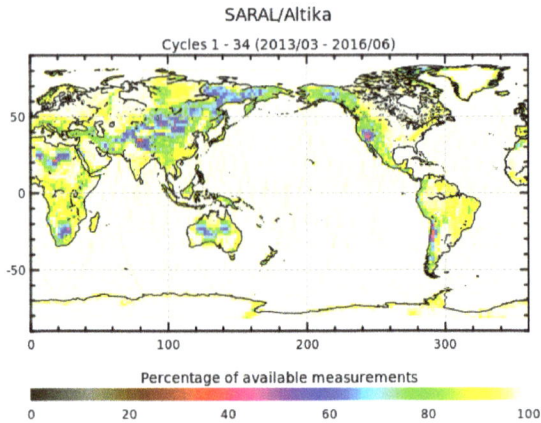

Figure 11. Map of the percentage of available measurements over land on SARAL/AltiKa's cycles 1 to 34.

The relative SSH bias compared to a zero reference of the TOPEX GMSL in 1993 (Global Mean Sea Level, CMEMS 2018 standards (Copernicus Marine Environment Monitoring Service)) is −82.3 mm for SARAL/AltiKa, whereas Sentinel-3A stands for −3.9 mm, Jason-2 for −18.5 mm and Jason-3 for −47.3 mm. Relative SSH bias and its evolution can also be derived from crossovers with a reference mission (such as Jason ones); even if not based exactly on the same length of the time series, results provided in [5] agree at the centimeter level with the ones presented here.

Its long-term stability as well, exceeds the mission requirements, featuring a very consistent mean sea level with respect to the reference record, derived from a combination of TOPEX/Jason-1/Jason-2 time series [29]. SARAL/AltiKa GMSL's shows a 4.8 mm/year evolution on this time period, thus keeping pace with Jason-2 (4.4 mm/year). It is a very satisfying statistic knowing that the time series for SARAL/AltiKa is still under 5 years which is the minimum period required to have a significant trend, see Figure 12.

Figure 12. SARAL/AltiKa's Global Mean Sea Level (red) compared to Jason-2 (blue) and the reference one (green).

As for crossovers analysis (see details in [5]), it demonstrates excellent performances, with 5.3 cm of standard deviation compared to 5.4 cm for Jason-2, using the radiometric correction [30]. This illustrates SARAL/AltiKa's very good skills to resolve mesoscale signatures.

Finally, SARAL/AltiKa's Ka-band frequency enables the resolution of small scales of ocean dynamics (mesoscale). Thanks to its higher rate of 40 Hz and a smaller and statistically more homogeneous footprint, its noise level is lower than Jason-2/3 (Figure 13). Provided the application of a fine editing (for 20 Hz measurements: based on the Sea Level Anomaly coherence of consecutives measurements) and a correction [31] dedicated to reduce correlated noise between altimeter range and Significant Wave Height (SWH), it does even better than delay-doppler altimetry missions such as Sentinel-3A. Smaller ocean scales are therefore observable notably around 35 km where the spectral "bump" [32] is clearly reduced compared to "standard" use of the data (see Figure 1 in [3] for comparison).

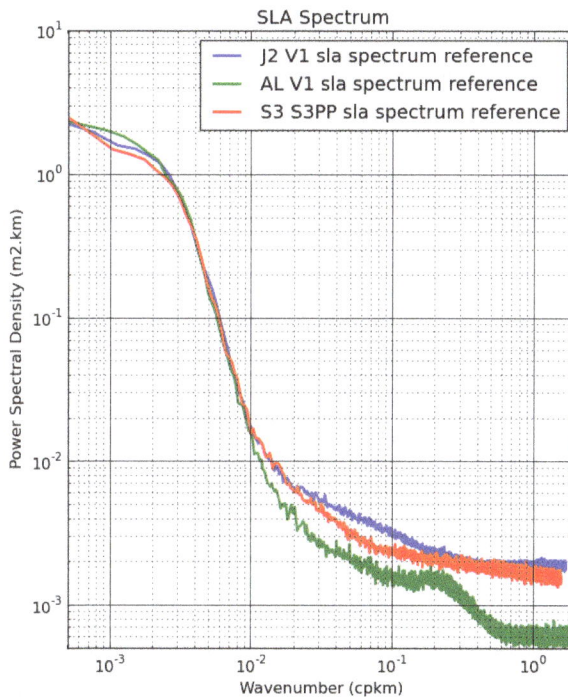

Figure 13. SARAL/AltiKa's SLA power spectrum (green) compared to Jason-2 (blue) and Sentinel-3A (red), with adapted editing and [31] correction for the three missions.

In conclusion, the SARAL/AltiKa mission performance remains excellent, compared to Jason-2, Jason-3 and Sentinel-3A, from fine scale ocean dynamics to long-term stability. And its quality will keep on improving, notably, thanks to several updates of geophysical corrections which are getting ready for the future reprocessing (GDR-E (Geophysical Data Record)) planned for 2018. The benefits of SARAL/AltiKa data to access the fine scale ocean dynamics are highlighted in our companion paper [3].

3. Unique Contributions of Ka-Band Information in AltiKa Data

The key feature of the altimetric payload has been the selection of Ka-band [27]. Using Ka-band avoids the need for a second frequency to correct for the ionosphere delay and eases the sharing of the antenna by the altimeter and the radiometer. The use of the Ka-band also allows the improvement of the range measurement accuracy in a ratio close to 2 (30 cm compared to 47 cm for Jason Ku-band) due to the use of a wider bandwidth and to a better pulse to pulse echo decorrelation. The higher pulse repetition frequency (4 KHz compared with 2 KHz on Jason-2) also permits a better along-track sampling of the surface. Moreover, the enhanced bandwidth (480 MHz compared with 320 MHz on Jason-2) enables a better resolution of the range. Finally, Ka-band antenna aperture is reduced, which limits the pollution within useful ground footprint. The known effect of rain on the Ka-band was such to put some uncertainty on the full availability of data in some regions. In fact, the larger sensitivity to small rates of rain was found to be less constraining than expected and in counterpart provides access to low rains climatology [28].

Apart from the improved accuracy and resolution for the liquid water surfaces (ocean and inland waters), the fact that AltiKa altimeter operates in Ka-band, which is higher than the previous frequencies, offer new paths of investigation. The penetration depth in snow is theoretically reduced from around 10 m in Ku-band to less than 1 m in Ka-band, such that the volume echo originates from the near subsurface. Second, the sharper antenna aperture leads to a narrower leading edge that reduces the impact of the ratio between surface and volume echoes of the height retrieval. Moreover, the volume echo in the Ka-band results from the near subsurface layer and is mostly controlled by ice grain size, unlike the Ku-band [33].

This section gives some illustrations of the unique characteristics of AltiKa instrument that offer unique contributions in fields that where not fully foreseen. Section 3.1 will give insights of the increase of resolution and data availability close to the coasts. Section 3.2 will focus on the Ka-band contribution for altimetry over snow and ice. Section 3.3 will study the rain sensitivity and finally Section 3.4 will give an original use of Ka-band measurements for studying the soil moisture.

3.1. Power Spectrum Density Analysis of SSH Signal from SARAL/AltiKa and Jason-2 (1 Hz) over Bay of Bengal

Indian coasts are vulnerable to high waves and rough sea conditions particularly during the pre- and post-monsoon seasons due to the presence of extreme atmospheric events such as cyclones. Unfortunately, wave models are still not mature enough to predict these waves due to the lack of accurate initial conditions and correct physical parameterizations. With the availability of altimeter data, assimilation of these data in numerical models is often found beneficial in improving the accuracies of the model predictions. One very significant promising development, particularly for the coastal regions, is the SARAL/AltiKa mission carrying onboard altimeter in the Ka-band and thus offering high spatial resolution suitable for coastal studies and assimilation in coastal wave models.

Tracks in the full Bay of Bengal region for the concurrent period of SARAL/AltiKa and Jason-2 (March 2013–July 2016) were used and 1 Hz Sea Surface Height (SSH) spectrum of both altimeters was studied. The Figure 14 shows the SSH spectrums of both the altimeters. Difference is noteworthy at wave length less than 50 km. The higher slopes in the wave lengths below 50 km shows the spectrum for SARAL/AltiKa have lesser noise than Jason-2 in this region. The power spectrum in the mesoscale region (50 km–250 km) is also fitted with power law a*k^b (k is wavenumber). The values of b for AltiKa is −1.96 as compared to b = −1.79 for Jason-2. Above 50 km wave length, both SARAL/AltiKa and Jason are comparable to each other. Also the approachability towards the coast has been studied using the SARAL/AltiKa based ISRO Space Applications Center (SAC) coastal product and the Coastal and Hydrology Altimetry product (PISTACH) product from Jason-2. Figure 15a shows the 20-Hz SWH from Jason-2 towards the coast and Figure 15b shows 40-Hz SWH from SARAL/AltiKa. Clearly the SARAL/AltiKa based coastal product provides more valid data in proximity to the coast that will help to improve the coastal models.

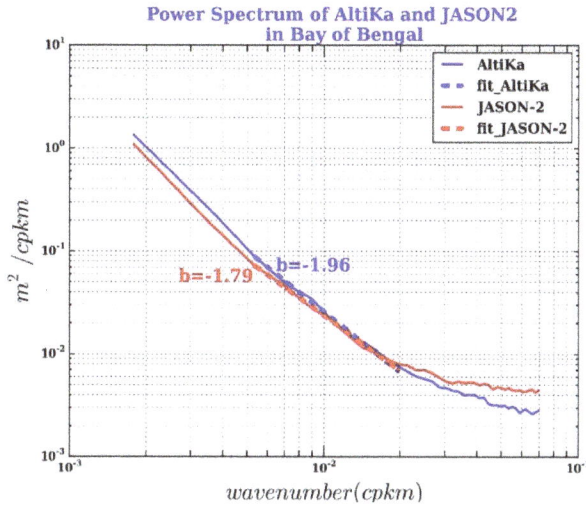

Figure 14. The power spectra of sea surface heights from SARAL/AltiKa and Jason-2 in Bay of Bengal. Green line represents (1/50 km).

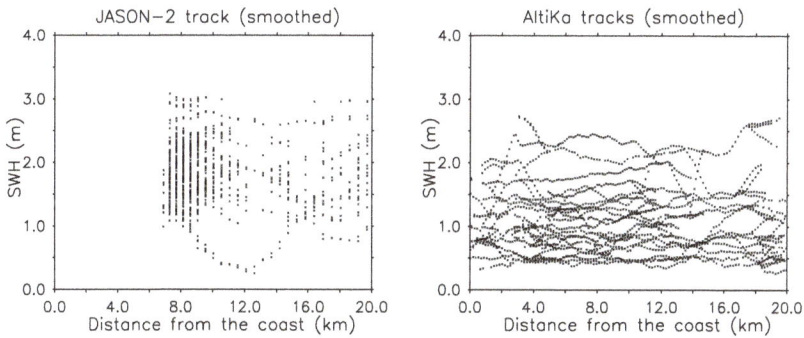

Figure 15. SWH from Jason-2 PISTACH product (**left**) and SAC coastal product (**right**) based on AltiKa data.

3.2. Complementarity between Ku and Ka Measurements over Ice Sheet Surfaces

Exploiting Ka frequency properties over land ice or sea ice is of particular interest to better assess the level of surfaces covered by ice or snow. Comparing measurements in Ku and Ka frequencies allows us to understand and quantify the penetration properties of the radar waves. Over water surfaces, Ku and Ka signals reflect at the air/water interface. It is commonly understood that in presence of snow or ice, the radar wave penetrates the ice pack if its wavelength is larger than the snow grain size (typically 0.5 mm but grain size depends on the temperature (snow metamorphism) and snow accumulation). According to [34], higher temperatures cause the snow grains to grow quickly, whereas higher accumulation rate causes them to slowly grow. Ku- and Ka-band wavelengths are respectively 2.21 cm and 0.84 cm. The factors between signal wavelength and grain size are respectively 44 and 17 in Ku- and Ka-bands. For a given wavelength, according to Mie theory [35], the scattering coefficient is conversely proportional to the radar wavelength at a power of 4. From Ku- to Ka-band, the scattering coefficient consequently increases by a factor 55. This leads to a penetration depth over snow surface between 0.1 m and 0.3 m in Ka-band [36]. Analysis performed on the altimeter signal acquired in Ku- and Ka-bands over the Antarctic ice sheet are consistent with this assumption.

To illustrate this theoretical introduction, we represent in Figure 16, the different penetration effects in Ku- and Ka-bands by comparing mean waveforms computed over ocean (SWH = 1 m ± 20 cm and similar epochs) and at a cross-over over Lake Vostok [37]. This lake in Antarctic is known to be quite perfectly flat, avoiding all effects induced by surface slopes (that can be different in Ku- and Ka-bands due to the different antenna gain patterns of the two missions). Over ocean, for SARAL/AltiKa (Ka-band) and Sentinel-3A (Ku-band in Pseudo-Low Resolution Mode), 1498 and 377 individual waveforms have been aggregated. Over Lake Vostok, for SARAL/AltiKa and Sentinel-3A, 1763 and 359 individual waveforms have been aggregated. We must note that waveforms have been artificially aligned to be compared.

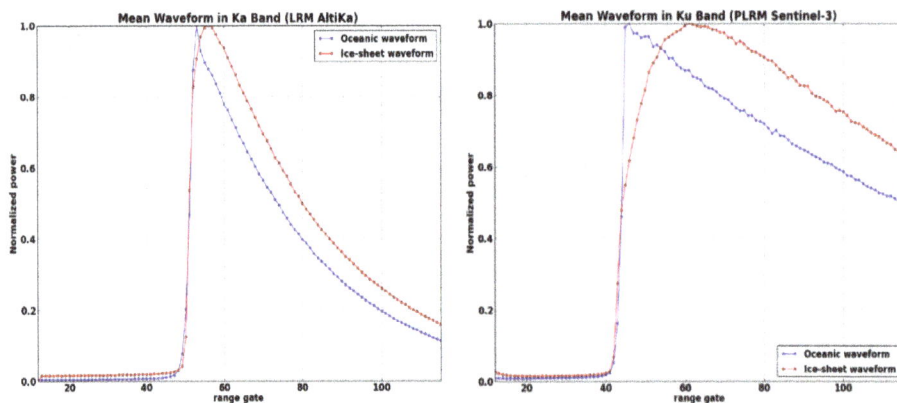

Figure 16. Mean SARAL/AltiKa (**left**) and Sentinel-3A (**right**) waveforms over ocean (blue) and Lake Vostok (red).

It appears on the left plot of Figure 16 that ocean and land ice leading edges are quite identical except for the very last points of the leading edge. Additional energy appears on the trailing edge samples for ice returns. This clearly indicates that the radar signal has been backscattered in the snow pack. However, considering that the distribution of energy in the leading edge is similar over ocean and over land ice, we can assume that an unique reflective surface is responsible for the main reflection (as for ocean) and that a retracking processing will provide the range corresponding to the distance between the satellite and this reflective surface. This surface can be assumed to be at the air/snow interface.

The plot on the right of Figure 16 shows that in Ku-band, returns are completely different over ocean and land ice. Clearly, the red echo can be considered as the summation of different contributions coming from different layers inside the snow pack, each contribution being attenuated depending on the absorption characteristics of each layer. In these cases, determining a range indicating the position of the main reflection (its distance from the radar) is very hard. If a threshold empirical retracker is used, the value of the threshold arbitrarily determines the reflective surface inside the snowpack.

For a better comparison of Ku and Ka measurements over Lake Vostok, waveforms in both bands are superimposed in Figure 17. The waveform acquired in Synthetic Aperture Radar (SAR) mode (Sentinel-3A, Ku-band) has been added (in red). As for SARAL/AltiKa, Sentinel-3A SAR mode mean echo over Lake Vostok presents a leading edge not impacted by penetration effects.

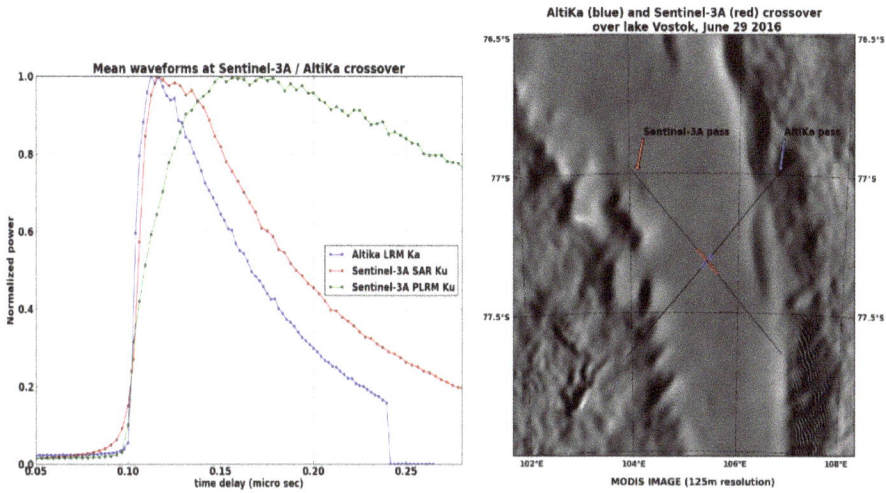

Figure 17. Left: Mean SARAL/AltiKa (blue) and Sentinel-3A (SAR in red, P-LRM in green) waveforms at a cross-over between SARAL/AltiKa and Sentinel-3A over Lake Vostok. **Right**: MODIS image of Lake Vostok with SARAL/AltiKa and Sentinel-3A tracks locations.

This analysis illustrates the complementarity between Ku and Ka measurements over ice sheet surfaces and the potential relevant information that could be retrieved by combining them. Of particular interest is the snow thickness that could be derived from the exploitation of the Ku and Ka waveform shapes, in particular their trailing edges.

Analysis of SARAL/AltiKa measurements over ice sheet have not only brought a new insight of the radar measurements of the snowpack but also encourage the development of new algorithms able to derive new parameters such as snow or ice pack properties. Scientific applications based on Ka-band unique characteristics and complementarity with Ku-band over ice sheet, icebergs and sea ice are developed in [3].

3.3. Impact of Rain on SARAL/AltiKa Measurements, a New Opportunity to Observe and Study the Rainfall Climatology

Since the first developments of the SARAL/AltiKa mission, one of the major concerns has been the Ka-band sensitivity to atmospheric conditions, and especially to atmospheric liquid water (both rain and clouds). Hence, the instrument power and gain have been optimized in order to maximize the Signal to Noise Ratio (SNR) and to minimize atmospheric attenuation issues [27]. As a result, the data loss over ocean is lower than anticipated (<0.1%) and the SNR is higher than the preflight expected value (14 dB) thanks to margins taken in the altimeter link budget (see details in [27]). An example of an AltiKa SNR map over ocean is provided in Figure 18 for cycle number 110, showing values varying from 18 to 32 dB. The altimeter link budget is thus excellent and allows a full data coverage over ocean.

Despite the very low number of lost data, the high sensitivity of the Ka-band to atmospheric liquid water (ten times larger than in Ku-band) impacts the altimeter waveform shape. AltiKa echoes are attenuated and distorted by rain events of heavy clouds [38–40], as shown in Figure 19a. The waveform amplitude is not only strongly attenuated but its shape is distorted, especially on the trailing edge part of the waveform. The example shown in Figure 19a is collocated to WindSat and SSMIS-F16/F17 rainfall rate measurements indicating a rain rate of about 4.5 mm/h during this event.

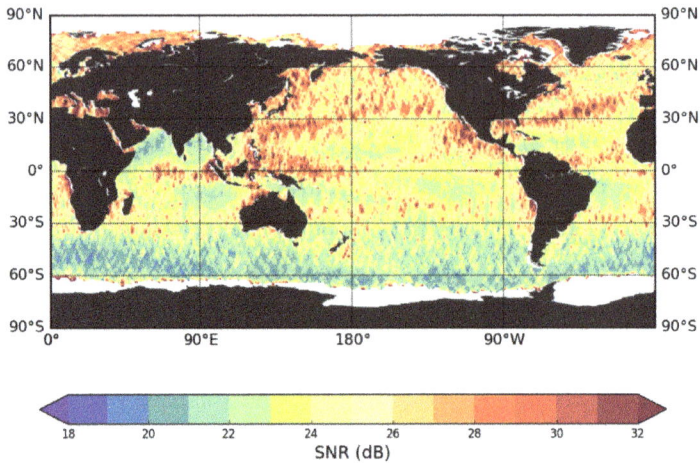

Figure 18. Map of Signal to Noise Ratio computed from standard MLE-4 estimates over ocean for cycle number 110 (from 2017/06/19 to 2017/07/24).

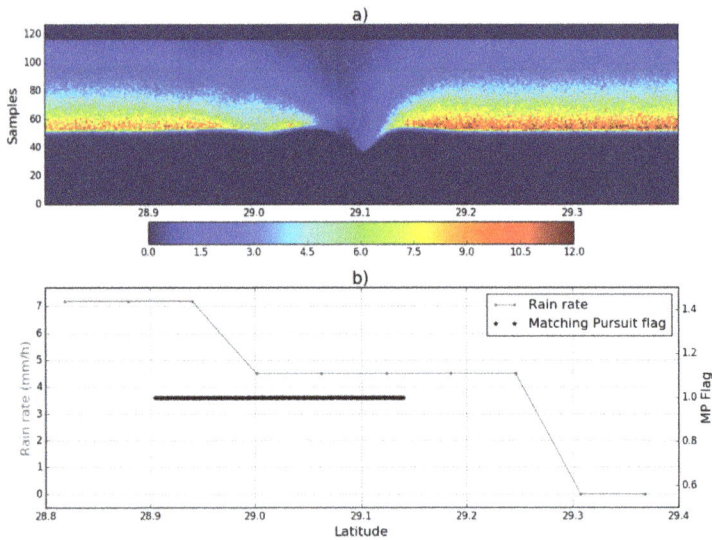

Figure 19. (**a**) AltiKa waveforms corrected from Automatic Gain Control during a rain event, pass 121 cycle 5; (**b**) collocated rainfall rate from SSMI and WindSat measurement in grey and Matching Pursuit flag in black [41].

Even if measurements are not lost during a rain event, the resulting waveform corruption may degrade the retracker performances and the geophysical estimates [42]. Indeed, operational ocean retrackers implemented in the ground segment are not designed to correctly process rain waveforms: these retrackers consider that the altimeter waveform footprint contains homogeneous backscattering properties. This is clearly not the case during rain events. As a result, the noise on each geophysical estimate increases when SARAL/AltiKa is overflying a rain cell. Impact of rain pollution on the range noise is illustrated in Figure 20. The collocation between SARAL/AltiKa and the WindSat/SSMIS measurements has been performed accounting for a maximum time delay of 1 hour. A clear rain rate dependency of the altimeter range noise (about 0.35 cm/mm/h) can be observed. The other geophysical estimates (significant wave height and backscattering coefficient) are impacted as well. Recent developments (in particular the "Adaptive" retracker) have shown very promising results to deal with such corrupted echoes [43].

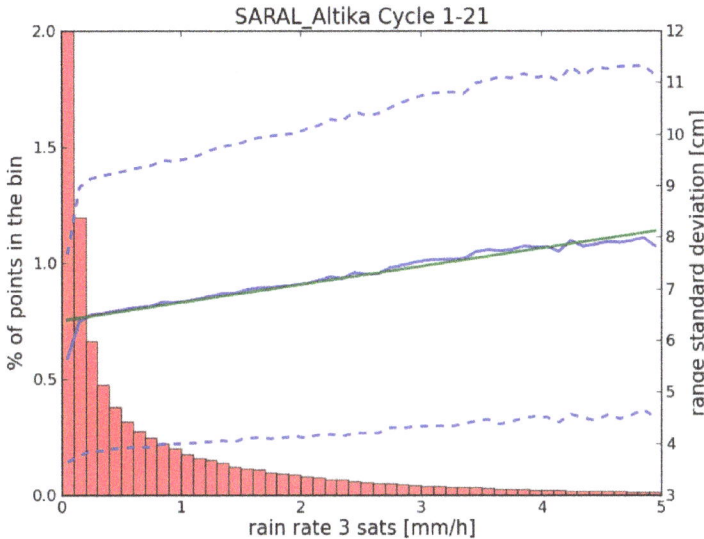

Figure 20. Variations of range std against rain rate. Please note that the first bin corresponding to dry measurements is truncated on the plot as it contains more than 70% of the measurement points. The plain blue line is the mean of the range standard deviation and the dashed blue lines represent the standard deviation for this mean. The green line is a linear regression of the mean (plain blue line).

The editing of altimeter data impacted by rain has been largely addressed in previous studies [44–46] and different methods exist to identify rain measurements. In the SARAL/AltiKa ground processing, the Matching Pursuit algorithm has been selected to flag rain data. This algorithm is based on the analysis of the short scale variations of the off-nadir angle estimate to identify rain cells in the SARAL/AltiKa measurements [41]. The performances of this algorithm have been assessed in [28] and show that the Matching Pursuit algorithm combined with the radiometer Integrated Liquid Water Content (ILWC) > 0.1 kg/m^2 provide a good rain flag. A map of 1 Hz flagged points is shown in Figure 21 illustrating the geographical distribution of the near 10% flagged measurements. As expected, most of the flagged points over ocean are located in the Intertropical Convergence Zone which reflects the global distribution of precipitations [47]. The comparison of the rain flag performances with WindSat and SSMIS-F16/F17 rainfall rates has shown acceptable performances but studies have been performed to improve and simplify rain flagging based on continuous wavelet transform. Results have been presented in [48].

Flagging SARAL/AltiKa measurements corrupted by rain presents a new opportunity to observe and study the rainfall climatology, seasonal and inter-annual variability. Moreover, the link between the signal attenuation and the rainfall rate can be exploited to extract valuable information on rain events. On-going studies seek to exploit Ka-band measurements to derive rain rate information, coupling them with the brightness temperature measurements provided by the radiometers for climatological monitoring.

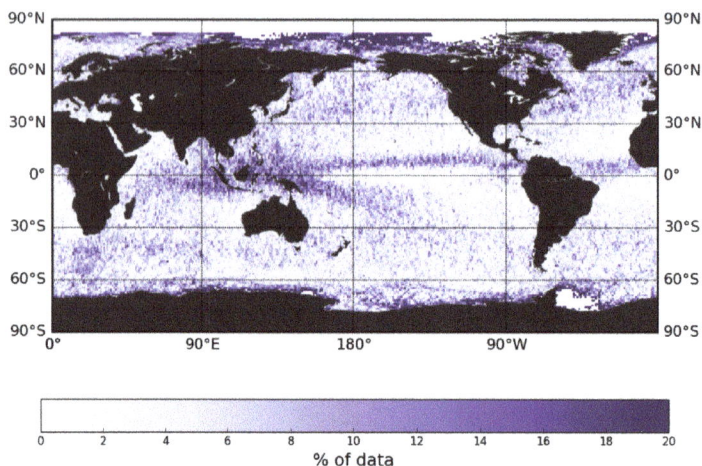

Figure 21. Percentage of 1 Hz flagged samples using the operational MP algorithm for which the ILWC >0.1 kg/m^2 for 1 year (cycles 100 to 110).

3.4. Relationship between Surface Soil Moisture and Radar Altimetry Backscattering Coefficients

Spatial and temporal variations of radar altimetry backscattering coefficients were related to the dynamics of surface properties. In semi-arid areas, soil moisture drives several surface processes including soil organic matter mineralization [49], vegetation productivity [50], land surface fluxes [51] and land surface atmosphere interactions [52]. A first inversion of the surface soil moisture (SSM) from the Ice-1 derived backscattering coefficients at Ku-band from Envisat RA-2 data was performed over Sahelian savannahs in the Gourma region of Mali [53]. Due to the lower sensitivity of nadir-looking altimeters to vegetation cover compared with side-looking SAR and scatterometers [54], higher correlation was found between in-situ SSM measurements and altimetry (R = 0.88, [53]) than with SAR (R = 0.85, [55]) and scatterometer (R = 0.63, [56]) backscattering coefficients over the sandy sites of the same semi-arid study area. Using the 16 first available cycles of the SARAL/AltiKa mission, the correlation is 0.88 between radar altimetry Ice-1 derived backscattering coefficients at Ka-band and level-3 SSM products derived from the Soil Moisture and Ocean Salinity satellite (SMOS) passive microwave observations over Sahelian savannahs [57]. The results obtained were generally better than the ones obtained using 3.5 times more numerous backscattering coefficients at Ku and C-bands from Jason-2 during the same observation period or the ones obtained using backscattering coefficients at Ku and S-bands from Envisat during its whole observation period (and comparing them to level-3 SSM products derived from AMSR-E) or passive microwave observations [57]. Time series of altimetry backscattering coefficients (obtained using Ice-1 retracking algorithm) at Ka-band from SARAL/AltiKa and Ku and C bands from Jason-2 are presented in Figure 22 from February 2013 to May 2016 over Sudano-Sahelian savanahs. This period corresponds to the 35 cycles of SARAL/AltiKa on its nominal orbit (and to cycles 169 to 291 of Jason-2). The backscattering level during the dry season decreases as the frequency increases—around 22.5 dB at C-band, 17 dB at Ku-band, and between 5 dB and 6 dB at Ka-band—whereas the amplitude of variations increases with the frequency—up to 10 dB at C-band, 20 dB at Ku-band and 30 dB at Ka-band. The larger dynamic of the backscattering coefficient at Ka-band during the rainy season is a strong advantage to accurately monitor the time-variations of SSM in semi-arid areas. Correlation coefficients between backscattering coefficients and SSM from SMOS (derived using the approach from [58]) were computed on the common period of availability of the two datasets, from February 2013 to May 2015. Due to the rapid response of SSM to rainfall—strong increase after a rainfall event followed by a fast decrease due to evaporation—in semi-arid environments, correlation were re-estimated applying a three-day smoothing window to limit the effect of the difference of acquisition time between altimeters and SMOS. Very similar results are obtained for both ascending and descending tracks:

(i)   R = 0.80 and 0.79 at Ka-band for SARAL/AltiKa tracks 0001 and 0846 respectively,
(ii) R = 0.58 and 0.54 at Ku-band for Jason-2 tracks 046 and 161 respectively,
(iii) R = 0.64 and 0.65 at C-band for Jason-2 tracks 046 and 161 respectively.

As the backscattering coefficient is linearly related to SSM over sand, better estimates of SSM can be expected from the inversion of altimetry backscattering coefficient at Ka-band than at other frequency bands commonly used in altimetry. In spite of the low temporal resolution of the SARAL/AltiKa mission with its repeat-period of 35 days, altimetry-derived SSM at Ka-band would be useful for the monitoring of the seasonal variations of this key variable along the altimeter tracks at high spatial resolution (∼165 m using the high-frequency 40 Hz data) and could be used for the calibration and validation of land-surface models and low resolution active (scatterometry) and passive (radiometry) microwave-based products. The characteristics of the backscattering coefficient at Ka-band illustrated in this study is also used to study the snow and ice state and important scientific results are given in [3].

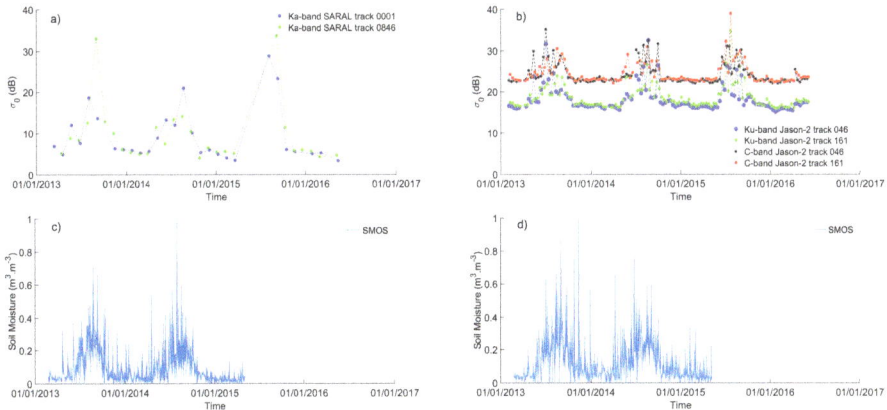

Figure 22. Time-series of backscattering coefficient (dB) for (**a**) SARAL/AltiKa (Ka-band) 0001 (blue) and 0846 (green) tracks and (**b**) Jason-2 (Ku and C-bands) 046 (blue and black) and 161 (light green and red), of volumetric soil moisture (m3.m-3), from SMOS (blue) at SARAL/AltiKa (**c**) and Jason-2 (**d**) sites.

4. Conclusions

Altimetry has been dominated by the Ku-band since the beginning and strong fears about a possible degradation by using Ka-band had limited the studies about launching a mission embarking on such technology. Vincent et al. [36] demonstrated the feasibility of such an altimeter but also importantly the need of higher resolution altimetry to address one of the most problematic features of ocean circulation, the mesoscale eddies and fronts. These features are essential to understanding the dynamics of ocean circulation on all space and time scales and theoretical studies of the cascade of energy over scales from 100–1000 km have always suffered from the lack of high-resolution observations. From [36] a Ka-band altimeter technology looked to be able to provide higher resolution to help answering these key issues. Selecting the Ka-band allows a larger bandwidth (480 MHz), which provides a vertical resolution of 0.3 m instead of 0.5 m in Ku-band. With such a resolution, the altimeter is close to a beam limited one: there is no 'plateau' in the echo as in Ku-band, since it strongly attenuates shortly after the leading edge, due to the small antenna aperture. This greatly reduces the pollution of 'land gates' into 'ocean gates', when considering land-sea transition areas. Moreover, due to the smaller antenna beamwidth, the Brown echo has a sharper shape in Ka-band than what is obtained with conventional Ku-band altimeters. Finally, the shorter decorrelation time of sea echoes at Ka-band also enables to double the number of independent echoes per second compared with Ku-band altimeters, allowing high frequency measurements (40 Hz, 20 Hz for Ku-band).

SARAL/AltiKa mission is the realization of the [36] study and in this paper we show that the quality of its measurements meets and generally exceeds the expectation of the visionary designers of this mission. Firstly, one of the main result is probably the fact that the AltiKa data quality and accuracy are met at a similar level whatever the water surfaces and whatever the scale (Section 2). This is demonstrated using global, regional or local methods and based on comparisons with various external data (in situ, other altimetric missions). The precision is at least at the same level or even better than for Ku-band for the open-oceans but improves substantially the along-track resolution (e.g., Figure 13 or Figure 14). These results will continue to improve with the reprocessing in 2018 with upgraded standards (GDR-E). Moreover, the characteristics of the AltiKa instrument allows to observe smaller water surfaces with equivalent precision and are of very great interest for coastal and inland waters studies [3]. The benefits of this increase of precision and resolution are highlighted through Science Applications of SARAL/Altika data [3]. Apart of these great achievements in term of precision and resolution, the characteristics of the altimeter and the Ka-band offer unique contributions in fields that where not fully foreseen. Before launch, concerns were raised about the sensitivity of the Ka-band to rain events, leading to missing and invalid measurements. However, in practice, the SARAL/AltiKa data return is remarkably high with few missing data (e.g., Figures 11 and 18). In fact, the sensitivity of Ka-band measurements to rain may also lead to a way of estimating very light rainfall over the oceans, for which we dramatically lack of information and which could lead to a great improvement of our knowledge of the oceanic rain climatology. Another specificity of the Ka-band is that electromagnetic wave penetration effects are minimized, which is particularly important over continental ice surfaces. From Ku- to Ka- bands, the scattering coefficient is increased by a factor 55: volume scattering is then being clearly dominant over surface scattering and gives information on snow or ice pack properties. Inversely, the radar wave extinction is also increased, leading to a penetration depth over the snow surface between 0.1 m and 0.3 m (versus 2 to 10 m in Ku-band). The altimetric observation and height restitution thus correspond to a thin subsurface layer and give complementary information than the one given historically by the Ku-band (see Section 3.2 and [3]). Finally, the spatial and temporal variations of Ka-band backscattering coefficients that are related to the dynamics of surface properties, show valuable information for studying snow and ice state but also surface soil moisture (e.g., Section 3.4). This last field of studies is at his infancy but results presented in this paper are clearly encouraging.

With the upcoming Surface Water & Ocean Topography (SWOT) mission embarking on a Ku-band nadir altimeter and a Ka-band interferometer (KaRIN) and the phase 0 of a new cryosphere satellite mission (bi-frequency altimeter in Ku- and Ka-band) that is currently studied by CNES and ESA, it is more than likely that SARAL/AltiKa Ka-band altimetric mission will continue to help preparing these missions but also help to understand Ku-band better.

Acknowledgments: The SARAL/AltiKa mission is an achievement of a very fruitful cooperation between ISRO and CNES. The contribution of EUMETSAT to data distribution is also strongly appreciated. We acknowledge the support of all the Investigators, PIs and Co-Is, of the SARAL/AltiKa Mission. Most of the studies performed by the French authors have been conducted and financed thanks to Centre National d'Etudes Spatiales (CNES), Centre National de la Recherche Scientifique (CNRS), and French Ministry of Research. The altimetry data used in Section 3.4 were made available by "Centre de Topographie des Océans et de l'Hydrosphère" (CTOH)—https://ctoh.legos.obs-mip.fr/. The SMOS data were obtained from the "Centre Aval de Traitement des Données SMOS" (CATDS), operated for the "Centre National d'Etudes Spatiales" (CNES, France) by IFREMER (Brest, France)—https://www.catds.fr/sipad/.

Author Contributions: Pascal Bonnefond and Jacques Verron coordinated the whole paper. K. N. Babu designed the study, analyzed the data and wrote the Section 2.1.1. Pascal Bonnefond and Olivier Laurain designed the study, analyzed the data and wrote the Section 2.1.2. Jean-François Crétaux and Muriel Bergé-Nguyen designed the study, analyzed the data and wrote the Section 2.1.3. Mathilde Cancet designed the study, analyzed the data and wrote the Section 2.2 while Pascal Bonnefond, Bruce J. Haines, Olivier Laurain and Christopher Watson provided the data and contributed to the writing and analysis. Annabelle Ollivier and Pierre Prandi designed the study, analyzed the data and wrote the Section 2.3. Aditya Chaudhary and Rashmi Sharma designed the study, analyzed the data and wrote the Section 3.1. Pierre Thibaut and Jérémie Aublanc designed the study, analyzed the data and wrote the Section 3.2. Jean-Christophe Poisson, Pierre Prandi and Pierre Thibaut designed the study, analyzed the data and wrote the Section 3.3. Frédéric Frappart designed the study, analyzed the data and wrote the Section 3.4. All the other co-authors helped in the analysis and paper writing.

Conflicts of Interest: The authors declare no conflict of interest.

References

1. Verron, J.; Sengenes, P.; Lambin, J.; Noubel, J.; Steunou, N.; Guillot, A.; Picot, N.; Coutin-Faye, S.; Sharma, R.; Gairola, R.M.; et al. The SARAL/AltiKa altimetry satellite mission. *Mar. Geod.* **2015**, *38*, 2–21, doi:10.1080/01490419.2014.1000471.
2. Verron, J.; Picot, N. Preface. *Mar. Geod.* **2015**, *38*, 1, doi:10.1080/01490419.2015.1052711.
3. Verron, J.; Bonnefond, P.; Aouf, L.; Birol, F.; Bhowmick, S.A.; Calmant, S.; Carret, A.; Conchy, T.; Crétaux, J.F.; Dibarboure, G.; et al. The benefits of the Ka-band as evidenced from the SARAL/AltiKa altimetric mission: Scientific applications. *Remote Sens.* **2017**, under review.
4. Church, J.; Clark, P.; Cazenave, A.; Gregory, J.; Jevrejeva, S.; Levermann, A.; Merrifield, M.; Milne, G.; Nerem, R.; Nunn, P.; et al. Climate change 2013: The physical science basis. Contribution of Working Group I to the fifth assessment report of the intergovernmental panel on climate change. In *Sea Level Change*; Cambridge University Press: Cambridge, UK; New York, NY, USA, 2013; Chapter 13.
5. Prandi, P.; Philipps, S.; Pignot, V.; Picot, N. SARAL/AltiKa global statistical assessment and cross-calibration with Jason-2. *Mar. Geod.* **2015**, *38*, 297–312, doi:10.1080/01490419.2014.995840.
6. Ablain, M.; Philipps, S.; Picot, N.; Bronner, E. Jason-2 global statistical assessment and cross-calibration with Jason-1. *Mar. Geod.* **2010**, *33*, 162–185, doi:10.1080/01490419.2010.487805.
7. Bonnefond, P.; Haines, B.; Watson, C. Coastal altimetry. In *In Situ Calibration and Validation: A Link from Coastal to Open—Ocean Altimetry*; Springer: Berlin/Heidelberg, Germany, 2011; Chapter 11, pp. 259–296, ISBN 978-3-642-12795-3.
8. Mertikas, S.P.; Daskalakis, A.; Tziavos, I.N.; Vergos, G.; Fratzis, X.; Tripolitsiotis, A. First calibration results for the SARAL/AltiKa altimetric mission using the gavdos permanent facilities. *Mar. Geod.* **2015**, *38*, 249–259, doi:10.1080/01490419.2015.1030052.
9. Bonnefond, P.; Desai, S.; Haines, B.; Leuliette, E.; Picot, N. Regional and Global CAL/VAL for Assembling a Climate Data Record Summary. In Proceedings of the Ocean Surface Topography Science Team Meeting, Miami, Florida, 2017. Available online: https://tinyurl.com/ycu32zeo (accessed on 9 January 2018).
10. Babu, K.N.; Shukla, A.K.; Suchandra, A.B.; Kumar, S.V.V.A.; Bonnefond, P.; Testut, L.; Mehra, P.; Laurain, O. Absolute calibration of SARAL/AltiKa in Kavaratti during its initial Calibration-validation phase. *Mar. Geod.* **2015**, *38*, 156–170, doi:10.1080/01490419.2015.1045639.
11. Bonnefond, P.; Exertier, P.; Laurain, O.; Guillot, A.; Picot, N.; Cancet, M.; Lyard, F. SARAL/AltiKa absolute calibration from the multi-mission Corsica facilities. *Mar. Geod.* **2015**, *38*, 171–192, doi:10.1080/01490419.2015.1029656.
12. Bonnefond, P.; Laurain, O.; Exertier, P.; Boy, F.; Guinle, T.; Picot, N.; Labroue, S.; Raynal, M.; Donlon, C.; Féménias, P.; et al. Calibrating SAR SSH of sentinel-3A and CryoSat-2 over the Corsica facilities. *Remote Sens.* **2018**, *10*, 92, doi:10.3390/rs10010092.
13. Crétaux, J.F.; Calmant, S.; Romanovski, V.; Shabunin, A.; Lyard, F.; Bergé-Nguyen, M.; Cazenave, A.; Hernandez, F.; Perosanz, F. An absolute calibration site for radar altimeters in the continental domain: Lake Issykkul in Central Asia. *J. Geod.* **2009**, *83*, 723–735, doi:10.1007/s00190-008-0289-7.
14. Crétaux, J.F.; Calmant, S.; Romanovski, V.; Perosanz, F.; Tashbaeva, S.; Bonnefond, P.; Moreira, D.; Shum, C.K.; Nino, F.; Bergé-Nguyen, M.; et al. Absolute Calibration of Jason Radar Altimeters from GPS Kinematic Campaigns over Lake Issykkul. *Mar. Geod.* **2011**, *34*, 291–318, doi:10.1080/01490419.2011.585110.
15. Crétaux, J.F.; Bergé-Nguyen, M.; Calmant, S.; Romanovski, V.; Meyssignac, B.; Perosanz, F.; Tashbaeva, S.; Arsen, A.; Fund, F.; Martignago, N.; et al. Calibration of Envisat radar altimeter over Lake Issykkul. *Adv. Space Res.* **2013**, *51*, 1523–1541, doi:10.1016/j.asr.2012.06.039.

16. Cretaux, J.F.; Bergé-Nguyen, M.; Calmant, S.; Djamangulova, N.; Satylkanov, R.; Lyard, F.; Perosanz, F.; Verron, J.; Samine Montazem, A.; Le Guilcher, G.; et al. Absolute calibration / validation of the altimeters on sentinel-3A and Jason-3 over the lake Issykkul. *Remote Sens.* **2017**, under review.

17. Arsen, A.; Crétaux, J.F.; del Rio, R.A. Use of SARAL/AltiKa over Mountainous Lakes, Intercomparison with Envisat Mission. *Mar. Geod.* **2015**, *38*, 534–548, doi:10.1080/01490419.2014.1002590.

18. Marty, J.C.; Loyer, S.; Perosanz, F.; Mercier, F.; Bracher, G.; Legresy, B.; Portier, L.; Capdeville, H.; Fund, F.; Lemoine, J.M.; et al. GINS: The CNES/GRGS GNSS scientific software. In Proceedings of the 3rd International Colloquium Scientific and Fundamental Aspects of the Galileo Programme, ESA Proceedings WPP326, Copenhagen, Denmark, 31 August–2 September 2011; Volume 31.

19. Jan, G.; Ménard, Y.; Faillot, M.; Lyard, F.; Jeansou, E.; Bonnefond, P. Offshore Absolute Calibration of space-borne radar altimeters. *Mar. Geod.* **2004**, *27*, 615–629, doi:10.1080/01490410490883469.

20. Cancet, M.; Bijac, S.; Chimot, J.; Bonnefond, P.; Jeansou, E.; Laurain, O.; Lyard, F.; Bronner, E.; Féménias, P. Regional in situ validation of satellite altimeters: Calibration and cross-calibration results at the Corsican sites. *Adv. Space Res.* **2013**, *51*, 1400–1417, doi:10.1016/j.asr.2012.06.017.

21. Cancet, M.; Lux, M.; Pénard, C. COMAPI: New Regional Tide Atlases and High Frequency Dynamical Atmospheric Correction. In Proceedings of the Ocean Surface Topography Science Team Meeting, Lisbon, Portugal, 2010. Available online: https://tinyurl.com/y98xedyq (accessed on 9 January 2018).

22. Carrère, L.; Lyard, F. Modeling the barotropic response of the global ocean to atmospheric wind and pressure forcing—Comparisons with observations. *Geophys. Res. Lett.* **2003**, *30*, 1275.

23. Lyard, F.; Lefevre, F.; Letellier, T.; Francis, O. Modelling the global ocean tides: Modern insights from FES2004. *Ocean Dyn.* **2006**, *56*, 394–415, doi:10.1007/s10236-006-0086-x.

24. Cancet, M.; Watson, C.; Haines, B.; Bonnefond, P.; Lyard, F.; Laurain, O.; Guinle, T. Regional CALVAL of Jason-2 and SARAL/AltiKa at Three Calibration Sites. In Proceedings of the Ocean Surface Topography Science Team Meeting, Reston, USA, 2015. Available online: https://tinyurl.com/ybm9huug (accessed on 9 January 2018).

25. Carrère, L.; Lyard, F.; Cancet, M.; Guillot, A.; Picot, N.; Dupuy, S. FES2014: A New Global Tidal Model. In Proceedings of the Ocean Surface Topography Science Team Meeting, Reston, USA, 2015. Available online: https://tinyurl.com/y9j9p4mf (accessed on 9 January 2018).

26. Valladeau, G.; Thibaut, P.; Picard, B.; Poisson, J.C.; Tran, N.; Picot, N.; Guillot, A. Using SARAL/AltiKa to improve Ka-band Altimeter measurements for Coastal Zones, Hydrology and ice: The PEACHI Prototype. *Mar. Geod.* **2015**, *38*, 124–142, doi:10.1080/01490419.2015.1020176.

27. Steunou, N.; Desjonquères, J.D.; Picot, N.; Sengenes, P.; Noubel, J.; Poisson, J.C. AltiKa Altimeter: Instrument description and in flight performance. *Mar. Geod.* **2015**, *38*, 22–42, doi:10.1080/01490419.2014.988835.

28. Tournadre, J.; Poisson, J.C.; Steunou, N.; Picard, B. Validation of AltiKa matching pursuit rain flag. *Mar. Geod.* **2015**, *38*, 107–123, doi:10.1080/01490419.2014.1001048.

29. Meyssignac, B.; Cazenave, A. Sea level: A review of present-day and recent-past changes and variability. *J. Geodyn.* **2012**, *58*, 96–109, doi:10.1016/j.jog.2012.03.005.

30. Picard, B.; Frery, M.L.; Obligis, E.; Eymard, L.; Steunou, N.; Picot, N. SARAL/AltiKa wet tropospheric correction: In-flight Calibration, retrieval strategies and performances. *Mar. Geod.* **2015**, *38*, 277–296, doi:10.1080/01490419.2015.1040903.

31. Zaron, E.D.; de Carvalho, R. Identification and reduction of retracker-related noise in altimeter-derived sea surface height measurements. *J. Atmos. Ocean. Technol.* **2016**, *33*, 201–210, doi:10.1175/JTECH-D-15-0164.1.

32. Dibarboure, G.; Boy, F.; Desjonqueres, J.D.; Labroue, S.; Lasne, Y.; Picot, N.; Poisson, J.C.; Thibaut, P. Investigating short-wavelength correlated errors on low-resolution mode altimetry. *J. Atmos. Ocean. Technol.* **2014**, *31*, 1337–1362, doi:10.1175/JTECH-D-13-00081.1.

33. Rémy, F.; Flament, T.; Michel, A.; Blumstein, D. Envisat and SARAL/AltiKa observations of the Antarctic Ice Sheet: A comparison between the Ku-band and Ka-band. *Mar. Geod.* **2015**, *38*, 510–521, doi:10.1080/01490419.2014.985347.

34. Legrésy, B.; Rémy, F. Using the temporal variability of the radar altimetric observations to map surface properties of the Antarctic ice sheet. *J. Glaciol.* **1998**, *44*, 197–206.

35. Wriedt, T. Mie Theory: A Review. In *The Mie Theory: Basics and Applications*; Hergert, W., Wriedt, T., Eds.; Springer: Berlin/Heidelberg, Germany, 2012; pp. 53–71.

36. Vincent, P.; Steunou, N.; Caubet, E.; Phalippou, L.; Rey, L.; Thouvenot, E.; Verron, J. AltiKa: A Ka-band Altimetry Payload and System for Operational Altimetry during the GMES period. *Sensors* **2006**, *6*, 208–234.

37. Aublanc, J.; Thibaut, P.; Lacrouts, C.; Boy, F.; Guillot, A.; Picot, N.; Rémy, F.; Blumstein, D. Altimetry Mission Performances over the Polar Ice Sheets: Cryosat-2, AltiKa and Sentinel-3A. In Proceedings of the Ocean Surface Topography Science Team Meeting, Miami, Florida, 2017. Available online: https://tinyurl.com/y7e7bckx (accessed on 9 January 2018).

38. Quartly, G.; Guymer, T.; Srokosz, A. The effects of rain on Topex radar altimeter data. *J. Atmos. Ocean. Technol.* **1996**, *13*, 1209–1229.

39. Tournadre, J.; Morland, J.C. The effects of rain on TOPEX/Poseidon altimeter data. *IEEE Trans. Geosci. Remote Sens.* **1997**, *35*, 1117–1135, doi:10.1109/36.628780.

40. Tournadre, J.; Lambin-Artru, J.; Steunou, N. Cloud and rain effects on AltiKa/SARAL Ka-band radar altimeter—Part I: Modeling and mean annual data availability. *IEEE Trans. Geosci. Remote Sens.* **2009**, *47*, 1806–1817, doi:10.1109/TGRS.2008.2010130.

41. Tournadre, J.; Lambin-Artru, J.; Steunou, N. Cloud and rain effects on AltiKa/SARAL Ka-band radar altimeter—Part II: Definition of a rain/cloud flag. *IEEE Trans. Geosci. Remote Sens.* **2009**, *47*, 1818–1826, doi:10.1109/TGRS.2008.2010127.

42. Prandi, P.; Debout, V.; Ablain, M.; Labroue, S. *SARAL/Altika Validation and Cross Calibration Activities: Annual Report 2016*; Technical Report; CLS: Ramonville Saint-Agne, France, 2016.

43. Thibaut, P.; Piras, F.; Poisson, J.C.; Moreau, T.; Aublanc, J.; Amarouche, L.; Picot, N. Convergent Solutions for Retracking Conventional and Delay Doppler Altimeter Echoes. In Proceedings of the Ocean Surface Topography Science Team Meeting, Miami, Florida, 2017. Available online: https://tinyurl.com/ybppndx6 (accessed on 9 January 2018).

44. Tournadre, J. Validation of Jason and envisat altimeter dual frequency rain flags. *Mar. Geod.* **2004**, *27*, 153–169, doi:10.1080/01490410490465616.

45. Tran, N.; Obligis, E.; Ferreira, F. Comparison of two Jason-1 altimeter precipitation detection algorithms with rain estimates from the TRMM Microwave Imager. *J. Atmos. Ocean. Technol.* **2005**, *22*, 782–794.

46. Tran, N.; Tournadre, J.; Femenias, P. Validation of envisat rain detection and rain rate estimates by comparing with TRMM data. *IEEE Geosci. Remote Sens. Lett.* **2008**, *5*, 658–662, doi:10.1109/LGRS.2008.2002043.

47. Adler, R.F.; Huffman, G.J.; Chang, A.; Ferraro, R.; Xie, P.P.; Janowiak, J.; Rudolf, B.; Schneider, U.; Curtis, S.; Bolvin, D.; et al. The Version-2 Global Precipitation Climatology Project (GPCP) monthly precipitation analysis (1979–Present). *J. Hydrometeorol.* **2003**, *4*, 1147–1167, doi:10.1175/1525-7541(2003)004<1147:TVGPCP>2.0.CO;2.

48. Poisson, J.; Thibaut, P.; Hoang, D.; Boy, F.; A. Guillot, A.N.P. Wavelet analysis of AltiKa measurements. In Proceedings of the Ocean Surface Topography Science Team, Constance, Germany, 28–31 October 2014.

49. Zech, W.; Senesi, N.; Guggenberger, G.; Kaiser, K.; Lehmann, J.; Miano, T.; Miltner, A.; Schroth, G. Factors controlling humification and mineralization of soil organic matter in the tropics. *Geoderma* **1997**, *79*, 117–161.

50. Hiernaux, P.; Mougin, E.; Diarra, L.; Soumaguel, N.; Lavenu, F.; Tracol, Y.; Diawara, M.; Jarlan, L. Rangeland response to rainfall and grazing pressure over two decades: Herbaceous growth pattern, production and species composition in the Gourma, Mali. *J. Hydrol.* **2009**, *375*, 114–127.

51. Brümmer, C.; Falk, U.; Papen, H.; Szarzynski, J.; Wassmann, R.; Brüggemann, N. Diurnal, seasonal, and interannual variation in carbon dioxide and energy exchange in shrub savanna in Burkina Faso (West Africa). *J. Geophys. Res. Biogeosci.* **2008**, *113*, doi:10.1029/2007JG000583.

52. Taylor, C.M.; Harris, P.P.; Parker, D.J. Impact of soil moisture on the development of a Sahelian mesoscale convective system: A case-study from the AMMA Special Observing Period. *Q. J. R. Meteorol. Soc.* **2010**, *136*, 456–470, doi:10.1002/qj.465.

53. Fatras, C.; Frappart, F.; Mougin, E.; Grippa, M.; Hiernaux, P. Estimating surface soil moisture over Sahel using ENVISAT radar altimetry. *Remote Sens. Environ.* **2012**, *123*, 496–507, doi:10.1016/j.rse.2012.04.013.

54. Fatras, C.; Frappart, F.; Mougin, E.; Frison, P.L.; Faye, G.; Borderies, P.; Jarlan, L. Spaceborne altimetry and scatterometry backscattering signatures at C- and Ku-bands over West Africa. *Remote Sens. Environ.* **2015**, *159*, 117–133, doi:10.1016/j.rse.2014.12.005.

55. Baup, F.; Mougin, E.; de Rosnay, P.; Hiernaux, P.; Frappart, F.; Frison, P.L.; Zribi, M.; Viarre, J. Mapping surface soil moisture over the Gourma mesoscale site (Mali) by using ENVISAT ASAR data. *Hydrol. Earth Syst. Sci.* **2011**, *15*, 603–616, doi:10.5194/hess-15-603-2011.

56. Gruhier, C.; de Rosnay, P.; Hasenauer, S.; Holmes, T.; de Jeu, R.; Kerr, Y.; Mougin, E.; Njoku, E.; Timouk, F.; Wagner, W.; et al. Soil moisture active and passive microwave products: Intercomparison and evaluation over a Sahelian site. *Hydrol. Earth Syst. Sci.* **2010**, *14*, 141–156, doi:10.5194/hess-15-603-2011.

57. Frappart, F.; Fatras, C.; Mougin, E.; Marieu, V.; Diepkilé, A.; Blarel, F.; Borderies, P. Radar altimetry backscattering signatures at Ka, Ku, C, and S bands over West Africa. *Phys. Chem. Earth Parts A/B/C* **2015**, *83*, 96–110, doi:10.1016/j.pce.2015.05.001.

58. Kerr, Y.H.; Waldteufel, P.; Richaume, P.; Wigneron, J.P.; Ferrazzoli, P.; Mahmoodi, A.; Al Bitar, A.; Cabot, F.; Gruhier, C.; Juglea, S.E.; et al. The SMOS soil moisture retrieval algorithm. *IEEE Trans. Geosci. Remote Sens.* **2012**, *50*, 1384–1403, doi:10.1109/TGRS.2012.2184548.

remote sensing

MDPI

Article

The Benefits of the Ka-Band as Evidenced from the SARAL/AltiKa Altimetric Mission: Scientific Applications

Jacques Verron [1,*], Pascal Bonnefond [2], Lofti Aouf [3], Florence Birol [4], Suchandra A. Bhowmick [5], Stéphane Calmant [4], Taina Conchy [6], Jean-François Crétaux [4], Gérald Dibarboure [7], A. K. Dubey [5], Yannice Faugère [8], Kevin Guerreiro [4], P. K. Gupta [5], Mathieu Hamon [9], Fatma Jebri [4], Raj Kumar [5], Rosemary Morrow [4], Ananda Pascual [10], Marie-Isabelle Pujol [8], Elisabeth Rémy [9], Frédérique Rémy [4], Walter H. F. Smith [11], Jean Tournadre [12] and Oscar Vergara [4,8]

[1]   Institut des Géosciences de l'Environnement (IGE), CNRS, 38041 Grenoble, France
[2]   SYRTE, Observatoire de Paris, PSL Research University, CNRS, Sorbonne Universités, UPMC Univ. Paris 06, LNE, 75014 Paris, France; pascal.bonnefond@obspm.fr
[3]   Météo-France, 31057 Toulouse, France; lotfi.aouf@meteo.fr
[4]   Laboratoire d'Etudes en Géophysique et Océanographie Spatiales (LEGOS), 31400 Toulouse, France; florence.birol@legos.obs-mip.fr (F.B.); stephane.calmant@ird.fr (S.C.); jean-francois.cretaux@legos.obs-mip.fr (J.-F.C.); guerrei@legos.obs-mip.fr (K.G.); fatma.jebry@gmail.com (F.J.); rosemary.morrow@legos.obs-mip.fr (R.M.); remy.omp@free.fr (F.R.); oscar.vergara@legos.obs-mip.fr (O.V.)
[5]   Space Applications Centre (ISRO), Ahmedabad 380015, India; suchandra@sac.isro.gov.in (S.A.B.); a_dubey@sac.isro.gov.in (A.K.D.); pkgupta@sac.isro.gov.in (P.K.G.); rajkumar_62@rediffmail.com (R.K.)
[6]   Universidade do Estado de Amazonas, Manaus 69020, Brazil; taina.conchy@inpa.gov.br
[7]   Centre National d'Etudes Spatiales (CNES), 31400 Toulouse, France; gerald.dibarboure@cnes.fr
[8]   Collecte Localisation Satellites (CLS), 31520 Ramonville Saint-Agne, France; yannice.faugere@cls.fr (Y.F.); mpujol@cls.fr (M.-I.P.)
[9]   Mercator Océan, 31520 Ramonville Saint-Agne, France; mathieu.hamon@mercator-ocean.fr (M.H.); elisabeth.remy@mercator-ocean.fr (E.R.)
[10]   Institut Mediterrani d'Estudis Avançats (IMEDEA) (CSIC-UIB), 07190 Esporles, Illes Balears, Spain; ananda.pascual@imedea.uib-csic.es
[11]   Laboratory for Satellite Altimetry, NOAA, College Park, MD 20740-3818, USA; walter.hf.smith@noaa.gov
[12]   Laboratoire d'Océanographie Physique et Spatiale (LOPS), 29280 Plouzané, France; jean.tournadre@ifremer.fr
*   Correspondence: jacques.verron@univ-grenoble-alpes.fr

Received: 29 November 2017; Accepted: 19 January 2018; Published: 24 January 2018

Abstract: The India–France SARAL/AltiKa mission is the first Ka-band altimetric mission dedicated primarily to oceanography. The mission objectives were firstly the observation of the oceanic mesoscales but also global and regional sea level monitoring, including the coastal zone, data assimilation, and operational oceanography. SARAL/AltiKa proved also to be a great opportunity for inland waters applications, for observing ice sheet or icebergs, as well as for geodetic investigations. The mission ended its nominal phase after three years in orbit and began a new phase (drifting orbit) in July 2016. The objective of this paper is to highlight some of the most remarkable achievements of the SARAL/AltiKa mission in terms of scientific applications. Compared to the standard Ku-band altimetry measurements, the Ka-band provides substantial improvements in terms of spatial resolution and data accuracy. We show here that this leads to remarkable advances in terms of observation of the mesoscale and coastal ocean, waves, river water levels, ice sheets, icebergs, fine scale bathymetry features as well as for the many related applications.

Keywords: altimetry; Ka-band; oceanography; hydrology; ice; geodesy

1. Introduction

The India–France SARAL/AltiKa mission is the first Ka-band altimetric mission dedicated to oceanography. SARAL was launched on 25 February 2013. Today, the SARAL/AltiKa mission is in its fifth year, which means that it has reached its design life. The mission will continue as long as the satellite and the ground segment continue to operate. The lifetime requirement of the AltiKa payload was just three years, which has clearly been exceeded. The instruments are still in perfect conditions.

The SARAL satellite is composed of a spacecraft bus developed by the Indian Space Agency (ISRO), and a payload developed by the French Space Agency (CNES). As its full name (SARAL stands for "Satellite with ARgos and ALtiKa") indicates, two missions are onboard the SARAL satellite: The AltiKa altimeter component and ARGOS-3, the new generation of ARGOS instrument. For the scientific applications discussed in this paper, we focus on AltiKa. The altimeter payload is composed of the AltiKa altimeter-radiometer in Ka-band, a Doris system for precise orbit determination (POD), and a Laser Retro-reflector Array instrument used for precise calibration of other POD instruments. AltiKa is a nadir-looking altimeter in line with Poseidon-3 on-board Jason-3 and other existing altimeters, but operates in a single frequency band: the Ka-band. It is the reduced ionosphere effects in Ka-band that makes it possible to use a mono-frequency altimeter. The altimeter shares the antenna with a bi-frequency radiometer required to correct the altimeter range for the wet troposphere path delay. The SARAL satellite orbit is almost polar, sunsynchronous and with a 35-day repeat cycle. This is the same orbit as Envisat with the initial motivation to continue the time series and to benefit from the existing mean sea surface. Since March 2015, technical issues have been encountered on the reaction wheels and this has led to the decision of relaxing the orbit. Then, after 4 July 2016, SARAL/AltiKa left its repetitive orbit by beginning a new phase named "SARAL Drifting Phase" (SARAL-DP). The data processing as well as data latency have not been changed. From this date, SARAL satellite does fly free of station keeping maneuvers which means there is a drift between successive ground tracks which are no more repetitive. The altimetry payload is operated nominally on the drifting orbit. The cycle duration of 35 days is no longer kept but the orbit has been chosen to preserve anyway subcycles within the range of 15 to 17 days quite relevant for mesoscale sampling.

For further information on SARAL/AltiKa, one can refer to introductory papers [1,2] and more widely to the Special Issue of Marine Geodesy dedicated to this satellite [3].

Since the beginning of the SARAL/AltiKa mission, performances proved to be compliant with nominal specifications with an overall observed performance for the sea surface height (SSH) RMS of 3.4 cm, lower than the mission requirement of 4 cm (see also [4]). In general, SARAL/AltiKa performances appear to be quite similar, and often better, than the Ku-band reference altimetric satellites such as Jason-2. Before launch, concerns were raised about the sensitivity of the Ka-band to rain events, leading to missing and invalid measurements. However, in practice, the SARAL/AltiKa data return is remarkably high with few missing data. Over the ocean, the data coverage is greater than 99.5%, which, once again, exceeds the mission requirements. SSH differences at crossovers is the main metric to assess the overall performance of satellite altimetry missions. In this regard, the results obtained from the SARAL/AltiKa mission are of the same order, or even better, as those obtained for Jason-2 mission. The same result is obtained for the sea level anomalies (SLA): SARAL/AltiKa data quality is as good (or even slightly better) as Jason-2. Another key factor for altimetry applications is the spectral content of the data. AltiKa altimeter provides a signal-to-noise ratio which has never been obtained before. Compared to Jason-2 and CryoSat-2 (SAR mode) data, the spectral content of SARAL/AltiKa observations is largely improved for all wavelengths below 70 km. This is explained by the much lower white noise, due to the excellent Ka-band retracking performances and also to the 40 Hz measurement rate (higher than the standard 20 Hz on Jason-class and CryoSat-2 missions). In conclusion, the mission performances widely confirm the nominal expectations in terms of accuracy, data quality and data availability.

The mission objectives are the observation of the oceanic mesoscale dynamics, including the coastal zone, but also the global and regional sea level monitoring, data assimilation and operational oceanography. It also includes the monitoring of the level of the main continental water bodies (lakes, rivers, enclosed seas), of large-scale sea level variations, the observation of polar oceans (as a result of the high inclination of its orbit), the analysis and forecast of wave and wind fields, the study of continental ice and sea ice, the access to low rains climatology and the marine biogeochemistry (mostly through improved mesoscale access).

Beyond the excellent quality of SARAL/AltiKa measurements, there are some specificities that emerge from the Ka-band altimetry. Different scientific results illustrate these specificities. The objectives of this paper are to give precise evidence of the main features of the SARAL/AltiKa that are associated with the Ka-band. In a companion paper [4], we looked at the quality assessment and specificities of AltiKa data. In the present paper, we focus more on the scientific aspects. Rather naturally, the paper is organized following the main scientific applications of SARAL/AltiKa: oceanography, hydrology, ice and geodesy. Some examples are given to illustrate some salient features of today's SARAL/AltiKa data with regard to standard altimetry: data accuracy allowing significant advances in the field of ocean mesoscale studies and for operational applications, in coastal areas, over ice, for inland waters and finally for geodesy.

2. The Ocean

As previously mentioned, SARAL/AltiKa's main scientific objective is to provide data products to the oceanographic community with the aim to improve our knowledge of the ocean mesoscale variability, mainly associated to eddies, meandering currents, fronts, filaments and squirts. The mesoscale variability refers to ocean signals with space scales of 50 km to 500 km and time scales from a few days to a few months. The associated kinetic energy is at least an order of magnitude larger than that of the mean circulation. It is now well recognized that the mesoscale processes are a key component of the ocean circulation, shaping the mean currents, controlling the exchanges with the atmosphere and the marine biogeochemistry, and clearly playing a major role in the way the ocean participate to the climatic system.

The sea level measurements of SARAL/AltiKa are used in theoretical and observational studies, modelling, data assimilation, etc. Because mesoscale and fine scale dynamical processes are particularly important in the coastal ocean, the observational capabilities of SARAL/AltiKa are also expected to bring an important contribution to the study of coastal dynamics. It is particularly important for many downstream applications including operational oceanography which is fed by a large amounts of in-situ and space-based data. Similar to previous altimeters, SARAL/AltiKa not only provide sea level measurements, but also ocean wave observations through the significant wave height (SWH) parameter. Ocean waves play a key role in ocean–atmosphere exchanges at the sea interface. It is essential to have the most accurate knowledge possible of the sea state for climate and meteorological studies.

2.1. Observability of the Fine-Scale Ocean Dynamics

Technological advances in all of the recent satellite altimeter missions have improved their signal-to-noise ratio, allowing us to observe finer-scale ocean processes with along-track altimeter data. By calculating wavenumber spectra from along-track SSH anomalies from the recent missions, we have been able to estimate the background noise levels of each mission, and the corresponding observable ocean scales, where the signal-to-noise is greater than 1. This statistical estimate of the altimetric noise and the observable scales has been performed over the global oceans using Jason-2 (Ku-band), SARAL/AltiKa (Ka-band) and CryoSat-2 in low-resolution mode (LRM) by [5], and from Jason-2, SARAL/AltiKa and Sentinel-3 in Synthetic Aperture Radar (SAR) mode by [6]. Regional analyses have also been performed over the western Mediterranean Sea by [7] using Jason-2, SARAL/AltiKa and CryoSat-2 SAR mode.

An example of the globally-averaged sea level anomaly (SLA) wavenumber spectra are shown in Figure 1 from three missions with different technologies : Jason-2, SARAL/AltiKa and Sentinel-3 in SAR mode [6]. All missions observe similar SLA spectral levels at scales larger than 100 km wavelength.

At shorter scales, the SLA spectra are dominated by the error level of each missions. Jason-2 has a higher mean noise level with a strong "bump" from 10 to 50 km wavelength due to surface roughness inhomogeneities within its relatively large footprint, mainly from surface waves. SARAL/AltiKa has the smallest noise at very short scales, but is also affected by the spectral "bump" at scales of 3–20 km wavelength (note that standard data processing is applied here; with more sophisticated treatment, these spectral bumps can be reduced without changing the hierarchy of satellite data quality for small ocean scales [4]). Sentinel-3 in SAR mode has a significant reduction in the spectral bump and the smallest error level at 10 km wavelength, with a different "red" noise spectrum. Both SARAL/AltiKa and Sentinel-3 SAR data observe similar SSH variability down to scales of 40 km wavelength.

Each altimeter's error field shows distinct seasonal and geographical variations, with higher noise in winter due to the rougher sea state [5]. We have recently extended this work to analyse SARAL/AltiKa over a longer three-year period when both missions were on a repeat groundtrack with low mean sea surface errors. SARAL/AltiKa has relatively low 1 Hz noise, but still exhibits strong seasonal variations in noise (Figure 2). The 1 Hz noise is higher in each hemisphere in the region of highest winter waves, but also increases in the boundary currents due to wave-current interactions, and in the tropics due to rain cells or calm water patches, all leading to inhomogeneities within the footprint. In the western Mediterranean Sea, similar seasonal patterns were found [7] (not shown).

These statistical noise levels impact on the ocean scales we can observe. In the global ocean, SARAL/AltiKa can observe scales of 30 km in the strong boundary currents and the tropics, reaching 70–80 km in the quieter eastern basin where the signal-to-noise ratio is low [5]. Conversely, the higher noise of Jason-2 limits the observation scales: we can only resolve processes of 40–50 km in the western boundary currents, and 80–100 km in the eastern basins. Note that these spectral wavelength limits correspond to individual feature diameters of half that size, i.e., with SARAL/AltiKa we can detect strong western boundary eddies down to 15 km diameter (30 km in wavelength). In the Mediterranean Sea, where the Rossby radius is small, the difference between SARAL/AltiKa and Jason-class altimeters is also striking [7]. In winter, when the mixed layers are deepest and the submesoscale is energetic, all of the altimeter missions can observe wavelengths down to 40–50 km. In summer, when the submesoscales are weaker, SARAL/AltiKa with its lower noise can detect ocean scales down to 35 km wavelength, whereas the higher noise from Jason-2 blocks the observation of scales less than 50–55 km wavelength.

The noise levels impacting on the observation of fine-scale dynamics depend on many factors, including surface roughness from different wind–wave–swell conditions, impact from rain or other geophysical corrections [5,8], and errors in the geoid or mean sea surface (MSS). Even with the new generation of MSS products, when a satellite drifts off the long-term mean repeat track with its accurate MSS, the error in SLA increases. Figure 3 shows the evolution of the SLA standard deviation, calculated with respect to different MSS products, during a period when the SARAL/AltiKa mission drifted from its long-term mean groundtrack (cycles 23–25, marked in gray, from [9]). The data have been filtered to retain only short wavelengths less than 200 km, which have the largest gridded MSS errors. The standard deviation for Jason-2 (in black) is calculated from the precise alongtrack MSS, whereas for SARAL/AltiKa it is calculated with respect to different MSS models. Figure 3 highlights the lower standard deviation of SARAL/AltiKa SLA compared to Jason, but even when the recent MSS models are used with the most geodetic data included (MSS CNES_CLS15), the error increases during the drifting orbit phase. Work is in progress to better understand and estimate these different geophysical signals, in order to reduce their effects on the SLA noise.

Although SARAL/AltiKa and Jason-2 show a similar pattern in seasonal and geographical variations of their noise, these studies quantify that the low background noise levels of SARAL/AltiKa enable this mission to make the most precise estimates of the fine-scale ocean dynamics, in the open ocean and regional and coastal seas. This is encouraging for further studies with SARAL/AltiKa , but also in the preparation for the future Ka-band wide-swath altimeter mission from NASA/CNES, the Surface Water Ocean Topography (SWOT) mission, whose lower noise should enable us to observed even smaller ocean scales.

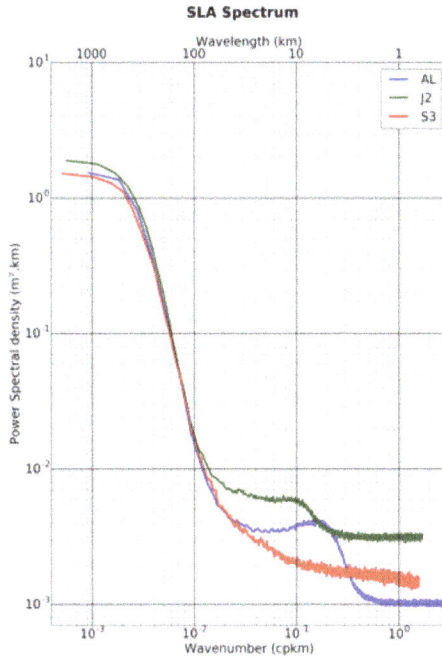

Figure 1. Globally-averaged wavenumber spectra of SSH anomalies (in m^2/cpkm) from the highest resolution along-track data for three different missions: Jason-2 20 Hz (J2 in green); SARAL/AltiKa 40 Hz (AL in blue); and Sentinel-3 20 Hz SAR (S3 in red). The spectral "bump" in Jason-2 and SARAL/AltiKa data is due to surface roughness inhomogeneities within the footprint, mainly from surface waves.

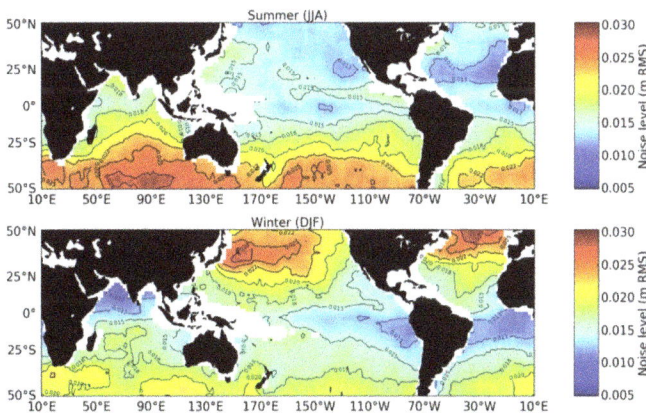

Figure 2. Geographical distribution of the 1 Hz noise level over the global oceans in: (**upper panel**) June–August; and (**lower panel**) December–February. The noise is estimated from wavenumber spectra averaged over $15° \times 15°$ boxes, as a constant value fit over the wavelengths 14–30 km.

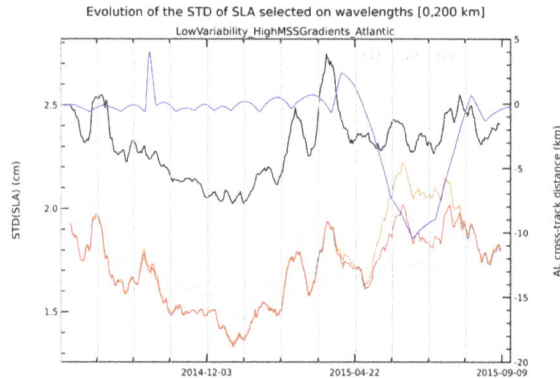

Figure 3. Temporal evolution of the standard deviation of the SLA, along Jason-2 tracks (calculated with respect to its alongtrack precise MSS in black) and along SARAL/AltiKa tracks (calculated with respect to different gridded MSS products: MSS CNES_CLS15 (in red), CNES_CLS11 (in orange) and DTU15 (in cyan). The SLA was high-pass filtered to keep only wavelengths shorter than 200 km. Statistics are computed over a low variability and intense bathymetric gradients region in the northeast Atlantic [20°S, 0°N; 30°W, 0°E]. The blue line shows the cross-track distance (in km; right axis) of the SARAL/AltiKa tracks compared to their long-term repeat groundtrack; this distance increases sharply during cycles C23, C24, and C25, marked in gray from 22 April 2015 onwards.

2.2. Observability of the Coastal Ocean Dynamics

Satellite altimetry represents a mature technology in the open ocean but its exploitation remains an issue in the coastal ocean for different reasons. Land contamination in the altimeter footprint lead to a signal which is much more complex to analyze than offshore and often lead to the rejection of the altimetry measurements during the standard processing step. Geophysical corrections are significantly less accurate in the coastal area. The radiometer measurements used to derive the wet-troposphere correction becomes perturbed by land 25–50 km from the coast (depending on the instrument). The coastal ocean dynamics have also shorter space and time scales than the scales of open ocean dynamics which are mostly resolved by altimetry. Consequently, coastal altimetry data are largely discarded in standard products distributed to users and then largely unexploited. However, having as many accurate observations of the coastal ocean as possible is crucial for improving our understanding and monitoring of physical ocean processes that impact a wide range of applications (coastal area management, risk prevention, fisheries, marine safety and shipping, pollution, etc). In the context of climate change, we need to observe and understand how the coastal sea level variations, in particular extreme events such as storm surges, but also those associated to much lower time scales, impact the coastal morphology (via erosion). In the last years, a lot of efforts have been dedicated in the coastal altimetry community in order to extend the observational capabilities of altimeters always closer to the coastline (e.g., http://www.coastalal.eu). The launch of the SARAL/AltiKa altimetry mission, with an instrument able to provide accurate sea level measurements at a higher spatial resolution and closer to the coastline, is a major step forward.

2.2.1. Observability of the Coastal OceanDynamics in the Central Mediterranean Sea

In a first illustration of the benefits of SARAL/AltiKa , we analyse if the improved altimetric capabilities of SARAL/AltiKa that has been shown in different studies (e.g., [5,7]) can lead to the detection of finer ocean scales in the Central Mediterranean Sea. In this area, the observation of the different current branches by altimetry is particularly challenging due to their short width and high space-time variability [10]. Figure 4 (left) shows an example of the wavenumber spectra calculated over the study area from both Jason-2 and SARAL/AltiKa along-track SSH. This very local result obtained from a data set including coastal altimetry observations (unlike the studies mentioned above) confirms what has already been obtained elsewhere. At wavelengths >80 km, SARAL/AltiKa and Jason-2 SLA spectra are very similar. Below 80 km, the spectral energy of SARAL/AltiKa decreases significantly more than for Jason-2. Using the same spectral analysis approach than [5,7] for different altimetry tracks crossing the Central Mediterranean Sea, we deduce that the mesoscale observational limit occurs around 46 km wavelength (for a 12–35 km estimated noise level) for Jason-2 and around 21 km wavelengths (for a 12–18 km estimated noise level) for SARAL/AltiKa. These numbers are slightly lower than the results obtained by [7] further north in the northwestern Mediterranean Sea, confirming the strong dependency of the observational capability of altimetry missions to the level of mesoscale energy. These results have been used to define for both altimetry missions the optimal spatial filtering of alongtrack SLA: a low-pass Lanczos filter with a 21/45 km cutoff frequency for SARAL/Jason-2.

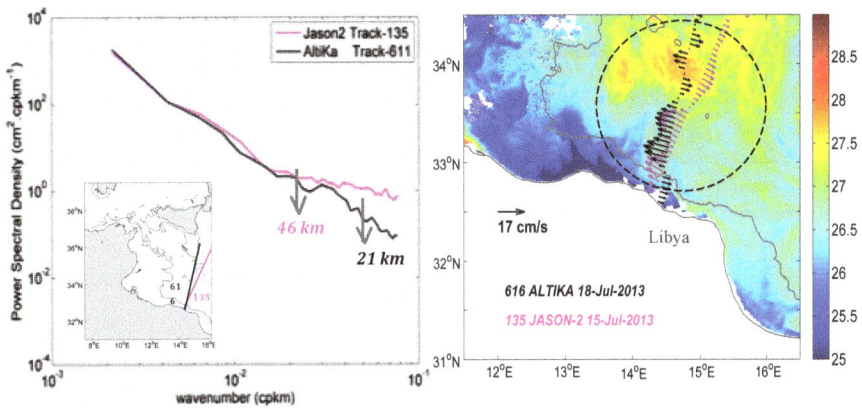

Figure 4. (**Left**) Wavenumber spectra from SSH observed in the Central Mediterranean Sea along the track 616 of SARAL/AltiKa (black) and along the track 135 of Jason-2 (pink); and (**Right**) comparisons of SST (in °C) with SARAL/AltiKa track 616 (black) and Jason-2 track 135 (pink) crosstrack velocities over the Tunisia-Lybian shelf for 18 July 2013. The SST pattern is derived from CMEMS products with 1/16° resolution. The 200 m isobath (in grey) is derived from the ETOPO2v1 database. The black dashed circle indicates the location of the Sidra Gyre.

To illustrate the impact of the improved observational capability of SARAL/AltiKa in terms of ocean circulation captured, the crosstrack absolute geostrophic velocities are computed from both the filtered SARAL/AltiKa and Jason-2 sea level data, adding the mean dynamic topography provided by AVISO to the SLA. They have then been compared to the information provided by high resolution Level 3 sea surface temperature (SST) 2D fields. Figure 4 (right) illustrates an example of this qualitative analysis. This zoom on the Tunisian-Lybian shelf for 18 July 2013 shows the presence of warmer waters north of the Lybian continental margin. From [10,11], it corresponds to the anticyclonic Sidra Gyre (black dashed circle), permanent feature of the regional circulation which advects warm and salty Ionian water through a surface current flowing northwestward offshore Libya. This feature is observed in both SARAL/AltiKa and Jason-2 current patterns but finer scales are derived from SARAL/AltiKa data and the undulations in the corresponding higher resolution surface currents perfectly correspond to the variations in the SST field.

Near the coast, on the Lybian shelf, the information provided by Jason-2 and SARAL/AltiKa diverge. In Jason-2, the surface currents still flows northwestward, while a southeastward current vein is observed in SARAL/AltiKa. For this date, SST data indicate the presence of colder waters flowing along the Libyan coast. They correspond to the AtlanticWaters advected Southeastward by the narrow Atlantic Lybian Current along the Tunisian-Libyan continental slope (see [10,11]). This very coastal current vein is captured by SARAL/AltiKa observations but not by Jason-2. This example shows how the better mesoscale resolution of SARAL/AltiKa allow to observe new signals, even near the coast, and then help to understand the fine-scale ocean dynamics.

2.2.2. SARAL-AltiKa Capabilities to Detect Coastal Currents: Comparisons With Jason-2 and HF Radar Data

In a second example, we refer to measurements of the surface velocity from radar. Among the various alternative sensors available, coastal High Frequency Radar is unique as it provides synoptic, high frequency and high resolution data at the boundary between the ocean and the atmosphere [12]. Here, we build up on a previous study [13] aimed at evaluating SARAL/AltiKa data in the coastal ocean (7–60 km from the shore). In that study, the altimeter performance was evaluated in the Ibiza Channel (Western Mediterranean) by estimating geostrophic currents and comparing them to high frequency (HF) radar surface velocities. Results showed that SARAL/AltiKa data were retrieved at a distance of only 7 km from the coast. The derived velocities revealed coherent mesoscale features with high temporal variability among the different cycles and with general reasonable agreement with HF radar fields (correlations of 0.54, which are significant at the 95% confidence level). In the present example, we also perform the comparison with Jason-2, whose track 187 crosses SARAL/AltiKa track 16 in the vicinity of Ibiza Island (Figure 5). The period of study is 18 April 2013 to 8 May 2014. The same processing as in [13] is applied for HF radar and as well as for both SARAL/AltiKa and Jason-2 missions. Surface geostrophic velocities from altimetry are computed by finite differences applied to the Absolute Dynamic Topography (ADT), which is obtained by adding SLA and the SMDT-MED2014 Mean Dynamic Topography [14]).

We then compare the mean across-track velocities for every satellite cycle with the equivalent interpolated HF radar vectors (3 days averages). As expected (Figure 5c,d) and due to the different repetitive cycle of each mission, the number of cycles available during the study period for SARAL/AltiKa is a factor of 3.5 less than for Jason-2. SARAL and HF radar mean velocities reveal a consistent pattern raising to positive significant correlations of 0.74 (Table 1). On the contrary, in the section sampled by Jason-2 (which is less than 10 km apart from SARAL/AltiKa), the time series are much noisier (both for radar and satellite) without a clear seasonal pattern. This translates in an insignificant correlation between Jason-2 and HF radar observations. However, RMS differences between satellite and HF radar velocities are of the same order (9 cm/s) for both missions. A more in depth investigation is needed to interpret and understand these results. Further works should span over a longer time period and interpret and extend the spatial coverage to an array of HF radars, taking advantage of the recent initiative to establish a pan-European HF radar network [12].

Table 1. Statistics of the comparisons between HF radar and altimetric across-track geostrophic velocities over the period comprised between 18 April 2013 and 8 May 2014. Twelve measurements are used for SARAL/AltiKa (repeat cycle of 35 days) and 36 measurements are considered for Jason-2 (repeat cycle of 9.9156 days; note that several cycles were missing for Jason-2).

	SARAL-AltiKa	Jason-2
Correlation	0.7382	-0.2481
Std (HF radar) (m/s)	0.0695	0.0518
Std (satellite) (m/s)	0.1055	0.0622
Rms Diff (HF radar − satellite)(m/s)	0.0942	0.0969

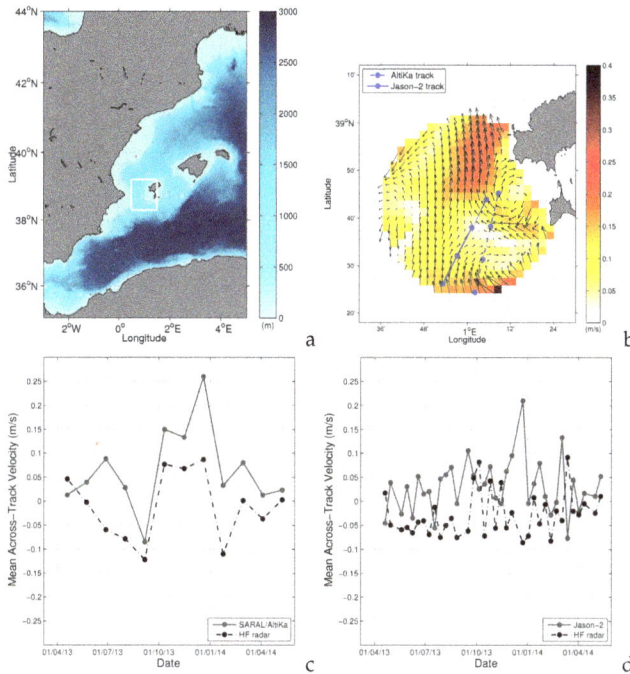

Figure 5. (**a**) Area of study in the Western Mediterranean. Colour contours represent bottom bathymetry and the white box corresponds to the area covered by the Ibiza HF radar shown in (**b**). (**b**) Coverage of SOCIB HF radar in the Ibiza Channel. The vectors correspond to surface velocities derived from HF radar (17 April 2013) field (three-day average). The colours are currents magnitude (in m/s). SARAL/AltiKa track 16 and Jason-2 track 187 are overlaid. The dots correspond to the position of the geostrophic velocity vectors derived from the altimeter missions. Note that the along-track delayed-time data provided at CMEMS is filtered with a 42 km filter and sub-sampled every 14 km (see details in [13]). (**c**) Comparison of the temporal variability of SARAL/AltiKa and HF radar mean cross-track velocity. X-axis is the number of cycles. (**d**) The same as (**c**) but for Jason-2.

2.3. SARAL/AltiK a Data Into Operational Systems

Altimetry is one of the most important data sources for assimilation into operational models. It provides global, real time, all-weather SSH measurements with high space and time resolution. Sea level is directly related to ocean circulation through the geostrophic approximation but sea level is also an integral of the ocean interior and is a strong constraint for inferring the 4D ocean circulation through data assimilation. Altimeters also measure SWH, which is essential for operational wave forecasting. High resolution from multiple altimeters is required to adequately represent ocean eddies and associated currents in models. Only altimetry can constrain the 4D mesoscale circulation in ocean models which is required for most operational oceanography applications.

There are very strong links between satellite oceanography and operational oceanography. The development of operational oceanography has been mainly driven by the development of satellite oceanography capabilities. The ability to observe the global ocean in near real time at high space and time resolution is indeed a prerequisite to the development of global operational oceanography and its applications. The first ocean parameter to be globally monitored from space was the SST on board meteorological satellites in the late 1970s. It is, however, the advent of satellite altimetry in the late 1980s that led the development of ocean data assimilation and global operational oceanography.

The satellite altimetry community was also keen to develop further the use of altimetry and this required an integrated approach merging satellite and in-situ observations with models.

2.3.1. SARAL/AltiK a Major Contributor to DUACS/CMEMS Multi-Mission Seal Level Products

For more than four years, SARAL/Altika products over ocean have been an important source of information for operational oceanography systems both in terms of assimilation and validation. Thanks to the efficiency of CNES and ISRO, the DUACS Sea Level system [15] was able to ingest SARAL/AltiKa data from 1 July 2013 onward in NRT, and merged them with the other flying satellites Jason-2, CryoSat-2 and HY-2A, to produce homogeneous set of Level 3 (along-track cross-calibrated SLA) and Level 4 (L4) products (multiple sensors merged as maps or time series).

Figure 6 shows the relative contribution of each mission in the multi-altimeter L4 products. This contribution (in %) is derived from a degrees of freedom of signal analysis described by [8]. During the first month, SARAL/AltiKa strongly contributes up to 40%, allowing us to mitigate the loss of Jason-1 few months earlier. Its contribution decreases by ~10% after the integration of HY-2A but rose again after HY-2A is removed of the system due to quality issues. During its whole repetitive period, SARAL/AltiKa was the second contributor, just after Jason-2 reference mission of the DUACS system because of an unexpected availability over the data over ocean as well as a very good timeliness of the OGDR and IGDR. With such a timeliness, and taking advantage of the "on the fly" DUACS system improvement implemented in February 2013, operational oceanography centres have been able since the beginning of the mission to access to homogeneous and cross calibrated DUACS SARAL/AltiKa products within a few hours. Another reason of this high contribution is the low noise level of SARAL/AltiKa enabled by its new Ka-band technology associated with its small footprint. Indeed, the energy removed from SARAL/AltiKa data by the DUACS filtering process is lower compared to what is removed from Jason-2 measurements by up to 20%. The good operational performances and data quality, routinely assessed by Calval teams, has allowed us to use SARAL/AltiKa in the fully operational Copernicus Marine Environment and Monitoring Service (CMEMS) from May 2015 onwards. The L3 and L4 Sea Level products are available on the CMEMS portal (http://marine.copernicus.eu/) for the whole SARAL/AltiKa period (2013–2017).

Regional CMEMS products (Mediterranean Sea, Arctic, European Shelves, etc.) particularly benefit from SARAL/AltiKa inputs. The availability of data above 82° latitude allowed us to increase the density of observation in the Arctic area, in complement to CryoSat-2. More measures have also been available near the coast. It is highlighted in Figure 7, showing the increase of measurement use to compute an Envisat/SARAL/AltiKa mean profile compared to an ERS-2/Envisat mean profile, both profiles having been estimated using the same period length, 14 years. One can notice the data gain in the coastal areas such as the Danish straits, where the availability of the previous missions was poor. The coming reprocessed data DUACS DT2018 (CMEMS V4) will benefit from this new mean profile, improving not only the products during the SARAL/AltiKa period, but also the full ERS-1/ERS-2/Envisat period through the use of this new mean profile allowing a notably better description of the coastal mesoscale.

After the end of the repetitive mission in July 2016, the degradation of the data quality due to the lack of a precise MSS was strongly mitigated by the use of MSS_CNESCLS2015 [9], and the contribution of SARAL/AltiKa in the DUACS system remain significant, though now lower than the new Jason-3 and Sentinel-3A missions. However, the data now produced with this new orbit become even still more interesting for the MSS studies as the sea level is now measured over uncharted tracks. Next versions of MSS will benefits from these dataset, notably at 10–50 km wavelength all over the ocean and on the coastal areas. This will eventually lower the error budget of the Sentinel-3-A and B mission relying on these MSS, still reinforcing the crucial contribution of SARAL/AltiKa to the Sea Level products.

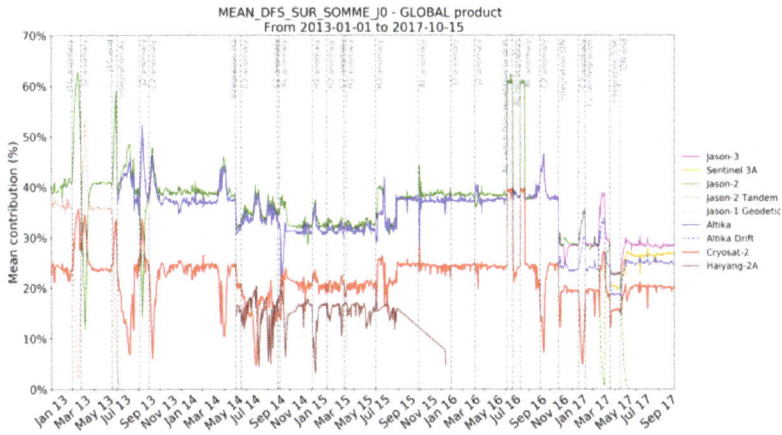

Figure 6. Relative contribution in percentage of each altimeter in the multi-mission NRT DUACS maps during the 2013–2017 period and link with altimeter events. The contribution is derived from the degrees of freedom of signal. The SARAL/AltiKa contribution is in blue, the plain curve corresponding to the repetitive period and the dotted curve to the drifting period.

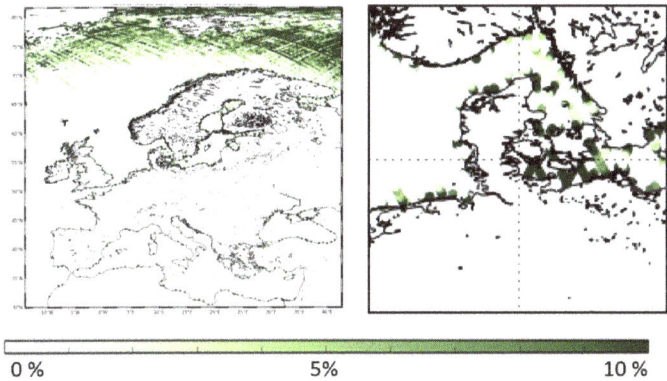

Figure 7. Relative increase in percentage of the number of point used to compute an Envisat/SARAL/AltiKa mean profile relatively to the previous a ERS-2/Envisat mean profile over: European Seas (**left**); and zooming over the Danish Straits (**right**).

2.3.2. Assimilation of SLA Data in the Mercator Ocean System

SARAL/AltiKa 35-day repeat orbit altimeter is well suited for the monitoring of mesoscale variability and to complement Jason-series observations (e.g., [16]). Its low noise level due to the use of Ka-band also allows a better resolution of along-track mesoscale signals [5]. SARAL/AltiKa thus provides essential altimeter observations for global ocean analysis and forecasting. The along track SARAL/AltiKa SLA observations has been successfully assimilated in the Mercator Ocean analysis and forecasting systems since 31 July 2013, only five months after the satellite launch. We analyse here its specific contribution in addition to other altimeters for global ocean analysis and forecasts. The new Copernicus global ocean system at 1/12° benefits from important improvements that lead to significant reduction of the analysis and forecast SLA errors. The observation errors for SST and SLA are now dynamically adjusted using an approach proposed by [17]. In 2015, SARAL/Altika diagnosed SLA observation errors have a global mean RMS of 3.6 cm, comparable to the Jason-2 error but lower than the one for HY-2A and CryoSat-2. This error estimate includes the mean dynamic topography, model representativity and processing error. The lower values are around 2 cm in the centre of the gyres but can reach more than 10 cm in high energetic regions, such as western boundary currents and the Antarctic circumpolar current.

Observing System Evaluation (OSE) experiments, covering the year 2015, have been carried out with this new system to assess the impact of an increase number of altimeters. Figure 8 shows the relative reduction of the SSH forecast Mean Square Error (MSE) due to the assimilation of SARAL/AltiKa ; Cryosat-2, Jason-2 and HY-2A being already assimilated. The impact is significant and highly regionally dependent. Innovation (observation minus model forecast) statistics are computed with Cryosat-2, Jason-2 and HY-2A data. The normalized SLA error difference between the experiment without and with SARAL/AltiKa SLA assimilated reaches 40% at mid-latitudes, whereas, in the same OSE but with the global 1/4° system, it reaches only 25% (not shown). This could be interpreted as a need of a higher density of observation to constrain the smaller scales of the 1/12° ocean model compared to the 1/4° system. The assimilation of SARAL/AltiKa SLA allows a gain of 0.8 day to get the same level of forecast error but without SARAL/AltiKa. The comparison of the estimated surface velocity fields in the different OSEs with subsurface drifters also shows an improvement with the addition of SARAL/AltiKa, even if three altimeter data sets have already been assimilated.

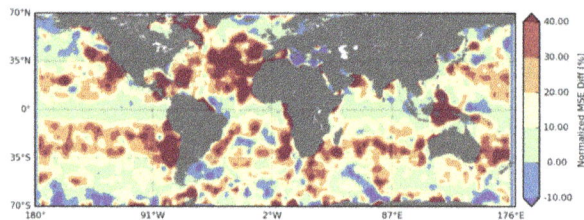

Figure 8. Normalized Mean Square forecast Error difference in percentage between the experiments assimilating three and four altimeter data sets under CryoSat-2, Jason-2 and HY-2A tracks with the 1/12° global system: (innov 3 alt)^2 − (innov 4 alt)^2 / (innov 3 alt)^2 − SARAL/AltiKa being the fourth altimeter.

2.3.3. Assimilation of SWH Data Into the Météo-France Operational Weather Forecast System

SARAL/AltiKa provides a very accurate SWH to improve the operational wave forecasting of Météo-France. During the last three years of operational use of SARAL/AltiKa, it has been demonstrated that the operational wave forecast is significantly improved for the global and regional scales. In addition, SARAL/AltiKa plays an important role in the correction of sea state in coastal zones. Figure 9 shows a snapshot of SWH from the operational Météo-France wave model (MFWAM) of a resolution of 10 km and dedicated to European seas during the storm happened in March 2016. The MFWAM model is routinely using SARAL/AltiKa, Jason-2 and CryoSat-2 altimeters wave data. The Jason-2 track overlaid on the snapshot indicates the accurate forecast at the peak of the event on 27 March 2016 at 21:00 UTC. The SARAL/AltiKa wave data were also used efficiently in the validation of high resolution coastal wave model of 200 m of resolution implemented at Météo-France. Figure 10 indicates a snapshot of SWH during the the storm of the 8 February 2016 which generated high waves in the English channel near the French coasts at Cherbourg. The SARAL/AltiKa track passing over the English channel and close to the French coast reveals the accurate forecast of the coastal model. Moreover with SARAL/AltiKa the assimilation of small wave heights (less than 1 m) in regional models MFWAM has been significantly improved for closed seas and lakes.

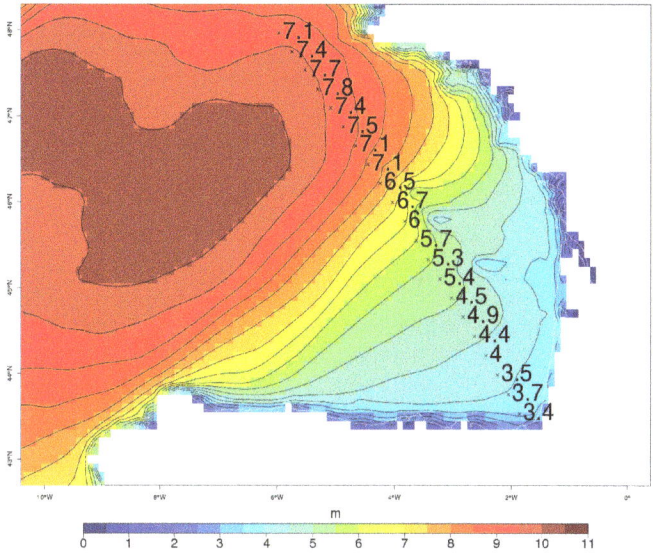

Figure 9. SWH (in m) from the regional wave model MFWAM dedicated to European seas on 27 March 2016 at 21:00 (UTC). The model MFWAM is driven by wind forcing from the atmospheric model of Météo-France ARPEGE (Action de Recherche pour Petite Echelle Grande Echelle) with a resolution of 10 km. The Jason-2 track in the gulf of Biscay indicates SWH (in m) during the peak of the event. SWH of more than 9 m was recorded off shore of the Brittany coast.

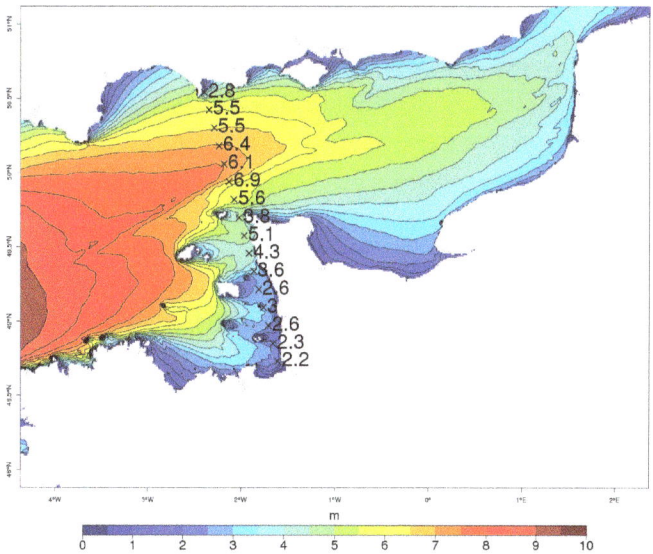

Figure 10. SWH (in m) in the English channel from the operational coastal model WaveWatch3 (WW3) with unstructured grid of a resolution of 200 m during the event of 8 February 2016 at 06:00 UTC. The model is driven by the wind forcing from the model ARPEGE. The SARAL/AltiKa track is overlaid and indicates SWH (in m) at the peak of the event near the coast.

2.4. Accuracy of SWH Data

The Comparative Analysis of Ku and Ka-Band Altimeters for Wave Observations

The measurements from both Ku- and Ka-band altimeters have been available concurrently. This case presented here aims to evaluate the suitability of both frequencies for wave and inland water studies. For this study the SARAL/AltiKa and Jason-2 data for year 2013 has been used. NOAA National Data Buoy Center (NDBC) buoy data have been used extensively to validate both SARAL/AltiKa and Jason-2 data individually for 2013. The SWH from both altimeters are collocated with NDBC buoy data at spatial and temporal intervals of 25 km and 30 min, respectively. Analysis has been carried out for various sea states and for the mid-latitude and tropical belts. The collocated points where difference of altimeter and observation is larger than three times the standard deviation of the difference are excluded from the analysis. The wave data for Cycle 4 from SARAL/AltiKa and Cycle 185 of Jason-2 have been analyzed over the Hudson's Bay area of Canada. This is the second largest bay in the world with average depth below 100 m and maximum depth of 270 m. The bathymetry of this region is shown in Figure 11a . Most of the Ku band altimeters does not show data availability at this shallow depth. Jason-2, for example, does not have data available over this region.

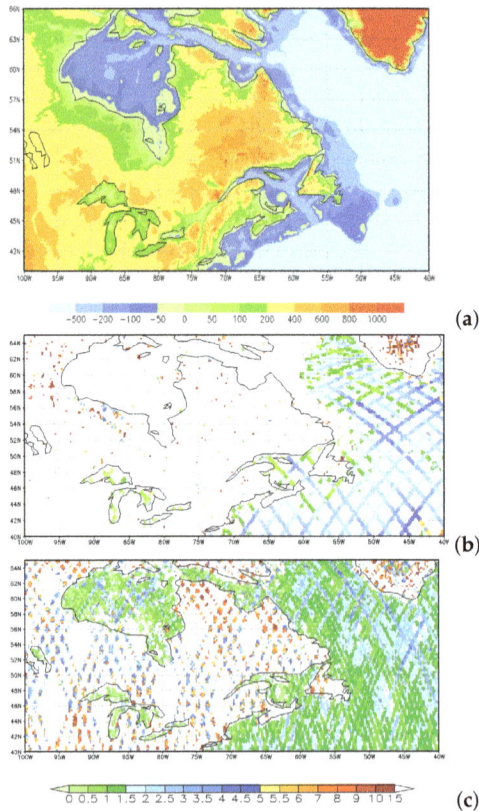

Figure 11. Availability of SWH data over very shallow depth from SARAL/Altika as compared to Jason-2. The example is Hudson Bay, with average depth of 100 m and maximum depth of 270 m: (**a**) bathymetry of Hudson Bay; (**b**) SWH from Jason-2; and (**c**) SWH from SARAL/AltiKa.

However, with emergence of SARAL/AltiKa, scientific fraternity is fortunate to get the valuable data from altimeter over this area. This opens the door to several applications for shallow waters including dams and reservoirs using Ka-band altimetry. Apart from being excellent in shallow water, Ka-band altimetry is known for its better approachability towards the coast. For example, over Indian subcontinent, the SWH data across a transect along 18°N has been shown in Figure 12 for Cycle 4 of SARAL/AltiKa and Cycle 185 of Jason-2. Very clearly, for the coast at about 83.5°E (marked by the yellow line), the SARAL/AltiKa has much better approachability, up to 84.2°E, as compared to Jason-2 (85.8°E). The comparison of SARAL/AltiKa and Jason-2 with NDBC buoys clearly shows that in calm and moderate sea conditions when SWH is between 0–2.5 m the performance of the SARAL/AltiKa is better than Jason-2 (Figure 13). The comparison of SWH with buoy observations for mid-latitudes and tropics clearly indicates that SARAL/AltiKa has excellent capabilities of observing waves for both the belts (Figure 14).

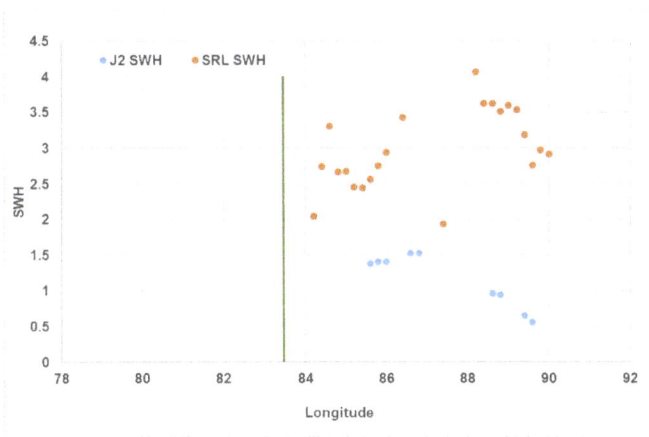

Figure 12. Availability of valid SWH data from SARAL/Altika and Jason-2 for transect along 18°N showing better available data near coast marked by green line.

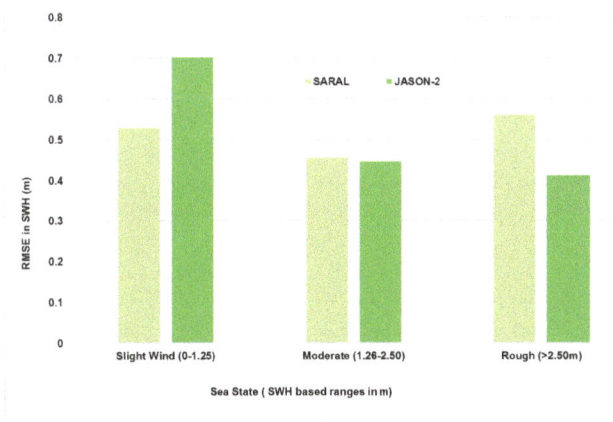

Figure 13. RMSE of SARAL/AltiKa and Jason-2 SWH with respect to buoy observed wave in various sea state categorized on basis of SWH.

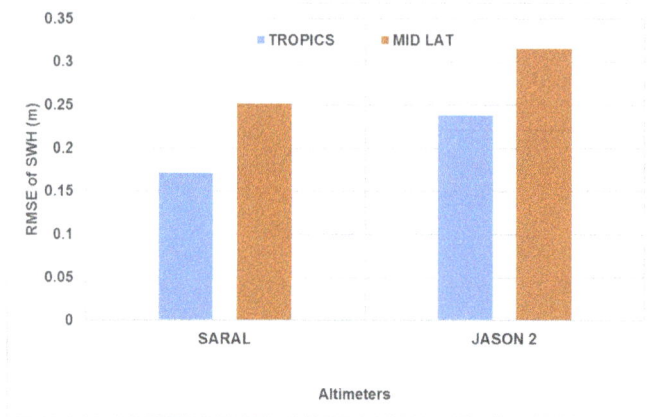

Figure 14. RMSE of SARAL/Jason-2 SWH with respect to buoy observed wave in mid-latitude (30–60°N) and Tropics (0–30°N).

3. Inland Waters

Worldwide, the number of operational ground stations for lake and river water levels and discharge monitoring has continuously decreased since the 1970s. Meanwhile, satellite capabilities for inland waters studies have emerged. In particular, many published studies have demonstrated that satellite altimetry allows measuring accurately water level along rivers and lakes [18,19]. For rivers, satellite altimetry has evolved from investigation of water height retrieval to the monitoring and discharge estimations. For lakes, it has evolved to the monitoring of water storage change [20]. Recent research focuses on integrating altimetry with other remote sensing techniques and hydraulic models to deal with key inland water resources issues such as flood (discharge, water spread and volume), water level monitoring (even in highly braided/fragmented river systems), rating curve generation for remote locations, reservoir operations, calibration of river/lake models, etc. Before it moves to a drifting orbit, SARAL/AltiKa contributes to meeting the growing demand of altimetry data for hydrological studies. Three examples of SARAL/Altika applications regarding rivers and lakes are provided in the following.

3.1. Rivers

3.1.1. Comparative Analysis of Water Level Retrieval Using Ku and Ka-Band Altimeters Over Brahmaputra River

Over inland water bodies, presence of sand bars, islands and vegetation contaminate radar return echo. Hence, with narrow footprint in Ka-band, altimeters enhance the possibilities of better data acquisition over the inland water bodies with multiple features. The decrease in time decorrelation between successive echoes has increased the pulse repetition frequency to 4000 Hz in Ka-band as compared to 2000 Hz with Ku-band altimeters. The limitation with higher frequency is that it has higher rainfall attenuation with increasing rainfall intensity, although total data loss due to this is less than 5%. In this study, river water levels were retrieved using Jason-2 and SARAL/AltiKa waveform datasets over Brahmaputra river for 2013. Three passes of Jason-2 (track number 242, 53 and 166) over the Brahmaputra river and nearby tracks of SARAL/AltiKa (track number 537, 810 and 438) were selected (see Figure 15). Time of pass of SARAL/AltiKa tracks and nearby pass data of Jason-2 were chosen. Beta parameter algorithm has been applied using the 20-Hz (Jason-2) and 40-Hz (AltiKa) waveform datasets to retrieve the river water levels with respect to mean sea level. A comparative analysis was carried out to assess the Ku-band vs. Ka-band altimetry performance for the river water level retrievals. Individual tracks were compared; it was found that SARAL/AltiKa performs slightly better for the upstream tracks, i.e., 537 vs. 242 and 810 vs. 53. For the downstream track where braiding pattern is high (track number 166, Jason-2 and 438, SARAL/AltiKa), Jason-2 performance decreased sharply (RMSE 1.19 m, Jason-2 vs. 0.41, SARAL/AltiKa) as compared to SARAL/AltiKa. This may be attributed due to higher band width (500 MHz) in Ka-band altimetry which provides shorter pulse duration; hence increase the vertical range resolution (0.3 m) as well as spatial resolution. Figure 16a,b presents the overall performance of the SARAL/AltiKa and Jason-2, respectively, taking into account all three tracks. The improved accuracy over the braided rivers highlight the high frequency Ka-band altimeter performance.

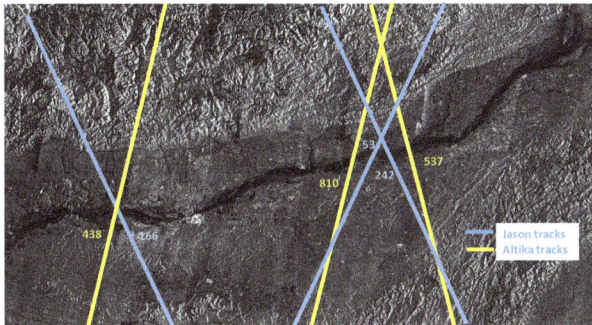

Figure 15. Brahmaputra river along with Jason-2 and SARAL/AltiKa tracks taken for the comparative analysis.

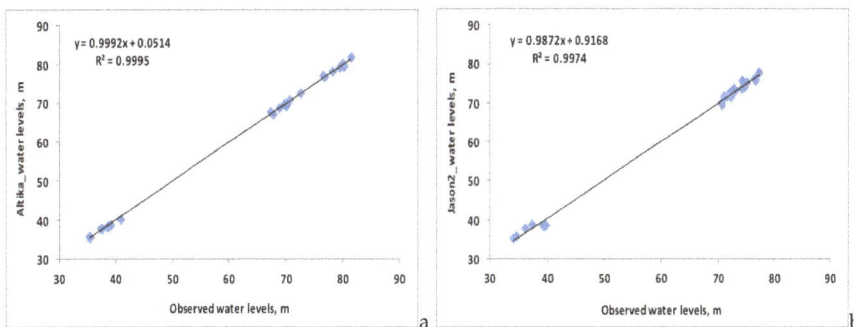

Figure 16. Overall relationship between observed versus retrieved water levels taking into account all study tracks: (**a**) SARAL/AltiKa ; and Jason-2 (**b**).

3.1.2. Discharge Estimates in the Congo Basin From Validated SARAL/AltiKa Measurements

The Congo basin, with an average discharge of 40,600 m^3/s and undergoing a 10% inter-annual variability [21,22], is the second largest hydrological basin on Earth. However, it is also one of the least monitored. Today, only a few tens of gauges still monitor the water level changes in the basin. In such a context, satellite altimetry provides a complement of information of highest interest [23]. A limited amount of works has been conducted to date to assess the quality of the SARAL/AltiKa data over rivers [24–26]. In the present study, we use the ranges issued at 40 Hz by the ICE1 retracking algorithm and applied the corrections for propagation in the atmosphere provided by global models such as ECMWF and Global Ionosphere Maps (GIM). The segment of height profiles at the crossing between the orbit ground track and a river reach (so called Virtual Stations) were subsequently processed with the median value of manually selected measurements being finally computed as the water height.

This procedure follows the results presented by [27,28] and has been widely used since then (see reviews in [18,20]). Two methods are usually used to assess the quality of the altimetry-derived series of water level: external comparison with in-situ gauge reading and internal comparison at cross-overs.

In the present study, we focus on the second method since the Congo basin is almost devoid of gauges. In addition, fortunately, the SARAL orbit forms crossovers at 1.5 day time difference over several rivers in the basin. Comparing water level measurements collected 1.5 days apart is not fully satisfactory but the positive side is that the difference is a conservative (over)estimate of the true measurement error since it includes the natural level variations. We selected 25 locations where the ascending and descending passes cross a river reach at less than 20 km, the maximum distance that we considered as relevant for a cross-over analysis. Within this dataset, the five cases correspond to reaches less than 250 m in width and 15 to reaches wider than 500 m (Figure 17). The colour code in Figure 17 stands for the standard deviation between the time series at the ascending and descending passes. The cumulative distribution of the cycle-by-cycle difference between all the measurement pairs (inset in Figure 17) shows that 90% of the measurement pairs present a difference in height less than 25 cm. The present result is consistent with the 10–30 cm errors found by [24,25] in the Ganga-Bramapoutra basin and by [26] in the Amazon basin. Noteworthy, there is no straightforward relationship with the width of the reach.

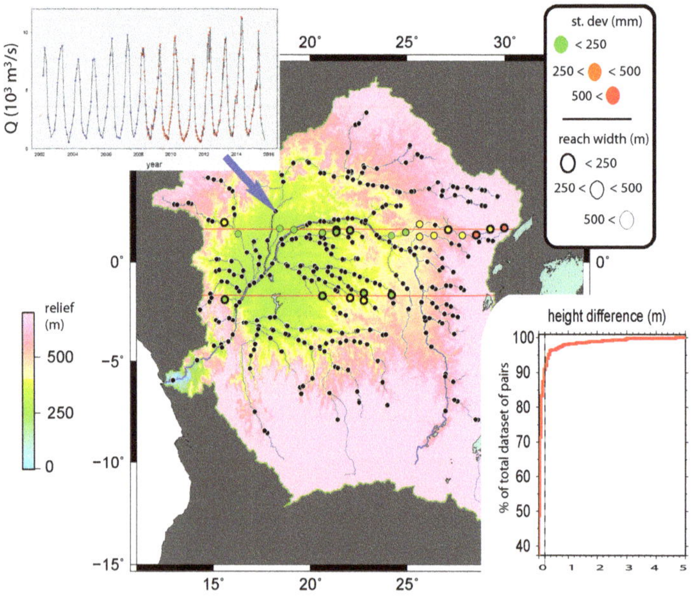

Figure 17. The Congo basin. The locations of the 350 SARAL/AltiKa Virtual Stations are shown with black dots. The +1.595° and −1.677° parallels in red are where the SARAL/AltiKa orbit forms crossover at 1.5 day temporal difference. The Virtual Stations used for the accuracy assessment at cross-overs are coded according the standard deviation between the two SARAL/AltiKa series of water level and the reach width, as recalled in the upper right inset. The cumulative histogram of the differences between measurement pairs is shown in the lower right inset. The dotted bar outlines the value of 20 cm in difference achieved by 90% of the population of height pairs. An example of Envisat + Jason-2 + SARAL/AltiKa discharge series on the Ubangui River derived from the rating curve established combining SARAL/AltiKa levels and MGB discharge is presented in the left hand inset.

One of the most important applications of water level measurements is their conversion into discharge estimates. Discharge estimates were obtained through the following steps: (1) same as first proposed by [29], compute distributed discharge between 2013 and mid-2014 by means of the rain-discharge model of the Institute of Hydraulic Research of Brazil (MGB-IPH) fed with rain fields from the JAXA/NASA Tropical Rainfall Measuring Mission (TRMM) mission; (2) compute height–discharge power laws compliant with the Manning equation; and (3) apply these rating curves to the SARAL/AltiKa series to derive discharge series. An example of such a discharge series is presented in the left hand inset of Figure 17. At this location on the Ubangui River, the Envisat/SARAL orbit and the Jason-2 orbit form a crossover right over the Ubangui River. Thus, a long continuous discharge series could be derived by applying the SARAL/AltiKa rating curve altogether to the Envisat, Jason-2 and SARAL/AltiKa times series of water level.

Overall, SARAL/AltiKa altimetry shows to be a powerful tool to estimate water levels with errors around 20 cm at more than 300 locations throughout the Congo basin, hence discharge in such a basin almost devoid of ground monitoring.

3.2. Lakes

A Case Study of the Tibetan Lakes

The Tibetan plateau is known by climatologists as the "Third Pole". It is located northeast of the Himalayan mountain range and has an average altitude of more than 4000 m. The Tibetan Plateau covers an area of 2.5 million square kilometres mainly made up of permafrost and covered by more than 1000 lakes of sizes larger than 1 km^2. Furthermore, the Tibetan Plateau is one of the regions of the world most sensitive to climate change since the average increase in temperature over the last few decades has been more than double the average for the rest of the Earth [30]. In such context, it has been demonstrated in many studies that lakes level changes in time are essential variables to track climate change impacts on water resources over the Tibetan plateau ([31], and many others). For example, an increase in temperature leads to an increase in evaporation but also accelerates glacier and permafrost thawing [32]. Changes in precipitation over long period of time also have direct impact on water contained in lakes [33]. Moreover it has been also shown in several studies that the lake level response time to climate change may be observable only over decades [20]. This has naturally placed the satellite altimetry as a useful tool for studying these links between lakes and climate changes. Due to its orbit the SARAL/AltiKa mission allows to monitor level changes on a large number of lakes over the Tibetan Plateau. It has moreover a very high accuracy over lakes, much better than that of Envisat or ERS-2 (placed on the same orbit) as shown in [34]. This is likely due to the choice of Ka band allowing a significant reduction of the footprint, and of the small ionospheric effect on the range measurement. The interest of this satellite is therefore significant since it allows extending the existing time series on the lakes over the Tibetan Plateau and may exhibit long term changes in an improved accuracy. Combining SARAL/Altika over lakes with other satellite altimeters is easily performed as a result of overlapping time of observations as shown in Figure 18. Inter-satellite instrumental biases are removed then to produce decadal level changes of lakes and thus observe and link the lake level with climate change (Figures 19–21). From these results we may draw two main conclusions:

- The responses of lakes to climate changes are not the same from one lake to another one and needs long term observations to be highlighted. For example, the Ziling lake which has grew up during more than 15 years seems over the last years to reach an equilibrium state which has been seen only with the SARAL/AltiKa measurements. It may be different for other lakes, as seen in Figures 20 and 21. It has been shown in [20] that long term changes of lakes over the Tibetan Plateau are highly variable and depend on regional climate change as well as the lake's bathymetry.

- Only SARAL/AltiKa allows extracting short term variability of water level of these lakes, since the other missions (ERS-2, Envisat, CryoSat-2, TOPEX/Poseidon, Jason-1 and Jason-2) were not precise enough to show these seasonal variabilities.

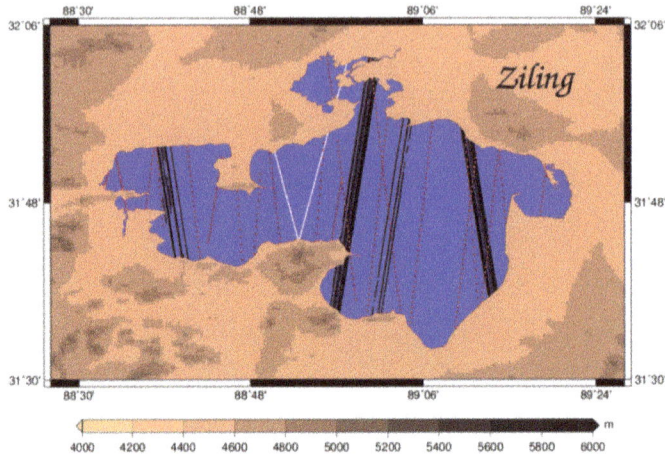

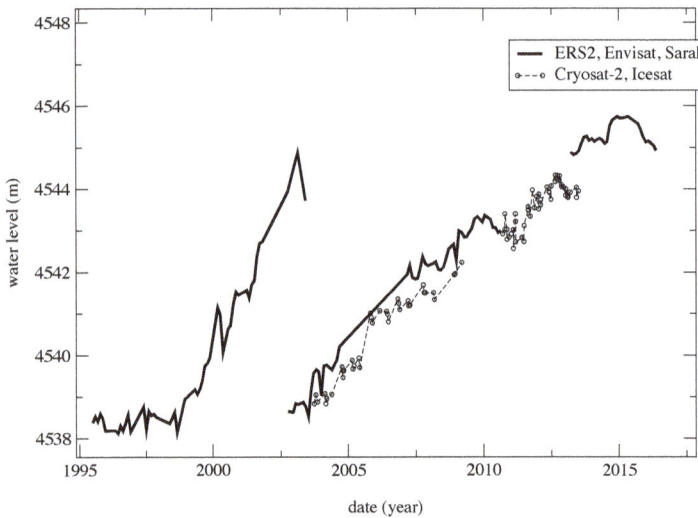

Figure 18. Tracks of satellite altimeters (black: ICESat; red: CryoSat-2; and white: ERS-2, Envisat and SARAL/AltiKa) over the Ziling Lake, one of the largest of the Tibetan Plateau (**top**); and water level changes from each satellite before correcting from inter-satellite tracks from 1995 to 2016 (**bottom**).

Lake Ziling

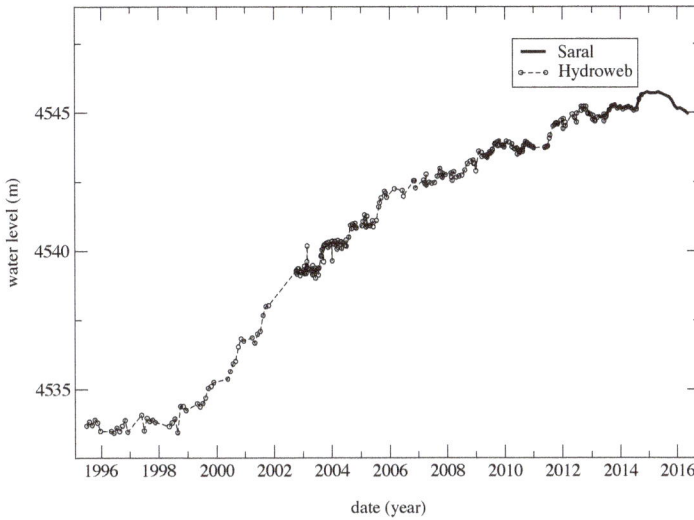

Figure 19. Water level change of the Ziling Lake from a multi-satellite data processing after removing the inter-satellite biases. After a long period of increasing starting in 1999, the lake level has stabilized in 2013/2014 (extracted from Hydroweb database http://hydroweb.theia-land.fr/ and from SARAL/AltiKa data processing for the period 2013–2016).

Lake Namco

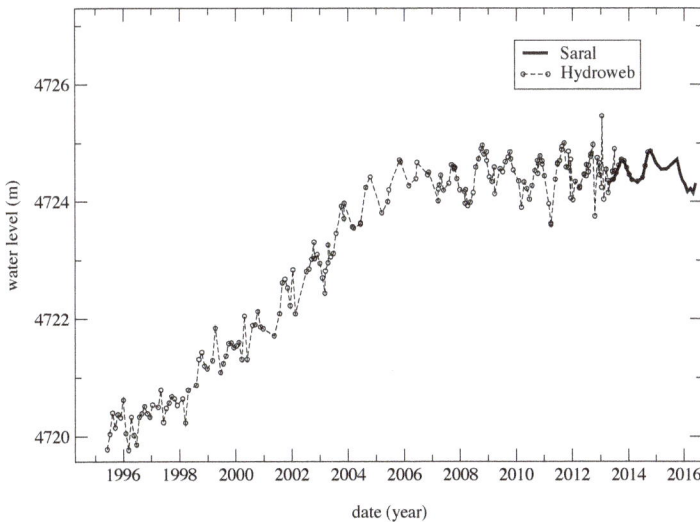

Figure 20. Water level change of the Namco lake from a multi-satellite data processing. After a long period of increasing, the lake level has stabilized in 2006/2007 (extracted from Hydroweb database http://hydroweb.theia-land.fr/ and from SARAL/AltiKa data processing for the period 2013–2016).

Lake Ngoring-Co

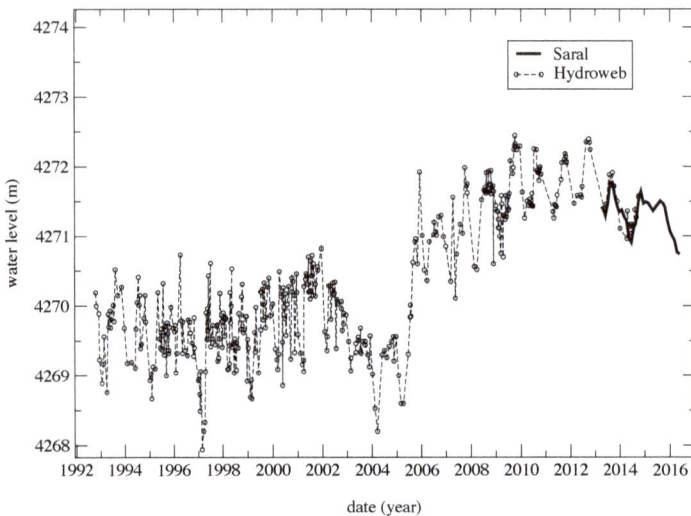

Figure 21. Water level change of Ngoring-Co Lake from a multi-satellite data processing. The lake was at equilibrium between 1992 and 2005 and has suddenly grown since 2006 to reach in a short time a new equilibrium, which slightly decreased after 2013 (extracted from Hydroweb database http://hydroweb.theia-land.fr/ and from SARAL/AltiKa data processing for the period 2013–2016).

4. Ice

For the last few decades, terrestrial ice has undergone large change and its survey is of a prime importance for climate study. Altimetry is one of the most powerful tools for ice sheet mass balance, sea ice thickness and iceberg detection studies. An altimeter operating in Ka- instead of Ku-band is far from being insignificant for these studies. The decrease of penetration depth within the snow, from 2 to 12 m for the Ku-band to less than 1 m for the Ka-band ensures the measurement of the snow surface. Moreover, the higher waveform rate and narrowed footprint of SARAL/AltiKa allows to better sample the small scale features and to better track the elevation transition. It is suitable to map the surface undulations or large crevasses above ice sheet or to detect leads on sea ice or small icebergs. Moreover, SARAL/AltiKa being on the same historical 35-day orbit than ERS-1-2 and Envisat allows to extend the survey for more than 20 years [35].

For sea ice, this phenomenon also seems to play a role but the major SARAL/AltiKa contribution yields from the reduced antenna aperture and in higher spatial resolution than previous altimeters. Indeed, the first echo is more likely to be at nadir decreasing influence of the off-nadir signal that often affects the retrieval measurement. Some studies in this issue confirm the ability of SARAL/AltiKa to detect leads and coastal polynyas as well as to represent spatial and temporal dynamics of water openings. Moreover, a meaningful contribution is the high sensitivity of the Ka-band backscatter to the snow and ice state, due to the altimeter wavelength. A few dB change may occur in a few weeks for any kind of snow or ice surface, even in the central part of the Antarctica ice sheet. It is observed in [36] a significant decrease of up to 15 dB at the end of ice season over Lake Baikal. The field measurements acquired along the SARAL/AltiKa tracks in Spring 2013 and Spring 2014 allow to suggest an important role of ice metamorphism and re-crystallisation.

4.1. Ice Sheet

Monitoring of the Antarctic Ice Sheet

Altimetry is probably one of the most powerful tools for ice sheet observation. Our vision of the Antarctic ice sheet has been deeply transformed since the launch of the ERS-1 satellite in 1991. With the launch of ERS-2, Envisat and SARAL/AltiKa, the series of altimetric observations now provides more than 20 years of continuous and homogeneous observations that allow monitoring of the shape and volume of ice sheets. The topography deduced from altimetry is one of the relevant parameters revealing the processes acting on ice sheet. Moreover, altimeter also provides other parameters such as backscatter and waveform shape [4] that give information on the surface roughness or snow pack characteristics.

The most surprising features of Ka-band observations above the Antarctica ice sheet is the very large space and time variability of the backscatter. While the Ku-band variability was about 0.5 dB over few cycles, it can reach 3 or 4 dB in Ka-band. This unexplained observation probably comes from the small wavelength (0.8 cm instead of 2.3 cm) that is very sensitive to snow metamorphism such as surface hoar. These structures are related to snow drift and sublimation and is one of the most unknown of the polar snow. Thus, one of the major limitation of radar altimeter above ice sheet is the penetration of the radar wave within the dry snowpack. For the classical Ku-band used by altimeters, the penetration reaches a few m. The slightest changes of surface echo or of volume echo affect the height retrieval. At the seasonal scale, the impact on height can be of few tens of centimetres. Moreover, this penetration may induce a long term artificial change in height caused by either change in the surface or snowpack properties. On the other hand, volume echo contains few interesting information on snow stratification and on ice grain size, the two volume components of Ku band. [37] show that the penetration depth in Ka-band is indeed between 50 cm and 1 m (see Figure 22). Moreover, they show that the stratification of the snow pack does not play a role in the volume component, the scattering by ice grain being the unique physics. The comparison between Ku and Ka-band may then correct from the penetration error and help in retrieve some snowpack parameter.

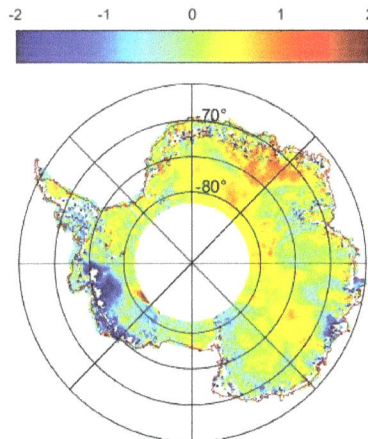

Figure 22. Difference (in m) between Ka and Ku-band height from the Envisat and SARAL/AltiKa altimeter over Antarctica. Due to the difference of penetration, in the central part, the Ku-band elevation is indeed below the Ka one. Near the coast, the smaller antenna aperture of Ka-band yields to a smaller slope error. In the West Antarctica sector, the negative pattern is due to the decrease of surface topography between the time of Envisat and of SARAL/AltiKa .

The characteristics of SARAL/AltiKa ensures a temporal stability of height retrieval that allows to obtain elevation trends with a good confidence. We well map the impressive loss of outlet glaciers of the West part of the continent. We are able to describe the large digging of Pine Island glacier from ERS-1 to Envisat and now its large enlargement during SARAL/AltiKa period [2].

4.2. Icebergs

Altimeters are powerful tools to study icebergs, either small (<3 km in length) or large (length > 16 km). Indeed, [38] demonstrated that small icebergs can produce a detectable signature in high resolution altimeter waveforms, at least in sea ice free ocean. Their method enables to detect, and to estimate the iceberg area from these signatures. The constraints on the method limit the detection to icebergs between 0.1 km^2 and 8 km^2. Nine satellite altimetry missions have already been processed to produce a 1992–present database of small iceberg location, area, volume and mean backscatter ([39,40]). Monthly mean probability of presence, area and volume of ice over a regular polar (100 × 100 km^2) or geographical (1° × 2°) grid are also available and are distributed on the Centre ERS d'Archivage et de Traitement (CERSAT) website. The archive of SARAL/Altika has of course been processed and is part of the database. Comparison of the detection results from classical Ku band altimeters (such as Jason-series) showed that the SARAL/AltiKa higher waveform frequency of 40 Hz as well as the smaller footprint significantly improves the detection of the smallest icebergs [40] and that SARAL/AltiKa could be used as a reference to inter-calibrate the volume of ice estimates from the different sensors.

Altimeter can also be used to measure the freeboard elevation profile of large icebergs [41,42]. Combining iceberg tracks from the National Ice Center (NIC) and the archives of three Ku band altimeters, Jason-1, Jason-2 and Envisat, a database of the daily position, free-board profile, length, width, area and volume of large icebergs covering the 2002–2012 period is presented in [42]. Compared to Ku-band altimeters, the SARAL/AltiKa smaller footprint and higher waveform rate allows a better estimation of the elevation and backscatter profiles over icebergs. Figures 23 and 24 compare Jason-2 and SARAL/AltiKa data over a large iceberg (C16) detected in the Weddell Sea. They present MODIS images of the iceberg (Figures 23 and 24a) as well as the raw altimeter waveforms (Figures 23 and 24b) and the waveforms repositioned using the altimeter tracker position to better show the iceberg signature (Figures 23 and 24c). The iceberg elevation profile is estimated using a simple tracker based on the detection for each waveform of the first occurrence (bin) of a power gradient larger than a given threshold [42]. As the altimeter approaches the iceberg from the north near 65.3°S, the tracker starts to move up (Figures 23 and 24c,d) mitigating the sea and iceberg surfaces elevation as it can be seen. As the tracker is not locked on the iceberg surface, the strong echo from the iceberg starts to appear in the first gate of the waveforms, and then moves toward the nominal track point (0), while the echo from the sea surface moves away from zero (Figures 23 and 24b). The length over which the altimeter footprint contains both ocean and iceberg depends on the angle between the ground track and the iceberg, and on the footprint size. SARAL/AltiKa smaller footprint (5.6 km vs. 8.6 km for the last range bin) significantly reduces the transition length (see Figures 23 and 24d) by almost a factor 2 compared to Jason-2, especially for slanted angles. The 40 Hz waveform rate also combined with the smaller footprint lead to a more rapid detection by the tracker of the elevation transition between ocean and iceberg and iceberg and ocean. The detection of the iceberg edges although not perfect for slanted angles is significantly better than that of Ku band altimeters. Furthermore, as SARAL/AltiKa footprint is smaller, the probability that a waveform is not contaminated by ocean, i.e., that the elevation and backscatter measurements are more precise, is higher than that for Ku band altimeters. This can be clearly seen in Figures 23 and 24; whereas the SARAL/AltiKa backscatter clearly shows a decrease associated to the iceberg slope (Figure 23d) and sharp jumps between ocean and icebergs, it is difficult to identify the transition between iceberg and ocean for Jason-2.

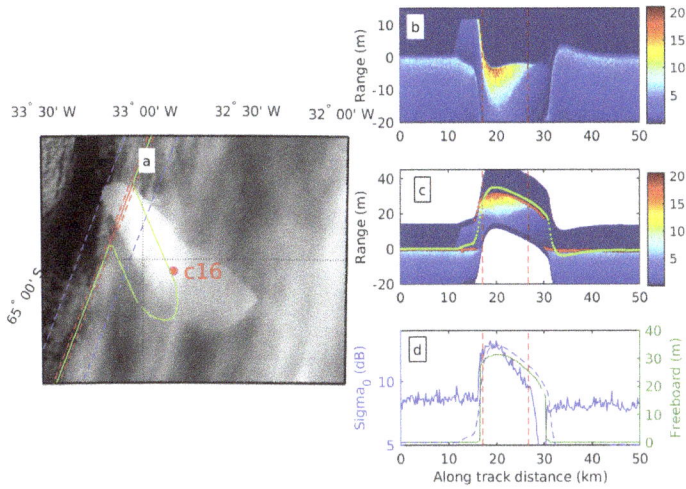

Figure 23. (a) MODIS image on 17 March 2013 17:34 UT and SARAL/AltiKa descending pass 102 Cycle 1 (17 March 2013 18:19UT). The red and blue dashed lines indicate the width of the first and last range bins, respectively, and the green line the elevation profile. (**b**) SARAL/AltiKa waveforms over C16. The dashed red lines indicate the iceberg limits from MODIS image. (**c**) Re-tracked waveforms using the tracker position, the green line represents the tracker position and the red line the elevation profile using a dedicated re-tracker. (**d**) Re-tracked elevation profile (green line), tracker position (dashed green line) and measured backscatter (blue line).

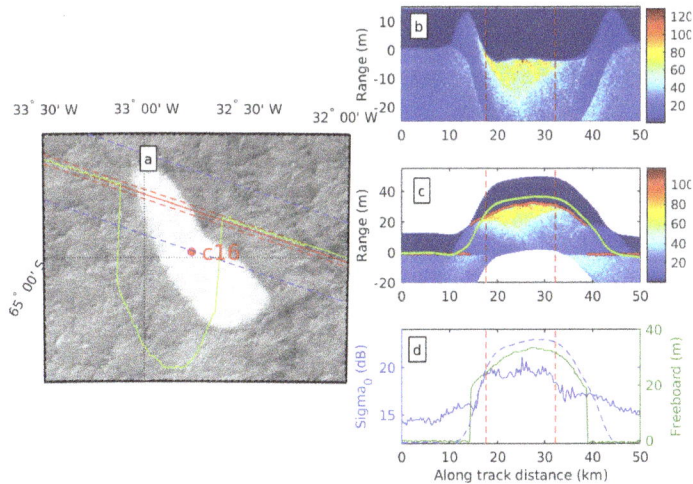

Figure 24. (a) MODIS image on 20 March 2013 16:45 UT and Jason-2 descending pass 180 Cycle 173 (20 March 2013 12:48UT). The red and blue dashed lines indicate the width of the fist range and last bins, respectively, and the green line the elevation profile. (**b**) Jason-2 waveforms over C16. The dashed red lines indicate the iceberg limits from MODIS image. (**c**) Re-tracked waveforms using the tracker position, the green line represents the tracker position and the red line the elevation profile using a dedicated re-tracker. (**d**) Re-tracked elevation profile (green line), tracker position (dashed green line) and measured backscatter (blue line).

4.3. Sea Ice

Regarding sea ice surfaces, SARAL/AltiKa offers mainly two advantages relatively to conventional Ku-band radar altimeters:

- First, its smaller footprint, higher vertical resolution (∼30 cm) and higher horizontal sampling (∼180 m) allow a better discrimination between ice floes and open sea ice fractures (generally referred to as leads), which strongly improves the measurement of sea level and freeboard height. These improvements are somehow counter-balanced by the technical issue of waveform saturation while the altimeter overflight surfaces with highly variable scattering. This effect tends to degrade the range estimation during the approach phase of specular surfaces like leads. However, the reactivity of the Attenuation Gain Controller can be easily corrected in future altimeter version.
- The second advantage of SARAL/AltiKa lies on its higher radar frequency: Unlike Ku-band radar altimeters, the Ka-band radar signal of AltiKa penetrates only part of the snowpack and possibly less than 3 cm [43]. The latter study uses this difference of penetration depth between Ku- and Ka-band altimeters to estimate a proxy of snow depth at the top of sea ice by combining SARAL/AltiKa (Ka-band) and CryoSat-2/SIRAL (Ku-band) (Figure 25). As the uncertainty related to the impact of snow depth on the freeboard-to-thickness conversion can be up to 100 %, this opportunity of measuring snow depth from Ka- and Ku-band radar altimetry represents a real breakthrough. In addition to the freeboard-to-thickness conversion, the snow depth measurement could be of hight interest for the quantification of energy, transfer between the atmosphere and the ocean as well as for the estimation of freshwater fluxes in the ocean.

The potential of combining Ka and Ku-band measurements to estimate snow depth at the top of sea ice is currently further investigated. In particular, the CryoVex airborne campaign carried, for the first time in March 2017, a Ka-band altimeter (Karen) in addition to the usual Ku-band altimeter (ASIRAS). The analysis of these highly accurate and colocated measurements will allow to provide more evidences on the interest of combining Ka and Ku band radar altimetry and could be a basis for the development of future polar altimetric missions.

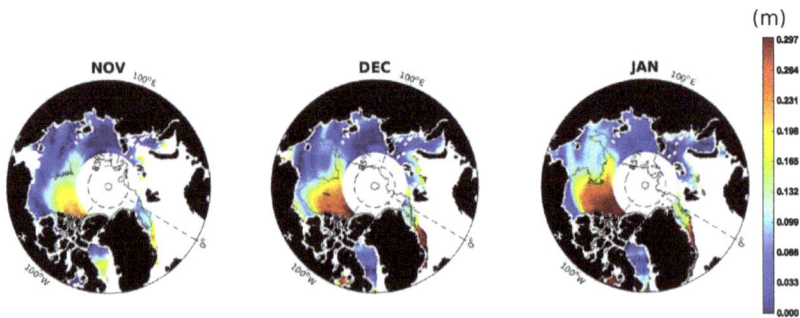

Figure 25. Maps of snow depth estimated from the combinaition of SARAL/AltiKa and CryoSat-2/SIRAL for November 2013, December 2013 and January 2014 [43].

5. Geodesy

Satellite altimetry has also become an effective remote sensing technique with important applications to geodesy [44]. Indeed, satellite altimetry has impact to geodesy by allowing mapping and monitoring a major part of the Earth surface and to make improvements on the measurements of the gravity field. The altimetric constellation with a number of successive and/or parallel missions with different orbit configuration and sampling characteristics are very supportive of geodetic applications.

The case of SARAL/AltiKa in its SARAL-DP configuration is of particular interest in this regard, as it is covering the earth with a very fine resolution well suited.

Ability to Find Uncharted Seamounts

Seamounts—undersea mountains of volcanic origin—are obstacles to flow, sites of enhanced mixing, habitat for fish, and possible hazards to submarines. Small seamounts are more common than large ones. Wessel et al. [45] found that all seamounts 2000 meters tall and taller appeared to have been found by the satellite altimeters of that time, but suggested that there might be as many as one hundred thousand seamounts between 1000 and 2000 m tall that had escaped detection. Since only about eight percent of the ocean floor has been sounded to one nautical mile resolution [46], the discovery of new seamounts will come mainly from new satellite altimetry. However, small seamounts remain difficult to find because they create only small anomalies in sea level (Figure 26). Smith [47] showed that the AltiKa altimeter on SARAL/AltiKa has much better precision in sea level measurement than previous Ku-band altimeters, and that this precision can be used to detect seamounts as small as 1 km tall. Marks [48] showed that a sea surface height profile made from the temporal median of SARAL/AltiKa repeat-track profiles can detect even smaller seamounts. It is now expected that SARAL's new geodetic mission will furnish data for finding many previously unknown seamounts. This will be a boon to scientists modelling ocean flow and mixing, to marine biologists and commercial fishermen, and perhaps also to navies that must navigate submarines.

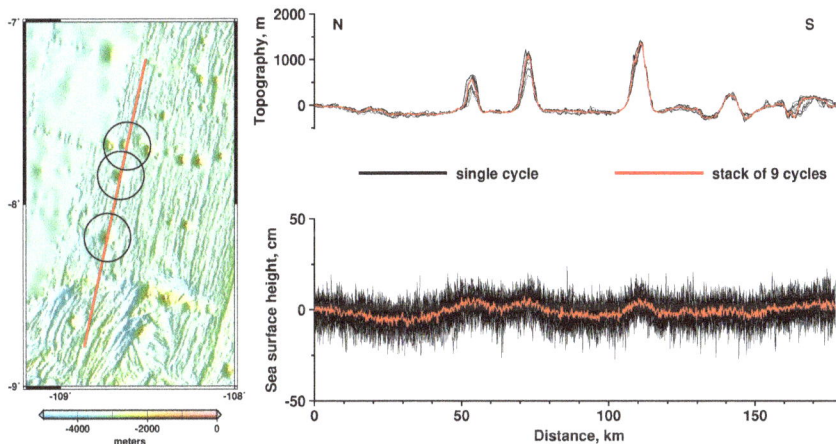

Figure 26. The red line shows the ground track of a SARAL/AltiKa repeat cycle (pass number 0396) over three small seamounts (in black circles). Seafloor topography is shown in the coloured image; the detailed portion is covered by a multi-beam echo sounder survey that confirms the position and height of the seamounts. Profiles along the red line in the previous image, showing: the sea floor topography under the satellite track (**top**); and the sea surface height anomaly (**bottom**). Black curves are individual results from each repeat cycle. The red curves are the median of all repeat cycles. The sea surface height anomaly produced by the seamounts is only 5 to 10 cm high, but is detectable, thanks to the better precision furnished by SARAL/AltiKa.

6. Conclusions

SARAL/AltiKa was launched on 25 February 2013. At the beginning, the SARAL/AltiKa mission was strongly motivated by a necessity by the research and operational oceanographic communities because of the need to fill the gap after the Envisat mission and before Jason-3 and Sentinel-3. Indeed, these objectives were completely fulfilled. However, the SARAL/AltiKa altimeter is also, and perhaps above all, the first oceanographic altimeter to use a high single frequency in Ka-band

(35.75 GHz). A little less than five years after launch, the satellite and the instruments onboard are performing well although the control of the reaction wheels appeared to be impossible at some stage, end of 2015 and beginning of 2016, leading the leave of the exact 35 repeat orbit and a shift to the Drifting Orbit Phase. Note this change had no impact of the quality and the availability of the SARAL/AltiKa data but only on the time/space sampling. As of today, all components of the altimetric system (i.e. AltiKa itself) are working properly.

Calibration and validation investigations have shown that the quality of the data meets the expectations and initial mission requirements and this is still correct [4]. Since the launch, the quality of all products appears to be continuously in line with the mission requirements. A large number of scientific investigations have been undertaken and SARAL/AltiKa has become a fully-fledged member of the altimetric constellation.

The advantages/drawbacks of the Ka-band summarize as follows:

- The higher frequency (35.75 GHz, to be compared to 13.5 GHz on Jason-2) leads to a smaller footprint (8 km diameter, to be compared to 20 km on Jason-2 and to 15 km for Envisat) and thus a better horizontal resolution.
- Ka-band allows to use a larger bandwidth (480 MHz to be compared to 320 MHz on Jason-2). This 480 MHz bandwidth provides a high vertical resolution (0.3 m) which is better with respect to other altimeters.
- The higher Pulse Repetition Frequency (4 kHz to be compared to 2 kHz on Jason-2) permits a decorrelation time of sea echoes at Ka-band shorter than at Ku-band, then allowing a better along-track sampling. This makes possible to increase significantly the number of independent echoes per second compared with Ku-band altimeters.
- Ka-band is much less affected by the ionosphere than one operating at Ku-band. This low ionospheric attenuation can even be considered as negligible, except for some exceptional ionospheric situations. It discards the need for a dual-frequency altimeter.
- Ka-band provides a better estimation of sea surface roughness than at Ku-band. The 8 mm wavelength in Ka-band is better suited to describing the slopes of small facets on the sea surface (capillary waves, etc.) and gives a more accurate measurement of the backscatter coefficient over calm or moderate seas, thus leading to a noise reduction of a factor of two compared to Jason-class altimeters for wave heights greater than 1 m. Moreover, the specificities of the Ka-band backscatter coefficient offer unique contributions in fields that where not foreseen (snow/ice morphology and its temporal variability (Section 4), soil moisture [4], etc.).
- With Ka-band, there is a lower radar penetration of snow and ice: penetration of snowpack is less than 3 cm for snow on sea ice and 1 m for continental ice, around ten times less than for Ku-band. The altimetric observation and height restitution thus correspond to a thin subsurface layer.
- A possible drawback of Ka-band was that the attenuation due to water or water vapour in the troposphere might affect Ka-band pulses in case of rain and increase significantlly the rate of missing data for strong rain rates. In fact, this was not found to be true in practice and rain had little influence on data availability and quality.

Regarding the oceans, SARAL/AltiKa provides an improved resolution of the SSH signal in particular of the mesoscales. Regarding the along-track SSH spatial resolution SARAL/AltiKa provides resolution down to 40–50 km wavelengths, whereas Jason and Envisat altimeters resolve down to 70–80 km wavelengths. Sentinel-3 SAR is still being analyzed, but it is suggested to also be resolved to 30–50 km wavelengths. **The coastal oceans**, directly benefit of the better discrimination in transition zones as well as this improved resolution of the measurements. Progress has been made in reducing the altimeter footprints and noise to approach closer to the coast, but also in processing algorithms, corrections, and products for coastal applications. The improved coastal altimeter data and their derived sea level and wind–wave data are being integrated into coastal observing systems. These data provide essential monitoring for both research and operational applications, in these coastal regions where in situ measurements are sparse.

Regarding inland waters, SARAL/AltiKa definitely improves the quality of hydrology products derived from satellite altimetry. Especially for water level time series over lakes and rivers it was shown that SARAL/AltiKa dramatically improved the quality due to its smaller footprint and its higher pulse repetition frequency. In particular, SARAL/AltiKa allows the access to smaller rivers and lakes that were not "visible" with standard altimeters. The impact of clouds or precipitation on the resulting water levels exists, but is not so strong that the data cannot be used for reliable and accurate estimation of water level time series. With regard to Envisat for example, the waveform disruption for SARAL/AltiKa for height estimation is better with SARAL/AltiKa. SARAL/AltiKa has the potential to provide better water level heights until 6 km near to the lake shore than Envisat. In addition, SARAL/AltiKa provides more homogenous water level heights over rivers than Envisat.

Regarding ice, the lesser radar penetration of snow is beneficial for various types of measurements. The difference between Ka and Ku-band height provides a good proxy of snow depth above sea ice that is a limitation to estimate ice thickness and is a climate indicator. Above ice sheet it ensures a mean to better constrain the waveform models and invert snowpack characteristics. Combined with a better spatial resolution, Ka-band in SAR mode for instance, would therefore allow closer monitoring of sea and continental ice. As for inland waters applications, ice applications gave evidence of the key importance of continuous observations along the same repeat orbit, as was possible with ERS-2, Envisat and SARAL/AltiKa (during the nominal phase).

Regarding geodesy, it appears that AltiKa is an excellent altimeter for resolving short-wavelength geoid anomalies. SARAL/AltiKa is now in its Drifting Phase, the dataset that would result would be a boon to marine geophysics, bathymetric estimation, and seamount mapping. Seamount size-frequency distribution models suggest that there may be as many as 105 seamounts between 1 and 2 km in height that are uncharted and were not detected by previous Ku-band altimeters. During its Drifting Phase, SARAL/AltiKa will surely find some of these.

A general observation is that the implementation of SARAL/AltiKa data into **operational systems** has been extremely easy both into operational oceanography centres for the SSH signal and into meteorological centres for the SWH in particular. It has also been the case for building databases of various levels of sophistication, for example for AVISO that integrates SARAL/AltiKa data as well as other altimeter data.

SARAL/AltiKa is in many respects a prototype of the altimetry of the future. SARAL/AltiKa represents the beginning of a new class of altimeters operating at Ka-band frequency with a small footprint and high pulse rate. The Ka-band is envisioned for several new altimetric satellite projects (i.e. AltiCryo from CNES, CryoSat-3 and SKIM from ESA) and the chosen band for the upcoming SWOT. It even opens more than previously the doors of interdisciplinarity. Indeed, the extended capabilities that are offered by the Ka-band allow to open even more widely some new frontiers of altimetry such as coastal oceanography, cryosphere, hydrology, beyond the traditional scope of the open ocean investigations. Ka-band altimetry, with SARAL/AltiKa as the today's most emblematic declination, can be seen also as a step towards improved resolution altimetry and a preparation for the NASA-CNES SWOT mission project and the KaRIN instrument (Ka-band Radar INterferometer).

Acknowledgments: The SARAL/AltiKa mission is an achievement of a very fruitful cooperation between ISRO and CNES. The contribution of EUMETSAT to data distribution is also strongly appreciated. We acknowledge the support of all the Investigators, PIs and Co-Is, of the SARAL/AltiKa Mission. Most of the studies performed by the French authors have been conducted and financed thanks to Centre National d'Etudes Spatiales (CNES), Centre National de la Recherche Scientifique (CNRS), and French Ministry of Research. The manuscript contents are solely the opinions of the authors and do not constitute a statement of policy, decision, or position on behalf of NOAA or the U. S. Government.

Author Contributions: Jacques Verron and Pascal Bonnefond coordinated the whole paper. For the ocean, Rosemary Morrow, Oscar Vergara, Marie-Isabelle Pujol and Gérald Dibarboure brought their contribution to Section 2.1; Florence Birol, Fatma Jebri and Ananda Pascual to Section 2.2; Yannice Faugère, Marie-Isabelle Pujol, Elisabeth Rémy, Mathieu Hamon and Lofti Aouf to Section 2.3; and Suchandra A. Bhowmick and Raj Kumar to Section 2.4. On inland waters, P. K. Gupta, A. K. Dubey, Taina Conchy and Stéphane Calmant brought their contribution to Section 3.1.1; and Jean-François Crétaux to Section 3.2. On ice, Frédérique Rémy brought her

Remote Sens. **2018**, *10*, 163

contribution to Section 4.1; Jean Tournadre to Section 4.2 ;and Kevin Guerreiro to Section 4.3. On geodesy, Walter H. F. Smith contributed to Section 5.

Conflicts of Interest: The authors declare no conflict of interest.

References

1. Vincent, P.; Steunou, N.; Caubet, E.; Phalippou, L.; Rey, L.; Thouvenot, E.; Verron, J. AltiKa: A Ka-band altimetry payload and system for operational altimetry during the GMES period. *Sensors* **2006**, *6*, 208–234.
2. Verron, J.; Sengenes, P.; Lambin, J.; Noubel, J.; Steunou, N.; Guillot, A.; Picot, N.; Coutin-Faye, S.; Gairola, R.; Raghava Murthy, D.V.A.; et al. The SARAL/AltiKa altimetry satellite mission. *Mar. Geodesy* **2015**, *38*, 2–21.
3. The SARAL/AltiKa Satellite Altimetry Mission. *Mar. Geodesy* **2015**, *38*. Available online: http://www.tandfonline.com/toc/umgd20/38/sup1 (accessed on 12 January 2018)
4. Bonnefond, P.; Verron, J.; Aublanc, J.; Babu, K.N.; Bergé-Nguyen, M.; Cancet, M.; Chaudhary, A.; Crétaux, J.F.; Frappart, F.; Haines, B.J.; et al. The benefits of the Ka-band as evidenced from the SARAL/AltiKa altimetric mission: Quality assessment and specificities of AltiKa data. *Remote Sens.* **2018**, *10*, 83, doi:10.3390/rs10010083.
5. Dufau, C.; Orsztynowicz, M.; Dibarboure, G.; Morrow, R.; Le Traon, P.Y. Mesoscale Resolution Capability of altimetry: Present & future. *J. Geophys. Res. Oceans* **2016**, *121*, 4910–4927.
6. Ablain, M.; Raynal, M.; Lievin, M.; Thibaut, P.; Dibarboure, G.; Picot, N. Improving altimeter sea level calculation at small ocean scales. In Proceedings of the 2016 Ocean Surface Topography Science Team (OSTST) Meeting, La Rochelle, France, 31 October–1 November 2016.
7. Morrow, R.; Carret, A.; Birol, F.; Nino, F.; Valladeau, G.; Boy, F.; Bachelier, C.; Zakardjian, B. Observability of fine-scale ocean dynamics in the Northwest Mediterranean Sea. *Ocean Sci.* **2017**, *13*, 13–29.
8. Dibarboure, G.; Pujol, M.I.; Briol, F.; Le Traon, P.Y.; Larnicol, G.; Picot, N.; Mertz, F.; Ablain, M. Jason-2 in DUACS: Updated System Description, First Tandem Results and Impact on Processing and Products. *Mar. Geodesy* **2011**, *34*, 214–241.
9. Pujol, M.-I.; Schaeffer, P.; Faugére, Y.; Raynal, M.; Dibarboure, G.; Picot, N. Gauging the improvement of recent mean sea surface models: A new approach for identifying and quantifying their errors. *J. Geophys. Res.* **2017**, submitted.
10. Jebri, F.; Birol, F.; Zakardjian, B.; Bouffard, J.; Sammari, C. Exploiting coastal altimetry to improve the surface circulation scheme over the central Mediterranean Sea. *J. Geophys. Res. Ocean.* **2016**, *121*, 4888–4909.
11. Sorgente, R.; Olita, A.; Oddo, P.; Fazioli, L.; Ribotti, A. Numerical simulation and decomposition of kinetic en e.g., in the Central Mediterranean: Insight on mesoscale circulation and en e.g., conversion. *Ocean Sci.* **2011**, *7*, 503–519.
12. Rubio, A.; Mader, J.; Corgnati, L.; Mantovani, C.; Griffa, A.; Novellino, A.; Quentin, C.; Wyatt, L.; Schulz-Stellenfleth, J.; Horstmann, J.; et al. HF Radar Activity in European Coastal Seas: Next Steps toward a Pan-European HF Radar Network. *Front. Mar. Sci.* **2017**, *4*, 8.
13. Pascual, A.; Lana, A.; Troupin, C.; Ruiz, S.; Faugère, Y.; Escudier, R. Assessing SARAL/AltiKa delayed-time data in the coastal zone: Comparisons with HFR observations. *Mar. Geodesy* **2015**, *38*, 260–276.
14. Rio, M.-H.; Pascual, A.; Poulain, P.-M.; Menna, M.; Barceló-Llull, B.; Tintoré, J. Computation of a new mean dynamic topography for the Mediterranean Sea from model outputs, altimeter measurements and oceanographic in-situ data. *Ocean Sci.* **2014**, *10*, 731–744.
15. Pujol, M.I.; Faugère, Y.; Taburet, G.; Dupuy, S.; Pelloquin, C.; Ablain, M.; Picot, N. DUACS DT2014: The new multi-mission altimeter dataset reprocessed over 20 years. *Ocean Sci.* **2016**, *12*, 1067–1090.
16. Le Traon, P.Y.; Antoine, D.; Bentamy, A.; Bonekamp, H.; Breivik, L.A.; Chapron, B.; Corlett, G.; Dibarboure, G.; Digiacomo, P.; Donlon, C.; et al. Use of satellite observations for operational oceanography: Recent achievements and future prospects. *J. Oper. Ocean.* **2015**, *8* (Suppl. 1), 12–27.
17. Desroziers, G.; Berre, L.; Chapnik, B.; Poli, P. Diagnosis of observation, background and analysis-error statistics in observation space. *Quart. J. R. Meteorol. Soc.* **2005**, *131*, 3385–3396.
18. Calmant, S.; Seyler, F. Continental surface waters from satellite altimetry. *Géosciences* **2006**, *338*, 1113–1122.
19. Crétaux, J.F.; Birkett, C. Lake studies from satellite radar altimetry. *Comptes Rendus Geosci.* **2006**, *338*, 1098–1112.
20. Crétaux, J.-F.; Abarca Del Rio, R.; Berge-Nguyen, M.; Arsen, A.; Drolon, V.; Clos, G.; Maisongrande, P. Lake volume monitoring from Space. *Surv. Geophys.* **2016**, *37*, 269–305.

21. Alsdorf, D.; Beighley, E.; Laraque, A.; Lee, H.; Tshimanga, R.; O'Loughlin, F.; Dinga, B.; Moukandi, G.; Spencer, R.G. Opportunities for hydrologica research in the Congo basin. *Rev. Geophys.* **2016**, *54*, 378–409.

22. Laraque, A.; Bellanger, M.; Adèle, G.; Guebanda, S.; Gulemvuga, G.; Pandi, A.; Paturel, J.E.; Robert, A.; Tathy, J.P.; Yambele, A. Recent evolution of Congo, Oubangui and Sangha Rivers flows. *Acad. R. Sci. Belgique Geogr. Ecol. Trop.* **2013**, *37*, 93–100.

23. Silva, J.; Calmant, S.; Seyler, F.; Medeiros Moreira, D.; Oliveira, D.; Monteiro, A. Radar Altimetry aids Managing gauge networks. *Water Resour. Manag.* **2014**, 28, 587–603.

24. Dubey, A.K.; Gupta, P.; Dutta, S.; Pratap Sing, R. Water Level retrieval Using SARAL/AltiKa Observations in the braided Brahmapoutra River, Eastern India. *Mar. Geodesy* **2015**, 38, 549–567.

25. Frappart, F.; Papa, F.; Marieu, V.; Malbeteau, Y.; Jordy, F.; Calmant, S.; Durand, F.; Bala, S. Preliminary assessment of SARAL/AltiKa observations over the Ganges-Brahmaputra and Irrawaddy Rivers. *Mar. Geodesy* **2015**, *38*, 568–580.

26. Schwatke, C.; Dettmering, D.; Borgens, E.; Bosch, W. Potential of SARAL/Altika for Inland Water Application. *Mar. Geodesy* **2015**, 38, 626–643.

27. Frappart, F.; Calmant, S.; Cauhopé, M.; Seyler, F.; Cazenave, A. Results of Envisat RA-2 Derived levels Validation over the Amazon basin. *Remote Sens. Environ.* **2006**, *100*, 252–264.

28. Silva, J.; Calmant, S.; Rotuono Filho, O.; Seyler, F.; Cochonneau, G.; Roux, E.; Mansour, J.W. Water Levels in the Amazon basin derived from the ERS-2 and Envisat Radar Altimetry Missions. *Remote Sens. Environ.* **2010**, *114*, 2160–2181.

29. Paris, A.; Santos da Silva, J.; Dias de Paiva, R.; Medeiros Moreira, D.; Calmant, S.; Collischonn, W.; Bonnet, M.-P.; Seyler, F. Global determination of rating curves in the Amazon basin. *Water Resour. Res.* **2016**, *28*, 3787–3814.

30. Wang, B.; Bao, Q.; Hoskins, B.; Wu, G.; Liu, Y. Tibetan Plateau warming and precipitation change in East Asia. *Geophys. Res. Lett.* **2008**, *35*, L14702.

31. Liu, J.; Wang, S.; Yu, S.; Yang, D.; Zhang, L. Climate warming and growth of high-elevation inland lakes on the Tibetan Plateau. *Glob. Planet. Chang.* **2009**, *67*, 209–217.

32. Lei, Y.; Yang, K.; Wang, B.; Sheng, Y.; Bird, B.W.; Zhang, G.; Tian, L. Response of inland lake dynamics over the Tibetan Plateau to climate change. *Clim. Chang.* **2014**, *125*, 281–290.

33. Phan, V.H.; Lindenb, R.C.; Menenti, M. Geometric dependency of Tibetan lakes on glacial runoff. *Hydrol. Earth Syst. Sci. Discuss.* **2013**, *10*, 729–768.

34. Arsen, A.; Cretaux, J.-F.; Abarca-Del-Rio, R. Use of SARAL/AltiKa over mountainous lakes, intercomparison with Envisat mission. *J. Adv. Space Res.* **2015**, *38*, 534–548.

35. Remy, F.; Flament, T.; Michel, A.; Verron, J. Ice sheet survey over Antarctica with satellite altimetry: ERS-2, Envisat, SARAL/AltiKa, the key importance of continuous observations along the same repeat orbit. *Int. J. Remote Sens.* **2014**, *35*, 5497–5512.

36. Kouraev, A.; Zakharova, E.; Remy, F. Study of Lake Baikal ice cover from radar altimetry and in situ observations. *Mar. Geodesy* **2015**, *38*, 477–486

37. Rémy, F.; Flament, T.; Michel, A.; Blumstein, D. Envisat and SARAL/AltiKa observations of the Antarctic ice sheet: A comparison between the Ku-band and the Ka-band. *Mar. Geodesy* **2015**, *38*, 510–521.

38. Tournadre, J. Signature of Lighthouses, Ships, and Small Islands in Altimeter Waveforms. *J. Atmos. Ocean. Tech.* **2007**, *24*, 1143–1149.

39. Tournadre, J.; Girard-Ardhuin, F.; Legresy, B. Antarctic iceb e.g., distributions, 2002–2010. *J. Geophys. Res.* **2012**, 117, C05004.

40. Tournadre, J.; Bouhier, N.; Girard-Ardhuin, F.; Remy, F. Antarctic iceb e.g., distributions 1992–2014. *J. Geophys. Res.* **2016**, *121*, 327–349.

41. McIntyre, N.F.; Cudlip, W. Observation of a giant Antarctic tabular iceberg by satellite radar altimetry. *Polar Rec.* **1987**, *145*, 458–462

42. Tournadre, J.; Bouhier, N.; Girard-Ardhuin, F.; Remy, F. Large iceb e.g., characteristics from altimeter waveforms analysis. *J. Geophys. Res.* **2015**, 120, 1954–1974.

43. Guerreiro, K.; Fleury, S.; Zakharova, E.; Rémy, F.; Kouraev, A. Potential for estimation of snow depth on Arctic sea ice from CryoSat-2 and SARAL/AltiKa missions. *Remote Sens. Environ.* **2016**, *186*, 339–349.

44. Smith, W.H.F.; Barale, V.; Gower, J.; Alberotanza, L. The Marine Geoid and Satellite Altimetry. In *Oceanography from Space*; Springer: Dordrecht, The Netherlands, 2010.

45. Wessel, P. Global distribution of seamounts inferred from gridded Geosat/ERS-1 altimetry. *J. Geophys. Res.* **2001**, *109*, 19431–19441.
46. Wessel, P.; Chandler, M.T. The spatial and temporal distribution of marine geophysical surveys. *Acta Geophys.* **2011**, *59*, 55–71.
47. Smith, W.H.F. The Resolution of Seamount Geoid Anomalies Achieved by the SARAL/AltiKa and Envisat RA2 Satellite Radar Altimeters. *Mar. Geodesy* **2015**, *38*, 644–671.
48. Marks, K.M.; Smith, W.H.F. Detecting small seamounts in AltiKa repeat cycle data. *Mar. Geophys. Res.* **2016**, *37*, 349–359.

remote sensing

MDPI

Article

Independent Assessment of Sentinel-3A Wet Tropospheric Correction over the Open and Coastal Ocean

Maria Joana Fernandes [1,2,*] **and Clara Lázaro** [1,2]

[1] Faculdade de Ciências, Universidade do Porto, 4169-007 Porto, Portugal; clazaro@fc.up.pt
[2] Centro Interdisciplinar de Investigação Marinha e Ambiental (CIIMAR/CIMAR), Universidade do Porto, 4050-123 Porto, Portugal
* Correspondence: mjfernan@fc.up.pt; Tel.: +351-220-402-452

Received: 25 January 2018; Accepted: 14 March 2018; Published: 20 March 2018

Abstract: Launched on 16 February 2016, Sentinel-3A (S3A) carries a two-band microwave radiometer (MWR) similar to that of Envisat, and is aimed at the precise retrieval of the wet tropospheric correction (WTC) through collocated measurements using the Synthetic Aperture Radar Altimeter (SRAL) instrument. This study aims at presenting an independent assessment of the WTC derived from the S3A MWR over the open and coastal ocean. Comparisons with other four MWRs show Root Mean Square (RMS) differences (cm) of S3A with respect to these sensors of 1.0 (Global Precipitation Measurement (GPM) Microwave Imager, GMI), 1.2 (Jason-2), 1.3 (Jason-3), and 1.5 (Satellite with ARgos and ALtika (SARAL)). The linear fit with respect to these MWR shows scale factors close to 1 and small offsets, indicating a good agreement between all these sensors. In spite of the short analysis period of 10 months, a stable temporal evolution of the S3A WTC has been observed. In line with the similar two-band instruments aboard previous European Space Agency (ESA) altimetric missions, strong ice and land contamination can be observed, the latter mainly found up to 20–25 km from the coast. Comparisons with the European Centre for Medium-Range Weather Forecasts (ECMWF) and an independent WTC derived only from third party data are also shown, indicating good overall performance. However, improvements in both the retrieval algorithm and screening of invalid MWR observations are desirable to achieve the quality of the equivalent WTC from Jason-3. The outcome of this study is a deeper knowledge of the measurement capabilities and limitations of the type of MWR aboard S3A and of the present WTC retrieval algorithms.

Keywords: Sentinel-3; satellite altimetry; microwave radiometer; wet tropospheric correction; wet path delay; sensor calibration

---

1. Introduction

Starting with European Remote Sensing 1 (ERS-1) in 1991 and TOPography Experiment (TOPEX)/Poseidon in 1992, satellite missions carrying radar altimeters have resulted in invaluable continuous measurements being obtained over the ocean, ice, rivers and lakes for more than 25 years. This has been made possible through the joint effort of various space agencies: the European Space Agency (ESA), the Centre National d'Études Spatiales (CNES), European Organisation for the Exploitation of Meteorological Satellites (EUMETSAT), the National Aeronautics and Space Administration (NASA), the Jet Propulsion Laboratory (JPL), the National Oceanic and Atmospheric Administration (NOAA), and the Indian Space Research Organization (ISRO). ESA supported the 35-day repeat missions ERS-1 (1991), ERS-2 (1995), and Envisat (2002), the geodetic mission CryoSat-2 (2010), and more recently, the 27-day repeat mission Sentinel-3A (2016), the latter in collaboration with EUMETSAT. Satellite with ARgos and ALtika (SARAL), a 35-day mission follow-up of Envisat, is a joint effort between the CNES and ISRO. Partly or in collaboration, NASA, NOAA, JPL, CNES, and EUMETSAT have been responsible for the so-called reference 10-day repeat missions: TOPEX/Poseidon (1991), Jason-1 (2002), Jason-2 (2008), and Jason-3 (2016) [1]. For a detailed and update review of satellite altimetry, see e.g., [2].

These missions have been designed to provide complementary spatial and temporal resolutions, allowing e.g., a better characterisation of the mesoscale oceanic circulation [3,4], the determination of accurate global and regional sea level trends [5–9], and unprecedented observations over river, lakes and reservoirs [10,11] as well as ice surfaces [12].

Satellite altimetry allows the determination of the height of e.g., the sea or lake surface above a reference ellipsoid by means of the following equation [13]:

$$h = H - R = H - R_{obs} - \Delta R \tag{1}$$

where H is the spacecraft height above a reference ellipsoid provided by precise orbit determination, referred to as an International Terrestrial Reference Frame, R_{obs} is the measured altimeter range corrected for all instrument effects, and R is the corresponding range corrected for all instrument, range, and geophysical effects. The term ΔR includes all corrections that need to be applied to the observed range due to the signal propagation delay through the atmosphere, its interaction with the sea surface and terms related with specific geophysical phenomena being given by Equation (2):

$$\Delta R = \Delta R_{dry} + \Delta R_{wet} + \Delta R_{iono} + \Delta R_{SSB} + \Delta R_{tides} + \Delta R_{DAC} \tag{2}$$

The first four terms in Equation (2) are the range corrections, accounting for the interaction of the altimeter radar signal with the atmosphere (dry, wet and ionospheric corrections) and with the sea surface (sea state bias). The last two terms refer to geophysical phenomena, dynamic atmospheric correction, and tides (ocean, load, solid earth and pole tides), which should be removed if they are not part of the signals of interest. A detailed description of the range and geophysical corrections can be found e.g., in [13,14].

Equation (1) evidences that the determination of an accurate surface height requires the knowledge of the spacecraft orbit, the range measurement, and of all correction terms in Equation (2) with the same accuracy. Amidst these terms is the wet tropospheric correction (WTC), due to the water vapour and liquid water content in the atmosphere. Since these variables have large space–time variability, if not properly modelled, the WTC is one of the major sources of uncertainty in many satellite altimetry applications [13,15,16].

Due to its variability, the most precise way to account for the WTC in satellite altimetry is by means of collocated measurements from a microwave radiometer (MWR), obtained on the same spacecraft. Two main types of radiometers have been deployed on the altimeter satellites: three-band radiometers on board the reference missions and two-band radiometers on all ESA missions and SARAL. Over the open ocean they provide accurate WTC retrievals within 1 cm [17].

The algorithms adopted in the WTC retrieval from the measured brightness temperatures (TBs) in the various spectral bands have been designed for the open ocean, assuming a constant ocean emissivity. In the presence of other surfaces such as land or ice, the measurements lay outside the predicted validity interval and the observations become unusable. Moreover, these instruments possess large footprints from 10 to 40 km depending on the frequency [18,19]. Therefore, in the coastal regions they sense land well before the altimeter, originating bands of invalid measurements around the coastline of a width of 10–40 km. In recent years, several methods have been developed to improve the MWR-derived WTC in the coastal zones, extending the validity of the correction up to the coast. An overview of these methods is given in [16].

Building directly on the proven heritage of ERS-1, ERS-2, and Envisat, launched on 16 February 2016, Sentinel-3A (S3A) carries a suite of innovative instruments that include [20]: (1) the Sea and Land Surface Temperature Radiometer (SLSTR) based on Envisat's Advanced Along Track

Scanning Radiometer (AATSR), to determine global sea surface temperatures to an accuracy of better than 0.3 K; the Ocean and Land Colour Instrument (OLCI), based on Envisat's Medium Resolution Imaging Spectrometer (MERIS), with 21 bands, (compared to 15 bands on MERIS), a design optimised to minimise Sun-glint, and a resolution of 300 m over all surfaces; a dual-frequency (Ku and C-band) advanced Synthetic Aperture Radar Altimeter (SRAL) developed from Envisat RA-2, CryoSat Synthetic Aperture Interferometric Radar Altimeter (SIRAL) and Jason-2/Poseidon-3, providing accurate surface topography measurements; and a microwave radiometer for accurate measurement of the wet pat delay of the SRAL observations [20].

SRAL and the MWR constitute the topography package, providing accurate measurements of sea surface height, significant wave height, and wind speed, essential for ocean forecasting systems and climate monitoring. Accurate SRAL measurements also extend to sea-ice, ice sheets, rivers, and lakes. Being the first altimeter mission to operate, globally, a radar altimeter in the SAR closed burst mode, with an improved along track resolution of about 300 m [21], Sentinel-3A is a pioneer and challenging mission in many aspects, pushing the experts of the various fields to the limit to tune the retrieval algorithms to its instruments and exploit the derived geophysical parameters.

As mentioned above, amongst its suite of instruments, Sentinel-3A carries a dual channel (23.8-GHz and 36.5-GHz) microwave radiometer aiming at the retrieval of the wet path delay, which provides important support for the quality of the SRAL measurements. Due to their instrumental characteristics and retrieval algorithms, the two-band MWRs deployed on ESA altimeter missions are known for their good performance in the open ocean [22]. However, when they approach the coast, the retrieval algorithm, which was designed for surfaces with ocean emissivity, generates very noisy values, as the footprint encounters surfaces with different levels of emissivity. The same happens at high latitudes in regions covered with ice [23,24].

This work aims at performing an independent assessment of S3A MWR-based WTC, in the open and coastal ocean, in support of S3A data improvement and exploitation.

The validation is performed by means of comparisons with independent data sets namely: wet path delays derived from the Global Precipitation Measurement (GPM) Microwave Imager (GMI); Global Navigation Satellite System (GNSS)-derived path delays determined at coastal stations; and wet path delays from the MWR on board Jason-2 (J2), Jason-3 (J3), and SARAL/AltiKa. In addition, the overall along-track performance is compared against independent estimates obtained from the GNSS-derived Path Delay Plus (GPD+) algorithm [24] and from atmospheric models. From this thorough analysis, a deeper knowledge of the measurement capabilities and limitations of the dual-frequency radiometer aboard S3A and of the present WTC retrieval algorithms is expected.

This paper is organised in five sections. Section 2 describes the data sets and the methodology used in the S3A WTC assessment. Section 3 describes the results obtained with the various datasets. Finally, Sections 4 and 5 present the discussions of the results and conclusions, respectively.

2. Dataset Description

2.1. Sentinel-3 Data and the Radar Altimeter Database System (RADS)

Sentinel-3A flies in a near-polar Sun-synchronous orbit with a Local Time of the Descending Node (LTDN) at 10:00 h. It has a high inclination of 98.65°, altitude of about 814.5 km, a 27-day repeat cycle (385 orbits per cycle), and 104-km inter-track spacing at the equator [20]. Sentinel-3B will be on a similar orbit, with an 180° phase difference, the two satellites ensuring twice the spatial and temporal coverage.

The Sentinel-3 data used in this study are from the Sentinel-3 Surface Topography Mission (STM) Level-2 Non-Time Critical (NTC) products from the so-called "Spring 2017 Reprocessing Campaign" [25,26]. The product reference is SRAL/MWR L2 Instrument Processing Facility (IPF) (SM-2): version 06.07 from Processing Baseline 2.15. These data have been available in the Radar Altimeter Database System (RADS, http://rads.tudelft.nl/rads/rads.shtml) [27] as reprocessed version

1 (rep1), since mid-August 2017. In addition to the native S3A products, the website includes the orbit and state-of-the-art range and geophysical corrections present in RADS for all other altimeter missions. The data span approximately 10 months, from 15 June 2016 (cycle 05) to 15 April 2017 (cycle 16).

As mentioned above, Sentinel-3A MWR operates in two frequencies. The first, located at a water vapour absorption line (23.8 GHz), is the primary water vapour sensing frequency; the second (36.5 GHz) is an atmospheric window sensitive to surface emissivity and to cloud liquid water. In the dual-frequency microwave radiometers, which do not include any low frequency channel, the effect of surface roughness is often taken into account through the altimeter-derived wind speed or the backscatter coefficient.

The relation between the measured brightness temperatures of the various MWR channels and the wet path delay (WPD) or WTC has been empirically established by means of statistical regression methods. Two main types of algorithms have been used in the retrieval of the WTC from the TB of the various MWR channels: (1) a parametric log-linear algorithm adopted in the reference missions [17]; and (2) a neural network algorithm first developed for Envisat, later used in SARAL/AltiKa and Sentinel-3 [28].

The function adopted in the retrieval of the WPD from the MWR brightness temperatures is usually derived from TB simulations, obtained from a database of atmospheric and sea-surface scenes using a radiative transfer model. Several databases can be used. Traditionally, in the ESA missions the database is built from European Centre for Medium-Range Weather Forecasts (ECMWF) analyses, while in the reference missions, it is usually built from radiosonde observations (for the atmospheric profiles) and from microwave imaging radiometers such as the Special Sensor Microwave Imager (SSM/I), the SSM/I Sounder (SSM/IS), or the Advanced Very High Resolution Radiometer (to provide sea-surface parameters) [29]. In summary, a WPD retrieval algorithm consists in the empirical establishment of the inverse function to get the WPD from the simulated TB.

In [29], the parametric log-linear algorithm is compared with the neural network formalism. These authors also analyse the advantages of the three-band MWR flying in the reference missions versus the two-band MWR on board the ESA missions, as well as how well additional parameters such as the backscatter coefficient can replace the missing third channel in the ESA missions.

The MWR-derived WTC present in this product has been computed by a neural network algorithm based on five inputs [30]: brightness temperatures at 23.8 GHz and 36.5 GHz, Ku-band ocean backscatter coefficient (not corrected for the atmospheric attenuation), sea surface temperature (four seasonal tables, 2° resolution), and the lapse rate (decreasing rate of atmospheric temperature with altitude, from a climatological table, 1° resolution) [30]. This algorithm is an evolution of the one developed for Envisat that only used the first three inputs described above [28].

The S3A products also include composite wet tropospheric correction [30], based on both radiometer and model-based corrections over areas where the radiometer WTC is missing or invalid due to the proximity of land (coastal areas and/or radiometer gaps in open oceans) [30,31]. A first assessment of this correction is also performed.

In the sea level anomaly (SLA) variance analyses performed in Section 3, S3A SRAL data and all required range and geophysical corrections to compute SLA were extracted from RADS.

2.2. The Global Precipitation Measurement (GPM) Microwave Imager (GMI)

The GMI is a dual-polarization, multi-channel, conical-scanning, passive microwave imaging radiometer on board the GPM satellite, launched on 27 February 2014. GPM has a non-Sun-synchronous orbit, at a mean altitude of 407 km and inclination of 65°, covering the latitude band between ±65°, allowing a full sampling of the Earth approximately every 2 weeks. The orbit plane completes half (180°) a rotation relative to the Sun every 41.1 days. The spacecraft undergoes yaw manoeuvres every ~40 days to compensate for the Sun's changing position and prevent the side of the spacecraft facing the Sun from overheating [32].

The GMI has been designed with a strict calibration accuracy requirement, enabling the instrument to serve as a microwave radiometric standard. The GMI features, ensuring its high calibration accuracy, include: protection of the hot load from Sun intrusion, noise diodes on the low-frequency channels for a dual calibration system, and a reflective antenna coating [33].

The GMI data used in this study are gridded products of total column water vapour (TCWV), in binary format provided by Remote Sensing Systems (RSS, http://www.remss.com/). Two $0.25° \times 0.25°$ global grids per day are provided, one containing the ascending and the other the descending GPM passes. The RSS products based on the Version-8.2 algorithm and the Radiative Transfer Model have been used [32].

As described in [34], the WTC has been computed from these TCWV products using the expression by [35], deduced from temperature and humidity profiles from ECMWF model fields:

$$\text{WTC} = -\left(a_0 + a_1 \text{WV} + a_2 \text{WV}^2 + a_3 \text{WV}^3\right) \text{WV} \times 10^{-2} \qquad (3)$$

with $a_0 = 6.8544$, $a_1 = -0.4377$, $a_2 = 0.0714$, and $a_3 = -0.0038$. In Equation (3), WV is in centimetres, as provided in the TCWV products, and WTC results are in metres.

In [24] the WTC derived from a previous version of these products (Version 7, V7) was compared against the WTC derived from the SSM/IS on board the Defense Meteorological Satellite Program satellite series F16 and F17. The SSM/IS sensors can be regarded as a reference, due to their stability and independent calibration [36,37]. This comparison shows that the scale factor and offset between SSM/IS and GMI are 0.99 and −0.26 cm, respectively, while the Root Mean Square (RMS) of the differences between the two data sets is 0.79 cm and 0.77 cm, before and after the adjustment to SSM/IS, respectively. According to RSS, the Version 8 (V8) brightness temperatures from GMI are slightly different from the V7 brightness temperatures; however there are no significant differences between the V7 and V8 ocean products. In the sequel, in this work the GMI V8.2 products have been adopted without applying any calibration parameters. It should be noted that, in spite of the fact that SSM/IS are considered stable sensors, the S3A WTC was not directly compared against the corresponding WTC derived from the SSM/IS TCWV products, also available from RSS, since both S3A, F16, and F17 are in Sun-synchronous orbits with a phase difference of 5–6 h (S3A/F16) and 3–5 h (S3A/F17).

2.3. Microwave Radiometers on Board Jason-2, Jason-3, and SARAL

Launched on 8 June 2008, the Ocean Surface Topography Mission (OSTM) Jason-2 carries the Advanced Microwave Radiometer (AMR) operating at 18.7, 23.8 and 34 GHz. In comparison with the dual-frequency MWR aboard the ESA missions, the additional low frequency channel improves the WTC retrieval, particularly by adding the ability to reduce land effects near the coast [38].

During the first years of the mission, like its predecessors TOPEX/Poseidon and Jason-1, Jason-2 flew in the so-called reference orbit: non-Sun-synchronous, inclination of 66°, altitude of 1336 km, 10-day repeat cycle (127 orbits per cycle), and 315 km inter-track spacing at the equator [39]. Jason-3 was launched to the reference orbit on 17 January 2016. During the calibration phase of approximately 6 months (20 repeat cycles) it shared the same orbit with Jason-2, with a time difference of about 1 min. In October 2016, Jason-2 moved to the interleaved orbit (same characteristics but with the ground tracks at half distance between those of the reference orbit) and later on, on 20 June 2017, to a long-repeat orbit, at an altitude of roughly 1309.5 km.

Jason-3 carries the Advanced Microwave Radiometer 2 (AMR-2), which is similar to the AMR but with improvements in instrument thermal control and stability [40]. Jason-3 is the first altimeter mission to implement special spacecraft cold sky calibration manoeuvres (CSCMs), combined with vicarious on-Earth ocean and land target references, for improving the long-term climate calibration of the radiometer [41]. Currently, Jason-2 also performs routine CSCMs.

Launched on 25 February 2013, SARAL/AltiKa flies on the same orbit as ERS-2 and Envisat: Sun-synchronous with an LTDN at 18:00 h and a Local Time of Ascending Node (LTAN) at 06:00 h.

It has an inclination of 98.5°, altitude of about 800 km, a 35-day repeat cycle (501 orbits per cycle), and 80 km inter-track spacing at the equator [42]. SARAL caries the AltiKa instrument, provided by CNES, consisting of a Ka-band altimeter and an embedded dual-frequency Microwave Radiometer (23.8 GHz/37 GHz) similar to that of Envisat but with a smaller footprint (8 km in Ka-band and 12 km in Ku-band) [19].

2.4. WTC from GNSS at Coastal Stations

Zenith total delays (ZTDs) from a network of 60 stations with a good global coverage have been used. The details of this computation are given in [43], with only the main features being outlined here. These stations have been chosen to cover the various regions of the world with different WTC variability conditions (Figure 1) and to ensure that all zenith wet delays (ZWDs) have been obtained using the same computation parameters, thus ensuring their long-term stability. All stations are located near the coast, up to 100 km from the coast and with an orthometric height <1000 m. The second condition aims at reducing the errors due to the height dependence of the WTC.

As illustrated in the background map of Figure 1, derived from ECMWF operational model fields, the RMS of the WPD has a clear zonal dependence, ranging from only a few cm near the poles up to 35 cm at low latitudes. The patterns of the RMS of WPD are dominated by the zonal dependence, associated with the dependence of the WPD on temperature and the increase in water vapour in the equatorial regions. The variance of the WPD (not shown) also has a strong dependency on latitude, reaching a maximum near the tropics, over regions where the conditions for strong monsoons prevail, and a minimum near the poles, where the content of water vapour in the atmosphere is minimal. Maximum values occur in the Northern Hemisphere, where the Indian and Pacific oceans meet; in the Southern Hemisphere, the highest variability is found northwards of Australia.

The GNSS provides accurate (4–6 mm) values of the ZTD, the sum of the dry and wet components of the tropospheric delay, at station height, while the quantity of interest in this study is the zenith wet delay (ZWD), symmetric with respect to the WTC, at sea (zero height) level. The latter is obtained from the ZTD at station level by subtracting the dry correction or zenith hydrostatic delay (ZHD) derived from the ECMWF sea-level pressure (SLP) field using the modified Saastamoinen model [44] and reducing ZHD and ZWD fields to sea level using the expression by [45], with modifications introduced by [46].

It has been shown [46] that the hydrostatic component of the tropospheric delay can be estimated from global grids of sea level pressure available from ECMWF models with an accuracy of 1 to 3 mm at a global scale, provided an adequate model for the height dependence of atmospheric pressure is adopted. Therefore, using ZHD computed from ECMWF model fields according to [46], the ZWD can be determined from the GNSS with an accuracy greater than 1 cm.

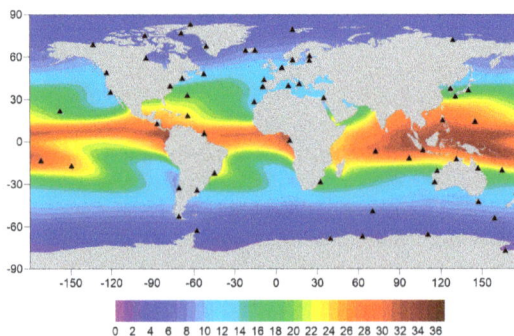

Figure 1. Location of a set of 60 Global Navigation Satellite System (GNSS) stations used in this study (adapted from [43]). The background map represents the Root Mean Square (RMS) of wet path delay (WPD) in cm.

2.5. GPD+ Wet Tropospheric Corrections

GNSS-derived Path Delay Plus (GPD+) provides wet tropospheric corrections derived by data combination, using space–time objective analysis, of all available wet path delays in the neighbourhood of a given point on the Earth. It has been designed: (1) to correct the WTC observations derived from the on-board MWR flagged as invalid due to various error sources: contamination by land, ice or rain, or instrument malfunction; and (2) to estimate improved WTC for satellites such as CryoSat-2, with no on-board MWR, for which only the model-derived WTC was available. In the first case, for all points with good MWR data, GPD+ keeps these values unchanged while for all invalid MWR measurements, a new estimate is obtained from the available observations. In the second case, a new estimate is obtained for all along-track points. Whenever the number of observations is null, GPD+ assumes the value of the first guess, which is the WTC from an atmospheric model: ECMWF operational for the most recent missions [47] and EMWF ReAnalysis (ERA) Interim for all missions with data prior to 2004 [23,24,48].

An important step in the GPD+ retrievals is the efficiency of the criteria to depict all invalid observations. This is done by using various flags provided with satellite data, when available, for example regarding the presence of land, ice, or rain in the radiometer footprint, or the distance from coast. Moreover, statistical criteria are established based on the comparison of MWR values and the corresponding model values, not only on the same point but also in the neighbouring along-track points. Figure 2 illustrates the invalid MWR points for Sentinel-3A cycle 06 (28.6% of the total ocean points). When only points with valid SLA are considered, the percentage of invalid MWR points varies, from cycle to cycle, from 11% to 17%.

For this study, two types of GPD+ WTC have been computed for S3A: (1) using only third-party data, i.e., wet path delays derived from a set of more than 800 GNSS stations and WTC from scanning imaging MWRs on board various remote sensing missions, hereafter designated GPD1; and (2) using all data including the S3A MWR measurements, provided they are valid, and only computing new estimates for the invalid MWR points, hereafter named GPD2.

The available WTC observations used in the GPD+ estimations, spanning the S3A mission period, include: valid observations from S3A´s on-board MWR; scanning imaging MWRs (SI-MWRs) from 11 different satellites; and GNSS-derived WTCs from more than 800 coastal stations.

Details on these data sets and on the GPD+ algorithm can be found in [24].

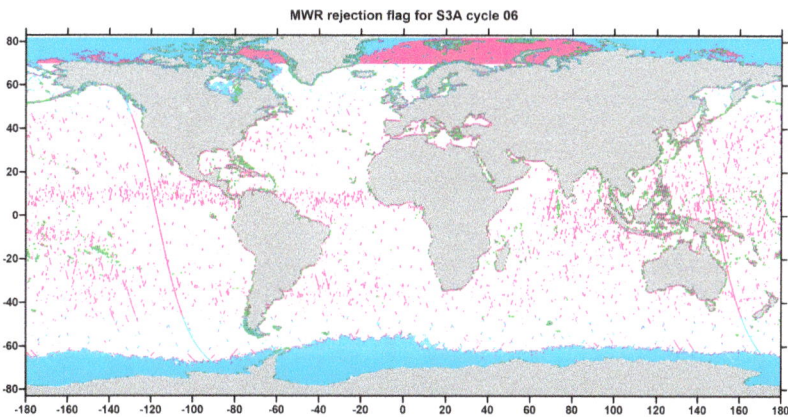

Figure 2. Sentinel-3A (S3A) points for cycle 06 with invalid microwave radiometer (MWR) observations: green—land contamination; blue—ice contamination; pink—rain, outliers, or additional condition such as all points above latitude 70°N or below 70°S.

2.6. The ECMWF Atmospheric Model

The atmospheric model used in this study is the ECMWF operational model. Global 0.125° × 0.125 grids of three single-level atmospheric parameters (sea level pressure (SLP), surface temperature (2-m temperature, 2T) and total column water vapour (TCWV)), available every 6 h have been used [49].

These model parameters are used both in the ZWD computations referred in Section 2.4 and to determine a model derived WTC for each S3A along-track position by space–time interpolation from the two closest grids, 6-h apart. In spite of its poor temporal sampling, the ECMWF model is used from here on as an additional point of comparison.

3. Results

Section 3 presents the various analyses performed aimed at a detailed assessment of the MWR-based wet tropospheric correction present on the Sentinel-3A products described in Section 2.1. In this section only the results are presented, their full discussion being described in Section 4.

3.1. Comparison with Other Microwave Radiometers

This section presents the comparison of Sentinel-3A MWR with other microwave radiometers on board the GMI, J2, J3 and SARAL. In these comparisons with other sensors, the wet path delay (WPD, symmetric with respect to the WTC) has been used. Unless otherwise mentioned, all statistical parameters detailed in this section (scale factor, offset, mean, and RMS of differences) refer to WPD. This is done to facilitate the representation and interpretation of the results.

3.1.1. Comparison with GMI

For the comparison with GMI, match points between GPM and S3A with a time difference ΔT < 45 min and within a distance ΔD < 50 km were computed. Only S3A points considered valid by the GPD+ algorithm [24] have been used (see Section 2.5). These criteria aim at removing S3A MWR observations contaminated by e.g., land, ice, rain, and outliers (Figure 2).

The use of match or collocated points is common practice in the comparison of a pair of sensors, thus comparing the measurements from both sensors in points with a given space–time difference. When analysing long periods of data, other approaches are possible such as e.g., in [50] where the authors organised the data sets in monthly bin boxes of three degrees, claiming that this approach is more favourable in e.g., the analyses of temporal variations of geophysical variables such as the WPD. Considering the short time span of the S3A dataset under inspection, match points are considered in this study.

Figure 3 illustrates the match points between GPM and S3A for the 10-month period of this study. The colour scale represents WPD differences between GMI and S3A in cm. Red colours mean that GMI measurements indicate wetter conditions than S3A, while blue colours indicate the opposite. The overall mean and RMS of the differences WPD(GMI)-WPD(S3A) are 0.17 cm and 0.95 cm, respectively (see Table 1).

Figure 4 illustrates the scattergram of the WPD from S3A against the WPD from the GMI (left) and against WPD difference between the GMI and S3A (right), using the whole set of match points (~219,000 points). The scale factor is 1.004 and the offset is 0.12 cm, indicating a very good overall agreement between the two radiometers, with the GMI measurements indicating slightly wetter conditions than S3A.

Figures 5 and 6 illustrate the time evolution of the WPD from both sensors and the respective differences. The top panel of Figure 5 represents the time evolution of the WPD from the GMI (blue) and S3A (pink) while the bottom panel shows the corresponding WPD differences between the GMI and S3A. A strong periodic signal of approximately 41 days (41.1 days) can be observed, due to GPM orbit plane rotation with respect to the Sun. It can be seen that this periodic pattern is due to the changes in the spatial sampling of the match points with time (cf. Figures 5 and 7 and discussion in Section 5).

When daily and 27-day RMS differences are computed (Figure 6), the same periodic pattern is shown, now at 82 days, and seem stable.

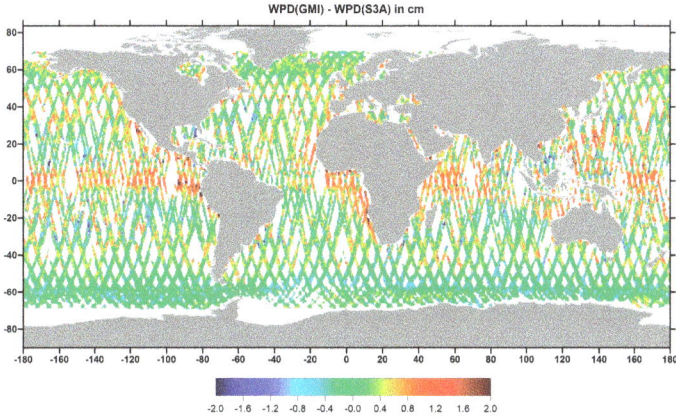

Figure 3. Spatial coverage of match points between S3A and the Global Precipitation Measurement (GPM) Microwave Imager (GMI) with time difference ΔT < 45 min and distance ΔD < 50 km, for S3A cycles 05–16, used in this study (~219,000 points). Colour scale indicates WPD differences between the GMI and S3A in cm.

Table 1. Statistical parameters of the comparison between the WPD from Sentinel-3A and other WPD sources. ECMWF: European Centre for Medium-Range Weather Forecasts; GPD: GNSS-derived Path Delay Plus; WTC: wet tropospheric correction; J2: Jason-2; J3: Jason-3; SARAL: Satellite with ARgos and ALtika.

WTC/S3A	Scale Factor	Offset (cm)	Mean Diff. (cm)	RMS Diff. (cm)
GMI/S3A	1.004	0.115	0.171	0.947
J2/S3A	1.010	−0.660	−0.556	1.213
J3/S3A	1.012	−0.831	−0.698	1.292
SARAL/S3A	0.955	0.557	0.029	1.536
ECMWF/S3A	1.013	−0.086	0.098	1.256
GPD1/S3A	1.005	0.173	0.251	0.967

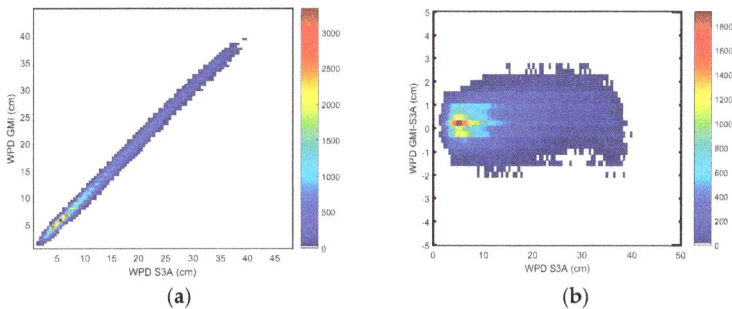

Figure 4. (a) Wet path delay from S3A versus WPD from the GMI; (b) WPD from S3A versus WPD differences between the GMI and S3A (~219,000 points).

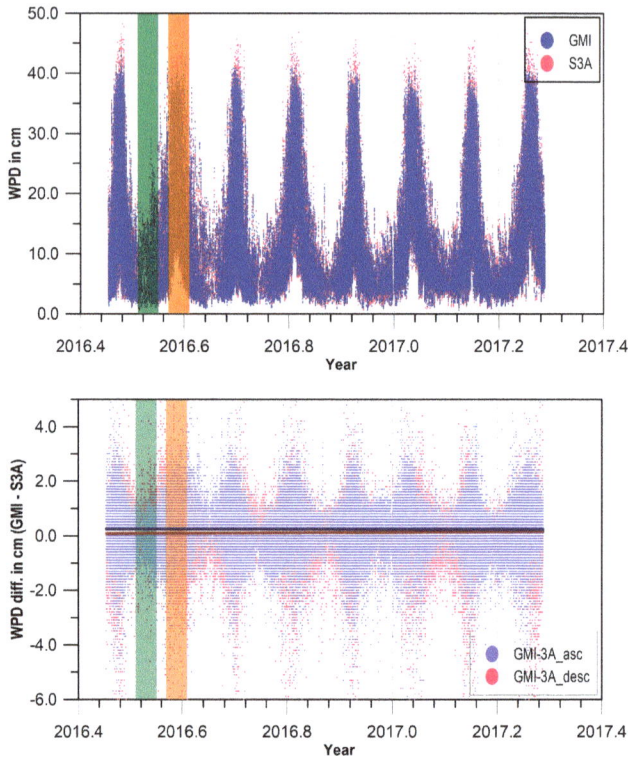

Figure 5. (**Top panel**): time evolution of the WPD from the GMI (blue) and S3A (pink); (**Bottom Panel**): time evolution of WPD differences between the GMI and S3A for ascending passes (blue) and descending passes (red). Colour bars refer to periods when the GMI/S3A match points are all located at high latitudes (green points in Figure 7) or low latitudes (orange points in Figure 7) to which correspond smaller or larger WPD variability.

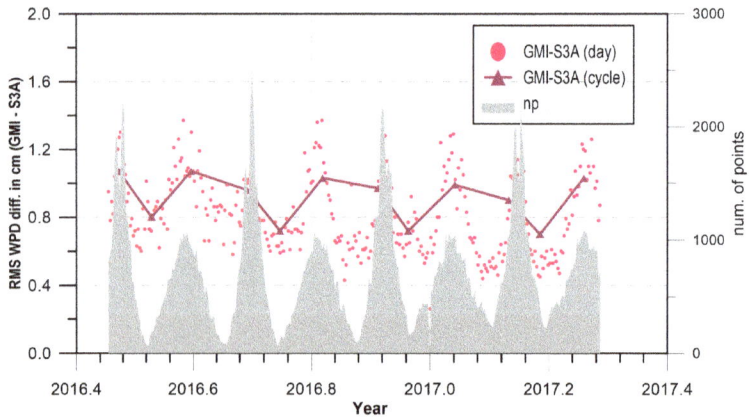

Figure 6. Time evolution of the daily RMS of the WPD differences between the GMI and S3A. The number of points is represented by "np".

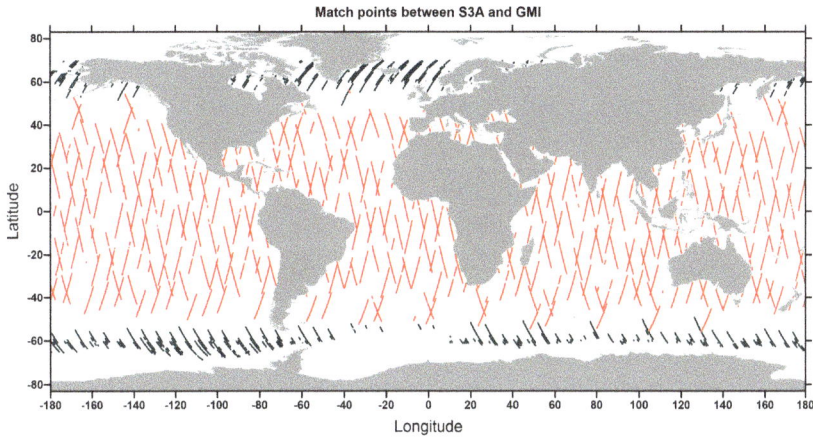

Figure 7. Match points between the GMI and S3A. Colours correspond to different time periods, indicated in Figure 5 by the corresponding colour bars.

3.1.2. Comparison with the MWR on Board Jason-2, Jason-3, and SARAL

To compare the WPD from S3A with that from Jason-2 (J2), Jason-3 (J3), and SARAL, the differences between the corresponding WPD at the crossovers between these missions have been computed. For this purpose, crossovers between S3A and Jason-2 and between S3A and Jason-3 with time difference $\Delta T < 180$ min have been computed. Note that both S3A and SARAL have Sun-synchronous orbits, with S3A LTAN at 22:00 h (LTDN at 10:00 h), and SARAL LTAN at 06:00 h (LTDN at 18:00). Thus, 240 min (4 h) is the minimum time interval to get crossovers between the two missions at low latitudes. With a 4-h difference, every ascending/descending S3A pass crosses a descending/ascending SARAL track. If a smaller period is considered, crossovers are all located at high latitudes, the corresponding values not being representative of the whole set of WTC variability conditions. For example, within a time interval of 180 min, all crossovers are above latitude 40° and below latitude −40°. In the sequel, for SARAL crossovers with time difference $\Delta T < 240$ min have been considered.

Figure 8 illustrates the spatial pattern of the WPD differences between J2, J3, SARAL, and S3A. The mean and RMS differences between J2, J3, SARAL, and S3A are presented in Table 1. Mean values are −0.56 cm, −0.70 cm and 0.03 cm, respectively, while the corresponding RMS are 1.21 cm, 1.29 cm, and 1.54 cm. The overall agreement with J2 and J3 is better than with SARAL, somehow expected since the crossovers with SARAL have a larger time difference.

Figure 9 shows the scattergrams of the WPD from S3A against the WPD from the various sensors (left panels) as well as against the difference between the WPD from each sensor and that from S3A (right panels). The scale factors and offsets for J2 and J3 are very similar: same scale factor (1.01) and offsets of −0.66 cm and −0.83 cm, respectively. The scale factor for SARAL is slightly different (0.96) with an offset of 0.56 cm. These scatterplots indicate that the best agreement is with J3 and the worst is with SARAL, due to the large time difference between the S3A/SARAL crossover points. Both Figures 8 and 9 indicate that J2 and J3 measure drier than S3A by about 0.6–0.8 cm, while SARAL measurements indicate wetter conditions than S3A by about 0.6 cm.

Figure 10 depicts the time evolution of the WPD differences between J2, J3, SARAL and S3A (daily and 27-day RMS values in centimetres). These results indicate that in spite of the small span of the data, there is an overall agreement between all analysed sensors.

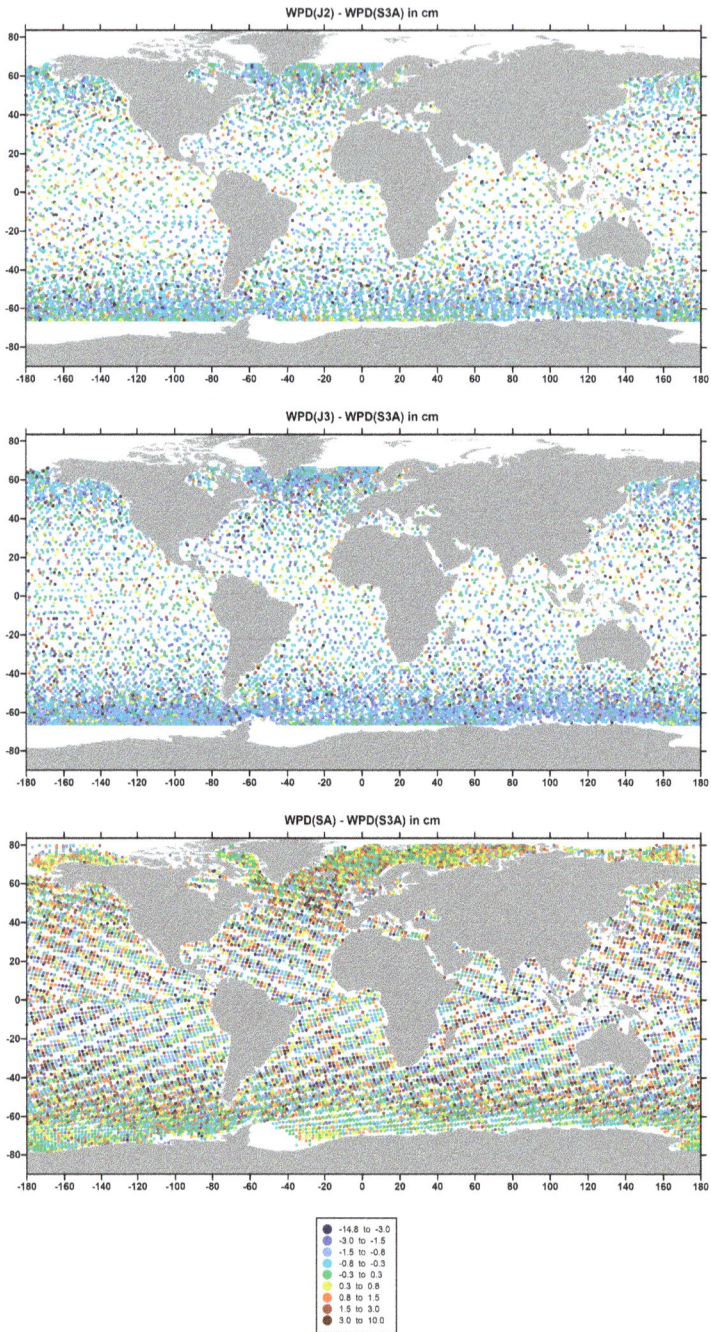

Figure 8. (**Top panel**): Crossover points between J2 and S3A with ΔT < 180 min (~11,600 points); (**Middle panel**): crossovers between J3 and the S3A with ΔT < 180 min (~12,600 points); (**Bottom panel**): crossovers between SARAL and S3A with ΔT < 240 min (~13,700 points).

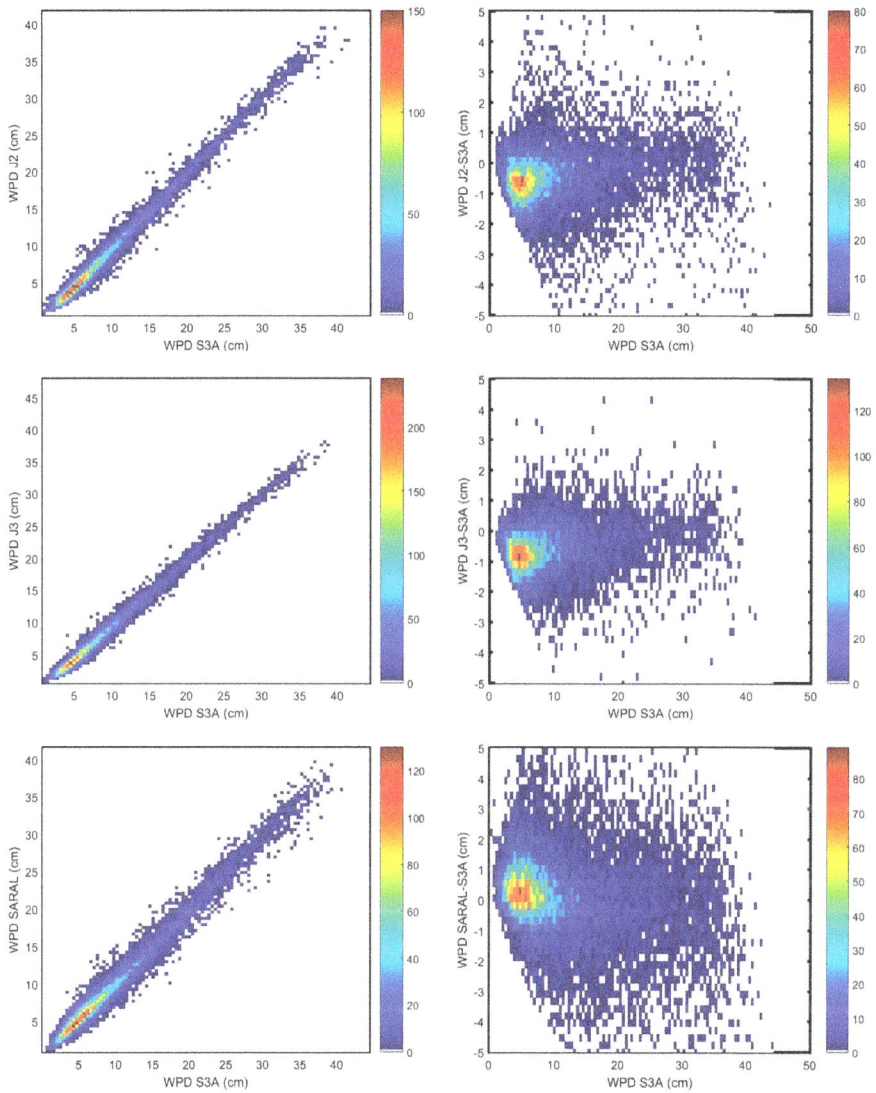

Figure 9. (**Left panels**): WPD from S3A versus WPD from J2 (**top**), J3 (**middle**), and SARAL (**bottom**), in cm; (**Right panels**): WPD from S3A versus WPD difference between J2 and S3A (**top**), between J3 and S3A (**middle**) and between SARAL and S3A (**bottom**).

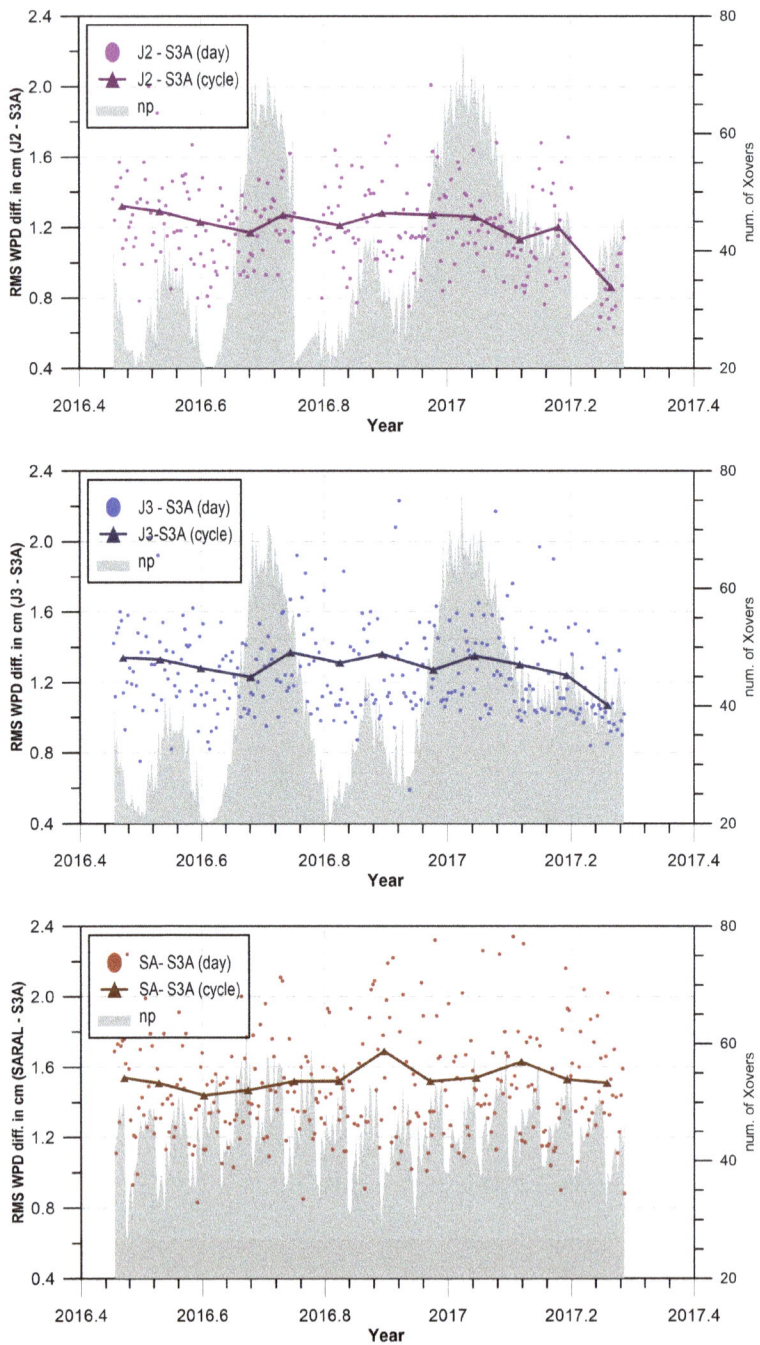

Figure 10. Time evolution of daily and 27-day RMS WPD differences between J2 and S3A (**Top**), between J3 and S3A (**Middle**), and between SARAL and S3A (**Bottom**), in cm.

3.2. Comparison with GNSS

Figure 11 shows the non-collocated comparison with the WTC derived at 60 GNSS stations, as described in Section 2.4. In the comparisons with S3A MWR, only valid MWR observations, except those related with the criterion for land contamination, have been selected. This means that the observations flagged as invalid due to all other error sources (ice and rain contamination and outliers) have been removed. In the comparisons with GPD+ all points have been used. For each epoch of the S3A measurement, a WTC from each GNSS station is linearly interpolated in time. Then, the WTC differences from all stations are binned in classes of distance from coast and the corresponding RMSs are computed.

As the GNSS stations are over land and the MWR measurement points are over the ocean, the distance from coast is also directly related with the distance between the points under comparison. Therefore, the RMS differences between the WTC from either the S3A MWR or GPD+ generally increase with the distance from coast, reaching a minimum at the maximum distance at which land contamination occurs in the MWR observations. Differences between the WTC computed at GNSS stations and derived from S3A MWR (or GPD+) are expected to increase as the distance to the GNSS station increases, as both measurements start to become decorrelated.

The RMSs of the differences between the WTC from GNSS and the WTC from S3A MWR (or the corresponding WPD) show that contamination is observed up to 20–25 km, in line with instrument specifications, the RMS of differences at these distances being about 1.8 cm. On the other hand, no land contamination is observed in the RMS of differences with respect to GPD+ (GPD1 mentioned above), being always smaller than those with respect to the MWR by about 0.3 cm, reaching 1.2 cm near the coast.

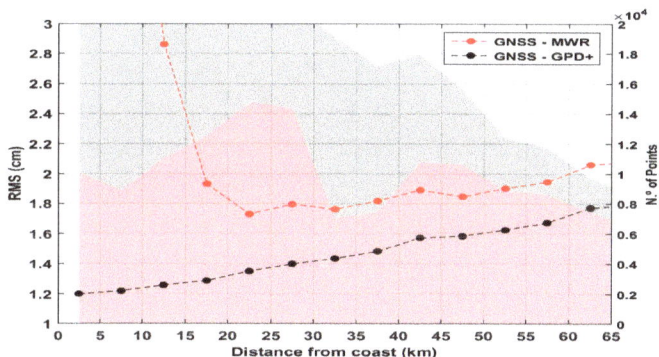

Figure 11. RMS differences between the WTC from the GNSS at coastal stations and the WTC from the S3A MWR, in cm. The grey and red coloured bars represent the number of points in each class of distance for GPD1 and the MWR, respectively.

3.3. Comparison with an Independent GPD+ WTC, ECMWF Operational Model and the Composite WTC

As mentioned in Section 2.5, two types of GPD+ WTC have been computed for S3A cycles 06 to 16: (1) GPD1—using only third-party data; and (2) GPD2—using all data including the S3A MWR. GPD1 WTC can be used as an external dataset, allowing an independent assessment of the S3A WTC. In the second case, GPD2 preserves the S3A MWR-derived WTC, whenever flagged as valid. Only new estimates are computed for the invalid points, using all available observations. The comparison between these two WTCs and between them and the S3A MWR-derived WTC (the latter comparison being performed only at valid MWR observations) gives further insight into the quality of the MWR observations, since it is performed using all S3A points and not only a subset (match points), as is the case of the comparison with the other MWR sensors.

Figure 12 shows an example where the different WTCs from S3A MWR, ECMWF operational model, GPD1 (top) and GPD2 (bottom) are compared for S3A pass 340, cycle 06. Figure 13 shows the corresponding plots for pass 462, same cycle. In these plots, only points with valid SLA are shown. Ice, land, and rain contamination can be observed in the MWR-derived WTC, more pronounced in pass 462 than in pass 340. Figure 14 illustrates the spatial distribution of the RMS of the WPD differences (cm) between GPD1 and S3A MWR for the whole period of study (cycles 05 to 16).

Figure 15 shows the time evolution of the RMS of the differences between the S3A MWR WTC and those from GPD1 and ECMWF, only for the valid MWR points. Note that for these points, GPD2 is equal to the MWR, so these differences are not shown.

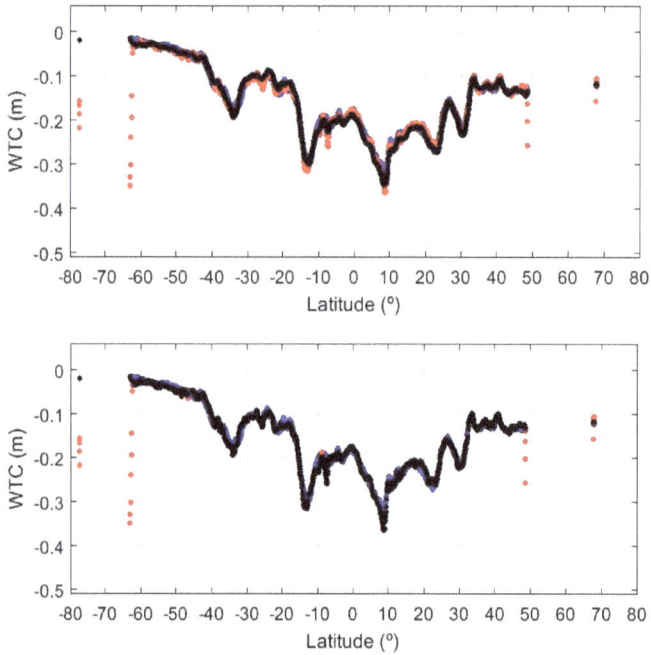

Figure 12. WTC (in metres) for S3A pass 340, cycle 06: ECMWF Operational (blue), MWR (red) and GPD (black, GPD1 in the top plot, GPD2 in the bottom plot) functions of latitude. The plot order is as mentioned in this caption. Thus, whenever the blue points cannot be seen, they are overlaid by the red and/or black points.

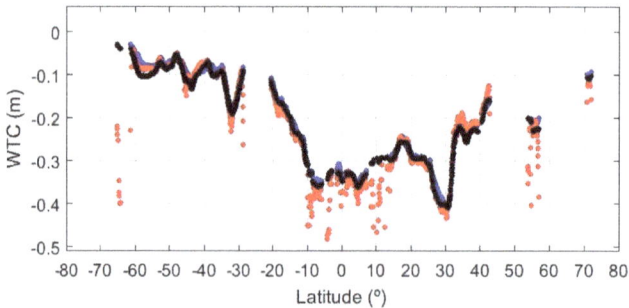

Figure 13. *Cont.*

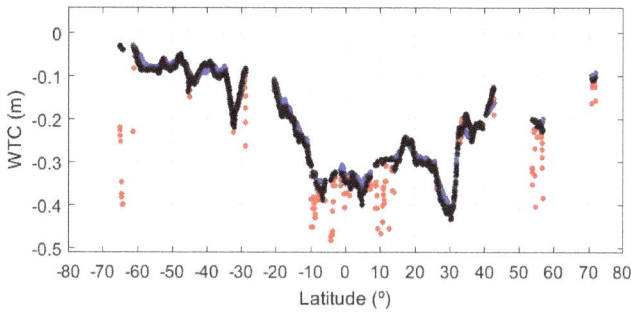

Figure 13. Same as in Figure 12 for S3A pass 462, cycle 06.

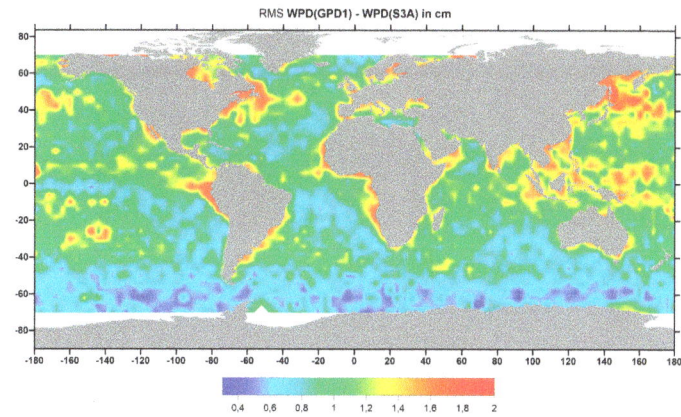

Figure 14. Spatial distribution of the RMS values of the WPD differences (cm) between GPD1 and the S3A MWR, for cycles 05 to 16.

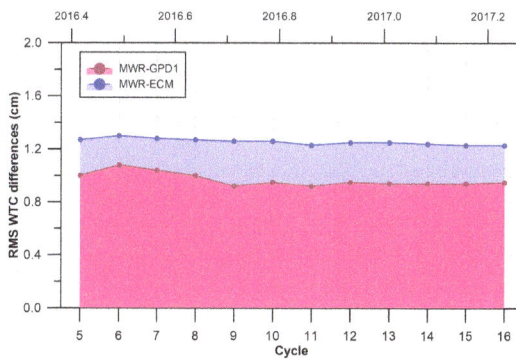

Figure 15. Time evolution of the RMS differences between the S3A MWR-based WTC and those from GPD1 and ECMWF.

The left panel of Figure 16 presents the scatterplots of the WPD from S3A versus the WPD from ECMWF (top) and GPD1 (bottom) in cm, while the right panel shows the corresponding scatterplots against the WPD differences between ECMWF and S3A (top) and between GPD1 and S3A (bottom), evidencing the good agreement between all WPD, better with respect to GPD1 than to the model, in line with the statistical parameters shown in Table 1.

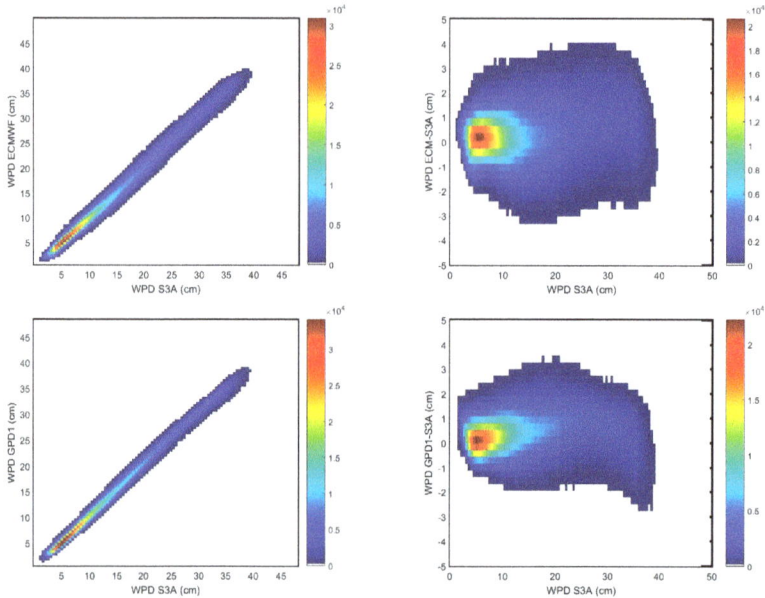

Figure 16. (**Left panels**): WPD from S3A versus WPD from ECMWF (**top**) and GPD1 (**bottom**), in cm; (**Right panels**): WPD from S3A versus WPD difference between ECMWF and S3A (**top**) and between GPD1 and S3A (**bottom**).

Aiming at inspecting the impact of the S3A WTC on the computation of SRAL-derived sea level anomalies in comparison with other WTC such as ECMWF, GPD1, and GPD2, various SLA variance analyses have been performed that can be divided into: (1) SLA along-track variance differences (weighted mean values per cycle and at collocated points, function of distance from coast and function of latitude); and (2) SLA analysis at crossovers (weighted mean cycle values and spatial pattern).

For this purpose, SLA datasets are first derived, for each S3A cycle, using the different WTC under comparison, a pair each time being compared. Then, for case (1), the difference between the weighted variance of each SLA dataset, computed using all along-track points, is estimated for each cycle. In addition, for the analysis of the SLA variance difference function of the distance from coast and function of latitude, the variance of co-located along-track SLA measurements for the whole study period and using each WTC is computed in bins of distance from the coast or latitude, respectively, and the differences are computed. For case (2), crossovers are first estimated using a pair of SLA datasets each time and the weighted variance of SLA differences is computed at cross-over points and estimated for each cycle. Moreover, the variances of the SLA differences at crossovers are computed in regular latitude × longitude grids (4° × 4°) and subtracted. Crossovers with a time difference less than 14 days, the minimum period to obtain S3A crossovers at all latitude bands, have been considered.

The results are shown in Figures 17–21, their discussion being described in Section 4.

The S3A products used in this study also include the composite WTC. Although [26] indicates that this correction is still not calibrated, a first assessment is also included here. Figure 22 depicts some representative examples of S3A passes where this WTC is compared against those from ECMWF, the MWR, and GPD2, showing that the present implementation of this correction evidences significant problems.

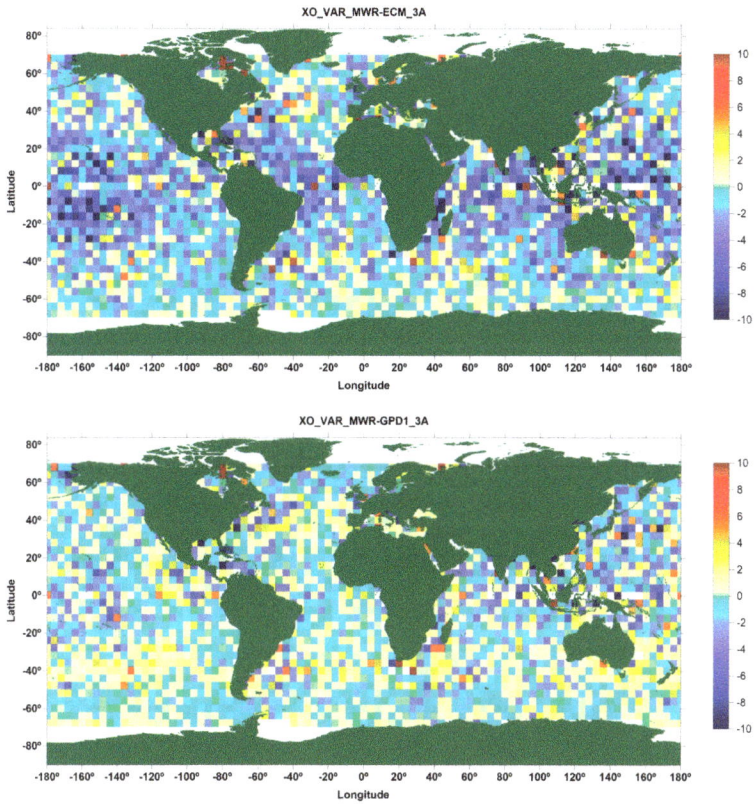

Figure 17. Spatial distribution of the weighted sea level anomaly (SLA) variance differences at crossovers, for SLA datasets computed using the MWR- and ECMWF-derived WTC (**Top**) and those using the WTC from the MWR and GPD1 (**Bottom**) for the period corresponding to S3A cycles 05 to 16. Only points with valid observations have been used.

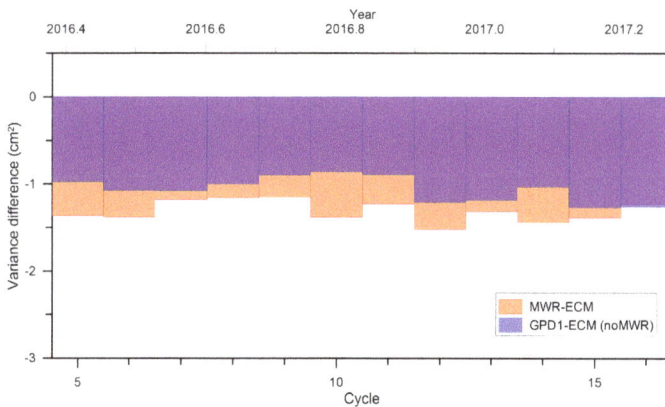

Figure 18. *Cont.*

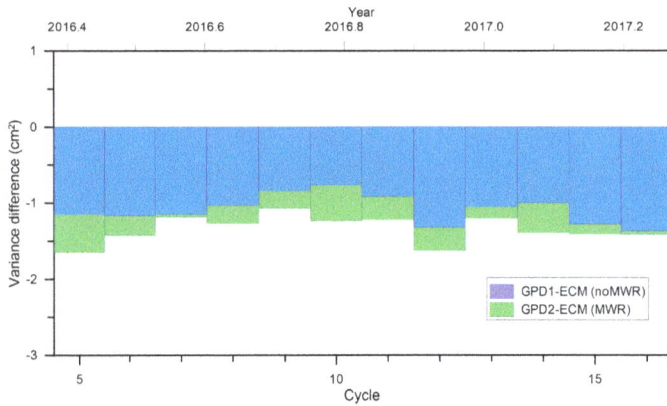

Figure 18. (**Top panel**): temporal evolution of weighted SLA variance differences at crossovers for SLA datasets computed using the WTC from the S3A MWR and ECMWF (orange) and the WTC from GPD1 and ECMWF (blue) using only points with a valid MWR; (**Bottom panel**): temporal evolution of weighted SLA variance differences at crossovers between SLA datasets computed using the WTC from GPD1 and that from ECMWF (blue) and between GPD2 and ECMWF (green), using all points with valid SLA.

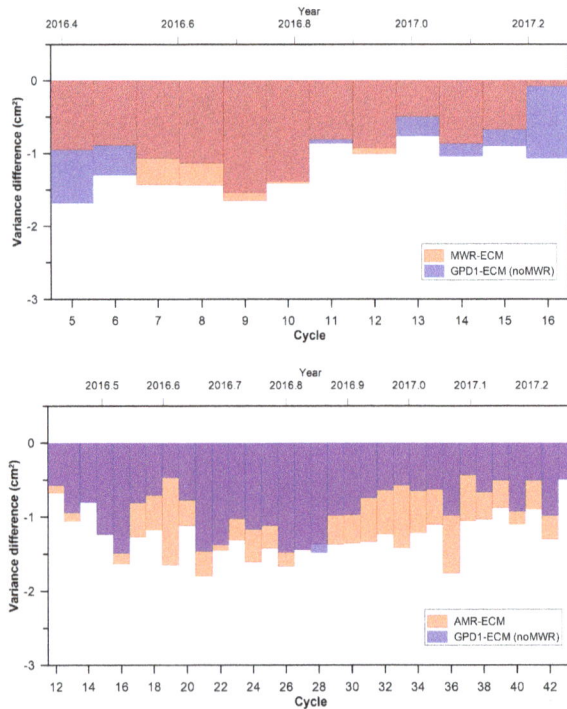

Figure 19. Temporal evolution of weighted along-track SLA variance differences for SLA datasets computed using the WTC from the on-board MWR and from ECMWF (orange) and those from GPD1 and ECMWF (blue) for S3A (**Top**) and J3 (**Bottom**). Only points with a valid MWR have been used.

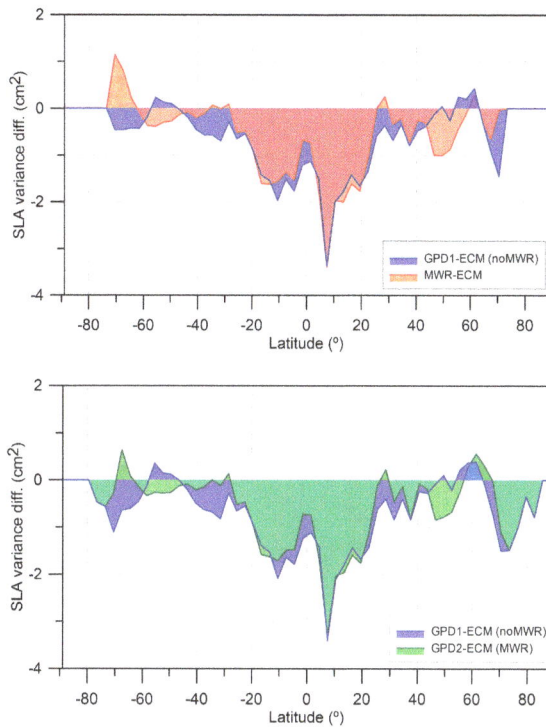

Figure 20. (**Top panel**): variance differences of SLA versus latitude, for SLA datasets computed using the WTC from the MWR and ECMWF (orange) and those from GPD1 and ECMWF (blue), over the period of S3A cycles 05 to 16 and using only points with a valid MWR; (**Bottom panel**): variance differences of SLA versus latitude between GPD1 and ECMWF (blue) and between GPD2 and ECMWF (green), over the period of S3A cycles 05 to 16 using all points with valid SLA.

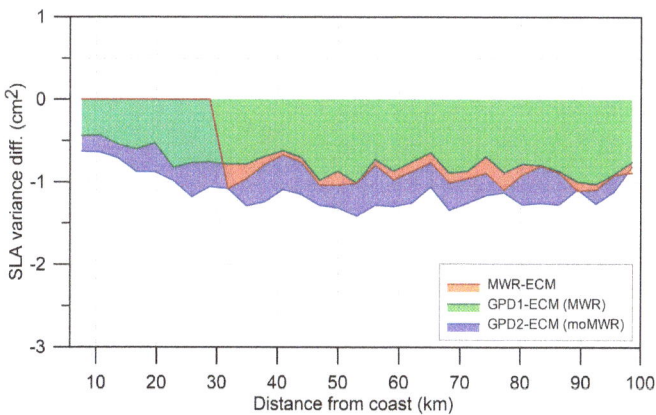

Figure 21. Variance differences of SLA versus distance from coast for SLA datasets computed using the WTC from MWR and ECMWF (orange), those from GPD1 and ECMWF (blue) and those from GPD2 and ECMWF (green) over the period of S3A cycles 05 to 16. In the first case only valid MWR points were selected while in the last two cases all points with valid SLA were considered

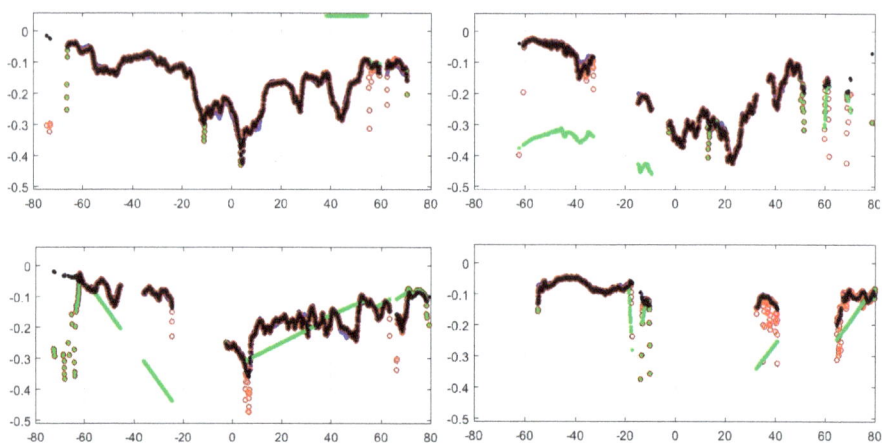

Figure 22. From top to bottom and left to right: WTC (in metres) for S3A cycle 06 passes 041, 646, 660, and 670, as a function of latitude (degrees). Shown are: ECMWF Operational (blue), MWR (red), GPD2 (black), and composite (green) WTC.

4. Discussion

This section presents a thorough discussion of the results presented in Section 3.

In terms of the global statistical parameters, GMI and Sentinel-3A agree very well, with mean and RMS of WPD differences (WPD(GMI)–WPD(S3A)) of 0.17 cm and 0.95 cm, respectively. The overall pattern of the WPD differences at match points shown in Figure 3 seems to indicate that at the low latitudes, where the WPD is larger, GMI measures wetter than S3A, while at the extreme southern latitudes, corresponding to regions of low WPD values, there is a tendency for GMI to measure drier than S3A. However, the scale factor (1.004) and small offset (0.12 cm) reveal a very good overall match between the two sensors, also demonstrated by the very small scatter of the scatter plot of the two WPDs illustrated in Figure 4.

The most curious feature of the comparison between GMI and S3A is the observed strong periodic pattern of 41 days. When analysing the way the spatial pattern of the match points changes with time (Figures 5 and 6), it can be observed that there are periods for which all match points between the two sensors are located only at high latitudes. An example of these periods is the one corresponding to the green bar in Figure 5, to which the points with the same green colour in Figure 7 (all above latitude 50°N or below latitude 50°S) correspond. During these periods, since the WPD at high latitudes is low (only a few cm) and with low variability, the differences between the two sensors are very small, with absolute values of less than 1 cm. On the contrary, during periods such as the one corresponding to the orange bar in Figure 5, the match points are located at the low latitudes, between ±50° (orange points in Figure 6). Over these periods, the WPD reaches higher values (up to 40–50 cm) and has larger variability, inducing larger differences between the two sensors. These results are a clear demonstration of the impact of data sampling in this type of study, which, if not properly accounted for, can lead to misinterpretations.

The authors of [26] report that the S3A MWR L1 brightness temperatures exhibit a difference of up to 1 K between ascending and descending tracks for the 23.8-GHz channel, with unknown origin. With a view to investigate if this potential bias can be observed in the current L2 products, the differences WPD (GMI)—WPD(S3A) have been represented in the bottom panel of Figure 5 separately for ascending (in blue) and descending (in red) tracks. No significant differences can be observed between the ascending and descending track differences. In terms of overall statistics, the S3A WPD differences with respect to GMI for the ascending tracks are about 1 mm larger than those for the descending tracks (compare dark blue with dark red lines in the bottom panel of Figure 5). This difference is too small, insignificant and cannot be related with the reported 1 K difference since the latter can lead to WPD differences ranging from 0.2 cm to 1 cm.

In spite of the small analysis period, the RMS differences (between GMI and S3A) computed for periods of the S3A cycles (27-day) evidence a long-term stable signal with an 82-day periodic pattern. This pattern occurs due to the fact that S3A orbit is Sun-synchronous, while GPM orbit is not. Due to its orbit characteristics, the GPM orbital plane takes 82 days to complete a full rotation with respect to the Sun and therefore with respect to the S3A orbital plane, explaining the observed periodic signal. In spite of these periodicities observed in the differences between GMI and the S3A MWR, due to the fact that GMI is on a non-Sun-synchronous orbit with a 65° inclination, the number of match points with S3A is larger than for example with any SSM/IS sensor, as these are on Sun-synchronous orbits, out of phase with respect to the S3A orbit by 3–6 h.

The comparison with the radiometers on board J2 and J3 indicate that on average these sensors measure drier than S3A by about 0.6–0.7 cm, while the RMS of the differences is within 1.2–1.3 cm. The scale factor (1.01), the negative offset of −0.7 and −0.8 cm and the blue predominant colour in the spatial pattern of the WPD differences shown in the top and medium panels of Figure 8 confirm this tendency.

For SARAL, the results are a bit different. The RMS of the differences is larger (1.5 cm) which can be explained by the fact that the crossover differences span a larger time interval (240 min versus 180 min used for J2 and J3). The scale factor of 0.96, offset of 0.6 cm, and the larger scatter observed in Figure 8 (bottom panel) and 9 (bottom panel) indicate a slightly larger difference between these two sensors, partly explained by the mentioned time difference between the corresponding crossovers.

The statistical parameters of the comparisons between S3A and the MWR aboard the other three altimeter missions are in agreement with the results presented in [24] and further work by these authors, comparing the same sensors with the SSM/I and SSM/IS sensors, which also indicate that J2 and J3 measure drier than the latter sensors by about 0.6 cm and 1 cm, respectively.

The time evolution of the daily and 27-day RMS differences between J2, J3, SARAL, and S3A illustrated in Figure 10 indicates a stable behaviour of the Sentinel-3A radiometer with respect to these sensors.

The comparisons with WTC derived at coastal GNSS stations show that land contamination can be observed in the S3A MWR observations up to 20–25 km from the coast. At this distance, the RMS of the differences between the WTC from S3A and the GNSS-derived ZWD is about 1.8 cm, increasing rapidly towards the coast, a clear indication that in the band of 0 km up to 25 km the MWR observations are not valid. On the contrary, no land contamination can be observed in the GPD1 WTC, derived solely from third party data. This stresses the need for robust criteria to remove the land-contaminated MWR observations and the important role of solutions such as GPD+ in providing valid WTC in the coastal regions.

Section 3.3 presents an extensive comparison between the S3A WTC and various corrections: GPD1 (from the GPD+ algorithm using only third party data), GPD2 (from the GPD+ algorithm, preserving the valid S3A MWR observations), the ECMWF operational model, and the composite WTC. In Figures 12 and 13, it can be observed that GPD1 and GPD2 are very similar, in spite of the fact that GPD1 does not use any observations from the S3A on-board MWR. It can also be concluded that, in regions with a valid MWR, the ECMWF operational model values are also very close, the main difference being that some, though important, small scale features are missing. These figures are clear illustration of typical S3A MWR-derived WTCs, where invalid observations can be observed near the coast, at high latitudes and at low latitudes, associated with heavy rain events.

The spatial distribution of the RMS of WPD differences between GPD1 and S3A MWR for the whole period of study (cycles 05 to 16), shown in Figure 14, indicates that the largest differences between these two WTCs are associated with regions with the largest WTC variability and ocean circulation patterns. Large differences also occur in some coastal regions, a possible indicator that some noisy MWR observations may still be present, in spite of the fact that all MWR measurements at distances from coast less than 30 km have been removed.

Both the statistical parameters in Table 1 and Figure 16 evidence that the S3A WPD is closer to GPD1 (RMS difference less than 1 cm) than to the ECMWF model (RMS difference 1.3 cm). The mean values of these differences are very small (0.1 cm and 0.2 cm for the comparison with GPD1 and ECMWF, respectively). The scale factors are both 1.01 and the offsets are also very small (0.2 cm, −0.1 cm). The time evolution of these differences, shown as 27-day RMS values in Figure 15, is stable.

The better agreement of S3A MWR-derived WTC with GPD1 in comparison with the ECMWF model is also observed in the scatterplots of Figure 16, where a smaller scatter can be observed for the former WTC pair. The better agreement of the S3A MWR with GPD1 than with the ECMWF model is explained by the fact that GPD1 WTC is also based on observations and the model has a poor temporal sampling of 6 h.

Figures 17–21 present the assessment of the S3A MWR-based WTC by means of SLA variance analysis. In this analysis, the WTC from S3A MWR, GPD1, and GPD2 are compared against the ECMWF model. In this way, both the performance of each WTC with respect to this model and their relative performance can be analysed.

The SLA variance analysis at crossovers presented in Figures 17 and 18 indicates that the S3A MWR reduces the variance both with respect to ECMWF and GPD1 (Figure 18, top panel), the first reduction being significant (mean cycle values of 1.1–1.5 cm^2), while the latter (not shown, inferred from Figure 18, top panel) is quite small (mean cycle values of 0.1–0.3 cm^2). The same is observed when comparing the performance of GPD1 and GPD2 against ECMWF (Figure 18, bottom panel). GPD1 reduces the mean cycle variance by 0.8–1.4 cm^2 while GPD2 reduces by 1.0–1.6 cm^2. Note that the comparisons involving the MWR only use the valid MWR points, while the comparisons involving only GPD1 and GPD2 use all points with valid SLA.

While the results from the variance analysis at crossovers indicate that the S3A MWR consistently reduces the SLA variance with respect to the WTC that only incorporates non-collocated radiometers (GPD1) and that the latter increases the variance with respect to the GPD+ WTC that makes use of the S3A MWR (GPD2), the same does not happen in along-track SLA variance analysis. In the along-track SLA variance differences, computed cycle by cycle (Figure 19), in most cycles GPD1 leads to a slightly larger SLA variance reduction with respect to ECMWF than the GPD WTC that incorporates the S3A MWR (GPD2). This is also demonstrated in Figures 20 and 21, which present the along-track variance differences in function of latitude and distance from coast, respectively. The top panel of Figure 20 shows that, in comparison with the MWR, GPD1 conducts to a slightly larger variance reduction with respect to ECMWF, except for some high-latitude bands. The bottom panel of the same figure shows that, in comparison with GPD2, with the exception of some high latitude bands, GPD1 also leads to a slightly larger SLA variance reduction with respect to ECMWF. In the first comparison only valid MWR points have been used, while in the latter all points with valid SLA have been analysed. This result is more evident in the SLA variance analysis function of distance from coast (Figure 21) where GPD1 consistently reduces the variance with respect to both MWR and GPD2.

This result is not expected since, in all previous analyses performed by the authors, the WTC derived from observations of the on-board MWR always reduced the SLA variance with respect to the WTC that only uses non-collocated observations, in all types of diagnoses and when only valid MWR points are selected. As an example, the bottom panel of Figure 19 illustrates for J3 the same variance analysis shown in the top panel for S3A, for the same period, proving that, contrary to S3A, the J3 AMR-2 WTC is consistently better than that of GPD1. Similar results have been obtained for Envisat, confirming that the GPD+ WTC that incorporates the valid observations from the Envisat radiometer consistently reduces the SLA variance with respect to the correction that only uses third party observations. These results indicate that although the overall performance of the S3A MWR seems good, improvements are still required to retrieve a WTC with the quality of that for J3.

In the latitude plots (Figure 20) some ice contamination can still be observed in the MWR-derived WTC and GPD2, suggesting that more robust editing is still desirable in the polar regions. We recall that a strong editing has already been performed by rejecting all MWR observations above latitude 70°N and below 70°S. Consequently, the preliminary version of the GPD+ WTC that makes use of the on-board MWR (GPD2) can still be improved, even with the present MWR dataset.

The examination of the composite correction present on the analysed S3A products shows that the implementation of this WTC still has serious problems, depicted in Figure 22. Moreover, the correction is not present in a large percentage of points, e.g., 21% of the points with valid SLA for cycle 06. Similar to the GPD+ WTC, the composite WTC aims at generating a continuous correction, valid everywhere, including the coastal zones and high latitudes. Also, similar to the GPD2 type of correction described in this paper, it also preserves the valid MWR values. The main difference between GPD+ and the composite WTC is that, on every invalid point, the former retrieves a new estimate by data combination of all available observations, while the latter uses the model values, adjusted to the closest valid MWR points. This requires robust criteria to detect valid/invalid MWR values, otherwise the model will be adjusted to spurious MWR observations, resulting in large biases as shown e.g., in the bottom left panel of Figure 22 or straight lines as shown in the right panels of the same figure. Similar behaviour has been observed, though in only a few occurrences, in the composite correction present in the Archiving, Validation and Interpretation of Satellite Oceanographic data (AVISO) Corrected Sea Surface Heights (CORSSH) products of TOPEX/Poseidon [23]. Due to potential implementation problems and the fact that GPD+ use observations while the composite is solely based on model values, it has been shown that the GPD type of WTC is a significant improvement with respect to the composite WTC, particularly in the coastal zone [9,23].

5. Conclusions

This study presents an independent assessment of the wet tropospheric correction derived from the two-band microwave radiometer deployed on Sentinel-3A, present on the L2 NTC products from the "Spring 2017" Reprocessing Campaign", Processing Baseline 2.15.

Studies such as this one play an important role in the improvement of altimeter-derived products, as satellite-derived data improvement is a long and continuous process. Sentinel-3 is a challenging mission in many aspects, with the research community having a long way to go to fully exploit the capabilities of its instruments, in particular SRAL. Six years after its end of mission, Envisat data are still being reprocessed and the reprocessing of TOPEX/Poseidon data has been delayed for several years.

Considering the relatively short time span of the dataset (10 months), the overall performance of the S3A MWR seems stable, with no drifts or irregular behaviour having been observed. A pronounced periodic signal has been observed in the differences with respect to GMI, due to the orbit configurations of the spacecraft housing these sensors.

When considering the linear adjustment of S3A MWR-derived WPD to those from four different sensors, scale factors very close to 1 have been obtained: GMI (1.00); J2 and J3 (1.11); and SARAL (0.96). The corresponding values for the WTC are the same since WTC is symmetric to WPD. All WPD offsets have absolute values of less than 1 cm: GMI (0.1 cm), J2 (−0.7 cm), J3 (−0.8 cm), and SARAL (0.6 cm). The corresponding offsets for the WTC are symmetric to those for WPD. The mean WPD differences have absolute values below 1 cm: GMI (0.2 cm), J2 (−0.6 cm), J3 (−0.7 cm) and SARAL (0.0 cm), the mean WTC differences being the corresponding symmetric values. RMS WPD (or WTC) differences (cm) of 0.95 (GMI), 1.3 (J2), 1.3 (J3), and 1.5 (SARAL) demonstrate a good agreement with all these sensors, in particular with GMI.

The comparison with GNSS shows land contamination in the S3A MWR observations up to 20–25 km and RMS differences between S3A MWR WTC and GNSS-derived ZWD of 1.8 cm at these distances. These results are in agreement with those found for the radiometers of other altimetric missions, for which the corresponding RMSs of differences are 1.7–1.8 cm [43].

Extensive comparison has been performed with the ECMWF model-derived WTC and two WTC versions from the GPD+ algorithm: one using only third party data (GPD1) and another preserving the valid S3A MWR points (GPD2). Direct WTC comparisons show small statistical parameters for the differences between all these corrections. As expected, smaller RMS values have been observed for the differences with respect to the GPD+ WTC (1 cm for GPD1) than for the differences with respect to ECMWF (1.3 cm). The scale factors with respect to both GPD1 and ECMWF are 1.01 and both offsets and mean differences have absolute values smaller than or equal to 0.2 cm.

As expected, all SLA variance analyses evidence a reduction of 1–2 cm^2 of the SLA computed with the S3A MWR WTC with respect to the ECMWF model, when mean cycle values are considered. The SLA variance analysis at crossover points indicates a small (mean cycle values less than 0.5 cm^2) but consistent variance reduction with respect to GPD1, however, the opposite is observed in the along-track variance analysis. This seems to suggest that improvements in both the retrieval algorithm and the criteria used to detect valid/invalid MWR observations are required to achieve a WTC with accuracy similar to the AMR-2 WTC for Jason-3. These results also highlight that the along-track comparison with the GPD1 type of WTC provides complementary and very important insight into the quality of the S3A MWR-derived WTC, not evident in the assessment against the other.

Similar to other dual-frequency radiometers on board the previous ESA missions, strong ice and land contamination is observed, the former in particular making the establishment of validation criteria for the MWR observations difficult at the high latitudes. Consequently, all observations with latitude absolute values larger than 70° have been rejected.

The composite WTC present in the analysed products was revealed to have implementation problems and was unsuitable for use, improvements being required in future versions of S3A data.

As a whole, this study contributes to a better knowledge of the wet path delay affecting satellite altimeter observations. In spite of the good overall performance of the S3A MWR when compared against other state-of-the-art radiometers, current limitations are also identified. Once a tuned retrieval algorithm is achieved, meeting the state-of-the art performances of current and past similar instruments, the inability of this type of radiometer to measure over non-ocean surfaces such as e.g., land, ice, or wetlands, prevent its use in these regions. Therefore, in regions where accurate satellite altimetry measurements are of crucial importance such as coastal or inland water zones, and accurate MWR-derived WTC retrieval is not possible, alternative methods such as the GPD+ type of corrections must be used.

The GPD+ WTC solely based on third-party observations, together with microwave radiometers on board other altimeter missions and GMI, were revealed to be very useful and independent tools to validate the Sentinel-3 radiometer-based wet tropospheric correction. Future work includes the monitoring of upcoming versions of this dataset and generation of updated versions of the GPD+ WTC making use of ameliorated MWR data, which will contribute to the generation of an accurate version of this important range correction for S3A SRAL measurements over the whole ocean, including coastal zones and at high latitudes.

In regions such as coastal and continental water zones, radiometers with additional high-frequency channels, such as the one being built for Jason-CS/Sentinel-6 [51] will provide smaller footprints, of great relevance for resolution of smaller scales of variability of the WTC.

Over inland water regions and to some extent in coastal regions, the availability of tropospheric corrections at high rate (frequency higher than 1 Hz), either from radiometers with smaller footprints or from high-resolution atmospheric models is also of great relevance, in particular in the exploitation of Sentinel-3 high-rate altimeter data. The advantage of having range corrections at a high rate applies not only to the WTC but also to the dry component of the tropospheric path delay, the dry tropospheric correction (DTC). In regions of sharp topography variations such as some steep coastal regions, lakes or rivers, one of the factors with most impact on the retrieval of accurate tropospheric corrections is their height dependence, particularly large for the DTC (1 cm per 40 m of height variation) [13].

The future of tropospheric corrections for satellite altimetry includes: WTC from MWRs with additional high-frequency channels and improved retrieval algorithms (over the open and coastal ocean and over lakes); and WTC and DTC from high-resolution atmospheric models, both computed at surface height using accurate Digital Elevation Models (DEM) or the altimeter-measured surface height (over inland water regions). In all cases, high-rate (e.g., 20 Hz) corrections are desirable to better account for spatial variability of km order or less, in particular the variability associated with the surface height variations.

Acknowledgments: This work was supported by the European Space Agency in the scope of the SAR Altimetry Coastal and Open Ocean–Performance Exploitation and Roadmap Study (SCOOP) project, Subcontract to SCOOP Contract N. 4000115385/15/I-BG. It is also a contribution to the Validation of Coastal ALtimetry from Sentinel-3 (VOCALS3) project. The authors would like to thank RADS for providing the S3A data, Remote Sensing Systems for providing the GMI products, and the European Centre for Medium-Range Weather Forecasts (ECMWF) for making the ECMWF operational model available.

Author Contributions: M.J.F. conceived, designed and performed the experiments and wrote the paper; C.L. contributed to data analysis editing and review of the paper.

Conflicts of Interest: The authors declare no conflicts of interest.

References

1. Benveniste, J. Radar altimetry: Past, present and future. In *Coastal Altimetry*; Vignudelli, S., Kostianoy, A., Cipollini, P., Benvensite, J., Eds.; Springer: Berlin/Heidelberg, Germany, 2011.
2. Stammer, D.; Cazenave, A. *Satellite Altimetry over Oceans and Land Surfaces*, 1st ed.; CRC Press: Boca Raton, FL, USA, 2017.
3. Le Traon, P.Y. From satellite altimetry to argo and operational oceanography: Three revolutions in oceanography. *Ocean Sci.* **2013**, *9*, 901–915. [CrossRef]
4. Le Traon, P.Y.; Dibarboure, G.; Jacobs, G.; Martin, M.; Remy, E.; Schiller, A. Use of satellite altimetry for operational oceanography. In *Satellite Altimetry over Oceans and Land Surfaces*; Stammer, D., Cazenave, A., Eds.; CRC Press: Boca Raton, FL, USA, 2017.
5. Ablain, M.; Cazenave, A.; Larnicol, G.; Balmaseda, M.; Cipollini, P.; Faugere, Y.; Fernandes, M.J.; Henry, O.; Johannessen, J.A.; Knudsen, P.; et al. Improved sea level record over the satellite altimetry ERA (1993–2010) from the climate change initiative project. *Ocean Sci.* **2015**, *11*, 67–82. [CrossRef]
6. Ablain, M.; Legeais, J.F.; Prandi, P.; Marcos, M.; Fenoglio-Marc, L.; Dieng, H.B.; Benveniste, J.; Cazenave, A. Satellite altimetry-based sea level at global and regional scales. *Surv. Geophys.* **2017**, *38*, 7–31. [CrossRef]
7. Quartly, G.D.; Legeais, J.F.; Ablain, M.; Zawadzki, L.; Fernandes, M.J.; Rudenko, S.; Carrère, L.; García, P.N.; Cipollini, P.; Andersen, O.B.; et al. A new phase in the production of quality-controlled sea level data. *Earth Syst. Sci. Data* **2017**, *9*, 557–572. [CrossRef]
8. Nerem, R.S.; Ablain, M.; Cazenave, A.; Church, J.; Leuliette, E. A 25-year satellite altimetry-based global mean sea level record: Closure of the sea level budget and missing components. In *Satellite Altimetry over Oceans and Land Surfaces*; Stammer, D., Cazenave, A., Eds.; CRC Press: Boca Raton, FL, USA, 2017.
9. Legeais, J.F.; Ablain, M.; Zawadzki, L.; Zuo, H.; Johannessen, J.A.; Scharffenberg, M.G.; Fenoglio-Marc, L.; Fernandes, M.J.; Andersen, O.B.; Rudenko, S.; et al. An accurate and homogeneous altimeter sea level record from the esa climate change initiative. *Earth Syst. Sci. Data* **2017**, *2017*, 1–35. [CrossRef]
10. Birkett, C.; Reynolds, C.; Beckley, B.; Doorn, B. From research to operations: The usda global reservoir and lake monitor. In *Coastal Altimetry*; Vignudelli, S., Kostianoy, A.G., Cipollini, P.J.B., Eds.; Springer-Verlag: Berlin/Heidelberg, Germany, 2011.
11. Cretaux, J.-F.; Nielsen, K.; Frappart, F.; Papa, F.; Calmant, S.; Benveniste, J. Hydrological applications of satellite altimetry: Rivers, lakes, man-made reservoirs, inundated areas. In *Satellite Altimetry in Coastal Regions*; CRC Press: Boca Raton, FL, USA, 2017.
12. Remy, F.; Memin, A.; Velicogna, I. Applications of satellite altimetry to study the antarctic ice sheet, satellite altimetry in coastal regions. In *Satellite Altimetry over Oceans and Land Surfaces*, 1st ed.; Stammer, D., Cazenave, A., Eds.; CRC Press: Boca Raton, FL, USA, 2017; p. 670.
13. Fernandes, M.J.; Lazaro, C.; Nunes, A.L.; Scharroo, R. Atmospheric corrections for altimetry studies over inland water. *Remote Sens.* **2014**, *6*, 4952–4997. [CrossRef]

14. Chelton, D.B.; Ries, J.C.; Haines, B.J.; Fu, L.L.; Callahan, P.S. Satellite altimetry. In *Satellite Altimetry and Earth Sciences: A Handbook of Techniques and Applications*; Fu, L.L., Cazenave, A., Eds.; Academic Press: San Diego, CA, USA, 2001.

15. Vieira, T.; Fernandes, M.J.; Lázaro, C. Analysis and retrieval of tropospheric corrections for cryosat-2 over inland waters. *Adv. Sp. Res.* **2017**, *46*. [CrossRef]

16. Cipollini, P.; Benveniste, J.; Birol, F.; Fernandes, M.J.; Obligis, E.; Passaro, M.; Strub, P.T.; Valladeau, G.; Vignudelli, S.; Wilkin, J. Satellite altimetry in coastal regions. In *Satellite Altimetry over Oceans and Land Surfaces*, 1st ed.; Stammer, D., Cazenave, A., Eds.; CRC Press: Boca Raton, FL, USA, 2017.

17. Keihm, S.J.; Janssen, M.A.; Ruf, C.S. Topex/poseidon microwave radiometer (TMR). III. Wet troposphere range correction algorithm and pre-launch error budget. *IEEE Trans. Geosci. Remote Sens.* **1995**, *33*, 147–161. [CrossRef]

18. Eymard, L.; Obligis, E. The Altimetric Wet Tropospheric Correction: Progress Since the ERS-1 Mission. In Proceedings of the15 Years of Progress in Radar Altimetry, Venice, Italy, 13–18 March 2006.

19. Tournadre, J.; Lambin-Artru, J.; Steunou, N. Cloud and rain effects on altika/saral ka-band radar altimeter-part i: Modeling and mean annual data availability. *IEEE Trans. Geosci. Remote Sens.* **2009**, *47*, 1806–1817. [CrossRef]

20. Sentinel-3 Team. *Sentinel-3 User Handbook*, 2nd ed.; GMES-S3OP-EOPG-TN-13-0001; European Space Agency: Paris, France, 2013.

21. Dinardo, S.; Fenoglio-Marc, L.; Buchhaupt, C.; Becker, M.; Scharroo, R.; Joana Fernandes, M.; Benveniste, J. Coastal sar and plrm altimetry in german bight and west baltic sea. *Adv. Sp. Res.* **2017**. [CrossRef]

22. Tournadre, J. Improved level-3 oceanic rainfall retrieval from dual-frequency spaceborne radar altimeter systems. *J. Atmos. Ocean. Technol.* **2006**, *23*, 1131–1149. [CrossRef]

23. Fernandes, M.J.; Lazaro, C.; Ablain, M.; Pires, N. Improved wet path delays for all esa and reference altimetric missions. *Remote Sens. Environ.* **2015**, *169*, 50–74. [CrossRef]

24. Fernandes, M.J.; Lazaro, C. Gpd+ wet tropospheric corrections for cryosat-2 and gfo altimetry missions. *Remote Sens.* **2016**, *8*, 851. [CrossRef]

25. EUMETSAT, S3a STM Reprocessing—"Spring 2017" (Level 0 to Level 2). Available online: eum/ops-sen3/rep/17/940906 (accessed on 31 August 2017).

26. EUMETSAT, Sentinel-3a Product Notice—STM l2 Marine ("Spring Reprocessing Campaign"). Available online: um/ops-sen3/doc/17/944329 (accessed on 15 September 2017).

27. Scharroo, R.; Leuliette, E.; Naeije, M.; Martin-Puig, C.; Pires, N. Rads Version 4: An efficient way to analyse the Multi-Mission altimeter database. In Proceedings of the ESA Living Planet Symposium, Prague, Czech Republic, 9–13 May 2016; ESA: Prague, Czech Republic, 2016.

28. Obligis, E.; Eymard, L.; Tran, N.; Labroue, S.; Femenias, P. First three years of the microwave radiometer aboard envisat: In-flight calibration, processing, and validation of the geophysical products. *J. Atmos. Ocean. Technol.* **2006**, *23*, 802–814. [CrossRef]

29. Thao, S.; Eymard, L.; Obligis, E.; Picard, B. Comparison of regression algorithms for the retrieval of the wet tropospheric path. *IEEE J. Sel. Top. Appl. Earth Obs. Remote Sens.* **2015**, *8*, 4302–4314. [CrossRef]

30. Collecte Localisation Satellites (CLS). *Surface Topography Mission (STM) Sral/Mwr L2 Algorithms Definition, Accuracy and Specification*; S3PAD-RS-CLS-SD03-00017; CLS: Ramonville St-Agne, France, 2011.

31. Mercier, F.; Rosmorduc, V.; Carrère, L.; Thibaut, P. Coastal and Hydrology Altimetry Product (PISTACH) Handbook. 2010. Available online: https://www.aviso.altimetry.fr/fileadmin/documents/data/tools/hdbk_Pistach.pdf (accessed on 19 March 2018).

32. Meissner, T.; Wentz, F.J.; Draper, D. *Gmi Calibration Algorithm and Analysis Theoretical Basis Document*; Report Number 041912; Remote Sensing Systems: Santa Rosa, CA, USA, 2012; 124p.

33. Draper, D.W.; Newell, D.A.; Wentz, F.J.; Krimchansky, S.; Skofronick-Jackson, G.M. The global precipitation measurement (GPM) microwave imager (GMI): Instrument overview and early on-orbit performance. *IEEE J. Sel. Top. Appl. Earth Obs. Remote Sens.* **2015**, *8*, 3452–3462. [CrossRef]

34. Fernandes, M.J.; Nunes, A.L.; Lazaro, C. Analysis and inter-calibration of wet path delay datasets to compute the wet tropospheric correction for cryosat-2 over ocean. *Remote Sens.* **2013**, *5*, 4977–5005. [CrossRef]

35. Stum, J.; Sicard, P.; Carrere, L.; Lambin, J. Using objective analysis of scanning radiometer measurements to compute the water vapor path delay for altimetry. *IEEE Trans. Geosci. Remote Sens.* **2011**, *49*, 3211–3224. [CrossRef]

36. Wentz, F.J. A well-calibrated ocean algorithm for special sensor microwave/imager. *J. Geophys. Res. Oceans* **1997**, *102*, 8703–8718. [CrossRef]

37. Wentz, F.J. *SSM/I Version-7 Calibration Report*; RSS Technical Report 011012; Remote Sensing Systems: Santa Rosa, CA, USA, 2013; p. 46.

38. Brown, S. A novel near-land radiometer wet path-delay retrieval algorithm: Application to the *Jason-2/OSTM* advanced microwave radiometer. *IEEE Trans. Geosci. Remote Sens.* **2010**, *48*, 1986–1992. [CrossRef]

39. Dumont, J.P.; Rosmorduc, V.; Picot, N.; Bronner, E.; Desai, S.; Bonekamp, H.; Figa, J.; Lillibridge, J.; Scharroo, R. OSTM/Jason-2 Products Handbook. 2011. Available online: http://www.ospo.noaa.gov/Products/documents/J2_handbook_v1-8_no_rev.pdf (accessed on 19 March 2018).

40. Dumont, J.P.; Rosmorduc, V.; Carrere, L.; Picot, N.; Bronner, E.; Couhert, A.; Guillot, A.; Desai, S.; Bonekamp, H.; Figa, J.; et al. Jason-3 Products Handbook. 2016. Available online: https://www.aviso.altimetry.fr/fileadmin/documents/data/tools/hdbk_j3.pdf (accessed on 19 March 2018).

41. Brown, S.; Islam, T. Jason-3 GDR calibration stability enabled by the cold sky maneuvers. In Proceedings of the Ocean Surface Topography Science Team Meeting, Miami, FL, USA, 23–27 October 2017.

42. Bronner, E.; Guillot, A.; Picot, N. SARAL/Altika Products Handbook. SALP-MU-M-OP-15984-CN. 2013. Available online: https://www.aviso.altimetry.fr/fileadmin/documents/data/tools/SARAL_Altika_products_handbook.pdf (accessed on 19 March 2018).

43. Vieira, T.; Fernandes, M.J.; Lázaro, C. Independent assessment of on-board microwave radiometer measurements in coastal zones using tropospheric delays from gnss. *IEEE Trans. Geosci. Remote Sens.* **2018**, under review.

44. Davis, J.L.; Herring, T.A.; Shapiro, I.I.; Rogers, A.E.E.; Elgered, G. Geodesy by radio interferometry—Effects of atmospheric modeling errors on estimates of baseline length. *Radio Sci.* **1985**, *20*, 1593–1607. [CrossRef]

45. Kouba, J. Implementation and testing of the gridded vienna mapping function 1 (VMF1). *J. Geod.* **2008**, *82*, 193–205. [CrossRef]

46. Fernandes, M.J.; Pires, N.; Lazaro, C.; Nunes, A.L. Tropospheric delays from gnss for application in coastal altimetry. *Adv. Space Res.* **2013**, *51*, 1352–1368. [CrossRef]

47. Dee, D.P.; Uppala, S.M.; Simmons, A.J.; Berrisford, P.; Poli, P.; Kobayashi, S.; Andrae, U.; Balmaseda, M.A.; Balsamo, G.; Bauer, P.; et al. The era-interim reanalysis: Configuration and performance of the data assimilation system. *Q. J. R. Meteorol. Soc.* **2011**, *137*, 553–597. [CrossRef]

48. Fernandes, M.J.; Lazaro, C.; Nunes, A.L.; Pires, N.; Bastos, L.; Mendes, V.B. Gnss-derived path delay: An approach to compute the wet tropospheric correction for coastal altimetry. *IEEE Geosci. Remote Sens. Lett.* **2010**, *7*, 596–600. [CrossRef]

49. Miller, M.; Buizza, R.; Haseler, J.; Hortal, M.; Janssen, P.; Untch, A. Increased resolution in the ecmwf deterministic and ensemble prediction systems. In *ECMWF Newsletter*; ECMWF: Reading, UK, 2010; Volume 124, pp. 10–16.

50. Thao, S.; Eymard, L.; Obligis, E.; Picard, B. Trend and variability of the atmospheric water vapor: A mean sea level issue. *J. Atmos. Ocean. Technol.* **2014**, *31*, 1881–1901. [CrossRef]

51. eoPortal:Satellite Missions Directory, Copernicus: Sentinel-6/Jason-CS (Jason Continuity of Service) Mission. Available online: https://directory.eoportal.org/web/eoportal/satellite-missions/content/-/article/jason-cs (accessed on 20 February 2018).

remote sensing

MDPI

Article

The Drifting Phase of SARAL: Securing Stable Ocean Mesoscale Sampling with an Unmaintained Decaying Altitude

Gérald Dibarboure [1],*, Alain Lamy [1], Marie-Isabelle Pujol [2] and Ghita Jettou [2]

[1] Centre National d'Etudes Spatiales, 31400 Toulouse, France; alain.lamy@cnes.fr
[2] Collecte Localisation Satellites, 31520 Ramonville Saint-Agne, France; mpujol@cls.fr (M.-I.P.); gjettou@cls.fr (G.J.)
*   Correspondence: gerald.dibarboure@cnes.fr

Received: 20 April 2018; Accepted: 2 July 2018; Published: 3 July 2018

Abstract: The French/Indian altimeter project Satellite with ARGOS and AltiKa (SARAL) completed its nominal 3-year mission on the historical European Remote-sensing Satellite (ERS) orbit in Spring 2016. In order to extend the lifetime of the satellite as much as possible, the agencies in charge of SARAL decided to initiate a so-called drifting phase where the satellite altitude is no longer maintained. In this paper we describe how the ocean mesoscale sampling capability of SARAL has been preserved during the drifting phase by initiating it at a specific altitude: the optimal starting point was approximately 1 km above the historical ERS/ENVIronment SATellite (ENVISAT) orbit. This strategy secured the ocean mesoscale sampling capability of SARAL drifting phase (DP) for 6 years or more. We also generalize these findings: any altimeter could follow SARAL's drifting phase strategy if their maneuvering capability is limited. Lastly, we explain how an altimetry mission or an entire altimeter constellation could be operated without any form of altitude control: some specific altitude bands (e.g., near 1230 km) guarantee a high-quality mesoscale sampling for years or decades even if the altitude is not maintained.

Keywords: altimetry; orbit decay; drifting orbit; geodetic orbit

1. Introduction and Context

The Satellite with ARGOS and AltiKa (SARAL) mission was developed by CNES (Centre National d'Etudes Spatiales) and ISRO (Indian Space Research Organization). Launched in 2013, the SARAL satellite was primarily a gapfiller for the Advanced Research and Global Observation Satellite (ARGOS) localization system and a Ka-band altimeter technology demonstrator (AltiKa). Verron et al. [1] and Vincent et al. [2] recall that SARAL/AltiKa was initially designed as a response to the requirements expressed by operational oceanography users to bridge the gap between the ENVIronment SATellite (ENVISAT) and the Sentinel-3 altimeter missions. An important objective of AltiKa was to complement the altimeter constellation and to help resolve large ocean mesoscale variability, i.e., scales ranging from 150 to 500 km and 15 to 50 days [3]. SARAL/AltiKa was instrumental in providing the ocean mesoscale sampling needed by operational systems such as AVISO [4] and the Copernicus Marine Environment Monitoring Service (CMEMS) [5]. Indeed, according to Chelton et al. [6] and Pascual et al. [7], two operational altimeters are required to monitor ocean mesoscale variability in delayed time, and up to four are needed in near real time.

Furthermore, Verron et al. [1,8] and Bonnefond et al. [9] recently reported that the Ka-band demonstrator exhibited unprecedented altimeter precision for topography over the ocean, rivers and lakes, and even over sea and land ice, as well as for wind and wave. This precision in turn led to innovative user-products and new scientific findings (e.g., [10–15]). Thanks to this demonstrated scientific value, CNES and ISRO decided to extend the SARAL mission beyond the 3-year nominal duration that was successfully completed in Spring 2016.

As reported by Bonnefond and Willis [16], SARAL exhibited an increasing need to reduce the stress on its momentum wheels in order to extend the lifetime of the mission. This could be achieved by limiting the orbit maintenance maneuvers. The agencies eventually decided that the best strategy was to stop all orbit control maneuvers and to let the altitude decay naturally. This strategy is also known as a drifting phase (DP). It was initiated in July 2016.

The main consequence of a DP was that the SARAL subsatellite track would rapidly drift away from the historical ground track formerly used by the European Remote-sensing Satellite (ERS) and ENVISAT missions. At this point it is important to highlight that the term "drift" refers to the evolution of the nadir track (i.e., position of the altimeter measurement profile): SARAL DP is therefore very different from the ENVISAT mission extension where the orbit was drifting but the nadir ground track was relatively stable and composed of 431 orbits per 30-day cycle.

Sandwell et al. [17] have shown that using a drifting ground track could be extremely useful for the marine geodesy community: the unprecedented precision of AltiKa could help resolve uncharted sea mounts and sea floor topography. Improving marine geodesy products requires a so-called "geodetic altimeter sampling", i.e., a sampling pattern where the subsatellite tracks over a long period (e.g., one year) create a very dense grid with a resolution of 8 km or less. In practice, this can be achieved by keeping precisely the satellite on a so-called "geodetic orbit", i.e., an orbit with a very long exact repeat cycle. This strategy was used for the Jason-1 Geodetic-Mission (GM) phase or the Jason-2 Long Repeat Orbit (LRO) phase. For SARAL DP, there is no longer an exact repeat because the orbit altitude decays continuously, yet the geodetic sampling of SARAL DP is naturally dense, albeit somewhat random (discussed in Sections 3.4 and 4.1).

However, putting SARAL on such a drifting ground track raised the question of the new mesoscale sampling capability (still a high priority for the mission). Dibarboure et al. [18] have shown that the geodetic orbit used by Cryosat-2 is very poor when it comes to mesoscale monitoring: the orbit has a monthly linear scanning pattern that leaves most of the ocean unobserved over a typical ocean mesoscale decorrelation period of 15 days. On the other hand, Dibarboure et al. [19] demonstrated that it could be possible to find geodetic orbits that were compatible with mesoscale monitoring: this was achieved by including intermediate sub-cycles (defined in Section 2.1) that maximize the ocean mesoscale sampling over a period of 15 to 20 days. But these studies were both carried out in the context of a well-maintained altitude, and the mesoscale sampling properties of decaying orbits has not been studied so far.

In this context, the objective of this paper is to demonstrate that it is possible to secure high-quality ocean mesoscale sampling and good marine geodesy sampling for 6 years or more even with an unmaintained decaying altitude. Sections 2 and 3 describe how the continuous evolution of the orbit sub-cycles could be predicted to find mesoscale-friendly altitude bands where a decaying altitude does not alter the altimeter sampling capability. In Section 4, the anticipated results are confirmed with in-flight data from the first 18 months of SARAL DP. Lastly Section 5 generalizes the SARAL DP findings to other altimeter missions.

2. Methodology Used to Prepare the Drifting Phase of SARAL

This section gives an overview of the methodology used in 2016 to define the drifting phase of SARAL: Section 2.1 describes how to infer the quality of mesoscale sampling for a given altitude, and Section 2.2 shows how we defined conservative scenarios to predict the altitude decay of SARAL DP. These two essential inputs are then used in Section 3 to predict the evolution of the ocean mesoscale sampling throughout the drifting phase, and to select the best starting point.

2.1. Quality of Mesoscale Sampling and Orbit Sub-Cycles

To measure the quality of the mesoscale sampling for a given altitude, we analyze the distribution of the satellite tracks in the longitude/time plane, following the methodology of Dibarboure et al. [19] for Jason-1: in essence, the quality of mesoscale sampling is inferred by gauging when altimeter tracks get too close in space or in time. Indeed, a single altimeter cannot resolve all the mesoscale space and time scales [6], so it is essential to avoid duplications between adjoining tracks: if the satellite tracks are too close, it means that a different region is not properly observed. Here, "too close" is measured in the longitude/time plane because all satellite tracks have the same geometry. Furthermore, the metric is based on mesoscale covariance models used by Ducet et al. [20] to build a multi-altimeter gridded map of the ocean mesoscale. The covariance models are a statistical approximation of the size of mesoscale features in space and time [21,22], i.e., a good description of when altimeter tracks are too close and correlated.

Note that because SARAL is the only satellite of the 2016+ altimeter constellation that is near the historical ERS/ENVISAT orbit, we only optimize its own sampling: we do not try to maximize the combined sampling of the entire altimeter constellation: the ground track geometry and repeat cycles of other altimeters are very different so the constellation-wide gain would be very small.

Figure 1 shows the distribution of the satellite tracks for the ERS/ENVISAT altitude originally used by SARAL (approximately 780 km). Each black dot is one satellite track. The vertical alignment of the black dots corresponds to the 35-day exact repeat cycle of this orbit. The grey circles are 150 km by 15 days. This is an approximation of the decorrelation scale of mesoscale eddies at mid-latitudes.

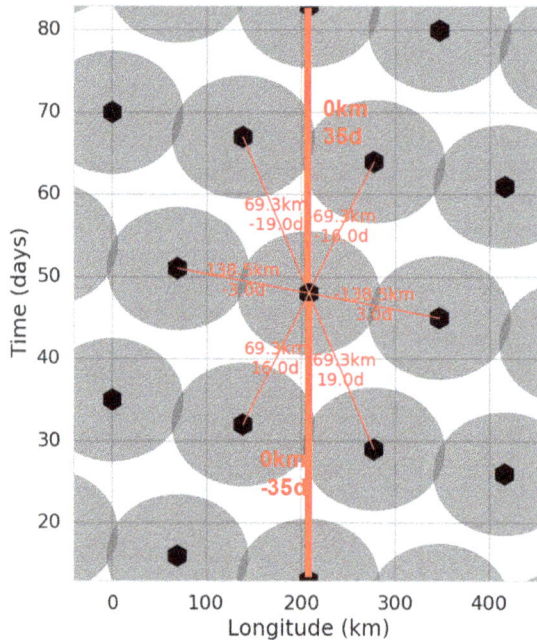

Figure 1. Longitude of Advanced Research and Global Observation Satellite (ARGOS) and AltiKa (SARAL/AltiKa) ascending tracks at 30°N as a function of time (black dots). The grey circles are 150 km by 15 days (typical decorrelation scale of ocean mesoscale). The red lines highlight the main orbit sub-cycles (major: 3 days, 16 days, minor: 19 days) as well as the 35-day repeat cycle of the European Remote-sensing Satellite (ERS)/ENVIronment SATellite (ENVISAT)/SARAL orbit.

In other words, if two grey circles overlap, then the corresponding satellite tracks are too close in space or in time: their measurements are highly correlated and in turn other regions of the space/time plane are completely unobserved. As expected from Le Traon and Dibarboure [23], the ERS/ENVISAT orbit is very good for mesoscale observation: the overlap between grey circles is almost non-existent as they are well-scattered in the longitude/time plane.

In addition to the true repeat cycle (highlighted with two thick vertical red lines), the diagonal alignment of grey circles in Figure 1 (thin red lines) shows the presence of orbital sub-cycles (SC). Orbital sub-cycles are defined by Rees [24] as a period of near-repeat for Earth remote-sensing satellites. The influence of SC in the context of altimetry missions is explained by Dibarboure et al. [19].

For the ERS/ENVISAT/SARAL altitude there are two dominant sub-cycles: 3 days and 16 days. In practice, dominant sub-cycles could also be defined by their global coverage: a sub-cycle corresponds to the time needed to collect a coverage that is globally homogeneous (e.g., 1000 km resolution after 3 days, or 180 km after 16 days). The 1002 tracks of the ERS/ENVISAT/SARAL repeat cycle are composed of two interleaved 16-day periods and one 3-day period. A 19-day sub-cycle is also visible in Figure 1: it is composed of a 16-day and a 3-day sub-cycle.

2.2. Predicting the Altitude Loss

Section 3.1 will show that the quality of the mesoscale sampling changes with the altitude and the associated sub-cycles. To determine how the sampling will evolve throughout the drifting phase, it is necessary to know how the altitude is likely to change in 3 years or more.

Like all low-earth orbit satellites, SARAL experiences an atmospheric drag force in a direction opposite to the direction of its motion. This force slows down the satellite, which in turn decreases its altitude. The drag force is given by Equation (1):

$$F_d = \frac{1}{2} \cdot \rho \cdot v \cdot C_b \tag{1}$$

where ρ is the atmospheric density, v is the satellite speed, and C_b the ballistic coefficient. Note that if the orbit is circular, the speed of the satellite is constant and it is a function of the orbit altitude. For the sake of simplicity, we here ignore that the satellite drag actually depends on the speed relative to the atmosphere which in turn creates a modulation of the drag force with latitude: the effect on the ground track geometry is extremely small and an analysis on the longitude of ascending node only requires us to know the mean satellite drag over a long period of time. The ballistic coefficient C_b is given by Equation (2) where C_x is the drag coefficient of the satellite, S is the area of the cross section surface in the direction of flight, and m is the mass of the satellite. In practice the coefficients used for SARAL in this study were estimated through an analysis of all the maneuvers performed by ISRO during the 3-year nominal mission.

$$C_b = \frac{C_x \cdot S}{m} \tag{2}$$

To get a projection of the altitude decay, it is necessary to have a prediction of the atmospheric density. The density ρ as a function of the altitude h around a reference altitude h_0 can be approximated by a simple exponential with variable scale height H (Equation (3)). Here, the scale height depends on the solar radio flux which is a proxy of the upper atmosphere density [25]. In practice, it is more accurate to use more sophisticated atmosphere models (e.g., MSIS2000). These models also use solar activity coefficient to parameterize the atmospheric density at all altitudes.

$$\rho = \rho_0 \cdot e^{\frac{h-h_0}{H}} \tag{3}$$

In this study, we used the solar flux predictions from the National Aeronautics and Space Administration (NASA) Marshall Space Flight Center [26]. The solar flux prediction used to prepare the SARAL DP in 2016 is shown in Figure 2a. Using the atmosphere model and the measured ballistic coefficient of SARAL, we obtain the altitude prediction in Figure 2b.

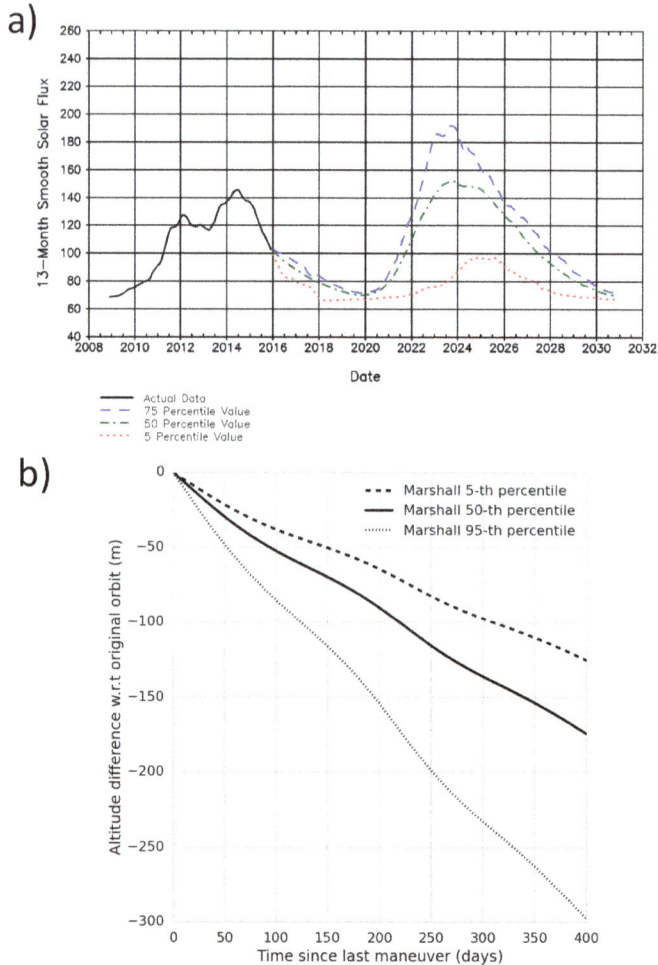

Figure 2. Panel (**a**) shows the solar flux prediction (unit: sfu, i.e., 10^{-22} W·m^{-2}·Hz^{-1}) used as an input (from the National Aeronautics and Space Administration (NASA) Marshall Space Flight Center). Panel (**b**) shows the altitude decay with respect to the original ERS/ENVISAT orbit for three solar flux scenarios: 5-th percentile (weak solar activity), 50-th percentile (median solar activity) and 95-th percentile (very intense solar activity).

The altitude decay is almost linear over one year, albeit modulated by the solar flux. To that extent, panel (b) shows three predictions: one for the 5-th percentile (weak solar activity, weak altitude decay), one for the 50-th percentile (median solar activity) and one for the 95-th percentile (strong solar activity).

The average altitude decay ranges from 150 m to 300 m per year. To prepare the SARAL drifting phase, we used a conservative value of 300 m/year (95-th percentile solar flux scenario): with this strategy it was very likely that the altitude would decrease more slowly. In other words, the drifting phase SARAL DP could be as long as three times our conservative estimate from Section 3.3 (the observed solar fluxes where indeed weaker and close to the 5-th percentile value, as discussed in Section 4.4).

3. Predictions Obtained to Define the Drifting Phase of SARAL

Based on the inputs from Section 2, the following sections explore the behavior of an unmaintained orbit strategy for SARAL, and the consequence for its mesoscale sampling capability. These predictions helped define the final strategy of the SARAL drifting phase. The observations after 18 months of SARAL DP are discussed in Section 4.

3.1. Diversity of the Mesoscale Sampling Near the ERS Altitude

The SARAL DP strategy was designed for a baseline of 3 years (i.e., as long as nominal lifespan of SARAL) with a backup scenario as long as 9 years (e.g., scenario where SARAL remains operational as long as the TOPEX/Poseidon mission). Because of the unmaintained decaying altitude and our conservative value of 300 m/year, the entire drifting phase could span over an altitude range of 1 km (baseline) to 3 km (backup).

But the mesoscale sampling properties over such a range of altitude can be very different: Figure 3 shows the position of SARAL tracks in the longitude/time plane for 5 different altitudes near the ERS orbit. Panel a in Figure 3 is 2 km above: it exhibits an exact 13-day repeat cycle and the 3-day sub-cycle that exists for all orbits in this altitude range. In panel (a), mesoscale sampling is not as good as for the ERS altitude because the 13-day repeat cycle is slightly too short: the grey circles are overlapping in the vertical direction. This weakness is the same as for the TOPEX orbit [23].

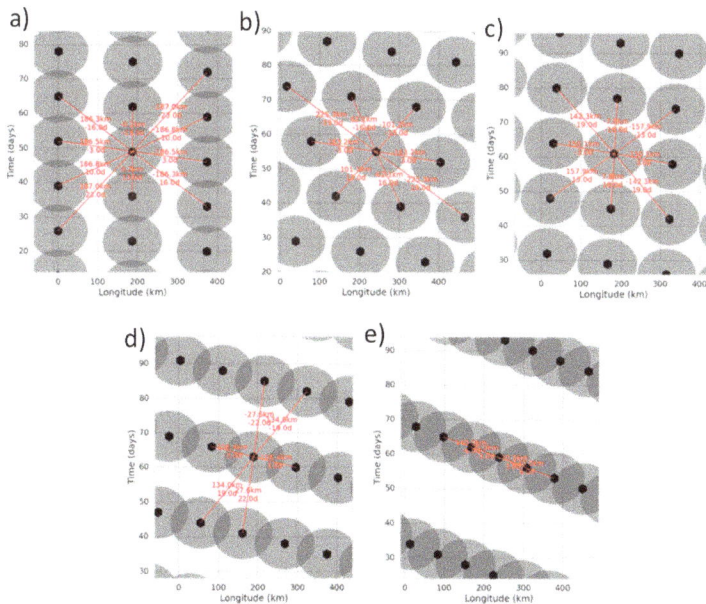

Figure 3. Same as Figure 1 for 5 altitudes near the historical ERS orbit. Panel (**a**) is approximately +2 km with respect to the ERS orbit. Panel (**b**) is approximately +1 km. Panel (**c**) is approximately +0.5 km. Panel (**d**) is approximately −1.5 km. Panel (**e**) is approximately −2.5 km. The alignment of gray circles (i.e., mesoscale-relevant orbit sub-cycles) is very different even though these 5 altitudes are relatively close to the ERS orbit.

Panel b in Figure 3 is 1 km above the ERS orbit. Here, the distribution of altimeter tracks in the longitude/time plane is different: the grey circles are rotated clockwise with respect to panel (a). The 13-day exact repeat of panel (a) has morphed into a 13-day sub-cycle and the 16-day sub-cycle has appeared (the 16-day SC is a composition of the 13-day SC and one 3-day SC). As a result, the mesoscale sampling is excellent at this altitude: the altimeter tracks are as far away from one another as possible, and there is no overlap between them.

Figure 3c is 500 m above the ERS altitude. The grey circles are rotated clockwise with respect to panel b in Figure 3, but the sampling capability is still excellent. The 16-day sub-cycle is very strong (almost vertical, i.e., a true repeat cycle), and the 19-day sub-cycle starts to appear due to the clockwise rotation (composition of a 16-day SC and a 3-day SC).

Panel d and panel e in Figure 3 are 1.5 km and 2.5 km below the SARAL altitude, respectively. At these altitudes, the sub-cycles are longer than 20-days, and the grey circles are largely overlapping: they are separated by less than 70 km and 3 days. The overlaps highlight the regions of the space and time plane that are correlated (as defined in Section 2.1) with two or more altimeter measurements in black. In panel (d) and panel (e), there is a large amount of overlap between adjoining circles; the visible consequence is that there are large bands of the space/time plane which are not observed whatsoever (not covered by grey circles). In essence, the nadir tracks are too close in space or in time and the resulting duplication between altimeter measurements creates large blind regions elsewhere.

The poor mesoscale sampling in panel (d) and (e) is due to the clockwise rotation of the grey circles in the longitude/time plane: they now align diagonally (very strong 3-day SC). This poor sampling pattern is typically observed on CryoSat-2 [18].

To summarize, the 3 examples above the ERS/ENVISAT orbit are good for mesoscale sampling, and it is preferable to stay between the current altitude and +1 km because the sampling at +2 km is not as good. The other 2 examples below the ERS/ENVISAT orbit are progressively worse and should be avoided because their SC are too long for mesoscale. But these are only a handful of examples, and the satellite altitude will decay continuously and change the sub-cycles.

3.2. The Continuous Evolution of Sub-Cycles (SC)

To understand how the orbit sub-cycles would change when there is no station-keeping maneuvers, we performed a long simulation starting 2-km above the ERS altitude and using the 300 m/year altitude decay obtained in Section 2.2. We chose to keep a constant altitude decay to show how a 15-year SARAL DP (starting in summer 2016) would look like at all altitudes.

Figure 4 shows the longitude of all subsatellite tracks as the altitude decays. We highlighted the altitudes studied in the 5 panels of Figure 3. The timeline goes from top to bottom (altitude decaying with time, 1 km of altitude is approximately equivalent to 3 years) so here the rotation of the grey circles discussed in Section 3.1 is counterclockwise: with this broader scale the rotation looks like horizontal parabolas (highlighted as colored arrows).

This continuous parabola-shaped effect is created by the modulo between the duration of a satellite revolution and the Earth rotation as seen by the satellite: a tiny change in the altitude also reduces the revolution period and the new modulo rotates the altimeter tracks in the longitude/time plane.

This result is important for SARAL DP: all sub-cycles appear, strengthen, become a true repeat cycle, weaken, and then disappear. More importantly this phenomenon occurs in a continuous way and the large parabolas are approximately 1-km large in altitude: in other words, if we find one orbit with good sampling properties, then there is also a 3-year time-frame where the sampling will be good because the sub-cycles change only progressively.

Incidentally, Figure 4 shows that larger sub-cycles are composed of smaller sub-cycles. Most SC can be decomposed as 1-day plus a given multiple of the 3-day sub-cycle (13, 16, 19, 22), but there are also larger sub-cycles composed of shorter sub-cycles (e.g., the 35-day repeat of ERS is 16 + 16 + 3 or 16 + 19). This rather complex composition effect is interesting because if two altitudes are attractive and separated by 1–2 km, then all intermediate altitudes will be good for mesoscale because the grey circle rotation seen in Figure 3 will ensure that the composition of intermediate sub-cycles is also good.

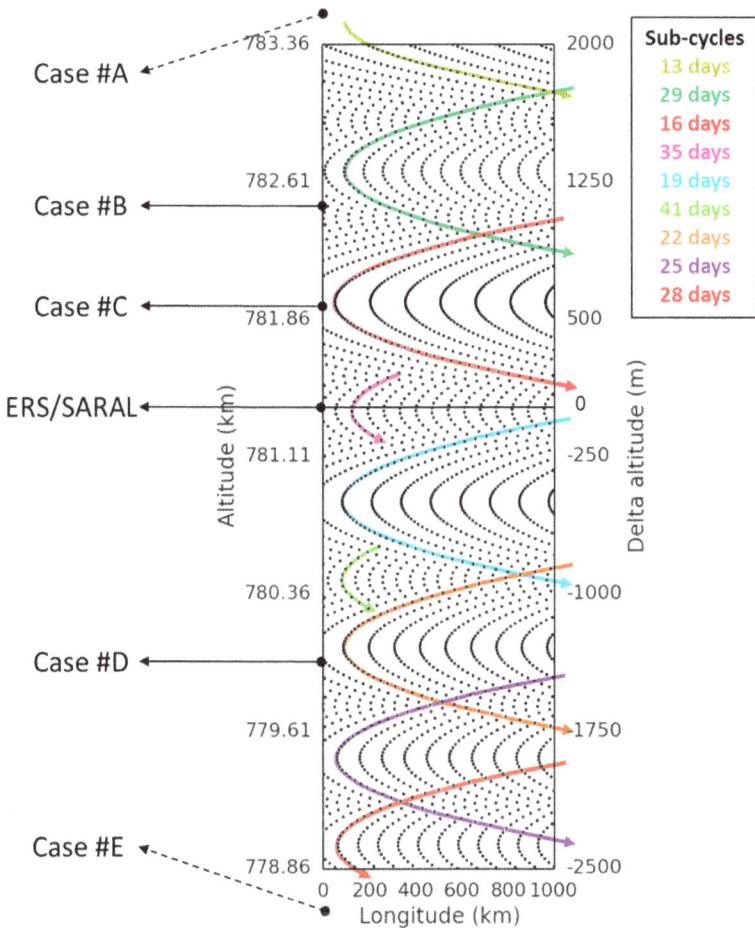

Figure 4. Simulated longitude of the AltiKa ascending tracks over a 15-year period. The simulation starts 2 km above the ERS orbit and it decreases by approximately 300 m per year. The continuously decreasing altitude transforms the orbit repeat cycle and sub-cycles (highlighted as colored parabolas): the 3-day sub-cycle of this altitude range generates composite sub-cycles increasing by 3 days (e.g., 13, 16, 19, 22, 25, 28), and the intermediate sub-cycles compose higher level sub-cycles (e.g., 29-day from 13 and 16, or 35 from 16 and 19). The circle alignment of the 5 panels in Figure 3 is explained by the colored parabolas.

As a result, the mesoscale-friendly altitudes aggregate in small bands that should be considered for SARAL DP. This aggregation was already observed but not explained by Dibarboure et al. [19] when they were looking for a good geodetic orbit for Jason-1.

3.3. Finding a Good Starting Point for the SARAL Drifting Phase (DP)

Assuming a baseline duration of 3 years, SARAL DP requires a 1-km altitude range where the sub-cycles are good for mesoscale. Figure 3 shows that sub-cycles ranging from 13 to 19 days are good although 15 to 17 days is better. Lastly the altitude range should be as close to the ERS/ENVISAT altitude as possible in order to avoid a strong maneuver when the drifting phase starts.

With all these constraints, Figure 4 shows that the best starting point for SARAL DP is approximately 1 km above the ERS/ENVISAT orbit: Figure 3b,c showed that the starting and intermediate altitudes yield an excellent sampling for mesoscale. After 3 years, SARAL/AltiKa is back at the ERS/ENVISAT altitude where the sampling capability is still good. From 3 to 6 years of drifting phase, the sampling is decent but sub-optimal due to the long 19-day SC. After 6 years, the sampling degrades because the main SC is too long. Figure 5 gives an overview of the dominant sub-cycles for all altitude as well as a qualitative index for mesoscale sampling.

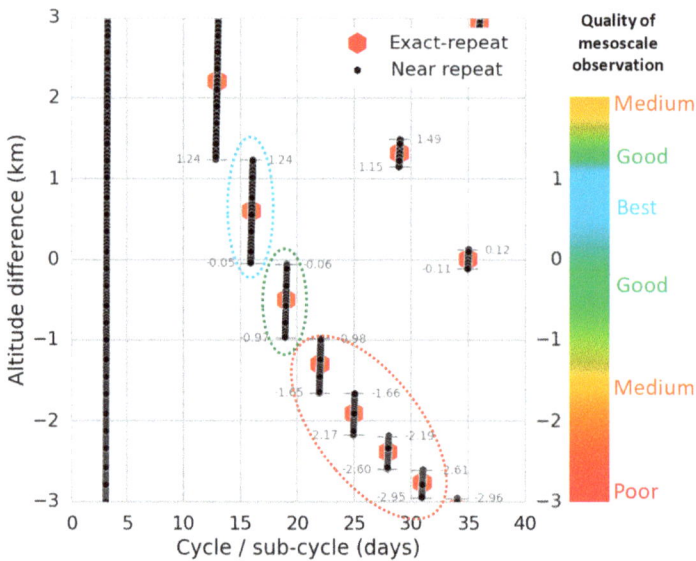

Figure 5. Overview of the dominant sub-cycles for a 6-km altitude range near the historical ERS orbit. The quality of mesoscale observation is controlled by some sub-cycles (15 to 19 days is better as shown in Figure 3) so some altitude ranges are better than others (colored ellipses and right-hand-side color bar).

3.4. Relaxed Altitude Control or No Control Whatsoever?

Assuming that SARAL/AltiKa is operated at least 3 years on the DP orbit, one might raise the following question: is it better to keep a relaxed altitude control or to have no altitude control whatsoever? E.g., one maneuver every 6–12 months to stay within 150–300 m of the altitude with an optimal sampling pattern for ocean mesoscale. As discussed above, such a yearly maneuver would be good but not essential for mesoscale for at least 3 years. But in this section we show that it would be very detrimental to the geodetic sampling.

Indeed, Zhang and Sandwell [27] recall that geodesists use altimeter data collected over a very long period (typically one year or more). They use these 1D altimeter datasets (subsatellite track) to compute a series of 2D gridded parameters such as geoid models, mean sea surface models, bathymetries, and gravity anomaly fields. For CryoSat-2 and Jason-1 GM (geodetic mission), the yearly cross-track resolution of altimeter tracks is of the order of 8 km at the equator. Thus, SARAL DP becomes exceedingly useful if it can collect a regular geodetic grid of 8 km after one year and a 4 km grid after two years. Zhang and Sandwell [27] emphasize the benefit of a drifting phase longer than 1.3 years, i.e., a cross-track resolution better than 6 km, to reveal a large number of uncharted seamounts, and important tectonic features.

Figure 6a shows a simulation of the zonal distance (in km) of all altimeter tracks (equator crossing, expressed as a distance with respect to the Greenwich meridian) after one year of SARAL DP: each track is a small black tick and the histogram shows how many tracks (i.e., how many black ticks) are located in each 8-km bin. Due to the uncontrolled altitude decay, the distribution is good albeit random with 0 to 3 satellite tracks per 8-km bin.

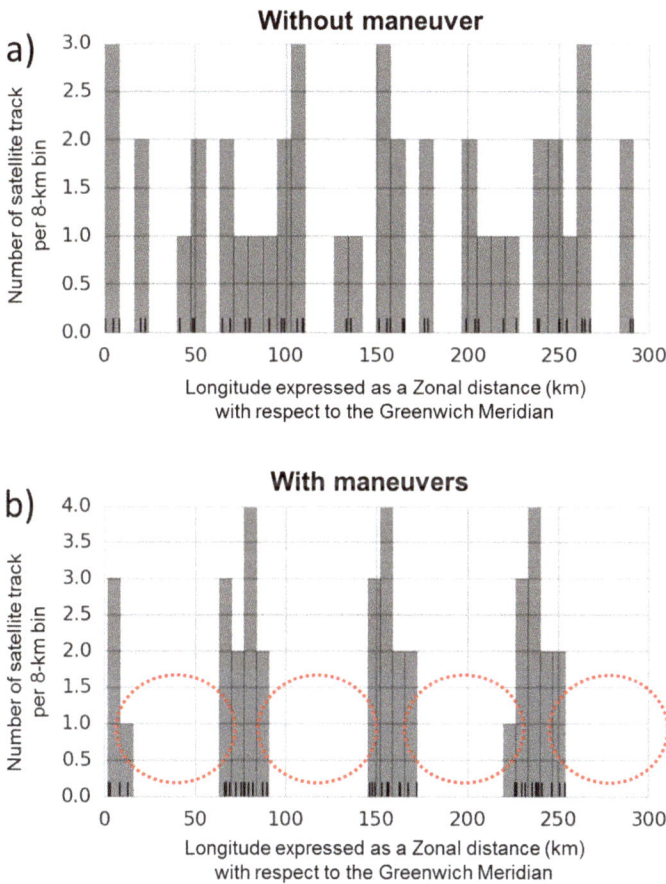

Figure 6. Simulated zonal distance (in km) of the equator crossing of ascending tracks after one year of AltiKa drifting phase (DP) (black ticks) over a 2.7° longitude region. The grey histogram shows the number of altimeter tracks in 8-km bins. Panel (**a**) is a simulation where the satellite altitude is not maintained: the distribution of altimeter tracks is relatively random with 0 to 3 tracks in each 8-km bin). Panel (**b**) is a simulation where the orbit altitude decay is mitigated with one maneuver in the middle of the drifting phase: this maneuver single-handedly aggregates all altimeter tracks in 30-km bands (typically 2 to 4 tracks per 8-km bin), leaving all the other regions completely unobserved (red circles).

In other words, this prediction shows that one year of SARAL DP does not yield a perfect 8-km geodetic grid since some grey bins are empty and other bins have up to 3 satellite track samples. But on average, all regions are reasonably well covered, albeit with a somewhat random and irregular pattern. This kind of random distribution was predicted for all regions and all solar flux scenarios.

In contrast, if a relaxed altitude control is performed, Figure 6b shows a very different picture. In this scenario a single maneuver is performed after 5 months of simulation in order to pull back SARAL to the altitude where the drifting phase was initiated. As a result, all the satellite tracks are put together in small bands of 30-km, leaving 50% of the globe unobserved even after one year. Assuming this pattern is repeated every 6 months, the geodetic dataset collected by AltiKa over the global ocean is less attractive than in Panel (a) of Figure 6, where the random distribution ensures that observation gaps are neither large nor systematic.

To summarize, as soon as the drifting phase starts, it is better for geodesy users to stop all maneuvers altogether and to let the satellite drift. Assuming that emergency debris avoidance must be performed, it is better to decrease the altitude than to increase it.

4. Observations after 18 Months of SARAL DP

The altitude of SARAL was successfully increased by 1 km in Summer 2016. The satellite has been operated on a slowly decaying altitude for approximately 18 months at the time of writing. This section gives an overview of the results obtained with in-flight data. Section 4.1 gives an overview of the geodetic sampling collected so far, and Sections 4.2 and 4.3 present some results on ocean mesoscale.

4.1. Geodetic Sampling

As expected from Section 3.4 and Figure 6a, the most important change of the drifting phase is the unprecedented cross-track resolution: Figure 7 shows the position of descending SARAL subsatellite tracks during the 35-day repeat phase (panel a) and during the first 15 months of SARAL DP (panel b). The longitude resolution of the historical ERS orbit was of the order 50 km at mid-latitudes. The cyclic sampling left very large uncovered regions, even after Spring 2015 when SARAL departed up to 10 km away from its nominal track for a brief period.

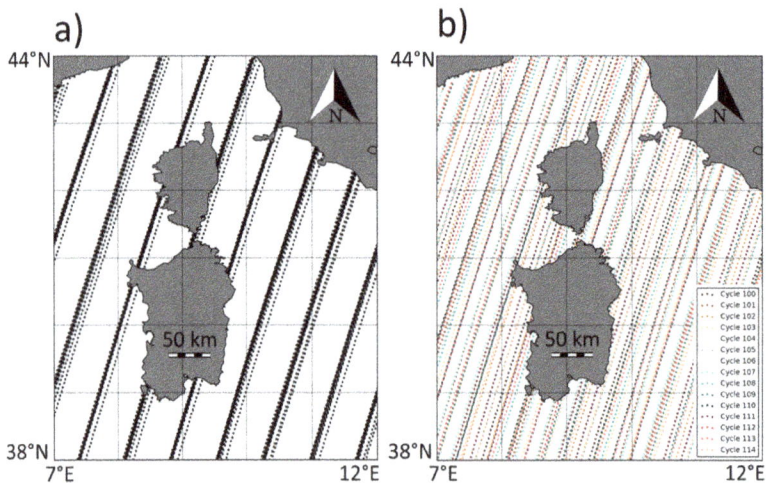

Figure 7. Position of the AltiKa measurements (1 Hz rate of the Geophysical Data Record product) from descending passes over a period of 15 months in the Mediterranean Sea. Panel (**a**) is for the 35-day repeat phase in 2015 and panel (**b**) is for the first 15 months of the drifting phase.

In contrast, the SARAL DP coverage is very dense (panel b, Figure 7): after 15 months, the grid resolution is approximately 4 km on average in this region. As discussed in Section 3.4, the geodetic grid is also somewhat random and very irregular: it can locally range from 1 km to 15 km.

Sandwell et al. [28] report that this grid is now used to enhance the smallest scales of geoid models and bathymetry fields in the open ocean. As more AltiKa measurements are collected, the local resolution will keep increasing, albeit still in an irregular way.

4.2. Mesoscale Sampling

Because the altitude of SARAL was increased at the beginning of the drifting phase, its mesoscale sampling has been modified as predicted in Section 3.3. Figure 8 shows the distribution of actual SARAL/AltiKa ascending nodes in the longitude/time plane (same metric as the simulations of Section 2 and Figure 1, but based on in-flight data). Panel (a) is for the 35-day repeat phase and panel (b) is for December 2017 after 18 months of SARAL DP. Each black dot is one satellite track, and the grey circles are their area of influence (high correlation) for large ocean mesoscale.

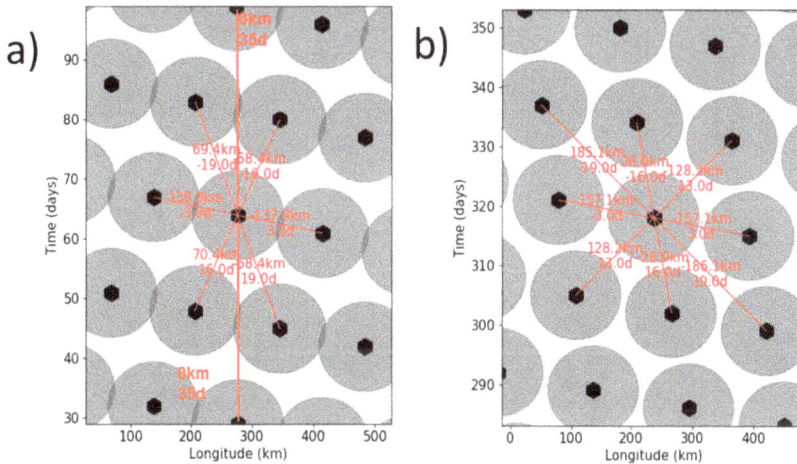

Figure 8. Same as Figure 1 for actual AltiKa data. Panel (**a**) is for the 35-day repeat phase in 2015 and panel (**b**) is December 2017 during the drifting phase.

Figure 8b shows that the distribution of SARAL DP tracks is consistent with the simulations from Figure 3, and more specifically with case #B. After 18-months the new phase of SARAL/AltiKa still provides a good sampling capability where each track/measurement minimizes the correlation with adjoining tracks.

More interestingly, the weight of SARAL/AltiKa in the Copernicus Marine Environment Monitoring Service [5] is routinely monitored in the frame of the operational production of multi-altimeter maps [22]: Figure 9 shows the evolution of the mean contribution of each altimeter used by the Optimal Interpolator. The interpolator is derived from Ducet et al. [20] and Le Traon et al. [21]. The global contribution of each satellite is measured using Degrees of Freedom of Signal following the methodology of Dibarboure et al. [29].

Figure 9a shows the contribution of SARAL/AltiKa in an idealized configuration where only the sampling change is taken into account. From September 2015 to June 2016, SARAL is on the historical 35-day orbit and its contribution is of the same order as Jason-2, i.e., substantially more than CryoSat-2. The weaker score of CryoSat-2 is consistent with findings from Dibarboure et al. [18]. Still, SARAL is slightly below Jason-2 by a couple of points on average. In contrast, after July 2016, SARAL DP is strictly equal to Jason-2: changing the altitude by +1 km slightly improved the sampling capability of AltiKa as predicted in Section 3.3.

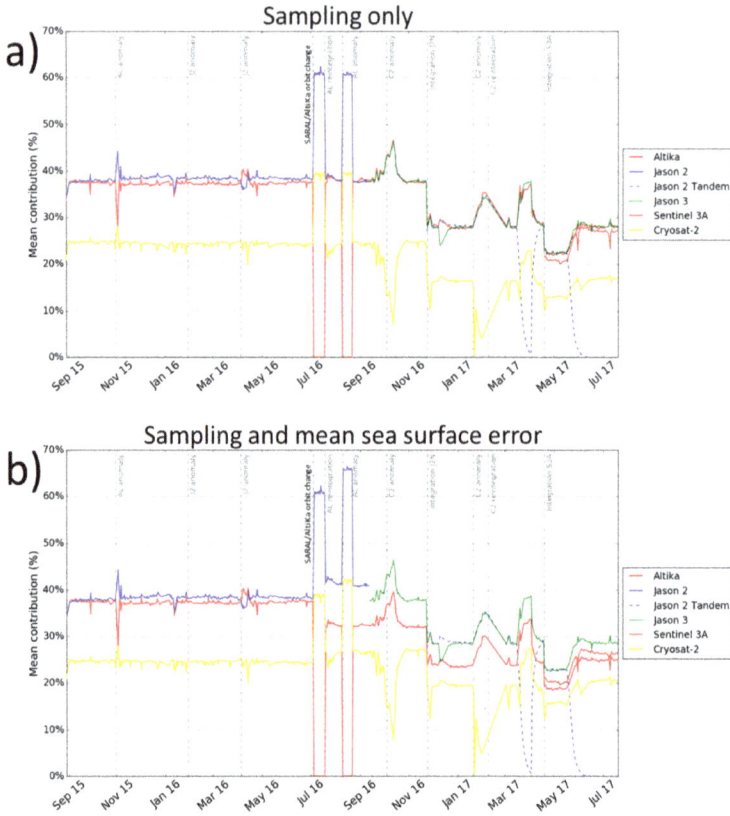

Figure 9. Relative contribution (%) of each altimeter to the multi-altimeter maps of the sea-level thematic assembly center of Copernicus Marine Environment Monitoring Service (CMEMS) from September 2015 to July 2017. Panel (**a**) measures the impact of SARAL DP sampling change only (the measurement error is assumed to be constant between the 35-day and the drifting phase). Panel (**b**) accounts for both the sampling change the increased error (sub-optimal mean sea surface model along the new uncharted orbit).

4.3. Impact of the Mean Sea Surface Error

An important component of the accuracy of altimetry products is the reference surface used to transform the Level-2 altimeter sea surface height (SSH) into Level-3 sea-level anomalies (SLA). Dibarboure et al. [18,19] have shown the difference between repeat orbits such as the 35-day orbit originally used for SARAL, and geodetic or non-repeat orbits such as SARAL DP.

The former can use a precise mean sea surface height (also known as mean profile) based on more than 2 decades of SSH from ERS-1 to SARAL: it is a very precise local estimate of the mean sea surface (MSS). In contrast, non-repeat missions rely on a global gridded MSS model which is slightly less accurate.

Pujol et al. [30] have measured the error of recent gridded MSS models using independent measurements from SARAL DP, HY-2A and Sentinel-3A. The error along uncharted tracks was shown to explain up to 30% of the total SLA variance for wavelengths ranging from 30 km to 100 km.

To that extent, Figure 9a is only an idealized view of the change between the 35-day phase and the drifting phase, because it does not account for the larger SLA error when a gridded MSS reference must be used in place of the precise 20-year mean profile. In contrast, Figure 9b takes into account both aspects of the orbit change: the slightly increased sampling capability and the additional MSS error.

The overall contribution of AltiKa decreases by approximately 6 points and in turn the relative contributions of other missions increase. Still, SARAL remains a strong contributor of the CMEMS system, including when both Jason-3 and Jason-2 plus Sentinel-3A are active. This is notably visible during the CryoSat-2 or Jason-2 data outages where the contribution of AltiKa exhibits some peaks as high a +5 to 7 points.

4.4. Evolution of the SARAL Altitude

After 18 months of drifting phase, the altitude of SARAL has decreased by less than 150 m. This is approximately the 5th percentile curve of the simulations from Figure 2, i.e., three times less than our conservative simulations of Figure 4 which was based on the 95th percentile scenario of the 2016 predictions. Thanks to a particularly weak solar activity, SARAL remains on the upper range of the desirable altitudes of Figure 5 (blue region, 16-day sub-cycle). In other words, the mesoscale sampling capability of SARAL/AltiKa will remain good for many years, and substantially longer than predicted by the conservative simulations.

5. Discussion: Using an Unmaintained Altitude with Other Altimeters

5.1. Generalizing the SARAL DP Strategy

SARAL/AltiKa is the first altimeter to be operated on an unmaintained decaying altitude. But other altimeters could use the same strategy in the future. There are essentially three reasons to use this kind of orbit in the context of radar altimetry:

- Platform ageing: if controlling the satellite's altitude becomes difficult or risky, it might be better to simply stop maneuvers altogether after the satellite is put on a well-chosen initial altitude.
- Simpler operations: keeping the satellite on its nominal orbit requires periodic operations (e.g., every 2 to 4 weeks for ERS and Jason altitudes). Letting the altitude decay naturally in a mesoscale-friendly altitude range might reduce the manpower needed for routine operations. While it is not significant for a single altimeter above 900 km, this strategy could be attractive for lower orbits (e.g., 500 km) or for altimeter constellations (e.g., 5+ satellites) because more manpower is required, and because each maneuver creates a small data gap or temporary degradation (e.g., discontinuity in precise orbit determination). Limiting the maneuvers to debris avoidance could reduce the operation cost of such a low-altitude constellation.
- Saving fuel: most of the satellite fuel is used during launch (to reach the nominal orbit) and for de-orbitation. Using a decaying altitude and avoiding station keeping maneuvers would save only a small amount of fuel. But for some satellites that experience major launch events, it is sometimes necessary to consume more fuel than anticipated (e.g., CryoSat-2 was injected relatively far away from its nominal altitude). In this situation, using a decaying altitude might be a backup scenario able to extend the satellite lifespan by saving fuel for the final de-orbitation.

To summarize, the SARAL DP decaying altitude strategy could be used for other altimeters in the future: as a new mission phase when the satellite suffers from ageing anomalies, as a backup orbit after launch anomalies, or for a dedicated mission/constellation put specifically at an altitude where the maintenance effort is limited.

In the following section, we expand the methodology and the results of Section 3 to determine automatically all the altitude ranges that are compatible with an SARAL DP strategy.

5.2. Methodology Addition

In Figure 3, we observed very different ocean mesoscale sampling properties. We used the overlap between grey circles as a way to detect poor orbits. This can be automated with the very simple correlation model (Equation (4)) used by Dibarboure et al. [19] for Jason-1.

$$C(dx, dt) = e^{-k \cdot \left(\frac{dx^2}{r_x^2} + \frac{dt^2}{r_t^2} \right)} \tag{4}$$

Here r_x and r_t are the spatial and temporal decorrelation radiuses of Section 2.1, and k is arbitrarily set to yield a correlation of 0.5 on the edge of the r_x/r_t circle. By computing $C(dx, dt)$ where dx and dt is the space and time distance between two altimeter tracks, it is possible to approximate the correlation between the datasets. Sensitivity tests show that the type of function and parameter k do not significantly impact the following results.

For a given altitude, we compute the position of all altimeter tracks in the space/time place over 50 days (i.e., complete mesoscale decorrelation), we pick one arbitrary track and we search for all possible neighbors in a space/time bubble. Then we keep the 5 non-aligned tracks with the highest correlation as they define the sub-cycles for this altitude. Pruning aligned tracks from this list is important: to illustrate, with a 3-day SC, we want to detect only the closest track aligned in this direction because 6, 9 and 12 days are only repeats of this SC (they are not separate sub-cycles because they all follow the same alignment).

This computation gives not only the list of dominant sub-cycles (e.g., 3 and 16 for the ERS/ENVISAT sub-cycles) but also all minor sub-cycles (e.g., 19-day SC visible in Figure 1) and the true repeat cycle (e.g., 35 days for ERS/ENVISAT). For each sub-cycle, we also have the correlation which indicates if this sub-cycle yields independent or correlated satellite tracks. If the correlation is too high for any SC, then the altitude is not attractive to observe mesoscale. If the correlation is low for all SC, then it is a good orbit candidate: it is worthy of more in-depth analyses not discussed in this paper.

5.3. Finding Good Altitude Ranges

We performed this computation for all altitudes from 300 km to 1500 km with an increment of 30 m. We used a threshold of $C = 0.5$ to reject mesoscale-unfriendly altitudes. Then we aggregated all valid orbits in altitude bands to determine how large these bands were.

Figure 10 shows the result near the ERS/ENVISAT altitude. Panel (a) of Figure 10, shows the correlation C for each altitude (each sub-cycle is displayed with a different color), and panel (b) of Figure 10, shows the valid (green) and invalid (red) regions based on our threshold (if any sub-cycle has a correlation $C > 0.5$ the altitude is in red). Panel (b) also highlights the 5 cases shown in Figure 3. This representation illustrates how the qualitative overlap discussed in previous sections is now quantified with this metric. The main green region in Panel (b) corresponds to the altitude selected by CNES and ISRO for SARAL DP.

5.4. Mesoscale-Friendly Altitude Bands for Unmaintained Altimeters

Figure 11 shows all the valid altitude bands found for the ERS/ENVISAT inclination. We performed sensitivity tests with the inclination of various altimetry missions (GEOSAT, Jason, Sentinel-3) that yielded similar results: the good/bad altitude ranges are translated but their properties are essentially the same.

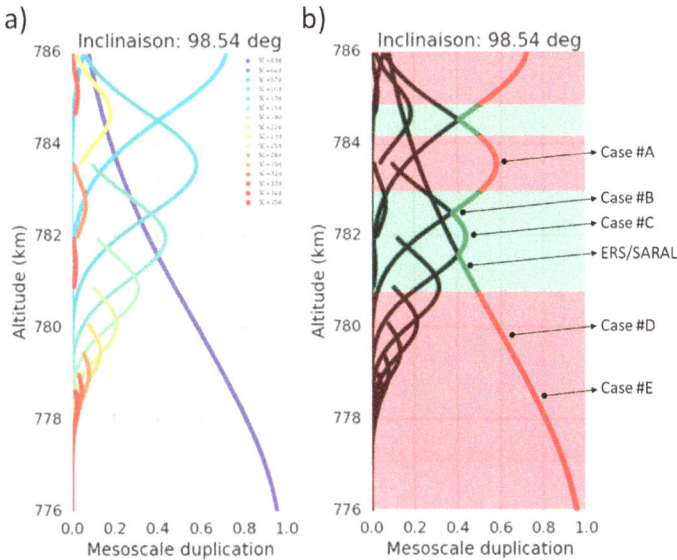

Figure 10. Quality of the mesoscale observation as a function of the altitude. For each altitude, we analyze the alignment of altimeter tracks in the space/time plane. Then we compute their correlation for ocean mesoscales. Each alignment is a sub-cycle and we can determine which sub-cycle contributes positively or negatively to the quality of mesoscale observation at this altitude (panel (**a**)). In panel (**b**), we separate favorable (green) from undesirable (red) altitudes by looking at the maximal correlation of each orbit sub-cycle. This metric corresponds to the overlap between the grey circles in Figure 1. The 5 orbit cases shown in Figure 3 and the ERS/ENVISAT altitude are highlighted in panel (**b**).

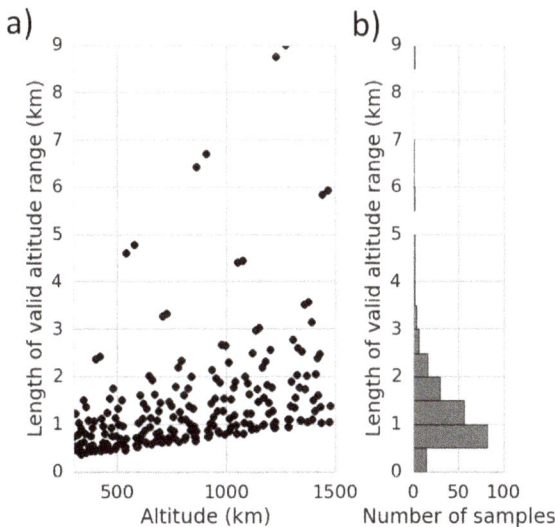

Figure 11. Length of the valid altitude bands (green regions from Figure 10b) for all altitudes and the inclination of the ERS orbit. Each dot in panel (**a**) is one green altitude range where the mesoscale correlation is always lower than 0.5. Panel (**b**) shows the associated distribution.

On average, they are rather thin (from 500 m to 1500 m) but a small fraction can be 2 to 3 km large. This finding is consistent with observations from Dibarboure et al. [19]: their orbit candidates for the so-called Extension-Of-Life phase of Jason-1 were found to be aggregated in thin altitude bands. In the case of SARAL DP there is a 2-km sweet spot just near the historical ERS/ENVISAT orbit: these larger bands are uncommon.

Furthermore Figure 11 also shows the existence of altitude bands as large as 9 km. These exceptional occurrences always come in pairs. This is explained by Figure 12: the longest valid altitude ranges are generally located symmetrically near the altitude of a very short repeat cycle. The black rectangles of panel (a) of Figure 12, on each side of the purple bell-shaped curve highlight are where valid altitude ranges are located. Panel (b) of Figure 12, is a zoom of the plain black rectangle of panel (a): it shows that the composition of a very short sub-cycle and multiple mesoscale-friendly sub-cycles minimizes ocean mesoscale duplication (i.e., the maximal correlation between adjoining satellite tracks). The validity of the altitude range is shown in green in panel (c) of Figure 12, the largest valid altitude band range from 1227 to 1236 km. The dashed black rectangle of panel (a) also contains a second valid range (essentially an upside-down mirror of panel (b)). The same dual/mirror pattern can be observed for most of the long valid ranges in Figure 11a.

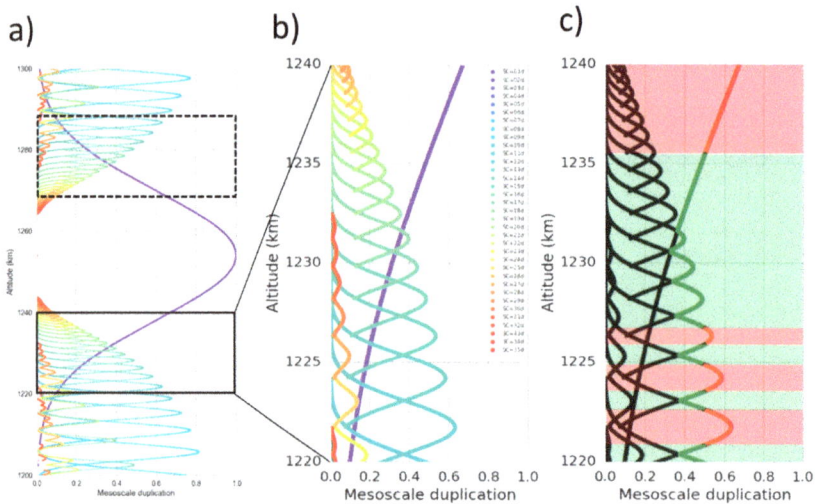

Figure 12. Same as Figure 10a from 1200 to 1300 km (panel (**a**)) and near 1230 km (panel (**b**)). In panel (**c**), we separate favorable (green) from undesirable (red) altitudes by looking at the maximal correlation of each orbit sub-cycle for the altitude range of panel (**b**). The altitude range in panel (**b**) and panel (**c**) corresponds to the plain rectangle of panel (**a**) and where the longest valid altitude range was found.

The 9-km bands are located above 1200 km where the atmospheric drag is very weak. To that extent, a dedicated altimeter mission parked at this altitude would have a great mesoscale sampling for decades. In practice, 1200 km could be too high for at least three reasons: aggressive radiation environment for the satellite, difficult to de-orbit at the end of the mission, unfavorable signal-to-noise ratio for the altimeter (e.g., possibly incompatible with a very small satellite).

In contrast, one could envision a low-cost, low-maintenance altimeter near 600 km for the following reasons:

- The lower altitude would better protect the satellite from radiations.
- The satellite de-orbitation would be easier (e.g., good for a small satellite).

- Altitudes lower than 700 km are attractive because the space debris density is lower: in turn one could anticipate fewer collision avoidance maneuvers (simpler operations)
- Lower altitudes yield a slightly better signal-to-noise ratio for the altimeter (less electrical power needed, i.e., also good for a small satellite).

Nevertheless, at 600 km the atmospheric drag increases exponentially (Figure 13). Even if there is a 4.5 km valid altimeter band that could host such as dedicated mission, the altitude of a SARAL-looking satellite would decay approximately 10 times faster than the 300 m/year of SARAL. To be viable, the platform should be specifically designed to minimize the atmospheric drag with an aerodynamic profile (e.g., smaller and compact solar arrays like Cryosat-2). The lifespan of the mission could also be extended by a handful of maneuvers (e.g., every couple of years) that would reset the altitude decay and start over a new maneuver-free cycle.

Such a low-cost and low-maintenance altimeter mission would be relevant in the context of low-cost constellations (e.g., 4 to 10 satellites): if all satellites are located at the same altitude in one sun-synchronous orbit plane, the altitude of all altimeters decays in a coherent way. With this scenario, the sampling capability of the entire constellation remains good throughout the entire altitude decay. In essence, this concept is a fire-and-forget constellation of small altimetry satellites whose orbit maintenance is limited to debris avoidance.

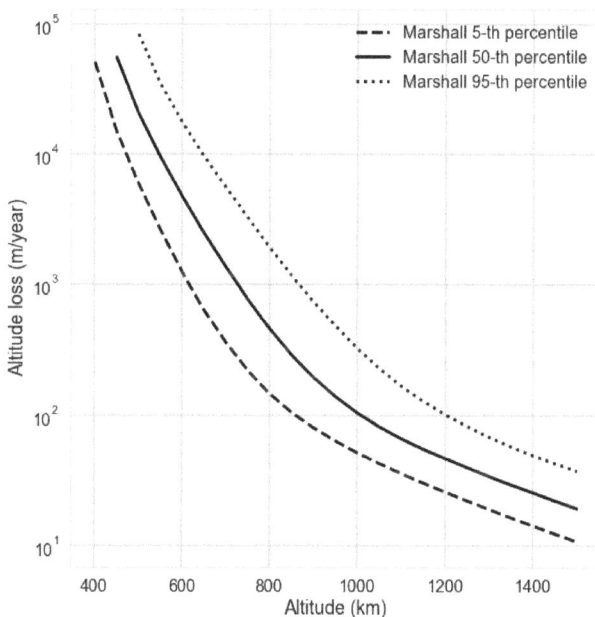

Figure 13. Approximate altitude loss (m/year) for an arbitrary satellite as a function of the initial altitude (km). Three scenarios are provided: three solar flux scenarios: 5-th percentile (weak solar activity, dashed), 50-th percentile (median solar activity, plain) and 95-th percentile (intense solar activity, dotted).

6. Conclusions

In this paper we showed that it is possible for altimeter missions to keep a good mesoscale sampling for years even if their altitude is not actively maintained and even if it decays due to the atmospheric drag. This strategy has been successfully applied on SARAL/AltiKa during the so-called drifting phase that was initiated in early July 2016 by ISRO and CNES.

Eighteen months later, SARAL DP has demonstrated the benefits of this strategy: reducing the maneuvers has made it possible to extend the satellite lifetime (less stress on momentum wheels), and the new DP orbit provides an enhanced sampling capability of ocean mesoscale, as well as precious measurements to improve marine geodesy references fields (e.g., bathymetry, geoid or mean sea surface).

For SARAL, the optimal starting point of the drifting phase was approximately 1 km above the historical ERS/ENVISAT orbit. Thanks to a continuously evolving combination of the 13, 16 and 19-day sub-cycles, the mesoscale sampling capability of SARAL DP has been secured for 6 years or more.

More generally, this strategy can be used for any altimeter (e.g., Jason-3 or Sentinel-3), for instance if the satellite maneuvering capability is limited by onboard ageing anomalies. As long as the satellite is put at the top of a favorable altitude range, the decaying altitude will not degrade the mesoscale sampling capability of the altimeter. It is possible to find mesoscale-friendly altitudes at all altitudes from 500 km to 1500 km.

We also identified a small list of larger mesoscale-friendly altitude bands where one could park a full-fledged drifting altimetry mission, or an entire altimeter constellation. In these sweet spots, the quality of mesoscale sampling is guaranteed for a very long time even if the altitude is not maintained. Near 1230 km, there are two very large altitude bands (9 km each) where an unmaintained altimeter would yield a good and stable mesoscale sampling for decades. At 600 km two large altitude ranges (4.5 km each) makes it possible to envision a low-cost/low-maintenance small altimeter satellite (good SNR, easy de-orbitation, orbit control limited to debris avoidance).

Author Contributions: G.D. carried out the mesoscale and sub-cycle analysis as well as mesoscale sampling simulations. A.L. performed the analysis of SARAL maneuvers and he computed the prediction of altitude decay. M.-I.P. performed the mesoscale sampling analysis on in-flight data using CMEMS/DUACS metrics. G.J. provided the geodetic coverage analysis of in-flight-data.

Funding: This work was sponsored by CNES as part of the SARAL cooperation with ISRO under contract SALP 104685/00.

Acknowledgments: The SARAL/AltiKa mission is an achievement of the fruitful cooperation between ISRO and CNES. We acknowledge the support of the SARAL/AltiKa science and project teams during the preparation of the SARAL drifting phase.

Conflicts of Interest: The authors declare no conflict of interest.

References

1. Verron, J.; Sengenes, P.; Lambin, J.; Noubel, J.; Steunou, N.; Guillot, A.; Picot, N.; Coutin-Faye, S.; Sharma, R.; Murthy, D.R.; et al. The SARAL/AltiKa altimetry satellite mission. *Mar. Geod.* **2015**, *38* (Suppl. 1), 2–21. [CrossRef]
2. Vincent, P.; Steunou, N.; Caubet, E.; Phalippou, L.; Rey, L.; Thouvenot, E.; Verron, J. AltiKa: A Ka-band Altimetry Payload and System for Operational Altimetry during the GMES Period. *Sensors* **2006**, *6*, 208–234. [CrossRef]
3. Morrow, R.; Seo, H.; Farrar, J.T.; Fu, L.L.; Le, P.-Y. Traon Ocean Eddies and Mesoscale Variability. In *Satellite Altimetry Over Oceans and Land Surfaces*; CRC Press: Boca Raton, FL, USA, 2017; pp. 315–342.
4. AVISO. SSALTO/DUACS User Handbook: (M)SLA and (M)ADT Near-Real Time and Delayed Time Products. CNES Document SALP-MU-P-EA-21065-CLS. 2016. Available online: https://www.aviso.altimetry.fr/en/data/product-information/aviso-user-handbooks.html (accessed on 2 July 2018).
5. CMEMS. Copernicus Marine Environment Monitoring Service–Sea Level Thematic Assembly Center Product User Manual. Available online: http://marine.copernicus.eu/documents/PUM/CMEMS-SL-PUM-008-017-036.pdf (accessed on 2 July 2018).
6. Chelton, D.B.; Schlax, M.G. The accuracies of smoothed sea surface height fields constructed from tandem altimeter datasets. *J. Atmos. Ocean. Technol.* **2003**, *20*, 1276–1302. [CrossRef]
7. Pascual, A.; Boone, C.; Larnicol, G.; Le Traon, P.Y. On the quality of real-time altimeter gridded fields: Comparison with in situ data. *J. Atmos. Ocean. Technol.* **2009**, *26*, 556–569. [CrossRef]

8. Verron, J.; Bonnefond, P.; Aouf, L.; Birol, F.; Bhowmick, S.A.; Calmant, S.; Conchy, T.; Crétaux, J.-F.; Dibarboure, G.; Faugère, Y. The benefits of the Ka-band as evidenced from the SARAL/AltiKa altimetric mission: Scientific applications. *Remote Sens.* **2018**, *10*, 163. [CrossRef]

9. Bonnefond, P.; Verron, J.; Aublanc, J.; Babu, K.N.; Bergé-Nguyen, M.; Cancet, M.; Chaudhary, A.; Crétaux, J.-F.; Frappart, F.; Laurain, O. The Benefits of the Ka-Band as Evidenced from the SARAL/AltiKa Altimetric Mission: Quality Assessment and Unique Characteristics of AltiKa Data. *Remote Sens.* **2018**, *10*, 83. [CrossRef]

10. Rémy, F.; Flament, T.; Michel, A. Envisat and SARAL/AltiKa observations of the Antarctic ice sheet: A comparison between the Ku-band and Ka-band. *Mar. Geod.* **2015**, *38*, 510–521. [CrossRef]

11. Smith, W.H.F. Resolution of Seamount Geoid Anomalies Achieved by the SARAL/AltiKa and Envisat RA2 Satellite Radar Altimeters. *Mar. Geod.* **2015**, *38*, 644–671. [CrossRef]

12. Valladeau, G.; Thibaut, P.; Picard, B.; Poisson, J.C.; Tran, N.; Picot, N.; Guillot, A. Using SARAL/AltiKa to improve Ka-band altimeter measurements for coastal zones, hydrology and ice: The PEACHI prototype. *Mar. Geod.* **2015**, *38* (Suppl. 1), 124–142. [CrossRef]

13. Arsen, A.; Crétaux, J.F.; Abarca del Rio, R. Use of SARAL/AltiKa over mountainous lakes, intercomparison with Envisat mission. *Mar. Geod.* **2015**, *38* (Suppl. 1), 534–548. [CrossRef]

14. Bhowmick, S.A.; Modi, R.; Sandhya, K.G.; Seemanth, M.; Balakrishnan Nair, T.M.; Kumar, R.; Sharma, R. Analysis of SARAL/AltiKa wind and wave over Indian Ocean and its real-time application in wave forecasting system at ISRO. *Mar. Geod.* **2015**, *38* (Suppl. 1), 396–408. [CrossRef]

15. Bhowmick, S.A.; Basu, S.; Sharma, R.; Kumar, R. Impact of Assimilating SARAL/AltiKa SWH in SWAN Model During Indian Ocean Tropical Cyclone Phailin. *IEEE Trans. Geosci. Remote Sens.* **2016**, *54*, 1812–1817. [CrossRef]

16. Bonnefond, P.; Willis, J. Minutes of the 2015 Ocean Surface Topography Science Team Meeting. Available online: http://www.aviso.altimetry.fr/fileadmin/documents/OSTST/OSTST_2015_Meeting_Report.pdf (accessed on 2 July 2018).

17. Sandwell, D.; Dietmar Müller, R.; Smith, W.; Garcia, E.; Francis, R. New global marine gravity model from CryoSat-2 and Jason-1 reveals buried tectonic structure. *Science* **2014**, *346*, 65–67. [CrossRef] [PubMed]

18. Dibarboure, G.; Renaudie, C.; Pujol, M.I.; Labroue, S.; Picot, N. A demonstration of the potential of Cryosat-2 to contribute to mesoscale observation. *Adv. Space Res.* **2012**, *50*, 1046–1061. [CrossRef]

19. Dibarboure, G.; Schaeffer, P.; Escudier, P.; Pujol, M.-I.; Legeais, J.F.; Faugère, Y.; Morrow, R.; Willis, J.K.; Lambin, J.; Berthias, J.P.; et al. Finding desirable orbit options for the "Extension of Life" phase of Jason-1. *Mar. Geod.* **2012**, *35*, 363–399. [CrossRef]

20. Ducet, N.; Le Traon, P.Y.; Reverdin, G. Global high resolution mapping of ocean circulation from the combination of TOPEX/POSEIDON and ERS-1/2. *J. Geophys. Res. (Oceans)* **2000**, *105*, 19477–19498. [CrossRef]

21. Le Traon, P.Y.; Faugère, Y.; Hernandez, F.; Dorandeu, J.; Mertz, F.; Ablain, M. Can we merge GEOSAT Follow-On with TOPEX/POSEIDON and ERS-2 for an improved description of the ocean circulation? *J. Atmos. Ocean. Technol.* **2003**, *20*, 889–895. [CrossRef]

22. Pujol, M.-I.; Faugère, Y.; Taburet, G.; Dupuy, S.; Pelloquin, C.; Ablain, M.; Picot, N. DUACS DT2014: The new multi-mission altimeter dataset reprocessed over 20 years. *Ocean Sci. Discuss.* **2016**, *12*, 1067–1090. [CrossRef]

23. Le Traon, P.Y.; Dibarboure, G. Mesoscale mapping capabilities from multiple altimeter missions. *J. Atmos. Ocean. Technol.* **1999**, *16*, 1208–1223. [CrossRef]

24. Rees, W.G. Orbital subcycles for Earth remote sensing satellites. *Int. J. Remote Sens.* **1992**, *13*, 825–833. [CrossRef]

25. Priester, W. Solar activity effect and diurnal variation in the upper atmosphere. *J. Geophys. Res.* **1961**, *66*, 4143–4148. [CrossRef]

26. Hathaway, D.H.; Wilson, R.M.; Reichmann, E.J. A synthesis of solar cycle prediction techniques. *J. Geophys. Res.* **1999**, *104*, 22375–22388. [CrossRef]

27. Zhang, S.; Sandwell, D.T. Retracking of SARAL/AltiKa radar altimetry waveforms for optimal gravity field recovery. *Mar. Geod.* **2017**, *40*, 40–56. [CrossRef]

28. Sandwell, D.; Müller, D.; Matthews, K.; Smith, W. Imaging Small-scale Seafloor and Sub-seafloor Tectonic Fabric Using Satellite Altimetry. In Proceedings of the 2017 AGU Fall Meeting, New Orleans, LA, USA, 11 December 2017.

29. Dibarboure, G.; Pujol, M.I.; Briol, F.; Traon, P.L.; Larnicol, G.; Picot, N.; Mertz, F.; Ablain, M. Jason-2 in DUACS: Updated system description, first tandem results and impact on processing and products. *Mar. Geod.* **2011**, *34*, 214–241. [CrossRef]

30. Pujol, M.-I.; Schaeffer, P.; Faugère, Y.; Raynal, M.; Dibarboure, G.; Picot, N. Gauging the improvement of recent mean sea surface models: A new approach for identifying and quantifying their errors. *J. Geophys. Res. Oceans* **2018**. [CrossRef]

remote sensing

MDPI

Article

HY-2A Altimeter Data Initial Assessment and Corresponding Two-Pass Waveform Retracker

Shengjun Zhang [1], Jiancheng Li [2,3], Taoyong Jin [2,3,*] and Defu Che [1]

[1] School of Resources and Civil Engineering, Northeastern University, Shenyang 110819, China;
    zhangshengjun@whu.edu.cn (S.Z.); chedefu@mail.neu.edu.cn (D.C.)
[2] MOE Key Laboratory of Geospace Environment and Geodesy, School of Geodesy and Geomatics,
    Wuhan University, Wuhan 430072, China; jcli@whu.edu.cn
[3] Collaborative Innovation Center of Geospatial Technology, Wuhan University, Wuhan 430072, China
*   Correspondence: tyjin@sgg.whu.edu.cn

Received: 30 January 2018; Accepted: 21 March 2018; Published: 23 March 2018

Abstract: The accuracy and resolution of the marine gravity field derived from multisatellite altimeter data sets mainly depend on the corresponding range precision and spatial distribution. Here, we preliminarily investigate the performance of HY-2A altimeter data by analyzing cross-mission sea surface height discrepancies with SARAL/AltiKa and calculating correlation coefficients with respect to tide gauge measurements. We also explore the improved range precision that can be achieved using a two-pass weighted least squares retracker which was proposed for the purpose of optimal gravity field recovery. Firstly, both the exact repetitive mission and the geodetic mission for HY-2A provide new track orientations and different data coverage for recovering the marine gravity field, and these dense geographical distributions are more greatly attributed to the geodetic mission in recent years. Secondly, HY-2A provides reliable sea surface height measurements based on exterior verifications by SARAL/AltiKa geophysical data records and tide gauge measurements, although the accuracy level is slightly lower than SARAL/AltiKa. Another more exciting finding is that the statistics of along-track sea surface heights in one-second intervals show that the two-pass retracking does further improve the range precision by a factor of 1.6 with respect to 20 Hz retracked results in sensor data records. In conclusion, the HY-2A mission can substantially improve the global accuracy and resolution of the marine gravity field and will reveal new tectonic features such as microplates, abyssal hill fabric, and new uncharted seamounts on the ocean floor.

Keywords: HY-2A; waveform retracking; range precision; marine gravity

1. Introduction

Satellite radar altimetry has been proven an effective tool for recovering the marine gravity field relying on its advantages of spatial scale, sustained operation at any time, and relatively low cost with respect to traditional shipboard instruments. Dozens of regional or global marine gravity field models have been published during the past 40 years based on the accumulated multisatellite altimeter data sets, especially those with nonrepetitive ground tracks [1–8]. The recovered marine gravity accuracy mainly depends on the data quality and the dense geographical distribution of satellite altimeters, while the range precision can be further improved by postprocessing techniques (e.g., waveform retracking, path delay correction modeling, and crossover adjustment). By contrast, the spatial distributions (e.g., adjacent intervals and orientations) of ground tracks are objectively determined in accordance with selected orbital parameters for multisatellite altimeter missions. Therefore, the resolution of a recovered marine gravity field is limited to the altimeter data density, especially along the cross-track directions.

A variety of particular processing techniques, such as the waveform retrackers and low-pass filters, are mostly discussed and adopted in previous marine gravity studies for obtaining improved altimeter measurements [9–11]. Moreover, the altimeter data sets from geodetic missions (GMs) provide high-density coverage that is crucial for recovering high-resolution marine gravity fields and thus play a more important role than data from exact repetitive missions (ERMs) [12], and the ERM data sets show remarkable advantages in oceanic operational applications. In addition, the ERM data sets should also be used for the construction of gravity fields as these repeated profiles have improved signal-to-noise by averaging nonunique, repetitive cycles. The diverse orientations of ground tracks are satisfied by four series of altimeter satellites with different orbit inclinations for 66°, 92°, 98.5° and 108°, respectively. The near polar orbital altimeter has a larger global coverage scope which is closer to the north and south poles, but has poor performances in obtaining the east components of vertical deflection at low-latitude zones. By contrast, the designed inclinations with bigger difference to 90° provide altimeter measurements with complementary features and all these different orbital inclinations make the solution of vertical deflection components more stable in marine gravity recovery studies [3].

HY-2A, China's first satellite for exploring the oceanic dynamic environment, was successfully launched on 16 August 2011 and the corresponding data sets were operationally distributed by China's National Satellite Ocean Application Service (NSOAS) since October 2011. The major payload of the HY-2A satellite is a dual-frequency radar altimeter operating at both Ku and C band for acquiring geophysical parameters such as sea surface height (SSH), significant wave height (SWH), and wind speed (WS) over sea surface. Firstly, HY-2A provides different orientations of ground tracks since its planned orbital inclination is 99.34° and the planned repetitive period is almost 14 days during the early stage of designed life, which are obviously different from other altimeter satellites [13]. Secondly, the geodetic mission phase (~168 days) with drifting orbits planned during the end stage of design life provides another gratifying aspect of denser spatial coverage for marine gravity recovery studies (Table 1). In summary, HY-2A has a great potential for enhancing the resolution and accuracy of the marine gravity field as long as the accuracy of SSH observations maintains an acceptable level for about several centimeters. Fortunately, the previous studies have verified that HY-2A has a similar performance to Jason-2 in foundation of cross calibration and crossover analysis, indicating that SSH observations have an accuracy of 6~8 cm or so [14–17], and meet the normal requirement of oceanic scientific researches and applications. However, most of the previous studies only focus on Interim Geophysical Data Record (IGDR) data, but the suitable waveform retrackers that can bring further improvement for SSH accuracy were poorly discussed before. Besides, HY-2A has a significantly lower application rate in the altimetry community compared to other typical missions, even lower than the afterward missions such as SARAL/AltiKa, Jason-3, and Sentinel-3A.

Under this circumstance, this paper will focus on two issues: (1) Whether the performance of HY-2A is reliable. The validation will be processed by comparing with the in situ time series of tide gauge measurements and by calculating the multimission crossover differences with contemporary missions; (2) Whether the range precision of HY-2A can be further improved by application of a two-pass weighted least squares retracker which was proposed by Sandwell and Smith (2005) for the purpose of optimal gravity field recovery. Previous studies have also shown that this two-pass approach successfully improves the range precision by a factor of ~1.5 for most conventional altimeters operating on both Ku-band and Ka-band [18–20]. If the HY-2A range precision also benefits from this two-pass retracking method, then there can be a potential dramatic enhancement in global marine gravity recovery relying on its geodetic mapping orbit.

Table 1. HY-2A mission specifications.

HY-2A	
Mission duration	16 August 2011–present
Size	8.56 m × 4.55 m × 3.185 m
Mass	≤1575 kg
Orbit type	Sun-synchronous orbit
Altitude	971~973 km
Inclination	99.34°
Repeat period	~14 days, ~168 days
Mode	Pulse-limited
Footprint size	2~10 km
Frequency	13.58 GHz. 5.2 GHz
Chirp bandwidth(Ku)	320 MHz/80 MHz/20 MHz
Chirp bandwidth(C)	160 MHz

2. HY-2A Data Description

The HY-2A, China's first oceanic satellite for exploring the dynamic environment with a radar altimeter on board, has been running operationally in orbit for more than 6 years since it was launched from Taiyuan on 16 August 2011. This satellite repeats its ground track as planned with a repetitive period of ~14 days for the first 4.5 years or so, while a geodetic mission with drifted ground tracks for each subcycle is operating currently since March 2016. All the level 2 products of HY-2A are administrated and distributed by NSOAS. Fortunately, we have successfully obtained most of the HY-2A Sensor Data Records (SDRs) including 20 Hz waveforms since January 2014 to April 2017 by official application from NSOAS (http://www.nsoas.gov.cn/). The specific information for the acquired HY-2A data sets is listed in Table 2.

Table 2. The general information for each cycle of acquired HY-2A altimeter data sets ('E' and 'G' indicates HY-2A data from ERM and GM, respectively).

Cycle	Time Scope	Data Quantity	Cycle	Time Scope	Data Quantity
E059	2013.12.21–2014.01.04	386 passes	E074	2014.07.19–2014.07.28	273 passes
E060	2014.01.04–2014.01.18	386 passes	E075	2014.08.03–2014.08.16	336 passes
E061	2014.01.18–2014.02.01	386 passes	E109	2015.11.21–2015.12.04	201 passes
E062	2014.02.01–2014.02.15	386 passes	E110	2015.12.07–2015.12.19	301 passes
E063	2014.02.15–2014.03.01	386 passes	E111	2015.12.19–2016.01.02	386 passes
E064	2014.03.01–2014.03.15	386 passes	E112	2016.01.02–2016.01.16	324 passes
E065	2014.03.15–2014.03.29	363 passes	E113	2016.01.16–2016.01.28	223 passes
E066	2014.03.29–2014.04.12	386 passes	E114	2016.01.30–2016.02.13	308 passes
E067	2014.04.12–2014.04.26	386 passes	E115	2016.02.13–2016.02.27	386 passes
E068	2014.04.26–2014.05.10	386 passes	E116	2016.02.27–2016.03.12	307 passes
E069	2014.05.10–2014.05.24	334 passes	E117	2016.03.12–2016.03.15	108 passes
E070	2014.05.24–2014.06.07	386 passes		Geodetic Mission	
E071	2014.06.07–2014.06.21	386 passes	G001	2016.03.24–2016.09.08	4542 passes
E072	2014.06.21–2014.07.05	376 passes	G002	2016.09.08–2017.02.23	4630 passes
E073	2014.07.05–2014.07.19	386 passes	G003	2017.02.23–2017.05.10	2030 passes

This paper focuses on the performance of HY-2A data, either the retracked measurements in level 2 products or the corresponding results of the two-pass waveform retracker. The spatial coverage is another crucial factor for evaluating altimetry data except for the common focus on range precision. Accordingly, the geographical distributions of HY-2A ground tracks were firstly analyzed over southeastern China seas by selecting the Cycle_E115 (from 13 February 2016 to 27 February 2016, 'E' represents ERM) and Cycle_G001 (from 24 March 2016 to 8 September 2016, 'G' represents GM) of HY-2A data. As shown in Figure 1, the HY-2A ERM data with a repetitive period of 14 days show significantly different ground tracks (red lines) over research regions compared with other typical altimeter missions, such as Jason-1/10 days (yellow lines), Jason-1 tandem with drifted intermediate orbit/10 days (orange lines), Geosat/17 days (purple lines), Envisat/35 days (blue lines), Envisat/30 days (green lines) and CryoSat-2/369 days (olive lines). The insert maps on the right side show magnified distributions within the black rectangular boxes for these indistinguishable tracks. The blue lines also represent the ground tracks of 35-day repeated orbit for ERS-1, ERS-2, and SARAL/AltiKa, while T/P, Jason-2, and Jason-3 have similar repetitive periods and geographical tracks with Jason-1 and its tandem mission. As shown in Figure 2, HY-2A GM data indeed has denser ground tracks (yellow lines) compared with ERM data (red lines) over the same research areas (subgraph), while a magnification of the local region within a blue rectangular box is also executed for giving better contrast effects.

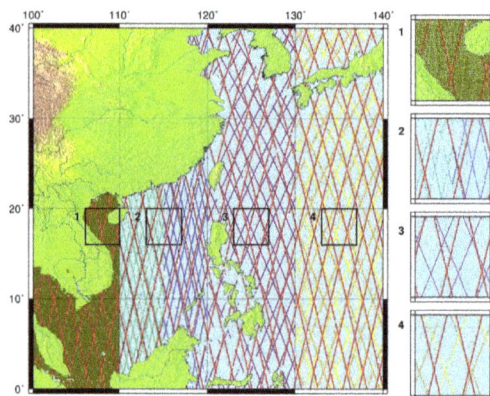

Figure 1. Distribution of multisatellite altimetry data over Research Region (HY-2A/ERM: red; Jason-1/ERM: yellow; Jason-1 Tandem/ERM: orange; Geosat/ERM: purple; Envisat/35 days: blue; Envisat/30 days: green; CryoSat-2/369 days: olive).

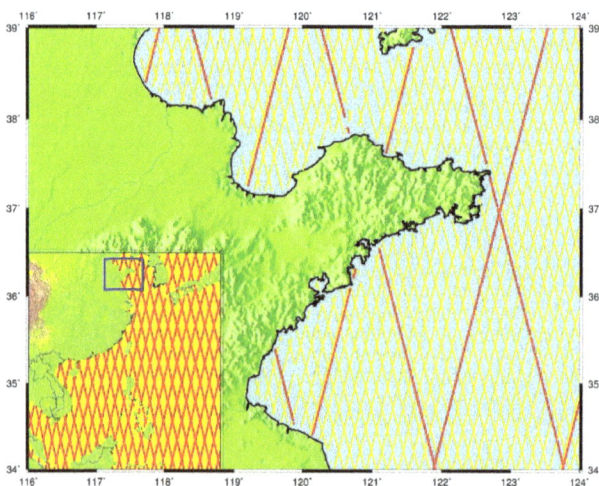

Figure 2. Distribution of HY-2A data over locally magnified region and research area (ERM: red; GM: yellow).

According to Sandwell and Smith's previous study [3], two or more satellites with different orbital inclinations and repetitive periods provide more stable computation of vertical deflections due to their diverse track orientations and intersections. Figure 1 proves that HY-2A ERM data is a great supplement to providing diverse track orientations of multisatellite materials for recovering marine gravity fields, while Figure 2 further verifies that the latest operation of HY-2A geodetic mission has brought denser data coverage. Under the premise of not considering the factor of range precision temporarily, there is enough reason to believe that HY-2A may potentially bring improvement of resolution for multisatellite, altimeter-derived marine gravity fields. Consequently, another crucial factor that directly relates to marine gravity recovery is the range precision of HY-2A and this study accordingly focuses on evaluating the performance of HY-2A measurements originally provided in level 2 products, which will be discussed in Section 3. Additionally, the two-pass retracker which was verified to be commonly effective for pulse-limited altimeter missions will be further investigated for HY-2A on evaluating whether the purpose of lower noise level is achieved. The procedures of estimating parameters and assessing performance are separately discussed in Sections 4 and 5.

3. Initial Assessment of HY-2A Measurement

3.1. Multi-Mission Crossover Analysis

Measurements from radar altimetry missions are classically validated at regional to global scales using relative calibration approaches based on statistical analysis at crossover points intra- and intermissions. Honestly, the smaller difference for intramission situation indicates better stability and internal conforming accuracy for onboard instruments while those crossover discrepancies between multimissions are more reliable for evaluating the range precision between altimetry measurements. Accordingly, we planned to firstly assess the HY-2A performance by analyzing crossover discrepancies with another typical high-precision mission. Meanwhile, the crossover difference for intramission situation is also considered for evaluating inner coincidence.

The satellite mission SARAL/AltiKa (hereafter referred as SARAL), which was launched in March 2013 by the Centre National d'Etudes Spatiales (CNES) and the Indian Space Research Organization (ISRO), shows outstanding performance in providing high-precision measurements due to the following improvements. SARAL operates at Ka-band frequency (35.75 GHz) and has a relatively small pulse-limited footprint and beam width in contrast to traditional Ku-band altimeters, which should decrease the contamination from bright off-nadir reflection. Furthermore, the relatively large bandwidth of SARAL allows a better vertical range sampling (~31 cm) than for Ku-band radar altimeters (~47 cm) which should strongly improve the determination of range [21–23]. Consequently, the SARAL geophysical data record (GDR) from cycle 1 to 12 was used as calibration data in this study, which is a homogeneous data set corresponding to more than one year (14 March 2013 to 8 May 2014).

Returning to the theory of crossover methodology, the validity requires that SSH measurements be entirely corrected from all ocean variability between the intersected passes of altimeters, while these SSH variations are caused by ocean tide, solid earth tide, barometer tide and wind effects. Therefore, the relevant environmental and geophysical corrections of the altimeter range measurements have been applied to calculating SSH. These corrections include dry and wet tropospheric path delay, ionospheric correction, sea state bias, ocean tide, solid earth tide, pole tide, high frequencies wind effect, and inverted barometer correction. All these corrections are provided in GDR products for HY-2A and SARAL, respectively, and no extra updating procedure is applied in this study. The specific information is listed in Table 3. Besides, the incompletely constrained time-varying effects of ocean surface will inevitably affect the crossover discrepancies for either intra- and intermissions. As a result, crossover differences with and without time limit need to be respectively considered during the statistical procedure. We selected the HY-2A data sets from E059 to E068 among the cross-mission validation based on the chosen SARAL data in order to ensure the overlap of measuring time.

Table 3. List of parameters used to estimate the SSH for both HY-2A and SARAL missions.

Contrastive Parameters	HY-2A	SARAL
Cycle range	59–68	1–12
Time scope	2013.12.21–2014.5.10	2013.3.14–2014.5.8
Dry troposphere correction	ECMWF	ECMWF
Wet troposphere correction	Radiometer	Radiometer
Ionospheric correction	Dual-frequency	GIM
Sea state bias	NSOAS empirical solution	NOAA empirical solution
Ocean tide	GOT00.2	GOT4.8
Solid earth tide	Cartwright and Tayler tables	Cartwright and Tayler tables
Pole tide	Wahr [24]	Wahr [24]
Inverted barometer correction	ECMWF	ECMWF
High-frequency fluctuations	Mog2D model	Mog2D model
Mean sea surface	NSOAS gridding solution	MSS_CNES_CLS_2011

Based on the procedures mentioned before, along-track SSH measurements are calculated for both HY-2A and SARAL missions. The HY-2A/HY-2A and SARAL/SARAL crossover points are respectively determined by fitting ground tracks of sample data, while the crossovers are defined as the positions where each satellite crosses its own ground position. In addition, the crossover points for intermissions are defined as the positions where HY-2A and SARAL cross the same sea surface position. To sum up, the geographical distributions of HY-2A data (red lines), SARAL data (blue lines), HY-2A/HY-2A (blue dots), SARAL/SARAL (red dots), and HY-2A/SARAL (green dots) crossovers are shown in Figure 3.

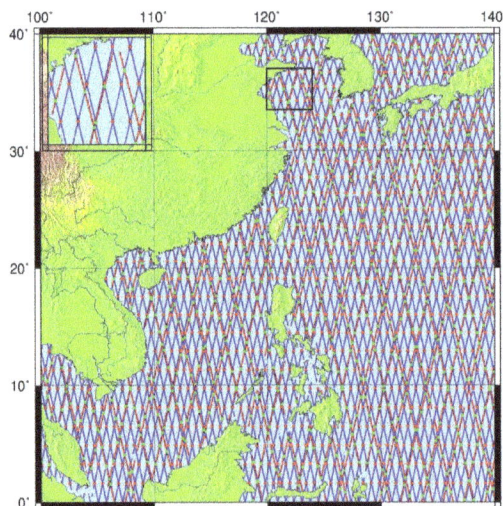

Figure 3. Distribution of HY-2A and SARAL altimetry data and corresponding crossover points intra- and intermissions (Red line: HY-2A data; Blue line: SARAL data; Blue dots: crossover points for HY-2A itself; Red dots: crossover points for SARAL itself; Green dots: crossover points for HY-2A and SARAL intermissions).

Following the calculated positions of crossover points, the minimum, maximum, mean, and root-mean-square (RMS) values of crossover discrepancies are computed between corresponding ascending and descending passes over ocean surfaces without and with time limits of 1, 2, 5 and 10 days, respectively. The specific results are listed in Table 4, which indicate that the measurements within the shortest time limits have the best coincidence under each situation. The differences between GOT00 and GOT4.8 might be as large as a centimeter level and influential in comparing SSH between HY-2A and SARAL. Consequently, we use the GOT-e 4.7 model (http://web.cecs.pdx.edu/~zaron/pub/GOT-e.html), which contains GOT-compatible grids and provides more valid values to cover the shoreline, for updating corresponding ocean tide corrections for both missions. The intermission crossover analysis is thus executed twice and the statistical information is listed in Table 5. HY-2A has a valid but slightly worse performance than SARAL according to crossover analysis for either intramission or intermission situations. Besides the close measuring time, the uniform GOT-e model also reduces the SSH discrepancies for intermissions. A preliminary conclusion is drawn that the accuracy of HY-2A measurements is around ~10 cm while the inherent uncertainty of SARAL is taken into account. This validation initially implies that HY-2A is capable of investigating marine gravity anomalies due to the new data coverage and reliable range precision.

Table 4. Statistical information of crossover differences under intramission situations. Unit: m.

Time Limit	HY-2A					SARAL				
	Num	Min	Max	Mean	RMS	Num	Min	Max	Mean	RMS
–	5136	−0.995	0.998	0.083	0.232	32146	−0.976	0.893	0.016	0.142
≤10 days	816	−0.979	0.841	0.096	0.182	2937	−0.607	0.732	0.017	0.082
≤5 days	442	−0.931	0.841	0.090	0.177	1468	−0.255	0.297	0.016	0.071
≤2 days	242	−0.899	0.841	0.086	0.170	557	−0.232	0.279	0.014	0.062
≤1 day	143	−0.664	0.608	0.095	0.148	371	−0.196	0.279	0.015	0.061

Table 5. Statistical information of crossover differences under intermission situation ('HY' and 'SA' represent HY-2A and SARAL, 'OTC' for ocean tide correction, while '_A' and '_D' denote ascending and descending passes, respectively). Unit: m.

Mission	Time Limit	OTC in GDR					Updated OTC with GOT-e				
		Num	Min	Max	Mean	RMS	Num	Min	Max	Mean	RMS
HY_A & SA_D	–	13,784	−0.998	0.958	0.191	0.250	13,216	−0.999	0.980	0.070	0.199
	≤1 day	131	−0.804	0.543	0.242	0.140	127	−0.942	0.406	0.097	0.139
HY_D & SA_A	–	11,590	−0.999	0.917	0.100	0.206	11,302	−0.999	0.782	−0.029	0.190
	≤1 day	100	−0.868	0.641	0.092	0.151	99	−0.999	0.321	−0.046	0.155
HY_A & SA_A	–	418	−0.993	0.650	0.209	0.226	396	−0.993	0.515	0.081	0.181
	≤1 day	5	0.041	0.439	0.222	0.162	5	−0.097	0.312	0.087	0.165
HY_D & SA_D	–	359	−0.992	0.951	0.124	0.224	349	−0.717	0.802	−0.0002	0.201
	≤1 day	3	0.047	0.419	0.175	0.211	3	−0.078	0.294	0.046	0.210
HY & SA	–	26,151	−0.999	0.958	0.150	0.235	25,263	−0.999	0.980	0.024	0.201
	≤1 day	239	−0.868	0.641	0.178	0.184	234	−0.999	0.406	0.035	0.175

3.2. Validation with Tide Gauge Measurements

Besides the classical cross-mission validation, another well-known method is that sea surface heights from radar altimetry missions and in situ sea level measurements from tide gauges serve as each other's buddy check. Consequently, we explored the performance of HY-2A observations in this section by comparing with time series of in situ sea level at tide gauge stations. Simultaneously, the 12 cycles of SARAL data were also evaluated during this procedure as a comparison. Nevertheless, the HY-2A time series were extended to 16 August 2014 (E075) because we wanted to involve the time series for as long as possible. This validation needed time overlap between altimetric measurements and in situ data series rather than between the two altimetric missions, while a significant data interruption happened to HY-2A after Cycle_E075.

Radar altimeters provide SSH observations, which need to be referenced and corrected from geophysical signals to provide sea level anomalies (SLA) compared with in situ measurements [25]. Therefore, the item of mean sea surface provided in level 2 products needs to be subtracted on the basis of calculating SSH as described in Section 3.1, after which along-track SLA series are obtained. We used valid-only satellite altimetry measurements relying on the quality flags and recommended data editing criteria in handbooks for both missions. The SLA series were then averaged onto a regular $1° \times 1°$ grid, with a temporal resampling corresponding to data intervals at half a month [26].

The University of Hawaii Sea Level Center (UHSLC) offers tide gauge data with two levels of quality control (QC), where more rigorous QC requires more effort and a greater time lag. The Research Quality Data (RQD) sets received thorough QC and were considered to be the final science-ready data set, while the RQD were released 1–2 years after data was received from the data originators by UHSLC. Consequently, this RQD obtained via ftp://ftp.soest.hawaii.edu/uhslc/rqds was used for evaluating the performance of HY-2A measurements in this study. Although the spatial coverage of tide gauges is limited to coastal areas or isolated islands in the open ocean, the tide gauges will sample the ocean every hour, while at a given point, the satellite altimetry sampling is higher than ten days. The RQD is maintained by Pat Caldwell [27] and archived as the NOAA Joint Archive for Mean Sea Level (JAMSL) and only available from the UHSLC at hourly and daily resolution in order to avoid confusion with monthly data provided by the Permanent Service for Mean Sea Level (PSMSL). Consequently, the daily RQD was used for validation and a preprocessing procedure was applied to high-frequency in situ measurements for resampling a time series at the low frequency of altimetry data in 15-day intervals.

Based on the acquired global RQD sets and time span for two altimeter missions, 292 tide gauge stations (red dots in Figure 4) were probably for validation as only these in situ data series span across the period from 2013 to 2015, while 6 stations (yellow stars in Figure 4) among them were initially chosen for validation. All these 6 stations are located at the isolated islands over the open ocean and their geographical distributions represent the most situations over open oceans. The specific information for the chosen tide gauges is listed in Table 6.

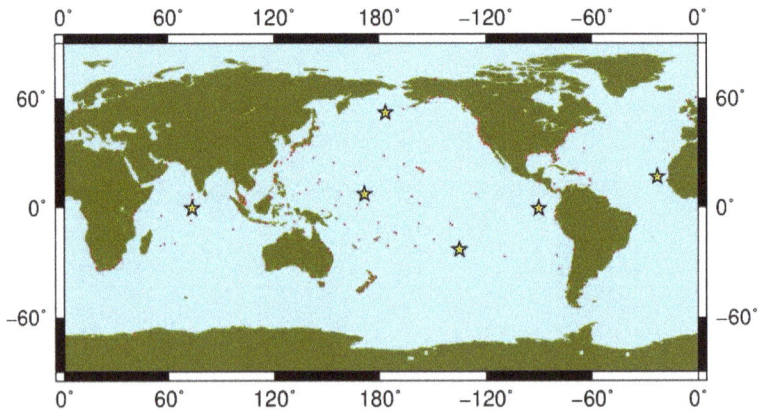

Figure 4. The 6 tide gauge stations (yellow stars) considered in this study. Red dots denote 292 tide gauge stations that span across the period from 2013 to 2015.

Table 6. Specific information of tide gauges in this study and corresponding correlation parameter (CI: Completeness index or percentage of data span without missing data; QC-YEARS: years which have received quality control).

Station	Latitude	Longitude	QC-YEARS	CI	Country	Contributor	Correlation Coefficient	
							SARAL	HY-2A
Majuro-B	07°–07°N	171°–22°E	1993–2016	98	Rep. Marshall I.	Nat. Tidal Ctr., BOM	0.8079	0.9572
Gan	00°–41°S	073°–09°E	1987–2015	93	Rep. of Maldives	UH Sea Level Center	0.8224	0.7606
Palmeira C.Verde	16°–45°N	022°–59°W	2000–2015	92	Portugal	UH Sea Level Cente	0.6278	0.5135
Adak Alaska	51°–52°N	176°–38°W	1950–2016	92	USA	National Ocean Service	0.4677	0.6054
Rikitea	23°–08°S	134°–57°W	1969–2015	93	French Polynesia	UH Sea Level Center	0.8859	0.4982
Santa Cruz	00°–45°S	090°–19°W	1978–2015	96	Ecuador	UH Sea Level Center	0.9016	0.9281
Los Angeles, CA	33°–43°N	118°–16°W	1923–2016	99	USA	National Ocean Service	0.6589	0.6276
Cape May, NJ	38°–58°N	074°–58°W	1965–2016	91	USA	National Ocean Service	0.8171	0.7386
Ko Lak	11°–48°N	099°–49°E	1985–2017	95	Thailand	Naval Hydro. Dept.	0.9157	0.9407
Hong Kong-B	22°–18°N	114°–13°E	1986–2016	99	China	HK Observatory	0.8045	0.1467

After resampling the tide gauge time series, a set of correlation coefficients between the altimetry grids and the in situ records were computed. The maximum correlation is found at four 1° × 1° grid points and corresponding satellite altimetry time series are then extracted for plotting collation maps. The maps are shown in Figure 5 and corresponding correlation coefficients for altimetry and in situ series are listed in Table 6. The average values of correlation coefficients are respectively, 0.71 and 0.75 for HY-2A and SARAL. Both the missions provide valid performance due to our numerical test and the agreement is slightly better for in situ/SARAL than for in situ/HY-2A.

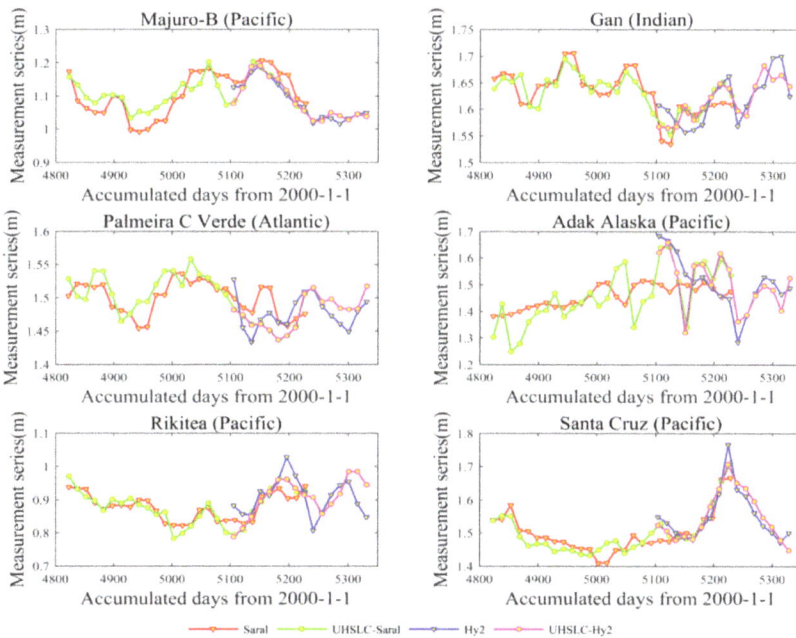

Figure 5. Comparison of HY-2A and SARAL SLA series with tide gauge sea level series at 6 stations (Red: SARAL; Green: TG for SARAL; Blue: HY-2A; Purple: TG for HY-2A).

We also calculated the correlation coefficients for the tide gauge sites used for validation in the recent published studies [28,29]. Four UHSLC sites (Los Angeles, CA; Cape May, NJ; Ko Lak; Hong Kong) with required data series were selected due to the time-scope repetition with HY-2A and SARAL. The specific information for these sites is also listed in Table 6. Slightly larger average values of correlation coefficients are obtained for HY-2A (0.76) and SARAL (0.79), respectively, while the Hong Kong station is excluded for statistics due to the serious data loss which occurred at surrounding areas for HY-2A. We can draw a similar conclusion that the HY-2A altimeter provides valid sea surface height measurements but at a slightly lower accuracy level than SARAL. Although HY-2A provides noisier measurements than SARAL according to cross-mission analysis and verification with in situ tide gauge measurements, we also found the reliable side of HY-2A data and the explorer procedure about the two-pass waveform retracker for the purpose of gravity recovery is accordingly reasonable and necessary.

4. Two-Pass Retracker for HY-2A

4.1. Theory of Two-Pass Retracker

Another major objective of our analysis is to evaluate whether the precision of HY-2A range measurements are further improved by using a two-pass waveform retracker which was proposed by Sandwell and Smith [9] initially and has already proven to be effective for other pulse-limited altimeter missions [19,20]. It is well known that waveform retracking is an effective and mature technique for solving limitations of data contamination by distribution of land or islands and diverse algorithms are developed either on empirical statistics or fitting functions for raw waveforms. Over the open ocean, the typical shapes of raw waveforms are well matched by the Brown [30] model with five parameters: arrival time, rise time, amplitude, square of the antenna mispointing angle, and thermal noise. The most concerning parameter for applications in marine gravity recovery is the arrival time, which provides an estimate of distance between the satellite and the instantaneous sea surface. In addition, the rise time and amplitude should also be set as variables for each raw waveform under any situations, while the thermal noise and square of the antenna mispointing angle can be treated as constant parameters for common waveforms obtained over open oceans.

Owing to the relationship between the errors in retracked estimates of arrival time and rise time [31], one effective approach to improving the precision of the arrival time is to fix the rise time parameter to a predetermined value based on surrounding waveforms. This can be accomplished with the two-pass retracking method where the waveforms are first fitted by the three-parameter Brown model (arrival time, rise time, and amplitude) with invariable constants describing thermal noise and antenna mispointing angles. The rise time parameter is then smoothed along-track, before retracking the waveforms a second time using a two-parameter Brown model (arrival time and amplitude) with the rise time being fixed to the along-track smoothed value [9]. Either the former three parameters or the latter two parameters will be solved during the weighted least squaring procedure, while the determination of the constants for describing thermal noise and antenna mispointing angles is critical before the two-pass model fitting procedure. The parameters and procedures involved in the two-pass retracker are deduced from the principle and power model of radar echoes over ocean surfaces.

The returned signal of conventional pulse-limited radar altimeters is proportional to the illuminated area scaled by the off-nadir roll-off of antenna gain pattern [30]. We are only interested in the slopes of SSH when using vertical deflection as an intermediate variable for recovering marine gravity field, and the constant offset with long-wave characteristics can be ignored accordingly. We also consider the case where the antenna bore sight is normally pointed to nadir. Under these simplifying assumptions, the power in the returned waveform can be modeled as following Equation (1), where A is the amplitude, t_0 is the arrival time, t is time for each sample gate, σ is the effective pulse length, and α is the trailing edge decay factor [32]. The effective pulse length σ can be calculated from two factors $\sigma^2 = \sigma_h^2 + \sigma_p^2$. The σ_h is the broadening of the pulse due to the interaction of ocean waves $\sigma_h = SWH/2c$, where c is the transferring speed of radar signals. The $\sigma_p = 0.513\tau_p$ is the standard deviation of the length of the outgoing pulse where $\tau_p = 1/B$ and B represents the chirp bandwidth for HY-2A (Ku: 320 MHz/80 MHz/20 MHz; C: 160 MHz). Three bandwidths for chirp signals are adopted in Model Compatible Tracker (MCT): the 320 MHz signal is arranged for oceanic measurement while

80 MHz is for the coastal zone, and the 20 MHz is for land or ice coverage. The trailing edge decay factor α mostly depends on the square of off-nadir pointing angle and can be treated as a constant value for HY-2A. Therefore, the fitting model has only three parameters $[A, t_0, \sigma_h]$ to be solved by using the weighted least squares approach, while the constant value α should be predetermined.

$$M(t) = \frac{A}{2}[1 + erf(\frac{t - t_0}{\sqrt{2\sigma}})]exp[-\alpha(t - t_0)] \tag{1}$$

$$X^2 = \sum_{i=1}^{N}[\frac{P_i - M(t_i - t_0 \cdot \sigma_h \cdot A)}{W_i}] \tag{2}$$

$$W_i = \frac{P_i + P_0}{\sqrt{K}} \tag{3}$$

The chi-square measure of the misfit between the returned power samples P_i and the model estimate $M(t_i)$ for the corresponding time t_i is formulated as Equation (2), where N is the number of gates (time samples) in the waveform, and W_i is the uncertainty assigned to each sample and given by Equation (3), where K is the number of statistically independent return echoes that were averaged to produce the waveform data records and P_0 is the power offset value reflecting the thermal noise level of the instrument. $K = 96$ is empirically determined for HY-2A as this mission collects ocean data at a pulse repetition frequency (PRF) of 1~4 kHz and similarly provides 20 Hz products in contrast to other altimeter missions [19]. P_0 will be given further discussion for selecting the optimal option in Section 4.2 although the results of the weighted least squares approach are largely insensitive to the numerical values for both K and P_0 [9].

In summary, the three parameters $[A, t_0, \sigma_h]$ can be firstly calculated through an iterative weighted least squares approach once the numerical values of α and P_0 are given previously. Besides, a threshold retracker is also introduced to provide an initial estimate of t_0 to accelerate the iterative process, while the corresponding threshold value is defined with respect to the cumulative sum of waveform power and should be predetermined as well. Secondly, the rise time parameter about SWH is smoothed along-track using a low-pass filter having a 0.5 gain at a full wavelength of 90 km [9,19]. The final step of the two-pass retracker is to process the raw waveforms again by fixing the rise time parameter to the smoothed value and only solving for the parameters of arrival time and amplitude. Based on the theory and procedures of the two-pass retracker mentioned before, we will discuss how to determine the corresponding parameters in the next section.

4.2. Parameter Determination

The typical shapes of HY-2A waveforms over the open ocean are also well matched by the Brown model similar to other Ku-band missions. By contrast, both HY-2A and Jason-1 waveforms have equally spaced gates or samples although the counts are 128 and 104, respectively. Both sets of waveforms have a low noise power level prior to the sharp arrival of the main pulse, which starts to show up around gate 30~32. Besides, HY-2A waveforms have relatively steeper trailing edge decays and corresponding parameters describing waveform tail should be adjusted. The special deformed waveforms over sea ice-covered areas, large lakes, or land areas are not discussed in this paper as we focus on the performance of the two-pass retracker for optimal gravity field recovery over open ocean areas. Consequently, we initially assume that the two-pass retracker is still effective for this China altimetry mission and accordingly determine the adjusted parameters for fitting HY-2A echoes.

As discussed in the theory section, we began the analysis for the two-pass retracker with five unknown parameters $[A, t_0, \sigma_h]$ and $[\alpha, P_0]$. The first three parameters need to be calculated by using the weighted least squares approach, while the latter items should be predetermined once the objective waveform samples are sufficient. In addition, a threshold value needs to be predetermined for providing an initial estimate of t_0 to accelerate the iterative process. The determination of $[\alpha, P_0]$ and the threshold value was discussed on the basis of HY-2A waveform samples from Cycle_E115.

Two descending passes labeled 222 and 260, respectively, were further selected for determining optimal parameters. Moreover, the screening criteria of valid results during the retracking procedure were also involved.

Firstly, α is a crucial parameter for fitting the waveforms, and the two-pass retracker is more sensitive to α than P_0 and threshold value [20]. The factor α depends mostly on the square of off-nadir pointing angle with a small order of magnitude and can be treated as a constant value for HY-2A. Based on the sampled passes, we gathered a number of successfully fitted waveform records and mean misfit values between original powers and modeled results respectively when using several typical values of α within the range of 0.005 and 0.015. Then we repeated this evaluation by sweeping through values of α at steps of 1.0×10^{-4} within range of 0.009 and 0.012. A best-fit α value of 0.0105 was approximately determined by considering both the valid records and misfit values for HY-2A. Meanwhile, the best-fit α values for different altimeter missions determined by previous studies are listed in Table 7 for summarization [18–20].

Table 7. The summarized best-fit α values for different satellite altimeter missions.

Mission	Geosat	ERS-1	Envisat	T/P Jason-1	CryoSat-2 LRM	CryoSat-2 SAR/SIN	SARAL	HY-2A
α value	0.006	0.022	0.09	0.0058	0.013	0.00744	0.0351	0.0105

Secondly, we investigated the relationship between waveform fitting results and the parameter P_0 describing the background noise level caused by temperature-dependent thermal noise in the receiver of the satellite altimeter. The magnitude of P_0 depends on the engineering characteristics of the altimeter and varies from the automatic gain control settings. Generally, P_0 can be either treated as a constant or empirically estimated through mean values or sum values of echo power for the first several sample gates [19,33]. 5 is used in this study for avoiding the false estimation due to complicated waveforms. Considering that the model-fitting results using the weighted least squares approach are largely insensitive to the numerical values of P_0, we treated P_0 as a constant similarly following the previous studies about ERS-1, Jason-1, and SARAL [19,20]. Besides, the power value for each gate is scaled by an arbitrary value of 3.3×10^6 to convert the original integers to floating point numbers and this scale value is completely irrelevant to the two-pass retracking procedure. Fortunately, the inheritance numerical value of P_0 from Jason-1 (5500) is still reliable for HY-2A waveforms under the selected scale value (3.3×10^6) because the ratios between the background noise level and the maximum power of echoes are similar for both missions.

Thirdly, a threshold value with respect to the cumulative sum of waveform power needs to be predetermined for providing an initial estimate of t_0 to accelerate the iterative process. Actually, the two-pass retracking results through iterative fitting are largely insensitive to the threshold value according to our experiments. Accordingly, we determined the optimal threshold value by minimizing the difference with respect to the results of the threshold retracker developed by Davis [34] for an exact initial objective gate, which leads to a valid threshold value of 0.015 for HY-2A.

Further discussion is needed about seriously deformed waveforms that will get unreliable results after the retracking procedure once the predetermined parameters [α, P_0] are given. Accordingly, we developed a waveform editing criteria based on estimated parameters [A, σ_h] as well as the fits to the waveforms executed. The editing threshold was established by constructing histograms of amplitude, chi-square misfit, and SWH versus standard deviation of the arrival time parameter. For HY-2A, valid waveforms should have model amplitudes within the range of 40,000–80,000, while the chi-squared misfit measurement should not exceed 800. The majority values of chi-squared misfit measurement are distributed within the range of 350~800 for valid retracked results. We also removed waveforms having SWH outside of the range 0.3–10 m for excluding observations over extremely unusual sea state conditions. Considering the robustness of the threshold retracker, we adopted the estimated results by the predetermined threshold instead when the HY-2A waveform records failed the editing criteria of the two-pass retracker.

While the predetermined parameters, threshold value, and waveform editing criteria were given, the results of the least squares fit of model waveforms and original waveform shapes are shown in Figure 6 for typical values of chi-squared misfit measurements. After examining thousands of waveforms over a range of sea states, we generally found that the fit to the HY-2A data is visually good, which suggested that the predetermined parameters were reliable. In accordance with the analysis by Garcia et al. [19], we also found a significant improvement (~5%) in the 20-Hz range precision if the least squares analysis was performed on three adjacent waveforms simultaneously, assigning the two outlier waveforms 1/2 the weight of the central waveform. Further enlargement of this three-waveform window did not reduce the range noise but increased processing time.

Once the two-pass retracker with three parameters $[A, t_0, \sigma_h]$ to be solved in the first step was optimized for selected constant parameters $[\alpha, P_0]$ and threshold value, we retracked the complete Cycle_E115 with the full three-parameter model and corresponding waveform editing criteria. Then the rise time parameter was low-pass filtered with its 0.5 gain at a full wavelength of 90 km. The final step was to retrack the waveforms by fixing the rise time parameter to the smoothed value and only solving the parameter of the arrival time and amplitude.

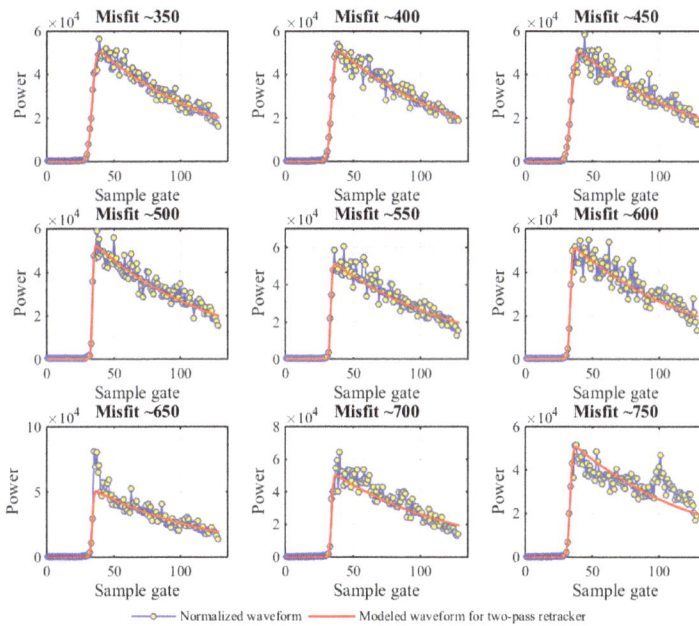

Figure 6. Least squares fit of model waveform of HY-2A for predetermined parameters (Blue: normalized waveform in SDR; Red: Modeled waveform of two-pass retracker).

5. Discussion

To assess the noise level of HY-2A data by the two-pass retracker, we performed a statistical analysis on the retracked SSH values (altitude minus range). All the corrections for path delay and geophysical environment were considered using items supplied by level 2 products. The noise was estimated as the standard deviation of the SSH with respect to the mean difference from the EIGEN6C4 model computed over a 1-second interval. The EIGEN6C4 model was used to account for areas of very steep geoid gradient where there could be a significant height variation (>20 mm) over the

7-km distance that the ground track moves in one second. This involved the provided 20 estimates corresponding to the HY-2A sampling rate along ground tracks.

Figure 7 shows the noise levels for the HY-2A sample pass labeled 260 in Cycle_E115. The noise level in the solved arrival time from the waveform leading edge should increase along with enhancing SWH and we plotted the noise level versus SWH thusly. Figure 7 also shows three independent estimates of noise level: the initial estimates derived from the retracked range measurement in SDR (red dots), the estimates from 3-parameter retracking (green dots), and ultimate 2-parameter retracking (blue dots), respectively. Additionally, the solid smoothed curves in the subgraph below with three corresponding colors are median averages of these estimates in 0.5-m SWH bins. The noise levels of three-parameter solutions and retracked results in SDR are at a similar level, while the latter is slightly superior. One more important finding is that the noise level of two-parameter retracked data is significantly lower than that for three-parameter solutions. Under the typical conditions of 2-m SWH, mean values of standard deviation for SDR, 3-parameter, and 2-parameter results are, respectively, 66.3, 67.1 and 40.9 mm for HY-2A, which is comparable to typical altimeter missions. The values for 3-parameter and 2-parameter retracked Jason-1 are, respectively, 75.9 and 46.4 mm, while CryoSat-2 LRM results are 64.7 and 42.7 mm [19]. This also demonstrates that the 2-parameter retracking decreases the noise level by a factor of 1.6 with respect to 3-parameter retracking, which is very close to the expected noise reduction of 1.57 based on the Monte Carlo simulation [9].

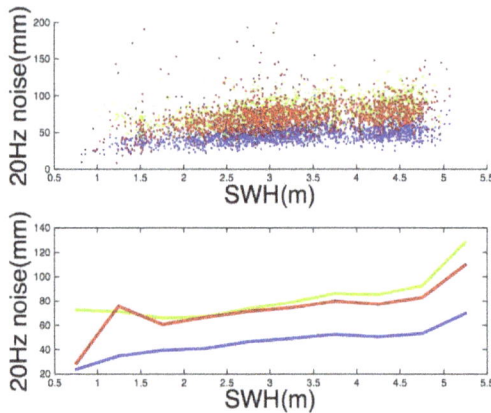

Figure 7. Standard deviation of retracked height with respect to EIGEN6C4 for HY-2A (Red—sensor data record; Green—3-parameter retracking; Blue—2-parameter retracking).

Spectral analysis is another effective way of analyzing the signal and noise ratio in the altimeter observations. On the basis of the Welch method [35], the power spectral density of along-track SSH for 3-parameter and 2-parameter retracked HY-2A data were respectively estimated using MATLAB and shown in Figure 8a. Both spectra show a rapid increase in power for wavelengths greater than 10 km and both the 3- and 2-parameter show a white noise level at wavelengths less than about 5 km. To highlight the differences between the two spectra, we calculated the spectra of the difference series between them as shown in Figure 8b. The increase in the difference spectra for wavelengths shorter than 90 km simply reflects the wavelength over which the SWH is smoothed between the two steps of iterative retracking procedures. At longer wavelengths, both the 3- and 2-parameter retrackers provide the same height measurement because the profiles contain the same SWH signal. The "hump" in the difference spectra between the wavelengths of 90 and ~5 km is also seen in spectra of two-pass retracked data from other pulse-limited altimeter missions [19,20]. This also demonstrates the band of wavelengths where two-pass retracking provides the most benefit [9].

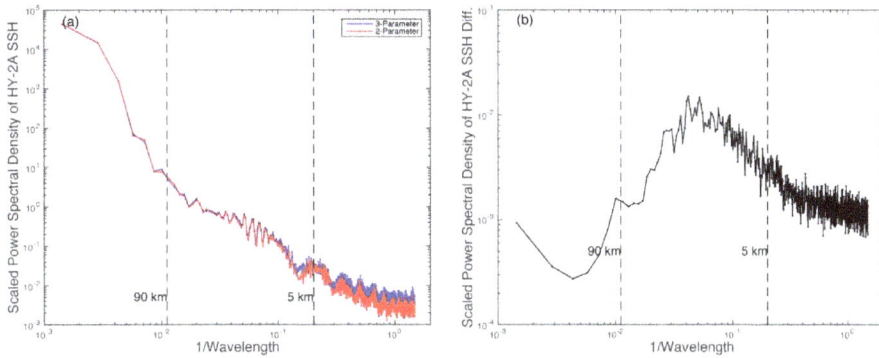

Figure 8. (a) Power spectral density of along-track retracked height from 3-parameter (blue) and 2-parameter (red) model for HY-2A data; (b) Power spectral density of along-track retracked height difference between 3-parameter and 2-parameter model for HY-2A. Wavelength range of spectral hump within 5–90 km is marked by vertical dashed lines.

The discrepancy at crossover points is another important index factor for assessing the accuracy of inner coincidence for altimeter SSH observations as mentioned before. Based on the HY-2A SDR data from Cycle_E115 over the southeastern China seas (100°~140°E, 0°~40°N), the statistic values of discrepancy at crossover points for three retracked results (SDR; 3-parameter and 2-parameter) are listed in Table 8, while the corrections for path delay and geophysical environment are also considered using items supplied by level 2 products. In addition, the corresponding results for Jason-1 GM data from Cycle500 are similarly calculated for comparison.

Table 8. Statistics of the SSH discrepancy at crossover points over southeastern China seas for HY-2A (Cycle_E115) and Jason-1 (Cycle500) sample data Unit: m.

Mission	Retracker Description	Crossover Points	Min	Max	MEAN	RMS
HY-2A	No retracker	125	−0.984	0.958	−0.287	0.412
	SDR—MLE4	125	−0.985	0.833	−0.280	0.391
	3-parameter	125	−0.975	0.916	−0.290	0.389
	2-parameter	125	−0.918	0.630	−0.272	0.377
Jason-1 GM	No retracker	126	−1.996	1.031	0.026	0.439
	SDR—MLE3	126	−0.713	0.547	0.048	0.191
	3-parameter	126	−0.531	1.286	0.072	0.230
	2-parameter	126	−0.446	1.006	0.066	0.208

Table 8 indicates that all the three retracked results show improvement with respect to results without any retracker, and more remarkable for Jason-1. Moreover, the ultimate retracked results derived from the 2-parameter model are slightly superior to that of the 3-parameter model for both missions, which proves that the second iterative fitting procedure with smoothed SWH values as priority information does bring improvement for range precisions. In addition, the larger discrepancies may lead to an initial conclusion that the precision of HY-2A SSH is slightly lower than Jason-1 data, which is consistent with the previous conclusion in Section 3. Nevertheless, the accuracy improvement by the two-pass retracker was verified for both missions and fortunately HY-2A can potentially calculate sea surface slopes at a similar accuracy level with respect to Jason-1 GM data due to the statistics of standard deviation over a 1-second interval.

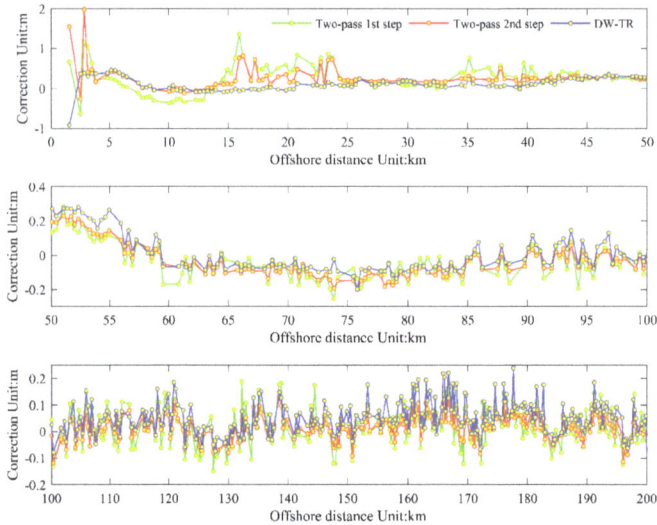

Figure 9. Comparison of range correction series for two-pass retracker and DW-TR retracker (Green: Two-pass retracker 1st step; Red: Two-pass retracker, 2nd step; Blue: DW-TR results).

We also introduced the DW-TR retracker proposed by Huang et al. [28] for comparison based on the same sequence of HY-2A waveforms. The sample data (Cycle_E075 Pass_001) is located at the Gulf of Alaska and contains 535 waveforms with an offshore distance of less than ~200 km. The range corrections for the sample data with respect to corresponding offshore distances are plotted in Figure 9. The comparison shows that the correlation coefficient is large (0.92) for typical waveforms with offshore distance larger than 50 km, and it dramatically declines to ~0.10 for contaminated waveforms with offshore distances less than 50 km. The two-pass retracker pays more attention to optimal gravity recovery over the open ocean, while the DW-TR retracker focuses mostly on the coastal applications. As a result, we can initially draw the conclusion that the two-pass waveform retracker is valid over the open ocean.

Moreover, the 2nd-step of the two-pass retracker (red lines) got a closer result with DW-TR retracker than the 1st step. It's hard to evaluate which retracker is more accurate relying on the current comparison. However, the accuracy improvement between the two steps of the retracker is detectable if we take the DW-TR result as reference. This improvement is also verified three other ways described before, STD calculation, power spectral density estimation, and crossover analysis. It's reasonable to draw the conclusion that the two-pass retracker can bring range precision enhancement for HY-2A.

6. Conclusions

The validation activities represent an essential part of describing HY-2A altimetry data quality assessment and allow a quick feedback to operational teams and users in the altimetry community. Based on the investigations performed in this study, HY-2A firstly provides unique data coverage over ocean regions within 80°N and 80°S according to its specific orbital parameters with respect to other altimetric missions. Secondly, HY-2A provides reliable measurements in level 2 products based on the analysis of the multimission crossover discrepancies and the exterior verification with tide gauge measurement series. Thirdly, the two-pass retracker does bring improvement of HY-2A range precision and significant decline of noise level by a factor of 1.6, similar to previous pulse-limited altimeter missions.

Remote Sens. **2018**, *10*, 507

In spite of the declined accuracy level of SSH observations with respect to SARAL based on our analysis, the along-track variances over a 1-second interval for retracked HY-2A results stay at a comparable level with respect to other missions (e.g., Jason-1/GM, CryoSat-2/LRM), which is an indicator for calculating reliable sea surface slopes. Additionally, the new data coverage due to HY2A's individual orbital parameters may provide detective signals over somewhere unreached previously. Therefore, the accumulation of HY-2A data, especially for the GM phase recently, provides a great opportunity to explore the finer structure of marine gravity fields.

Almost 5 years ago, marine gravity fields having accuracies of 3~5 mGal (e.g., S&S V18 [18] and DNSC08 [11]) were based primarily on dense track coverage from 18 months of Geosat data from geodetic missions collected in the early 1990s and 12 months of European Remote-Sensing Satellite-1 (ERS-1/GM) data collected in 1995~1996. Over the last five years, the latest global mainstream marine gravity model such as DTU13 [36] and V23.1 [37] have got precision breakthrough at 1~3 mGal which mainly benefits from new missions with nonrepetitive ground tracks such as Jason-1/GM and CryoSat-2/369d. Moreover, both the orbits of SARAL and HY-2A satellites began to drift by up to 10 km from their initial 35-day track and 14-day track, respectively, and resulted in several cycles of altimeter data for new ocean coverage. Consequently, there are enough reasons to believe that the objective of 1-mGal accuracy and 1-minute resolution for the global marine gravity field is accessible in the recent future. Besides, the next big step in gravity field improvement may come with the higher range precision and finer resolution provided by the planned SWOT mission, which is scheduled for launch in 2020 [38].

Acknowledgments: The multi-satellite altimeter data were provided by CNES and European Space Agency, while the HY-2A data and tide-gauge data were respectively distributed by NSOAS and UHSLC. The authors also thank Wang Haihong for providing DW-TR results for HY-2A sample waveforms. The manuscript was considerably improved through constructive comments from three anonymous reviewers and the assigned academic editor, which are gratefully acknowledged. This work was supported by the National Natural Science Foundation of China (No. 41721003, 41210006), by the National Key Research and Development Program of China (No. 2016YFB0501702), by the 973 Project of China (No. 2013CB73301, 2013CB73302), and by the DAAD Thematic Network Project under Grant 57173947.

Author Contributions: Jiancheng Li and Taoyong Jin conceived and designed the experiments. Shengjun Zhang performed the experiments and wrote the first draft, and all authors analyzed the data and wrote the final draft.

Conflicts of Interest: The authors declare no conflict of interest.

References

1. Rapp, R.H. Geos 3 data processing for the recovery of geoid undulations and gravity anomalies. *J. Geophys. Res.* **1979**, *84*, 3784–3792. [CrossRef]
2. Haxby, W.F.; Karner, G.D.; Labrecque, J.L.; Weissel, J.K. Digital images of combined oceanic and continental data sets and their use in tectonic studies. *EOS Trans. Am. Geophys. Union* **1983**, *64*, 995–1004. [CrossRef]
3. Sandwell, D.T.; Smith, W.H.F. Marine gravity anomaly from Geosat and ERS1 satellite altimetry. *J. Geophys. Res.* **1997**, *102*, 10039–10054. [CrossRef]
4. Andersen, O.B.; Knudsen, P. Global marine gravity field from the ERS-1 and Geosat geodetic mission altimetry. *J. Geophys. Res.* **1998**, *103*, 8129–8137. [CrossRef]
5. Hwang, C. Inverse Vening Meinesz formula and deflection-geoid formula: Applications to the predictions of gravity and geoid over the South China Sea. *J. Geodesy* **1998**, *72*, 304–312. [CrossRef]
6. Sandwell, D.T.; Garcia, E.S.; Soofi, K.; Wessel, P.; Chandler, M.; Smith, W.H. Towards 1-mGal accuracy in global marine gravity from Cryosat-2, Envisat and Jason-1. *Lead. Edge* **2013**, *32*, 892–898. [CrossRef]
7. Andersen, O.B.; Jain, M.; Knudsen, P. The Impact of Using Jason-1 and Cryosat-2 Geodetic Mission Altimetry for Gravity Field Modeling. *IAG 150 Years Symp.* **2016**, *143*, 205–210.
8. Zhang, S.; Sandwell, D.T.; Jin, T.; Li, D. Inversion of marine gravity anomalies over southeastern China seas from multi-satellite altimeter vertical deflections. *J. Appl. Geophys.* **2017**, *137*, 128–137. [CrossRef]
9. Sandwell, D.T.; Smith, W.H.F. Retracking ERS-1 altimeter waveforms for optimal gravity field recovery. *Geophys. J. Int.* **2005**, *163*, 79–89. [CrossRef]

10. Hwang, C.; Guo, J.; Deng, X.; Hsu, H.Y.; Liu, Y. Coastal gravity anomalies from retracked Geosat GM altimetry: Improvement, limitation and the role of airborne gravity data. *J. Geodesy* **2006**, *80*, 204–216. [CrossRef]

11. Andersen, O.B.; Knudsen, P.; Berry, P. The DNSC08GRA global marine gravity field from double retracked satellite altimetry. *J. Geodesy* **2010**, *84*, 191–199. [CrossRef]

12. Sandwell, D.T. Antarctic marine gravity field from high-density satellite altimetry. *Geophys. J. Int.* **1992**, *109*, 437–448. [CrossRef]

13. Zhao, G.; Zhou, X. Precise orbit determination of Haiyang-2 using satellite laser ranging. *Chin. Sci. Bull.* **2013**, *58*, 589–597. [CrossRef]

14. Legeais, J.-F.; Ablain, M.; Faugère, Y.; Mertz, F.; Soussi, B.; Vincent, P. HY-2A and DUACS (Data Unification and Altimeter Combination System) altimeter products. In Proceedings of the OSTST 2012 (Ocean Surface Topography Science Team) Meeting, Venice, Italy, 22–29 September 2012.

15. Bao, L.; Gao, P.; Peng, H.; Jia, Y.; Shum, C.K.; Lin, M.; Guo, Q. First accuracy assessment of the HY-2A altimeter sea surface height observations: Cross-calibration results. *Adv. Space Res.* **2015**, *55*, 90–105. [CrossRef]

16. Peng, H.; Lin, M.; Mu, B. Global statistical evaluation and performance analysis of HY-2A satellite radar altimeter data. *Haiyang Xuebao* **2015**, *37*, 54–66. (In Chinese)

17. Yang, L.; Zhou, X.; Lin, M.; Lei, N.; Mu, B.; Zhu, L. Global statistical assessment of HY-2A altimeter IGDR data. *Prog. Geophys.* **2016**, *31*, 629–636. (In Chinese) [CrossRef]

18. Sandwell, D.T.; Smith, W.H.F. Global marine gravity from retracked Geosat and ERS-1 altimetry: Ridge segmentation versus spreading rate. *J. Geophys. Res.* **2009**, *114*, B014411. [CrossRef]

19. Garcia, E.S.; Sandwell, D.T.; Smith, W.H.F. Retracking CryoSat-2, Envisat, and Jason-1 radar altimetry waveforms for improved gravity field recovery. *Geophys. J. Int.* **2014**, *196*, 1402–1422. [CrossRef]

20. Zhang, S.; Sandwell, D.T. Retracking of SARAL/AltiKa Radar Altimetry Waveforms for Optimal Gravity Field Recovery. *Mar. Geodesy* **2017**, *40*, 40–56. [CrossRef]

21. Prandi, P.S.; Philipps, S.; Pignot, V.; Picot, N. SARAL/AltiKa global statistical assessment and cross-calibration with Jason-2. *Mar. Geodesy* **2015**, *38*, 297–312. [CrossRef]

22. Vu, P.; Frappart, F.; Darrozes, L. Multi-Satellite Altimeter Validation along the French Atlantic Coast in the Southern Bay of Biscay from ERS-2 to SARAL. *Remote Sens.* **2018**, *10*, 93. [CrossRef]

23. Salameh, E.; Frappart, F.; Marieu, V. Monitoring Sea Level and Topography of Coastal Lagoons Using Satellite Radar Altimetry: The Example of the Arcachon Bay in the Bay of Biscay. *Remote Sens.* **2018**, *10*, 297. [CrossRef]

24. Wahr, L.W. Deformation of the earth induced by polar motion. *J. Geophys. Res. (Solid Earth)* **1985**, *90*, 9363–9368. [CrossRef]

25. Valladeau, G.; Legeais, J.F.; Ablain, M.; Guinehut, S.; Picot, N. Comparing altimetry with tide gauges and Argo profiling floats for data quality assessment and mean sea level studies. *Mar. Geodesy* **2012**, *35*, 42–60. [CrossRef]

26. CalVal In-Situ Altimetry/tide Gauges. *Validation of Altimeter Data by Comparison with Tide Gauge Measurements*; CLS: Ramonville St-Agne, France, 2014; CLS.DOS/NT/13-285, Issue 1.1, Nomenclature: SALP-RP-MA-EA-22294-CLS.

27. Caldwell, P.C.; Merrifield, M.A.; Thompson, P.R. *Sea Level Measured by Tide Gauges from Global Oceans—The Joint Archive for Sea Level Holdings (NCEI Accession 0019568)*; Version 5.5; NOAA National Centers for Environmental Information: Silver Spring, MD, USA, 2015.

28. Huang, Z.; Wang, H.; Luo, Z.; Shum, C.K.; Tseng, K.-H.; Zhong, B. Improving Jason-2 Sea Surface Heights within 10 km Offshore by Retracking Decontaminated Waveforms. *Remote Sens.* **2017**, *9*, 1077. [CrossRef]

29. Xu, X.; Birol, F.; Cazenave, A. Evaluation of Coastal Sea Level Offshore Hong Kong from Jason-2 Altimetry. *Remote Sens.* **2018**, *10*, 282. [CrossRef]

30. Brown, G. The average impulse response of a rough surface and its applications. *IEEE Trans. Antennas Propag.* **1977**, *25*, 67–74. [CrossRef]

31. Maus, S.; Green, C.M.; Fairhead, J.D. Improved ocean-geoid resolution from retracked ERS-1 satellite altimeter waveforms. *Geophys. J. Int.* **1998**, *134*, 243–253. [CrossRef]

32. Rodriguez, E. Altimetry for non-Gaussian oceans: Height biases and estimation of parameters. *J. Geophys. Res. Oceans* **1988**, *93*, 14107–14120. [CrossRef]

33. Martin, T.V.; Zwally, H.J.; Brenner, A.C.; Bindschadler, R.A. Analysis and retracking of continental ice sheet radar altimeter waveforms. *J. Geophys Res.* **1983**, *88*, 1608–1616. [CrossRef]
34. Davis, C.H. A surface and volume scattering retracking algorithm for ice sheet satellite altimetry. *IEEE Trans. Geosci. Remote Sens.* **1993**, *31*, 811–818. [CrossRef]
35. Welch, P.D. The use of fast Fourier transforms for the estimation of power spectra: A method based on time averaging over short, modified periodograms. *IEEE Trans. Audio Electroacoust.* **1967**, *15*, 70–73. [CrossRef]
36. Andersen, O.B.; Knudsen, P.; Kenyon, S.; Holmes, S. Global and arctic marine gravity field from recent satellite altimetry (DTU13). In Proceedings of the 76th EAGE Conference and Exhibition, Amsterdam, The Netherlands, 16–19 June 2014.
37. Sandwell, D.T.; Müller, R.D.; Smith, W.H.F.; Garcia, E.; Francis, R. New global marine gravity model from CryoSat-2 and Jason-1 reveals buried tectonic structure. *Science* **2014**, *346*, 65–67. [CrossRef] [PubMed]
38. Fu, L.L.; Ubelmann, C. On the transition from profile altimeter to swath altimeter for observing global ocean surface topography. *J. Atmos. Ocean. Technol.* **2013**, *31*, 560–568. [CrossRef]

remote sensing

MDPI

Article

Using Satellite Altimetry to Calibrate the Simulation of Typhoon Seth Storm Surge off Southeast China

Xiaohui Li [1], Guoqi Han [2,*], Jingsong Yang [1,*], Dake Chen [1], Gang Zheng [1] and Nan Chen [3,4]

[1] State Key Laboratory of Satellite Ocean Environment Dynamics, Second Institute of Oceanography, State Oceanic Administration, Hangzhou 310012, China; lixiaohui1991@live.cn (X.L.); dchen@sio.org.cn (D.C.); zhenggang@sio.org.cn (G.Z.)
[2] Fisheries and Oceans Canada, Northwest Atlantic Fisheries Centre, St. John's, NL A1C 5X1, Canada
[3] State Key Laboratory of Marine Environmental Science, Xiamen University, Xiamen 361005, China; chennan@udel.edu
[4] Center for Remote Sensing, College of Earth, Ocean and Environment, University of Delaware, Newark, DE 19716, USA
* Correspondence: guoqi.han@dfo-mpo.gc.ca (G.H.); jsyang@sio.org.cn (J.Y.); Tel.: +1-709-772-4326 (G.H.)

Received: 26 February 2018; Accepted: 13 April 2018; Published: 23 April 2018

Abstract: Satellite altimeters can capture storm surges generated by typhoons and tropical storms, if the satellite flies over at the right time. In this study, we show TOPEX/Poseidon altimeter-observed storm surge features off Southeast China on 10 October 1994 during Typhoon Seth. We then use a three-dimensional, barotropic, finite-volume community ocean model (FVCOM) to simulate storm surges. An innovative aspect is that satellite data are used to calibrate the storm surge model to improve model performance, by adjusting model wind forcing fields (the National Center for Environment Prediction (NCEP) reanalysis product) in reference to the typhoon best-track data. The calibration reduces the along-track root-mean-square (RMS) difference between model and altimetric data from 0.15 to 0.10 m. It also reduces the RMS temporal difference from 0.21 to 0.18 m between the model results and independent tide-gauge data at Xiamen. In particular, the calibrated model produces a peak storm surge of 1.01 m at 6:00 10 October 1994 at Xiamen, agreeing with tide-gauge data; while the peak storm surge with the NCEP forcing is 0.71 m only. We further show that the interaction between storm surges and astronomical tides contributes to the peak storm surge by 34% and that the storm surge propagates southwestward as a coastally-trapped Kelvin wave.

Keywords: storm surge; satellite altimetry; calibration; numerical modelling; FVCOM

1. Introduction

Storm surges, generated by extreme wind stress acting on shallow, continental shelf seas, can lead to severe coastal floods, particularly when they coincide with a high astronomic tide [1]. They can result in devastating economic and social impacts, including loss of life, damage to property and disruption of essential services [2–5]. As climate changes, coastal areas are becoming increasingly more vulnerable to storm surges [4]. To mitigate loss of life and damage to property by storm surges, great efforts have been made in monitoring and forecasting storm surges, such as enhancing tide-gauge networks, integrating satellite observations and improving numerical prediction of storm surge and inundation [6].

In recent years, some studies have indicated that satellite altimetry is capable of observing and studying storm surge features. Scharroo et al. showed that Geosat follow-on sea level anomalies reached 90 cm at the coast of the Gulf of Mexico during Hurricane Katrina [7]. Han et al. clearly showed the utility of satellite altimetry in observing and understanding storm surges, complementing tide-gauge observations for the analysis of storm surge characteristics and for the validation and improvement of storm surge models [4]. Lillibridge et al. reported that the storm surge caused by Hurricane Sandy was captured by the HaiYang-2A (HY-2A) satellite [8]. Recently, Chen et al. showed a detailed analysis of HY-2A satellite observations combined with tide-gauge data during the passage of Sandy. For this event, Montauk's tide-gauge recorded a maximum surge of 173 cm, and HY-2A showed positive sea-level anomalies of about 183 cm during the storm event. Further analysis suggests that the continental shelf wave generated during the passage of Sandy and observed by altimetry and tide-gauges has a propagating speed of 6.5 m/s [6]. The SARAL/Altika altimeter captured a storm surge event in the North Sea during Cyclone Xaver in December 2013 [9]. Han et al. studied storm surge features in the Gulf of Mexico during Hurricane Isaac, as observed by Jason-1 and Jason-2 altimeters and tide-gauge data, showing the utility of the constellation of altimeter missions and prospects of the upcoming Surface Water and Ocean Topography (SWOT) mission [10]. These studies have shown that satellite altimetry is very useful for observing and understanding features of storm surges. They have also indicated that it is highly opportunistic for a single satellite altimetry mission to capture storm surges due to infrequent sampling. Satellite altimetry does not provide along-coast sea surface height distribution. On the other hand, there are a few recent studies on using satellite altimetry data to improve storm surge simulation through data assimilation [11–13]. These studies [11–13] showed the positive impacts and challenges of assimilating altimetry data on storm surge hindcasts and forecasts in the Gulf of Venice.

Typhoons often cause storm surges off the eastern and southern coasts of China in summer and fall, for example along the coast of Fujian. Taiwan Strait, connecting the South China Sea and the East China Sea (Figure 1), is a shallow water area between Fujian and Taiwan, with an average depth of about 60 m [14]. The tide is very strong and dominated by semi-diurnal constituents [14–16]. Storm surge models have been developed for Taiwan Strait and adjacent waters [16,17]. According to previous studies, the interaction between tide and storm surge is notable, especially in shallow waters where tidal range is large [18–20]. In spite of the fact that the region is often hit by typhoons and storm surges, there has been little literature reporting on storm surges observed by satellite altimetry.

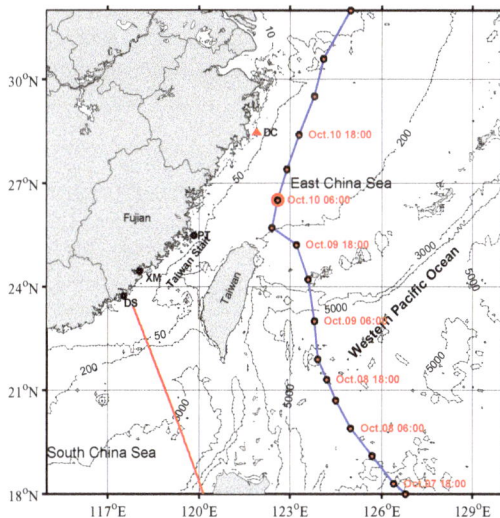

Figure 1. Map showing the study area off Southeast China with bathymetric contours in meters. Typhoon Seth's track and locations at specific times are shown as blue lines and red dots. The red line is the TOPEX/Poseidon (T/P) satellite ground track. Pingtan (PT), Xiamen (XM) and Dongshan (DS) tide stations are marked by circles. The Dachen (DC) weather station is also depicted (red triangle).

In this study, we use TOPEX/Poseidon (T/P) satellite altimetry observations to show cross-shelf variation of Typhoon Seth storm surge off Southeast China on 10 October 1994. The T/P satellite ground track had a pass nearby Xiamen at 05:46 UTC on 10 October 1994 (Figure 1), when Typhoon Seth was located close to the southeastern coast of China. We then apply a state-of-the-art finite-volume community ocean model (FVCOM) [21] to simulate and understand the Typhoon Seth storm surge off Southeast China. The novel aspect of this study is using satellite altimetry observations to calibrate the storm surge model by adjusting the model wind forcing fields. We further integrate the T/P satellite altimetry with the simulated results from the calibrated model to investigate features of the storm surge. Note that storm surges are not only the sea level rises directly forced by wind stress and/or low atmospheric pressure, but also coastally-trapped free propagating signals generated remotely by storms. On 1 October 1994, Seth started in an area near the Marshall Islands and then strengthened into a strong typhoon on 7 October moving northwestward east of Taiwan. As it moved through the Ryukyu Islands, winds gusted to 110 knots (200 km/h). At 00:00 on 10 October, it turned northeastward.

This paper is organized as follows. In Section 2, we describe altimetry data, tide-gauge data, the FVCOM model and its setup, as well as the model calibration procedure. Section 3 evaluates and calibrates the model results against observations. Section 4 discusses tide-surge interactions and the mechanisms of storm surge propagation. We provide conclusions in Section 5.

2. Data and Methods

2.1. T/P Data

The T/P satellite was launched on 10 August 1992. We have used 1-Hz altimetric sea surface height anomalies along a T/P satellite ground track (Track 88) nearby Xiamen, Fujian province (Figure 1). The T/P data were obtained from the Centre for Topographic studies of the Ocean and Hydrosphere (CTOH) X-TRACK product. The altimetric sea surface height anomalies are corrected for ionospheric and tropospheric delays, sea state bias, as well as ocean, solid Earth and pole tides. Inverse barometer or high-frequency dynamic corrections are not applied. The X-TRACK product starts from Geophysical Data Record (GDR) data, with enhancement in geophysical corrections (e.g., the ionospheric correction and the wet tropospheric correction) and thus has more valid data points than a standard GDR product in coastal regions [22]. The X-TRACK sea surface height anomalies are referenced to the mean sea surface calculated from the along-track T/P, Jason-1 and Jason-2 data from 1993 to 2015.

2.2. Tide-Gauge Data

Hourly sea level data at Xiamen (Figure 1) for 1994 are obtained from the University of Hawaii Sea Level Center (UHLSC, http://uhslc.soest.hawaii.edu/). Inverse barometer correction is not applied, since we are interested in storm surge. Harmonic analysis including 59 tidal constituents is carried out to retrieve tidal constants [23]. We subtract the tide height predicted using the retrieved tidal constants from the hourly sea level data to produce non-tidal sea level anomalies.

2.3. FVCOM (3.2.1) Ocean Circulation Model

The model used in this study is FVCOM, which has the advantages of horizontal grid flexibility and computational efficiency [21]. Thus, this model is highly suitable for the present study area with an irregular complex coastline, reaching from shallow waters to the steep topography of the shelf break.

The model domain covers the eastern and southern coast of China and the adjacent deep ocean (Figure 2). With 29,235 unequally-spaced nodes and 56,534 elements, the horizontal grid has a typical resolution of 10 km over the shelf and 2–5 km along the coast and shelf edge. Vertically, 11 levels are uniformly distributed. We use bottom topography from the Earth Topography 1-arc-minute gridded global relief (ETOPO1) dataset [24]. The bathymetry was smoothed to improve the model stability. This technique limits the depth difference for three vertices at each triangle. On the basis of the Courant–Friedrichs–Levy (CFL) numerical stability condition, model equations are solved with an integration time step of 2 s for the external mode and an internal to external mode ratio of 10. We use the quadratic bottom friction formulation, with a minimum drag coefficient of 0.0025 [21].

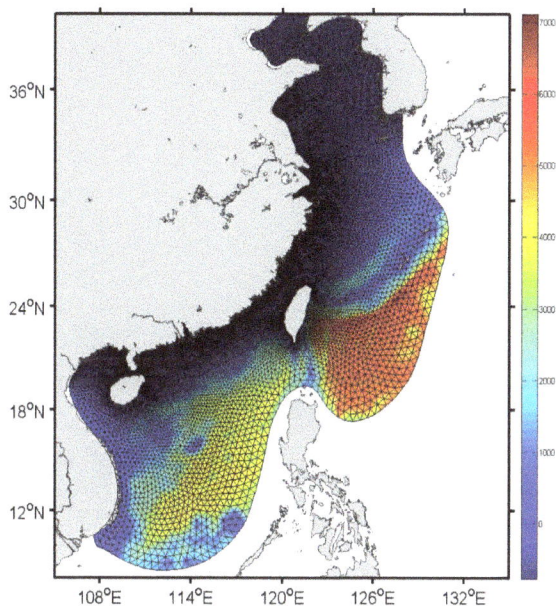

Figure 2. Mesh with 29,235 nodes in the waters off the eastern and southern coast of China. The fine grid resolution is around 2 km along the Taiwan Strait coast.

The model is forced by winds and air pressure at the sea surface. We used 6-hourly, 10-m, 0.25° by 0.25° wind stress and sea level pressure of the National Center for Environmental Prediction (NCEP), from NCAR's RAD (National Center for Atmospheric Research, Research Data Archive, http://rda.ucar.edu/datasets) over the entire computational domain. Tidal heights for the eight main tidal constituents (semi-diurnal (M2, S2, N2 and K2) and diurnal (K1, O1, P1 and Q1)) predicted by OTIS (OSU Tidal Inversion Software) were specified along the open boundaries [25].

The model sea level and velocity were initialized from zero. The model reaches an approximate dynamic equilibrium after running for 15 days. The results from 26 September to 18 October 1994 were analyzed to examine the storm surge.

We carry out the model runs for tide forcing only, for wind forcing only and for tide and wind forcing simultaneously. The non-tidal sea level anomalies were obtained by subtracting the tide level of the model run for tide forcing only from the total sea level of the model run for tide and wind forcing simultaneously.

2.4. Modification of NCEP Winds

A good quality wind field is essential for storm surge calculations [11–13]. However, the wind field from the atmospheric reanalysis product usually underestimates the maximum sustainable wind of hurricanes and typhoons. Various approaches have been used to adjust reanalysis winds in the vicinity of the storm center, for example, by fitting the analytical cyclone model of Holland [26,27] within the radius of maximum wind (*RMW*) and then blending the analytical model wind with the reanalysis wind field. In this study, we follow their approach, except for not fitting the analytical model. Ignoring the asymmetry of the typhoon, we modify the NCEP wind fields as follows:

$$V_w = (\frac{r}{RMW} \times ntimes + \frac{RMW - r}{RMW}) \times V_{NCEP}, 0 \leq r \leq RMW \qquad (1)$$

$$V_w = (\frac{r - RMW}{3 \times RMW} + \frac{4 \times RMW - r}{3 \times RMW} \times ntimes) \times V_{NCEP}, RMW < r \leq 4 \times RMW \qquad (2)$$

$$V_w = V_{NCEP}, r > 4 \times RMW \qquad (3)$$

where r is the radial distance from the typhoon center, V_{NCEP} is the background wind speed and *ntimes* is a parameter to be modified. When *ntimes* equals 1, the formulas give the original NCEP wind. We calculate the maximum wind radial distance using the NCEP wind data with the center location provided by the Typhoon Online of China (http://www.typhoon.gov.cn/). Then, the maximum wind radial distance was treated as the *RMW* to construct the horizontal wind fields, neglecting asymmetry. We modify the wind forcing using Equations (1) to (3) for the period from 9 October to 11 October, by choosing the *ntimes* = 1.1, 1.2, 1.3, 1.4 and 1.5, respectively.

2.5. Calibration of the Stom Surge Model against Altimetry Data

De-tided model sea surface heights forced by the NCEP and modified winds are interpolated on to the satellite ground track at the time (05:46 UTC on 10 October 1994) of the satellite passing, one cycle before and one cycle after. To eliminate the impacts of the mean sea surface height difference between the altimetric and model results on the storm surge comparison, we first calculate an average from Cycle 75 (07:48 UTC, 30 September 1994) and 77 (03:44 UTC, 20 October 1994) and then subtract the average from sea surface height of Cycle 76. This procedure is applied to the model and altimetric results, respectively. We calculate the root-mean–square (RMS) difference between the altimetric and model results for each *ntimes*. The run with the smallest RMS difference is chosen as the baseline case.

3. Results

3.1. Simulated Tides

The simulated co-tidal chart for the dominant constituent M2 constituent (Figure 3) in the study region agrees generally with the observation-based co-tidal chart in Fang et al. (2004). As can be seen, the M2 tide enhances significantly in Taiwan Strait, with an amplitude greater than 2 m. At the Xiamen (XM) tide-gauge station, the simulated major semi-diurnal and diurnal tidal constituents and shallow-water constituents agree well with the observations (Table 1). The good agreement of the model tides with observations (especially for the shallow-water constituents) suggests that the bottom friction formulation used in the present model is reasonable.

Table 1. Comparison of the amplitude (m) and phase (°) between that simulated and observed at XM.

Tide	Simulated Amplitude	Observed Amplitude	Simulated Phase	Observed Phase
Q1	0.06	0.05	120	125
O1	0.30	0.28	128	140
P1	0.11	0.11	162	172
K1	0.35	0.34	163	175
N2	0.30	0.37	112	129
M2	1.67	1.85	132	148
S2	0.52	0.54	169	194
K2	0.14	0.15	156	192
MN4	0.03	0.03	266	289
M4	0.09	0.08	282	309
MS4	0.06	0.05	319	353

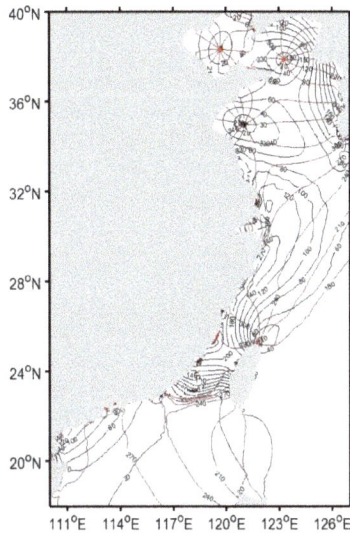

Figure 3. M2 co-tidal and co-phase charts from FVCOM. The red and black lines show the phase lag (in degrees relative to Beijing local time (UT + 8 h)) and amplitude (in centimeters), respectively.

3.2. Comparison of Model Storm Surge under NCEP Winds with Tide-Gauge Data

The de-tided model sea surface heights are interpolated to the Xiamen tide-gauge station. Forced by the NCEP wind only, the model significantly underestimates the peak storm surge (Figure 4). The peak storm surge is 1.01 m from the tide-gauge data, but 0.71 m only from the model result, an underestimate of 0.3 m (30%). On the other hand, the model captures the timing of the peak storm surge well, suggesting that both the timing of the NECP wind evolution and the model storm surge propagation speed are realistic. Note that the pole tide is not removed from the tide-gauge storm surge. However, the pole tide is small in magnitude (<1 cm) at the time, and its impact is negligible in the present study.

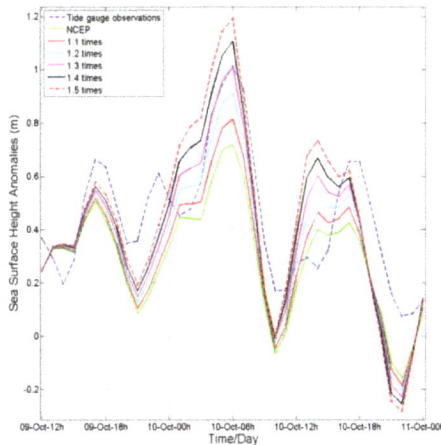

Figure 4. Sea level anomalies comparison between observations and simulations at Xiamen.

3.3. Storm Surge from T/P Observations

The altimetric sea surface height anomalies during Typhoon Seth (Cycle 76, 10 October 1994) increase rapidly from the shelf break (at 22.3°N where the 200-m isobath is located) toward the coast of Fujian (Figure 5). Quality altimetric data are unavailable within about 50 km from the coast. The altimetric sea surface height anomalies reach 0.7 m at the location closest to the coast. In contrast, the altimetric sea surface height anomalies have much smaller cross-shelf variations before (Cycle 75) and after (Cycle 77) Typhoon Seth.

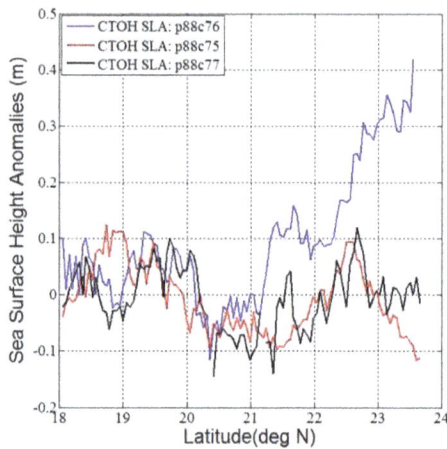

Figure 5. Altimetric sea surface height anomalies along Track 88 observed by T/P. Blue: Cycle 76 during Typhoon Seth. Red: Cycle 75. Black: Cycle 77. CTOH, Centre for Topographic studies of the Ocean and Hydrosphere. SLA, Sea Level Anomalies.

3.4. Comparison of Model Storm Surge under NCEP Winds with T/P Observations

We can see a significant difference between the altimetric and model results, with the model underestimation (Figure 6). Over the continental shelf inshore of 22.3°N, the difference represents the model underestimation of storm surge. The root-mean-square (RMS) difference is 0.15 m (Table 2). In the deep waters offshore of 22.3°N, the difference may in part be attributed to missing baroclinic ocean processes in the present barotropic model, which are present in the altimeter observations. The deep-water region is dynamically complex, with seasonally-varying slope currents, mesoscale eddies and Kuroshio intrusion currents [28]. There may be trapped waves propagating along the shelf break.

Table 2. RMS difference of non-tidal sea level anomalies between the model and T/P observations.

	RMS (22.3 to 23.55°N)
NCEP	0.15 m
modified by 1.1	0.13 m
modified by 1.2	0.11 m
modified by 1.3	0.10 m
modified by 1.4	0.11 m
modified by 1.5	0.12 m

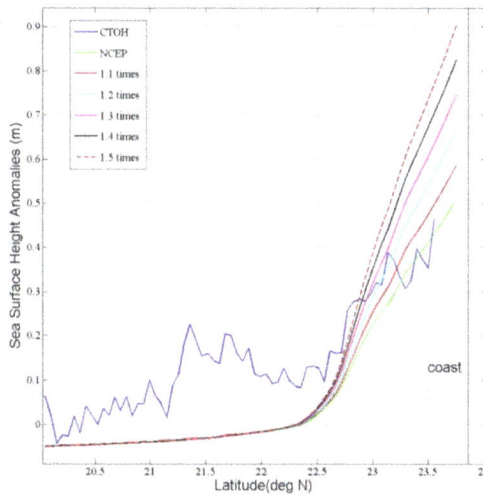

Figure 6. De-tided altimetric and model surface height anomalies for Cycle 76 during Hurricane Seth, relative to those averaged for Cycle 75 and 77.

3.5. Calibration against Altimetric Observations

Because the model forced by the NCEP wind and tide is able to well reproduce major tidal constituents and the shallow-water tidal constituents (see Section 3.1), we can conclude that the model underestimation of the coastal storm surge magnitude is not caused by the bottom friction parameterization. Also because the model captures the timing of the peak storm surge well, we can consider both the timing of the NECP wind evolution and the model storm surge propagation speed realistic. On the other hand, upon examining the NCEP wind speed, we find that the maximum NCEP wind is much lower than the maximum sustained wind of the best-track data. For example, at 6:00 10 October 1994, the maximum NCEP wind is 30.33 m/s. Therefore, we modify the wind forcing using Equations (1)–(3) for the period from 9 October to 11 October, by choosing the *ntimes* = 1.1, 1.2, 1.3, 1.4 and 1.5, respectively. Table 3 shows the differences in the maximum wind speed between the modified results and best-track data. The RMS difference of the maximum wind speed between the modified wind fields and best-track data is smallest at *ntimes* = 1.3 (about 2 m/s), while it is about 10 m/s between the NCEP and observed wind fields (Table 4). Figure 7 shows the wind patterns before and after the modification at 6:00 10 October 10 1994. A further evaluation of the NCEP and modified wind fields against independent weather station data was made at Dachen (Table 5). The results show the smallest mean and RMS differences at *ntimes* = 1.3 − 1.4.

Table 3. Comparison of the maximum wind speed (m/s) between the NCEP, modified and best-track wind fields.

	9 October, T12	9 October, T18	10 October, T00	10 October, T06	10 October, T12	10 October, T18	11 October, T00
NCEP	33	31	29	30	27	22	22
Modified by 1.1	36	34	32	33	30	24	25
Modified by 1.2	40	38	34	36	32	27	27
Modified by 1.3	43	41	37	39	35	29	29
Modified by 1.4	46	44	40	42	38	31	31
Modified by 1.5	49	47	43	45	40	33	33
best-track	45	40	40	40	35	30	30

Table 4. RMS difference in maximum wind speed (m/s) between best-track and modified winds from 12:00 9 October to 00:00 11 October. For NCEP winds, *ntimes* = 1.0.

ntimes	NCEP	1.1	1.2	1.3	1.4	1.5
RMS	9.6	6.7	3.9	1.6	2.1	4.5

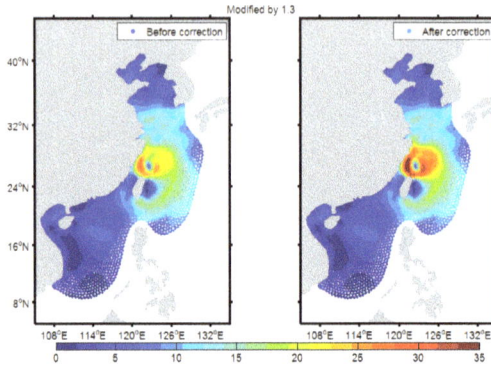

Figure 7. Comparison between the modified and NCEP wind field at 06:00, 10 October.

Table 5. NCEP, modified (*ntimes* = 1.1 to 1.5) and observed wind speed (m/s) at Dachen from 12:00 9 October to 00:00 11 October.

ntimes	9 October, T12	9 October, T18	10 October, T00	10 October, T06	10 October, T12	10 October, T18	11 October, T00	Mean
NCEP	8.9	13.1	15.0	19.1	13.6	14.9	11.9	13.8
1.1	8.9	13.3	15.9	20.5	14.8	16.3	12.8	14.6
1.2	8.9	13.3	16.7	21.9	15.9	17.7	13.7	15.4
1.3	8.9	13.5	17.5	23.4	17.1	19.1	14.6	16.3
1.4	8.9	13.7	18.3	24.9	18.2	20.6	15.5	17.2
1.5	8.9	13.8	19.1	26.3	19.4	21.9	16.4	18.0
Dachen	9.3	12.0	14.0	22	25.0	22	12	16.6

The model results with the modified wind forcing show improved agreement of the model storm surge over the continental shelf inshore of 22.3°N with altimetric data as the *ntimes* increases from one (Figure 6). The best agreement reaches when *ntimes* is 1.3, with the RMS difference reduced to 0.10 m (Table 2).

Finally, we use tide-gauge data as an independent verification of the calibration. When forced by the modified wind (1.3-times the NCEP wind fields) that produces the nearly best agreement with the T/P observations, the model storm surge achieves the best agreement with tide-gauge data at Xiamen (Table 6, Figure 4), in terms of the RMS difference and the peak surge magnitude. Therefore, we choose the model run forced by both the tide and the 1.3-times the NCEP wind fields as the baseline run for further discussion in the next section.

Table 6. RMS difference of non-tidal sea level anomalies between the model and tide-gauge observations at Xiamen.

	RMS
NCEP	0.21 m
modified by 1.1	0.19 m
modified by 1.2	0.18 m
modified by 1.3	0.18 m
modified by 1.4	0.19 m
modified by 1.5	0.21 m

4. Discussion

4.1. Effect of Tide-Surge Nonlinear Interactions

Both model results and tide-gauge data show a significant 12-h oscillation in non-tidal sea level anomalies at the Xiamen tide-gauge station. To understand this oscillation, we have run a model simulation without tides. The model run without tides does not show the 12-h oscillation at all (Figure 8). Therefore, the 12-h oscillation is associated with the nonlinear interactions between the wind-driven storm surge and the tides. The tide-surge interactions in Taiwan Strait were discussed by Zhang et al. (2010) [17], attributable mainly to the bottom friction and advection. In addition, the highest surge of 0.70 m in the simulation without tides occurs about 1 h earlier. At the time of the peak storm surge, the non-linear interactions contribute by 0.34 m (Figure 8) and the pure wind effects by 0.67 m.

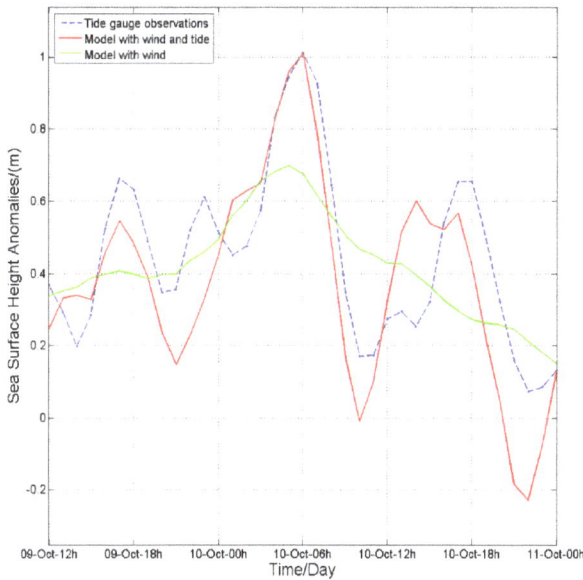

Figure 8. Non-tidal sea surface height anomalies at Xiamen from tide-gauge observations, from the model run forced by the National Center for Atmospheric (NCAR) wind modified by 1.3-times and from the model run forced by both the tide and the NACR wind modified by 1.3-times.

The sea level anomalies due to the nonlinear interactions are negatively correlated with the tide height, that is high tides tend to reduce storm surge and vice versa (Figure 9). The peak positive and negative nonlinear effects are behind the low and high tides, respectively, suggesting that the nonlinear interactions are mainly associated with the bottom friction. The correlation coefficients from 06:00 8 October to 06:00 12 October are -0.46 ($p < 0.001$) at both XM and Pingtan (PT). A numerical study by Rego and Li. [2010] showed a similar relationship between surge and tides in the Gulf of Mexico [29].

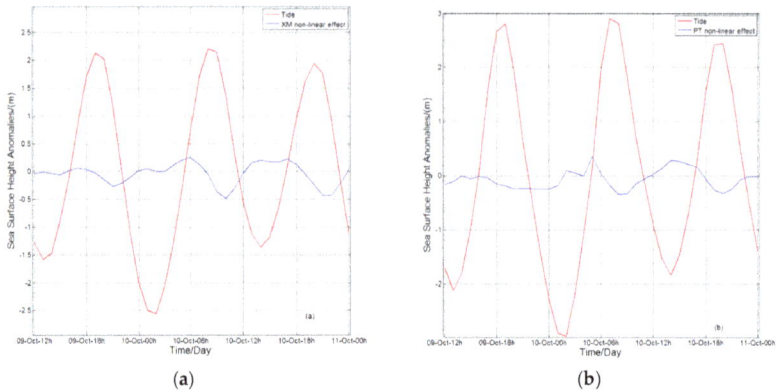

(a)   (b)

Figure 9. Sea surface height anomalies due to the tide-surge nonlinear interactions and tidal height at (a) XM and (b) PT. The model is forced by both the tide and the NCEP wind modified by 1.3-times.

4.2. Storm Surge as a Coastally-Trapped Kelvin Wave

The de-tided model sea level anomalies from 03:00 to 08:00 on 10 October 1994 (Figure 10) indicate a storm surge generated along the coast of Fujian, when the typhoon was located northeast of Taiwan. They also showed a southwestward propagation of storm surge. The sea level anomalies were relatively high near the Pingtan tide-gauge station at 03:00 on 10 October (Figure 10a). The high sea level anomalies propagated southwestward, arriving at the Xiamen tide-gauge station at 06:00 (Figure 10d) and at the Dongshan tide-gauge station 2 h later.

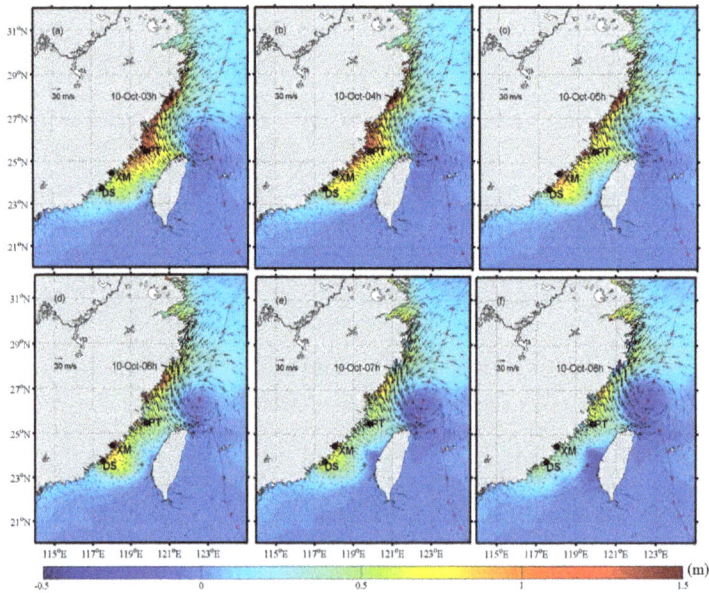

Figure 10. (a–f) show temporal change of the model non-tidal sea surface height anomalies and the forcing wind fields at 03:00, 04:00, 05:00, 06:00, 07:00 and 08:00 on 10 October. The model run is forced by both the tide and the NCEP wind modified by 1.3-times.

The coastally-trapped waves are often generated by a passing storm, which propagate along the coast with the coastline on its right in the Northern Hemisphere. We analyze the correlation of the simulated sea level anomalies at Pingtan, Xiamen and Dongshan. Xiamen is used as the reference. The lagged correlation coefficients are calculated and shown in Figure 11. The correlation results indicate that it takes 3 h for the storm surge wave to travel from Pingtan to Xiamen and 2 h from Xiamen to Dongshan (Figure 11). The distance between Pingtan and Xiamen is 217 km and between Xiamen and Dongshan is 145 km. Therefore, the propagation speed of the coastally-trapped wave is ~20 m/s, which is consistent with the propagation speed (23 m/s) of the Kelvin wave based on an averaged depth of Taiwan Strait.

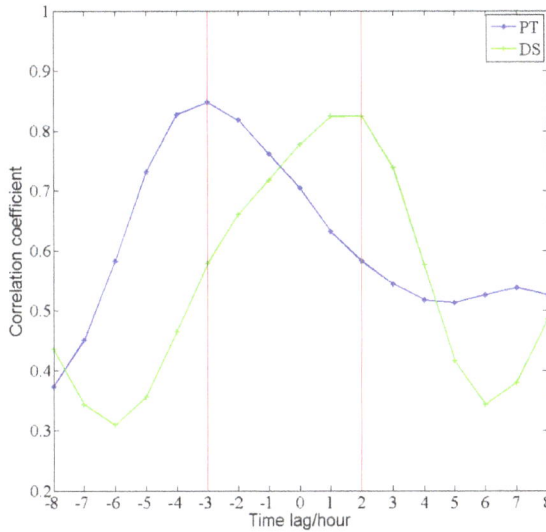

Figure 11. The lagged correlation coefficients of the model non-tidal sea surface height anomalies between PT and XM (blue) and between DS and XM (green). Negative values in the time lag mean the former station leading the latter for each pair. The model run is forced by both the tide and the NCEP wind modified by 1.3-times.

Several previous studies [4,6,10,30,31] have shown that storm surges propagate as first-mode barotropic continental shelf waves. In our case, we calculate the dispersion relationship for a continental shelf wave across a section aligned with the T/P track (Figure 1), based on the barotropic continental shelf wave theory [32]. A continental shelf wave must have a period longer than the local inertial period, which is 29 h at 24.37°N. The dispersion relationship shows that the phase speed of the first-mode continental shelf wave for this area must be less than 8 m/s (Figure 12). The storm surge signal during Seth has a dominant period of 12 h (shorter than the shortest possible period of 29 h for a continental shelf wave) and a propagating phase speed of ~20 m/s (greater than the greatest possible phase speed of 8 m/s for a continental shelf wave); therefore, it is impossible for it to be a continental shelf wave.

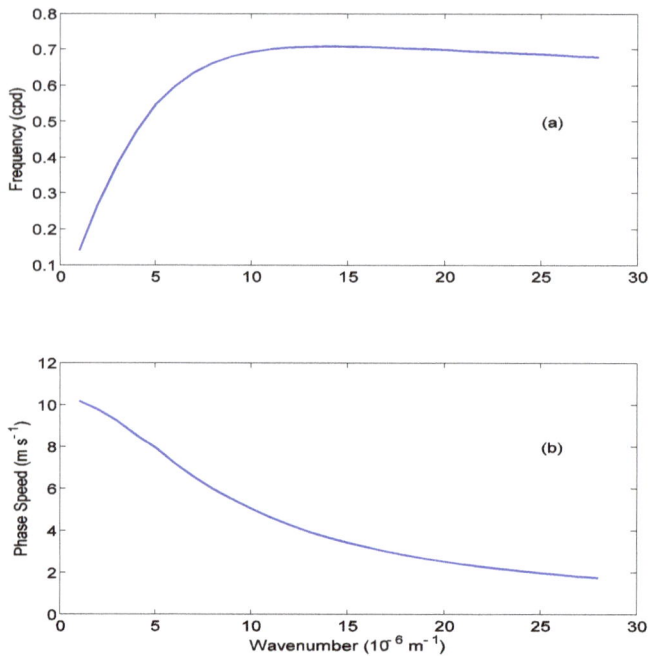

Figure 12. (a) The dispersion relationship estimated for the first-mode continental shelf wave in the coastal area nearby the T/P ground track; (b) the corresponding wave speed curve.

4.3. Strength and Limitation

In the present study, the calibrated model produced the storm surge in good agreement with the tide-gauge observations. The study points to the usefulness of altimetric data in improving storm surge modelling, with its cross-shore information making up for the limitation of a one-time single profile. Such an altimetric profile would be particularly useful for regions lacking tide-gauge observations. The model results are used to explain the impacts of tide-surge interactions and the alongshore propagation of storm surge. However, the altimeter data are used to adjust the wind forcing in an ad hoc way, instead of being dynamically assimilated into the storm surge model. Though the underestimation of typhoon winds in the reanalysis product is a general issue, the degree of underestimation varies storm by storm and is difficult to generalize. We also recognize that the present study provides one single case only. More cases are needed to demonstrate the improvements from a statistical perspective in the future.

The present study adjusts the wind forcing field in reference to the typhoon best-track data to minimize model-altimetry mismatch. This approach is simple and is proven to work well with the pre-determination that the underestimation of Typhoon Seth wind speed by the NCEP product is the main cause of the model deficiency. Some recent studies [11–13] have used another approach that uses altimetry to improve initial conditions and scatterometry to correct wind fields. Their approach corrects the bias of the global model wind field in reference to scatterometer winds and dynamically assimilates altimetric sea level data to provide a realistic initial state to begin the model simulation. Therefore, the latter approach, though methodologically complex and computationally more expensive, is of more general use and thus is to be pursued in the future.

Remote Sens. **2018**, *10*, 657

5. Conclusions

A 3D, barotropic, finite-volume coastal ocean model was developed to simulate and study the storm surge along the southeastern China during Typhoon Seth in October 1994. A novel aspect is that the model storm surge was calibrated against a TOPEX/Poseidon observed cross-shelf storm surge profile, by adjusting the model wind forcing fields in reference to the typhoon best-track data. Through the calibration process, we determined the baseline wind forcing fields, i.e., 1.3-times the NCEP wind fields, which agree approximately with the best-track data for the maximum sustained speed while have a spatial structure similar to the NCEP wind fields.

The model results from the baseline wind fields reduce the along-track RMS difference between the model and altimetric data from 0.16 m to 0.09 m. It reduces the RMS temporal difference from 0.21 m to 0.18 m between model and tide-gauge data at Xiamen. In particular, the baseline model produces a peak storm surge of 1.01 m at 6:00 10 October 1994 at Xiamen, agreeing with tide-gauge data. The model results also show that the nonlinear interactions between the storm surge and astronomical tides contribute to the peak storm surge by 34% and that the storm surge propagates southwestward as a coastally-trapped Kelvin wave.

Acknowledgments: The X-TRACK sea surface height product was obtained from CTOH, France. This work was supported by the National Programme on Global Change and Air-Sea Interaction (GASI-IPOVAI-04), the National 863 Program of China (2013AA09A505), the National Science Foundation (41621064, 41306191, 41306192 and 41406203) of China, the European Space Agency and Ministry of Science and Technology Dragon 4 Cooperation Programme (32249) and the Project of State Key Laboratory of Satellite Ocean Environment Dynamics, Second Institute of Oceanography (No. SOEDZZ1804). Helpful comments were received from two anonymous reviewers.

Author Contributions: G.H., J.Y., and D.C. conceived the idea and designed the approach. X.L. implemented and ran the model, as well as analyzed the data, under the close guidance of G.H. and J.Y. G.Z. and N.C. contributed to the data analysis. X.L. and G.H. wrote the paper. J.Y. and D.C. contributed to the paper writing.

Conflicts of Interest: The authors declare no conflict of interest.

References

1. Brown, J.D.; Spencer, T.; Moeller, I. Modeling storm surge flooding of an urban area with particular reference to modeling uncertainties: A case study of Canvey Island, United Kingdom. *Water Resour. Res.* **2007**, *43*, W6402. [CrossRef]
2. Price, J.F. Upper Ocean Response to a Hurricane. *J. Phys. Oceanogr.* **1981**, *2*, 153–175. [CrossRef]
3. Sanford, T.B.; Black, P.G.; Haustein, J.R.; Feeney, J.W.; Forristall, G.Z. Ocean Response to a Hurricane. Part I: Observations. *J. Phys. Oceanogr.* **1987**, *11*, 2065–2083. [CrossRef]
4. Han, G.; Ma, Z.; Chen, D.; Deyoung, B.; Chen, N. Observing storm surges from space: Hurricane Igor off Newfoundland. *Sci. Rep.* **2012**, *2*, 1010. [CrossRef] [PubMed]
5. Shutler, J.D.; Quartly, G.D.; Donlon, C.J.; Sathyendranath, S.; Platt, T.; Chapron, B.; Johannessen, J.A.; Girard-Ardhuin, F.; Nightingale, P.D.; Woolf, D.K.; et al. Progress in satellite remote sensing for studying physical processes at the ocean surface and its borders with the atmosphere and sea ice. *Prog. Phys. Geogr.* **2016**, *40*, 215–246. [CrossRef]
6. Chen, N.; Han, G.; Yang, J.; Chen, D. Hurricane Sandy storm surges observed by HY-2A satellite altimetry and tide-gauges. *J. Geophys. Res. Oceans* **2014**, *119*, 4542–4548. [CrossRef]
7. Scharroo, R.; Smith, W.; Lillibridge, J.L. Satellite altimetry and the intensification of Hurricane Katrina. *Eos Trans. Am. Geophys. Union* **2005**, *40*, 366. [CrossRef]
8. Lillibridge, J.; Lin, M.; Shum, C.K. Hurricane Sandy Storm Surge Measured by Satellite Altimetry. *Oceanography* **2013**, *2*, 8–9. [CrossRef]
9. Fenoglio-Marc, L.; Scharroo, R.; Annunziato, A.; Mendoza, L.; Becker, M. Cyclone Xaver seen by geodetic observations. *Geophys. Res. Lett.* **2015**, *22*, 9925–9932. [CrossRef]
10. Han, G.; Ma, Z.; Chen, N.; Chen, N.; Yang, J. Hurricane Isaac storm surges off Florida observed by Jason-1 and Jason-2 satellite altimeters. *Remote Sens. Environ.* **2017**, *198*, 244–253. [CrossRef]

11. De Biasio, F.; Bajo, M.; Vignudelli, S.; Umgiesser, G.; Zecchetto, S. Improvements of storm surge forecasting in the Gulf of Venice exploiting the potential of satellite data: The ESA DUE eSurge-Venice project. *Eur. J. Remote Sens.* **2017**, *50*, 428–441. [CrossRef]

12. Bajo, M.; De Biasio, F.; Umgiesser, G.; Vignudelli, S.; Zecchetto, S. Impact of using scatterometer and altimeter data on storm surge forecasting. *Ocean Model.* **2017**, *113*, 85–94. [CrossRef]

13. De Biasio, F.; Vignudelli, S.; Della Valle, A.; Umgiesser, G.; Bajo, M.; Zecchetto, S. Exploiting the Potential of Satellite Microwave Remote Sensing to Hindcast the Storm Surge in the Gulf of Venice. *IEEE J. Sel. Top. Appl. Earth Obs. Remote Sens.* **2016**, *9*, 5089–5105. [CrossRef]

14. Ye, A.; Chen, Z.; Yu, Y. Numerical inverstigation of three-dimensional semidiurnal tidal waves in Taiwan Strait and its adjacent areas. *Oceanol. Limnol. Sin.* **1985**, *6*, 439–450.

15. Li, Y.; Cai, W.; Li, L.; Lin, M. The tide characteristics of the seas adjacent to Fujian and Taiwan derived from TOPEX/POSEIDON altimeter data. *Acta Oceanol. Sin.* **2002**, *24*, 154–162.

16. Zhang, W.Z.; Hong, H.S.; Shang, S.P.; Chen, D.W.; Chai, F. A two-way nested coupled tide-surge model for the Taiwan Strait. *Cont. Shelf Res.* **2007**, *27*, 1548–1567. [CrossRef]

17. Zhang, W.Z.; Shi, F.; Hong, H.S.; Shang, S.P.; Kirby, J.T. Tide-surge Interaction Intensified by the Taiwan Strait. *J. Geophys. Res.* **2010**. [CrossRef]

18. Prandle, D.; Wolf, J. Surge-Tide Interaction in the Southern North Sea. In *Elsevier Oceanography Series*; Elsevier: New York, NY, USA, 1978; pp. 161–185.

19. Zhen, W.; Chen, F.; Chen, X. Tides and tidal currents in the Taiwan Strait. *J. Oceanogr. Taiwan Strait* **1982**, *2*, 1–4.

20. Pan, H.; Liu, F. A numerical study of the tide-surge interaction in the East China Sea and the South China Sea. *Chin. J. Oceanol. Limnol.* **1994**, *1*, 13–21.

21. Chen, C.; Liu, H.; Beardsley, R.C. An Unstructured Grid, Finite-Volume, Three-Dimensional, Primitive Equations Ocean Model: Application to Coastal Ocean and Estuaries. *J. Atmos. Ocean. Technol.* **2003**, *20*, 159–186. [CrossRef]

22. Birol, F.; Fuller, N.; Lyard, F.; Cancet, M.; Nino, F.; Delebecque, C.; Fleury, S.; Toublanc, F.; Melet, A.; Saraceno, M.; et al. Coastal Applications from Nadir Altimetry: Example of the X-TRACK Regional Products. *Adv. Space Res.* **2016**. [CrossRef]

23. Pawlowicz, R.; Beardsley, B.; Lentz, S. Classical tidal harmonic analysis including error estimates in MATLAB using T_TIDE. *Comput. Geosci.* **2002**, *28*, 929–937. [CrossRef]

24. Amante, C.; Eakins, B.W. ETOPO1 1 Arc-Minute Global Relief Model_Procedures, Data Sources and Analysis. *NOAA Tech. Memo. NESDIS NGDC-24* **2008**. [CrossRef]

25. Egbert, G.D.; Erofeeva, S.Y. Efficient Inverse Modeling of Barotropic Ocean Tides. *J. Atmos. Ocean. Technol.* **2002**, *19*, 183–204. [CrossRef]

26. Weisberg, R.H.; Zheng, L. Circulation of Tampa Bay driven by buoyancy, tides and winds, as simulated using a Finite Volume Coastal Ocean Model. *J. Geophys. Res.* **2008**, *111*, C1005. [CrossRef]

27. Ma, Z.; Han, G.; de Young, B. Oceanic responses to hurricane Igor over the Grand Banks: A modelling study. *J. Geophys. Res. Oceans* **2015**, *120*. [CrossRef]

28. Su, J.; Yuan, Y. (Eds.) *Hydrography in the Chinese Seas*; Ocean Press: Beijing, China, 2005; p. 367. (In Chinese)

29. Rego, J.O.L.; Li, C. Nonlinear terms in storm surge predictions: Effect of tide and shelf geometry with case study from Hurricane Rita. *J. Geophys. Res.* **2010**, *115*. [CrossRef]

30. Tang, C.L.; Gui, Q.; De Tracey, B.M. Barotropic response of the Labrador/Newfoundland Shelf to a moving storm. *J. Phys. Oceanogr.* **1998**, *17*, 1152–1172. [CrossRef]

31. Thiebaut, S.; Vennell, R. Observation of a fast continental shelf wave generated by a storm impacting Newfoundland using wavelet and cross-wavelet analyses. *J. Phys. Oceanogr.* **2010**, *40*, 417–428. [CrossRef]

32. Brink, K.H.; Chapman, D.C. *Programs for Computing Properties of Coastal-Trapped Waves and Wind-Driven Motions over the Continental Shelf and Slope*; WHOI Tech. Rep. WHOI-87-24; Woods Hole Oceanographic Institution: Woods Hole, MA, USA, 1987; 119p.

remote sensing

MDPI

Article

Evaluation of Satellite-Altimetry-Derived Pycnocline Depth Products in the South China Sea

Yingying Chen [1,2], Kai Yu [1,3], Changming Dong [1,4], Zhigang He [5], Yunwei Yan [3] and Dongxiao Wang [2,*]

[1] School of Marine Science, Nanjing University of Information Science and Technology, Nanjing 210044, China; cyy@scsio.ac.cn (Y.C.); yukai041@nuist.edu.cn (K.Y.); cmdong@gmail.com (C.D.)
[2] State Key Laboratory of Tropical Oceanography, South China Sea Institute of Oceanology, Chinese Academy of Sciences, Guangzhou 510000, China
[3] State Key Laboratory of Satellite Ocean Environment Dynamics, Second Institute of Oceanography, State Oceanic Administration, Hangzhou 310000, China; yanyunwei@sio.org.cn
[4] Department of Atmospheric and Oceanic Sciences, University of California, Los Angeles, CA 90095, USA
[5] College of Ocean and Earth Science, Xiamen University, Xiamen 361000, China; zghe@xmu.edu.cn
* Correspondence: dxwang@scsio.ac.cn; Tel.: +86-18-664-692-695

Received: 27 June 2017; Accepted: 8 August 2017; Published: 12 August 2017

Abstract: The climatological monthly gridded World Ocean Atlas 2013 temperature and salinity data and satellite altimeter sea level anomaly data are used to build two altimeter-derived high-resolution real-time upper layer thickness products based on a highly simplified two-layer ocean model of the South China Sea. One product uses the proportional relationship between the sea level anomaly and upper layer thickness anomaly. The other one adds a modified component (η'_M) to account for the barotropic and thermodynamic processes that are neglected in the former product. The upper layer thickness, in this work, represents the depth of the main pycnocline, which is defined as the thickness from the sea surface to the 25 kg/m^3 isopycnal depth. The mean upper layer thickness in the semi-closed South China Sea is ~120 m and the mean reduced gravity is ~0.073 m/s^2, which is about one order of magnitude larger than the value obtained in the open deep ocean. The long-term temperature observations from three moored buoys, the conductivity-temperature-depth profiles from three joint cruises, and the Argo measurements from 2006 to 2015 are used to compare and evaluate these two upper layer thickness products. It shows that adding the η'_M component is necessary to simulate the upper layer thickness in some situations, especially in summer and fall in the northern South China Sea.

Keywords: upper layer thickness; satellite altimeter; two-layer ocean model; South China Sea

1. Introduction

The development of satellite remote sensing technology has enabled the retrieval of high-resolution, real-time, global information on the ocean. However, remote sensors can only "see" the sea surface. To infer underwater information using remote sensing data, one of the simplest and most used ways is to use the two-layer ocean model [1]. This model simplifies the stratified ocean to a two-layer homogeneous fluid, regarding the main pycnocline as the interface and the main pycnocline depth as the upper layer thickness (ULT). With the two-layer ocean model, one can easily derive the ULT, and the barotropic mode and first baroclinic mode of the circulation using satellite altimeter sea level anomaly (SLA) data.

The altimeter-derived ULT and circulation products are widely used in various areas for different purposes. Garzoli et al. [2] monitored the upper layer transport in the southeastern Atlantic Ocean; Sainz-Trapaga et al. [3] identified the Kuroshio Extension, its bifurcation, and its northern branch;

Goni and Wainer [4] investigated the variability of the Brazil Current front. After adding satellite sea surface temperature (SST) data to the two-layer ocean model, Shay et al. [5] estimated upper layer thermal structure and upper layer heat content, and then studied the effects of a warm oceanic feature associated with Hurricane Opal. Following Shay et al. [5], Pun et al. [6] improved typhoon intensity forecasts and analyzed the spatial and temporal errors in the western North Pacific Ocean with the satellite-derived upper layer heat content. Vertical thermal structure and upper-ocean heat content from satellite remote sensing data had also been obtained by combining climatological hydrographic data and the two-layer ocean model (e.g., [7–9]). Most of the studies mentioned above were for the open ocean. Lin et al. [10] were the first to use altimeter data and the two-layer ocean model to study the spatial and temporal variation of ULT in the semi-closed South China Sea.

The SLA changes through mainly three processes: barotropic motions, the vertical Ekman pumping of the main pycnocline and the near-surface density flux above the seasonal pycnocline [11]. The Ekman pumping of the main pycnocline is the most important process, and the other two are usually one order of magnitude smaller than that of the main pycnocline fluctuations in the open ocean. The altimeter-derived ULT product based on the two-layer ocean model is mainly dependent on the proportional relationship between the SLA and main pycnocline fluctuations. As with previous studies in the open ocean, the effects of barotropic motions and near-surface density flux on the variability of SLA were not discussed in the work of Lin et al. [10]. However, these factors may play a more important role in the semi-closed South China Sea. The South China Sea is the largest marginal sea in the Northwest Pacific. Its mean depth is ~1200 m and maximum depth is ~5000 m, with a diamond-shaped deep basin oriented along the northeast-southwest direction (Figure 1). The SLA and upper layer circulation in the South China Sea have strong seasonal variability that is primarily driven by the monsoon wind [12–14]. The coastal regions show a higher seasonal sea level cycle than deep water (e.g., [15,16]). Liu et al. [17] found that the seasonal variability of the SLA is forced mainly by surface wind curl, and secondarily (about 20%) by surface net heat flux. They concluded that the surface net heat flux can obviously change the seasonal variability of the SLA, especially in the central South China Sea, by expanding/contracting the water column in the mixed layer, and the influence of the surface net heat flux is much stronger than for the open ocean. Cheng and Qi [18] argued that on seasonal time scales the baroclinic component to a great extent explains the SLA over the deep part of the South China Sea basin, whereas the barotropic component has a significant contribution to the SLA over shallow water areas in the South China Sea.

In the present study, instead of the historical ocean profiles, the gridded data is used to derive the altimeter-derived ULT products in the South China Sea. We first get the high-resolution ($0.25° \times 0.25°$) altimeter-derived ULT products and give a comprehensive assessment of the ULT products with multifarious in situ observations. A monthly varied modified component is first introduced into the two-layer ocean model to examine the effects of the barotropic component and near-surface density flux on the ULT products in the South China Sea. This paper is organized as follows: Section 2 describes the data. Section 3 introduces the methodology. Section 4 determines the product parameters. Section 5 evaluates the products with different in situ datasets. Section 6 discusses the physical mechanisms of the modified component. Finally, we give our conclusions and discussions in Section 7.

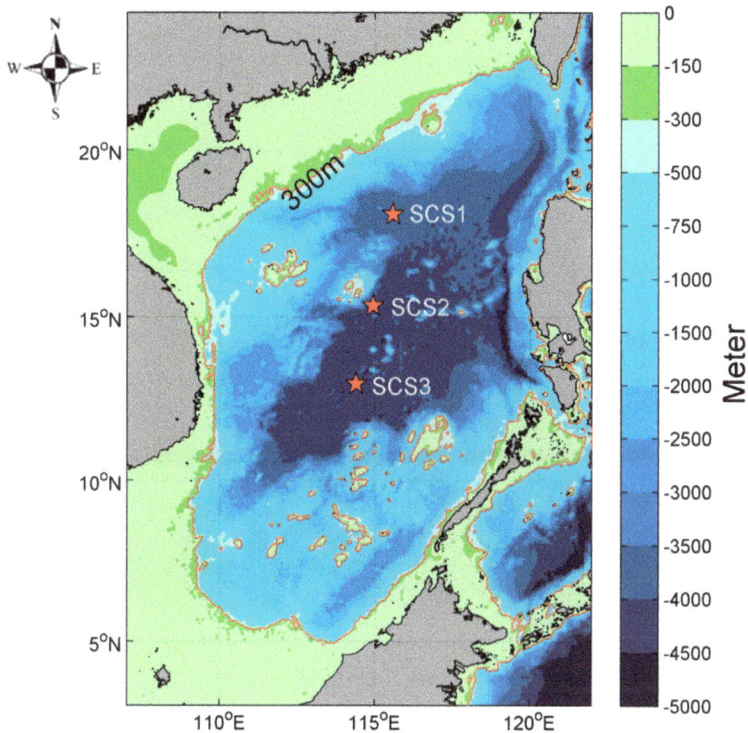

Figure 1. Bottom topography of the South China Sea. The locations of three ATLAS buoys SCS1 (18°5.9′N, 115°35.8′E), SCS2 (15°20.6′N, 114°57.3′E), SCS3 (12°58.5′N, 114°24.5′E) are indicated by red stars.

2. Data

Two datasets are used to derive the ULT product in the South China Sea. The first is the climatological monthly temperature and salinity data at standard depth levels with a horizontal resolution of 0.25° both in longitude and latitude taken from the World Ocean Atlas 2013 [19]. The second is the daily and climatological monthly delayed-time level 4 gridded SLA products from multi-satellite observation provided by Archiving, Validation and Interpretation of Satellite Oceanographic Data (AVISO [20], version 5.1). The SLA data have the same spatial resolution as the World Ocean Atlas 2013 and have been modified for various atmospheric, tidal and instrumental corrections [21,22].

Three different kinds of in situ datasets are used to validate and evaluate the ULT product. The first dataset consists of measurements from three autonomous temperature line acquisition system (ATLAS) buoys (named SCS1, SCS2 and SCS3), deployed by the Institute of Oceanography, Taiwan University during the South China Sea Monsoon Experiment, whose locations are shown in Figure 1. The temporal spans are from 17 April 1997 to 9 April 1999 for SCS1, from 11 April 1998 to 10 April 1999 for SCS2, and from 12 April 1998 to 10 April 1999 for SCS3. The temporal resolution is 10 min, and the measurement depth is 1 m, 25 m, 50 m, 75 m, 100 m, 125 m, 150 m, 200 m, 250 m, 300 m and 500 m.

The Argo measurements available in the South China Sea from 2006 to 2015 (Figure 2a) are provided by the US Global Ocean Data Assimilation Experiment (USGODAE [23]). The conductivity-temperature-depth (CTD) probe data from three joint cruises in July 1998 (from 4 June to 21 July 1998), August 2000 (from 2 August to 3 September 2000) and May 2004 (from 29 April to 25 May 2004) are provided by the South China Sea Institute of Oceanology of the Chinese Academy of Sciences. These three cruises include 133, 189 and 143 different CTD casts, respectively (Figure 2b). These data are widely used in South China Sea studies (e.g., [24,25]).

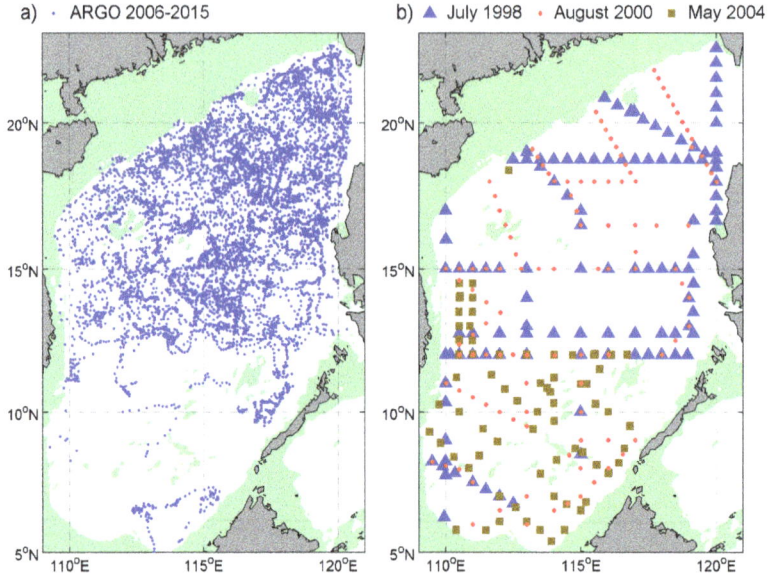

Figure 2. Distribution of (**a**) Argo measurements from 2006 to 2015 and (**b**) CTD data of three joint cruises investigation in July 1998 (blue triangles), August 2000 (red points) and May 2004 (golden squares).

3. Methods

The ocean bottom pressure (P_o) is the sum of the sea surface atmospheric pressure (P_a) and the weight per unit area of the water column:

$$P_o = P_a + g \int_{-D}^{\eta} \rho dz \tag{1}$$

where D is the mean ocean depth, η represents the sea surface height above geoid, g is the gravitational acceleration, and ρ is sea water density. According to the Boussinesq approximation, Equation (1) can be further approximated by

$$P_o = P_a + g \int_{-D}^{0} \rho dz + g\rho_0 \eta \tag{2}$$

where ρ_0 is the mean density of the ocean, which is equal to 1025 kg/m^3 in this study. The fluctuating ocean bottom pressure (P_o') is caused by the variability of atmospheric pressure, the density fluctuations under the mean sea surface and the SLA (η'):

$$P_o' = P_a' + g \int_{-D}^{0} \rho' dz + g\rho_0 \eta' \tag{3}$$

Now η' can be partitioned into barotropic (η_b') and steric components (η_s'):

$$\eta' = \eta_b' + \eta_s'. \tag{4}$$

The barotropic component changes as a result of internal mass redistribution in the ocean or water-mass flux, which are directly correlated with ocean bottom pressure, or changes as a result of the variability of atmospheric pressure:

$$\eta_b' = \frac{P_a' - P_o'}{g\rho_0}. \tag{5}$$

The steric component, also called the baroclinic component, changes depending on the seawater density of the water column but corresponds neither to ocean bottom pressure nor to sea surface atmospheric pressure:

$$\eta_s' = -\frac{1}{\rho_0} \int_{-D}^{0} \rho' dz. \tag{6}$$

According to Gill and Niiler [11], the steric height anomaly, η_s', is produced mainly by the variability of vertical Ekman pumping of the main pycnocline and near-surface density flux above the seasonal pycnocline.

The ocean is continuously stratified, with a sharp vertical density gradient in the pycnocline. The two-layer model, proposed by Goni et al. [1], simplifies the stratified ocean as a two-layer homogeneous fluid, regarding the main pycnocline depth as the interface. Let the thicknesses and densities of the upper and lower layer be h_1, ρ_1, h_2, and ρ_2, respectively. The sea surface height is then

$$\eta = h_1 + h_2 - D, \tag{7}$$

and the ocean bottom pressure is then

$$P_o = P_a + g\rho_1 h_1 + g\rho_2 h_2 \tag{8}$$

Combining Equations (7) and (8) yields:

$$\eta = \varepsilon h_1 + B, \tag{9}$$

where

$$\varepsilon = (\rho_2 - \rho_1)/\rho_2, \tag{10}$$

and

$$B = \frac{P_o - P_a}{g\rho_2} - D = \frac{\rho_1}{\rho_2}h_1 + h_2 - D. \tag{11}$$

The fluctuating part of Equation (9) is

$$\eta' = \varepsilon h_1' + B'. \tag{12}$$

As one can see, B' has the same meaning as η_b' in Equation (4), and varies with the variabilities of the ocean bottom pressure and sea surface atmospheric pressure. On the other hand, the $\varepsilon h_1'$ has the same meaning as η_s' in Equation (4), and this illustrates that the steric component of the SLA in the two-layer ocean model is only caused by the fluctuations of the interface (main pycnocline). However, the effect of near-surface density flux is not considered in this model.

Generally speaking, the lower layer thickness in the deep ocean is much larger than that of the upper layer. If it is further assumed that the lower layer is infinitely deep and the fluid in the lower layer is stagnant, the model can be referred to as a 1.5-layer reduced gravity model. In the 1.5-layer reduced gravity model, the barotropic component, B', is negligible and the SLA is equal to the baroclinic component and proportional to the ULT anomaly (ULTA), h_1' [3]. Letting η_{ULT}' denote the part of SLA that is proportional to the ULTA, then we have

$$\eta'(t) = \eta_{ULT}'(t) = \varepsilon h_1'(t). \tag{13}$$

As a result, the real-time ULT product can be derived from the corresponding altimeter data:

$$h_1(t) = \overline{h}_1 + h_1'(t) = \overline{h}_1 + \frac{1}{\varepsilon}\eta'(t), \tag{14}$$

where $\overline{h}_1$ represents the climatological mean ULT. Both $\overline{h}_1$ and ε can be calculated from the World Ocean Atlas 2013 data. The 1.5-layer reduced gravity model is widely used in simulating the upper-ocean circulation in the South China Sea (e.g., [26–28]).

Equation (14) only considers the proportional relationship between the SLA and ULTA. However, the barotropic and sea surface thermodynamic processes will change the SLA, while having little effect on the main pycnocline fluctuations. They will break the proportional relationship between the SLA and ULTA and bring error into the altimeter-derived ULT product based on Equation (14). To consider the effects of these processes, we add a modified component η_M', and then η' can be divided into two parts η_{ULT}' and η_M':

$$\eta'(t) = \eta_{ULT}'(t) + \eta_M'(t). \tag{15}$$

The climatological monthly η_{ULT}' can be derived from h_1' based on Equation (13). The h_1' is calculated from the World Ocean Atlas data. Then the climatological monthly η_M' can be produced by subtracting the World Ocean Atlas-derived η_{ULT}' from the η', which is obtained from the climatological monthly altimeter data. Then the real-time ULT product can be obtained with the daily altimeter data by regarding the climatological monthly η_M' as a background field:

$$h_1(t) = \overline{h}_1 + \frac{1}{\varepsilon}\eta_{ULT}'(t) = \overline{h}_1 + \frac{1}{\varepsilon}(\eta'(t) - \eta_M'(t)). \tag{16}$$

4. Parameter Determination

To get the altimeter-derived ULT products, the first step is to determine the interface of the two-layer ocean model. Given that the effect of salinity on the density is smaller than the effect of temperature, previous investigators usually choose the depth of a certain isotherm as the interface to correspond to the main thermocline depth (e.g., [1–4]). After analyzing the in situ hydrographic data from the World Ocean Database 2005, Lin et al. [10] found the 16 °C isotherm has a better linear correlation with the SLA than those at 12 °C, 14 °C, 18 °C, or 20 °C, and thus used the depth of the 16 °C isotherm as the interface of the South China Sea. In our study, from the climatological mean temperature and density profiles over the South China Sea region derived from the World Ocean Atlas 2013 (Figure 3a), one can see that the thermocline and pycnocline depths are nearly the same in the South China Sea. Both are mainly located between 50 m and 200 m. Around this depth range, different isotherms and isopycnals depths (Figure 3a) are adopted to calculate the climatological monthly ULT and ULTA using the World Ocean Atlas 2013 data. All available spatial and temporal ULTA values in the South China Sea are compared with the corresponding SLA (Figure 3b,c). The spatial grid pixels are excluded when the water depth is shallower than 300 m (Figure 1) because of the inapplicability of the two-layer ocean model in the continental shelf area. The correlation coefficients between the SLA and ULTA reach a maximum when choosing the 20 °C isotherm or 25 kg/m³ isopycnal. Both the maximum correlation coefficients are ~0.77 and the corresponding depths of these two choices are ~120 m (Figure 3a). The strong similarity suggests that the salinity effect is weak. The 25 kg/m³ isopycnal depth will be used to represent the ULT in the remainder of this paper. The 20 °C isotherm depth is adopted to represent the ULT only when the ULT product is validated with the ATLAS buoys' data, because the ATLAS buoys lack salinity data.

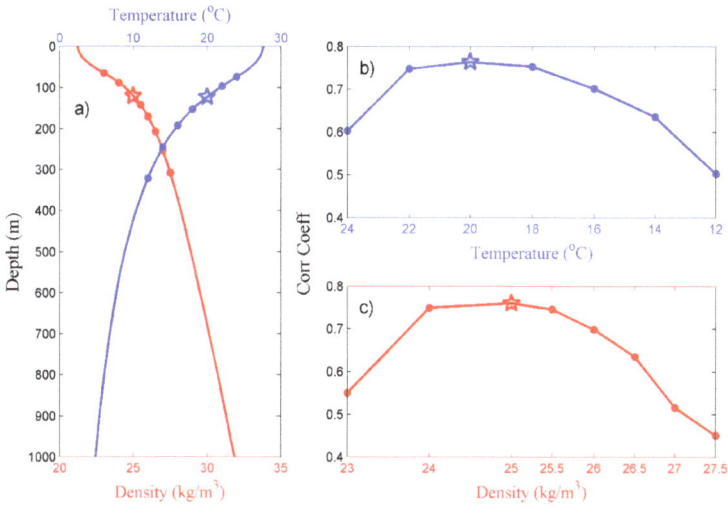

Figure 3. (**a**) The mean temperature (blue) and density (red) profile in the South China Sea derived from the World Ocean Atlas 2013. The correlation coefficient between the SLA and ULTA calculated by different (**b**) temperature and (**c**) density. The stars indicate where the maximum correlation coefficients appear.

Figure 4 shows the comparison of the correlation between the SLA and ULTA in different seasons. The winter, spring, summer and fall values correspond here to January, April, July and October, respectively. The correlation coefficient is high (beyond 95% t-test confidence interval) throughout the year. The mean correlation coefficient is 0.76 ± 0.07. The maximum correlation coefficient is 0.86 which happens in spring (Figure 4b), and the minimum is 0.72 in summer (Figure 4c).

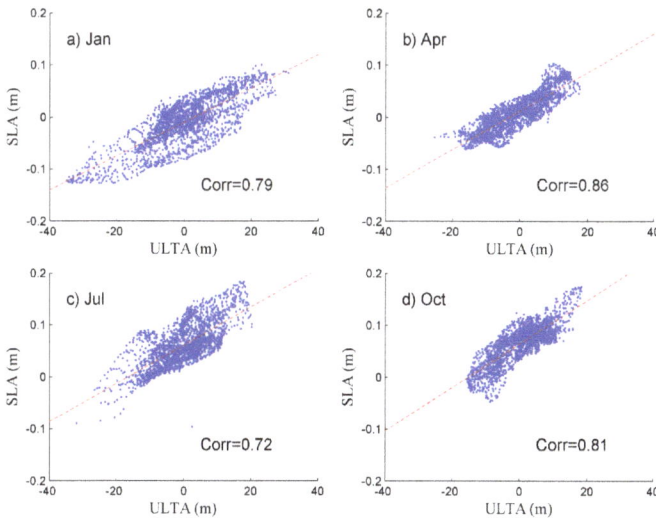

Figure 4. The scatter plots of SLA and ULTA calculated by 25 kg/m^3 isopycnal in January, April, July and October (**a–d**). The line of best fit is represented by the red dashed line.

After determining the interface of the two-layer ocean model, the climatological mean $\bar{h}_1, \rho_1, \rho_2$ and ε can be calculated from the World Ocean Atlas 2013 data. Figure 5a illustrates the climatological mean ULT, $\bar{h}_1$, in the South China Sea. The mean ULT is ~120 m and becomes thicker toward the east. There are two eddy-like thinner ULT areas at 18°N, 118°E and 14°N, 111°E, corresponding to the cold eddy northwest of the Luzon Island [13,29] and the cold eddy off the Vietnamese coast [27,30], respectively. Figure 5b shows that the distribution of ε in the South China Sea. ε increases gradually from the northwest to southeast South China Sea, with the values between 7.0×10^{-3} and 7.8×10^{-3} in most of our study area. For the regions near the continental shelf edge, the ε rapidly decreases to ~5.0×10^{-3}. Around the Nansha Islands, ε also decreases rapidly. The reduced gravity g' can be estimated by multiplying ε by the gravitational acceleration. This gives g' ~0.073 m/s^2 in the South China Sea, which is about one order of magnitude larger than the value obtained in the open deep ocean (e.g., [1,2]).

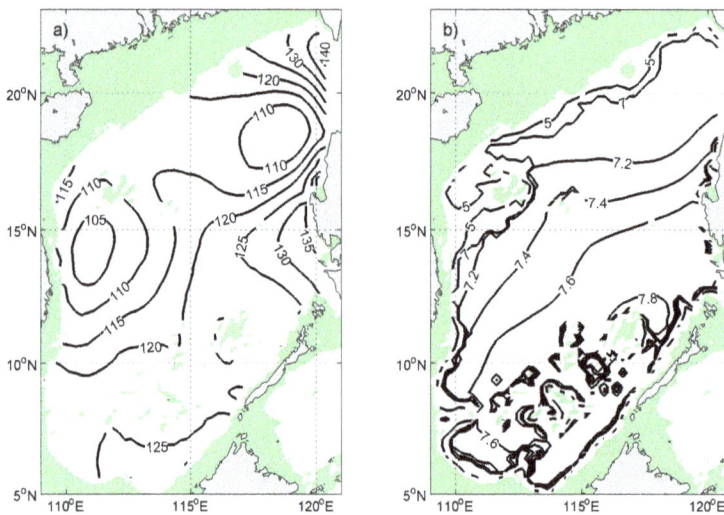

Figure 5. Climatological mean (**a**) ULT (CI: 5 m) and (**b**) ε ($\times 10^{-3}$) in the South China Sea derived from the World Ocean Atlas 2013.

The first row of Figure 6 shows the climatological seasonal variation of η', SLA. In winter, η' is negative in most of the South China Sea, and reaches a minimum in the northeastern South China Sea. This is mainly due to the reinforcement of the cold eddy northwest of the Luzon Island. η' gradually increases in spring and reaches a maximum in the northeastern South China Sea in summer. Then η' decreases in fall and starts another cycle. The maximum seasonal variability of η' in the northeastern South China Sea suggests that the cold eddy northwest of the Luzon Island dominates the seasonal variability of the η' in the South China Sea. The monsoon is thought to be the primary force for the seasonal shift of the cold eddy and η' [16]. The seasonal variability of η'_{ULT} (second row in Figure 6) shows great similarity to η', except that η'_{ULT} is smaller than η' in the northeastern South China Sea and larger than η' in the north central part of the South China Sea in winter. In summer and fall, η'_{ULT} is smaller than η' in the north central, northwestern and southern South China Sea. These differences can be clearly illustrated by the seasonal variability of η'_M (third row in Figure 6). Figure 7a shows the correlation coefficient between the monthly η' and η'_{ULT}. A significant positive correlation exists between 10°N and 15°N in the northeastern South China Sea, while in the north central, northwestern and southern South China Sea, the positive correlation is weak and even turns into a negative correlation in the north central and northwestern South China Sea. On the other hand, the correlation coefficient between η' and η'_M (Figure 7b) is relatively high where the positive correlation between η' and η'_{ULT} is lower. Interestingly, the region where the higher positive correlation between η'_M and η'_M occurs is exactly the area where one can find the larger ratio of the standard deviation between η'_M and η'_{ULT} (Figure 7c), which shows the relative importance of these two components in the seasonal variability amplitude of η'. It can be concluded that the proportional relationship between η' and h'_1 is stronger and one can get an accurate ULT product based on Equation (14) between 10°N and 15°N and in the northeastern South China Sea. However, in the north central, northwestern and southern South China Sea, η'_M becomes important, and one must derive the ULT product according to Equation (16) instead of Equation (14).

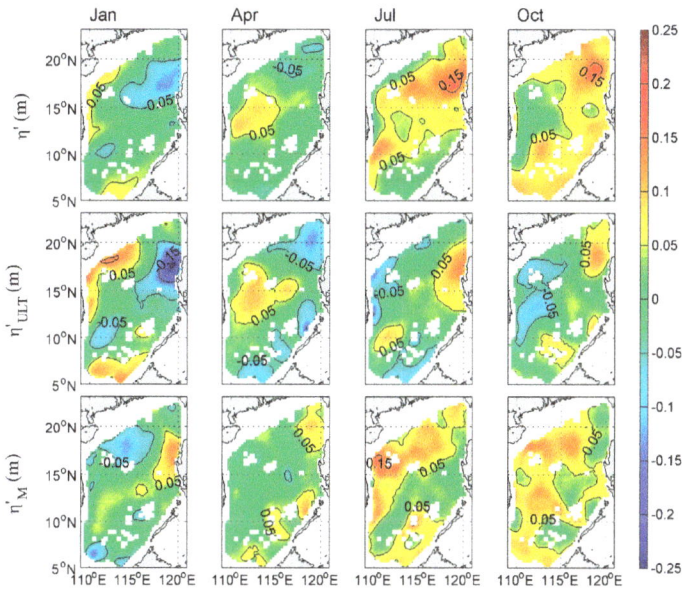

Figure 6. Seasonal variability of η', η'_{ULT} and η'_M (CI: 0.1 m).

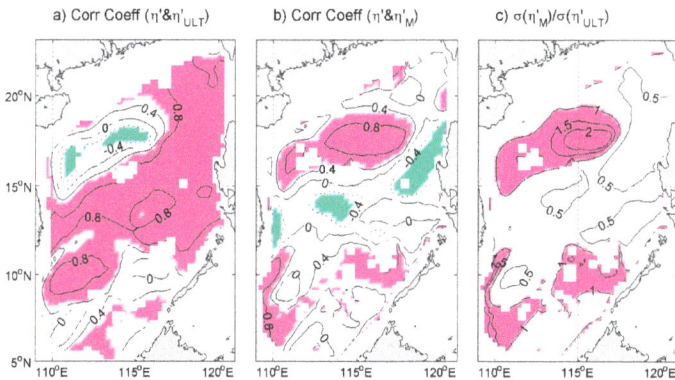

Figure 7. Spatial distributions of correlation coefficient between the monthly η' (**a**) with η'_{ULT} and (**b**) with η'_M. Shading in (**a**,**b**) indicates correlation beyond 95% confidence level; (**c**) The ratio of the standard deviation between η'_M and η'_{ULT}. The ratio larger than 1 is shaded in (**c**).

5. Product Evaluation

After identifying the parameters, one can easily get the altimeter-derived real-time ULT products based on Equations (14) and (16), respectively. These two products are compared and evaluated with different in situ observations.

The Argo-data-derived ULTs are directly obtained through computing the 25 kg/m^3 isopycnal depth at each Argo profile. Then the gridded altimeter-derived ULTs were interpolated to Argo profile locations to compare with the Argo-data-derived ULTs. The Argo measurements in the South China Sea are separated into the part north of 15°N and the part between 10°N and 15°N, because the relative importance of η'_{ULT} and η'_M is quite different in these two areas and the Argo data are very scarce in the southern South China Sea (Figure 2a). Figure 8 shows the comparison of the mean bias and root-mean-square error (RMSE) between the altimeter-derived and Argo-data-derived ULT in these two areas based on the two methods. Overall, the biases are positive, which indicate that the altimeter-derived products always overestimate the ULT directly calculated by the Argo observations. The overestimation can be somewhat reduced after considering η'_M. At the same time, the RMSE also achieves a certain degree of reduction. The bias and RMSE for the region north of 15°N are obviously larger than those between 10°N and 15°N. Specifically, without considering η'_M, the maximum bias (~12 m) and maximum RMSE (more than 15 m) can be found in summer and fall for the region north of 15°N. After adding η'_M, the bias is reduced to less than 3 m and the RMSE decreases by more than 27%. The situations are the same for the region between 10°N and 15°N, except that the bias and RMSE and the improvements after considering η'_M are relatively smaller than those north of 15°N. These results correspond to the large positive η'_M in summer and fall (third row in Figure 6) and prove the relative importance of η'_M in summer and fall, especially in the northern South China Sea. However, in winter and spring, the improvements are very limited after considering η'_M.

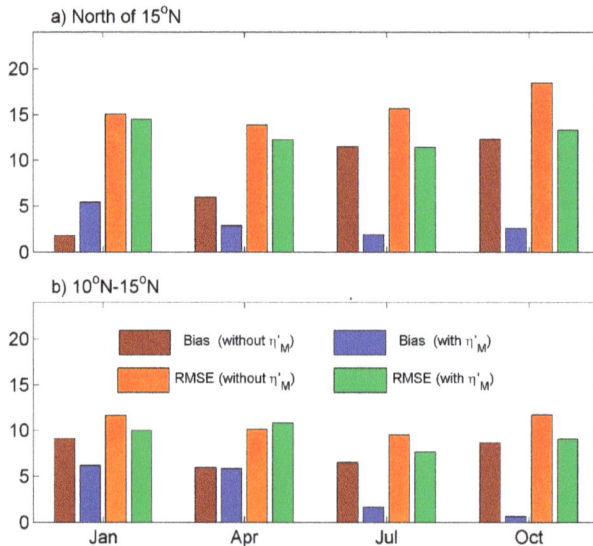

Figure 8. Bias and RMSE (Unit: m) between the altimeter-derived and Argo-data-derived ULT for the South China Sea (**a**) north of 15°N and (**b**) between 10°N and 15°N.

The CTD data collected during three joint cruises are used to validate the altimeter-derived ULT products. Figure 9 shows the differences between altimeter-derived and cruise-data-derived ULTs for these three cruises. The ULT differences at each station are derived in the same way as the

Argo profile. Results for the first two cruises are shown in the first two columns of Figure 9. They are during the summer and fall and cover nearly all the South China Sea. The altimeter-derived products, which neglect η'_M, evidently overestimate the cruise-data-derived ULT in these two cruises, especially in the north central, northwestern and southern South China Sea. After considering η'_M, the overestimations are substantially reduced, and the RMSE are reduced from 11.8 m and 14.5 m to 8.6 m and 11.7 m, respectively for the cruise in July 1998 and August 2000. The reduction rate of the RMSE for these two cruises is about 27% and 19%, respectively. However, both the altimeter-derived products underestimate the cruise-data-derived ULT off the southeast coast of Vietnam in August 2000 (second column in Figure 9). That may be because of the effects of an anticyclonic eddy, which is part of a dipole eddy pair associated with the summer eastward jet (e.g., [14,26,27,31,32]). The details still need further study. Furthermore, η'_M has a very limited influence on ULT products for the cruise in May 2004 (third column in Figure 9), and this agrees with previous studies that η'_M is very small (Figure 6) and the improvements are very limited after considering η'_M in spring (Figure 8b).

Next, we validate the two satellite-derived products with the data of the three ATLAS buoys in the South China Sea (Figure 10). The SCS1 is located in the northern South China Sea (Figure 1) where the relative importance of η'_M reaches its maximum (Figure 7c). The correlation coefficient and RMSE between the time series of ULT derived from the altimeter without considering η'_M and from the SCS1 is 0.54 m and 15.2 m. The altimeter-derived product obviously overestimates the buoy-derived ULT during the summer and fall. Consistent with our expectations, these overestimations are markedly reduced after considering η'_M. The new product fits very well with the buoy-derived ULT, especially in summer and fall. The correlation coefficient rises to 0.81, and the RMSE decreases to 9.6 m. These results confirm the important role of η'_M in simulating the ULT in the northern South China Sea in summer and fall. For the buoys SCS2 and SCS3, when η'_M is neglected, the altimeter-derived products correspond well with the buoy-derived ULT. Their correlation coefficient is 0.91 and 0.69, and their RMSE is 9.7 m and 8.1 m for SCS2 and SCS3, respectively. After considering η'_M, the time series are basically the same (Figure 10b,c), which is consistent with the result that the effects of η'_M are smaller in the middle of the South China Sea (Figure 7c).

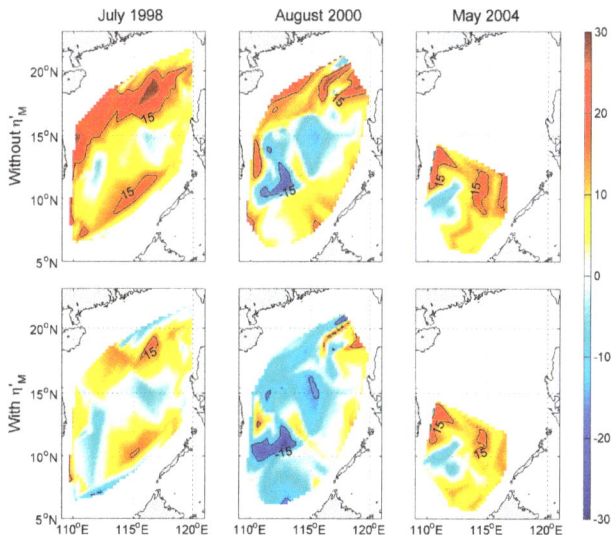

Figure 9. Spatial distribution of differences (Units: m) between altimeter-derived and cruise-data-derived ULT in July 1998 (**left**), August 2000 (**middle**) and May 2004 (**right**) without η'_M (**up**) and with η'_M (**down**).

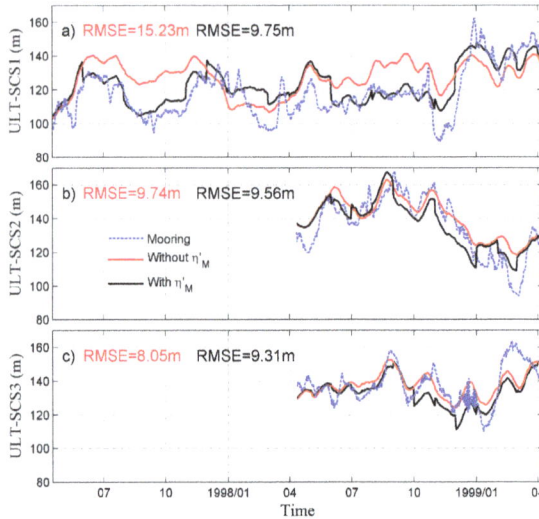

Figure 10. (**a–c**) Time series of buoy-data-derived ULT (blue dashed line), altimeter-derived ULT without η'_M (red solid line) and with η'_M (black solid line). The locations of three ATLAS buoys are marked in Figure 1.

6. Physical Mechanisms of the Modified Component

It is worth discussing the physical mechanisms that control η'_M. According to Gill and Niiler [11], the wind can affect η' in two ways: by changing the bottom pressure through barotropic motions and by changing the main thermocline through vertical Ekman pumping. In addition, the surface density flux can affect the near-surface steric height anomaly by changing the density in the mixed layer. Therefore, η'_{ULT} looks to be mainly under the control of baroclinic dynamics and η'_M is dominated by barotropic processes and the near-surface baroclinic thermodynamic processes. By comparing the 1.5-layer baroclinic planetary wave model and the barotropic Sverdrup model, Liu et al. [17] noted that the seasonal variability of η' in the South China Sea is mainly determined by surface wind curl on baroclinic Rossby waves and the barotropic contribution is very small. They also find that the surface steric height anomaly in the upper 50 m, which is controlled by surface heat flux forcing, contributes ~20% of the total steric height anomaly. According to Liu et al. [17], the surface steric height anomaly in the upper 50 m can be calculated as:

$$\eta'_{50} = -\frac{1}{\rho_0} \int_{-50\,m}^{0} \rho' dz, \tag{17}$$

in the same way as Equation (6). The differences between η'_s and η'_{50} represent the baroclinic dynamics component of η', which is mainly correlated with the main thermocline variability. Liu et al. [17] also mentioned that the barotropic component of η' can be calculated by subtracting η'_s from the altimeter η' if both are exact.

Figure 11 shows the seasonal variability of the total steric height anomaly (η'_{WOA}), the surface steric height anomaly (η'_{WOA50}), and the baroclinic dynamics component of η' (η'_{WOAULT}) measured by the World Ocean Atlas 2013. Figure 11 also illustrates the seasonal variability of the barotropic component of η' (η'_{BT}) calculated by subtracting η'_{WOA} from η' (first row in Figure 6). The major seasonal features of η'_{WOA} and η'_{WOAULT} agree well with η' and η'_{ULT} (Figure 6), which suggests that the seasonal variability of the SLA is dominated by the baroclinic part, especially by the baroclinic dynamic processes. The seasonal variability of η'_{WOA50} is relatively small, and mainly manifests as negative anomalies (smaller than −0.02 m) in winter and spring and positive anomalies (larger than 0.02 m) in summer and fall in the northern South China Sea. In the same way as Figure 7c, the relative importance of η'_{WOA50}, η'_{BT}, and η'_{WOAULT} is shown by comparing their standard deviation (Figure 12). From Figure 12a, η'_{WOA50} is found to be more important than η'_{WOAULT} only in the north central part of the South China Sea. This position is exactly where the buoy SCS1 is located and the maximum ratio of the standard deviation between η'_M and η'_{ULT} is found. This demonstrates that, in this region, the variability of η'_M can be partly explained as the effects of surface thermodynamic processes. The surface heat flux expands/contracts the water column in the mixed layer and brings a positive/negative η'_M in summer/winter, especially in the northern South China Sea. Comparing with η'_{WOA50}, η'_{BT} plays a more important role. η'_{BT} shows high positive values (larger than 0.05 m) in the northeastern South China Sea in winter, in the southern South China Sea in spring, and in the western South China Sea in summer and fall. The seasonal variability of η'_{BT} is the same as, but weaker than, η'_M. Besides the north central part of the South China Sea, the ratio of the standard deviation between η'_{BT} and η'_{WOAULT} is more than 1 in the northwestern and southern South China Sea (Figure 12b). After combining η'_{WOA50} and η'_{BT}, the ratio of the standard deviation between the combined effect and η'_{WOAULT} (Figure 12c) fits very well with that between η'_M and η'_{ULT} (Figure 7c). This confirms that η'_M is mainly under the control of the combined effect of surface thermodynamic and barotropic processes. However, whether the differences between η'_{WOA} and satellite-derived η' can represent the barotropic component is still open to discussion. The differences may also include the effects of the time inconsistency of these two datasets. The Gravity Recovery and Climate Experiment (GRACE) [33,34] will be included to discuss the seasonal variability of the barotropic component in future work.

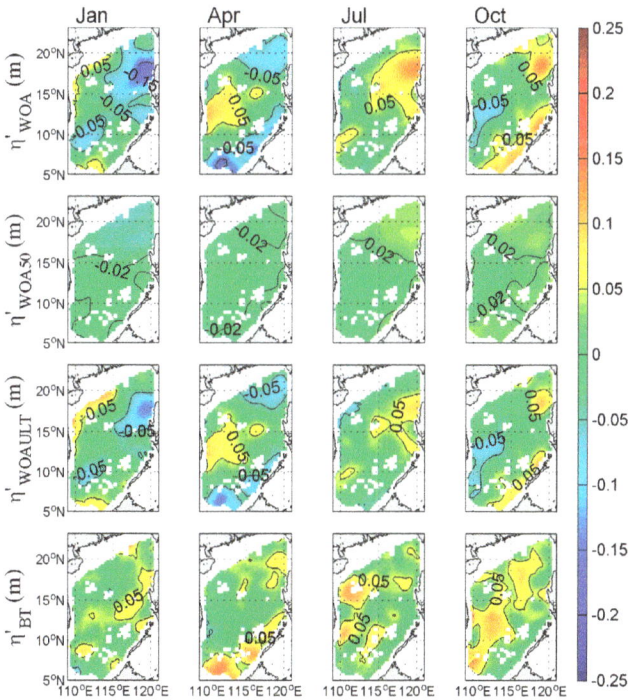

Figure 11. Seasonal variability of η'_{WOA}, η'_{WOA50}, η'_{WOAULT} and η'_{BT}.

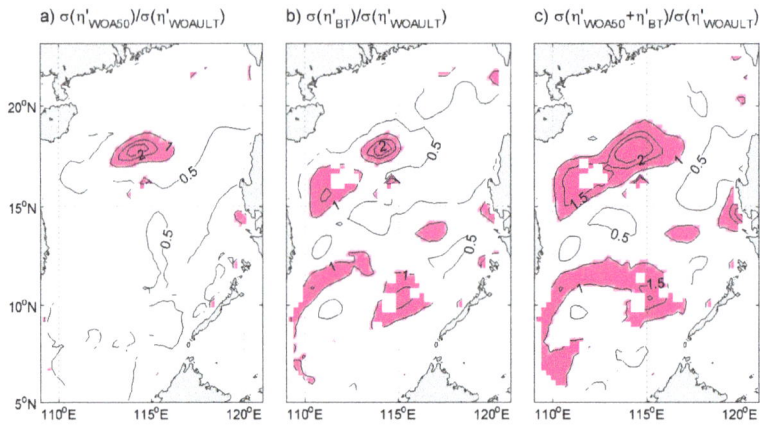

Figure 12. The ratio of the standard deviation (**a**) between η'_{WOA50} and η'_{WOAULT}; (**b**) between η'_{BT} and η'_{WOAULT}; and (**c**) between $\eta'_{WOA50} + \eta'_{BT}$ and η'_{WOAULT}. The ratio larger than 1 is shaded.

7. Conclusions and Discussion

The climatological monthly gridded temperature and salinity and SLA data are used to quantify the background field parameters of Equations (14) and (16) based on a highly simplified two-layer ocean model. The first method considers that the SLA is directly proportional to the ULTA, and the other separates SLA into two parts: the component proportional to ULTA, η'_{ULT}, and the modified component η'_M, which can correct the difference between η' and η'_{ULT} introduced by other physical processes. With these equations, one can derive two real-time ULT products using the SLA data. As far as we know, this is the first time to have obtained the altimeter-derived ULT products with gridded World Ocean Atlas 2013 data. Compared to the former products ($2° \times 2°$) derived from the site data (e.g., [10]), the new products have higher resolution ($0.25° \times 0.25°$) and thus can identify the ULT structures at smaller scales.

The 25 kg/m^3 isopycnal depth is chosen as the representation of the ULT after comparing different choices. The alternative choice is the 20 °C isotherm depth due to the salinity effect being weak. In the South China Sea, the mean ULT is ~120 m, and there are two obvious eddy-like thinner ULT areas corresponding to two cold eddies. The mean reduced gravity is ~0.073 m/s^2, which is about one order of magnitude larger than the value obtained in the open deep ocean. Both η' and η'_{ULT} display a clear seasonal variability and their variability shows great similarity, except in the north central, northwestern and southern South China Sea, where the η'_M has higher correlation with η' and larger standard deviation relative to η'_{ULT}.

Three different in situ observations are used to compare and evaluate these two ULT products. Both the high-resolution altimeter-derived products satisfactorily simulate the ULT in the South China Sea. The range of the RMSE is from less than 10 m to ~15 m in different evaluations. The monthly varied η'_M is firstly introduced into the two-layer ocean model. We found that η'_M is really necessary to correct the overestimation in summer and fall reported by the altimeter-derived ULT product without considering η'_M, especially in the northern South China Sea. However, in winter and spring, and in the middle South China Sea, the corrective action of η'_M is limited. Nevertheless, it is comforting to see that the bias and RMSE is relatively small in these situations even if without η'_M. The in situ observations in the South China Sea south of 10°N were too few to evaluate the satellite-derived ULT product. Finally, the physical mechanisms of η'_M are found to be mainly under the control of the combined effect of barotropic and surface thermodynamic processes.

Acknowledgments: This research was supported by the National Key R&D Program of China (grant 2016YFA0601804 and 2016YFA0601803), National Science Foundation of China (grant 41606021, 41606017, and 41476022), Open Fund of State Key Laboratory of Satellite Ocean Environment Dynamics (grant QNHX1610), the Startup Foundation for Introducing Talent of NUIST (grant 2014r017 and 2014r072), the National Program on Global Change and Air-Sea Interaction (grant GASI-03-IPOVAI-05), and the Priority Academic Program Development of Jiangsu Higher Education Institutions (PAPD). The authors would like to thank Gregory P. King for language editing. The authors also thank South China Sea Institute of Oceanology of the Chinese Academy of Sciences for supplying the ATLAS buoys' data during the South China Sea Monsoon Experiment and the CTD probe data of the joint cruise investigations.

Author Contributions: D.W. and C.D. conceived and designed the experiments; Y.C., K.Y, Z.H. and Y.Y. performed the experiments; Y.C. and K.Y, analyzed the data; Y.C. wrote the paper. K.Y and Y.Y. helped modify the manuscript.

Conflicts of Interest: The authors declare no conflict of interest.

References

1. Goni, G.J.; Kamholz, S.; Garzoli, S.L.; Olson, D.B. Dynamics of the Brazil-Malvinas confluence based on inverted echo sounders and altimetry. *J. Geophys. Res.* **1996**, *101*, 16273–16289. [CrossRef]
2. Garzoli, S.L.; Goni, G.J.; Mariano, A.J.; Olson, D.B. Monitoring the upper southeastern Atlantic transports using altimeter data. *J. Mar. Res.* **1997**, *55*, 453–481. [CrossRef]
3. Sainz-Trapaga, S.; Goni, G.J.; Sugimoto, T. Identification of the Kuroshio Extension, its bifurcation and Northern Branch from altimetry and hydrographic data during October 1992–August 1999: Spatial and temporal variability. *Geophys. Res. Lett.* **2001**, *28*, 1759–1762. [CrossRef]
4. Goni, G.J.; Wainer, I. Investigation of the Brazil current front variability from altimeter data. *J. Geophys. Res.* **2001**, *106*, 31117–31128. [CrossRef]
5. Shay, L.K.; Goni, G.J.; Black, P.G. Effects of a warm oceanic feature on Hurricane Opal. *Mon. Weather Rev.* **2000**, *128*, 1366–1383. [CrossRef]
6. Pun, I.F.; Lin, I.I.; Wu, C.R.; Ko, D.S.; Liu, W.T. Validation and application of altimetry derived upper ocean thermal structure in the western North Pacific Ocean for typhoon intensity forecast. *IEEE Trans. Geosci. Remote Sens.* **2007**, *45*, 1616–1630. [CrossRef]
7. Polito, P.S.; Sato, O.T. Patterns of sea surface height and heat storage associated to intraseasonal Rossby waves in the tropics. *J. Geophys. Res.* **2003**, *34*, L09603. [CrossRef]
8. Lentini, C.A.D.; Goni, G.J.; Olson, D.B. Investigation of Brazil current rings in the confluence region. *J. Geophys. Res.* **2006**, *111*. [CrossRef]
9. Momin, I.M.; Sharma, R.; Basu, S. Satellite-derived heat content in the tropical Indian Ocean. *Remote Sens. Lett.* **2011**, *2*, 269–277. [CrossRef]
10. Lin, C.Y.; Ho, C.R.; Zheng, Z.W.; Kuo, N.J. Validation and variation of upper layer thickness in South China Sea from satellite altimeter data. *Sensors* **2008**, *8*, 3802–3818. [CrossRef] [PubMed]
11. Gill, A.E.; Niiler, P.P. The theory of seasonal variability in the ocean. *Deep Sea Res.* **1973**, *20*, 141–177. [CrossRef]
12. Wyrtki, K. Scientific Results of Marine Investigations of the South China Sea and the Gulf of Thailand. In *Physical Oceanography of the Southeast Asian Waters*; Scripps Institution of Oceanography: La Jolla, CA, USA, 1961; Volume 2, p. 195.
13. Qu, T. Upper-Layer Circulation in the South China Sea. *J. Phys. Oceanogr.* **2000**, *30*, 1450–1460. [CrossRef]
14. Liu, Y.; Weisberg, R.H.; Yuan, Y. Patterns of upper layer circulation variability in the South China Sea from satellite altimetry using the self-organizing map. *Acta Oceanol. Sin.* **2008**, *27*, 129–144. [CrossRef]
15. Amiruddin, A.; Haigh, I.; Tsimplis, M.; Calafat, F.; Dangendorf, S. The seasonal cycle and variability of sea level in the South China Sea. *J. Geophys. Res.* **2015**, *120*, 5490–5513. [CrossRef]
16. Cheng, Y.; Hamlington, B.; Plag, H.-P.; Xu, Q. Influence of ENSO on the variation of annual sea level cycle in the South China Sea. *Ocean Eng.* **2016**, *126*, 343–352. [CrossRef]
17. Liu, Z.; Yang, H.; Liu, Q. Regional dynamics of seasonal variability in the South China Sea. *J. Phys. Oceanogr.* **2001**, *31*, 272–284. [CrossRef]
18. Cheng, X.; Qi, Y. On steric and mass-induced contributions to the annual sea-level variations in the South China Sea. *Glob. Planet Chang.* **2010**, *72*, 227–233. [CrossRef]
19. World Ocean Atlas 2013. Available online: http://apdrc.soest.hawaii.edu (accessed on 4 March 2017).

20. Archiving, Validation and Interpretation of Satellite Oceanographic Data. Available online: ftp.aviso. altimetry.fr (accessed on 4 March 2017).
21. Le Traon, P.Y.; Ogor, F. ERS-1/2 orbit improvement using TOPEX/POSEIDON: The 2 cm challenge. *J. Geophys. Res.* **1998**, *103*, 8045–8057. [CrossRef]
22. Le Traon, P.Y.; Nadal, F.; Ducet, N. An improved mapping method of multisatellite altimeter data. *J. Atmos. Ocean. Technol.* **1998**, *15*, 522–534. [CrossRef]
23. US Global Ocean Data Assimilation Experiment. Available online: http://www.usgodae.org/cgi-bin/argo_ select.pl (accessed on 4 March 2017).
24. Liu, Y.; Yuan, Y.; Su, J.; Jiang, J. Circulation in the South China Sea in summer of 1998. *Chin. Sci. Bull.* **2000**, *45*, 1648–1655. [CrossRef]
25. Zeng, L.; Wang, D.; Chen, J.; Wang, W.; Chen, R. SCSPOD14, a South China Sea physical oceanographic dataset derived from in situ measurements during 1919–2014. *Sci. Data* **2016**, *3*, 160029. [CrossRef] [PubMed]
26. Metzger, E.J.; Hurlburt, H. Coupled dynamics of the South China Sea, the Sulu Sea, and the Pacific Ocean. *J. Geophys. Res.* **1996**, *101*, 12331–12352. [CrossRef]
27. Wang, G.; Chen, D.; Su, J. Generation and life cycle of the dipole in the South China Sea summer circulation. *J. Geophys. Res.* **2006**, *111*, C06002. [CrossRef]
28. Zhuang, W.; Qiu, B.; Du, Y. Low-frequency western Pacific Ocean sea level and circulation changes due to the connectivity of the Philippine Archipelago. *J. Geophys. Res.* **2013**, *118*, 6759–6773. [CrossRef]
29. He, Y.; Cai, S.; Wang, D.; He, J. A model study of Luzon cold eddies in the northern South China Sea. *Deep Sea Res. Part I* **2015**, *97*, 107–123. [CrossRef]
30. Gan, J.; Qu, T. Coastal jet separation and associated flow variability in the southwest South China Sea. *Deep Sea Res. Part I* **2008**, *55*, 1–19. [CrossRef]
31. Wu, C.R.; Shaw, P.T.; Chao, S.Y. Assimilating altimetric data into a South China Sea model. *J. Geophys. Res.* **1999**, *104*, 29987–30005. [CrossRef]
32. Chu, X.; Xue, H.; Qi, Y.; Chen, G.; Mao, Q.; Wang, D.; Chai, F. An exceptional anticyclonic eddy in the South China Sea in 2010. *J. Geophys. Res.* **2014**, *119*, 881–897. [CrossRef]
33. Chambers, D.P.; Bonin, J.A. Evaluation of Release 05 time-variable gravity coefficients over the ocean. *Ocean Sci.* **2012**, *8*, 859–868. [CrossRef]
34. Chambers, D.P.; Willis, J.K. A Global Evaluation of Ocean Bottom Pressure from GRACE, OMCT, and Steric-Corrected Altimetry. *J. Atmos. Ocean Technol.* **2010**, *27*, 1395–1402. [CrossRef]

remote sensing

MDPI

Article

3D Geostrophy and Volume Transport in the Southern Ocean

María Isabel Vigo [1,*], David García-García [1], María Dolores Sempere [1] and Ben F. Chao [2]

[1] Department of Applied Mathematics, University of Alicante, E-03080 Alicante, Spain; d.garcia@ua.es
    (D.G.-G.); m.sempere@ua.es (M.D.S.)
[2] Institute of Earth Sciences, Academia Sinica, Taipei 11529, Taiwan, ROC; bfchao@earth.sinica.edu.tw
*   Correspondence: vigo@ua.es

Received: 14 March 2018; Accepted: 3 May 2018; Published: 5 May 2018

Abstract: The 3D geostrophic currents and the associated volume transport (VT) can be estimated from the GOCE and Altimetry satellite data and in-situ temperature and salinity profiles measured by the Argo floats. We do so for the Southern Ocean between 20°S and 65°S with their time variability down to the depth of 1975 m (in 58 layers) over the 11-year period of 2004–2014. The results depict the Southern Ocean circulation where a zonal Antarctic Circumpolar Current (ACC) interacts with a meridional thermohaline circulation. The VT reproduces the polar front and the subantarctic front of the ACC, as well as the large scale and mesoscale currents in the Southern Ocean. Our estimates for the Agulhas current and the East Australia currents are also quantitatively comparable with results from other approaches in the literature based on in-situ data. For ACC, the estimated VT at the Drake Passage is 185 Sv for the norm of the time average VT, or 202 Sv for the mean of the norms of the monthly VT, which are larger than previous estimations (ranging from 134 to 175 Sv). The estimate is potentially reconciled when only the zonal transport is considered (181 Sv). The Drake Passage total VT appears to be quite stable during the studied period, unlike its (dominant) zonal and meridional components which show higher variability that mostly compensate each other. The spatially averaged ACC VT shows per 1° width in the main stream a mean value of 29.6 Sv or 35.8 Sv (depending on the method used), an annual signal with an amplitude of 0.33 ± 0.06 Sv that peaks in early April, with no significant semi-annual signals nor linear trend. Water transports of barotropic and baroclinic origin have been isolated in the VT series showing that 75% of transport is barotropic and the remaining 25% baroclinic, while the variability and annual signal in the ACC is fully barotropic.

Keywords: ocean geostrophy; water volume transport; satellite geodesy; space gravity; altimetry; Argo; Southern Ocean; ACC

---

1. Introduction

Physical forces acting on a body of water in the Navier-Stokes equation of motion include gravity, friction, Coriolis, and pressure gradient. In the oceans, dependent on the forces involved in the particular situation under consideration, different horizontal currents arise when wind stress is exerted on the sea surface. That balanced by the Coriolis force include the geostrophic and the Ekman flows. The geostrophic currents arise when the pressure gradient is balanced by the Coriolis force. It acts to the left (right) to the velocity direction in the southern (northern) hemisphere. The Ekman currents occur when the frictional force plays a significant role; its motion at the surface is 45° to the left (right) of the wind, while the total mass transport including all depths moves 90° to the left (right) on average. Ekman transport moves water mass to a side of the wind, which generates a pressure gradient that ultimately forms a geostrophic flow. Thus the two currents are not completely independent.

This study focuses on the geostrophic currents of the Southern Ocean. Historically, surface currents have been studied from ship drifter data [1], whereas the associated volume transport were inferred from in situ measurements of bottom [2], buoy or moored instruments [3], drifts and current meter records [4], hydrographic measurements [5], shipboard acoustic Doppler current profiler [6], and temperature (T) and salinity (S) climatological profiles [7]. In modern times, remote sensing instruments on board satellites opened new methodologies to observe currents. Wunsch and Gaposchkin [8] estimated the global geostrophic currents from hydrographic and altimetric sea level data, along with a modeled geoid. However, the necessary measurements were not accurate enough until decades later. The required quality of sea surface height (SSH) measurements was obtained in the early 1990s with the launch of the TOPEX/Poseidon (T/P) altimetric satellite [9]. The hydrographic measurements started to become continuous and globally monitored with the deployment of free-drifting profiling floats of the Argo project since 2000 [10]. In the meantime, the space gravity missions as GRACE and GOCE greatly improved the determination of the Earth's geoid [11]. Presently, the combined processing of these new datasets allows the estimation of not only the global ocean surface geostrophic currents but also the three-dimensional (3D) structures with depth.

Cadden et al. [12] obtained promising results by combining climatological T/S profiles and altimetric SSH measurements with a GRACE-derived geoid. Mulet et al. [13] studied the Atlantic Meridional Overturning Circulation by the Wunsch and Gaposchkin's method combining altimetric SSH, sea surface temperature, climatological Argo data, and a GRACE-based geoid. Kosempa and Chambers [14] obtained 3D geostrophic velocity fields combining satellite altimetry, a GRACE-based geoid and Argo T/S profiles. Their results were favorably compared to estimated currents at 1000 dbar based on Argo float trajectories. For the first time, geostrophic volume transport fields of the Southern Ocean have been obtained from their 3D geostrophy, where a significant correlation is evident between the average zonal transport over the entire Indian Ocean basin and the Antarctic Annual Oscillation at low frequencies. In this work we will use a GOCE-derived geoid, which is a notable advancement in precision and resolution over the GRACE based geoids .

The Antarctic Circumpolar Current (ACC) arises as the principal current when studying the geostrophic currents of the Southern Ocean. The mightiest and longest current not blocked by continents and connecting the Pacific, Atlantic and Indian oceans, it consists of two varying main fronts, which from north to south are the Subantarctic Front, and the Polar Front (Figure 1). Closer to Antarctica there is a third and weaker Southern Front. Adjacent to the coast there are two deep-reaching cyclonic gyres located at the Weddell and Ross Seas (Figure 1), and a third gyre between 80°E and 90°E of longitude (not shown in Figure 1). They are considered to bring ACC water near the cold continental coast to form deep waters [15]. It is worth noting that the ACC presents a net loss of heat to the atmosphere, that thermally isolates the cold Antarctic waters from the subtropical warm ones.

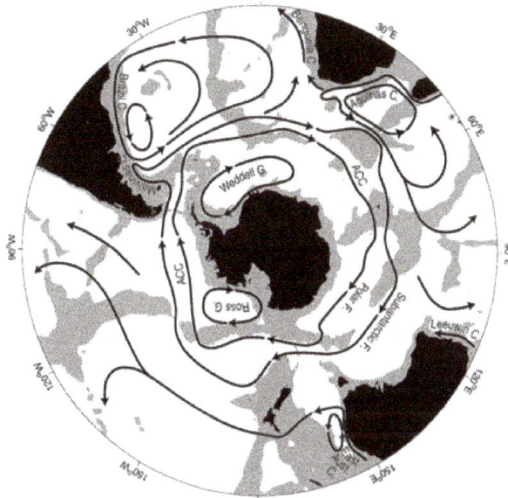

Figure 1. Schematic map of major currents in the southern hemisphere oceans south of 20°S, showing the two major cores of the Antarctic Circumpolar Current (ACC), the Subantarctic Front and the Polar Front. (F is short for front, C for current and G for gyre). Depths shallower than 3500 m are shaded. Reproduced with permission from Rintoul et al. [16].

Without global satellite coverage, the study of the ACC has been limited to the analysis of the currents at two passages. One is located south of Tasmania, where volume transports around 147 sverdrups (Sv = 10^6 m^3/s or 10^9 kg/s of flux) have been reported [17,18]. The second is the Drake Passage to the south of South America. The mean flow there varies between 134 to 140 Sv [2,5,19,20], while recent studies have considerably upped these estimates to 173–175 Sv [3,7].

In this study we shall calculate the 3D geostrophic currents and the associated volume transport (VT) using the SSH from a monthly merged solution from several altimetric missions, a geoid model from the GOCE mission, and the Argo monthly T/S profiles. We shall do so for the Southern Ocean between 20°S and 65°S at 58 depths from near surface to 1975 m depth for the 11-year period of 2004–2014.

2. Methodology

The geostrophic flow in the ocean can be estimated based on the combination of space and in situ data [8]. We do so using altimetric and gravity measurements from satellites, and in situ data from the Argo floats. To this end, we first define the Absolute Dynamic Topography (ADT) and the Relative Dynamic Topography (RDT) as follows:

$$ADT(x,y,t) = SSH(x,y,t) - N(x,y), \tag{1}$$

$$RDT(x,y,z,t) = \frac{1}{g(y)} \int_{P(z)}^{0} \frac{dP}{\rho(x,y,z,t)}, \tag{2}$$

where N represents a time-averaged geoid, x denotes the longitude, y the latitude, z depth, t time, g the latitude-dependent gravitational acceleration, $P(z)$ the pressure at depth z (in Pascal units), and ρ the density. We obtain the density from the ocean T, S, and pressure via the state Equation of Seawater from the Gibbs Seawater Oceanography Toolbox [21].

For the surface geostrophic currents, zonal speed (positive eastward) u_s and the meridional speed (positive northward) v_s follow from the geostrophic equation, i.e., the balance between the pressure gradient force and the Coriolis force at the surface:

$$u_s(x,y,t) \quad = \quad -\frac{g(y)}{f}\frac{\partial ADT}{\partial y}(x,y,t),$$

$$v_s(x,y,t) \quad = \quad \frac{g(y)}{f}\frac{\partial ADT}{\partial x}(x,y,t),$$

(3)

where $f = 2\omega \sin y$ is the Coriolis parameter depending on ω, the angular rate of Earth's rotation. The surface geostrophic current provides the boundary value for the geostrophic current at any depth $z = z_i$ according to:

$$u_s(x,y,t) \quad = \quad -\frac{g(y)}{f}\frac{\partial RDT}{\partial y}(x,y,z_i,t) + u(x,y,z_i,t),$$

$$v_s(x,y,t) \quad = \quad \frac{g(y)}{f}\frac{\partial RDT}{\partial x}(x,y,z_i,t) + v(x,y,z_i,t).$$

(4)

Substitution and expansion of Equations (3) and (4) allow one to calculate the geostrophic current at depth z_i from the geostrophic surface current and the spatial gradients of ADT and RDT (at depth z_i):

$$u(x,y,z_i,t) \quad = \quad -\frac{g(y)}{f}\left(\frac{\partial ADT}{\partial y}(x,y,t) - \frac{\partial RDT}{\partial y}(x,y,z_i,t)\right),$$

$$v(x,y,z_i,t) \quad = \quad \frac{g(y)}{f}\left(\frac{\partial ADT}{\partial x}(x,y,t) - \frac{\partial RDT}{\partial x}(x,y,z_i,t)\right).$$

(5)

Estimating the geostrophic currents at several depths yields the three dimensional geostrophic flow, hereafter referred to as 3D geostrophy.

The volume of water transport at a cell of a regular grid from the surface to a depth D by the geostrophic flow can be estimated integrating vertically the 3D geostrophy from $z = -D$ to $z = 0$, and multiplying the result by the width of the cell perpendicular to the transport as follows:

$$VT_u(x,y,t) \quad = \quad w_{NS} \cdot \int_{-D}^{0} u(x,y,z,t)dz,$$

$$VT_v(x,y,t) \quad = \quad w_{EW}(y) \cdot \int_{-D}^{0} v(x,y,z,t)dz.$$

(6)

where w_{NS} (w_{EW}) is the North-South (East-West) width of the grid cell. Note that in a regular grid w_{EW} depends on latitude y, unlike w_{NS}. Besides, both components of the VT depend on the depth of integration D. Units of volume transport are Sverdrups.

3. Data

3.1. Sea Surface Height

Sea level maps are provided by the CCI-Sea Level Project (http://www.esa-sealevel-cci.org) as a monthly merged solution from several altimetry satellites (Jason 1&2, Topex/Poseidon, Envisat, ERS1&2, GFO) for the time span 1993/01/01 to 2014/12/31, with a spatial resolution of 0.25 degrees. These maps are given as anomalies with respect to the DTU10-MSS mean sea surface where all recommended geophysical and atmospheric corrections are applied. This version of the product is Version 1.1. The product can be found in "ESA Sea Level CCI project team (2016): ESA Sea Level Climate Change Initiative (Sea Level CCI): Time series of gridded Sea Level Anomalies (SLA). Centre for Environmental Data Analysis", downloaded in 06/2016. http://catalogue.ceda.ac.uk/uuid/682e9b455aae4f8fb2275580e8e21f1f (See [22] for further details).

3.2. Mean Sea Surface, Geoid, and Mean Dynamic Topography

A high-resolution mean sea surface model, DTU13MSS, computed by the Danish National Space Center [23] based on two decades of multi-mission satellite altimeters from nine different satellites has been used. By combining it with the geoid EIGEN-6C3, a high resolution global combined gravity field model based on the 4th release of the GOCE Direct Approach (see [24] for details), the DTU13MDT mean dynamic topography model was derived. Note that EIGEN-6C3 is a combined geoid model that also uses GRACE data and to reduce the applied filter it also assimilates surface gravity data based on satellite altimetry (mainly DTU10GRA). Andersen et al. [23] demonstrated that the Mean Dynamic Topography (MDT) DTU13MDT combined with DTU13MSS can be used to derive realistic geostrophic currents comparable to oceanographic derived MDT. Both data sets are available via ftp.space.dtu.dk/DTU13.

3.3. Temperature and Salinity Profiles from Argo Data

We use the Roemmich - Gilson Argo Climatology data [25] (data provided by SCRIPPS available at http://sio-argo.ucsd.edu/RG_Climatology.html, downloaded in April 2018). Data are $1° × 1°$ regular monthly grids at 58 depths from near surface to 1975 m. They are estimated via weighted least-squares fit to the nearest 100 Argo-only T and S profiles. We take the time period 2004–2014 since at 2004 the number of Argo profiling floats reached 1000, and the southern limit of 65°S because that is the Argo spatial coverage limit.

3.4. Simulated Geostrophic Currents from ECCO Model

We shall compare our results against the output of the ECCO Model Version 4 Release 3 [26,27], for the overlapping period 2004–2014. This version of the model assimilates both in situ (T and S profiles, including Argo's), and satellite remote sensing measurements including sea surface height from altimetry missions, sea surface T and S, sea-ice concentration, and ocean bottom pressure from GRACE mission. Further details can be found at https://ecco.jpl.nasa.gov/products/all/. One of the outputs is the simulated surface currents as monthly maps with a spatial resolution of $0.5° × 0.5°$, which are reduced to $1° × 1°$ for comparison with our estimate. To obtain the gestrophic component of the ECCO Surface Currents we subtracted the Ekman component of the Surface Currents as obtained from GEKCO2 product [28]. The Ekman component was computed from remote-sensing altimeter and scatterometer data sets, and is distributed by the CERSAT as daily Ekman Surface Currents maps with a spatial resolution of $0.25° × 0.25°$. Maps are reduced to monthly $1° × 1°$ maps for comparison. In order to derive the RDT at different depths (see Equation (2)) we use ECCO T and S profiles at 37 depths down to 1975 m. From the RDT we derive the 3D geostrophic currents following the methodology of Section 2.

3.5. In-Situ Observations from Drifters

To validate the estimated surface geostrophic currents, their mean will be compared with in-situ measurements provided by drifting buoys. Here we use an annual climatology for the near-surface geostrophic currents from drifter buoys provided by the Global Drifter Program (www.aoml.noaa.gov, Version number: 3.1. Created on: 9 March 2018). A more detailed description of the data can be found in [29,30].

4. Results and Discussion

4.1. 3D Geostrophic Currents

The 3D geostrophic currents are computed according to Equation (5) at 58 depths from near surface to 1975 m depth for the period 2004–2014. We have computed the ADT and the velocities at the surface with a spacial resolution of 0.25°, whereas, for the RDT, the maximum spatial resolution is 1°. Thus, the 3D geostrophy is provided at 1° spatial resolution, but surface currents velocities have been computed at 0.25°. Note, that since the geostrophic velocity is based on the spatial gradient, its determinable speed at each point is affected by the spatial resolution of the data sets. The time-averaged zonal and meridional velocities at different depths are shown in Figures 2 and 3, while Figure 4 shows the mean of geostrophic current speed at the same depths. The latter is estimated as the Euclidean norm of the zonal and meridional components of the current. Note that the blank spots increase by depths due to the bathymetry. Furthermore, in Figure 3 we can discern approximately every three degrees meridional stripes that seem to be an artefact due to some of the processing of the data to produce the ADT. It likely comes from the assimilation of GRACE data in the used geoid, since GRACE has a polar orbit and meridional stripes are characteristic of unfiltered GRACE data (see e.g., [31]). Note that this noise is not observable in the geoid itself but comes out when it is combined with the SHH and we compute the partial derivatives to obtain the surface geostrophic currents. In any case, the noise only affects the meridional component, which in most of the cases is residual with respect to the zonal component in the studied region. This figures clearly illustrate geostrophic currents being strongest near the surface, whose strength decays with depth. The Subantartic and Polar fronts are clearly identified, showing a primary eastward component. The alternation of northward and southward components of the currents reveals the presence of eddies. Considering the 'ACC region' as contoured in Section 4.3 (see further details below), we obtain a mean (latitude weighted) speed at the region at different depths of 22.4 ± 0.2 cm/s (2.5 m), 19.7 ± 0.1 cm/s (500 m), 17 ± 0.1 cm/s (1000 m), 15.2 ± 0.1 cm/s (1500 m), and 14 ± 0.1 cm/s (1975 m). A mean decrease from the surface to 1975 m of 3.8 cm/s is observed, but in areas of maximum intensity this decrease reaches values over 25 cm/s, from which more than half (around 13 cm/s) takes place within the first 500 m depth, in both zonal and meridional components. The ACC also extends bellow the surface with decreasing strength, nevertheless it still reaches values as large as 30 cm/s at 1975 m, which is in good agreement with previous studies [32]. Note that besides ACC, the 3D geostrophy estimated here agrees well with the general current pattern of the Southern Ocean depicted in Figure 1, showing comparable current systems. For example, the Brazil-Malvinas Confluence region in the southwestern Atlantic basin can be observed with values ranging from 30 cm/s to 15 cm/s at 1975 m. The Agulhas Current, which is the warm shallow western boundary current of the southwest Indian Ocean, and the Benghela Current, which is the cold eastern boundary current of the southeast Atlantic Ocean, are very clear at all depths: a westward current can be seen from south Madagascar, becoming south-westward along the east coast of South Africa, and then moving eastward leaving the coast, while a west-northern current transports cold water from Antartica along the western South Africa coast. The northern part of the Ross Sea Gyre between 180°E and 120°W can also be recognized.

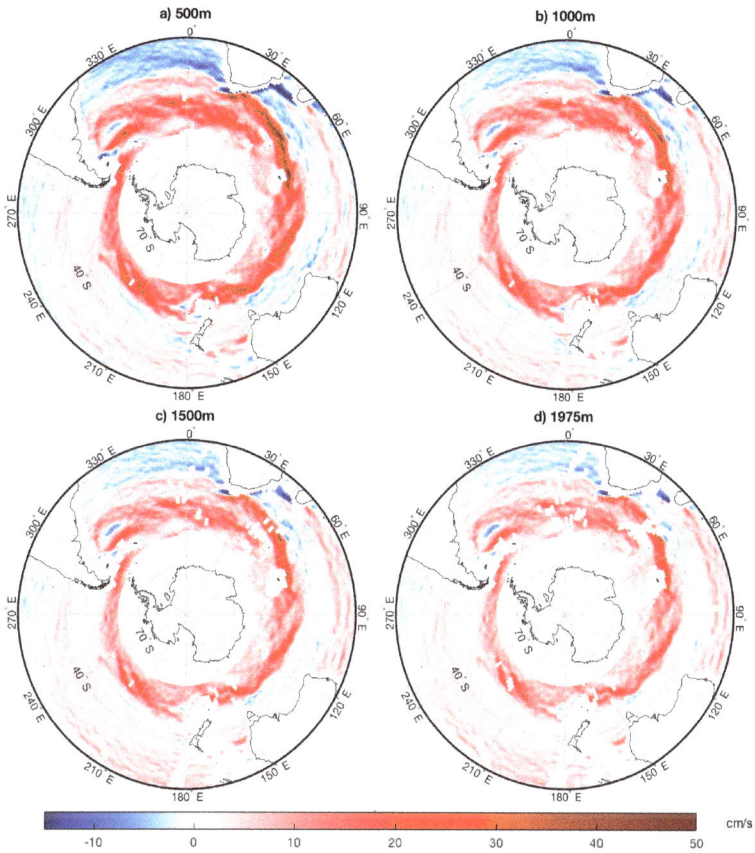

Figure 2. 2004–2014 Mean zonal geostrophic current at depths of: (**a**) 500 m, (**b**) 1000 m, (**c**) 1500 m, and (**d**) 1975 m. Positive (red) values represent eastward direction while negative (blue) westward. The zonal geostrophic currents reach values up to 60 cm/s (the color scale is saturated at 50 cm/s to better resolve the interrelation between different depths).

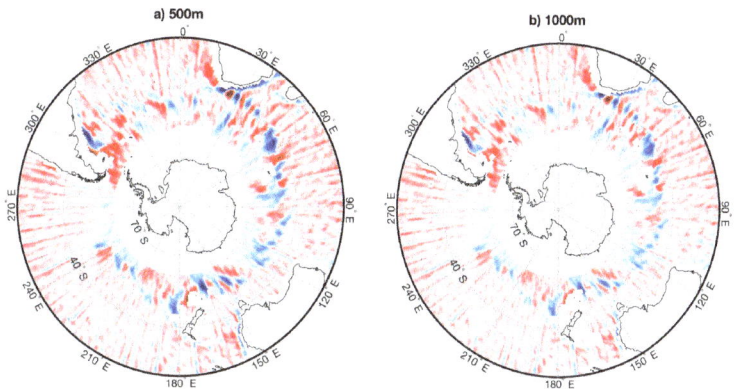

Figure 3. *Cont.*

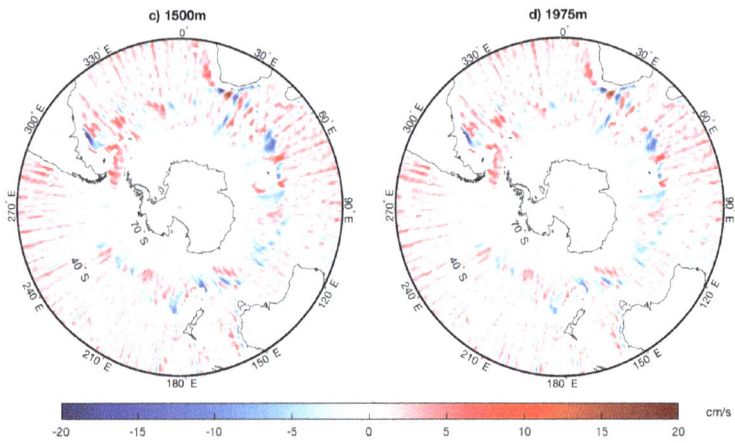

Figure 3. Same as Figure 2, but for the mean meridional Geostrophic current (color saturated at 20 cm/s).

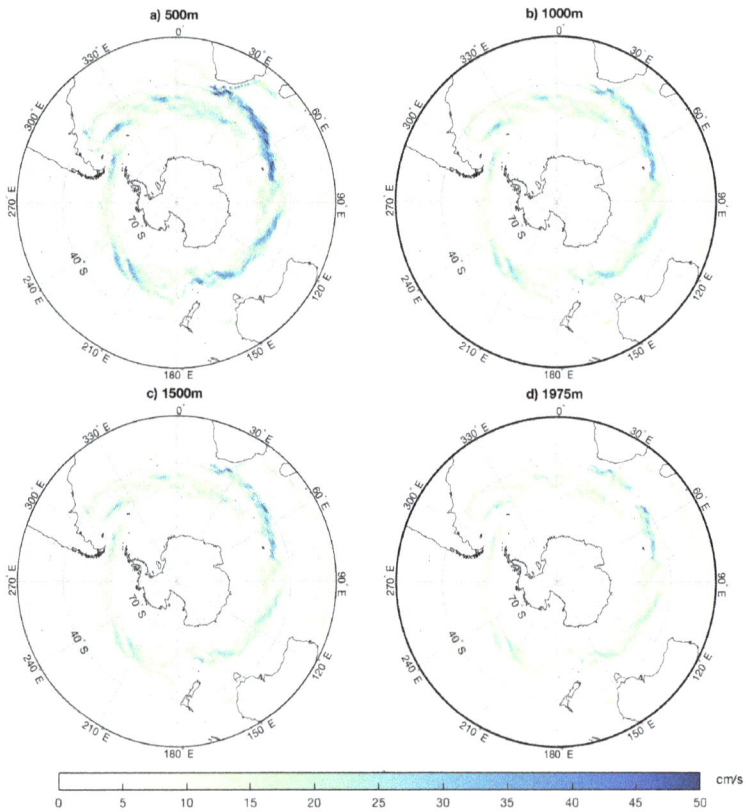

Figure 4. Same as Figure 2, but for the mean speed of the geostrophic current (color saturated at 65 cm/s).

4.2. Comparison with Simulated ECCO geoStrophic Currents and In-Situ Surface Currents Observations

We compare the 3D geostrophic currents derived from GOCE, altimetry, and Argo data for the region [65°S, 20°S] with the 3D Geostrophic currents from the ocean circulation model ECCO that assimilates in situ and satellite measurements (see Section 3.4). Note that, unlike ECCO, in this study we use a GOCE geoid which is independent from altimetry. For validation purposes, we also compare the two products with in situ observations derived from satellite-tracked surface drifting buoy observations [29] from which we remove the Eckam component in the same way as before (see Section 3.4). We will limit our validation to mean surface currents as only the time averaged velocities are available although the drifters currents uses data for the period 1979–2015.

Figure 5 shows the mean speed of the surface geostrophic currents from the three data sets: this study (GOCE and altimetry based), ECCO products, and Drifters. Our results and drifters show a remarkable good agreement whereas the ECCO value over the Southern Ocean presents much lower speeds (around 50% lower, note the different ranges of the color bar for Figure 5b). Considering the complex representation of the velocities $u + iv$, where u and v are the zonal and meridional components, we can obtain the complex correlation coefficient as described by Kundu [33] whose magnitude gives the overall measure of correlation and whose phase gives the average relative displacement or veering between the two velocity maps. For the velocities from this study and the drifters, the complex correlation coefficient is 0.78 with a mean angular offset of veering of 3°. While for the surface currents from ECCO and drifters, the complex correlation coefficient is slightly lower, at 0.69, though with a smaller relative shift of 0.9°, but notice that the angular offset of veering is more significant the higher the correlation is. Quantitatively our study suggests that the ECCO based approach significantly under-estimates the current variabilities while our approach produces a more realistic estimate closer to the direct drifter observations at least for the mean values. The two indirect methods use similar altimetric data, thus the improvement in our study illustrates the utility of an independent GOCE-derived geoid.

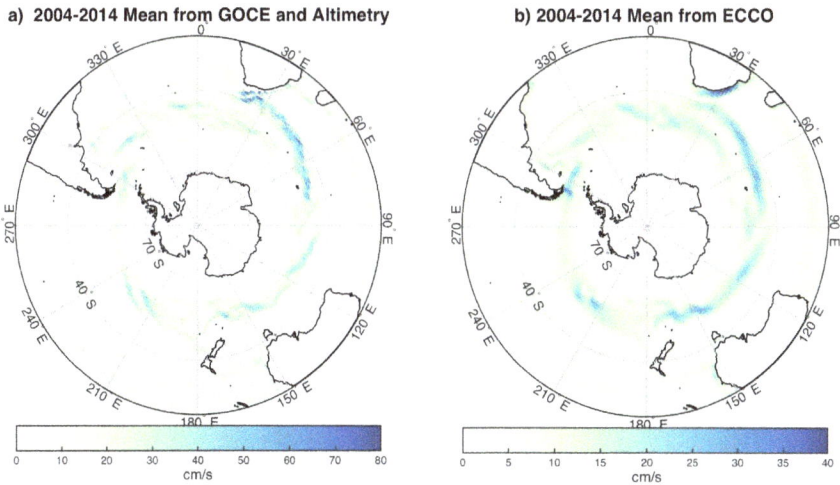

a) 2004-2014 Mean from GOCE and Altimetry   b) 2004-2014 Mean from ECCO

Figure 5. *Cont.*

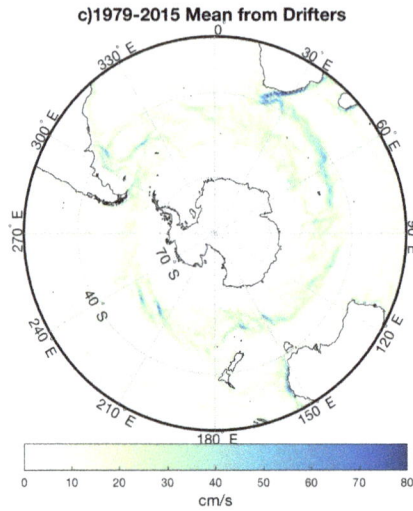

c)1979-2015 Mean from Drifters

Figure 5. Mean surface geostrophic currents from (**a**) this study (GOCE and altimetry based), (**b**) ECCO, and (**c**) drifters. Note the different ranges of the color bar.

Figure 6 shows the mean speed variations from surface down to 1975 m depth for the 3D geostrophic currents from this study (GOCE, Altimetry, and Argo based), compared to those from ECCO products. Again the patterns are quite similar, but with lower intensities in the ECCO about 30%; the latter produces a lower intensity variation also by depths. Figure 6a also shows higher values in certain regions of the ACC, particularly south of Africa. Note that these variations with depth with respect to the surface depend on the RDT and not on the reference surface current.

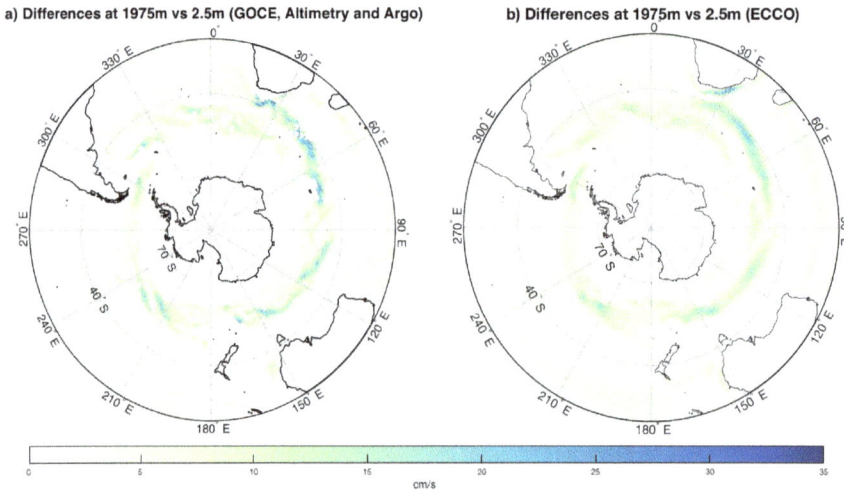

a) Differences at 1975m vs 2.5m (GOCE, Altimetry and Argo)
b) Differences at 1975m vs 2.5m (ECCO)

Figure 6. Differences of mean speed from geostrophic currents at 1975 m vs. 2.5 m of depth: (**a**) this study (GOCE, Altimetry and Argo based) ; (**b**) from ECCO model.

For comparison, Figure 7 shows the time series of the zonal geostrophic current speeds at 1000 m depth at two different locations characterized by very different current intensities. The first location is

(40°S, 138°E), south of Australia and north of the subtropical front, where zonal currents are small, and shows a mean zonal speed of 0.1 ± 0.2 cm/s, with standard deviation of 2.4 cm/s. The second location is (51°S, 148°E), south of Tasmania close to the polar front, where the currents are expected to be large. The mean zonal speed there is estimated to be 12 ± 0.8 cm/s with a standard deviation of 8.8 cm/s. The zonal speed at these two locations has also been studied by [14], following a similar methodology but using a GRACE derived geoid. For the overlapping time period (2004–2012), the time series variability is similar to that reported by [14], but intensities obtained by our approach at the second location are considerably larger. That is in agreement with previous studies (e.g., [34–38]) that reported how GOCE geoid enhances the estimate of geostrophic circulation over the GRACE geoid in providing more detailed patterns and higher intensities closer to in-situ observations. The dotted lines are the zonal speed at 1000 m as estimated from the ECCO model, which again under-estimates it.

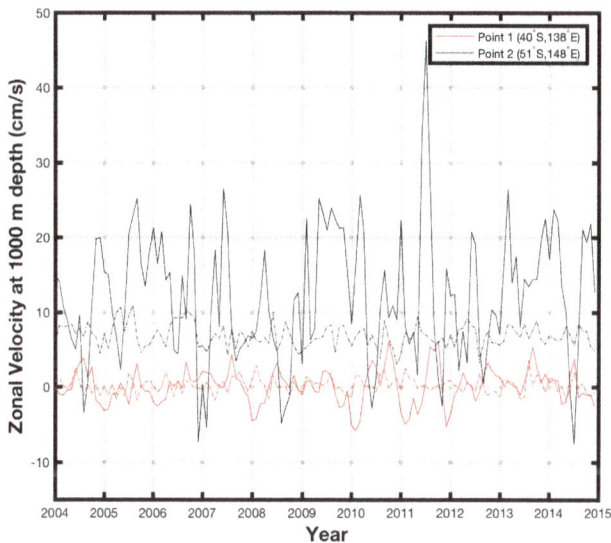

Figure 7. Time series of zonal speed at 1000 m depth at two localities as a function of time (2004–2014): Red curve for (40°S, 138°E) in South Australia (Point1), and black curve for (51°S, 148°E) south of Tasmania (Point 2). Solid lines are our results, dotted lines are those from the ECCO Model.

4.3. Volume Transports

The 3D integrated zonal and meridional VT are evaluated according to Equation 6 (see Section 2), with the limit of integration going from 1975 m depth to the surface. Following [39] we define the barotropic transport as the portion of VT due to a water column moving uniformly as fast as the bottom current, and baroclinic transport as the part of VT with respect to the barotropic transport, so that the geostrophic VT is the sum of the two. The time averaged VT for the period 2004–2014 is shown in Figure 8 in two different ways. Arrows in Figure 8a are the mean VT vectors, while the colored background shows their Euclidean norm, where the color scale is saturated at 80 Sv (maximum values can reach 129 Sv) for better visualization. On the other hand, Figure 8b gives the mean of the monthly VT vector norms map. In this case the maximum values reach 141 Sv, and the color scale saturated at 100 Sv. The 'ACC region' is defined as the grid points in Figure 8b with values above 20 Sv (some isolated points, likely outliers, have been edited out).

As expected, the VT patterns are similar to those from the geostrophic currents. In particular, ACC can be clearly identified as the dominant current. The eastward component is driving the transport, with some alternating north-south departures produced by eddies. In some areas, as in the south of Africa, the subantarctic and polar fronts can be discerned. The mean of the monthly VT vector norms in 1° grid cell width reaches values from 10.5 Sv at a few points to 141 Sv, with a mean value 35.8 ± 0.5 Sv. On the other hand, the norm of the mean vectors is 29.6 ± 0.4 Sv per 1° grid cell width, from which 22.6 ± 0.3 Sv is of barotropic origin and 7.1 ± 0.1 Sv is baroclinic transport. Hereafter, we will call the Euclidean norm of the mean VT vectors the total VT. Figure 9 shows the time evolution in the ACC region of the zonal, meridional, and total VT, for both the barotropic and the baroclinic components and their sum. Note that these values refer to the VT per 1° grid cell. So, for any section along a meridian, the mean zonal VT at a given time can be estimated multiplying the zonal value at Figure 9 at such time by the number of grid points at the given section. A time average of the volume transport at those sections can be found later in this section. Besides, as the mean number of grid points per meridional section is 10, it can be inferred that the mean VT for this kind of section is 296 Sv. The total VT as shown in Figure 9 is indistinguishable from the zonal component. Approximatively 75% of the mean VT is barotropic, while the remaining 25% is baroclinic. However, the variability of the VT is entirely driven by the barotropic transport. In particular, the VT (and the barotropic component) shows an annual signal of 0.33 ± 0.06 Sv fully driven by the barotropic transport, reaching the maximum on about 8 April. The annual signal explains a 32% of the standard deviation of the signal for total VT. The semiannual signal and the linear trends are not significantly different from zero. Temporally, the total VT and the total barotropic transport show an oscillation with a period of 2–3 years. On the other hand, the mean meridional VT in the ACC (see Figure 9b) shows a mean value of -0.6 ± 0.2 Sv, which means a southward transport. Around 70% of the meridional mean ACC transport is barotropic and 30% is baroclinic, its variability driven by the barotropic transport. None of them display significant annual, semiannual or linear trends, although a decadal oscillation seems to be discernible.

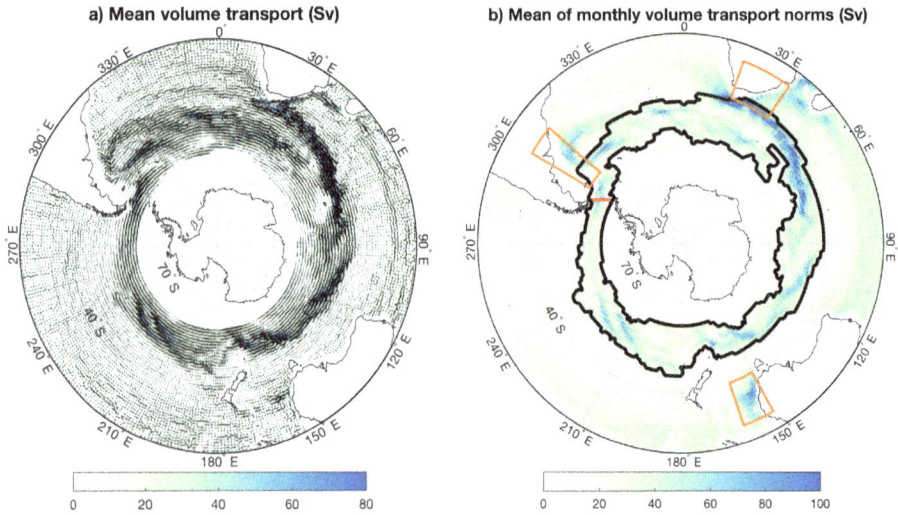

Figure 8. 2004–2014 mean geostrophic volume transport: (**a**) Arrows are the mean vectors, and the color represents their norms. (The color scale is saturated at 80 Sv, maximum values can reach 129 Sv); (**b**) Mean of the monthly vector norms. The ACC region is outlined in black, the Drake Passage in red, and the regions considered for the Agulhas, Brazil-Malvinas and East Australia currents are outlined in orange. The color scale is saturated at 100 Sv, maximum values can reach 141 Sv. Units are Sv.

Of great interest in the ACC is the Drake Passage (red line in Figure 8b), the narrowest part of the current, which is also one of the most studied areas in the Southern Ocean with a large amount of in situ data. The VT in the Drake Passage is estimated by simply adding the VT vectors at the

Drake Passage grid points (red line in Figure 8b, a total of 6 grid cells). Our estimation considering the mean of the monthly VT vector norms is 202.7 Sv, with a standard deviation of 32.4, while considering the total VT (norm) of the mean vector is 185.2 Sv. This estimate is quite a bit larger than those reported in the literature, which range from 134 to 141 Sv (e.g., [5,19,20]), and more recently between 173.3 ± 11 Sv [3] and 175 ± 5 Sv [7]. The differences are apparently reduced when only the zonal component is considered, which shows a mean value of 181.5 Sv with a standard deviation of 43.9 Sv. The origin of this discrepancy is not clear, and will be discussed later. Figure 10a shows the temporal variation of the total VT through the Drake Passage. It shows a big discrepancy from the mean total VT of the ACC (Figure 9); the latter shows an annual signal that is absent in the Drake Passage, whose interannual variability shown by the 12-month running means is not similar either. The total VT at the Drake Passage is quite stable during the period; only a slight increase of around 15 Sv is shown between 2011–2012. Yet the signal varies from year to year; for example, the mean signal is 193 Sv in 2007, and 220 Sv in 2011. In any case, the total VT at Drake Passage is quite stable while considerable variability can be observed in the (dominant) zonal and (relatively small) meridional components (Figure 10a). The meridional component is much smaller at a mean value of 36.5 Sv and a standard deviation of 22.2 Sv. The two components' variability are highly correlated with a (negative) coefficient of −0.86, as can be seen in Figure 10b. They largely add to each other to form a relatively steady total VT. When the total VT is decomposed in the perpendicular and parallel directions of the Drake Passage, the mentioned variability exists only in the parallel direction, which could be interpreted as a small oscillation on the main direction of the total VT.

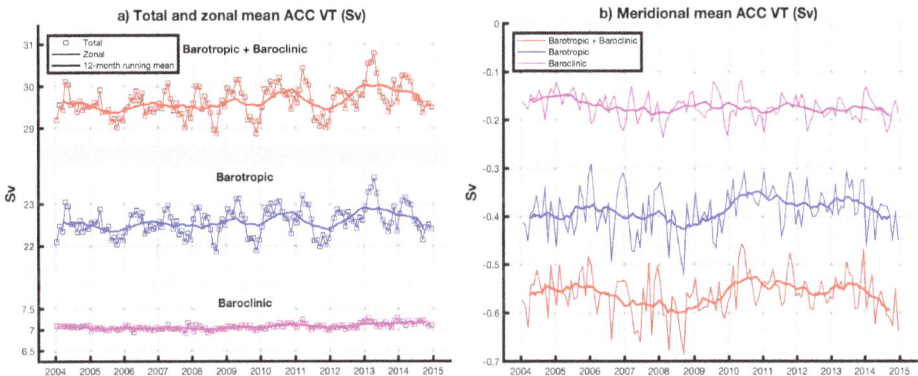

Figure 9. Volume transport in the ACC region per 1° grid cell: (**a**) zonal (thin curves) and total (squares) VT; (**b**) mean meridional (thin curves) VT. Barotropic transport in blue, baroclinic transport in magenta, and their sum (barotropic+baroclinic) in red. Thick curves are 12-months running means. Note that these values refer to the VT per unit cross-section area. Units are Sv.

The barotropic and baroclinic components of the VT at Drake Passage is shown in Figure 10c,d. The former accounts for about 80% of the VT, while the latter accounts for the remaining 20%. As in the ACC, the variability of VT at Drake Passage is entirely related to barotropic transport. Note that the VT and the barotropic transport show similar standard deviations for all components, and a significant correlation coefficient of 0.98. The baroclinic transport shows standard deviations 10 times smaller, and correlation coefficients between 0.3 and 0.4.

Other major current systems apart from the ACC are clearly observed in Figure 8b. Three of them are outlined in the orange box in the map: the Brazil-Malvinas current ([53.5°S, 30.5°S] × [300.5°E, 310.5°E]), the Agulhas current ([38.5°S, 25.5°S] × [19.5°E, 35.5°E]), and the East Australian current [40.5°S, 25.5°S] × [150.5°E, 160.5°E]. In all cases the currents are selected as the grid points with speeds above 20 Sv. The Brazil-Malvinas current shows a mean of the VT vector norms of 39.4 ± 0.7 Sv, and a total VT of 23.3 ± 0.8 Sv. The Agulhas current shows a mean of the VT vector norms of 60.1 ± 2.6 Sv, and a total VT of 35.5 ± 2.8 Sv. In the southern part of the Agulhas system, near the coast in the region delimited by 31.5°S–32.5°S, and 30.5°E–31.5°E the water flows southward along the coast, and we have a southward meridional transport of 68.9 Sv, which is close to the 69.7 Sv reported in [40] or the 76 Sv in [41]. Note the differences with both approaches: results in Bryden et al. [40] are based on one year data (1995) from an array of six moorings, which are extrapolated to the whole area and integrated from 0 to 2400 m depth, and in Donohue et al. [41] are based on data from only the first half of 1995, while our results combine 11 years of satellite data (not includeding 1995) in 1° × 1° fields integrated from 0 to 1975 m. The East Australian current shows a mean of the VT vector norms of 47.9 ± 1.7 Sv, and a total VT of 13.4 ± 0.8 Sv. In the smaller region located at 27°S from 153.4–155°E a southward transport of 15.8 ± 10.5 Sv above 2000 m was reported by Sloyan et al. [42], based on data from a mooring array between April 2012 and August 2013. Since the time period overlaps with ours, we computed the mean meridional VT for the same region and time period, and we obtain a southward VT of 14.7 ± 7 Sv. The two studies yield very similar VT estimates, which is remarkable keeping in mind the different nature of the data sets. The agreements corroborate that the estimated VT is robust for the Agulhas and East Australian currents.

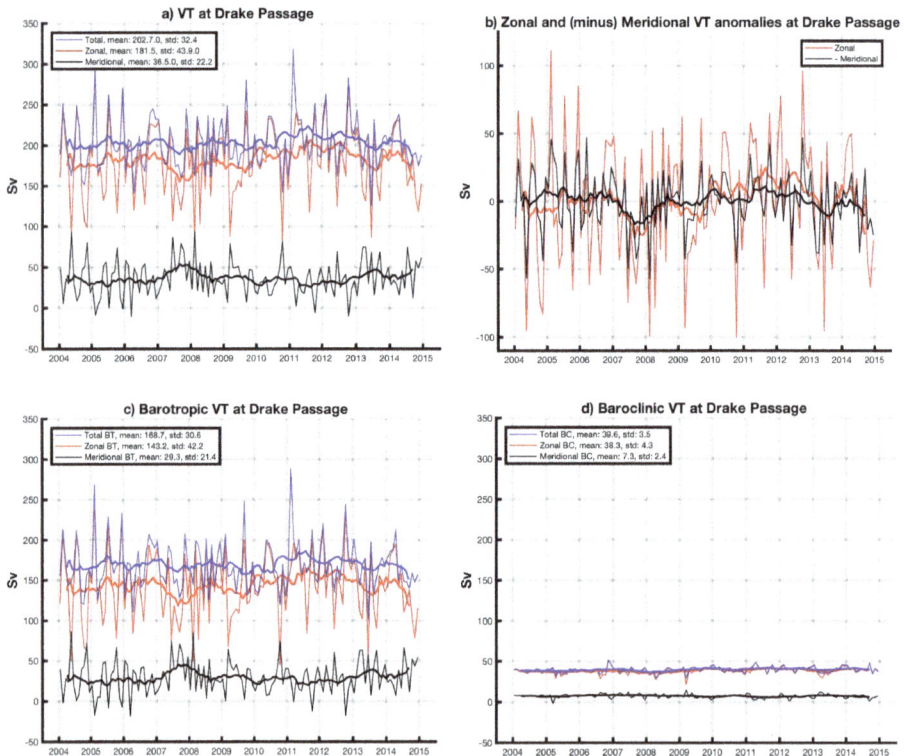

Figure 10. VT at Drake Passage. (**a**) Mean zonal (red), meridional (black), and total VT (blue). (**b**) Anomalies with respect to the mean of zonal and negative meridional signals in (**a**). (**c**) Same as (**a**) but for the barotropic transport. (**d**) Same as (**a**) but for the baroclinic transport. Thick curves are 12-months running means. The sign of the meridional component is reversed to better show its correlation with the zonal component.

In Figure 11, by time averaging and integration latitudinally for the whole region (65°S–20°S, Figure 11a), and the ACC region (Figure 11b), we obtain a longitudinal series of VT and the zonal and meridional components for the Southern Ocean and the ACC. Dotted curves represent the same analysis for barotropic VT. The VT in both regions is clearly dominated by its zonal component, and it is mainly due to barotropic transport. We can clearly identify the three more prominent peaks in the zonal and total VT in Figure 11a with the major local currents, and those roughly delimit the three ocean basins. Around 30°E we find the first peak corresponding to the Agulhas current, from there we are in the Indian Ocean, where we can observe the highest VT until around 115°E. The Australian continent goes from that point to 150°E. In that section the Southern Ocean narrows and the transport remains lower. Then, VT suddenly picks up at the location of the East Australia current around 150°E. From that point we are in the Pacific Ocean until around 285°E, where the Southern Ocean narrows even more at Drake Passage, and that is more or less where the Pacific Ocean connects to the Atlantic Ocean. Around 300°E the peak observed corresponds to the Malvinas current, and the Atlantic Ocean goes from there to 30°E. As expected, at each basin, the VT transport decays from west to east. Thus, the three prominent minima in the zonal and total VT correspond to three choke points of the ACC: the south of Australia between 120°E and 150°E, New Zealand between 170°E and 180°E, and the Drake Passage between 280°E and 300°E. The longitudinal structure of the ACC is similar to that of the whole region, although less pronounced, with the exception of the maximum VT south of Africa, where an almost maximum meridional VT occurs from north to south (Figures 11b and 8a). In any case, it seems that the zonal VT drives the low frequency (in latitude) variability of the total VT, and the meridional VT drives the high frequency (also in latitude). This high variability in the meridional component is due to eddies structures.

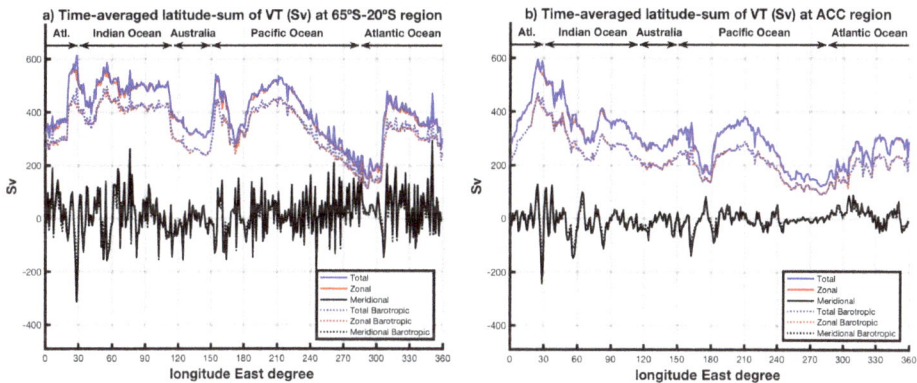

Figure 11. The time-averaged total VT (solid blue curve) and the zonal (solid red curve) and meridional (solid black curve) components of VT. (**a**) Whole region (65°S to 20°S), and (**b**) ACC region (as outlined in Figure 8b). Dotted curves are the same but for barotropic transport.

5. Conclusions

The 3D geostrophic currents and the associated volume transport in the Southern Ocean can be estimated based on data from satellites (altimetric SSH, and a GOCE-based geoid) and Argo floats. The use of an independent geoid from the altimetric SSH is of paramount importance for to better ocean circulations, and the utility of a geoid derived from GOCE is a notable advancement in precision and resolution over those based on the GRACE geoid in previous studies [12–14]. We directly combine the enhanced knowledge of the surface circulation provided by the latest release of GOCE data in combination with altimetry, with in-situ observations of the sub-surface Argo floats to obtain more accurate estimates of the 3D sub-surface geostrophic currents. The results are given as monthly maps from the surface down to 1975 m depth containing 58 layers. The analyzed region comprises the Southern ocean between 20°S and 65°S, and the time period is 2004–2014. The obtained results clearly depict the currents structure in the Southern Ocean. They are quantitatively compared with previous studies, but it is unable to reproduce the Ross and Weddel Gyres due to the limitation of the Argo data. Surface currents are in remarkably good agreement with those from the drifter observations. Comparisons with output from the ECCO model that also assimilates altimetry SSH and Argo data (among others) but that lacks an independent geoid, evidences that the surface geostrophic currents and their decrease in strength by depths, are greatly underestimated in ECCO.

From the estimated 3D geostrophy we obtain the VT that we analyze for the whole region, and in particular for the ACC, showing quite similar behaviour in terms of variability and components contribution. Global estimates of the VT allow us to analyze the ACC as a whole, which is not possible when studying a section of the ACC as the Drake Passage. For the ACC, the signal shows a mean value for each 1° cell width, depending on the method used, of 29.6 (the norm of mean VT) or 35.8 Sv (the mean of VT norms). The baroclinic and barotropic components of the VT are such that 75% of the total transport is barotropic which supports the importance of an independent precise geoid in its determination. Nevertheless, in this study we show the dominance of the barotropic component for the ACC which does not necessary hold in other current systems. The ACC VT presents an annual signal with 0.33 ± 0.06 Sv amplitude that peaks in early April ($97° \pm 11°$), and no significant semi-annual signals or linear trend. The annual signal is of barotropic origin, while the baroclinic annual signal is essentially absent. At Drake Passage, our mean geostrophic VT estimate is 181 Sv for the zonal component, 185 Sv for the norm of the time averaged VT vector, and 202 Sv for the mean of the norms of the monthly VT. These estimates are quite a bit larger than the usual mean values reported in the literature. The methodology and data used being different, assessing possible systematic errors in the VT estimation awaits further research. However, the agreement observed in the Agulhas and the East Australia currents suggests that an overestimation, if it existed, would not have a uniform effect everywhere. Griesel et al. [43] evaluated four different MDT products for the ACC transport, that make use of GRACE satellite gravity for the geoid. They concluded that they all imply an unrealistic lack of mass conservation, which would lead to an overestimation for the ACC transport. As such, the effect of accounting for an ocean mass balance constraint requires further research. Understanding these discrepancies will help to improve the future methodologies. The total VT at the Drake Passage is relatively stable during the studied period, unlike its zonal and meridional components that show variabilities that cancel each other in the total VT, implying that the transport vector oscillates around its mean direction, remaining quasi constant in magnitude.

In summary, the combination of altimetry, space gravimetry missions, and in situ Argo measurements, which are in a state of maturity, has confirmed their capability to yield realistic geostrophic current and VT estimates up to 1975 m depth. This new approach enables the possibility of new studies in oceanography at global scales.

Author Contributions: M.I.V. conceived and designed the study. M.I.V. and D.G.-G. processed and analyzed the data. M.D.S. processed all data to obtain the 3D geostrophy and VT under the supervision of M.I.V. and B.F.C. in an earlier approach of the work. All authors were involved in the analysis of the results and contributed to the discussion of results and the preparation of the manuscript.

Funding: This research was funded by Taiwan MoST grant number106-2116-M-001-013.

Acknowledgments: We thank the three anonymous reviewers whose constructive comments/suggestions helped improve and clarify this manuscript. We acknowledge the support of all data providers: ESA in the frame of the CCI Sea Level Project for Altimetry data; DTU SPACE from the Danish National Space Center for MDT and MSS products; SCRIPPS for Roemmich and Gilson Temperature and Salinity profiles from ARGO data; Global Drifter Program for the near-surface geostrophic currents from drifter buoys; LEGOS and CERSAT for the Ekman Surface Currents. M.D.S. is supported by the phD grant UAFPU2014-5884 from the University of Alicante.

Conflicts of Interest: The authors declare no conflict of interest.

Remote Sens. **2018**, *10*, 715

References

1. Richardson, P.L. Worldwide ship drift distributions identify missing data. *J. Geophys. Res.* **1989**, *94*, 6169–6176. [CrossRef]
2. Whitworth, T.; Peterson, R.G. Volume Transport of the Antarctic Circumpolar Current from Bottom Pressure Measurements. *Am. Meteorol. Soc.* **1984**, *15*, 810–816. [CrossRef]
3. Donohue, K.A.; Tracey, K.L.; Watts, D.R.; Chidichimo, M.P.; Chereskin, T.K. Mean Antarctic Circumpolar Current transport measured in Drake Passage. *Geophys. Res. Lett.* **2016**, *43*, 11760–11767. [CrossRef]
4. Frankignoul, C.; Bonjean, F.; Reverdin, G. Interannual variability of surface currents in the tropical Pacific during 1987–1993. *J. Geophys. Res.* **1996**, *101*, 3629–3647. [CrossRef]
5. Cunningham, S.A.; Alderson, S.G.; King, B.A.; Brandon, M.A. Transport and variability of the Antarctic Circumpolar Current in Drake Passage. *J. Geophys. Res.* **2003**, *108*, 8084. [CrossRef]
6. Firing, Y.; Chereskin, T.K.; Mazloff, M.R. Vertical structure and transport of the Antarctic Circumpolar Current in Drake Passage from direct velocity measurements. *J. Geophys. Res.* **2011**, *116*, C08015. [CrossRef]
7. Colin de Verdière, A.; Ollitrault, M. A direct determination of the World Ocean barotropic circulation. *J. Phys. Oceanogr.* **2016**, *46*, 255–273. [CrossRef]
8. Wunsch, C.; Gaposchkin, E.M. On using satellite altimetry to determine the general circulation of the oceans with application to geoid improvement. *Rev. Geophys.* **1980**, *18*, 725–745. [CrossRef]
9. Fu, L.L.; Cazenave, A. (Eds.) *Satellite Altimetry and Earth Science*; Academic Press: New York, NY, USA, 2001; p. 463.
10. Roemmich, D.; Owens, W.B. The Argo Project: Global ocean observations for the understanding and prediction of climate variability. *Oceanography* **2000**, *13*, 45–50. [CrossRef]
11. Rummel, R.; Balmino, G.; Johannessen, J.; Visser, P.; Woodworth, P. Dedicated gravity field missions—Principles and aims. *J. Geodyn.* **2002**, *33*, 3–20. [CrossRef]
12. Cadden, D.D.H.; Subrahmanyam, B.; Chambers, D.P.; Murty, V.S.N. Surface and subsurface geostrophic current variability in the Indian Ocean from altimetry. *Mar. Geod.* **2009**, *32*, 19–29. [CrossRef]
13. Mulet, S.; Rio, M.H.; Mignot, A.; Guineut, S.; Morrow, R. A new estimate of the global 3D geostrophic ocean circulation based on satellite data and in-situ measurements. *Deep Sea Res. II* **2012**, *77–80*, 70–81. [CrossRef]
14. Kosempa, M.; Chambers, D.P. Southern Ocean Velocity and Geostrophic transport fields estimated combining Jason altimetry and ARGO data. *J. Geophys. Res. Oceans* **2014**, *119*, 4761–4776. [CrossRef]
15. Carter, L.; McCave, I.N.; Williams, M.J.M. Circulation and water masses of the Southern Ocean: A review. *Dev. Earth Environ. Sci.* **2008**, *8*, 85–114. [CrossRef]
16. Rintoul, S.R.; Hughes, C.W.; Olbers, A. The Antarctic Circumpolar Current system. In *Ocean Circulation and Climate: Observing and Modelling the Global Ocean*; Siedler, G., Church, J., Gould, J., Eds.; International Geophysics 77; Academic Press: New York, NY, USA, 2001; pp. 271–302, ISBN 978-0-12-641351-9.
17. Rintoul, S.R.; Sokolov, S. Baroclinic transport variability of the Antarctic Circumpolar Current south of Australia (WOCE repeat section SR3). *J. Geophys. Res.* **2001**, *106*, 2815–2832. [CrossRef]
18. Rintoul, S.R.; Sokolov, S.; Church, J. A 6 year record of baroclinic transport variability of the Antarctic Circumpolar Current at 140° E derived from expendable bathythermograph and altimeter measurements. *J. Geophys. Res.* **2002**, *107*, 3155. [CrossRef]
19. Ganachaud, A.; Wunsch, C. Improved estimates of global ocean circulation, heat transport and mixing from hydrographic data. *Nature* **2000**, *408*, 453–457. [CrossRef] [PubMed]
20. Koenig, Z.; Provost, C.; Ferrari, R.; Sennéchael, N.; Rio, M.H. Volume transport of the Antarctic Circumpolar Current: Production and validation of a 20 year long time series obtained from in situ and satellite observations. *J. Geophys. Res. Oceans* **2014**, *119*, 5407–5433. [CrossRef]
21. McDougall, T.J.; Barker, P.M. Getting started with TEOS-10 and the Gibbs Seawater (GSW) Oceanographic 423 Toolbox, version 3.0 (R2010a). *SCOR/IAPSO WG* **2011**, *127*, 1–28; ISBN 978-0-646-55621-5.
22. Ablain, M.; Cazenave, A.; Larnicol, G.; Balmaseda, M.; Cipollini, P.; Faugére, Y.; Fernandes, M.J.; Henry, O.; Johannessen, J.A.; Knudsen, P.; et al. Improved sea level record over the satellite altimetry era (1993–2010) from the Climate Change Initiative project. *Ocean Sci.* **2015**, *11*, 67–82. [CrossRef]
23. Andersen, O.; Knudsen, P.; Stenseng, L. The DTU13 MSS (Mean Sea Surface) and MDT (Mean Dynamic Topography) from 20 years of Satellite Altimetry. In *IAG Symposia*; Springer: Cham, Switzerlands, 2015; doi:10.1007/1345_2015_182.

24. Förste, C.; Bruinsma, S.; Shako, R.; Marty, J.-C.; Flechtner, F.; Abrikosov, O.; Dahle, C.; Lemoine, J.-M.; Neumayer, K.H.; Biancale, R.; et al. EIGEN-6—A new combined global gravity field model including GOCE data from the collaboration of GFZ-Potsdam and GRGS-Toulouse. In Proceedings of the EGU General Assembly, Vienna, Austria, 3–8 April 2011.
25. Roemmich, D.; Gilson, J. The 2004–2008 mean and annual cycle of temperature, salinity, and steric height in the global ocean from the Argo Program. *Prog. Oceanogr.* **2009**, *82*, 81–100. [CrossRef]
26. Fukumori, I.; Wang, O.; Fenty, I.; Forget, G.; Heimbach, P.; Ponte, R.M. ECCO Version 4 Release 3. 2017, Available online: ftp://ecco.jpl.nasa.gov/Version4/Release3/doc/v4r3_estimation_synopsis.pdf (accessed on April 2018).
27. Forget, G.; Campin, J.-M.; Heimbach, P.; Hill, C.N.; Ponte, R.M.; Wunsch, C. ECCO version 4: An integrated framework for non-linear inverse modeling and global ocean state estimation. *Geosci. Model Dev.* **2015**, *8*, 3071–3104. [CrossRef]
28. Sudre, J.; Maes, C.; Garçon, V.C. On the global estimates of geostrophic and Ekman surface currents. *Limnol. Oceanogr. Fluids Environ.* **2013**, *3*, 1–20. [CrossRef]
29. Laurindo, L.; Mariano, A.; Lumpkin, R. An improved near-surface velocity climatology for the global ocean from drifter observations. *Deep-Sea Res. I* **2017**, *124*, 73–92. [CrossRef]
30. Laurindo, L. On the Air-Sea Exchange of Mechanical Energy: A Meso to Large-Scale Assessment Using Concurrent Drifter and Satellite Observations. Ph.D. Dissertation, University of Miami, Coral Gables, FL, USA, **2018**.
31. Zhang, Z.-Z.; Chao, B.F.; Lu, Y.; Hsu, H.-T. An effective filtering for GRACE time-variable gravity: Fan filter. *Geophys. Res. Lett.* **2009**, *36*, L17311. [CrossRef]
32. Johnson, G.C.; Bryden, H.L. On the size of the Antarctic Circumpolar Current. *Deep-Sea Res.* **1989**, *36*, 39–53. [CrossRef]
33. Kundu, P.J. Ekman Veering Observed near the Ocean Bottom. *J. Phys. Oceanoghrapy* **1975**, *6*, 238–242. [CrossRef]
34. Bingham, R.J.; Knudsen, P.; Andersen, O.; Pail, R. An initial estimate of the North Atlantic steady state geostrophic circulation from GOCE. *Geophys. Res. Lett.* **2011**, *38*, L01606. [CrossRef]
35. Feng, G.; Jin, S.; Sanchez-Reales, J.M. Antarctic circumpolar current from satellite gravimetric models ITG-GRACE2010, GOCE-TIM3 and satellite altimetry. *J. Geodyn.* **2013**, *72*, 72–80. [CrossRef]
36. Knudsen, P.; Bingham, R.; Andersen, O.; Rio, M.H. A global mean dynamic topography and ocean circulation estimation using a preliminary GOCE gravity model. *J. Geod.* **2011**, *85*. [CrossRef]
37. Sánchez-Reales, J.M.; Vigo, M.I.; Jin, S.G.; Chao, B.F. Global Surface Geostrophic Currents from Satellite Altimetry and GOCE. In Proceedings of the 4th International GOCE User Workshop, Munich, Germany, 31 March–1 April 2011, ESA SP-696.
38. Sánchez-Reales, J.M.; Vigo, M.I.; Jin, S.G.; Chao, B.F. Global Surface Geostrophic Currents Derived from Satellite Altimetry and GOCE Geoid. *Mar. Geod.* **2012**, *35*, 175–189. [CrossRef]
39. Fofonoff, N.P. Dynamics of ocean currents. In *The Sea. Volume 1: Physical Oceanography*; Wiley-Interscience: Hoboken, NJ, USA, 1962; pp. 323–395.
40. Briden, H.L.; Beal, L.M.; Duncan, L.M. Structure and Transport of the Agulhas Current ant its temporal variability. *J. Oceanogr.* **2005**, *61*, 479–492. [CrossRef]
41. Donohue, K.A.; Firing, E.; Beal, L. Comparison of three velocity sections of the Agulhas Current and Agulhas Undercurrent. *J. Geophys. Res.* **2000**, *105*, 28585–28593. [CrossRef]
42. Sloyan, B.M.; Ridgway, K.R.; Cowley, R. The East Australian Current and Property Transport at 27°S from 2012 to 2013. *J. Phys. Oceanogr.* **2016**, *46*. [CrossRef]
43. Griesel, A.; Mazloff, M.R.; Gille, S.T. Mean dynamic topography in the Southern Ocean: Evaluating Antarctic Circumpolar Current transport. *J. Geophys. Res.* **2011**, *117*, C01020. [CrossRef]

remote sensing

MDPI

Article

Multi-Satellite Altimeter Validation along the French Atlantic Coast in the Southern Bay of Biscay from ERS-2 to SARAL

Phuong Lan Vu [1,*], Frédéric Frappart [1,2], José Darrozes [1], Vincent Marieu [3], Fabien Blarel [2], Guillaume Ramillien [1], Pascal Bonnefond [4] and Florence Birol [2]

[1] GET-GRGS, UMR 5563, CNRS/IRD/UPS, Observatoire Midi-Pyrénées, 14 Avenue Edouard Belin, 31400 Toulouse, France; frederic.frappart@get.omp.eu (F.F.); jose.darrozes@get.omp.eu (J.D.); Guillaume.Ramillien@get.omp.eu (G.R.)

[2] LEGOS-GRGS, UMR 5566, CNES/CNRS/IRD/UPS, Observatoire Midi-Pyrénées, 14 Avenue Edouard Belin, 31400 Toulouse, France; florence.birol@legos.obs-mip.fr (F.B.); fabien.blarel@legos.obs-mip.fr (F.B.)

[3] UMR CNRS 5805 EPOC—OASU—Université de Bordeaux, Allée Geoffroy Saint-Hilaire CS 50023, 33615 Pessac CEDEX, France; vincent.marieu@u-bordeaux.fr

[4] SYRTE, Observatoire de Paris, PSL Research University, CNRS, Sorbonne Universités, UPMC Univ. Paris 06, LNE, 75014 Paris, France; pascal.bonnefond@obspm.fr

* Correspondence: phuonglan.vu@get.omp.eu; Tel.: +33-7-8232-1136

Received: 5 November 2017; Accepted: 28 December 2017; Published: 11 January 2018

Abstract: Monitoring changes in coastal sea levels is necessary given the impacts of climate change. Information on the sea level and its changes are important parameters in connection to climate change processes. In this study, radar altimetry data from successive satellite missions, European Remote Sensing-2 (ERS-2), Jason-1, Envisat, Jason-2, and Satellite with ARgos and ALtiKa (SARAL), were used to measure sea surface heights (SSH). Altimetry-derived SSH was validated for the southern Bay of Biscay, using records from seven tide gauges located along the French Atlantic coast. More detailed comparisons were performed at La Rochelle, as this was the only tide gauge whose records covered the entire observation period for the different radar altimetry missions. The results of the comparison between the altimetry-based and in-situ SSH, recorded from zero to five kilometers away from the coast, had root mean square errors (RMSE) ranging from 0.08 m to 0.21 m, 0.17 m to 0.34 m, 0.1 m to 0.29 m, 0.18 m to 0.9 m, and 0.22 m to 0.89 m for SARAL, Jason-2, Jason-1, ENVISAT, and ERS-2, respectively. Comparing the missions on the same orbit, ENVISAT had better results than ERS-2, which can be accounted for by the improvements in the sensor mode of operation, whereas the better results obtained using SARAL are related to the first-time use of the Ka-band for an altimetry sensor. For Jason-1 and Jason-2, improvements were found in the ocean retracking algorithm (MLE-4 against MLE-3), and also in the bi-frequency ionosphere and radiometer wet troposphere corrections. Close to the shore, the use of model-based ionosphere (GIM) and wet troposphere (ECMWF) corrections, as applied to land surfaces, reduced the error on the SSH estimates.

Keywords: sea surface height; coastal altimetry; validation; tide gauge

1. Introduction

Satellite altimetry is a radar technique detailing the topography of the earth's surface based on the measurement of the distance between the satellite and the surface, derived from the two-way travel time of an electromagnetic wave emitted by the altimeter, or altimeter range and the precise knowledge of the satellite orbit [1,2]. The primary objectives of satellite radar altimetry are to measure the marine geoid, ocean currents, and sea level variability. In the coastal zones, sea level data retrieval and interpretation is particularly complex. This is due to the interaction of the radar signal with land topography [3–6], geophysical corrections becoming inaccurate or incorrect when land surfaces are encompassed by the footprint of altimeters and radiometers [7], and to the short time scales and small spatial scales of the coastal dynamics [8] that are only partly sampled by the different altimeter missions, even when used in conjunction. Advances were made in altimeter instrument design with the development and launch of the Ka-band AltiKa instrument with on-board Satellite with ARgos and ALtiKa (SARAL). The use of this higher frequency reduces noise and has a smaller radar footprint in the images than in those obtained with the classical Ku-band instruments [9]. As a result, the coastal band is reduced where echoes are impacted by the presence of land. The launches of the first altimeters using the Synthetic Aperture Radar (SAR) technique [10], SIRAL on-board Cryosat-2 and SRAL on-board Sentinel-3, allowed more reliable observations in coastal areas. In parallel, advances were made in the development of new processing algorithms, corrections, and products for coastal applications of radar altimetry [11–14].

In this study, a comparison was completed of the most-used altimetry missions for monitoring the dynamic topography of the ocean surface, Jason-1, Jason-2, ERS-2, Envisat, and SARAL on their nominal tracks, operating in Low Resolution Mode (LRM) along the Atlantic coast of the Bay of Biscay (42–48°N, 0–10°W) from 1993 to 2015. Compared to the permanent calibration facilities or Calibration/Validation (Cal/Val) sites, parts of the ESA Earth Observation ground segment are located in Corsica, the Western Mediterranean Sea at the Harvest platform [15], the Eastern Pacific Ocean in Gavdos [16], the Eastern Mediterranean Sea [17], and the Bass Strait, between Australia and Tasmania [18]. Our goal was to evaluate the quality of altimeter-based sea surface heights (SSH). Our study had two main interests: to be well-instrumented with the presence of seven tide gauges in close proximity to eight altimeter tracks from five altimetry missions (Figure 1), and to investigate different tidal conditions than the classical altimetry Cal/Val sites. In contrast to the Corsica and Gavdos Mediterranean sites that are in semi-diurnal micro-tidal environments, the Harvest platform on the Pacific coast of California and the Bass Strait site between Australia and Tasmania are in mixed semi-diurnal macro and micro-tidal environments, respectively. The Bay of Biscay study area is in a semi-diurnal macro-tidal environment [19–21].

The error in the SSH retrieval from radar altimetry is caused by two main components: the ionosphere and wet troposphere corrections. For all satellites, the Ionosphere Correction (IC), due to electron content in the ionosphere, is obtained either by the bi-frequency altimeters, in Ku and C bands for T/P, Jason-1/2, and in Ku and S bands for Envisat, or from the electronic content of the ionosphere, such as the Global Ionospheric Maps (GIM) [22], available after 1998, or the New Ionospheric Climatology 2009 (NIC 09) [23] for mono-frequency altimeters such as ERS-2 or SARAL. The Wet Troposphere Correction (WTC), due to liquid water in the atmosphere, is derived either from radiometer measurements present on altimetry satellites or from meteorological model outputs, such as the ones made available by the European Center for Medium Range Weather Forecasting (ECMWF). Bi-frequency and radiometer-based corrections are used over the ocean even in coastal areas.

This study assessed the altimetry-based SSH in terms of number of observations, bias, root mean square error (RMSE), and correlation coefficient, for almost all the missions that were in orbit between 1993 and 2015, using records from seven tide gauges located along the French Atlantic coast in the southern Bay of Biscay. Contrary to what is commonly performed when using radar altimetry in coastal areas, the performance was assessed for different combinations of corrections applied to the altimeter range available in the Geophysical Data Records (GDR). A detailed analysis of the impact of these corrections on the accuracy of the altimetry-based SSH was performed at the La Rochelle tide gauge because this tide gauge is the only one that covers the entire observation period for the different radar altimetry missions.

2. Datasets and Study Area

2.1. Radar Altimetry Data

The radar altimetry data used in this study was obtained from the high precision altimetry missions launched after 1992 with the exception of Topex/Poseidon and Geosat Follow-On. Figure 1 shows a timeline of the different missions. The data were made available by the Centre de Topographie de l'Océan et de l'Hydrosphère (CTOH [24]). Altimetry data from all satellites were corrected for the classical propagation and geophysical corrections needed from the coast, including instrumental, ionosphere, wet and dry troposphere, sea state bias, solid earth, and pole tide corrections.

Timeline for RADAR Altimeters with short periods used in our study

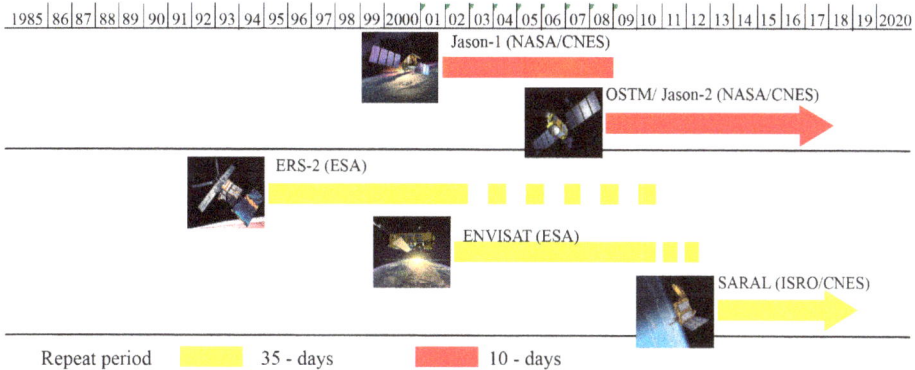

Figure 1. Timeline for radar altimeters used in our study (modified from [25]).

2.1.1. Jason-1

This mission was launched in December 2001 by the National Aeronautics and Space Administration (NASA) and the Centre national d'études spatiales (CNES) to ensure continuity of the oceanographic observations provided by the Topex/Poseidon mission. The Jason-1 payload is composed of the Poseidon-2 altimeter operating at Ku (13.575 GHz) and C (5.3 GHz) bands, a microwave radiometer, and a triple system for precise orbit determination [26] Jason-1 orbits at an average altitude of 1336 km, with an inclination of 66°, on a 10-day repeating cycle. Observations along the ground track of all ocean and continental surfaces from these altimeters are collected from 66°N to 66°S.

2.1.2. Jason-2

The Jason-2 mission was launched on 20 June 2008 as a co-operation between CNES, EUMESAT, NASA, and The National Oceanic and Atmospheric Administration (NOAA). Its payload is mostly composed of the Poseidon-3 radar altimeter from CNES, the Advanced Microwave Radiometer (AMR) from JPL/NASA, and a triple system for precise orbit determination: the real-time tracking system Détermination Immédiate d'Orbite par Doris Embarqué (DIODE) of the Doppler Orbitography by Radiopositioning Integrated on Satellite (DORIS) instrument from CNES, a Global Positioning System (GPS) receiver and a Laser Reflector Array (LRA) from NASA. Jason-1 and Jason-2 share the same orbit of 1336 km above the earth, with a cycle time of 10 days. Poseidon-3 radar altimeter is a two-frequency solid-state altimeter that accurately measures the distance between the satellite and the surface (range). It also provides ionospheric delay corrections over the ocean with a precision of about 2 cm, operating at Ku (13.575 GHz) and C (5.3 GHz) bands [25,27].

2.1.3. ERS-2

European Remote Sensing-2 (ERS-2) was launched in 1995 by the European Space Agency (ESA) as an ERS-1 follow-up mission, designed to study the earth's environment. The satellite carries, among other instruments, a radar altimeter (RA) operating at Ku-band (13.8 GHz), developed for measuring the height of the ocean, land, and ice caps. ERS-2 orbits at an average altitude of 790 km, with an inclination of 98.54°, on a sun-synchronous orbit with a 35-day repeating cycle. It provides observations of the earth's surface, including ocean, land, and ice caps, from 82.4°N to 82.4°S. This orbit was formerly used by the ERS-1 mission, with an equatorial ground-track spacing of about 85 km. ERS-2 data are available from 17 May 1995 to 9 August 2010. After 22 June 2003, the dataset coverage is limited to ground station visibility.

2.1.4. Envisat

The Envisat mission was launched on 1 March 2002 by ESA, carrying 10 instruments including the advanced radar altimeter (RA-2). Envisat was based on the on-board sensor on the ERS-1 and 2 satellites. RA-2 was a nadir-looking pulse-limited radar altimeter operating at two frequencies at the Ku (13.575 GHz), as ERS-1 and 2, and S (3.2 GHz) bands. The goal was to collect radar altimetry data over ocean, land, and ice caps [26,28] Envisat remained on its nominal orbit until October 2010 and its mission ended 8 April 2012. RA-2 stopped operating correctly at S-band in January 2008. Its initial orbital characteristics were the same as for ERS-2.

2.1.5. SARAL/AltiKa

The SARAL mission was launched on 25 February 2013. SARAL is collaboration between CNES and Indian Space Research Organization (ISRO). Its payload included the AltiKa radar altimeter and bi-frequency radiometer, and a triple system for precise orbit determination: the real-time tracking system DIODE of DORIS instrument, a Laser Reflector Array (LRA), and the Advance Research and Global Observation Satellite (ARGOS-3). The AltiKa radar altimeter is a solid-state mono-frequency altimeter that provides accurate range measurements and is the first altimeter to operate Ka-band (35.75 GHz). Formerly, the SARAL orbit was used by ERS-1 and 2 and Envisat, with an equatorial ground-track spacing of about 85 km [9], and a 35-day repeating cycle. It was placed on a drifting orbit as of 7 April 2016.

2.2. Tide Gauge Records

The tidal data used in this study was obtained from records of tide gauges SSH from the French Réseaux de référence des observations marégraphiques (REFMAR) networks [29]. We used data from 7 tide gauge stations located along the southwestern coast of France (Table 1; Figure 1). The SSH data were provided at 10-min intervals and were referenced to the Institut Géographique National 1969 (IGN69) ellipsoid. For comparison with the altimetry data, the difference between the two reference systems (ZH/GRS80 in Table 1) was calibrated at the 7 station tide gauges. We used daily Mean Sea Surface (MSS) from 1995 to 2016, available on Système d'Observation du Niveau des Eaux Littorales (SONEL) networks (www.sonel.org), obtained from different tidal filters (e.g., Doodson filter and Demerliac filter) to reduce the tidal effects. Several such linear filters compute the daily mean sea levels [27,30].

2.3. Study Area

The study area is located along the Atlantic coast in the Bay of Biscay (42–48°N, 0–10°W), an extension of the Atlantic Ocean, off the western coast of Europe, bordered by France and Spain. In this area, the continental shelf is narrow in the south (~30 km) and extends to 180 km in the north [31]. The oceanic circulation is characterized by the presence of a large-scale gyre, the Iberian Poleward Current (IPC) [29,32] These characteristics, combined with a seasonal wind regime [30,33] and a large amount of fresh water from river discharges [31,34], drive a complex system of coastal currents, mesoscale eddies, local upwellings, and internal tide [33,35].

In the study area, five passes of ERS-2/Envisat/SARAL were available (Figure 2a), three descending passes (#0360, #0818, #0274), and two ascending passes (#0945, #0859). Three Jason-1 and Jason-2 passes, two descending (#70, #248), and an ascending pass (#213) were available.

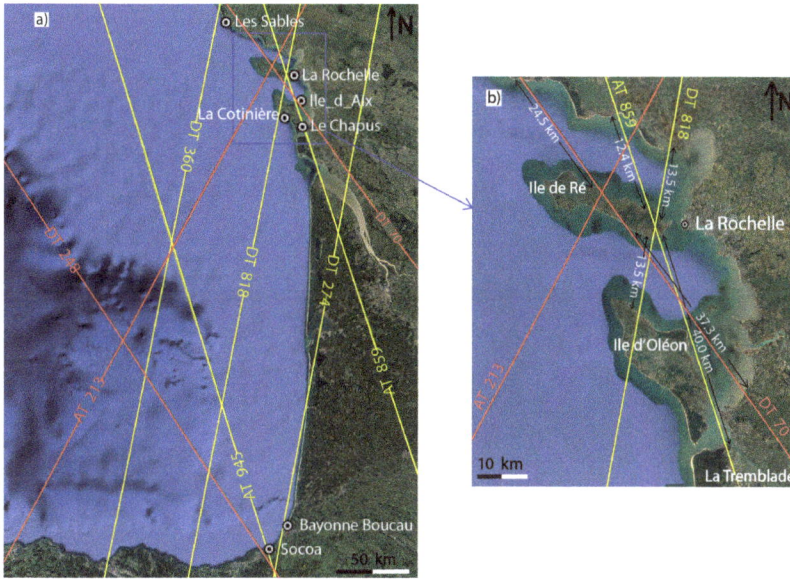

Figure 2. Google Earth image of the study area (**a**) with superposition of theoretical traces altimetry. Jason-1 and Jason-2 tracks are represented with red lines and ERS-2, Envisat, and SARAL tracks with yellow lines; (**b**) detail near La Rochelle tide gauge with Jason-1 and Jason-2 tracks in red lines and ERS-2, Envisat, and SARAL tracks in yellow.

Envisat and ERS-2 altimetry heights are referenced to the WGS 84 ellipsoid, and Jason-1 and 2 and SARAL are relative to the Topex/Poseidon ellipsoid. The corrections applied to the different sensors, contained in the Envisat, ERS-2, Jason-1 and 2, and SARAL GDR standard data products are listed in Table 2.

Table 1. Locations of the tide gauge stations and the available time period in the study area. Columns 6 and 7 show the satellite track number and the distance track–tide gauge. The green highlights the tracks close to the coastline (<15 km). The yellow tracks are located in a mixed domain between the coastal area and open ocean (15–25 km), and the red tracks the open ocean (>25 km).

Tide Gauge Station	Longitude (°)	Latitude (°)	Validation Period	ZH/GRS80 (m)	Jason-1 and 2		ERS-2, Envisat, SARAL	
					Track	Distance (km)	Track	Distance (km)
Bayonne Boucau	43.5273	−1.5148	1999–2015	46.41	248	24.3	274, 945	3.0, 18.6
Bourceranc le Chapus	45.8534	−1.1778	2012–2013	43.49	70	7.0	818, 859	17.8, <1
Ile d'Aix	46.0074	−1.1743	2011–2015	43.49	70, 213	2.88, 29.0	818, 859	13.7, 4.7
La Rochelle la Pallice	46.1585	−1.2206	1996–2015	43.46	70, 213	8.8, 16.8	818, 859	5.8, 5.4
Les Sables d'Olonne	46.4974	−1.7935	1995–2015	44.54	70	7.0	360	6.6
La Cotinière	45.9136	−1.3278	2006–2015	43.45	213	24.8	818	4.7
Saint Jean de Luz Socoa	43.3952	−1.6816	2002–2015	46.64	248	5.4	274, 945	6.2, 2.1

Table 2. Ocean ranges available and corrections for each altimetry mission.

Altimetry Mission	Period	Format	Ocean Retracking Algorithm	Corrections to the Range			
				Ionosphere	Dry Troposphere	Wet Troposphere	Sea State Bias
ERS-2	Cycle 1 (May 1995) to cycle 085 (July 2003)	REAPER GDR	MLE-3	NIC 09 GIM	ECMWF	Radiometer ECMWF	Empirical
Jason-1	Cycle 001 (January 2002) to cycle 259 (January 2009)	GDR E	MLE-3	Bi-frequency (C-Ku) GIM model	ECMWF model derived	Jason-1 radiometer (JMR) and ECMWF model	Empirical model derived
Envisat	Cycle 006 (May 2002) to cycle 094 (October 2010)	GDR C v2.1	MLE-3	Bi-frequency (C-Ku) GIM model	ECMWF model derived	Microwave Radiometer (MWR) and ECMWF model	Empirical model derived
Jason-2	Cycle 001 (July 2008) to cycle 250 (April 2015)	GDR D	MLE-4 MLE-3	Bi-frequency (C-Ku) and GIM model	ECMWF atmospheric pressures	Jason-2 radiometer (AMR) and ECMWF model	Empirical model derived from Jason-2 data
SARAL	Cycle 001 (March 2013) to cycle 028 (November 2015)	GDR T	MLE-4	GIM- derived	ECMWF model derived	Radiometer and ECMWF model	Empirical model derived

3. Methods

3.1. SSH from Altimetry

The absolute calibration technique is required in the determination of the absolute altimeter bias. This technique requires simultaneous measurements of $SSH_{\text{in situ}}$ by an altimeter in the same terrestrial reference frame at the exact same location or comparison point. The absolute altimeter bias ($Bias_{altimeter}$) is estimated as follows [33,36]:

$$Bias_{altimeter} = SSH_{altimeter} - SSH_{\text{in situ}} - \Delta h_{geoid} \tag{1}$$

where $SSH_{altimeter}$ and $SSH_{\text{in situ}}$ are the SSH estimated from the altimeter and in situ measurements, respectively, and Δh_{geoid} represents the difference in geoid heights between the tide gauge and the altimeter measurement locations. A negative bias indicates that the measure of the SSH by the altimeter is underestimated; either the altimeter range is being overestimated, or the orbit is biased downwards [6]. The SSH from the altimeter is given as [2,37]:

$$SSH_{altimeter} = H - \left(R + \sum \Delta R_{environmental} + \sum \Delta R_{geophysical} \right) \tag{2}$$

where H is the height of the center of the mass of the satellite above the ellipsoid, estimated using the Precise Orbit Determination (POD) technique; R is the nadir altimeter range from the center of the mass of the satellite to the sea surface while considering instrumental corrections; and $\sum \Delta R_{environmental}$ and $\sum \Delta R_{geophysical}$ are the sum of the environmental and geophysical corrections applied to the range, respectively. The environmental corrections are composed of the following contributions:

$$\sum \Delta R_{environmental} = \Delta R_{ion} + \Delta R_{dry} + \Delta R_{wet} + \Delta R_{SSB} \tag{3}$$

where ΔR_{ion} is the atmospheric refraction range delay due to the free electron content associated with the dielectric properties of the ionosphere, ΔR_{dry} is the atmospheric refraction range delay due to the dry gas component of the troposphere, ΔR_{wet} is the atmospheric refraction range delay due to the water vapor and the cloud liquid water content of the troposphere, and ΔR_{SSB} is the range correction caused by the various effects of the sea surface roughness within the radar footprint. This is known as sea state bias (SSB) and is the sum of electromagnetic (EM), skew, and tracker biases. The geophysical corrections are composed of the following contributions:

$$\sum \Delta R_{geophysical} = \Delta R_{load} + \Delta R_{solid\ Earth} + \Delta R_{pole} + \Delta R_{atm} \tag{4}$$

where ΔR_{load}, $\Delta R_{solid\ Earth}$ ΔR_{pole}, and ΔR_{atm} are the corrections accounting for crustal vertical motions due to the solid earth, pole tides, and rapid fluctuations of the atmosphere, respectively.

The leveling of the tide gauge precisely determines the ellipsoidal height (H_0) of the gauge zero as:

$$SSH_{\text{in situ}} = h + H_0 \tag{5}$$

where h is the instantaneous stage value in the gauge record at time t. H_0 values are provided by REFMAR (Table 1).

3.2. SSH Processing and Editing

Altimetry SSH was obtained using the Multi-mission Altimetry Processing Software (MAPS) [35,38]. This software computes high frequency SSH referenced to the WGS84 ellipsoid along the altimeter tracks from the altimeter range and a set of corrections chosen by the user. The range can be any of the ranges available in the GDR from different retracking algorithms. MAPS allows a refined selection of the valid altimetry data to build several virtual stations where a SSH time series over the ocean [39] or water levels over rivers and estuaries [38,40,41] can be constructed. The altimetry-based SSH were automatically computed along the track by MAPS at a 20-Hz frequency rate (i.e., ~0.35 km along the track) for Envisat, ERS-2, Jason-1, and Jason-2, and at 40 Hz (i.e., ~0.18 km along the track) for SARAL. Many different retracking algorithms are used to measure the waveforms applied to the Brown model. Range values suitable for deriving SSH were retracked using the Maximum Likelihood Estimator (MLE) retracking algorithm for all altimeters [42] The MLE3 algorithm [43] estimates with three degrees of freedom: range, significant wave height, and power, whereas the MLE4 algorithm [41,44] estimates four degrees of freedom including the three previous plus the slope of the waveform trailing edge. We used both algorithms and compared them in this study. Data processing involved three main steps: a coarse delineation of the study area using Google Earth, a refined selection of the valid altimetry data through visual inspection, and the computation of the SSH time series using all the valid points selected by the user. The average altimetry-based SSH was computed for each cycle using the median and the mean of the selected altimetry heights, along with their respective deviation, meaning their mean absolute and standard deviations, respectively.

3.3. Comparisons between Altimetry-Based and In Situ-Based SSH

Based on the availability of the altimetry data, the comparison with in-situ SSH, including number of cycles, bias, root means square errors (RMSE), and correlation coefficient (R), was performed using satellite data with open ocean conditions at distances between 5 km and 50 km from the coast. Depending on the frequency of sampling of each tide gauge, the maximum time-lag between in situ records and altimetry data was lower than 10 min for any measurement considered (Section 2.2).

3.4. Comparisons of the Atmosphere Delays over the Open Ocean

Bias and RMSE were compared between model-based and bi-frequency ionosphere corrections. Model-based and radiometer-derived wet troposphere corrections were performed over the entire Bay of Biscay. According to a prior study [42,45] the comparisons were performed in along-track cells 7 km in length (i.e., 1-Hz). More details about their definition during the complete observation period of each altimetry mission, whose data were used in this study, are available in previous reports [40,46].

4. Results

Comparisons between altimetry-based SSH and tide gauge records were performed at seven locations (Table 1) along the French Atlantic coast between La Rochelle and Bayonne under eight altimetry tracks (Figure 2). We will present the results for the La Rochelle tide gauge as the in situ data records exhibit a longer period of common availability with the altimetry missions than the other tide gauges, and two altimeter tracks, ascending and descending, were available for each altimetry mission. Similar results were obtained from the other tide gauges.

4.1. Multi-Missions Comparisons over the French Southern Part of the Bay of Biscay

The metrics compared between the altimetry-based SSH, estimated zero and five kilometers from the coast, and SSH measured by the tide gauge were: number of cycle observations (N), bias, standard deviation (SD), root mean square error (RMSE), and correlation coefficient (R) for the different SSH values. The results are outlined in Table 3 for all the tide gauges present along the French Atlantic coast of the southern part of the Bay of Biscay. The best results were obtained with SARAL (R > 0.99 and RMSE < 0.20 m) among all the tide gauges. Excellent agreement was found using Jason-1 (R > 0.9 and RMSE < 0.15 m for the majority of the time) and Jason-2 (R > 0.9 and RMSE < 0.20 m for the majority of the time). Lower but good agreement was found using Envisat (R > 0.7 for the majority of the time and RMSE < 0.21 m). For ERS-2, the quality of the SSH retrieval was variable.

Table 3. Comparisons of distance to the gauge, number of cycle observations (N), bias, root mean square error (RMSE), SD, and the correlation coefficient (R) from zero to five kilometers away from the coast for all tide gauge stations present along the French Atlantic coast of the southern part of the Bay of Biscay for ERS-2, Jason-1, Envisat, Jason-2, and SARAL. The ionosphere and wet troposphere corrections were derived from the Global Ionospheric Maps (GIM) and European Center for Medium Range Weather Forecasting (ECMWF) models, respectively.

Altimeter	Station	Altimetry Track	Distance Alt/Gauge (km)	N	Bias (m)	RMSE (m)	SD (m)	R
	La Rochelle	818	5.8	13	−0.6	1.4	1.05	0.41
ERS-2	La Rochelle	859	5.4	25	−0.65	0.22	1.45	0.99
	Bayonne Boucau	274	3.0	21	−1.00	0.89	1.05	0.51
	La Rochelle	70	8.8	97	−0.81	0.29	0.98	0.93
Jason-1	La Rochelle	213	16.8	61	−0.73	0.21	0.71	0.97
	Bayonne Boucau	248	24.3	25	−0.94	0.15	1.00	0.99
	Saint Jean de Luz Socoa	248	5.4	35	−0.77	0.10	1.00	0.996
	La Rochelle	818	5.8	77	0.37	0.90	1.00	0.73
	La Rochelle	859	5.4	76	0.49	0.18	1.15	0.99
Envisat	Bayonne Boucau	274	3.0	42	−0.07	0.77	0.99	0.67
	Saint Jean de Luz Socoa	274	6.2	51	−0.17	0.63	0.99	0.8
	Les Sables d'Olonne	360	6.6	38	0.25	0.84	0.97	0.7
	La Rochelle	70	8.8	150	−0.30	0.19	0.99	0.99
	La Rochelle	213	16.8	37	−1.20	0.21	1.27	0.91
	Bayonne Boucau	248	24.3	118	−0.65	0.34	1.03	0.93
Jason-2	Saint Jean de Luz Socoa	248	5.4	84	−0.78	0.33	1.03	0.92
	Les Sables d'Olonne	70	7.0	111	−0.67	0.17	0.99	0.98
	Ile d'Aix	213	29.0	14	−1.1	0.27	1.27	0.83
	Ile d'Aix	70	2.88	58	−0.37	0.21	1.44	0.98
	La Rochelle	818	5.8	25	−0.29	0.19	0.98	0.99
	La Rochelle	859	5.4	26	−0.28	0.13	1.22	0.99
	Bayonne Boucau	274	3.0	26	−0.83	0.12	0.94	0.99
SARAL	Saint Jean de Luz Socoa	274	6.2	14	−0.80	0.12	0.94	0.996
	Les Sables d'Olonne	360	6.6	14	−0.35	0.08	1.12	0.998
	Ile d'Aix	818	13.7	22	−0.37	0.21	0.98	0.995
	Ile d'Aix	859	4.7	22	−0.47	0.14	1.22	0.99

4.2. Multi-Mission Comparisons at La Rochelle Tide Gauge

The site of La Rochelle was chosen for comparisons between tide gauge records and altimetry-based SSH because the tide gauge records cover the entire observation period for the different radar altimetry missions. For all missions in this study, we considered one ascending track and one descending track. These data were compared with the La Rochelle data in terms of N, bias, SD, RMSE, and R for the different SSH values computed using all possible combinations of ocean ranges varying from zero to five kilometers to the coast (Table 4). The considerable bias between altimetry and the tide gauge five kilometers from the coast is explained, in Section 3.1, by the slope of the geoid relative to the ellipsoid. To highlight the differences between the sensors, we present the results for the SSH estimates obtained from the GIM model for the ionosphere correction and the model-based wet troposphere correction. Therefore, the complete ERS-2 record was not used; only the data acquired from 30 November 1998 (cycle 37) to 11 August 2003 (cycle 85) was used since the release of the GIM-based ionosphere correction.

The comparison results between the tide gauge records and the altimetry-based SSH from five kilometers from the coast are presented in Figure 3 for Jason-1 and Jason-2 data, and in Figure 4 for ERS-2, Envisat, and SARAL data. The quality of the results is strongly impacted by the direction of the satellite groundtrack. The results differ between the tracks where the acquisitions were performed first on the ocean and then on the land and the opposite situation. Descending track 70 passes over the Bay of Biscay for 24.5 km between the mainland and Ré Island, and then for another 37.3 km between Ré Island and the mainland, whereas ascending track 213 passes from the ocean to Ré Island and then for 12.4 km from Ré Island to the mainland (Figure 2b). For the Jason-2 data, the number of valid cycles is higher along the descending track, with 60% of 250, cycles than along the ascending track with only 15%. Therefore, even though the descending track 70 passes from land to ocean, the results obtained for this track, the RMSE of 0.19 m and R of 0.99, are better than those of the ascending track with a RMSE of 0.21 m and a R of 0.91

(Figure 3a,b). For the Jason-1 data, the number of valid cycles is higher along the descending track, with 37% of 259 cycles, than along the ascending track with only 24% valid. Accurate results were obtained for the ascending track with a RMSE of 0.21 m and a R of 0.97, which are better than those of the descending track, with a RMSE of 0.29 m and a R value of 0.93. Both were statistically representative with more than 50 samples or cycles (Figure 3c,d). The differences between the satellites (Figure 3) could account for the disparity in tracking mode between the closed-loop Poseidon-2 tracker onboard Jason-1 and the open loop (DIODE/digital elevation model) Poseidon-3 tracker mode onboard Jason-2 [44,47].

The impact of the retracking algorithm was also analyzed. The results of the comparisons between the tide gauge records and the Jason-2 track 213 altimetry-based SSH at La Rochelle, with the two available ocean retracking algorithms (MLE-4 and MLE-3), are presented in Figure S1 for distances varying from 0 to 50 km from the coast. The results show that within the zero to five kilometer zone, only 15% of cycles are valid. Furthermore, the SSH derived from altimetry over five kilometers from the coast is affected not only by less reliable geophysical and environmental corrections, but it is also influenced by the land. The RMSE and R obtained were 0.33 m and 0.8, respectively, using MLE-3, and 0.21 m and 0.91, respectively, using MLE-4.

Figure 4 shows the comparison results for ERS-2, Envisat, and SARAL virtual stations at La Rochelle. Large differences were observed between ascending track 859 and descending track 818 for all missions. As seen in Figure 1b, descending track 818 passes over the Bay of Biscay for 13.5 km between the mainland and Ré Island, and then for another 13.5 km between Ré Island and Oléron Island. Ascending track 859 passes over the Bay of Biscay for 40 km from la Tremblade over the mainland to Ré Island. For ERS-2 data, more valid cycles were found along the ascending track with 51% of 49 cycles, compared to the descending track with only 27% (Figure 4a,b). These results indicate that many cycles are affected by tracking loss effects. Much better results were obtained for the ascending track, with a RMSE of 0.22 m and a R of 0.99, than for the descending track with a RMSE of 1.4 m and a R of 0.41. This poor result can be accounted for by the low cycle number (N = 13), which is not statistically representative.

Table 4. Comparisons between altimetry-based SSH and in-situ SSH in La Rochelle from zero to five kilometers away from the coast (N, bias, RMSE, SD, and R) for ascending and descending tracks of sERS-2, Jason-1, Jason-2, Envisat, and SARAL satellites, and for various ionospheric and tropospheric corrections including GIM, bi-frequency, and model-based wet troposphere.

Altimeter	Correction Iono-Wet Tropo	Descending Track					Ascending Track				
		N	Bias (m)	RMSE (m)	SD (m)	R	N	Bias (m)	RMSE (m)	SD (m)	R
ERS-2	Nic09-Rad	9	0.32	1.3	1.12	0.54	33	−0.40	0.21	1.41	0.99
	GIM-Rad	8	0.25	1.3	1.08	0.52	25	−0.42	0.24	1.43	0.99
	Nic09-Mod	14	−0.53	1.4	1.11	0.42	37	−0.68	0.22	1.40	0.99
	GIM-Mod	13	−0.6	1.4	1.05	0.41	25	−0.65	0.22	1.45	0.99
Jason-1	Bi-Rad	144	−0.36	0.46	0.98	0.92	56	−0.78	0.29	1.01	0.94
	Bi-Mod	97	−0.81	0.46	1.01	0.82	54	−0.75	0.28	1.05	0.94
	GIM-Rad	154	−0.28	0.37	1.00	0.95	84	−0.73	0.18	0.70	0.98
	GIM-Mod	97	−0.81	0.29	0.98	0.93	61	−0.73	0.21	0.71	0.97
Envisat	Bi-Rad	77	0.6	0.9	1.00	0.73	76	0.63	0.19	1.17	0.99
	Bi-Mod	77	0.49	0.9	1.00	0.73	76	0.58	0.18	1.14	0.99
	GIM-Rad	77	0.59	0.9	0.99	0.73	76	0.63	0.19	1.19	0.99
	GIM-Mod	77	0.37	0.9	1.00	0.73	76	0.49	0.18	1.15	0.99
Jason-2 (MLE-3)	Bi-Rad	119	−0.61	0.2	1.02	0.97	22	−1.20	0.26	0.75	0.86
	Bi-Mod	119	−0.61	0.2	1.01	0.97	22	−1.20	0.26	0.75	0.85
	GIM-Rad	132	−0.54	0.22	0.99	0.98	42	−1.30	0.33	0.91	0.81
	GIM-Mod	132	−0.54	0.21	0.99	0.98	42	−1.30	0.33	0.91	0.80
Jason-2 (MLE-4)	Bi-Rad	142	−0.31	0.36	0.99	0.95	31	−1.20	0.23	0.90	0.89
	Bi-Mod	142	−0.31	0.36	0.99	0.95	31	−1.20	0.23	0.90	0.89
	GIM-Rad	150	−0.30	0.19	0.99	0.99	37	−1.20	0.2	1.27	0.91
	GIM-Mod	150	−0.30	0.19	0.99	0.99	37	−1.20	0.21	1.27	0.91
SARAL	GIM-Rad	25	−0.29	0.19	0.99	0.99	26	−0.28	0.13	1.22	0.99
	GIM-Mod	25	−0.29	0.19	0.98	0.99	26	−0.28	0.13	1.22	0.99

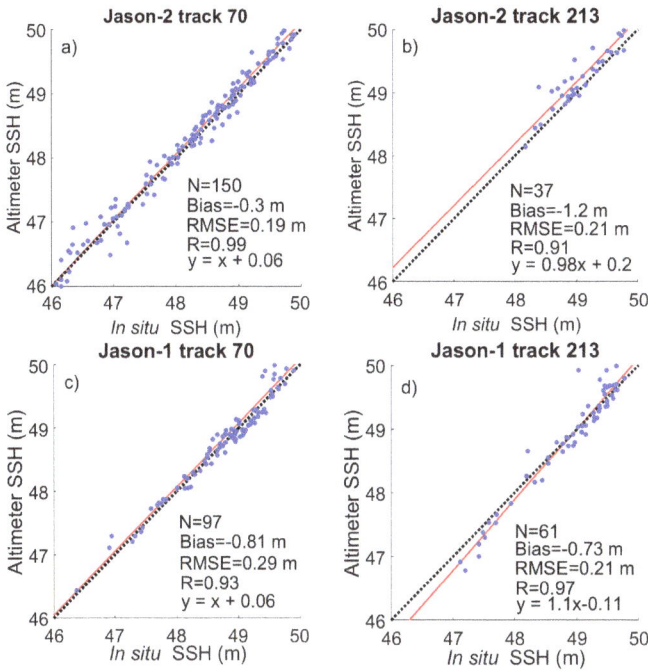

Figure 3. Comparisons between in situ and altimeter-based sea level anomalies in La Rochelle for (**a,b**) Jason-2 and (**c,d**) Jason-1 for descending track 70 and ascending track 213. The ionosphere and wet troposphere corrections were derived from GIM and ECMWF models, respectively.

More data were acquired in the 330 MHz Ku chirp bandwidth acquisition mode (ocean mode with a 30-m window size) than in the 82.5 MHz Ku chirp bandwidth acquisition mode (ice mode with a 116-m window size) along the 818 ERS-2 descending track and the 859 ERS-2 ascending track. This could account for the low accuracy of the ERS-2 SSH measurement and the considerable loss of data. The number of valid cycles obtained for Envisat on the ascending and descending tracks were close to 77 cycles accounting for 82% of the total. However, the results obtained for the 859 Envisat ascending track, with a RMSE of 0.18 m and a R of 0.99, were better than those for the 818 Envisat descending track, with a RMSE of 0.9 m and a R of 0.73 (Figure 4c,d). Along the 859 Envisat ascending track, the data were acquired in the 320 MHz Ku chirp bandwidth acquisition mode with a 64-m window size. Along the 818 Envisat descending track, the data were acquired in both the 320 and 80 MHz Ku chirp bandwidth acquisition modes with 64- and 256-m window sizes, respectively, over land, and in the 320 MHz bandwidth over open ocean. Over the study area, the altimeter switched between these two modes, accounting for both the data loss and the decrease in accuracy. Similar results were found in the Gironde Estuary [38,41]. Better results were obtained using SARAL data, with a RMSE of 0.19 m and a R value of 0.99 for the 818 descending track, and a RMSE of 0.13 m and a R of 0.99 for the 859 ascending track. The number of valid cycles obtained for SARAL was 92% of the total for both the ascending and descending tracks (Figure 4e,f). This result can be accounted for by the larger effective footprint radius of ERS-2 and Envisat at Ku-band than the one of SARAL at Ka-band, and the surrounding lands along the track are encompassed in the scene observed by the altimeter. Therefore, the number of valid cycles for the SARAL altimeter are greater than the number of valid cycles with ERS-2 and Envisat. If these results are confirmed, we can attribute this accuracy to the smaller effective footprint at Ka-band. Radar echoes are less affected by the presence of land surrounding the study area.

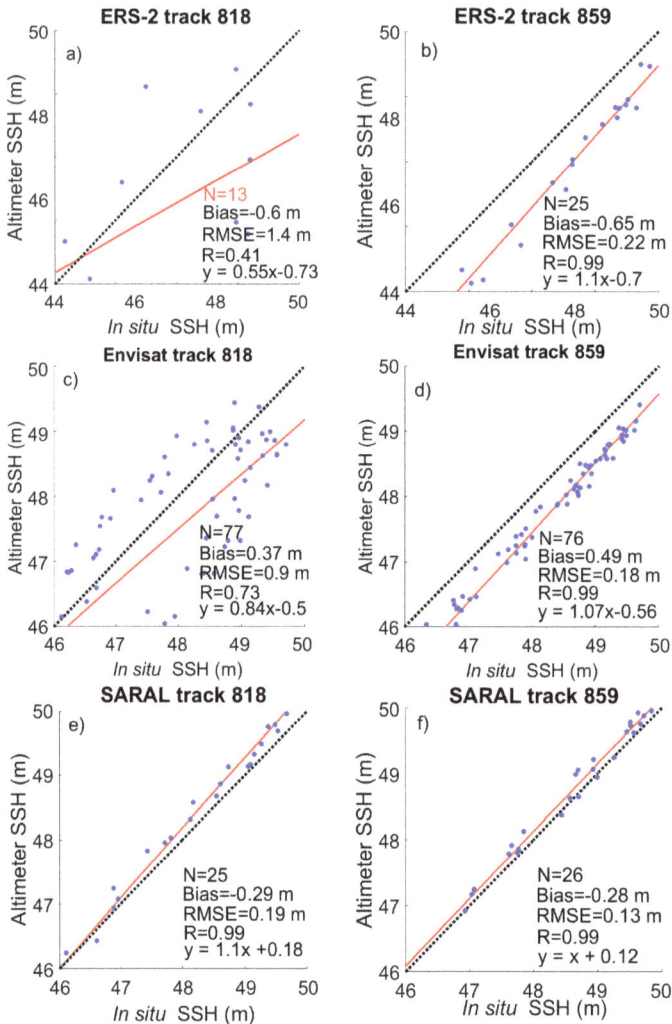

Figure 4. Comparisons between in situ and altimeter-based sea level anomalies in La Rochelle for (**a,b**) ERS-2; (**c,d**) Envisat; and (**e,f**) SARAL for descending track 0818 and ascending track 0859. The ionosphere and wet troposphere corrections were derived from GIM and ECMWF models, respectively.

4.3. Comparisons of Atmosphere Corrections over the Open Ocean

4.3.1. Comparisons of Bi-Frequency and Model-Based Ionospheric Corrections

The bias and RMSE for the ionosphere corrections from bi-frequency measurements and GIM model for Jason-1, Envisat, and Jason-2 using the MLE-3 and MLE-4 retracking algorithms are presented in Table 4 and Figures 5 and 6, Figures S2 and S3, respectively, over the entire southern Bay of Biscay. Low biases and RMSE, lower than 0.03 m and 0.05 m, respectively, are found when the distance to the coast is more than 25 km, except for a limited number of locations along the track where the bias reaches 0.1 to 0.2 m, and RMSE reaches up to 0.3 to 0.5 m for Envisat. The Jason-1 and Jason-2 missions exhibit very low variation differences in the ionosphere corrections along the tracks far from the shore. To reduce this variability, ionosphere corrections were smoothed on distances of several tenths of kilometers. Conversely, a larger variability was observed along the Envisat tracks. Biases and RMSE increased when closer to the shore, reaching several tenths of centimeters. Envisat stopped operating properly at S-band in January 2008, meaning that only the GIM-based ionosphere correction was available until the end of Envisat's scientific mission in October 2010 along the nominal orbit and April 2012 on the drifting orbit.

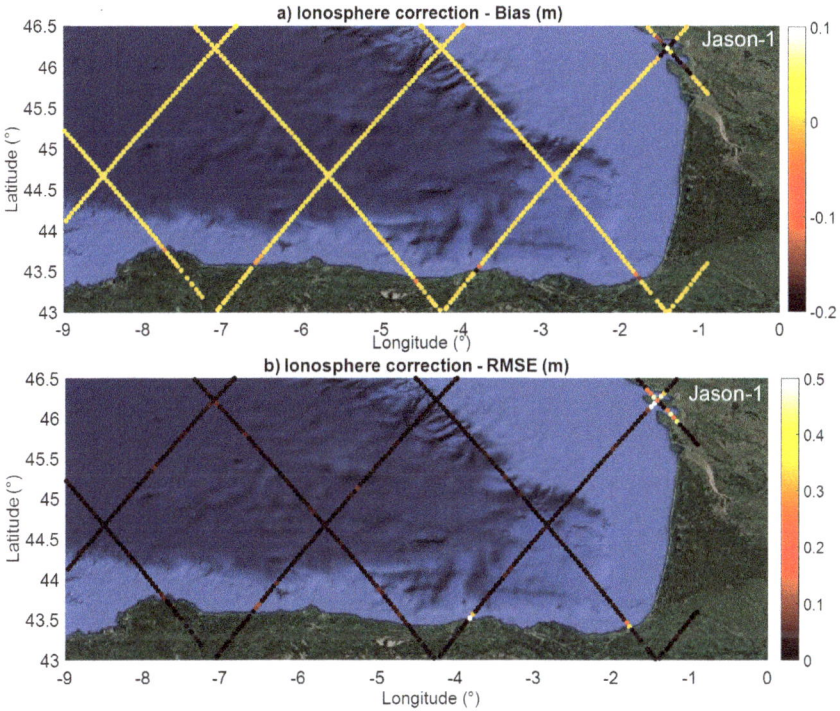

Figure 5. Comparison between ionosphere corrections from bi-frequency measurements and the GIM model for Jason-1: (**a**) bias and (**b**) RMSE.

Bi-frequency along-track profiles and model-based ionospheric corrections (GIM for all the missions and NIC09 for ERS-2) within 50 km of the coast are presented in Figure 7 for Jason-2 and Jason-1 (track 70), and Envisat and ERS-2 (track 859). Larger discrepancies were observed closer to the coast, reaching 0.25 m for Jason-1 when averaged over the entire observation period. Lower differences, less than 0.10 m, were observed using Jason-2 and less than 0.03 m for Envisat on average. Nevertheless, individual differences up to one meter were observed for distances from the coast, to less than 10 km between model-based and bi-frequency ionosphere corrections. Minute differences were observed between the two ionosphere correction models when compared over their common period of availability.

Due to the footprint size of the altimeter and the different frequencies used, land areas were encompassed in the image. As the radar electromagnetic wave can penetrate land surfaces with a different penetration depth, depending on the frequency and other parameters such as soil type, moisture, and roughness. For Jason-2, Jason-1 (track 213), and Envisat (track 818), within 50 km of the coast near La Rochelle, the bi-frequency-based ionospheric correction cannot be used within five kilometers of the coast, due to land contamination [45,48] (Figure 7). On the contrary, the model-based ionosphere correction presents quite stable values, in the range of a few centimeters, for the entire observation period. For ERS-2, both GIM- and NIC09-based ionospheric corrections can be used within five kilometers of the coast (Figure 7).

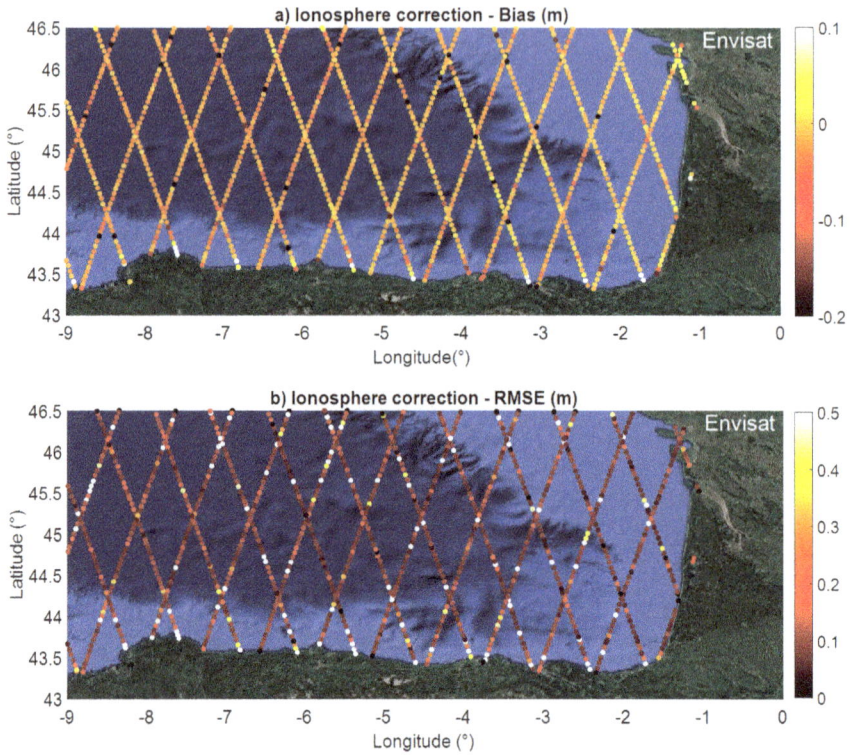

Figure 6. Comparison between the ionosphere corrections from bi-frequency measurements and the GIM model for Envisat: (**a**) bias and (**b**) RMSE.

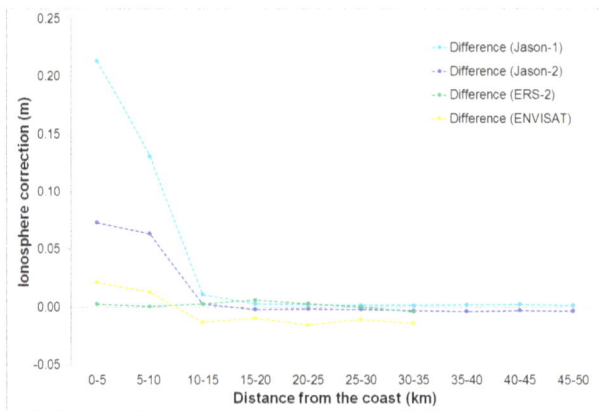

Figure 7. Difference between the ionosphere corrections from GIM and bi-frequency measurements for Jason-1 (light blue), Jason-2 (blue), and Envisat (yellow), and between GIM and Nic09 for ERS-2 (green) over La Rochelle within 50 km of the coast.

4.3.2. Wet Tropospheric Correction Comparison between Radiometers and ECMWF Model

Bias and RMSE for the wet troposphere corrections from radiometer measurements and the ECMWF model for Jason-1, Envisat, Jason-2, and SARAL are presented in Figures 8 and 9 and Figures S4–S6, respectively, for the entire southern Bay of Biscay. Over the open ocean, the difference between WTC derived from the ECMWF model and radiometers is negligible. Within a short distance from the coast (<25 km), both bias and RMSE increase. They remain quite small for Jason-1 (Figure 8) and Jason-2 (Figure S5) missions (lower than 0.05 m), and increase up to −0.4 m and 0.2 m for ERS-2 (Figure S4), −0.3 m and 0.2 m for Envisat (Figure 9), and −0.15 m and 0.05 m for SARAL (Figure S6), respectively.

As mentioned previously, due to the footprint size in the different frequencies used by the radiometers, land areas are encompassed in the tracks. The brightness temperatures measured in the different frequencies are affected by the presence of land in the footprint of the radiometer [11]. Corrections based on the deconvolution technique have allowed the removal of the land contribution from the signal measured by the radiometer [46,49] to provide reliable corrections up to five kilometers from the coast, as with Jason-2 (Figure S5).

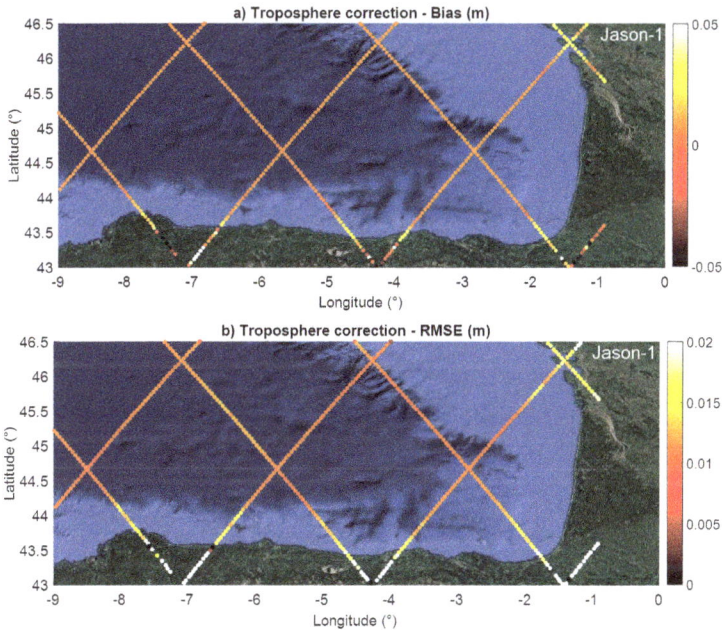

Figure 8. Comparison between wet troposphere corrections from radiometer measurements and the CMWF model for Jason-1b (**a**) Bias and (**b**) RMSE.

Considering the tracks of Jason-1 and Jason-2 (track 213), and ERS-2, Envisat, and SARAL (track 818) are within 50 km of the coast near La Rochelle, the improvement in the quality of the wet troposphere correction is visible (Figure 10). The wet troposphere correction derived from the ECMWF model output, considered as a reference, is quite stable against time and along the tracks. A clear improvement is observed between Jason-1, with an almost constant bias of five centimeters along the 50 km stretch, and Jason-2, with almost no bias up to 10–15 km from the coast that reaches 0.03 m at the coast, and also between ERS-2 with a small negative bias that increases from 20–25 km from the coast to reach −0.20 m at the coast, Envisat with almost no bias up to 10–15 km from the coast that reaches −0.13 m at the coast, and SARAL with almost no bias up to 10–15 km from the coast that reaches −0.06 m at the coast.

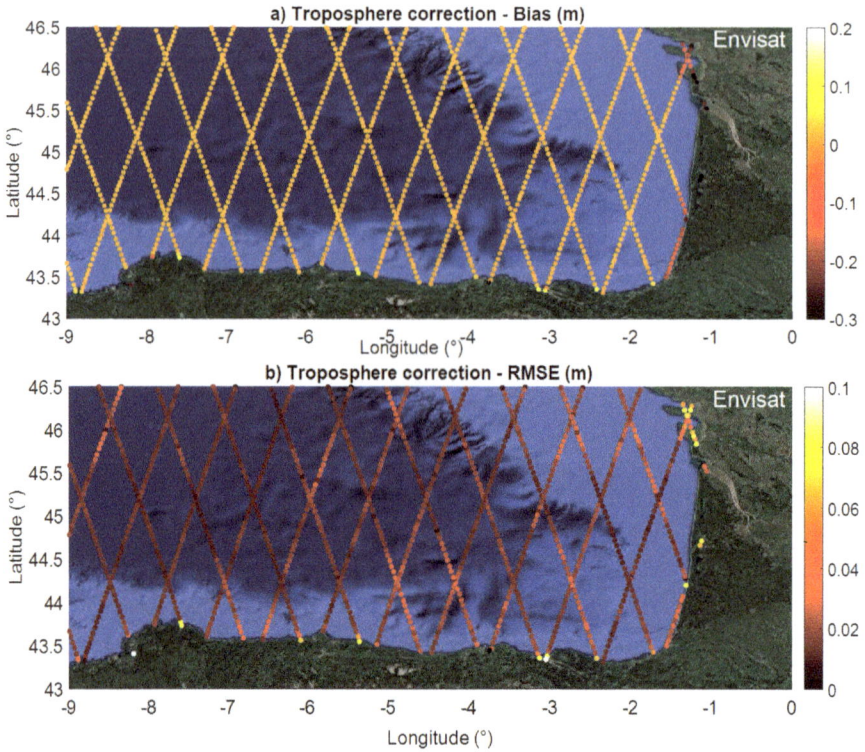

Figure 9. Comparison between wet troposphere corrections from radiometer measurements and the ECMWF model for Envisat: (**a**) bias and (**b**) RMSE.

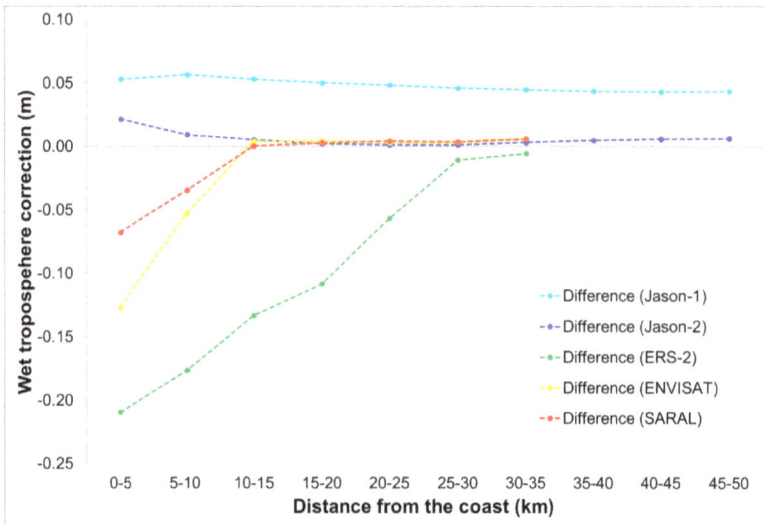

Figure 10. Difference between wet troposphere corrections from ECMWF and radiometer measurements for Jason-1 (light blue), Jason-2 (blue), ERS-2 (green), Envisat (yellow), and SARAL (red).

5. Discussion

Despite a decrease in the accuracy when approaching the coast, radar altimetry measurements can be used for monitoring the sea surface height close to the shore. The comparisons made with tide gauge records from stations located along the French Atlantic coast in the southern Bay of Biscay showed that the accuracy of the SSH estimates increased for the more recent missions on the 35-day repeating orbit from ERS-2 to SARAL, whereas similar results were observed between Jason-1 and Jason-2 on the 10-day repeating period orbit (Tables 3 and 4). Compared to ERS-2, Envisat benefits from having more acquisitions in ocean mode, at a higher bandwidth frequency mode, that allows a more accurate determination for the middle of the leading edge, and therefore for the altimeter range. SARAL, the first mission to operate at Ka-band, benefits from its smaller footprint (~8 km) [9,50] compared to all the previous altimetry missions (i.e., ~20 km for Jason-2 and ~15 km for Envisat) to obtain better estimates in terms of R (higher than 0.99), and the majority of the time, in terms of RMSE (generally lower than 0.2 m) (Table 3). In this study, if the RMSE values were large, the results presented here were obtained very close to the coast in a macro-tidal environment where the standard deviation of the SSH was generally higher than one meter (Tables 3 and 4). As seen in Figure S1, for a distance more than 10 km from the coast, the results obtained were similar to what was found at the Cal/Val sites. Better results in terms of R and RMSE were obtained when the altimeter ground-track passed from the ocean to land (e.g., track 859 of ERS-2, Envisat, and SARAL). Due to the complexity of the environment under the Jason-1 and Jason-2 tracks around the La Rochelle tide gauge, with the presence of Ré and Oléron islands, no similar conclusion was obtained. Comparisons were performed not only using classical bi-frequency-based ionosphere corrections and radiometer-derived wet troposphere corrections, as measurements were performed over open ocean, but also using corrections derived from model outputs over land, including GIM for the ionosphere and ECMWF for the wet troposphere. The analysis of these different types of corrections showed very good agreement over open ocean but more discrepancies for distances less than 25 km from the coast. However, these results were variable depending on the mission. This latter result shows that model-based ionosphere and wet troposphere corrections could be used for homogeneous long-term altimetry-based SSH from multiple missions, especially given the loss of the secondary frequency, as for Envisat. The deconvolution technique used to filter out the land contribution in the radiometer measurements enables the use of this correction up to the coast in the case of Jason-2 (Figure 10). Efforts to homogenize the corrections are needed to build long term records of altimetry-based SSH. In the near future, the Pertuis Charentais region, encompassed in a few tenths of kilometers by four tide gauge stations (La Rochelle, l'Ile d'Aix, La Cotinière, and Le Chapus) could be the first Cal/Val site located along the Atlantic coast. This group of stations benefit from numerous ascending and descending tracks from all the high-precision altimetry missions including Cryosat-2, Sentinel-3A, and the future Sentinel-3B. These stations also have a large diversity in their sea state due to their configuration in the open ocean, away from Ré and Oléron islands and the coastal part between these islands and the shore.

6. Conclusions

This study provides a thorough assessment of the performance of radar altimetry-based SSH for all the missions in orbit from ERS-2 to SARAL with the exception of Geosat Follow-On. From all the retracking algorithms, MLE-4 provided the most accurate estimates, especially for distances less than 10 km from the coast. The most recent missions, whose data were considered in this study, showed a clear improvement in the quality of the altimetry-derived SSH data. Results of the comparisons between in situ and Jason-2-based SSH within five kilometers of the coast had a good RMSE, lower than 0.34 m. As the descending track of ERS-2 and Envisat passes from land to ocean, the RMSE obtained was higher compared to the same in situ tide gauges, 1.4 m and 0.9 m, respectively. Due to the large footprint size of Envisat and ERS-2 (18 km), land areas were encompassed, thereby reducing accuracy. The best RMSE was less than 0.21 m for the SSH derived from the SARAL altimeter and the in situ tide gauge at the same nine locations. The comparison statistics showed that the SARAL altimeter provides better results than other missions over the coastal area. This is due to both improvements in the processing algorithms (MLE-4 vs. MLE-3) and the use of the Ka-band for AltiKa onboard SARAL that is characterized by a smaller footprint that enables obtaining valid range estimates closer to the shore than when using the classical Ku-band.

This study showed that model-based corrections of the ionosphere (GIM) and wet troposphere (ECMWF) delays provide more realistic estimates close to the shore than the bi-frequency-based ionosphere and radiometer-based wet troposphere corrections that are generally used. Besides, these model-based corrections provide very similar values as the instrumental corrections over the open ocean, demonstrating their potential for building homogenous long term datasets.

Better estimates of SSH are likely to be provided by Jason-3, launched 17 January 2016, which can operate in open-loop mode, taking into account a priori ranges from a digital elevation model (DEM) and Sentinel-3, launched 16 February 2016. Sentinel-3 is the first altimeter with a high-resolution mode, also known as Synthetic Aperture Radar mode (SAR), on a repetitive orbit and that can also operate in open-loop. This can increase the number of independent measurements over coastal areas, ice sheet margins, land, and inland waters. The new-era coastal altimetry satellites are becoming an important part of coastal ocean observing systems [47,51].

Supplementary Materials: The following are available online at www.mdpi.com/2072-4292/10/1/93/s1. Figure S1: Comparisons between in situ and *SSH* Jason-2 trace 213 in La Rochelle as a function of the distance from the coast with method MLE-3 and MLE-4. (a) Bias, (b) RMSE, (c) Number of observations, (d) Correlation coefficients R, Figure S2: Comparison between ionosphere corrections from bi-frequency measurements using MLE-4 retracking algorithm and GIM model for Jason-2 (a) Bias, (b) RMSE, Figure S3. Comparison between ionosphere corrections from bi-frequency measurements using MLE-3 retracking algorithm and GIM model for Jason-2 (a) Bias, (b) RMSE , Figure S4: Comparison between wet troposphere corrections from radiometer measurements and ECMWF model for ERS-2 (a) Bias, (b) RMSE, Figure S5: Comparison between wet troposphere corrections from radiometer measurements and ECMWF model for Jason-2 (a) Bias, (b) RMSE, Figure S6: Comparison between wet troposphere corrections from radiometer measurements and ECMWF model for SARAL (a) Bias, (b) RMSE.

Acknowledgments: This study was supported by Centre National de la Recherche Scientifique (CNRS) through the framework of MISTRAL project, Centre National d'Etudes Spatiales through OSTST project "FOAM" and TOSCA project "CTOH". PLV gratefully acknowledge the Vietnamese Government's 911 project for funding during my thesis, without which the present study could not have been completed. We thank three anonymous Reviewers for their very helpful comments that helped us in improving the quality of the manuscript.

Author Contributions: All of the authors of the present work contributed to the discussion of the results, as well as the writing of the manuscript. Phuong Lan Vu and Frédéric Frappart designed the study. Phuong Lan Vu, Frédéric Frappart and José Darrozes analyzed and discussed the preliminary results. Frédéric Frappart, Vincent Marieu and Fabien Blarel provided the altimetry from Multi—satellite dataset and MAPS software.

Conflicts of Interest: The authors declare no conflicts of interest.

References

1. Fu, L.L.; Cazenave, A. *Satellite Altimetry and Earth Sciences: A Handbook of Techniques and Applications*; International Geophysics Series; Academic Press: San Diego, CA, USA, 2001; Volume 69, 463p.
2. Frappart, F.; Blumstein, D.; Cazenave, A.; Ramillien, G.; Birol, F.; Morrow, R.; Rémy, F. Satellite Altimetry: Principles and Applications in Earth Sciences. In *Wiley Encyclopedia of Electrical and Electronics Engineering*; Webster, J., Ed.; John Wiley & Sons: Hoboken, NJ, USA, 2017; pp. 1–25.
3. Deng, X.; Featherstone, W.E.; Hwang, C.; Berry, P.A.M. Estimation of contamination of ERS-2 and POSEIDON satellite radar altimetry close to the coasts of Australia. *Mar. Geodesy* **2002**, *25*, 249–271. [CrossRef]
4. Deng, X.; Featherstone, W.E. A coastal retracking system for satellite radar altimeter waveforms: Application to ERS-2 around Australia. *J. Geophys. Res.* **2006**, *111*. [CrossRef]
5. Gommenginger, C.; Thibaut, P.; Fenoglio-Marc, L.; Qyartly, G.; Deng, X.; Gomez-Enri, J.; Challenor, P.; Gao, Y. Retracking altimeter waveforms near the coasts: A review of retracking methods and some applications to coastal waveforms. In *Coastal Altimetry*; Vignudelli, S., Kostianoy, A., Cipollini, P., Benveniste, J., Eds.; Springer: Berlin/Heidelberg, Germany, 2011; pp. 61–101.
6. Bonnefond, P.; Exertier, P.; Laurain, O.; Thibaut, P.; Mercier, F. GPS-based sea level measurements to help the characterization of land contamination in coastal areas. *Adv. Space Res.* **2013**, *51*, 1383–1399. [CrossRef]

7. Andersen, O.B.; Scharroo, R. Range and geophysical corrections in coastal regions: And implications for mean sea surface determination. In *Coastal Altimetry*; Vignudelli, S., Kostianoy, A., Cipollini, P., Benveniste, J., Eds.; Springer: Berlin/Heidelberg, Germany, 2011; pp. 103–146.

8. Robinson, A.R.; Brink, K.H. The Global Coastal Ocean: Multiscale Interdisciplinary Processes. In *The Sea*; Harvard University Press: Cambrige, UK, 2005; Volume 13, p. 1062.

9. Verron, J.; Sengenes, P.; Lambin, J.; Noubel, J.; Steunou, N.; Guillot, A.; Picot, N.; Coutin-Faye, S.; Gairola, R.; Raghava Murthy, D.V.A.; et al. The SARAL/AltiKa altimetry satellite mission. *Mar. Geodesy* **2015**, *38*, 2–21. [CrossRef]

10. Wingham, D.J.; Francis, C.R.; Baker, S.; Bouzinac, C.; Brockley, D.; Cullen, R.; de Chateau-Thierry, P.; Laxon, S.W.; Mallow, U.; Mavrocordatos, C.; et al. CryoSat: A mission to determine the fluctuations in Earth's land and marine ice fields. *Adv. Space Res.* **2006**, *37*, 841–871. [CrossRef]

11. Obligis, E.; Desportes, C.; Eymard, L.; Fernandes, M.J.; Lázaro, C.; Nunes, A. Tropospheric Corrections for Coastal Altimetry. In *Coastal Altimetry*; Vignudelli, S., Kostianoy, A., Cipollini, P., Benveniste, J., Eds.; Springer: Berlin/Heidelberg, Germany, 2011; pp. 147–176.

12. Passaro, M.; Cipollini, P.; Vignudelli, S.; Quartly, G.D.; Snaith, H.M. ALES: A multi-mission subwaveform retracker for coastal and open ocean altimetry. *Remote Sens. Environ.* **2014**, *145*, 173–189. [CrossRef]

13. Fernandes, M.J.; Lázaro, C.; Nunes, A.L.; Scharroo, R. Atmospheric Corrections for Altimetry Studies over Inland Water. *Remote Sens.* **2014**, *6*, 4952–4997. [CrossRef]

14. Birol, F.; Fuller, N.; Lyard, F.; Cancet, M.; Niño, F.; Delebecque, C.; Fleury, S.; Toublanc, F.; Melet, A.; Saraceno, M. Coastal applications from nadir altimetry: Example of the X-TRACK regional products. *Adv. Space Res.* **2017**, *59*, 936–953. [CrossRef]

15. Bonnefond, P.; Exertier, P.; Laurain, O.; Ménard, Y.; Orsoni, A.; Jan, G.; Jeansou, E. Absolute Calibration of Jason-1 and TOPEX/Poseidon Altimeters in Corsica. *Mar. Geodesy* **2003**, *26*, 261–284. [CrossRef]

16. Haines, B.J.; Desai, S.D.; Born, G.H. The Harvest experiment of the climate data record from TOPEX/Poseidon, Jason-1 and the ocean surface topography mission. *Mar. Geodesy* **2010**, *33*, 91–113. [CrossRef]

17. Mertikas, S.P.; Ioannides, R.T.; Tziavos, I.N.; Vergos, G.S.; Hausleitner, W.; Frantzis, X.; Tripolitsiotis, A.; Partsinevelos, P.; Andrikopoulos, D. Statistical models and latest results in the determination of the absolute bias for the radar altimeters of Jason satellites using the Gavdos facility. *Mar. Geodesy* **2010**, *33*, 114–149. [CrossRef]

18. Watson, C.S.; White, N.C.; Church, J.A.; Burgette, R.J.; Tregoning, P.; Coleman, R. Absolute calibration in Bass Strait, Australia: TOPEX, Jason-1 and OSTM/Jason-2. *Mar. Geodesy* **2011**, *34*, 242–260. [CrossRef]

19. Pingree, R.D.; Mardell, G.T.; New, A.L. Propagation of internal tides from the upper slopes of the Bay of Biscay. *Nature* **1986**, *321*, 154–158. [CrossRef]

20. New, A.L. Internal tidal mixing in the Bay of Biscay. *Deep Sea Res. Part A Oceanogr. Res. Pap.* **1988**, *35*, 691–709. [CrossRef]

21. Pairaud, I.L.; Lyard, F.; Auclair, F.; Letellier, T.; Marsaleix, P. Dynamics of the semi-diurnal and quarter-diurnal internal tides in the Bay of Biscay. Part 1: Barotropic tides. *Cont. Shelf Res.* **2008**, *28*, 1294–1315. [CrossRef]

22. Iijima, B.A.; Harris, I.L.; Ho, C.M.; Lindqwister, U.J.; Mannucci, A.J.; Pi, X.; Reyes, M.J.; Sparks, L.C.; Wilson, B.D. Automated daily process for global ionospheric total electron content maps and satellite ocean altimeter ionospheric calibration based on Global Positioning System data. *J. Atmos. Sol.-Terr. Phys.* **1999**, *61*, 1205–1218. [CrossRef]

23. Scharroo, R.; Smith, W.H.F. Global positioning system-based climatology for the total electron content in the ionosphere. *J. Geophys. Res.* **2010**, *115*. [CrossRef]

24. CTOH: Centre for Topographic Studies of the Oceans and Hydrosphere. Available online: http://ctoh.legos. obs-mip.fr (accessed on 4 December 2017).

25. United States Department of Agriculture. Available online: https://ipad.fas.usda.gov/cropexplorer/global_ reservoir (accessed on 4 December 2017).

26. Carayon, G.; Steunou, N.; Courrière, J.; Thibaut, P. Poseidon-2 radar altimeter design and results of in-flight performances. *Mar. Geodesy* **2003**, *26*, 159–165. [CrossRef]

27. Desjonquères, J.-D.; Carayon, G.; Steunou, N.; Lambin, J. Poseidon-3 Radar Altimeter: New Modes and In-Flight Performances. *Mar. Geodesy* **2010**, *33*, 53–79. [CrossRef]

28. Zelli, C. ENVISAT RA-2 advanced radar altimeter: Instrument design and prelaunch performance assessment review. *Acta Astronaut.* **1999**, *44*, 323–333. [CrossRef]
29. Service Hydrographique et Océanographique de la Marine. Available online: http://data.shom.fr/donnes/refmar (accessed on 4 December 2017).
30. Besser, G. *Marées*; Service Hydrographique et Océanographique de la Marine: Paris, France, 1985.
31. Charria, G.; Lazure, P.; le Cann, B.; Serpette, A.; Reverdin, G.; Louazel, S.; Batifoulier, F.; Dumas, F.; Pichon, A.; Morele, Y. Surface layer circulation derived from Lagrangian drifters in the Bay of Biscay. *J. Mar. Syst.* **2013**, *109*, 60–76. [CrossRef]
32. Frouin, R.; Fiúza, A.F.G.; Ambar, I.; Boyd, T.J. Observations of a Poleward Surface Current off the Coasts of Portugal and Spain During Winter. *J. Geophys. Res.* **1990**, *95*, 679–691. [CrossRef]
33. Le Cann, B.; Pingree, R.D. Structure, strength, and seasonality of the slope current in the Bay of Biscay region. *J. Mar. Biol. Assoc.* **1990**, *70*, 857–885.
34. Lazune, P.; Jegou, A.-M.; Kerdreux, M. Analysis of salinity measurements near islands on the French continental shelf of the Bay of Biscay. *Sci. Mar.* **2006**, *70*, 7–14.
35. Pichon, A.; Correard, S. Internal tides modelling in the Bay of Biscay. Comparisons with observations. *Sci. Mar.* **2006**, *70*, 65–88. [CrossRef]
36. Ménard, Y.; Jeansou, E.; Vincent, P. Calibration of the TOPEX/POSEIDON Altimeters Additional Results at Harvest. *J. Geophys. Res.* **1994**, *99*, 24487–24504. [CrossRef]
37. Chelton, D.B.; Ries, J.C.; Haines, B.J.; Fu, L.-L.; Callahan, P.S. Satellite Altimetry. In *Satellite Altimetry and Earth Sciences*; Fu, L.-L., Cazenave, A., Eds.; Academic Press: San Diego, CA, USA, 2001; pp. 1–131.
38. Frappart, F.; Papa, F.; Marieu, V.; Malbéteau, Y.; Jordy, F.; Calmant, S.; Durand, F.; Bala, S. Preliminary assessment of SARAL/AltiKa observations over the Ganges-Brahmaputra and Irrawaddy Rivers. *Mar. Geodesy* **2015**, *38*, 568–580. [CrossRef]
39. Frappart, F.; Roussel, N.; Biancale, R.; Martinez Benjamin, J.J.; Mercier, F.; Pérosanz, F.; Garate Pasquin, J.; Martin Davila, J.; Perez Gomez, B.; Gracia Gomez, C.; et al. The 2013 Ibiza calibration campaign of Jason-2 and SARAL altimeters. *Mar. Geodesy* **2015**, *38*, 219–232. [CrossRef]
40. Frappart, F.; Legrésy, B.; Niño, F.; Blarel, F.; Fuller, N.; Fleury, S.; Birol, F.; Calmant, S. An ERS-2 altimetry reprocessing compatible with ENVISAT for long-term land and ice sheets studies. *Remote Sens. Environ.* **2016**, *184*, 558–581. [CrossRef]
41. Biancamaria, S.; Frappart, F.; Leleu, A.S.; Marieu, V.; Blumstein, D.; Boy, F.; Sottolichio, A.; Valle-Levinson, A. Satellite radar altimetry water elevations performance over a hundred meter wide river: Evaluation over the Garonne River. *Adv. Space Res.* **2017**, *59*, 128–146. [CrossRef]
42. Amarouche, L.; Thibaut, P.; Zanife, O.Z.; Dumont, J.-P.; Vincent, P.; Steunou, N. Improving the Jason-1 ground retracking to better account for attitude effects. *Mar. Geodesy* **2004**, *27*, 171–197. [CrossRef]
43. Rodríguez, E. Altimetry for non-Gaussian oceans: Height biases and estimation of parameters. *J. Geophys. Res.* **1988**, *93*, 14107–14120. [CrossRef]
44. Brown, G. The average impulse response of a rough surface and its applications. *IEEE Trans. Antennas Propag.* **1977**, *25*, 67–74. [CrossRef]
45. Astudillo, O.; Dewitte, B.; Mallet, M.; Frappart, F.; Rutllant, J.; Ramos, M.; Bravo, L.; Goubanova, K.; Illig, S. Surface winds off Peru-Chile: Observing closer to the coast from radar altimetry. *Remote Sens. Environ.* **2017**, *191*, 179–196. [CrossRef]
46. Blarel, F.; Frappart, F.; Legrésy, B.; Blumstein, D.; Rémy, F.; Fatras, C.; Mougin, E.; Papa, F.; Prigent, C.; Niño, F.; et al. Altimetry backscattering signatures at Ku and S bands over land and ice sheets. In *Remote Sensing for Agriculture, Ecosystems, and Hydrology XVII*; Neale, C.M.U., Maltese, A., Eds.; Society of Photo-Optical Instrumentation Engineers (SPIE): Bellingham, WA, USA, 2015.
47. Martin-Puig, C.; Leuliette, E.; Lillibridge, J.; Roca, M. Evaluating the Performance of Jason-2 Open-Loop and Closed-Loop Tracker Modes. *Am. Meteorol. Soc.* **2016**. [CrossRef]
48. Vignudelli, S.; Kostianoy, A.G.; Cipollini, P.; Benveniste, J. *Coastal Altimetry*; Springer-Verlag: Berlin/Heidelberg, Germany, 2011; p. 578.
49. Brown, S. A Novel Near-Land Radiometer Wet Path-Delay Retrieval Algorithm: Application to the Jason-2/OSTM Advanced Microwave Radiometer. *Geosci. Remote Sens.* **2010**, *48*, 1986–1992. [CrossRef]

50. Bonnefond, P.; Verron, J.; Aublanc, J.; Babu, K.N.; Bergé-Nguyen, M.; Cancet, M.; Chaudhary, A.; Crétaux, J.-F.; Frappart, F.; Haines, B.J.; et al. The Benefits of the Ka-Band as Evidenced from the SARAL/AltiKa Altimetric Mission: Quality Assessment and Unique Characteristics of AltiKa Data. *Remote Sens.* **2018**, *10*, 83. [CrossRef]
51. Liu, Y.; Kerkering, H.; Weisberg, R.H. *Coastal Ocean Observing Systems*; Elsevier: Amsterdam, The Netherlands, 2015; p. 461. ISBN 978-0-12-802022-7.

remote sensing

MDPI

Article

Calibrating the SAR SSH of Sentinel-3A and CryoSat-2 over the Corsica Facilities

Pascal Bonnefond [1,*], Olivier Laurain [2], Pierre Exertier [2], François Boy [3], Thierry Guinle [3], Nicolas Picot [3], Sylvie Labroue [4], Matthias Raynal [4], Craig Donlon [5], Pierre Féménias [6], Tommaso Parrinello [6] and Salvatore Dinardo [7]

[1] SYRTE, Observatoire de Paris, PSL Research University, CNRS, Sorbonne Universités, UPMC University Paris 06, LNE, 75014 Paris, France
[2] Géoazur, Observatoire de la Côte d'Azur, 06905 Sophia-Antipolis CEDEX, France; Olivier.Laurain@oca.eu (O.L.); Pierre.Exertier@oca.eu (P.E.)
[3] Centre National d'Etudes Spatiales (CNES), 31401 Toulouse CEDEX, France; Francois.Boy@cnes.fr (F.B.); Thierry.Guinle@cnes.fr (T.G.); Nicolas.Picot@cnes.fr (N.P.)
[4] Collecte Localisation Satellites (CLS), 31520 Ramonville Saint-Agne, France; Sylvie.Labroue@cls.fr (S.L.); mraynal@cls.fr (M.R.)
[5] European Space Agency/European Space Research and Technology Centre (ESA/ESTEC), 2201 AZ Noordwijk, The Netherlands; Craig.Donlon@esa.int
[6] European Space Agency/European Space Research Institute (ESA/ESRIN), 00044 Frascatti, Italy; Pierre.Femenias@esa.int (P.F.); Tommaso.Parrinello@esa.int (T.P.)
[7] HeSpace, 64293 Darmstadt, Germany; Salvatore.Dinardo@eumetsat.int
*   Correspondence: Pascal.Bonnefond@obspm.fr; Tel.: +33-1-4051-2229

Received: 20 November 2017; Accepted: 10 January 2018; Published: 11 January 2018

Abstract: Initially developed to monitor the performance of TOPEX/Poseidon and to follow the Jason legacy satellite altimeters at Senetosa Cape, Corsica, this calibration/validation site has been extended to include a new location at Ajaccio. This addition enables the site to monitor Envisat and ERS missions, CryoSat-2 and, more recently, the SARAL/AltiKa mission and Sentinel-3A satellites. Sentinel-3A and CryoSat-2 carry altimeters that use a synthetic aperture radar (SAR) mode that is different to the conventional pulse-bandwidth limited altimeters often termed "low resolution mode" (LRM). The aim of this study is to characterize the sea surface height (SSH) bias of the new SAR altimeter instruments and to demonstrate the improvement of data quality close to the coast. Moreover, some passes of Sentinel-3A and CryoSat-2 overfly both Senetosa and Ajaccio with only a few seconds time difference, allowing us to evaluate the reliability and homogeneity of both ground sites in term of geodetic datum. The Sentinel-3A and CryoSat-2 SSH biases for the SAR mode are respectively +22 ± 7 mm and −73 ± 5 mm (for CryoSat-2 baseline C products). The results show that the stability of the SAR SSH bias time series is better than standard LRM altimetry. Moreover, compared to standard LRM data, for which the measurements closer than ~10 km from the coast were generally unusable, SAR mode altimeters provide measurements that are reliable at less than few hundred meters from the coast.

Keywords: altimetry; SAR; calibration; validation

---

1. Introduction

The Corsica geodetic facilities that are located both at Senetosa Cape for the TOPEX/Poseidon (T/P) and Jason satellite ground tracks and near Ajaccio for the Envisat satellite ground tracks have been developed to calibrate successive satellite altimeters in an absolute sense. Since 1998, the successful calibration process used to calibrate many oceanographic satellite altimeter missions has been regularly updated, see details in [1–3].

One of the main issues in comparing offshore altimetric sea surface height (SSH) measurements together with in-situ tide gauge data is taking into account the geoid slope (of several cm/km) over distances of tens of kilometers between both locations. As discussed in [4], other factors including ocean dynamics and tidal differences, impact the comparisons with independent coastal in situ measurements, but are negligible in such areas and within the short distance in our study (less than 20 km). A specific Global Positioning System (GPS) campaign was conducted in 1999 in order to determine a geoid map of about 20 km long and 5.4 km wide centered on the T/P and Jason satellite's ground track #085 at the Senetosa Cape site [4]. In addition, a second campaign was conducted in 2005 in the Ajaccio area under the Envisat ground track #130 (Figure 1) [5,6]. Both local geoids are the key datum for the absolute calibration of the satellite altimeters. Indeed, any distortion or systematic errors between offshore and tide gauge locations will cause a systematic error in the absolute SSH bias. From this point of view, the calibration process also requires accurate vertical offsets to be determined for each individual tide gauge in use. In order to achieve and maintain accuracy at the level of just a few millimeters or even less, geodetic campaigns must be carried out regularly in order to check or to eventually update such offsets.

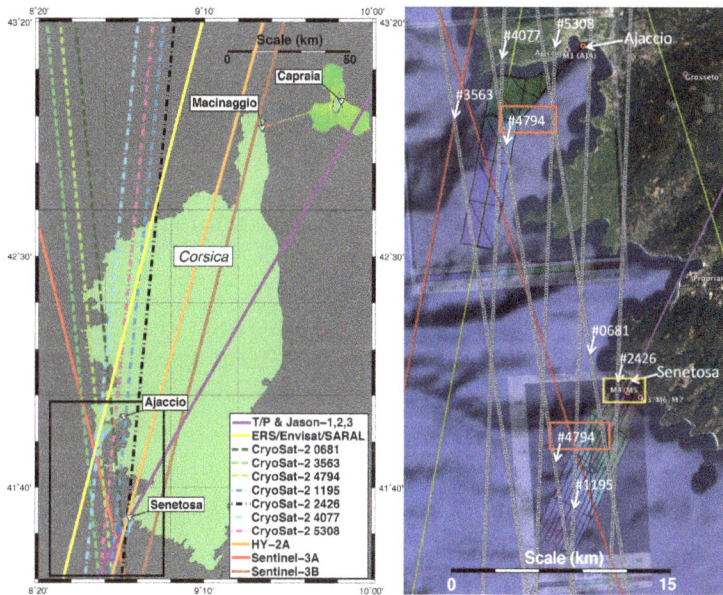

Figure 1. Left: General configuration of the Corsica calibration site with all of the satellite altimeter missions ground tracks that have been monitored since 1998 (the black frame on the left bottom corresponds to the zoom on the right map). **Right**: zoom on Ajaccio and Senetosa sites with Sentinel-3A (red) and CryoSat-2 (grey) tracks. The red rectangles highlight the CryoSat-2 pass #4794 that crosses the Senetosa and Ajaccio sites (see Section 0). The yellow rectangle shows the area that is discussed in Section 0 for pass #2426 (see Figure 8c).

The present paper focuses on the calibration of both the Sentinel-3A and CryoSat-2 missions using the same facilities and similar data processing. Sentinel-3A has a 27-day repeat cycle, where CryoSat-2 has a repeat cycle of 369 days and a 30-day sub-cycle; this repetition allows them to perform the absolute calibration over the Senetosa and Ajaccio area on a monthly basis for both satellites. Due to the configuration of Sentinel-3A repeat ground track and some CryoSat-2 passes, each altimeter overflies both the Senetosa and Ajaccio sites with a time delay of about five seconds corresponding to a distance of about 37 km (Figure 1). As a consequence, it allows us to determine two SSH biases for each mission that should be equivalent considering that, most of the time, the sea state conditions at both sites are almost identical due to the short time delay and the short distance between sites. From the analysis of all available calibration passes, systematic differences can be related to errors in the geodetic references or in the situ data. For T/P-Jason and ERS-Envisat-SARAL/AltiKa this was difficult to determine as each altimeter mission overflew only a single site, Senetosa or Ajaccio respectively. Since July 2016, the SARAL/AltiKa mission is in a drifting orbit phase allowing a similar study to the one presented here that is discussed in [7].

In this paper we first present the data and the methodology (Section 0). In the results section, Section 0, we present a re-processing of the Ajaccio tide gauge data following a re-analysis of the Global Positioning System (GPS) surveys that are used to monitor the location and any vertical movement of the tide gauges (Section 0). This resulted in an updated datum and a −30 mm vertical correction to the tide-gauge-derived SSH that is used in the calibration process of all the missions calibrated using the Ajaccio tide gauge data. The two following sections, 0 and 0, are then dedicated to the analysis of the Sentinel-3A [8] and CryoSat-2 [9] SSH biases, respectively.

2. Data and Methodology

2.1. Data

The data sets used for Sentinel-3 and CryoSat-2 are described below.

Sentinel-3A:

- Data sources: S3-PDGS (payload data ground segment) reprocessing 2017, Non-Time Critical (NTC) products: cycle 5–19, processing baseline 2.15 [10]. The processing baseline (PB) 2.15 consists of the following processors: SRAL L1 IPF V6.11, MWR L1 IPF V6.04, SRAL L2 IPF V6.07.
- Sentinel-3A SRAL (Ku/C Radar Altimeter) processing for SAR and Pseudo LRM (PLRM) and applied corrections

  ○ Standard processing: dual-frequency ionosphere measurement and wet troposphere radiometry. In this study, we also compared the dual frequency ionosphere measurements with the global ionosphere maps model (GIM), together with a comparison of the wet troposphere radiometer measurements with the wet troposphere model from European Centre for Medium-Range Weather Forecasts (ECMWF)). The chosen wet correction for the model corresponds to the field that is computed at sea level (MOD_WET_TROPO_COR_ZERO_ALTITUDE)
  ○ Dry troposphere
  ○ Sea state bias (SSB)
  ○ Solid, loading and pole tides

- In situ data

  ○ Ajaccio: Service Hydrographique et Océanographique de la Marine (SHOM) radar tide gauge data in real time
  ○ Senetosa: pressure tide gauges (data retrieved end of July 2017)

CryoSat-2:

- Data sources: from SARvatore service at ESA G-POD (https://gpod.eo.esa.int/services/CRYOSAT_SAR/) using baseline B products with two different retrackers. It is important to note that this product does not correspond to the current baseline C products processed by the CryoSat-2 payload data ground segment (PDGS), the main difference being the correction of a known range bias of 673 mm:

○   SAMOSA2 retracker (default): The SAMOSA (SAR Altimetry MOde Studies and Applications) model [11] has been implemented in its first-order formulation (SAMOSA2) in a retracker as in [12]. This retracker uses a look-up table to mitigate the effect of the model's approximation of the squared PTR (point target response) with a Gaussian curve, it implements a stack masking for the Doppler beams padded to zero and has in principle no enhancement in the coastal zone.

○   SAMOSA+ retracker (the SAMOSA2 model tailored for inland water, sea ice and the coastal zone domain): in the case of SAMOSA+, the open ocean SAMOSA [11] model and retracker [13] has been implemented in the retracking scheme with two significant additions [12]: The first one concerns the selection of the first-guess epoch. This is not selected as the position of the waveform peak but as the position of the moving correlation peak in 20 consecutive waveforms (after aligning them for tracker shift). The rationale behind this choice is to attempt to mitigate the typical off-ranging effect in coastal data. The second concerns the treatment of land-contaminated waveforms. In case waveforms are not contaminated by land, the SAMOSA model was used with the mean square slope set to zero, i.e., as in [13], whereas in the case of land contaminated waveforms, we used a dual step retracking: in the first step, the SWH is still estimated as in [13], while in the second step, the SWH was set to zero and the third free parameter in the retracking becomes the mean square slope. The output of this second step is the range and the amplitude Pu (retracked waveform amplitude, see Equation (22) in [12]).

- Cryosat-2 SIRAL (synthetic aperture interferometric radar altimeter) data processing for SAR and applied corrections

    ○   Dry troposphere
    ○   Wet troposphere from ECMWF model
    ○   Ionosphere from GIM
    ○   Sea state bias = 3.5% of SWH
    ○   Solid, loading and pole tides

- In situ data

    ○   Ajaccio: SHOM radar tide gauge data in real time
    ○   Senetosa: pressure tide gauges (data retrieved end of July 2017)

2.2. Methodology

The main difference in the processing in this study, compared to previous studies [1–3,6], is the difference in the footprint area in over which the geoid height is interpolated. In the case of LRM (low resolution mode, e.g., Jason missions), the footprint is defined as a circle using the significant wave height to define the diameter [14], while in the case of SAR (synthetic aperture radar, e.g., Sentinel-3A) the footprint is defined as a rectangle with a width of 300 m along-track and a length across-track that is also a function of SWH with the same formula as LRM (see Figure 5 or Figure 8 for an illustration).

Our standard processing of the SSH bias uses (when available) the wet tropospheric correction derived from the radiometer in addition to the ionospheric correction derived from the dual frequency of the altimeter. We performed a comparison with numerical weather prediction (NWP) model corrections that are available in the products for the wet troposphere and the ionosphere corrections, from ECMWF and GIM respectively (see Section 0).

3. Results

3.1. Solving the Vertical Offset in the Ajaccio Tide-Gauge-Derived SSH

In [3] we showed that there was an anomalous SSH bias of -30.5 ± 4.5 mm between the Ajaccio tide gauge and our GPS-based sea level measurements. We thus retrieved the GPS data from recent surveys performed in 2013, 2014 and 2015 and computed new height differences between the antenna reference point (ARP) of the permanent receiver (AJAC) and the fundamental marker close to the tide gauge (G). This processing was done in baseline mode with the GAMIT software [15] and the results are shown in Table 1. The standard errors of the individual determinations range from 0.2 to 1.1 mm mainly due to the number of days of observation. It provides a much more coherent weighted mean value of 50.0371 m, rather than the 2005 reference, with a standard error of the weighted mean of 0.4 mm (Table 1), thus changing our historical reference by 18.3 mm.

Additionally, during the deployment of the Ajaccio tide gauge in 2012, the Service Hydrographique et Océanographique de la Marine (SHOM) performed SSH observations with a contact gauge and determined a 13.0 mm bias (standard deviation of 3.6 mm) which was introduced as a "calibration offset" into the tide gauge system. We decided to remove this quantity from the instrument and by adding it to the new correction (of 18.3 mm); the total systematic error of the tide gauge SHH has been established to be 31.4 mm. This is in very good agreement with the SSH bias observed between the Ajaccio tide gauge and our GPS-based sea level measurements (of −30.5 ± 4.5 mm [3]). We then changed our datum to a rounded offset of −30 mm in the tide gauge SSH. The results presented in this paper take this correction into account.

Another geodetic survey is planned for next year to confirm the height difference between the ARP of the permanent receiver (AJAC) and the fundamental marker close to the tide gauge (G), but we are already very confident in the current result given the 0.4 mm standard error.

Table 1. Height differences between permanent Global Positioning System (GPS) (AJAC) and tide gauge (G) at Ajaccio (in m).

Year	Reference Used	AJAC-G	Number of Days
2005	50.0188 ± 0.0100		
2013		50.0386 ± 0.0011	2
2014		50.0360 ± 0.0010	3
2015		50.0370 ± 0.0002	7
Average		50.0371 ± 0.0004	

3.2. Sentinel-3A SSH Calibration and Corrections Validation

The Sentinel-3A ascending pass #741 overflies the Senetosa site and ~5 s later the Ajaccio site. For the first time this allows us to compare the SSH biases, which can be independently determined at both locations, and then to assess geodetic references together with in-situ measurements. Launched on February 16th 2016, Sentinel-3A was in a ground-segment ramp-up phase during the first months with different versions of the IPF (instrument processing facilities) making the different cycles difficult to compare. Thanks to a reprocessing campaign [10] completed in September 2017 by the respective ESA and EUMETSAT agencies for cycles 5–19 (June 2016 to July 2017) with the same Processing Baseline 2.15, it is now possible to analyze a homogeneous set of data using the Non-Time Critical (NTC) product. We thus analyzed and compared the two modes which are used to derive the sea surface height from the radar measurements: (i) the SAR (synthetic aperture radar) mode and (ii) the PLRM (pseudo LRM) mode that mimics the classic LRM. A complete description of the products used and the processing options is given in Section 0.

Concerning the ionospheric correction, both the dual-frequency altimeter derived and the GIM model were averaged over a distance of ~140 km centered on the geoid area of Senetosa and Ajaccio (slightly shifted by ~14 km for Senetosa to avoid measurements over Sardinia). The results presented in Figure 2 show good consistency between the instrument and GIM model in terms of bias (+6 mm) and stability (5–6 mm). Moreover, the results from either the Senetosa or Ajaccio sites show good consistency.

Concerning the wet tropospheric correction deduced from the satellite radiometer, we have reduced the land contamination (as the radiometer has a large footprint) by filtering to keep the non-contaminated signal within our area of study. The selected corrections are linearly interpolated with an outlier rejection at 3-σ up to the edges of the study area. They are then considered as a constant for both Senetosa and Ajaccio processing. The Senetosa interpolation area was used for both the Senetosa and Ajaccio sites (Figure 3) because in this area the radiometer is almost not contaminated by land. The permanent GPS receiver at Senetosa and Ajaccio also allows us to compute the wet tropospheric corrections at these locations. Recently, Collecte Localisation Satellites (CLS) have developed an algorithm to reduce the land contamination in coastal conditions (available for cycle five to 15) [16]. The three wet tropospheric corrections were compared and the differences are shown in Figure 2. We note that the radiometer wet tropospheric correction provided in the products appears to be biased on average (Senetosa and Ajaccio) by −17.5 mm compared to the model and −20.5 mm compared to the GPS. The differences between the two sites are very small (3–5 mm). The ~−20 mm bias is probably due to a remaining land contamination in the radiometer footprint. The analysis of the new radiometer correction developed by CLS to limit this effect shows better agreement with the correction derived from GPS. On the same time series (cycle 5–15), the bias is not only reduced by 15 mm in average but the standard deviation is also improved by 2 mm in average (Table 2).

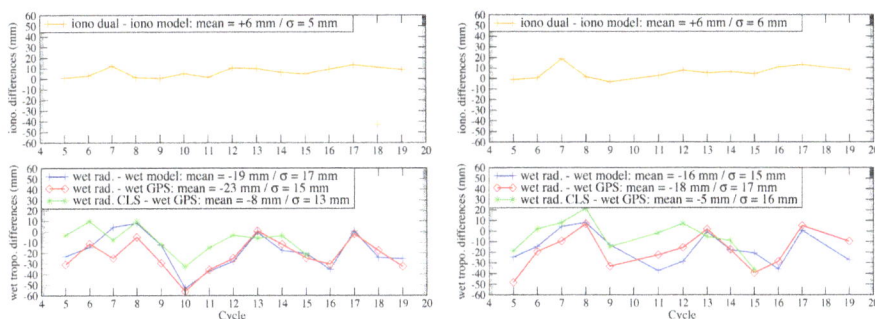

Figure 2. Sentinel-3A corrections differences at the Senetosa (**left**) and Ajaccio (**right**) calibration sites. Ionospheric correction differences in the upper panels. Wet tropospheric correction differences in the lower panels.

Figure 3. Map of the Sentinel-3A configuration: the shaded areas correspond to the locations where the wet tropospheric correction (radiometer and model) is interpolated.

A second study of the wet tropospheric correction was conducted using the Ajaccio interpolation area (Figure 3) that should be more sensitive to land contamination compared to Senetosa. All the other parameters are identical to the standard processing performed for Ajaccio. Figure 4 shows that, in this case, the wet tropospheric correction is biased by −79 mm (light blue) using the Ajaccio GPS wet tropospheric correction as a reference. The 59 mm increase in the difference between using the Ajaccio and Senetosa interpolation areas illustrates the challenge to fully remove land contamination in the radiometer footprint in such complex coastal areas. In this case, the CLS correction clearly provides a great improvement even though a residual bias of −14 mm still exists (orange). In addition, the differences between the wet tropospheric corrections deduced from both the Senetosa and Ajaccio GPS data (~37 km separation distance and only five seconds time lag for a typical satellite overpass) are given. It is worth noting that, if the corrections are equivalent (+1 mm) on average, the dispersion of 12 mm certainly reflects different meteorological conditions at both sites at the same time. We conclude and confirm that it is very important to use GPS measurements as close as possible to satellite radiometer data when computing wet tropospheric correction terms.

Table 2. Wet tropospheric correction differences on the common cycles (5–15) (in mm).

	Radiometer-GPS	
	Mean	σ
Senetosa		
Standard product	−23	16
CLS	−8	13
Ajaccio		
Standard product	−20	17
CLS	−5	16

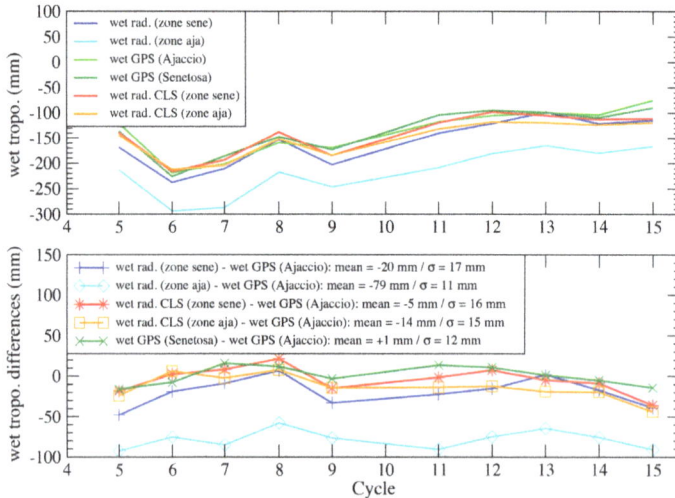

Figure 4. Sentinel-3A wet tropospheric correction differences at Ajaccio (using the Ajaccio interpolation area). Wet tropospheric correction in the upper panel: derived from satellite microwave radiometer using the Senetosa interpolation area (dark blue), radiometer using the Ajaccio interpolation area (light blue), GPS from Ajaccio (light green), GPS from Senetosa (dark green), radiometer (from Collecte Localisation Satellites (CLS)) using the Senetosa interpolation area (red) and radiometer (from CLS) using the Ajaccio interpolation area (orange). Wet tropospheric correction differences in the lower panel: reference is always the wet tropospheric correction from the Ajaccio GPS and same color code is used for the corrections being compared.

Finally, we also compared the corrections derived from the satellite instruments (radiometer and altimeter) using either SAR or PLRM processing. We found very small differences (<1 mm) that are caused by range differences between these two processing schemes. Finally, we compared the sea state bias corrections for SAR and PLRM that, for the moment, use the same correction tables: there are few millimeter differences originating from significant wave height (SWH) and wind speed differences, as the waveform shapes are different for SAR and PLRM.

The SSH calibration was performed independently at Ajaccio and Senetosa using respectively a radar tide gauge and three pressure tide gauges. Figure 5 illustrates the process used that is detailed in [3]. We recall that by convention the SSH bias corresponds to the difference ($SSH_{altimetry}$ − $SSH_{tide\ gauge}$). This process was first carried out without taking into account the Ajaccio tide gauge SSH offset (see Section 0); a 26 mm offset clearly appears in the averaged results (−5 mm for Ajaccio and +21 mm for Senetosa) highlighting that the Ajaccio offset is clearly detectable by our absolute SSH calibration process. As a result, the Ajaccio tide gauge SSH time series were corrected by −30 mm, definitively. In Figure 5, the insert in green illustrates the improvement brought by SAR processed data. Indeed, when overflying the small Sanguinaires islands, only 2–3 data look bad and can be easily detected and rejected using a simple 3-σ criteria (circled crosses). For comparison with LRM in the same area, all the Envisat data that were at a distance less than 10 km from the coast were eliminated [6]. However, it is important to note that the areas used to perform the calibration (blue boxes in Figure 5) are far enough (>10 km) to avoid any land contamination for the altimeter, except maybe in PLRM mode in the Ajaccio area.

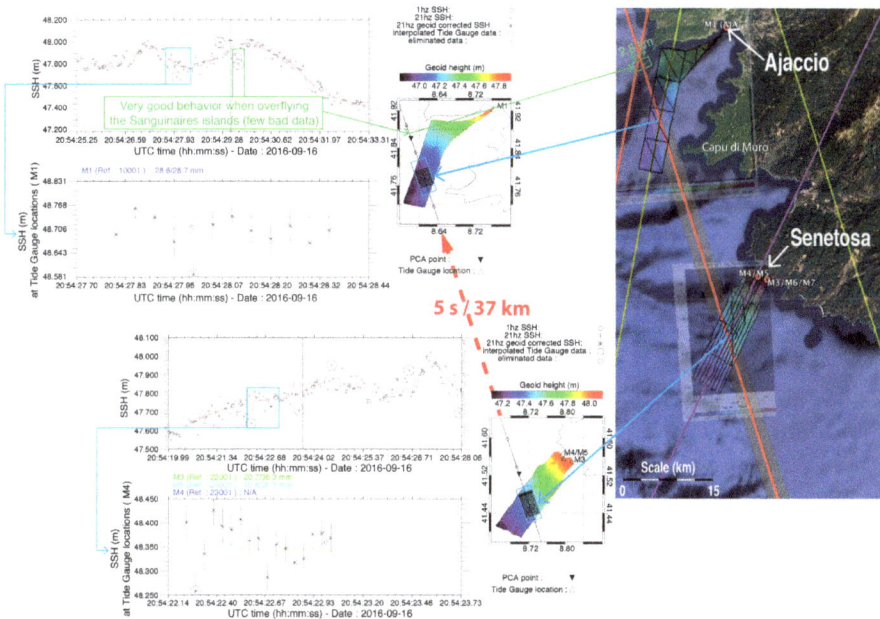

Figure 5. Example of sea surface height (SSH) bias processing for Sentinel-3A in synthetic aperture radar (SAR) mode (pass 741, cycle 8). For each independent processing (Ajaccio, top/Senetosa, bottom) the upper plot shows ±4 s of altimeter SSH series around the point of closest approach (PCA). The lower plots show the selected SSH (inside the geoid area) that are corrected from the geoid height differences between the SSH location and the tide gauge locations.

The Figure 6 upper plots shows the SSH bias time series for SAR and PLRM respectively at Senetosa (left) and Ajaccio (right), while the lower plots show the differences (SAR-PLRM) either for "orbit-range" only or total SSH bias ("orbit-range + corrections"). The Ajaccio time series looks slightly noisier for SAR and much more so for PLRM compared to the Senetosa time series. This can be explained by more possible land contamination in the radar footprint in the Ajaccio area (e.g., the Capu di Muro and Sanguinaires islands, see Figure 5). The very good consistency of the "orbit-range" and "SSH bias" differences indicates that the correction differences between SAR and PLRM are very small, as explained previously. They are mainly due to SSB differences because SWH and wind speed estimations are different between SAR and PLRM processing. For the moment, the SSB model in the product is the same for SAR and PLRM. An improvement in terms of consistency is foreseen by the further tuning of SSB models for both SAR and PLRM, instead of using the same model. Thus, the main difference is due to range estimation that is different in SAR and PLRM processing.

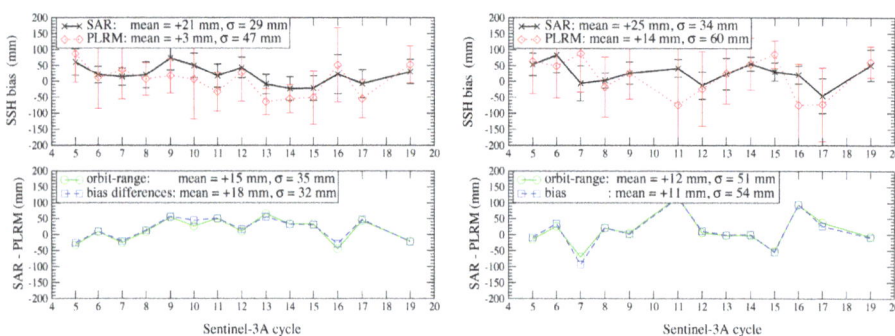

Figure 6. Sentinel-3A SSH bias time series from Senetosa (**left**) and Ajaccio (**right**). The upper panels show the time series for SAR (black crosses) and Pseudo LRM (PLRM) (red diamonds). The lower panels show the differences between SAR and PLRM for the "orbit-range" (green circles) and the total SSH bias (blue squares).

For each cycle, we have computed the averaged SSH bias from both Senetosa and Ajaccio independently (see the time series in Figure 7). First, it shows that the SSH bias for SAR is very stable with a standard deviation of 24 mm. Concerning the PLRM, as expected due to data processing issues (reconstruction of pseudo LRM waveforms from original SAR measurements that make the PLRM measurements noisier than real LRM), the standard deviation of 42 mm is higher than the classical LRM missions (e.g., 38 and 37 mm for Jason-1 and Jason-2 respectively [2]). Second, the 13 mm difference in the SSH bias between SAR and PLRM comes mainly from the differences in the altimeter range. From the entire data set (13 cycles), our best estimate of the SSH bias for Sentinel-3A is then:

- Sentinel-3A SAR: +22 ± 7 mm
- Sentinel-3A PLRM: +9 ± 12 mm

Recent studies using the transponder installed in Crete show that the SAR range bias is 0 ± 12 mm [17] and 8 ± 12 mm [18], so our results are in very good agreement because most of the SSH bias (+22 mm) originates from the radiometer wet tropospheric correction.

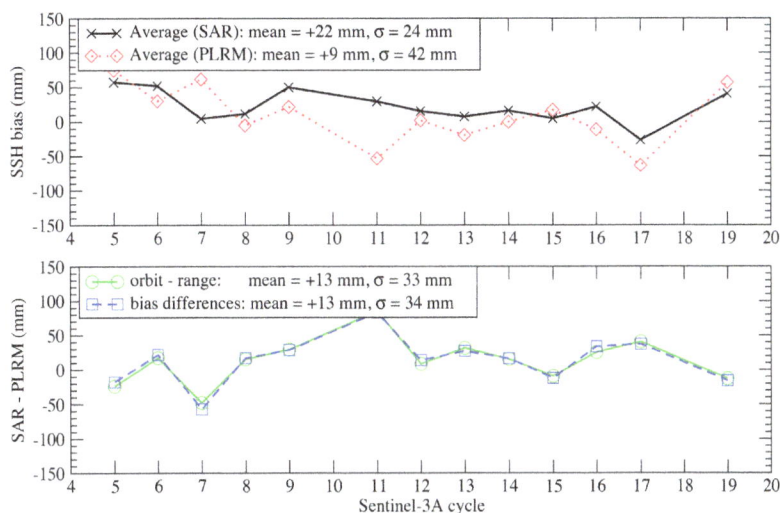

Figure 7. Average of Senetosa and Ajaccio Sentinel-3A SSH biases for both SAR and PLRM (upper panel). The lower panel shows the differences between SAR and PLRM for the "orbit-range" (green circles) and the total SSH bias (blue squares).

3.3. CryoSat-2 SSH Calibration

Some of the CryoSat-2 altimeter passes overfly the Senetosa or Ajaccio facilities (or both). These passes are listed on Figure 1 (with their relative orbit number). In this study, we analyzed all these overflights with the same methodology as used for Sentinel-3A. In terms of our data set, we used the SARvatore service [19], which is part of the European Space Agency (ESA) grid processing on demand (G-POD) service. G-POD allows us to select an area and then process all of the L1a (FBR, full bit rate) data available to provide L2 products. It is possible to choose different options, but we have left all the options as default values. G-POD uses the SAMOSA2 retracker as default. However, another retracking approach tailored for coastal zones [11,12] was also available (SAMOSA+, SAR altimetry mode studies and applications) and we generated a separate set of data using this option. The different processing schemes were then used to analyze their impact on the SSH bias: SAMOSA2 for one set and SAMOSA+ for the other. The details on the data sets are given Section 0.

The configurations of the overflights are different from one CryoSat-2 ground track pattern to another (see Figure 1 right) but most of them have data very close to the coast (a few hundred meters, see Table 3). Descending pass #2426 is particularly interesting because the closest data is at ~1400 m from the M4/M5 tide gauges at Senetosa. Figure 8 illustrates the quality of the data for cycle 1 and the improvement for some data when using SAMOSA+ retracking. For the data before the point of closest approach (PCA), the SSH from SAMOSA+ decreases and appears more homogeneous with the rest of the time series (marked by green arrows in Figure 8a,b): one measurement at only 400 m across-track from the coast has been clearly improved even with about 37% of the footprint on land (marked by red arrows in Figure 8). However, most importantly, all the data immediately after the coastline (in time after the PCA, because it is a descending track) are of very good quality (marked by blue arrows in Figure 8) and do not look contaminated by land effects even with the classic SAMOSA2 retracker. This is clearly linked to the small area of the footprint for SAR in comparison with LRM. In this example, the footprint is a rectangle with a width of 300 m along-track and a length across-track of 3000 m (see white rectangle area in Figure 8c).

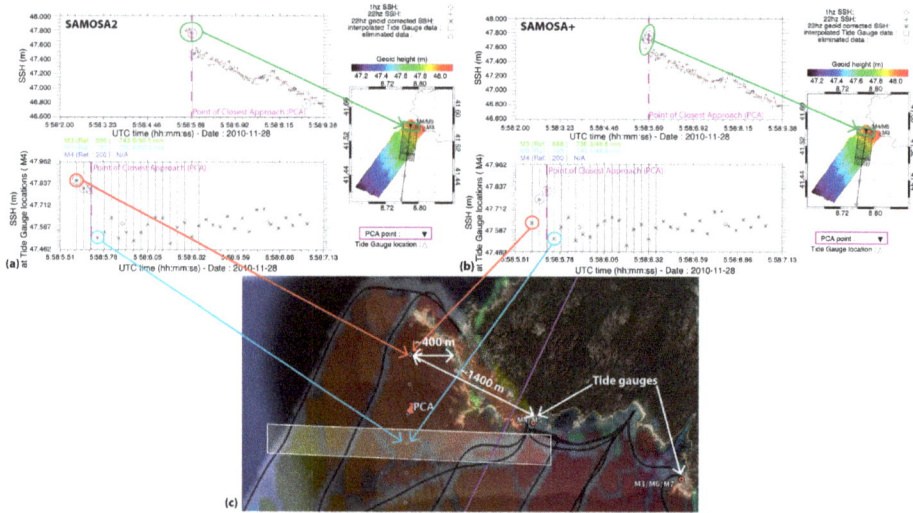

Figure 8. Example of retracking improvement for pass #2426 cycle 001. The upper panels of (**a**,**b**) show the altimeter SSH (crosses) without geoid correction, while the lower ones show the altimeter SSH after geoid correction (stars with error bars). An automatic outlier detection is performed and the rejected data are circled. (**a**) Corresponds to the G-POD standard retracking (SAMOSA2) while (**b**) corresponds to the improved retracking (SAMOSA+ is the SAMOSA2 model tailored for inland water, sea ice and coastal zone domains). (**c**) Shows a zoom of the data locations (circles) very close to the coast; the white rectangle represents the SAR footprint for one data; the black lines correspond to the Catamaran-GPS tracks [4] that were used to map the Senetosa geoid (colored map in background).

Table 3 provides the details of the SSH calibration for each pass and for the two retrackings used, and Figure 9 illustrates the time series of these SSH biases. Two passes in the Ajaccio area were not used because too few data were in the geoid area (#3563 and #5308, see Figure 1). The standard deviations of the individual mean SSH bias were respectively 14.6 mm and 14.0 mm for SAMOSA2 and SAMOSA+. This shows a small improvement for SAMOSA+, but if the same computation is done only for the "coastal" passes (#4077, #4794 and #2426), the standard deviations are this time respectively 15.2 mm and 12.5 mm. SAMOSA+ clearly improved the SSH when very close to the coast, avoiding some outliers even in the time series (see data circled in red in Figure 9 for pass #4794 at Ajaccio). The mean SSH bias for Senetosa and Ajaccio was −746 mm for both SAMOSA2 and SAMOSA+, but the overall standard deviation of the time series was improved by ~13 mm with SAMOSA+ ($\sqrt{(29^2-26^2)}$). The CryoSat-2 SSH bias is then:

- CryoSat-2 SAMOSA2: −746 ± 5.5 mm (−73 ± 5.5 mm for baseline C)
- CryoSat-2 SAMOSA+: −746 ± 5.0 mm (−73 ± 5.0 mm for baseline C)

We were unable to find in the literature another value of the CryoSat-2 absolute SSH bias (using sea level in situ measurements), but our results look very coherent with the updated Svalbard transponder result [20]: 722 ± 6 mm (standard deviation of 29 mm, baseline B, before correcting the 673 mm known range bias). It is worth noting that the 24 mm remaining difference between these two determinations can be due to the geophysical corrections, and notably that the SSB is not affecting the transponder processing.

For pass #4794, overflying both the Senetosa and Ajaccio facilities, we computed the averaged time series in the same way as for the Sentinel-3A pass shown in Figure 7. This is presented in Figure 10, where the average SSH bias from the Ajaccio and Senetosa overflights (5 s time lag and 37 km distance) showed a very low standard deviation (15 mm for SAMOSA+ and 17 mm for SAMOSA2) and a bias very close to the global average: −740 mm for SAMOSA+ and −741 mm for SAMOSA2 vs. −746 mm for the global average (for both SAMOSA+ and SAMOSA2). For comparison, with a similar configuration, the standard deviation over 13 cycles for Sentinel-3A was 24 mm.

Table 3. Details of the SSH biases for each pass and for the two retrackers (SAMOSA2 and SAMOSA+).

Passes	SAMOSA2			SAMOSA+			Comment
	Mean (mm)	σ (mm)	Number of Cycles	Mean (mm)	σ (mm)	Number of Cycles	
Ajaccio							
3563	NA	NA	NA	NA	NA	NA	Coastal *
4077	−746	32	5	−743	30	5	Coastal *
4794	−727	36	5	−731	24	5	Coastal *
5308	NA	NA	NA	NA	NA	NA	Coastal *
Average	−736	33	10	−737	26	10	
Senetosa							
0681	−766	17	5	−768	14	4	d > 3 km
1195	−733	23	3	−733	23	3	d > 8 km
2426	−757	36	4	−756	35	4	Coastal *
4794	−742	26	6	−746	23	6	d > 13 km
Average	−751	27	18	−751	26	17	
Total	−746	29	28	−746	26	27	

* Coastal means that we take the measurements up to the coast; for the others the distance "d" is the closest measurement to the coast.

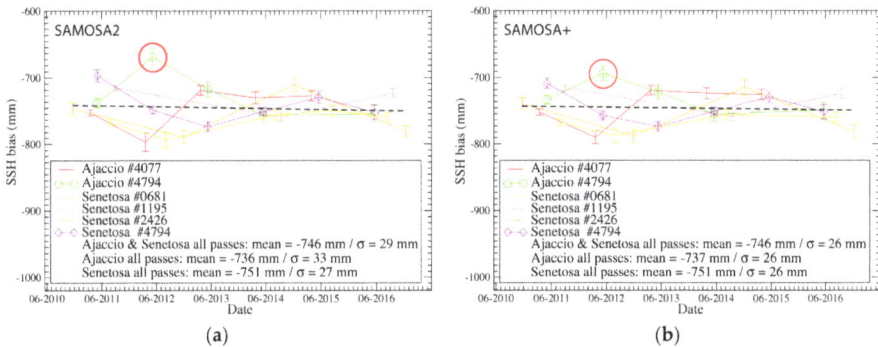

(a) (b)

Figure 9. Time series of the CryoSat-2 SSH biases for each pass and the averages from Ajaccio, Senetosa and both (dash black line): (**a**) corresponds to the G-POD standard retracking (SAMOSA2) while (**b**) corresponds to the improved retracking (SAMOSA+) tuned for coastal, inland waters and ice.

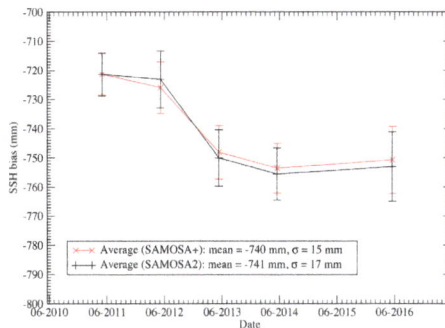

Figure 10. Example of pass #4794 with a configuration « à la » Sentinel-3A from which the SSH biases from Ajaccio and Senetosa were averaged (using SAMOSA+).

4. Discussion and Conclusions

The configuration of the Sentinel-3A and CryoSat-2 passes allowed us to cross-compare the results obtained independently from both the Senetosa and Ajaccio sites. Our results confirm previous results that were obtained from SARAL/AltiKa together with GPS-based sea level measurements [3]. The SSH measured by the Ajaccio tide gauge is biased by ~30 mm and our present study shows the origin of this offset. By taking into account this offset, we compared results from the Senetosa and Ajaccio sites for both the Sentinel-3A and CryoSat-2 altimeters. For the Sentinel-3A the site-difference between Senetosa and Ajaccio was −4 mm (21–25 mm, see Figure 6) while for CryoSat-2 the difference was −14 mm ((−751)–(−737)) mm, see Figure 9 from SAMOSA+ results). These comparisons show that no significant systematic error remains in the geodetic datum or in situ measurements. This is a very important result that gives us confidence also for the historical calibrations performed independently at each site. However, differences still exist at the millimeter level between the SSH bias obtained at the Senetosa and Ajaccio sites for Sentinel-3A and CryoSat-2 (respectively −4 and −14 mm). This probably reflects some remaining errors in the altimeter measurement system (coastal contamination, sea state bias, …) and/or in the local geodetic datum of both the Senetosa and Ajaccio sites. Finally, our analysis of the wet tropospheric corrections available for the Sentinel-3A calibration showed that the effect of land contamination on the radiometer was ~−20 mm, which is actually the main part of the SSH bias (+22 ± 7 mm). In terms of the range itself, we can consider that SAR measurement system on Sentinel-3A as "unbiased". For CryoSat-2, it is important to recall that the product used from G-POD is based on baseline B, so to be comparable to the current baseline C, one must remove the known range bias of 673 mm to the SSH biases determined in this study: when using baseline C products, the CryoSat-2 SSH bias is −73 ± 5 mm. Last but not least, the SSH biases for both Sentinel-3A and CryoSat-2 were compared to the range biases determined with transponders located respectively in Crete and Svalbard, and the remaining differences (respectively 22 mm and 24 mm) between these two determinations can be due to geodetic datum errors or geophysical corrections and, notably, the sea state bias that is not affecting transponder processing. This is probably the first time in altimetry that the results from transponders and in situ SSH calibration agree at this level of accuracy.

Acknowledgments: This study was conducted and financed thanks to Centre National d'Etudes Spatiales (CNES), Centre National de la Recherche Scientifique (CNRS), and the French Ministry of Research.

Author Contributions: Pascal Bonnefond and Olivier Laurain designed the study, analyzed the data and wrote the paper. All the other co-authors helped in paper writing and analysis.

Conflicts of Interest: The authors declare no conflict of interest.

References

1. Bonnefond, P.; Exertier, P.; Laurain, O.; Menard, Y.; Orsoni, A.; Jan, G.; Jeansou, E. Absolute Calibration of Jason-1 and TOPEX/Poseidon Altimeters in Corsica Special Issue on Jason-1 Calibration/Validation. *Mar. Geod.* **2003**, *26*, 261–284. [CrossRef]
2. Bonnefond, P.; Exertier, P.; Laurain, O.; Jan, G. Absolute Calibration of Jason-1 and Jason-2 Altimeters in Corsica during the Formation Flight Phase. *Mar. Geodesy* **2010**, *33*, 80–90. [CrossRef]
3. Bonnefond, P.; Laurain, O.; Exertier, P.; Guillot, A.; Picot, N.; Cancet, M.; Lyard, F. SARAL/AltiKa absolute calibration from the multi-mission Corsica facilities. *Mar. Geodesy* **2015**, *38*, 171–192. [CrossRef]
4. Bonnefond, P.; Haines, B.; Watson, C. In Situ Calibration and Validation: A Link from Coastal to Open-ocean altimetry, chapter 11. In *Coastal Altimetry*; Vignudelli, S., Kostianoy, A., Cipollini, P., Benveniste, J., Eds.; Springer: Berlin/Heidelberg, Germany, 2011; pp. 259–296. ISBN 978-3-642-12795-3.
5. Bonnefond, P.; Exertier, P.; Laurain, O.; Menard, Y.; Orsoni, A.; Jeansou, E.; Haines, B.; Kubitschek, D.; Born, G. Leveling Sea Surface using a GPS catamaran. *Mar. Geodesy* **2003**, *26*, 319–334. [CrossRef]
6. Bonnefond, P.; Exertier, P.; Laurain, O.; Lyard, F.; Bijac, S.; Cancet, M.; Chimot, J.; Féménias, P. Corsica: An experiment for long-term Altimeter calibration and sea level monitoring. In Proceedings of the ESA Living Planet Symposium, Bergen, Norway, 28 June–2 July 2010.

7. Bonnefond, P.; Verron, J.; Babu, K.N.; Bergé-Nguyen, M.; Cancet, M.; Chaudhary, A.; Crétaux, J.F.; Frappart, F.; Haines, B.J.; Laurain, O.; et al. The Benefits of the Ka-Band as Evidenced from the SARAL/AltiKa Altimetric Mission: Quality Assessment and Unique Characteristics of AltiKa Data. *Remote Sens.* **2017**, *10*, 83. [CrossRef]
8. Donlon, C.; Berruti, B.; Buongiorno, A.; Ferreira, M.-H.; Féménias, P.; Frericka, J.; Goryl, P.; Kleina, U.; Laur, H.; Mavrocordatos, C.; et al. The Global Monitoring for Environment and Security (GMES) Sentinel-3 mission. *Remote Sens. Environ.* **2012**, *120*, 37–57. [CrossRef]
9. Bouffard, J.; Webb, E.; Scagliola, M.; Garcia-Mondéjar, A.; Baker, S.; Brockley, D.; Gaudelli, J.; Muir, A.; Hall, A.; Mannan, R.; et al. CryoSat Instrument Performance and Ice Product Quality Status. *Adv. Space Res.* **2017**. [CrossRef]
10. Sentinel-3A SRAL Reprocessed Dataset Available on CODA from 16 October. Available online: https://www.eumetsat.int/website/home/News/DAT_3648215.html (accessed on 6 November 2017).
11. Ray, C.; Martin-Puig, C.; Clarizia, M.P.; Ruffini, G.; Dinardo, S.; Gommenginger, C.; Benveniste, J. SAR altimeter backscattered waveform model. *IEEE Trans. Geosci. Remote Sens.* **2015**, *53*, 911–919. [CrossRef]
12. Dinardo, S.; Fenoglio-Marc, L.; Buchhaupt, C.; Becker, M.; Scharroo, R.; Fernandez, J.M.; Benveniste, J. Coastal SAR and PLRM Altimetry in German Bight and West Baltic Sea. *Adv. Space Res.* **2017**, in press. [CrossRef]
13. Fenoglio-Marc, L.; Dinardo, S.; Scharroo, R.; Roland, A.; Dutour Sikiric, M.; Lucas, B.; Becker, M.; Benveniste, J.; Weiss, R. The German Bight: A validation of CryoSat-2 altimeter data in SAR mode. *Adv. Space Res.* **2015**, *55*, 2641–2656. [CrossRef]
14. Chelton, D.B.; Walsh, E.J.; MacArthur, J.L. Pulse compression and sea level tracking in satellite altimetry. *J. Atmos. Ocean. Technol.* **1989**, *6*, 407–428. [CrossRef]
15. King, R.; Bock, Y. Documentation for the GAMIT GPS Analysis Software. Release 10, Massachusetts Institute of Technology, Cambridge. 2000. Available online: http://www-gpsg.mit.edu/~simon/gtgk/GAMIT.pdf (accessed on 5 October 2017).
16. Picard, B.; Fréry, M.; Pardé, M.; Bonnefond, P.; Laurain, O.; Crétaux, J.-F. A Wet Tropospheric Correction Dedicated to Hydrological and Coastal Applications. In Proceedings of the Ocean Surface Topography Science Team Meeting, Miami, FL, USA, 23–27 October 2017; Available online: https://meetings.aviso.altimetry.fr/fileadmin/user_upload/tx_ausyclsseminar/files/OSTST-2017_PICARD_NewCoastalNN_POSTER.pdf (accessed on 6 November 2017).
17. Mertikas, S.; Donlon, C.; Mavrocordatos, C.; Féménias, P.; Galanakis, D.; Tziavos, I.; Boy, F.; Vergos, G.; Andersen, O.; Frantzis, X.; et al. Multi-mission calibrations results at the Permanent Facility for Altimetry Calibration in west Crete, Greece attaining Fiducial Reference Measurement Standards. In Proceedings of the Ocean Surface Topography Science Team Meeting, Miami, FL, USA, 23–27 October 2017; Available online: https://meetings.aviso.altimetry.fr/fileadmin/user_upload/tx_ausyclsseminar/files/OSTST_CalVal_20_Oct_2017_Mertikas_AAA.pdf (accessed on 6 November 2017).
18. Garcia-Mondejar, A.; Mertikas, S.; Galanakis, D.; Labroue, S.; Bruniquel, J.; Quartly, G.; Féménias, P.; Mavrocordatos, C.; Wood, J.; Garcia, G.; et al. Sentinel-3 Transponder Calibration Results. In Proceedings of the Ocean Surface Topography Science Team Meeting, Miami, FL, USA, 23–27 October 2017; Available online: https://meetings.aviso.altimetry.fr/fileadmin/user_upload/tx_ausyclsseminar/files/S3_OSTST_Poster_20171012_V8.pdf (accessed on 6 November 2017).
19. Dinardo, S. Guidelines for the SAR (Delay-Doppler) L1b Processing. 2013. Available online: http://wiki.services.eoportal.org/tiki-download_wiki_attachment.php?attId=2540 (accessed on 5 October 2017).
20. Garcia-Mondejar, A.; Fornari, M.; Bouffard, J.; Féménias, P.; Roca, M. CryoSat-2: Range, Datation and Interferometer Calibration with Svalbard Transponder. *Adv. Space Res.* **2017**, under review.

remote sensing

MDPI

Article

Coastal Improvements for Tide Models: The Impact of ALES Retracker

Gaia Piccioni *, Denise Dettmering, Marcello Passaro, Christian Schwatke, Wolfgang Bosch and Florian Seitz

Deutsches Geodätisches Forschungsinstitut der Technischen Universität München (DGFI-TUM), Arcisstrasse 21, 80333 München, Germany; denise.dettmering@tum.de (D.D.); marcello.passaro@tum.de (M.P.); christian.schwatke@tum.de (C.S.); wolfgang.bosch@tum.de (W.B.); florian.seitz@tum.de (F.S.)
* Correspondence: gaia.piccioni@tum.de; Tel.: +49-(89)-23031-1214

Received: 9 April 2018; Accepted: 3 May 2018; Published: 3 May 2018

Abstract: Since the launch of the first altimetry satellites, ocean tide models have been improved dramatically for deep and shallow waters. However, issues are still found for areas of great interest for climate change investigations: the coastal regions. The purpose of this study is to analyze the influence of the ALES coastal retracker on tide modeling in these regions with respect to a standard open ocean retracker. The approach used to compute the tidal constituents is an updated and along-track version of the Empirical Ocean Tide model developed at DGFI-TUM. The major constituents are derived from a least-square harmonic analysis of sea level residuals based on the FES2014 tide model. The results obtained with ALES are compared with the ones estimated with the standard product. A lower fitting error is found for the ALES solution, especially for distances closer than 20 km from the coast. In comparison with in situ data, the root mean squared error computed with ALES can reach an improvement larger than 2 cm at single locations, with an average impact of over 10% for tidal constituents K_2, O_1, and P_1. For Q_1, the improvement is over 25%. It was observed that improvements to the root-sum squares are larger for distances closer than 10 km to the coast, independently on the sea state. Finally, the performance of the solutions changes according to the satellite's flight direction: for tracks approaching land from open ocean root mean square differences larger than 1 cm are found in comparison to tracks going from land to ocean.

Keywords: ocean tides; coastal altimetry; ALES retracker

1. Introduction

The ability to predict tides in coastal areas is of crucial importance for our society. In certain regions, tidal events combined with extreme meteorological conditions are responsible for severe flooding and consequent environmental issues. Another critical function of tide models is related to ocean satellite altimetry: altimetric measurements need to be corrected for tidal signal in order to separate the tidal-related variability of sea level from the anomalies coming from the ocean dynamic topography. Therefore, more accurate tide models result in more reliable altimetric sea level retrievals. During the last decades, improvements in oceanographic models and observation techniques brought remarkable results in tide monitoring and prediction. A fundamental benefit comes from satellite altimetry, which provides global-scale sea-level observations with an accuracy of few centimeters [1]. These measurements are mainly exploited in modern tide models as constraint for hydrodynamic modeling, or to empirically derive tidal information from satellite sea-level time-series. As described by [2], after the exploitation of satellite data TOPEX/Poseidon (launched in 1992) tide models showed an enhancement of approximately 5 cm over the previous models. However, significant errors for the major constituents M_2 and S_2 were found at high latitudes [3]. Also, low accuracy was observed in shallow waters, where tidal constituents are highly dependent on bathymetry and the shape of the oceanic shelf [4]. Major efforts in these areas brought a dramatic progress for shallow-water tides, with a consequent larger agreement among different models, and a clear improvement on the single constituents [5]. However, lower performances were observed in coastal regions, resulting in large discrepancies among the models. For models assimilating satellite measurements such situation may be due to a poor availability and quality of altimetric data, highly influenced by the presence of land [6], patches of water at very low sea state within the altimeter footprint [7], or ice [8]. In these areas, the returned echo assumes shapes that are considerably different from the typical open ocean radar return and therefore the signal needs to be fitted with a dedicated algorithm (called retracker). Exploiting these recent advances in data pre-processing, some dedicated coastal products are currently available [9].

The purpose of this paper is to assess the influence of a tailored coastal retracking method on the quality of an ocean tide model, which is an important step towards more oceanographic applications of coastal altimetry [10]. In other words, we want to quantify the difference at the coast between tidal constituents estimated with a dedicated coastal retracker and the same constituents derived with an ordinary open ocean retracker. The coastal retracker used for this experiment is the Adaptive Leading Edge Subwaveform (ALES) retracker. The reliability of this retracker has been proven in a number of applications such as the regional estimation of the seasonal cycle and trend of the sea level [11,12]. Moreover, improvements in areas with a complex macrotidal regimes were validated in [13]. The approach applied to derive the tidal constituents represents a prototype of the new Empirical Ocean Tide (EOT) model. This model takes advantage of the most recent altimetric products, with focus on coastal performances. The model scheme follows the former EOT11a approach [3]: residual tidal constituents are derived on a least-squares-based harmonic analysis applied to Sea Level Anomalies (SLA). In this case, an along-track solution was preferred compared to the classical grids in order to study the evolution of the performances and the impact of the retrackers with respect to the distance to the coast. The data used for the model are illustrated in Section 2 together with a brief description of the ALES retracker and the in situ dataset used for the comparison of the models (Sections 2.1 and 2.2). A more detailed explanation of the tide model approach can be found in Section 3. In Section 4 the methods used for the model comparison are shown, and in Section 5 the results are presented and discussed. Finally, in Section 6 the conclusions and future work are described.

2. Dataset Description

2.1. Altimeter Dataset

In this study, high-rate observations from Jason-1 and Jason-2 missions were used. The high-rate (20 Hz) data allow a ground spatial resolution of circa 350 m along-track, which was preferred over low-rate (1-Hz) products for this dedicated investigation over coastal areas. The data were extracted from the DGFI-TUM's Open Altimeter Database (OpenADB: https://openadb.dgfi.tum.de), which contains the original Sensor Geophysical Data Records (SGDR) and derived high-level products. Version SGDR-E is available in OpenADB for Jason-1, while for Jason-2 version SGDR-D was used. For the two missions only data provided during the reference orbit phase are included, obtaining a continuous time-series of 14 years, from January 2002 until February 2016. In order to compute the tidal constants, values of Sea Level Anomalies (SLA) are needed. At each point, SLA are calculated according to [14]:

$$SLA = H - R - h_{MSS} - h_{geo} \tag{1}$$

where H is the orbital height of the satellite, R is the range, h_{MSS} is the height of the Mean Sea Surface (**MSS**), and h_{geo} is the sum of the heights of all the geophysical corrections. The **MSS** and the geophysical corrections applied for both missions are listed in Table 1. SLA values are additionally flagged with the following criteria:

- $-2.5\,\text{m} \leq \text{SLA} \leq 2.5\,\text{m}$ [3]
- $\text{SWH} < 11\,\text{m}$ [15]
- $7\,\text{dB} < \text{BS} < 30\,\text{dB}$ [15]
- Distance to coast >3 km

where SWH is the Significant Wave Height and BS is the backscatter coefficient. Note that the backscatter coefficient is commonly defined in literature as σ_0, however in this case BS is used to avoid ambiguities with the unit-weight variance (see Section 4). The tidal correction plays an important role in this investigation. The rationale behind the EOT approach consists of the following steps:

1. Application of a pre-existing tide model to correct the SLA
2. Estimation of residual periodic components associated with tides in the corrected SLA
3. Estimation of a new tide correction to adjust and improve the original FES2014 solution

Details on this procedure are given in Section 3. The pre-existing tide model used to correct SLAs is the Finite Element Solution 2014 (FES2014) and it is characterized by new high-resolution and coastal features, essential basis for a coast-dedicated tide model. According to the range used, two experiments are defined in this study: SGDR and ALES. The first uses the range obtained from the ocean retracker of the standard product. This is based on the MLE4 algorithm which adopts the Brown-Hayne (BH) functional form [16,17]. The BH models the expected reflected radar signal from the ocean surface and is considered to be suboptimal in the coastal zone. The second is based on the ALES retracker which restricts the application of BH to only a portion of the fitted radar echo (selected according to a first estimation of the sea state) in order to guarantee the precision of the measurement also in the open ocean, while avoiding spurious reflections typical of the coastal zone. The SSB correction applied depends on the range used. For both the retracking algorithms, the estimates of SSB are also provided. In particular, the ALES SSB is computed using the same SSB model of the SGDR [18] applied to the 20-Hz estimations of SWH and Wind Speed from ALES. This strategy has already been validated with in situ data in [19].

Table 1. List of corrections used to compute Sea Level Anomalies for this study.

Correction	Model	Reference
Mean Sea Surface	DTU15MSS	Andersen et al. [20]
Inverse barometer	Dynamic Atmospheric Correction (DAC)	Carrère et al. [21]
Wet and Dry troposphere	ECMWF	ECMWF [22]
Ionosphere	NOAA Ionosphere Climatology 2009 (NIC09)	Scharroo and Smith [23]
Ocean and Load tide	FES2014	Carrère et al. [24]
Solid Earth and Pole Tide	IERS Conventions 2003	McCarthy and Petit [25]
ALES Sea State Bias	ALES	Passaro et al. [19,26]
SGDR Sea State Bias	SGDR	AVISO/PODAAC [15]

2.2. Tide Gauge Dataset

The harmonic constants resulting from the along-track model are compared against in situ data at the coast. These data were taken from the Global Extreme Sea Level Analysis (GESLA) dataset, which is a unique-format collection of different datasets containing high-frequency (every one hour) sea-level measurements [27]. The harmonic constants used for the comparison were computed via the least-squares method, following e.g., [13]. Within this dataset, tide gauges were selected according to the following criteria:

- Maximum distance to satellite track: 50 km.
- GESLA data already assimilated in FES2014 model (Cancet, personal communication) are discarded.
- Stations near estuaries are discarded. Exceptions for fjords (e.g., Finnish and Canadian coasts).
- Final manual screening on the selected stations: tide gauges with timeseries shorter than one year are discarded while part of the timeseries containing doubtful offsets are not considered.

For each site, one or two crossing tracks were found, obtaining a total of 85 tracks for 70 tide gauges. Their locations are shown in Figure 1.

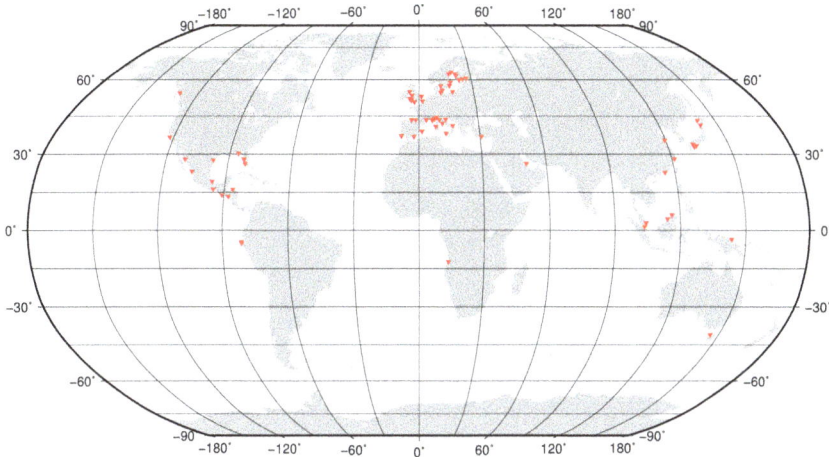

Figure 1. Location of the in situ data used in this work.

3. Tide Model Approach

The method used to compute the tidal constants is based on version 11a of DGFI-TUM's EOT model. For this work, the tidal analysis was based on Jason-1 and Jason-2 missions only, and an along-track solution was chosen. The approach is described in detail hereafter.

3.1. Selection of the Nodes

In the first step, the tracks of interest were selected according to the position of the tide gauges. The points along track at which the tidal constants were computed (also called nodes) are placed on the reference points belonging to the CTOH Topex/Poseidon nominal path (see acknowledgements), with a distance of circa 7 km between two nodes. Each node represents the center of a circular area of influence with radius (ψ_{max}) 15 km. All the SLA observations located within this area are selected for the tidal analysis. In order to account for the different behavior of SLAs, every observation i is weighted with a Gaussian function inversely proportional to its distance from the node ψ_i [3]:

$$w_i = e^{-\beta \psi_i^2} \tag{2}$$

where w_i is the value of the weight, ψ is the distance between the observation and the node, and β is defined as:

$$\beta = \frac{\ln 2}{\tau^2} \tag{3}$$

with $\tau = 0.4\psi_{max}$. The quantity τ is called half-weight width and determines the steepness of the Gaussian function. Namely, it defines the distance from the node for which the weight has value 0.5; in this case the value of τ is 6 km. In Figure 2, the node configuration together with the weighting representation is shown.

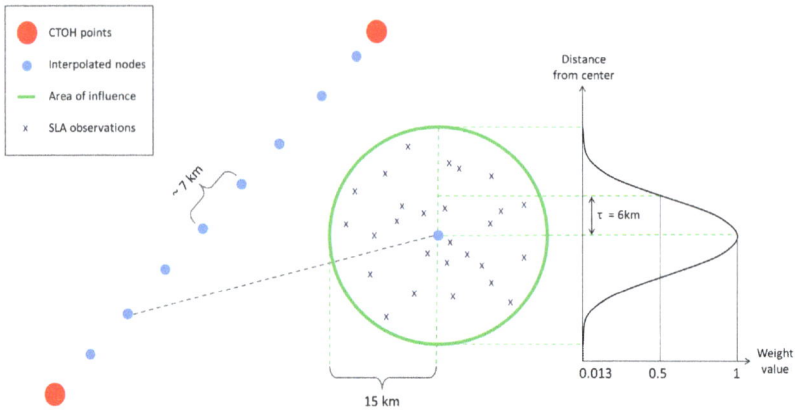

Figure 2. Scheme of the nodes and weighting process. From left to right: the nodes (pale blue) are located over the CTOH track points (red) with a distance of 7 km. For each node a circular area of interest is defined (green), and all the SLA observations (crosses) within this area are selected. The observations are weighted with a Gaussian function dependent on the distance from the node.

3.2. Computation of Tidal Constituents

After collecting the SLA observations, the components of the residual tide signal are computed for the main constituents: M_2, N_2, S_2, K_2, K_1, O_1, P_1, and Q_1. The solution is estimated by a weighted least-squares approach. The unweighted fitting equation for an observation i can be written as [28]:

$$SLA_i = m_j + a \cdot t_i + \sum_{k=1}^{n} \left(A_k \cos P_k \cdot f_k \cos \left(\theta_k + u_k \right) + A_k \sin P_k \cdot f_k \sin \left(\theta_k + u_k \right) \right) \qquad (4)$$

Together with the tidal elements, also the j-th mission offset m_j and the slope a—coming from the SLA time series at the node—are calculated [29]. The summation represents the sum of the n tidal constituents which are defined by the amplitude A_k, the phase P_k, and the given astronomical arguments summarized by the symbol θ_k, dependent on the time of the observation t_i. The nodal corrections f_k and u_k are also given, and can be obtained according to [30]. $A_k \cos P_k$ and $A_k \sin P_k$ are respectively the unknown coefficients of the in-phase and quadrature components and characterize the residual signal of constituent k. All the known right-hand-side elements of Equation (4) are used to form the design matrix of the least-squares approach. In this study, the weight matrix is diagonal and is filled with the weights of the SLAs computed with Equation (2). In combination with the least-squares estimation, a Variance Component Estimation (VCE) is also applied [3]. The VCE is used to weight the contribution of the different missions following the iterative procedure described in e.g., [31]. Finally, the residuals are added to the FES2014 constituents in order to obtain a full tidal signal, from which the amplitude and phase are derived.

4. Evaluation Methods

The different performances between the derived tide models are compared against the GESLA in situ data by computing the Root-Mean-Square difference (*RMS*) of their harmonic constants. For a k-th constituent, the RMS is computed as:

$$RMS_k = \sqrt{\frac{(A_M \cos P_M - A_T \cos P_T)^2 + (A_M \sin P_M - A_T \sin P_T)^2}{2}} \qquad (5)$$

where A_M and P_M are the amplitude and phase of one solution (*SGDR* or *ALES*) for that constituent, and A_T and P_T the ones given by the tide gauge observations. The absolute *RMS* difference between the *SGDR* and the *ALES* solution is written as $\Delta RMS_k = RMS_{k,SGDR} - RMS_{k,ALES}$ and is measured in cm. The relative *RMS* difference is also shown, which is described as:

$$\Delta RMS_k [\%] = \frac{\Delta RMS_k}{RMS_{k,SGDR}} \cdot 100 \tag{6}$$

This difference is expressed in percentage and indicates the relative improvement or worsening of the ALES solution with respect to the SGDR results. For an overall performance, the Root-Sum Squared (*RSS*) of the available n constituents is also calculated:

$$RSS = \sqrt{\sum_{k=1}^{n} RMS_k^2} \tag{7}$$

Both ΔRMS_k and RSS quantities are estimated for all the nodes along-track. In particular, results for the closest node to the tide gauge of interest (henceforth CNTG) are also considered, as they would represent the accuracies with respect to the coastal true values. To highlight the discrepancies among the ALES and SGDR solutions, it was chosen to express the results in terms of absolute differences, such as:

$$\Delta RSS = RSS_{SGDR} - RSS_{ALES} \tag{8}$$

A positive ΔRSS corresponds to higher RSS for the SGDR solutions, and therefore an improvement of ALES solutions with respect to a model using an ordinary retracker. Finally, the internal quality of the models is compared using σ_0, the unit-weight variance of the least-squares fit, and is inversely proportional to the number of observations [32]. The larger is σ_0, the higher is the uncertainty of the fitting.

5. Results and Discussion

5.1. Number of Observations

One of the most advantageous features of ALES retracker is the large amount of valid coastal measurements available along track. In this work this benefit is shown in terms of observations available for each node. In Figure 3 the difference between the number of observations of ALES and the ones retrieved with SGDR are displayed. This difference is expressed as: Δ_{obs}, i.e., observations of ALES minus observations of SGDR. Each dot represents a node along the tracks, plotted against the distance to the coast. The red markers highlight the positive values, that is, the nodes for which ALES provides a larger amount of data with respect to SGDR. The blue dots are used for the negative values. An interesting, yet expected behavior is observed for values below 20 km from the coast: far more observations are available with ALES while approaching the coast, with some exceptions for few points.

5.2. Fitting Uncertainty

An analogous comparison is shown for the variable σ_0, that represents the quality of the least squares fit. In Figure 4a the difference at each node between the σ_0 computed for the SGDR solutions and σ_0 obtained from ALES is shown. A positive value on the Y the change. -axis (red dots) corresponds to a larger fitting error for the SGDR solutions, and negative values (blue dots) for the contrary. From the plot it is clear that in most cases an improvement for σ_0 is achieved with ALES, with exception for few coastal points. The dependence of σ_0 on the number of observations may explain the smaller errors for ALES. However, from Figure 4b one can notice that large improvements in σ_0 are reached also for a lower amount of data. On the other hand, the few cases with larger internal errors may be found at nodes with more data availability. These special cases, which accounts for only the 1.5% of the cases, may be justified by residual erroneous estimations in the ALES data, which were not identified by the outliers analysis.

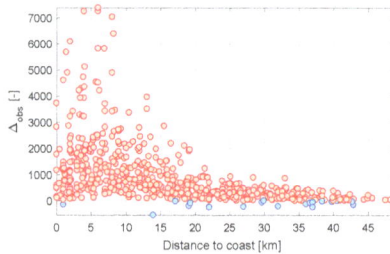

Figure 3. Difference in the number of observations between ALES and SGDR at each node against the distance to coast. The blue dots show the cases for which less observations are available for ALES, while the red dots correspond to a larger amount of data for ALES.

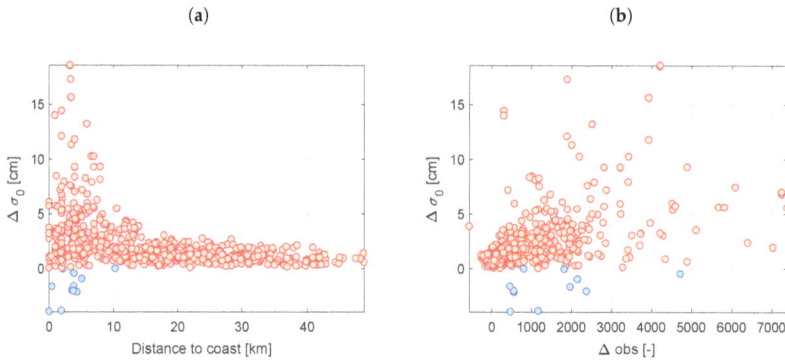

Figure 4. Difference of σ_0 values between SGDR minus ALES at each node against the distance to coast (**a**), and against difference in number of observations between ALES and SGDR (**b**).

5.3. Comparison Against In Situ Data

In this Section the results are compared in terms of RMS and RSS differences against in situ data. The first paragraph gives an overview of the results; the second paragraph discusses the dependency of the results on distance to coast, distance to tide gauge, track direction, and SWH.

5.3.1. General Results

In Figure 5, the spatial distribution of the absolute differences ΔRSS are shown. The differences are computed for the CNTG, for all the 85 tracks. In general, improvements are found for 66 tracks, with an average of 0.4 cm and a maximum value of 1.9 cm. The red dots indicate the highest improvements for the ALES solutions, which are located unevenly between Europe and the American continent.These higher values may be due to improvements to only few single constituents. This can be observed in Figure 6, where the absolute ΔRMS of the closest nodes to the tide gauge of interest are plotted against the longitude of the in situ site. The plot is divided in three rows for an easier visualization, and the ΔRMS of each tidal constant is color-coded according to the legend. An example of large improvements for single constituents can be seen at Prince Rupert, western Canada, or Ringhals, Sweden (respectively longitudes: $-130.32°$ and $12.11°$), where the values of ΔRMS for M_2 and S_2 for Prince Rupert, or Q_1 and O_1 are larger than 2.5 cm. In contrast, there are locations such as La Union, El Salvador, and Swansea, UK (longitudes $-87.82°$ and $-3.98°$) where the ALES solution shows a loss in performance—again differences larger than 2.5 cm—for constituent M_2. The RMS differences for all the 85 tracks are summarized in Table 2. The average values were computed using the single RMS values obtained at each site, at the CNTG. A mean improvement of 2 mm can be measured for the ALES solutions with respect to SGDR. It is important to stress that in the global average, results based on ALES are superior to results based on SGDR for every constituent. For K2, O1 and P1 the improvement is over 10%. For Q1, the improvement is over 25%. For larger RMS (such as M_2 and S_2) a minor effect of ALES is observed for the relative differences.

Table 2. Average of RMS for major constituents for the closest points to the tide gauges. The values are expressed in cm. The last column shows the relative difference between the two solutions.

Constituents	RMS_{ALES} (cm)	RMS_{SGDR} (cm)	ΔRMS (%)
M2	8.0	8.2	2.4
N2	2.1	2.3	8.7
S2	3.5	3.7	5.4
K2	1.4	1.6	12.5
K1	2.1	2.2	4.5
O1	1.4	1.6	12.5
Q1	0.8	1.1	27.3
P1	1.2	1.4	14.3

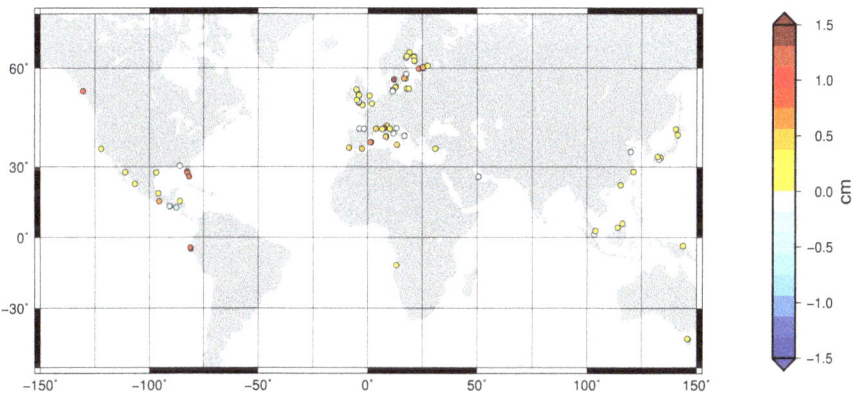

Figure 5. Geographical distribution of the ΔRSS (in cm) for the closest nodes to the tide gauge of interest.

5.3.2. Study of the Dependencies

Also in this section, the CNTGs are used to study the performances of the two retracker solutions. It must be pointed out that the CNTGs may not coincide with the closest points to the coast, as they depend on the position of the track with respect to land. For this reason, it was chosen to analyze the ΔRSS values against the distance to coast (Figure 7a) as well as against the distance to the tide gauge of interest (Figure 7b). The first plot shows not only that the nodes are mostly concentrated within 10 km to the coast, but also that improvements with ALES larger than 0.5 cm occur for nodes closer than 5 km. On the other hand, no visible dependency is observed between the values of ΔRSS and their distance from the tide gauge: in fact, the same improvements over 0.5 cm appear also for distances above 20 km. The dependency on the distance to the coast is also shown for the ΔRMS of the single constituents, Figure 8. Within 10 km from the coast, improvements below 2 cm can be found for all constituents. Larger variability is observed for the major constituents, and single values can reach e.g., ±5 cm for M_2 and ±3 cm for S_2.

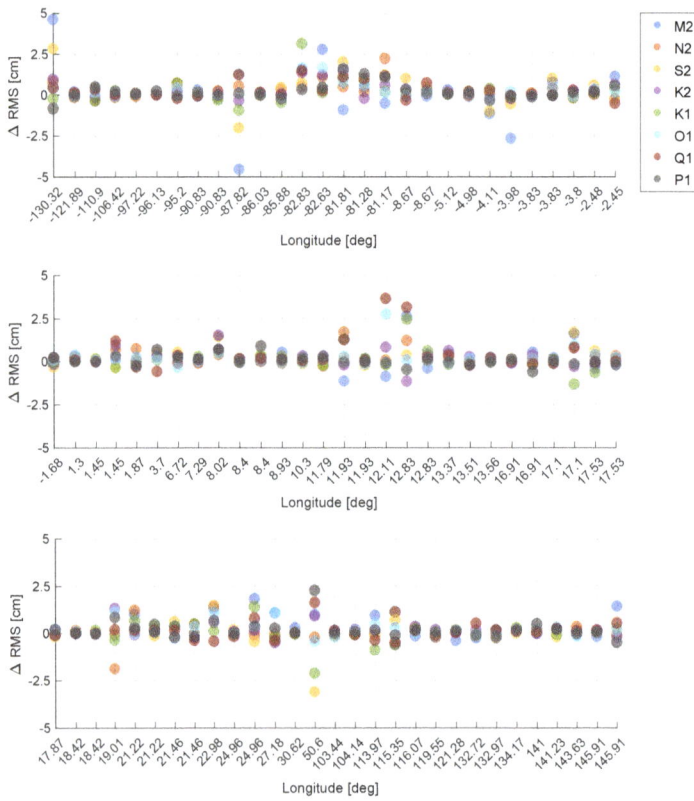

Figure 6. Difference of RMS for major constituents at the CNTG. The values are plotted against the longitude of their location.

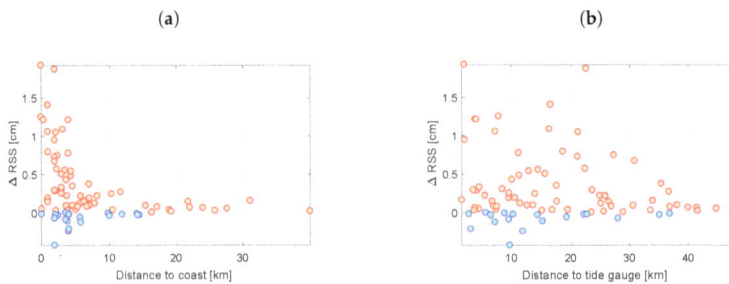

Figure 7. Difference of RSS against distance to coast (**a**) and tide gauge (**b**) in km. The values are shown for the closest nodes to the tide gauge of interest.

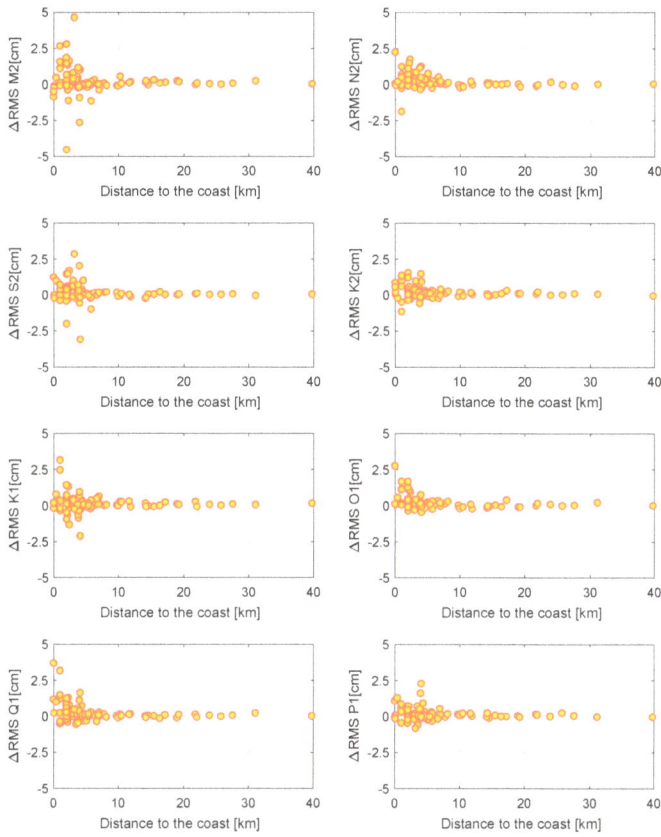

Figure 8. Absolute ΔRMS of major constituents at the CNTG. The values are plotted against the distance to the coast (in km).

Another aspect analyzed is the influence of the track direction on the results, because the performance of a retracker may change whether the satellite approaches land from ocean or flies from ocean to land, as well as if there is a bay (case: land-ocean-land) or the coast is parallel to the track (case: parallel to land). These results are shown in Table 3. The four main headers indicate the track position and the number of tracks used for the RMS average. The RSS computed from the RMS averages for each case are also displayed in the last row of the table. In general, lower values are found for the ALES solution, with exception of few constituents. An interesting result regards the transition land/ocean (i.e., the first four columns): both SGDR and ALES solutions show a higher performance for all constituents for the case ocean-to-land, against the land-to-ocean results, reaching differences larger than 1 cm for single constituents. This can be seen also from the averaged RSS, which show discrepancies of 26 mm between the ALES solutions and 21 mm for SGDR. This situation may be justified by a different behavior of the on-board tracker according to the flight direction, which may consequently influence the performance of the retrackers [7]. A clear example is presented in Figure 9, where the RMS of constituent M_2, computed for the ALES solution, is plotted for the nodes of tracks 111 and 92. Track 111 is ascending, and it goes from ocean to land, while track 92 is descending, going from land to ocean. It can be observed, that even though few nodes at track 92 are closer to the tide gauge, they still show a larger RMS with respect to nodes belonging to track

111. Moreover, larger discrepancies are found between ALES and SGDR for the case ocean-to-land, for which RSS values differ of 6 mm against 1 mm in the land-to-ocean case. Unfortunately, for the cases land-ocean-land and parallel-to-land only few tracks were available. However, from both the single RMS and the RSS values similar performance is found between ALES and SGDR solutions.

Table 3. Average of RMS computed for major constituents at the closest points to the tide gauges. The averages are computed after dividing the tracks according to their position with respect to the coast. The values are in cm.

Constituents	Land to Ocean: 30		Ocean to Land: 34		Land-Ocean-Land: 15		Parallel to Land: 6	
	RMS_{ALES}	RMS_{SGDR}	RMS_{ALES}	RMS_{SGDR}	RMS_{ALES}	RMS_{SGDR}	RMS_{ALES}	RMS_{SGDR}
M2	6.6	6.9	4.8	5.0	19.3	19.2	4.6	4.7
N2	1.7	1.8	1.3	1.6	4.8	5.2	1.4	1.4
S2	3.1	3.2	2.1	2.4	7.7	7.8	2.6	2.5
K2	1.2	1.3	1.0	1.3	2.8	2.9	1.7	1.7
K1	1.9	1.9	1.4	1.5	3.8	4.2	2.2	2.2
O1	1.2	1.3	1.0	1.3	2.5	2.7	1.6	1.6
Q1	0.8	0.9	0.7	1.0	1.3	1.8	0.9	1.0
P1	1.5	1.7	0.7	0.9	1.9	1.9	1.1	1.2
RSS	8.4	8.5	5.8	6.4	22.1	22.8	6.5	6.6

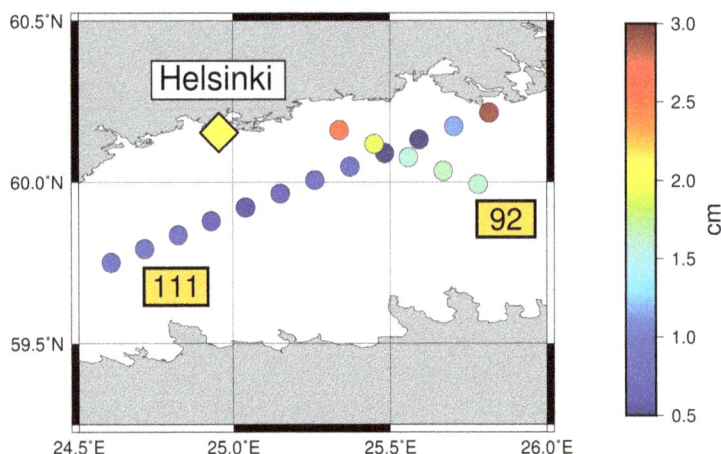

Figure 9. RMS values for M_2 constituent computed with ALES solutions for tracks 111 (**ascending**) and 92 (**descending**). The tracks face the tide gauge station of Helsinki (**diamond-shape marker**). The nodes of each track are represented by the round markers and the color shows the value of the RMS with respect to the tide gauge, in cm.

Finally, the sea state dependency is shown for the absolute ΔRSS. It was chosen to represent the sea state as the average of the SWH at each node, plus its standard deviation. The SWH values are taken from the ALES product. While the improvement of the ALES data for calm sea states (<2.5 m) is expected [33], the available literature concerning data quality in comparison with SGDR for wavy seas is still scarce. Indeed, from Figure 10, relevant improvements (>0.5 cm) are observed for sea states within 2.5 m, while only few examples are available for high states. However, ΔRMS > 1 cm are found above 3 m, showing no sensitive relation between the sea state and the data analyzed.

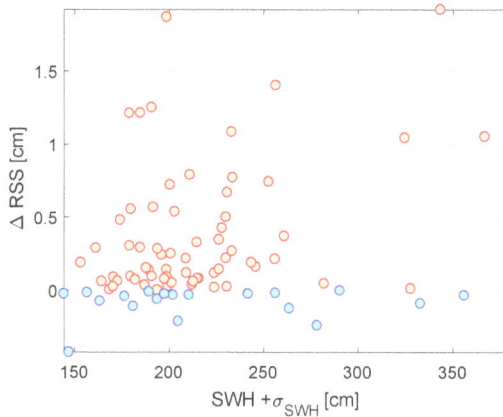

Figure 10. ΔRSS absolute values for the closest node to the tide gauge against the sea state, represented as the averaged SWH computed at each node plus its standard deviation.

6. Conclusions and Outlook

In this work, we have tested the impact of a coast-dedicated retracker on the estimation of ocean tidal constituents. The experiment aimed to compare tidal constants computed with the standard SGDR product against the ALES coastal retracker. The tidal constituents were derived on nodes defined along Jason satellites' tracks, applying the method of weighted least-squares on SLA which were previously corrected for the FES2014 tide model. The results were compared with in situ observations in terms of RMS and RSS values. It was shown that with ALES an increased number of sea level retrievals were available at each node, especially at distances closer than 20 km from the coast. The largest improvements are detected at distances within 10 km to the coast, independently from the geographical location and the sea state. A similar behavior was detected for the least-squares fitting error, which also show no clear dependency on the number of observations. In addition, no evident dependency is found for the RMS improvements to the distance to the tide gauges. The general ΔRSS results over the track nodes showed an average improvement of 0.4 cm for 66 tracks. However, the averaged RMS suggest a mean impact of few mm for all tidal constants. The ΔRMS highlighted a positive impact of ALES for single constituents, which can reach values >2.5 cm.

The RMS averages were presented after dividing the results according to the satellite's flight direction. From this experiment it was possible to see that the performance of both solutions change according to the track direction. The transition ocean-to-land shows smaller RMS for both ALES and SGDR solutions, and for all constituents. Single differences may exceed 1 cm when compared with land-to-ocean solutions.

In conclusion, improvements independent from the node position, together with a lower fitting error and a large data availability, make ALES a favorable choice for coastal tidal analysis. Indeed, the retracker will be exploited within the the new version of the EOT model. However, future research should be extended to minor tidal constituents, as well as dedicated regional analyses.

The improvements shown in this study were found despite the ALES retracking strategy was only applied in the residual analysis of the EOT procedure, while the original FES2014 model, which corrects for most of the tidal variability, is still based on SGDR data. We expect therefore that the use of ALES data could bring a decisive improvement in coastal tide modeling if used as a data source to estimate the full tidal component of the sea level variability. Finally, we recommend further investigations aimed to quantify the impact of additional altimetric corrections on the tidal estimation at the coast and to promote more oceanographic applications of coastal altimetry data.

Remote Sens. **2018**, *10*, 700

Author Contributions: G.P. and D.D. conceptualized the research work; G.P. performed the experiments and analyzed the data with D.D. and M.P.; The manuscript was written by G.P. and M.P. contributed in writing the ALES fundamentals. C.S. provided special OpenADB tools used in the processing algorithm and insights on data structure; all the authors revised the manuscript and gave constructive comments.

Acknowledgments: The authors thank NOVELTIS, LEGOS, CLS Space Oceanography Division, CNES and AVISO for providing the FES2014 model. The TOPEX/Poseidon nominal tracks are available at: http://ctoh.legos.obs-mip.fr/products/altimetry/topex-poseidon-nominal-path/view distributed by the CTOH observational service. The tracks were selected by using the AVISO pass locator, accessible on: https://www.aviso.altimetry.fr/en/data/tools/pass-locator.html. The tide-gauge data used to compare our results were taken from the GESLA public dataset, available at: http://gesla.org/.

Conflicts of Interest: The authors declare no conflict of interest.

References

1. Bonnefond, P.; Haines, B.J.; Watson, C. In situ absolute calibration and validation. In *Coastal Altimetry*; Vignudelli, S., Kostianoy, A.G., Cipollini, P., Benveniste, J., Eds.; Springer: Berlin/Heidelberg, Germany, 2011; pp. 259–296, ISBN 978-3-642-12795-3.
2. Shum, C.K.; Woodworth, P.L.; Andersen, O.B.; Egbert, G.D.; Francis, O.; King, C.; Klosko, S.M.; Provost, C.L.; Li, X.; Molines, J.-M.; et al. Accuracy assessment of recent ocean tide models. *J. Geophys. Res.* **1997**, *102*, 25173–25194. [CrossRef]
3. Savcenko, R.; Bosch, W. *EOT11a—Empirical Ocean Tide Model From Multi-Mission Satellite Altimetry*; DGFI Report No. 89; Deutsches Geodätisches Forschungsinstitut: Munich, Germany, 2012. [CrossRef]
4. Andersen, O.B. Shallow water tides in the Northwest European shelf region from TOPEX/POSEIDON altimetry. *J. Geophys. Res.* **1999**, *104*, 7729–7741. [CrossRef]
5. Stammer, D.; Ray, R.D.; Andersen, O.B.; Arbic, B.K.; Bosch, W.; Carrère, L.; Cheng, Y.; Chinn, D.S.; Dushaw, B.D.; Egbert, G.D.; et al. Accuracy assessment of global barotropic ocean tide models. *Rev. Geophys.* **2014**, *52*, 243–282. [CrossRef]
6. Gommenginger, C.; Thibaut, P.; Fenoglio-Marc, L.; Quartly, G.; Deng, X; Gómez-Enri, J.; Challenor, P.; Gao, Y. Retracking altimeter waveforms near the coasts. In *Coastal Altimetry*; Vignudelli, S., Kostianoy, A.G., Cipollini, P., Benveniste, J., Eds.; Springer: Berlin/Heidelberg, Germany, 2011; pp. 61–101, ISBN 978-3-642-12795-3.
7. Passaro, M.; Cipollini, P.; Vignudelli, S.; Quartly, G.D.; Snaith, H.M. ALES: A multi-mission adaptive subwaveform retracker for coastal and open ocean altimetry. *Remote Sens. Environ.* **2014**, *145*, 173–189. [CrossRef]
8. Andersen, O.B.; Piccioni, G. Recent Arctic sea level variations from satellites. *Front. Mar. Sci.* **2016**, *3*, 76. [CrossRef]
9. Cipollini, P.; Benveniste, J.; Birol, F.; Fernandes, M.J.; Passaro, M.; Vignudelli, S. Satellite altimetry in coastal regions. In *Satellite Altimetry over Oceans and Land Surfaces*; Stammer, D., Cazenave, A., Eds.; CRC Press: Boca Raton, FL, USA, 2017; pp. 343–380.
10. Liu, Y.; Weisberg, R.H.; Vignudelli, S.; Roblou, L.; Merz, C.R. Comparison of the X-TRACK altimetry estimated currents with moored ADCP and HF radar observations on the West Florida Shelf. *Adv. Space Res.* **2012**, *50*, 1085–1098. [CrossRef]
11. Passaro, M.; Fenoglio-Marc, L.; Cipollini, P. Validation of Significant Wave Height from improved satellite altimetry in the German Bight. *IEEE Trans. Geosci. Remote Sens.* **2015**, *53*, 4. [CrossRef]
12. Passaro, M.; Dinardo, S.; Quartly, G.D.; Snaith, H.M.; Benveniste, J.; Cipollini, P.; Lucas, B. Cross-calibrating ALES Envisat and CryoSat-2 Delay-Doppler: A coastal altimetry study in the Indonesian Seas. *Adv. Space Res.* **2016**, *58*, 289–303. [CrossRef]
13. Lago, L.S.; Saraceno, M.; Ruiz-Etcheverry, L.A.; Passaro, M.; Oreiro, F.A.; D'Onofrio, E.E.; González, R. Improved sea surface height from satellite altimetry in coastal zones: A case study in Southern Patagonia. *IEEE J. Sel. Top. Appl. Earth Obs. Remote Sens.* **2017**, *10*, 3493–3503. [CrossRef]
14. Andersen, O.B.; Scharroo, R. Range and geophysical corrections in coastal regions—And implications for mean sea surface determination. In *Coastal Altimetry*; Vignudelli, S., Kostianoy, A.G., Cipollini, P., Benveniste, J., Eds.; Springer: Berlin/Heidelberg, Germany, 2011; pp. 103–145, ISBN 978-3-642-12795-3.

15. Picot, N.; Case, K.; Desai, S.; Vincent, P.; Bronner, E. *AVISO and PODAAC User Handbook. IGDR and GDR Jason Products*; SALP-MU-M5-OP-13184-CN (AVISO), JPL D-21352 (PODAAC); Physical Oceanography Distributed Active Archive Center: Pasadena, CA, USA, 2012.
16. Brown, G. The average impulse response of a rough surface and its applications. *IEEE Trans. Antennas Propag.* **1977**, *25*, 67–74. [CrossRef]
17. Hayne, G.S. Radar altimeter mean return waveforms from near-normal- incidence ocean surface scattering. *IEEE Trans. Antennas Propag.* **1980**, *28*, 687–692. [CrossRef]
18. Tran, N.; Thibaut, P.; Poisson, J.-C.; Philipps, S.; Bronner, E.; Picot, N. Impact of Jason-2 wind speed calibration on the sea state bias correction. *Mar. Geod.* **2011**, *34*, 3–4. [CrossRef]
19. Gómez-Enri, J.; Cipollini, P.; Passaro, M.; Vignudelli, S.; Tejedor, B.; Coca, J. Coastal altimetry products in the strait of Gibraltar. *IEEE Trans. Geosci. Remote Sens.* **2016**, *54*, 99. [CrossRef]
20. Andersen, O.B.; Stenseng, L.; Piccioni, G.; Knudsen, P. The DTU15 MSS (Mean Sea Surface) and DTU15LAT (Lowest Astronomical Tide) reference surface. In Proceedings of the ESA Living Planet Symposium 2016, Prague, Czech Republic, 9–13 May 2016.
21. Carrère, L.; Faugère, Y.; Bronner, E.; Benveniste, J. Improving the dynamic atmospheric correction for mean sea level and operational applications of atimetry. In Proceedings of the Ocean Surface Topography Science Team (OSTST) Meeting, San Diego, CA, USA, 19–21 October 2011.
22. Persson, A. *User Guide to ECMWF Forecast Products*; ECMWF: Reading, UK, 2015.
23. Scharroo, R.; Smith, W.H.F. A global positioning system-based climatology for the total electron content in the ionosphere. *J. Geophys. Res.* **2010**, *115*, 16. [CrossRef]
24. Carrère, L.; Lyard, F.; Cancet, M.; Guillot, A.; Picot, N. FES 2014, a new tidal model—Validation results and perspectives for improvements. In Proceedings of the ESA Living Planet Symposium 2016, Prague, Czech Republic, 9–13 May 2016.
25. McCarthy, D.D.; Petit, G.; (Eds.) *IERS Conventions (2003)*; IERS Technical Note 32; Verlag des Bundesamts für Kartographie und Geodäsie: Frankfurt, Germany, 2004.
26. Passaro, M.; Kildegaard Rose, S.; Andersen, O.B.; Boergens, E.; Calafat, F.M.; Dettmering, D.; Benveniste, J. ALES+: Adapting a homogenous ocean retracker for satellite altimetry to sea ice leads, coastal and inland waters. *Remote Sens. Environ.* **2018**, in press. [CrossRef]
27. Woodworth, P.L.; Hunter, J.R.; Marcos, M.; Caldwell, P.; Menendez, M.; Haigh, I. Towards a global higher-frequency sea level data set. *Geosci. Data J.* **2017**, *3*, 50–59. [CrossRef]
28. Savcenko, R.; Bosch, W. Residual tide analysis in shallow water-contribution of ENVISAT and ERS altimetry. In *ESA-SP636 (CD-ROM), Proceedings of the Envisat Symposium 2007, Montreux, Switzerland, 23–27 April 2007*; Lacoste, H., Ouwehand, L., Eds.; ESA: Noordwijk, The Netherlands, 2007.
29. Bosch, W.; Savcenko, R.; Flechtner, F.; Dahle, C.; Mayer-Gürr, T.; Stammer, D.; Taguchi, E.; Ilk, K.H. Residual ocean tide signals from satellite altimetry, GRACE gravity fields, and hydrodynamic modelling. *Geophys. J. Int.* **2009**, *178*, 1185–1192. [CrossRef]
30. Doodson, A.T.; Warburg, H.D. *Admiralty Manual of Tides*; H. M. Stationery Off.: London, UK, 1941.
31. Teunissen, P.; Amiri-Simkooei, A. Variance component estimation by the method of least-squares. In *VI Hotine-Marussi Symposium on Theoretical and Computational Geodesy, Proceedings of the IAG Symposium, Wuhan, China, 29 May–2 June 2006*. Xu, P, Liu, J., Dermanis, A., Eds.; Springer:Berlin/Heidelberg, 2008; ISBN 978-3-540-74583-9.
32. Koch, K.-R. *Parameter Estimation and Hypothesis Testing in Linear Models*, 2nd ed.; Springer: Berlin/Heidelberg; Bonn, Germany, 1999; ISBN 978-3-540-65257-1.
33. Passaro, M.; Cipollini, P.; Benveniste, J. Annual sea level variability of the coastal ocean: The Baltic Sea-North Sea transition zone. *J. Geophys. Res. Oceans* **2015**, *120*, 3061–3078. [CrossRef]

remote sensing

MDPI

Article

Improving Jason-2 Sea Surface Heights within 10 km Offshore by Retracking Decontaminated Waveforms

Zhengkai Huang [1,2], Haihong Wang [1,3,*], Zhicai Luo [4], C. K. Shum [5,6], Kuo-Hsin Tseng [7,8,9] and Bo Zhong [1,3]

[1] School of Geodesy and Geomatics, Wuhan University, Wuhan 430079, China; zhkhuang@whu.edu.cn (Z.H.); bzhong@sgg.whu.edu.cn (B.Z.)
[2] Guangxi Key Laboratory of Spatial Information and Geomatics, Guilin University of Technology, Guilin 541004, China
[3] Key Laboratory of Geospace Environment and Geodesy, Ministry of Education, Wuhan University, Wuhan 430079, China
[4] MOE Key Laboratory of Fundamental Physical Quantities Measurement, School of Physics, Huazhong University of Science and Technology, Wuhan 430074, China; zcluo@hust.edu.cn
[5] Division of Geodetic Science, School of Earth Sciences, Ohio State University, Columbus, OH 43210, USA; ckshum@osu.edu
[6] State Key Laboratory of Geodesy and Earth's Dynamics, Institute of Geodesy and Geophysics, Chinese Academy of Sciences, Wuhan 43077, China
[7] Department of Civil Engineering, National Central University, Taoyuan 32001, Taiwan; khtseng@csrsr.ncu.edu.tw
[8] Center for Space and Remote Sensing Research, National Central University, Taoyuan 32001, Taiwan
[9] Institute of Hydrological and Oceanic Sciences, National Central University, Taoyuan 32001, Taiwan
* Correspondence: hhwang@sgg.whu.edu.cn; Tel.: +86-27-6877-1756

Received: 7 June 2017; Accepted: 19 October 2017; Published: 23 October 2017

Abstract: It is widely believed that altimetry-derived sea surface heights (SSHs) in coastal zones are seriously degraded due to land contamination in altimeter waveforms from non-marine surfaces or due to inhomogeneous sea state conditions. Spurious peaks superimposed in radar waveforms adversely impact waveform retracking and hence require tailored algorithms to mitigate this problem. Here, we present an improved method to decontaminate coastal waveforms based on the waveform modification concept. SSHs within 10 km offshore are calculated from Jason-2 data by a 20% threshold retracker using decontaminated waveforms (DW-TR) and compared with those using original waveforms and modified waveforms in four study regions. We then compare our results with retracked SSHs in the sensor geophysical data record (SGDR) and with the state-of-the-art PISTACH (Prototype Innovant de Système de Traitement pour les Applications Côtières et l'Hydrologie) and ALES (Adaptive Leading Edge Subwaveform) products. Our result indicates that the DW-TR is the most robust retracker in the 0–10 km coastal band and provides consistent accuracy up to 1 km away from the coastline. In the four test regions, the DW-TR retracker outperforms other retrackers, with the smallest averaged standard deviations at 15 cm and 20 cm, as compared against the EGM08 (Earth Gravitational Model 2008) geoid model and tide gauge data, respectively. For the SGDR products, only the ICE retracker provides competitive SSHs for coastal applications. Subwaveform retrackers such as ICE3, RED3 and ALES perform well beyond 8 km offshore, but seriously degrade in the 0–8 km strip along the coast.

Keywords: coastal altimetry; sea surface height; Jason-2; waveform retracking

1. Introduction

Satellite radar altimetry is a powerful technology for remotely sensing physical properties of global oceans. Sea surface heights (SSHs) are the primary product of altimetry, which have greatly benefited studies of extensive scientific issues in geodesy, oceanography, geophysics and many other disciplines [1]. Despite the remarkable success of altimetry in the open ocean, which can provide SSHs with an accuracy of several centimeters, altimetry-derived SSHs in coastal zones are severely degraded due to inherent limitations of this technique [2–4]. The Altimeter emits pulse-limited radar signals and receives echoes from illuminated Earth surfaces in nadir. The distance from satellite to the sea surface covered by the radar footprint, referred to as the range, is measured by tracking the received waveform. Hence, SSH can be derived by subtracting the range from satellite altitude. In coastal zones, altimeter waveforms show diversely complex shapes because of land in the footprint and complicated sea state conditions [5,6]. In this case, erroneous range measurements might be delivered by on-board retrackers, which are designed for normal waveforms in the open ocean. In addition, inaccurate media and geophysical corrections such as atmospheric delay corrections, tides, sea state bias (SSB), etc., are non-negligible factors for the poor quality of coastal SSHs [7–9]. These corrections are more complex near the coast and it is considerably more difficult to improve them.

Coastal regions play a very important role in the geographical and economical support of human society, and therefore the coastal environment is subject to great pressure from human activities [10]. High quality and dense sampling of geospatial and environmental information is expected not only for coastal management but also for scientific research, e.g., [11]. The increasing demands of coastal applications have motivated excellent efforts in the altimetry community to improve coastal applicability, including proposing new altimetry concepts (Ka-band altimetry [12], delay-Doppler altimetry [13,14], wide-swath altimetry [15], GNSS altimetry [16], constellations of altimeters [17]) and reprocessing conventional altimeter data [18–20]. In the last decade, a series of projects were supported by some space agencies and research institutions, aiming to retrieve valid altimeter data as close as possible to the coast. Hereafter a few data sets have been available for coastal applications, such as PISTACH (Prototype Innovant de Système de Traitement pour les Applications Côtières et l'Hydrologie) as well as coastal and hydrology altimetry data products [21], COASTALT [22] and X-TRACK [23]. There is increasing consensus that coastal altimetry is a critically important yet challenging discipline.

The endeavor in coastal altimetry consists of refining auxiliary corrections and retracking waveforms [4,20], which are performed to tackle the aforementioned factors degrading conventional altimeter data near the coasts. A summary of the significant achievements was given in [24], including improvements in waveform retracking, tropospheric corrections, tide models, and dynamic atmosphere corrections. There are also some studies on SSB correction, aiming at better modelling the error induced by ocean surface waves, whitecaps and foam [25–29]. Among these improvements, waveform retracking is getting considerable attention in recent years, because of its evident effect on the enhancement of altimeter measurements.

The waveform retracking technique was originally designed for post-processing open ocean radar waveforms. The Brown mathematical model [30] is used to fit ocean waveforms in order to refine parameters related to reflective surfaces, referred to as OCEAN retracker. For waveforms over mixed surfaces, a threshold retracker (TR) was developed [31] for specular or multi-peak shapes. This approach determines retracked gate using a preset percentage of maximum echo power. Later, several revised algorithms based on TR were proposed for coastal applications, and successfully validated in some case studies. For instance, Hwang et al. [32] presented an improved threshold retracker (ITR) and verified better performance than the OCEAN and TR methods using Geosat geodetic mission (GM) altimeter data around Taiwan. Another modified threshold retracker (MTR) was applied to Jason-2 data over California coastal ocean, which could filter the pre-leading edge bumps [33]. This kind of retracker extracts subwaveform in order to avoid superimposed signals from non-marine surfaces by detecting the apparent leading edge. Similarly, assuming waveforms contain a portion unaffected by land effects, a series of subwaveform retrackers have been developed [34–38]. Some other approaches that use refined model-based retrackers, such as Brown plus specular peak model [39] and the Brown with asymmetric Gaussian peak (BAGP) model [6], were developed to conform peaky altimetric waveforms that occupy a large percentage of coastal areas. In addition, some researchers developed adaptive retracking systems by applying an optimal retracker for various characteristic shapes, on the basis of waveform classification [36,38,40–42]. Although these retrackers contribute enormously to coastal SSH retrievals with better accuracy and shortened coastal gaps, handling coastal waveforms within 10 km of the shore is still a thorny issue [2,38,43].

The major obstacle for coastal waveform retracking is due to the diversity and complexity of coastal waveforms. According to the classification of the coastal waveforms within PISTACH processing [21], the percentage of typical ocean waveforms starts decreasing around 10 km away from the coast, whilst the quantity of waveforms with the shape of the Brown plus anomalous peaks rapidly increases. More than a quarter of the waveforms do not obey the Brown model at the distance of 8 km off the shoreline [6]. Within 5 km away from the shoreline, the peaky waveforms outnumber the ocean waveforms. Anomalous peaks in waveforms might cause an overestimation of some waveform parameters for empirical retrackers or make model-based retracker failure. In order to purify waveforms, a waveform modification (WM) concept was presented by Tseng et al. [43]. In their study, they proposed an algorithm to modify polluted coastal waveforms before retracking. By retracking modified waveforms, it was demonstrated in four cases in North America that SSHs in 1–7 km coastal zone had smaller root mean square error (RMSE) than those from retracking original waveforms, comparing with the in-situ tide gauge data. However, shifted leading edges might be misjudged as spurious peaks and then distorted in their procedure, if the selected reference waveform is not representative (further discussion given in Section 3.1). Another issue is that an interpolation for amending outlier gates might induce noise from adjacent gates.

In this paper, an alternative strategy for decontaminating waveforms is presented in order to overcome these defects of the WM method. We revise the criterion for selecting reference waveforms, and directly set the outliers as null value rather than amending them by interpolation, i.e., the detected spurious gates would not be taken into consideration in the retracking procedure. The performance of this new approach is evaluated in four coastal regions within 10 km offshore using Jason-2 data. The article is structured as follows. The study regions and dataset used in the present study are described in Section 2. In Section 3, the WM method is reviewed and a modified algorithm is proposed. The strategy for computation and evaluation of the retracked SSHs is also introduced in this section. The retracked SSHs by various retrackers are validated using geoidal heights and tide gauge data in Section 4. Further discussions about our results are presented in Section 5. Finally, some conclusions and recommendations are summarized in Section 6.

2. Materials and Study Regions

2.1. Jason-2 Altimeter Data

The Ocean Surface Topography Mission (OSTM)/Jason-2 satellite, launched in June 2008, is an international mission between the National Aeronautics and Space Administration (NASA), the Central National d'Etudes Spatiales (CNES), the National Oceanic and Atmosphere Administration (NOAA) and the European Organization for the Exploitation of Meteorological Satellites (EUMETSAT). One vital objective of Jason-2 is to extend the time series of SSH observations beyond TOPEX/Poseidon (T/P) and Jason-1 by more than two decades. As a successor to T/P and Jason-1, Jason-2 not only inherited the characteristics of its predecessors: a similar payload (Poseidon altimeter) and an identical orbit (inclination: ~66°; revisit period: ~10 days), but also made some improvements in instruments, data processing and algorithms. For instance, the Poseidon-3 altimeter on Jason-2 has a lower instrumental noise and adopts a better tracking algorithm for land and ice surface than preceding generations of Poseidon altimeters [44]. These improvements are expected to increase the accuracy of sea surface height measurement to about 2.5 cm [45]. A detailed description of Jason-2 mission is referred to in [46].

The Jason-2 sensor geophysical data record (SGDR) product version d (JA2-GPS-PdP) provided by AVISO (Archiving, Validation and Interpretation of Satellite Oceanographic data)/CNES was used in this study. This dataset is a reprocessed delivery product, containing waveforms and all relevant corrections needed to retrieve SSHs. The waveform information is required for ground post-processing (i.e., retracking). Originally, the Jason-2 altimeter receives about 2060 elementary echoes per second. In order to reduce the thermal and speckle noise, waveforms are produced at a 20 Hz rate by averaging elementary echoes over 50 ms on board [44]. Detailed description of waveform sampling is suggested to refer [47]. Each waveform in SGDR comprises 104 sampling gates with an on-board tracking gate at #32.5. In addition to the operating tracker ranges, three types of retracked ranges calculated by OCEAN (MLE4), MLE3, and ICE retrackers, are included in SGDR [48].

For comparison, we employed the coastal and hydrology altimetry products released by the Collecte Localisation Satellites (CLS) in the frame of the PISTACH project. The PISTACH project is funded by CNES, aiming to improve Jason-2 altimetry data over coastal areas and continental waters. The PISTACH products are divided into two categories: one for coastal applications and the other for hydrology. Coastal products cover the whole ocean and a 25-km fringe over land. Hydrology products cover all continents plus a 25-km band over coastal ocean. Four alternative ranges are available in PISTACH products, namely, ICE1, ICE3, RED3 and OCE3 [21]. Therein, the ICE1 retracker, as well as the ICE retracker in SGDR, is based on the Offset Center of Gravity (OCOG) method [49], which determines range epoch by the first sample in which power is larger than 30% of the COG amplitude. The ICE3 retracker has the same principle as the ICE1 retracker, but only part of the samples around the main leading edge is taken into computation. The RED3 method retracks parameters with a maximum likelihood estimator in a small window selected in the same way as ICE3 retracker. The OCE3 range is obtained by applying the classical MLE3 retracker to filtered waveforms by singular value decomposition filtering.

2.2. Study Regions

Four study regions with different relief and sea state conditions were selected in order to evaluate the SSHs retrieved by various retrackers (Figure 1). Characteristics of the coast have a great effect on the shape of altimeter waveforms [50]. Complex topography will decrease the amount of recognizable waveforms in the coast area. Region (a) is located near the NOAA gauge station in Los Angeles, west coast of US. Ocean bottom elevation has a sudden drop near to 5 km offshore, with a relatively smooth slope between 0–4 km and 6–10 km. Region (b), neighboring to Cap May at the east coast of US, has the most gradual variation in topography and the shallowest water depth among the four study regions. Region (c) is at the south of Madeira Island, Portugal. In this region, the coastal topography is composed of a mountain ridge and the ocean depth sharply drops by 2000 m within 4 km. In contrast, water depth is also very shallow (<22 m) in region (d). Different from other regions, region (d) has a small angle between satellite track and coastline, which lengthens the effect of land contamination and results in a great number of waveforms deviating from the Brown model.

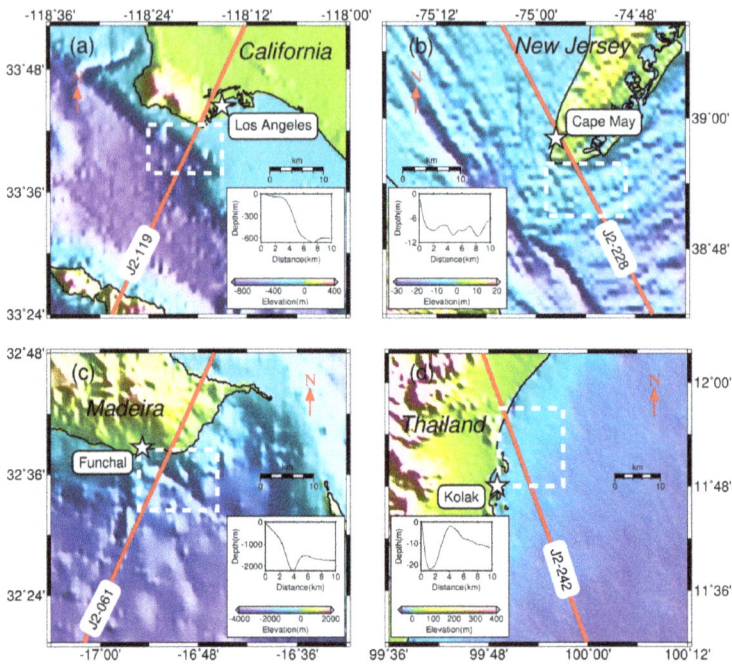

Figure 1. Study regions: (**a**) Los Angeles, CA, USA; (**b**) Cape May, NJ, USA; (**c**) Funchal, Madeira Island, Portugal; (**d**) Ko Lak, Prachuap Khiri Khan, Thailand. Red line denotes ground track of Jason-2 satellite. Orientation of the pass tag over the red line implies flight direction of the satellite. Evaluated data are located within the rectangle with the white dashed line. Gauge station is marked as the white pentacle with black edge. Topography is extracted from the digital elevation model SRTM30-plus [51]. Interpolated along-track ocean depth within 0–10 km offshore is plotted in the sub-panel.

In each study region, there is a Jason-2 pass and a vicinal tide gauge station. These four track segments used in this study consist of two approaching and two leaving land, corresponding to ascending and descending passes respectively. The flight direction of the satellite is denoted by the orientation of the pass tag in Figure 1. A brief introduction to the Jason-2 data used in the four study locations is given in Table 1. Cycles of altimeter data used in the study are selected according to the time span of available gauge data. The tide gauge data are downloaded from the University of Hawaii Sea Level Center (UHSLC). The hourly research quality data are used, which are the final science-ready dataset released 1–2 years after raw data are received by UHSLC [52].

Table 1. Brief introduction to Jason-2 data in each study region.

Region	Pass	Cycle [1]	Time Span	Direction
Los Angeles, USA	119	2-238(234)	July 2008–December 2014	Ascending, approaching land
Cape May, USA	228	2-238(234)	July 2008–December 2014	Descending, leaving land
Funchal, Portugal	061	2-202(196)	July 2008–December 2013	Ascending, approaching land
Ko Lak, Thailand	242	2-262(260)	July 2008–December 2014	Descending, leaving land

[1] The figures in parentheses refer to the number of available cycles.

3. Methodologies

3.1. Recapitulation of the WM Technique

WM is a statistical approach to trim off anomalous peaks of land contamination in the stacked waveforms [43]. An empirical criterion is exploited to detect the spurious peaks, and then the corresponding gates are amended by a 2-D interpolation from neighboring nodes. Firstly, the averaged waveform over deeper oceans is calculated as a reference waveform for each pass. Secondly, the residual waveform is defined as the following Equation (1), i.e., subtracting the reference from the original waveform.

$$\Delta P_n = P_n - P_{ref} \tag{1}$$

where P_n, ΔP_n are the n-th original waveform and the corresponding residual, P_{ref} is the reference waveform. If the i-th gate value $\Delta P_n(i)$ in the residual waveform fulfills the criterion given in Equation (2), the original waveform sampling $P_n(i)$ related to the i-th gate will be judged as an outlier.

$$|\Delta P_n(i)| > 2\sigma_n \tag{2}$$

where σ_n is the standard deviation of the n-th residual waveform. Finally, the outlier is replaced by the interpolated value from the neighboring samples in the current waveform and its adjacent waveforms.

The crucial point of the WM method is the selection of the reference waveform. An averaged waveform within 20–30 km of the coast is treated as an ocean waveform, of which the midpoint of the leading edge is very close to the nominal tracking gate. Tseng et al. [43] used it as the reference to detect spurious gates with contaminations from non-marine surfaces and achieved good results in most cases. However, there is one possibility when the leading edge of the waveform migrates far away from the nominal tracking gate. It often happens in the transition zone between land and sea. In this case, the reference waveform might lose its representativeness and bring about worse results. It can explain why the WM method had no improvement for SSHs within 0.5–1 km of the shoreline. Figure 2 demonstrates a failed case of the WM method near Cape May, NJ, USA, which is also the study region of Case 1 in [43]. It can be seen that original waveforms (Figure 2a) begin to shift at a latitude of 38.83° compared to waveforms over deeper ocean. These waveforms form an apparent bump between 38.85° and 38.90°. The shift causes large residuals near the nominal tracking gate for these waveforms (see Figure 2b) and subsequently misleads the judgment of contaminated gates. The leading edge in modified waveforms would be seriously distorted as a result (Figure 2c).

There is also a potential problem in repairing outliers. Nearby gates of the detected outliers have a high probability of being polluted by land effect, but their amplitudes are not enough to be judged as anomalies. Therefore, interpolation from nearby gates might not completely remove the anomalous peaks.

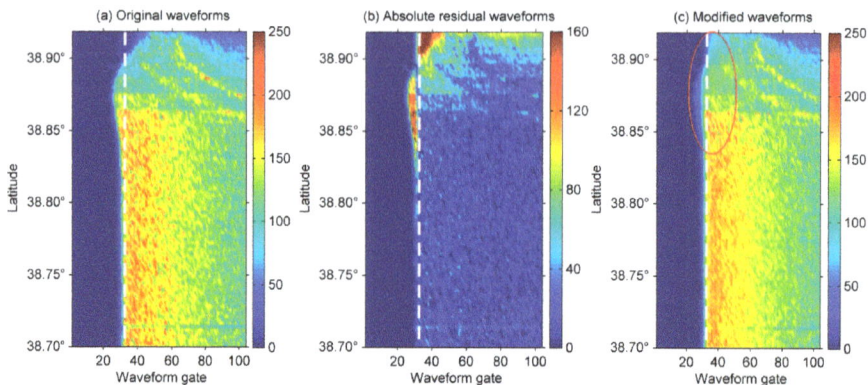

Figure 2. Failed case of coastal waveform modification illustrated by Jason-2 pass #228, cycle #50, near Cape May, NJ, USA. White dashed line shows the nominal tracking gate. The upper boundary in latitude indicates the coastline. Waveforms along with latitude are color-coded by waveform sampling power. (**a**) Original waveforms; (**b**) Absolute residual waveforms relative to the mean waveform over 20–30 km offshore; (**c**) Modified waveforms.

3.2. Waveform Decontamination

To avoid the abovementioned problems, three revisions for the WM method were made in this study. Here, the revised procedure is referred to as waveform decontamination (WD), to distinguish it from the WM. Firstly, we substituted the averaged waveform within 20 km of the coast as the reference waveform for the mean waveform between 20–30 km. Secondly, the standard deviation (σ_n) of the individual waveform was replaced with the root mean square (RMS) of all residual waveforms in the study region

$$\text{RMS} = \sqrt{\frac{\sum_{n=1}^{N} \sum_{i=1}^{M} \Delta P_n^2(i)}{\text{NM}}} \tag{3}$$

where N is the total number of waveforms in the study region, M is the number of waveform sampling gates (104 for Jason-2). By applying the new criterion, spurious gates were detected as in the WM method. Finally, outliers were directly set to null value in our procedure, instead of being repaired in the WM.

The first two revisions make the criterion for outlier detection more tolerant of the waveform shift. This is because the reference waveform and RMS used in the WD contain information of coastal waveforms to be modified, and therefore are more representative of these waveforms than those in the WM. The third revision we made is aiming to avoid errors induced by interpolation. These modifications are helpful to reduce the misjudgment of outliers and the risk of waveform distortion after amending. An example in Figure 3 shows the improvement after these revisions. The leading edge of the observed waveform (black curve with dots in Figure 3a) deviates from that of the reference waveform for the WM (blue curve in Figure 3a). This phase shift causes extreme residuals at two sampling gates on the leading edge (see Figure 3b). Therefore, these gates will be misjudged as anomalous samples by the WM. In contrast, the misjudgment does not happen in the case of the WD. This is because the reference waveform for the WD (red curve in Figure 3a) has a much gentle leading edge, which intersects the leading edge of the observed waveform. Meanwhile, the RMS computed using Equation (3) is larger than σ_n used in Equation (2), since the RMS contains contribution from shifted waveforms in the study area.

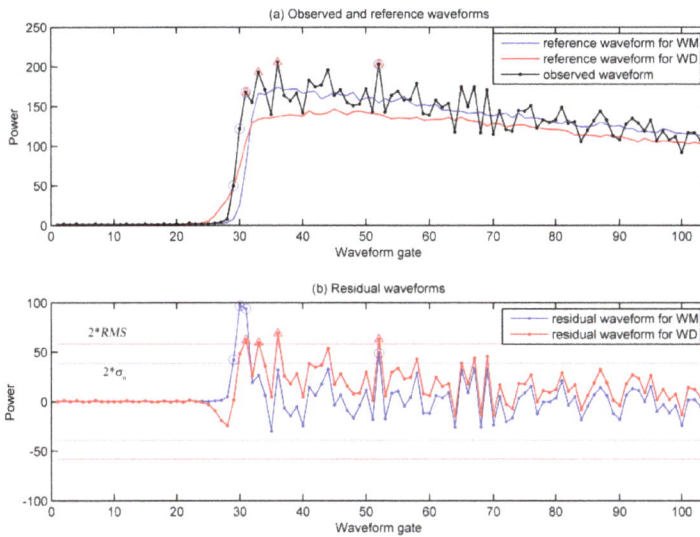

Figure 3. Comparison between waveform modification (WM) and waveform decontamination (WD) in the case of Figure 2. Reference waveforms for the two methods are shown in (**a**). The blue one is used in the WM, averaged within 20–30 km offshore. The red one is used in the WD, averaged between 0–20 km offshore. The black curve with dots is the observed waveform at latitude 38.84°. Its corresponding residual waveforms are shown in (**b**). The blue circles and the red triangles denote anomalous samples detected by WM and WD, respectively.

3.3. Computation and Evaluation of the Retracked SSHs

In order to test the efficiency of the WD method, 20% TR was used to retrack the original waveforms (Ori-TR), modified waveforms (MW-TR) and decontaminated waveforms (DW-TR), respectively. Because the decontaminated waveform may contain null values, traditional formula of TR to determine the retracked gate should be slightly reformed as

$$G_r = G_l + \frac{T_h - P_l}{P_k - P_l} \cdot (G_k - G_l) \tag{4}$$

where G_r is the retracked gate, T_h is the threshold level, G_k is the first gate with power exceeding T_h, G_l is the first non-null gate before G_k, P_k and P_l are the power at gate G_k and G_l, respectively. The threshold level T_h is computed using

$$T_h = T_0 + th \cdot (P_{max} - T_0) \tag{5}$$

where T_0 is the thermal noise which is determined by averaging the power in the first five gates, *th* is the threshold (20% in the present study), P_{max} is the maximum power of a waveform. Then the retracked range correction is calculated using the offset between the retracked gate and the nominal tracking gate multiplied by the one way range corresponding to the sampling interval.

After waveform retracking correction, the media and geophysical corrections were included to correct range estimations. All atmosphere corrections used in this study were model-derived values. The dry and wet troposphere corrections were calculated using climate models from the European Centre for Medium-range Weather Forecasts (ECMWF). The ionosphere corrections were derived from the Global Ionospheric Maps (GIM). The ocean/solid earth/pole tide corrections were computed using FES2004, Cartwright and Taylor tidal potential, Equilibrium model, respectively [45]. These corrections were provided both in SGDR and in PISTACH. The SSB corrections and the dynamic atmosphere corrections we used in this study were retrieved from the PISTACH product. The SSB corrections associated with OCEAN retracker in PISTACH had the smallest number of default values, and hence were applied to the other retracked SSHs.

Two methods were employed to evaluate the retracked SSHs. First, along-track SSHs were compared to their corresponding geoidal heights derived from EGM2008 model [53]. The standard deviation (SD) of difference between along-track SSHs and geoidal heights was computed, which can be used to assess the variability of the along-track SSHs during each cycle. Second, the SSHs were compared to nearby tide gauge data. In order to evaluate the altimeter-derived SSHs using in-situ sea level, tidal corrections were removed from the SSHs. Selecting the gauge station as the reference point, the SSH time series were determined by correcting the geoid differences between locations of measurements and gauge station. To avoid datum differences between altimeter measurements and in-situ observations, the temporal mean value in the time series was removed. Since the sampling interval of the altimeter-derived SSH time series is only ~10 days, the in-situ sea level was linearly interpolated at each epoch of altimetry observations from the hourly sampling of gauge level data. Finally, the SD and correlation coefficient between altimeter and in-situ time series were computed.

4. Results

In order to retain as many measurements as possible, no data editing was performed on SGDR and PISTACH products. Nine retrackers were compared in this study, including three retrackers (ICE, OCEAN, MLE3) from SGDR, three retrackers (ICE3, OCE3, RED3) from PISTACH, and three retrackers (Ori-TR, MW-TR, DW-TR) performed in this study. Table 2 lists the 20 Hz data availability of various retrackers within 10 km of the shoreline in four study regions. It should be noted that the distance to the coastline was approximately represented by the along-track distance to the point where the satellite ground track crosses the coastline. The data availability rates of OCEAN and OCE3 are very low in the coastal area. OCE3 only successfully retracked 54% waveforms, which is equivalent to the percentage of ocean waveforms classified in PISTACH. In the following discussion, the OCE3 retracker was not taken into account due to its low data availability. It should be clarified that all retrackers were applied to the 0–20 km coastal band, although this study focuses on an area within 10 km offshore.

Table 2. Data availability of various retrackers within 10 km offshore in each study region.

Region	20 Hz Data (pt)	Ocean Waveform (%)	Retracker (%)								
			ICE	OCEAN	MLE3	ICE3	OCE3	RED3	Ori-TR	MW-TR	DW-TR
Los Angeles	8082	57	100	92	100	99	57	97	100	100	100
Cape May	8265	74	100	97	100	100	74	99	100	100	100
Funchal	6746	53	100	70	100	100	52	100	100	100	100
Ko Lak	8569	30	100	66	100	99	30	88	100	100	100
Mean	7916	54	100	81	100	99	54	96	100	100	100

4.1. Comparison to Geoid

The SDs of the difference between various retracked along-track SSHs and geoid within 10 km away from the coastline are computed for all cycles after discarding crude measurements by a 3σ de-outlier process. Figure 4 shows histograms of the SDs in the four study regions. Table 3 presents the mean SDs before and after removing outliered cycles (>3σ) and corresponding improvement percentage (IMP) [32] relative to the non-retracked SSHs. The SD after removing outliered cycles is denoted as Cal. SD in Table 3 and the following table. The highest precision and the maximum IMPs are indicated by underlined numbers. The percentages of valid data and numbers of invalid cycles are also tabulated in Table 3.

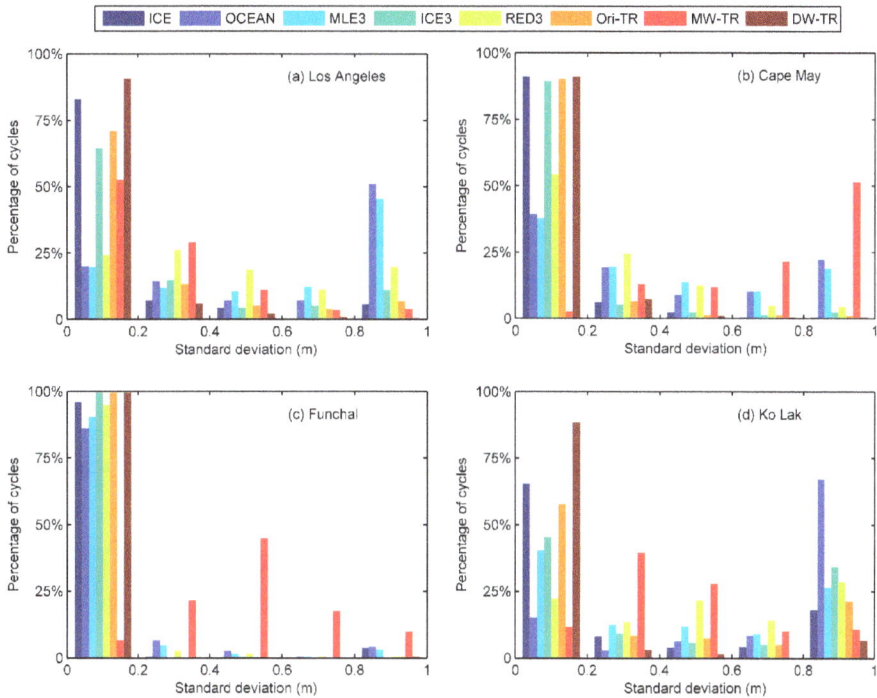

Figure 4. Histograms for standard deviations of various retracked sea surface heights (SSHs) relative to geoid in four study regions: (**a**) Los Angeles, (**b**) Cape May, (**c**) Funchal, (**d**) Ko Lak. All available cycles are included for statistical analysis. Only frequencies of standard deviations less than 1 m are plotted with the step of 0.2 m. Retrackers are represented by different colors shown in the upmost legend.

It is notable that DW-TR exhibits better performance than the other retrackers. In overall average, DW-TR preserves most of data points (97%) and achieves the largest improvement, which is about 82% and 89% before and after the calibration, respectively. It exceeds two 30% threshold retrackers, ICE and ICE3, by 5% and 15% in accuracy, respectively. For the same 20% threshold retrackers, DW-TR outperforms Ori-TR by 9% and MW-TR by 34%. It indicates that the waveform decontamination procedure can efficiently reduce the superimposed noise in coastal waveforms. Overall, three model-based retrackers (OCEAN, MLE3 and RED3) have more data loss and less improvement in accuracy than threshold retrackers. In region (a), OCEAN and MLE3 retracked results are even worse than the non-retracked SSHs. Obviously, the redundant peaks in waveforms will cause the fitted waveform using the Brown model to deviate seriously from the actual waveform. Although RED3 applies a window to filter the waveform, its improvement is not significant.

Histograms in Figure 4 also illustrate that DW-TR is the most robust retracker. For the DW-TR retracked SSHs, more than 85% of cycles have the SD <0.2 m in all the four cases. The OCEAN retracker is less robust compared to the other retrackers. It can be found over region (a) and (d) that the precision of OCEAN retracked measurements is larger than 0.8 m for more than 50% of cycles. Furthermore, frequency of high precision (<0.2 m) cycles for MW-TR is relatively less than the other retrackers. It is particularly evident in region (b). This result indicates that the precision of MW-TR retracked measurements in many cycles deteriorate after waveform modification.

Table 3. Mean standard deviation of the difference between retracked along-track SSHs and geoid within 10 km offshore in four study regions.

Region (a): Los Angeles								
Retracker	ICE	OCEAN	MLE3	ICE3	RED3	Ori-TR	MW-TR	DW-TR
SD (cm)/IMP (%)	32/68	128/−28	120/−20	56/44	94/6	46/54	39/61	25/75
Cal. SD (cm)/IMP (%)	20/79	109/−15	103/−8	35/63	63/34	29/69	35/63	18/81
Valid data (%)	96	77	97	94	92	95	99	97
Invalid cycle	5	2	5	9	7	8	1	1
Region (b): Cape May								
Retracker	ICE	OCEAN	MLE3	ICE3	RED3	Ori-TR	MW-TR	DW-TR
SD (cm)/IMP (%)	36/82	85/58	64/68	31/85	59/71	32/84	117/42	31/85
Cal. SD (cm)/IMP (%)	15/92	64/67	56/71	16/92	34/82	17/91	94/51	17/91
Valid data (%)	97	85	98	96	93	97	97	97
Invalid cycle	1	3	2	5	2	2	1	1
Region (c): Funchal								
Retracker	ICE	OCEAN	MLE3	ICE3	RED3	Ori-TR	MW-TR	DW-TR
SD (cm)/IMP (%)	30/68	26/73	21/78	10/89	17/82	14/85	64/33	14/85
Cal. SD (cm)/IMP (%)	10/89	18/81	14/85	9/90	13/86	9/90	61/34	9/90
Valid data (%)	96	69	96	99	97	99	99	99
Invalid cycle	6	6	6	1	5	1	1	1
Region (d): Ko Lak								
Retracker	ICE	OCEAN	MLE3	ICE3	RED3	Ori-TR	MW-TR	DW-TR
SD (cm)/IMP (%)	77/60	135/30	71/63	118/39	107/45	86/55	71/63	35/82
Cal. SD (cm)/IMP (%)	49/74	127/33	60/68	93/51	72/62	56/70	57/70	16/92
Valid data (%)	95	53	97	96	84	94	96	94
Invalid cycle	9	3	6	4	11	10	8	14
Average over Four Regions								
Retracker	ICE	OCEAN	MLE3	ICE3	RED3	Ori-TR	MW-TR	DW-TR
SD (cm)/IMP (%)	44/70	94/33	69/47	54/64	69/51	45/70	73/50	26/82
Cal. SD (cm)/IMP (%)	24/84	80/42	58/54	38/74	46/66	28/80	62/55	15/89
Valid data (%)	96	71	97	96	92	96	98	97
Invalid cycle	5.3	3.5	4.8	4.8	6.3	5.3	2.8	4.3

4.2. Comparison to Tide Gauge Data

Figure 5 shows an example of coastal (0–10 km) SSH variation from each retrackers in region (a). Temporal mean is subtracted from each time series, referring to as height anomaly. It is obvious that the measurements from SGDR and PISTACH products are much noisy than those of DW-TR with respect to tide gauge. In Figure 5a, the OCEAN and MLE3 retracked SSHs are apparently more dispersed than the ICE results. Figure 5b illustrates that the SSHs from ICE3 and RED3 in many cycles are largely deviated from the gauge data. Figure 5c presents the results from three 20% threshold retrackers using original, modified and decontaminated waveforms, respectively. The retracked results by DW-TR are most consistent with the gauge data. It is evident that a number of deviated cycles exist in the time series estimated by MW-TR.

The statistical results of comparison over the four study regions are summarized in Table 4. The results show, on average, DW-TR yields the smallest SDs and the highest IMPs. It also obtains the largest number of valid data and the smallest number of invalid cycles. Furthermore, DW-TR presents the highest correlation between height anomaly and tide gauge. It is the only retracker with a mean correlation coefficient higher than 0.9. Compared to Ori-TR, DW-TR achieves 40% (50–30 cm) improvement before calibration and 33% (30–20 cm) after calibration, whilst a decrease in accuracy from MW-TR is observed. MW-TR is better than Ori-TR only in region (d). Concerning retrackers in SGDR and PISTACH products, ICE performs with greater stability than the other retrackers. In some cases (e.g., near Cape May coast), ICE provides comparable results to DW-TR. Negative IMPs appear in some cases for OCEAN, MLE3, ICE3 and RED3, indicating that these retrackers are unreliable. Although the PISTACH retrackers are developed for coastal applications, their performance should be evaluated case by case. For instance, the results of ICE3 are better in region (c), but worse in the other regions than those of the ICE retracker.

Table 4. Statistics of various retracked SSHs within 10 km offshore compared with tide gauge in each region.

Region (a): Los Angeles								
Retracker	ICE	OCEAN	MLE3	ICE3	RED3	Ori-TR	MW-TR	DW-TR
SD (cm)/IMP (%)	34/65	69/31	103/−4	55/44	58/41	41/59	46/53	22/78
Cal. SD (cm)/IMP (%)	21/76	55/37	76/13	35/60	34/61	24/72	37/57	17/81
Valid data (%)	98	76	97	95	92	98	98	97
Invalid cycles	3	4	6	6	7	6	6	1
Correlation	0.89	0.6	0.43	0.77	0.75	0.82	0.7	0.95

Region (b): Cape May								
Retracker	ICE	OCEAN	MLE3	ICE3	RED3	Ori-TR	MW-TR	DW-TR
SD (cm)/IMP (%)	33/56	88/−19	82/−10	41/45	71/4	36/51	50/32	35/53
Cal. SD (cm)/IMP (%)	22/63	46/23	44/26	24/60	26/57	23/62	39/35	22/63
Valid data (%)	97	85	97	97	93	97	97	97
Invalid cycles	1	1	3	2	1	1	1	1
Correlation	0.92	0.65	0.69	0.9	0.89	0.92	0.73	0.92

Region (c): Funchal								
Retracker	ICE	OCEAN	MLE3	ICE3	RED3	Ori-TR	MW-TR	DW-TR
SD (cm)/IMP (%)	51/36	39/51	24/69	16/80	17/78	27/66	41/48	14/82
Cal. SD (cm)/IMP (%)	18/68	19/65	15/72	16/71	16/71	13/76	35/37	13/76
Valid data (%)	98	76	97	95	92	98	98	97
Invalid cycles	25	24	24	22	23	22	24	22
Correlation	0.9	0.93	0.93	0.97	0.97	0.98	0.83	0.98

Region (d): Ko Lak								
Retracker	ICE	OCEAN	MLE3	ICE3	RED3	Ori-TR	MW-TR	DW-TR
SD (cm)/IMP (%)	89/6	143/−51	94/1	133/−41	122/−29	94/0	56/41	49/49
Cal. SD (cm)/IMP (%)	57/36	113/−27	75/15	104/−17	56/36	58/34	36/60	27/70
Valid data (%)	91	51	94	91	81	91	92	92
Invalid cycles	20	15	17	17	20	21	18	18
Correlation	0.67	0.47	0.61	0.58	0.72	0.64	0.76	0.84

Average over Four Regions								
Retracker	ICE	OCEAN	MLE3	ICE3	RED3	Ori-TR	MW-TR	DW-TR
SD (cm)/IMP (%)	52/41	85/3	76/14	61/32	67/24	50/44	48/44	30/66
Cal. SD (cm)/IMP (%)	30/61	58/25	53/32	45/44	33/56	30/56	37/47	20/73
Valid data (%)	96	72	96	95	90	96	96	96
Invalid cycles	12.3	11	12.5	11.8	12.8	12.5	12.3	10.5
Correlation	0.85	0.66	0.67	0.81	0.83	0.84	0.76	0.92

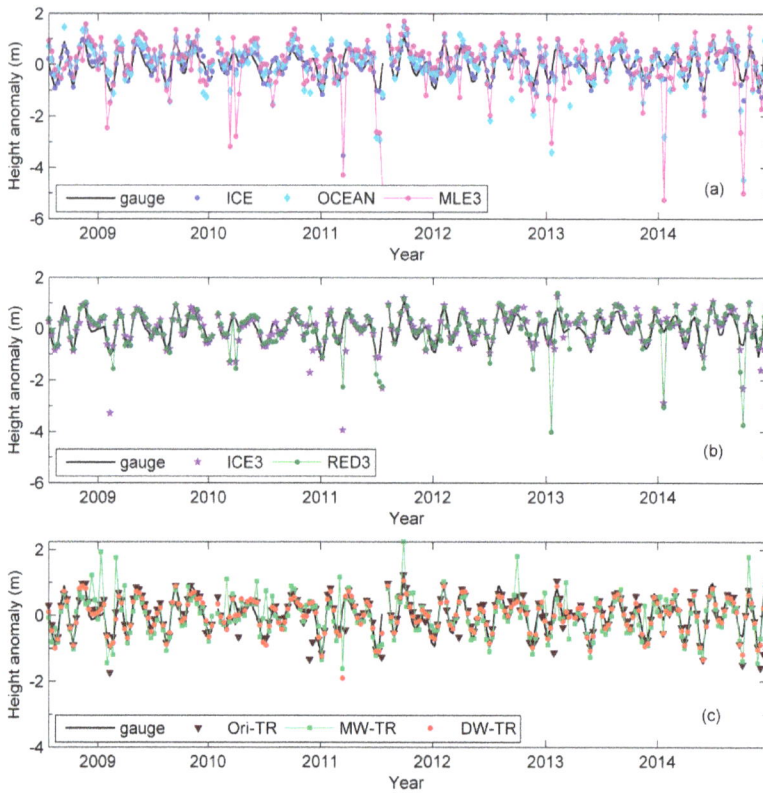

Figure 5. Coastal (0–10 km) SSH variation from different retrackers: (**a**) Sensor geophysical data record (SGDR) retrackers; (**b**) Prototype Innovant de Système de Traitement pour les Applications Côtières et l'Hydrologie (PISTACH) retrackers; (**c**) 20% threshold retrackers using original, modified and decontaminated waveforms near Los Angeles gauge station. Black line denotes tide gauge data.

In order to assess how close each retracker can extend precise measurements to the coastline, the SDs and correlation coefficients over each 1-km band within 20 km of the shoreline are computed, which are shown in Figure 6 in terms of along-track distance. It can be seen that DW-TR achieves the smallest SDs and the highest correlation in all cases within 10 km of the shoreline, whilst performing as well as OCEAN, ICE, and Ori-TR in the 10–20 km zone. It can extend measurements to 1 km offshore with nearly consistent accuracy and data gaps as well as high correlation, except for in region (d). There is a peninsula extruding toward the middle segment of the selected satellite track in region (d), which results in large error and low correlation at an along-track distance of 4–6 km. As expected, the performance of OCEAN gets worse when approaching the coast. Data gaps in each band within 10 km for different retrackers are illustrated in Figure 7. From 7 km, the number of invalid cycles for OCEAN is significantly larger than those of other retrackers. Although the OCEAN retracker achieves high accuracy in area very near close to the coast sometimes, it is based on very huge data gaps as displayed in Figure 7.

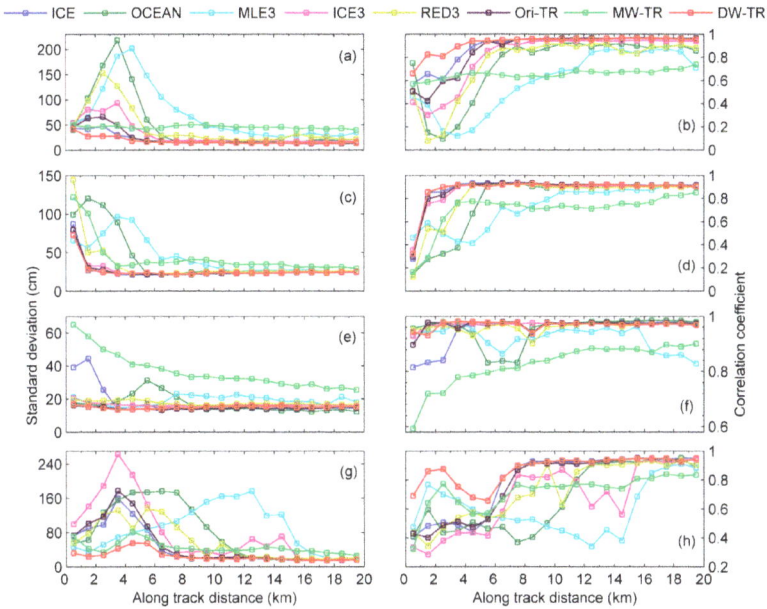

Figure 6. Along-track standard deviations (left column) and correlation coefficients (right column) between various retracked SSHs and tide gauge within 20 km offshore. Each node is computed using the SSHs within the corresponding zone of 1 km in width. Study region is arranged in row: (**a,b**) Los Angeles, (**c,d**) Cape May, (**e,f**) Funchal, (**g,h**) Ko Lak.

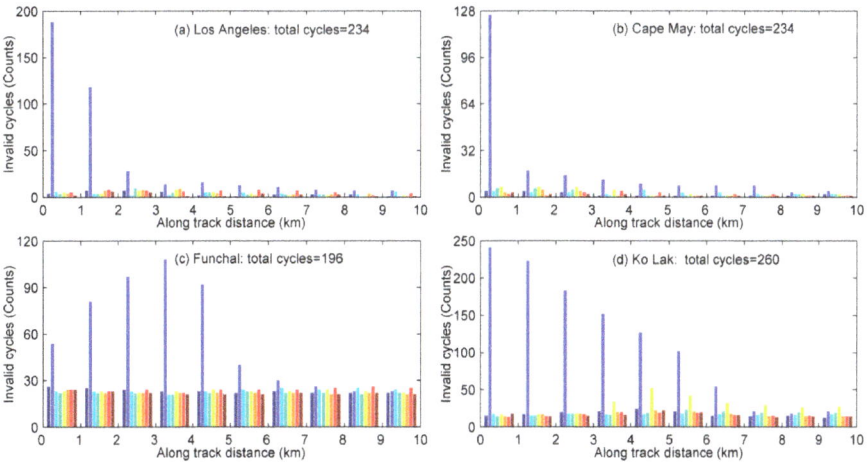

Figure 7. Invalid cycles for different retrackers over each 1-km band within 10 km offshore in four study regions: (**a**) Los Angeles, (**b**) Cape May, (**c**) Funchal, (**d**) Ko Lak. Color legend is the same as Figure 4.

5. Discussion

5.1. Negative Effects of Contaminations in Coastal Waveforms on Retrackers

Comparison results in Section 4 demonstrate the inefficiency of Brown-model-based retrackers (OCEAN, MLE3, RED3 and OCE3) for coastal altimetry. Obviously, the estimates from the maximum likelihood estimator applied to the Brown model cannot accurately reproduce the variations of the coastal waveform, due to the superimposed echoes from non-marine surfaces [6]. On the other hand, the contamination in coastal waveform may disturb the estimation of waveform amplitude, and then lead to unreliable retracked SSHs from empirical retrackers [43]. In general, SSHs will be underestimated by OCOG and threshold retrackers due to the anomalous peaks. It can explain why the empirical retrackers (ICE, ICE3 and Ori-TR) using original waveforms have inconsistent accuracy along the distance to the coast (see Figure 6), and have a relatively large number of invalid cycles (see Tables 3 and 4).

The subwaveform technique may improve the precision of altimeter observations. But it is difficult to accurately extract the subwaveform, when there are multiple peaks or an excessive peak in the waveform. Anomalous peaks may mislead the judgment of the leading edge, thus resulting in seriously deviated height estimations. This is the reason that the ICE3 and RED3 retrackers, which retrack portion samples of the waveform, show worse performance and less valid data in some cases. To further discuss the performance of subwaveform retrackers in coastal area, we analyzed the state-of-the-art ALES (Adaptive Leading Edge Subwaveform) product in the four study regions. ALES picks out subwaveform including the leading edge using an adaptive window, and models it with the classic Brown model by means of least square estimation [38]. According to the result in Figure 8, ALES works effectively to about 8 km offshore. The best case is near Ko Lak, Thailand, where SSHs from ALES have much higher accuracy than those from DW-TR up to 4 km offshore. However, the ALES outputs show significant degradation over areas very close to land. The same result was found around the Tsushima Island in Japan [54]. Apparently, ALES does not completely overcome the difficult problem of accurate subwaveform extraction in coastal ocean.

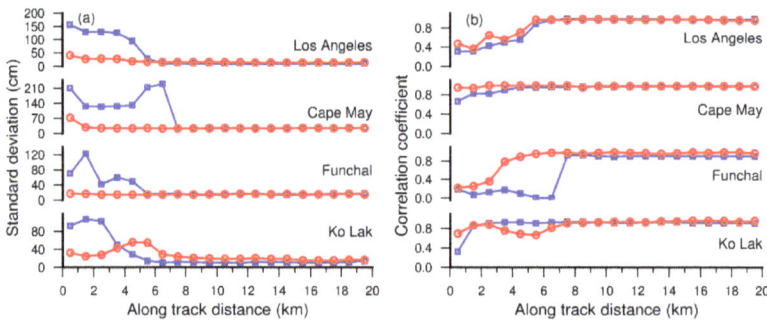

Figure 8. *Cont.*

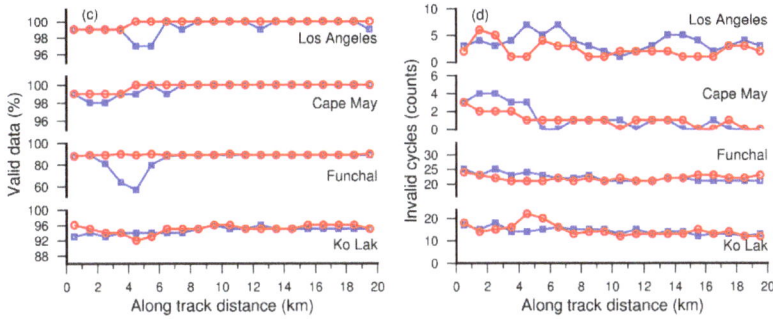

Figure 8. Standard deviation (**a**) correlation coefficient (**b**) valid data percentile (**c**) and invalid cycle number (**d**) of ALES product (**blue**) in the four study regions. For comparison, the results of DW-TR (**red**) are also shown in the figure.

5.2. Improvement of the WD with Respect to the WM

In nature, the effect of the WM and the WD is not to improve the retracking algorithm, but to purify waveforms. It is similar to the subwaveform technique to some extent. The subwaveform approach excludes polluted samples by selecting part of each waveform, while the modification procedure is designed to detect and remedy spurious samples by some empirical methods.

As pointed out in Section 3.1, shifts of the leading edges of the waveforms will cause the failure of the WM procedure. In ocean-land or land-ocean transition zones, these shifts will be frequently introduced by the tracker algorithm, which is based on the determination of the center of gravity of the waveform energy [55]. Most of sampling gates in the shifted leading edges will be misjudged as outliers by the WM method (see Figure 9). Therefore, retracking modified waveforms does not achieve similar average accuracy compared to retracking original waveforms over the four test regions. From Figure 9, it can be seen that outliers detected by the WD are much less than those by the WM. The WD method presented in this paper successfully overcomes the insufficiency of WM. All results validate the highest performance of the DW-TR retracker which is based on purified waveforms by the WD.

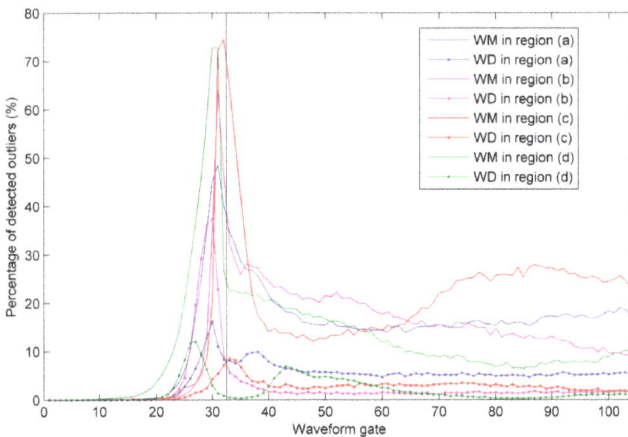

Figure 9. Percentage of detected outliers at each sampling gate in the four study regions for the WM (solid curves) and for the WD (solid curve with dot). The black dashed line shows the location of the nominal tracking gate.

5.3. Bias Analysis

Biases between DW-TR, Ori-TR and OCEAN retrackers were estimated for each region, presented in Table 5. Biases were computed by averaging differences between two retracked SSH sets within 13–20 km where correlation coefficients were higher than 0.9 for the three retrackers and for the four considered regions. Biases of DW-TR and Ori-TR relative to OCEAN are of the same order of 65 to 78 cm, with standard deviation of the order of 6 to 11 cm. Biases between DW-TR and Ori-TR are of the order of 1 cm with standard deviation less than 2 cm. It implies that the WD method proposed in this paper will not induce a significant bias into the SSH estimations. The bias between DW-TR and OCEAN is mainly a result of the difference between the 20% threshold retracker and the OCEAN retracker. Therefore, for operational applications covering a larger area, the WD method can still improve the accuracy of SSHs without significant biases, if the same retracker is employed for retracking decontaminated waveforms in coastal regions and original waveforms in open oceans.

Table 5. Mean biases with standard deviation between retracked SSHs by DW-TR, Ori-TR and OCEAN. Biases were estimated using along-track points within the 13–20 km band where correlation coefficient with tide gauge was higher than 0.9 for the three retrackers and for the four regions.

	Region (a)	Region (b)	Region (c)	Region (d)
DW-TR–OCEAN (cm)	72.6 ± 11.9	73 ± 6.5	78 ± 11.8	65.7 ± 6.9
Ori-TR–OCEAN (cm)	72.9 ± 11.7	71.9 ± 6.3	77.6 ± 10.8	65 ± 8.2
DW-TR–Ori-TR (cm)	0.2 ± 1.1	1.4 ± 1.2	0.5 ± 1.7	1.3 ± 1.9

5.4. Test for Tracks Paralleling or Crossing Intricate Coastlines

To test the ability of our proposed method in more complicated situations, experiments were performed for three more tracks passing through areas with different types of coastlines. The first track with a length of about 100 km is taken from pass 229, which is nearly parallel to the northwest coast of Indonesia (the bottom in Figure 10). There is an island in the south end of this track. Maximum distance to land is less than 20 km. The second track from pass 228 is located near the Delta Amacuro in Venezuela (the middle in Figure 10). A 65 km long segment is truncated in which the longest distance to land is about 11 km. Many small islands distribute on the side of the track where the river estuary is. The third one is a 45-km-long track from pass 242, located in the Singapore Strait. There are lots of islands on both sides of the track. The farthest distance away from land is less than 5 km, and then almost every waveform on this track contains land contamination.

Since tide gauge data are not available, we compared the sea level anomaly (SLA) from DW-TR to those from ICE, ALES and Ori-TR retrackers. Figure 10 shows examples of SLA profiles from different retrackers. For all cases, reference waveform used in the WD was averaged using all waveforms in the selected track. It can be seen that DW-TR shows robust performance in three cases. For pass 229, the performance of ALES and DW-TR is better than those of ICE and Ori-TR retrackers. ICE and Ori-TR suffer from large errors in the region close to the tongue at a latitude of about 1.2°. Obviously, anomalous peaks cause the sea level to be underestimated by ICE and Ori-TR. The subwaveform technique and the WD process can exclude or remove anomalous peaks well in this case. For pass 228, four retrackers work well except at both ends of the track where the satellite nadirs are much closer to coastline. Compared to other retrackers, the DW-TR obtains a relatively larger number of reasonable SLAs where the distance to land is less than 5 km. In the case of pass 242, the performance of ICE, Ori-TR and ALES becomes worse, while the DW-TR still keeps high effectivity and reliability. Comparisons of SLAs by the DW-TR with those by the Ori-TR further confirm that the WD may not produce a significant bias.

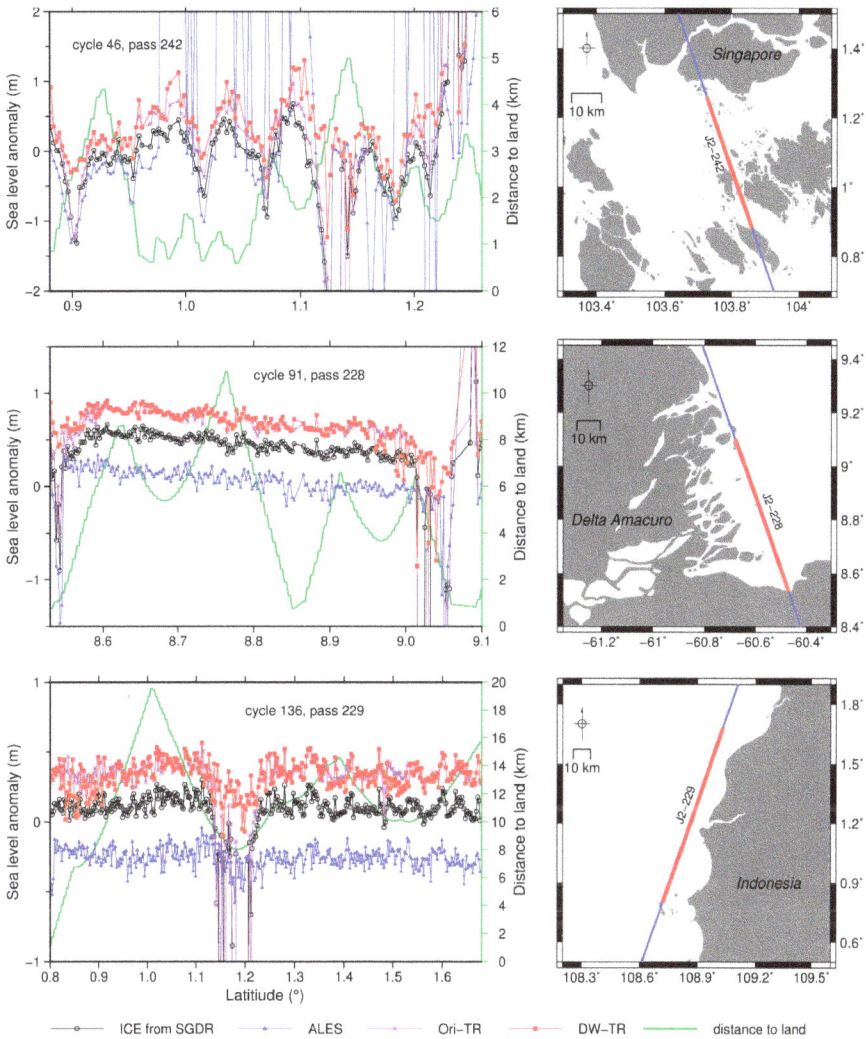

Figure 10. Sea level anomaly profiles of Jason-2 altimetry along pass 242 from cycle 46 (**upper**), pass 228 from cycle 91 (**middle**), and pass 229 from cycle 136 (**bottom**), which cross different types of coastlines. Tracks (the red parts are used here) and coastlines are shown in the right.

5.5. Issues for Further Research

Although the WD method was validated in several regions in this study, more validations are needed for future global coastal applications since coastal situations are so complicated and quite different everywhere. It is worth mentioning the important role of the reference waveform for the WD. The impact of the reference waveform and it optimal determination might be worthy of further study. No strict rule for the selection of the reference waveform is given in this study. But it should be kept in mind that the reference waveform has to contain information of coastal waveforms to be modified.

It is difficult to accurately detect all contaminated samples in a waveform in a simple way. Underestimation of SLA at both ends of track 228 in Figure 10 implies that there are still some anomalous peaks remained after the WD. Therefore, there is certainly room for improvement of outlier detection. Realigning along-track waveforms [39,54] before the WD might be a good way to further improve outlier detection, which is under investigation.

It should be emphasized that we used the same SSB corrections for all the retracked SSHs, which might induce some errors. The SSB is the sum of electromagnetic bias, skewness bias and tracker bias [25,45]. The accuracy of the SSB is mainly up to the accuracy of estimated significant wave height (SWH), varying with different retrackers. It is expected that the retracked SSHs can be further improved by applying SSB correction corresponding to the specific retracker. Focusing on the performance of various retrackers, SWH is not estimated in this study. It remains a topic of research to assess the impact of SSB on the retracked results.

6. Conclusions and Recommendations

It is proved that removing the superimposed signals from non-marine surfaces by empirical or model-fitting methods is helpful to improve the performance of waveform retracking [6,39,43]. In this paper, an improved empirical method for waveform decontamination was developed based on the WM algorithm. According to the experiments over four coastal regions with different situations, retracked SSHs using the decontaminated waveforms are superior to those using original waveforms and modified waveforms. Comparison results also suggest the DW-TR retracker outperforms SGDR, PISTACH and ALES retrackers in terms of accuracy and data coverage within 10 km offshore. It should be emphasized that the WD method is appropriate for other altimeter data, although it was only validated here using Jason-2 altimeter data. Moreover, only the 20% threshold retracker was applied on the decontaminated waveforms as SSHs were concerned in the present study. In order to extract more parameters, such as significant wave height, model-based retrackers should be used. Theoretically, all retrackers can be exploited to retrack the decontaminated waveforms. Their performance, however, should be further evaluated.

For the SGDR products, ICE provides more accurate measurements than OCEAN and MLE3 in coastal areas, which is consistent with previous studies [33,43]. It is reasonable since in some degree, the ICE retracker moderates the influence of excessive energy in the trailing edge by combining threshold and OCOG retrackers. The performance of ICE3 and RED3 in PISTACH products is not stable. Generally speaking, the two retrackers have high accuracy beyond about 8 km offshore, and become worse in the 0–8 km coastal band. The same performance can be found for ALES in the four study regions. It indicates that these subwaveform retrackers encounter a barrier to accurately extract noise-free subwaveforms in coastal oceans very close to land.

The main purpose of this study is to improve SSHs within 10 km offshore. Experiments in the four study regions were performed using data in the 0–20 km coastal band. The subsequent test in three particular cases shows the new retracking strategy based on the WD proposed here is not limited to a 20 km wide strip along the coast, and can adapt to diverse complicated situations. So, it has great potential for operational application in the global coastal zone. Another advantage of the WD method is that it will not induce a bias between coastal regions and open oceans, as long as the same retracker is applied on decontaminated waveforms and original waveforms. This feature may extend applications of the new method to a larger area. With the help of the WD, the retracking process can achieve more accurate coastal altimetry data with the ability to compare them to the data from open oceans. It will greatly benefit coastal applications connected to sea level, such as coastal circulation studies, coastal tide modelling and refining gravity field in coastal zones. By combining the WD with sophisticated retrackers, significant wave height and backscatter coefficient are expected to be acquired with a higher accuracy, which is helpful for monitoring coastal sea state. This work will be performed in the future study.

Acknowledgments: We acknowledge AVISO for the Jason-2 SGDR data, CLS for the PISTACH data, and UHSLC for the tide gauge data used in this study. The ALES data were obtained from JPL Physical Oceanography DAAC and developed by the UK National Oceanography Centre. This work was supported by the National Natural Science Foundation of China (Grant No. 41774010, 41174021, 41429401), by the Open Fund of Guangxi Key Laboratory of Spatial Information and Geomatics (Grant No. 15-140-07-26), and by the Belmont Forum/IGFA G8 Coastal Vulnerability Program via the United States National Science Foundation (Grant No. CER-1342644).

Author Contributions: Haihong Wang and Zhicai Luo conceived and designed the experiments; Zhengkai Huang and Haihong Wang performed the experiments and wrote the first draft; all authors analyzed the data and wrote the final draft.

Conflicts of Interest: The authors declare no conflict of interest. The founding sponsors had no role in the design of the study; in the collection, analyses, or interpretation of data; in the writing of the manuscript, and in the decision to publish the results.

References

1. Fu, L.L.; Cazenave, A. *Satellite Altimetry and Earth Sciences: A Handbook of Techniques and Applications*; Academic Press: San Diego, CA, USA, 2001.
2. Deng, X.; Featherstone, W.E.; Hwang, C.; Berry, P.A.M. Estimation of contamination of ERS-2 and POSEIDON satellite radar altimetry close to the coast of Australia. *Mar. Geodesy* **2002**, *25*, 249–271. [CrossRef]
3. Vignudelli, S.; Cipollini, P.; Roblou, L.; Lyard, F.; Gasparini, G.P.; Manzella, G.; Astraldi, M. Improved satellite altimetry in coastal systems: Case study of the Corsica Channel (Mediterranean Sea). *Geophys. Res. Lett.* **2005**, *32*, L07608. [CrossRef]
4. Gómez-Enri, J.; Cipollini, P.; Passaro, M.; Vignudelli, S.; Tejedor, B.; Coca, J. Coastal altimetry products in the Strait of Gibralta. *IEEE Trans. Geosci. Remote Sens.* **2016**, *54*, 5455–5466. [CrossRef]
5. Deng, X. Improvement of Geodetic Parameter Estimation in Coastal Regions from Satellite Radar Altimetry. Ph.D. Thesis, Curtin University of Technology, Perth, Australia, 2004.
6. Halimi, A.; Mailhes, C.; Tourneret, J.Y.; Thibaut, P.; Boy, F. Parameter estimation for peaky altimetric waveforms. *IEEE Trans. Geosci. Remote Sens.* **2013**, *51*, 1568–1577. [CrossRef]
7. Anderson, O.B.; Scharroo, R. Range and geophysical corrections in coastal regions: And implications for mean sea surface determination. In *Coastal Altimetry*; Vignudelli, S., Kostianoy, A.G., Cipollini, P., Benveniste, J., Eds.; Springer: Berlin, Germany, 2011; pp. 103–145.
8. Feng, H.; Vandemark, D. Altimeter data evaluation in the coastal gulf of Maine and Mid-Atlantic bight regions. *Mar. Geodesy* **2011**, *34*, 340–363. [CrossRef]
9. Idris, N.H.; Deng, X.L.; Anderson, O.B. The importance of coastal altimetry retracking and detiding: A case study around the Great Barrier Reef, Australia. *Int. J. Remote Sens.* **2014**, *35*, 1729–1740. [CrossRef]
10. Vignudelli, S.; Snaith, H.M.; Lyard, F.; Cipollini, P.; Venuti, F.; Birol, F.; Bouffard, J.; Roblou, L. Satellite radar altimetry from open ocean to coasts: Challenges and perspectives. In Proceedings of the 5th Asia-Pacific Remote Sensing Symposium, Goa, India, 13–17 November 2006; Volume 6406, pp. 1–12.
11. Reale, F.; Dentale, F.; Pugliese Carratelli, E.; Torrisi, L. Remote sensing of small-scale storm variations in coastal seas. *J. Coast. Res.* **2014**, *30*, 130–141. [CrossRef]
12. Vincent, P.; Steunou, N.; Caubet, E.; Phalippou, L.; Rey, L.; Thouvenot, E.; Verron, J.; AltiKa, A. Ka-band altimetry payload and system for operational altimetry during the GMES period. *Sensors* **2006**, *6*, 208–234. [CrossRef]
13. Raney, R.K. The delay/Doppler radar altimeter. *IEEE Trans. Geosci. Remote Sens.* **1998**, *36*, 1578–1588. [CrossRef]
14. Halimi, A.; Mailhes, C.; Tourneret, J.Y.; Boy, F.; Moreau, T. Including antenna mispointing in a semi-analytical model for delay/Doppler altimetry. *IEEE Trans. Geosci. Remote Sens.* **2015**, *53*, 598–608. [CrossRef]
15. Lowe, S.T.; Zuffada, C.; Chao, Y.; Kroger, P.; Young, L.E.; LaBrecque, J.L. 5-cm-precision aircraft ocean altimetry using GPS reflections. *Geophys. Res. Lett.* **2002**, *29*, 131–134. [CrossRef]
16. Allan, T. The story of GANDER. *Sensors* **2006**, *6*, 249–259. [CrossRef]
17. Enjolras, V.; Vincent, P.; Souyris, J.C.; Rodriguez, E.; Phalippou, L.; Cazenave, A. Performances study of interferometric radar altimeters: From the instrument to the global mission definition. *Sensors* **2006**, *6*, 164–192. [CrossRef]

18. Anzenhofer, M.; Shum, C.K.; Rentsh, M. *Coastal Altimetry and Applications*; Report No. 464; Geodetic Science and Surveying, The Ohio State University: Columbus, OH, USA, 1999.

19. Bouffard, J.; Vignudelli, S.; Cipollini, P.; Menard, Y. Exploiting the potential of an improved multimission altimetric data set over the coastal ocean. *Geophys. Res. Lett.* **2008**, *35*, L10601. [CrossRef]

20. Cipollini, P.; Calafat, F.M.; Jevrejeva, S.; Melet, A.; Prandi, P. Monitoring sea level in the coastal zone with satellite altimetry and tide gauges. *Surv. Geophys.* **2016**, *38*, 33–57. [CrossRef]

21. Mercier, F.; Rosmorduc, V.; Carrere, L.; Thibaut, P. Coastal and Hydrology Altimetry Product (PISTACH) Handbook. CLS-DOS-NT-10-246, SALP-MU-P-OP-16031-CN 01/00. 2010. Available online: https://www.aviso.altimetry.fr/fileadmin/documents/data/tools/hdbk_Pistach.pdf (accessed on 23 October 2017).

22. Cipollini, P.; Susana Barbosa, S.; Caparrini, M.; Challenor, P.; Coelho, H.; Dinardo, S.; Fernandes, S.; Gleason, S.; Gómez-Enri, J.; Gommenginger, C.; et al. COASTALT Project's contribution to the development and dissemination of coastal altimetry. In Proceedings of the 20 Years of Progress in Radar Altimetry Symposium, Venice, Italy, 24–29 September 2012.

23. Birol, F.; Fuller, N.; Lyard, F.; Cancet, M.; Nino, F.; Delebecque, C.; Léger, F. Coastal applications from nadir altimetry: Example of the X-TRACK regional products. *Adv. Space Res.* **2017**, *59*, 936–953. [CrossRef]

24. Vignudelli, S.; Kostianoy, A.G.; Cipollini, P.; Benveniste, J. *Coastal Altimetry*; Springer: Berlin, Germany, 2011.

25. Tran, N.; Vandemark, D.; Chapron, B.; Labroue, S.; Feng, H.; Beckley, B.; Vincent, P. New models for satellite altimeter sea state bias correction developed using global wave model data. *J. Geophys. Res.* **2006**, *111*, C09009. [CrossRef]

26. Feng, H.; Yao, S.; Li, L.; Tran, N.; Vandemark, D.; Labroue, S. Spline-based nonparametric estimation of the altimeter sea-state bias correction. *IEEE Geosci. Remote Sens. Lett.* **2010**, *7*, 577–581. [CrossRef]

27. Tran, N.; Vandemark, D.; Labroue, S.; Feng, H.; Chapron, B.; Tolman, H.L.; Lambin, J.; Picot, N. Sea state bias in altimeter sea level estimates determined by combining wave model and satellite data. *J. Geophys. Res.* **2010**, *115*, C03020. [CrossRef]

28. Pires, N.; Fernandes, M.; Gommenginger, C.; Scharroo, R. A Conceptually Simple Modeling Approach for Jason-1 Sea State Bias Correction Based on 3 Parameters Exclusively Derived from Altimetric Information. *Remote Sens.* **2016**, *8*, 576. [CrossRef]

29. Reale, F.; Dentale, F.; Pugliese Carratelli, E. Numerical simulation of whitecaps and foam effects on satellite altimeter response. *Remote Sens.* **2014**, *6*, 3681–3692. [CrossRef]

30. Brown, G. The average impulse response of a rough surface and its applications. *IEEE Trans. Antennas Propag.* **1977**, *25*, 67–74. [CrossRef]

31. Davis, C.H. A robust threshold retracking algorithm for measuring ice-sheet surface elevation change from satellite radar altimeters. *IEEE Trans. Geosci. Remote Sens.* **1997**, *35*, 974–979. [CrossRef]

32. Hwang, C.; Guo, J.Y.; Deng, X.L.; Hsu, H.Y.; Liu, Y.T. Coastal gravity anomalies from retracked Geosat/GM altimetry: Improvement, limitation and the role of airborne gravity data. *J. Geodesy* **2006**, *80*, 204–216. [CrossRef]

33. Lee, H.; Shum, C.K.; Emery, W.; Calmant, S.; Deng, X.; Kuo, C.Y.; Roesler, C.; Yi, Y.C. Validation of Jason-2 altimeter data by waveform retracking over California coastal ocean. *Mar. Geodesy* **2010**, *33*, 304–316. [CrossRef]

34. Guo, J.; Gao, Y.; Hwang, C.; Sun, J. A mutli-subwaveform parametric retracker of the radar satellite altimetric waveform and recovery of gravity anomalies over coastal oceans. *Sci. China Earth Sci.* **2010**, *53*, 610–616. [CrossRef]

35. Yang, Y.; Hwang, C.; Hsu, H.J.; Dongchen, E.; Wang, H. A subwaveform threshold retracker for ERS-1 altimetry: A case study in the Antarctic Ocean. *Comp. Geosci.* **2012**, *41*, 88–98. [CrossRef]

36. Yang, L.; Lin, M.; Liu, Q.; Pan, D. A coastal altimetry retracking strategy based on waveform classification and subwaveform extraction. *Int. J. Remote Sens.* **2012**, *33*, 7806–7819. [CrossRef]

37. Idris, N.H.; Deng, X. The retracking technique on multi-peak and quasi-specular waveforms for Jason-1 and Jason-2 missions near the coast. *Mar. Geodesy* **2012**, *35*, 217–237. [CrossRef]

38. Passaro, M.; Cipollini, P.; Vignudelli, S.; Quartly, G.D.; Snaith, H.M. ALES: A multi-mission adaptive subwaveform retracker for coastal and open sea altimetry. *Remote Sens. Environ.* **2014**, *145*, 173–189. [CrossRef]

39. Gómez-Enri, J.; Vignudelli, S.; Quartly, G.D.; Gommenginger, C.P.; Cipollini, P.; Challenor, P.G.; Benveniste, J. Modeling ENVISAT RA-2 waveforms in the coastal zone: Case study of calm water contamination. *IEEE Geosci. Remote Sens. Lett.* **2010**, *7*, 474–478. [CrossRef]
40. Deng, X.; Featherstone, W. A coastal retracking system for satellite radar altimeter waveforms: Application to ERS-2 around Australia. *J. Geophys. Res.* **2006**, *111*, C06012. [CrossRef]
41. Wang, H.H.; Luo, Z.C.; Yang, Y.D.; Zhong, B.; Zhou, H. An adaptive retracking method for coastal altimeter data based on waveform classification. *Acta Geodesy Cartogr. Sin.* **2012**, *41*, 729–734.
42. Khakia, M.; Forootan, E.; Sharifi, M.A. Satellite radar altimetry waveform retracking over the Caspian Sea. *Int. J. Remote Sens.* **2014**, *35*, 6329–6356. [CrossRef]
43. Tseng, K.H.; Shum, C.K.; Yi, Y.C.; Emery, W.J.; Kuo, C.Y.; Wang, H.H. The improved retrieval of coastal sea surface heights by retracking modified radar altimetry waveforms. *IEEE Trans. Geosci. Remote Sens.* **2014**, *52*, 991–1001. [CrossRef]
44. Desjonquères, J.D.; Carayon, G.; Steunou, N.; Lambin, J. Poseidon-3 radar altimeter: New modes and in-flight performances. *Mar. Geodesy* **2010**, *33*, 53–79. [CrossRef]
45. AVISO+ CNES. Available online: https://www.aviso.altimetry.fr/en/missions/current-missions/jason-2/index.html (accessed on 20 May 2017).
46. Lambin, L.; Morrow, R.; Fu, L.L.; Willis, J.; Bonekamp, H.; Lillibridge, J.; Perbos, J.; Zaouche, G.; Vaze, P.; Bannoura, W.; et al. The OSTM/Jason-2 mission. *Mar. Geodesy* **2010**, *33*, 4–25. [CrossRef]
47. Chelton, D.B.; Ries, J.C.; Haines, B.J.; Fu, L.-L.; Callahan, P.S. Satellite Altimetry. In *Satellite Altimetry and Earth Sciences: A Handbook of Techniques and Applications*; Fu, L.-L., Cazenave, A., Eds.; Academic Press: San Diego, CA, USA, 2001; pp. 1–131.
48. Dumont, J.-P.; Rosmordue, V.; Picot, N.; Desai, S.; Bonekamp, H.; Figa, J.; Lillibridge, J.; Sharroo, R. OSTM/Jason-2 Products Handbook. CNES: SALP-MU-M-OP-15818-CN; EUMETSAT: EUM/OPS-JAS/MAN/08/0041; JPL: OSTM-29–1237; NOAA/NESDIS: Polar Series/OSTM J400; 2011; Available online: https://www.aviso.altimetry.fr/fileadmin/documents/data/tools/hdbk_j2.pdf (accessed on 23 October 2017).
49. Wingham, D.J.; Rapley, C.G.; Griffiths, H. New techniques in satellite altimeter tracking systems. In Proceedings of the 1986 International Geoscience and Remote Sensing Symposium on Remote Sensing: Today's Solutions for Tomorrow's Information Needs, Noordwijk, The Netherlands, 8–11 September 1986; Volume 3, pp. 1339–1344.
50. Gommenginger, C.; Thibaut, P.; Fenoglio-Marc, L.; Quartly, G.; Deng, X.; Gómez-Enri, J.; Challenor, P.; Gao, Y. Retracking altimeter waveforms near the coasts. In *Coastal Altimetry*; Vignudelli, S., Kostianoy, A.G., Cipollini, P., Benveniste, J., Eds.; Springer: Berlin, Heidelberg, 2011; pp. 61–102.
51. Becker, J.J.; Sandwell, D.T.; Smith, W.H.F.; Braud, J.; Binder, B.; Depner, J.; Fabre, D.; Factor, J.; Ingalls, S.; Kim, S.H.; et al. Global bathymetry and elevation data at 30 arc seconds resolution: SRTM30_PLUS. *Mar. Geodesy* **2009**, *32*, 355–371. [CrossRef]
52. Caldwell, P.C.; Merrfield, M.A.; Thompson, P.R. Sea Level Measured by Tide Gauges from Global Oceans—The Joint Archive for Sea Level Holdings (NCEI Accession 0019568), Version 5.5. NOAA National Centers for Environmental Information, Dataset. 2015. Available online: https://uhslc.soest.hawaii.edu/datainfo/ (accessed on 23 October 2017).
53. Pavlis, N.K.; Holmes, S.A.; Kenyon, S.C.; Factor, J.K. The development and evaluation of the Earth Gravitational Model 2008 (EGM2008). *J. Geophys. Res.* **2012**, *117*, B04406. [CrossRef]
54. Wang, X.; Ichikawa, K. Coastal waveform retracking for Jason-2 altimeter data based on along-track echograms around the Tsushima Islands in Japan. *Remote Sens.* **2017**, *9*, 762. [CrossRef]
55. Thibaut, P.; Poisson, J.C.; Bronner, E.; Picot, N. Relative performance of the MLE3 and MLE4 retracking algorithms on Jason-2 altimeter waveforms. *Mar. Geodesy* **2010**, *33*, 317–335. [CrossRef]

remote sensing

MDPI

Article

SWOT Spatial Scales in the Western Mediterranean Sea Derived from Pseudo-Observations and an Ad Hoc Filtering

Laura Gómez-Navarro [1,2,*]**, Ronan Fablet** [3]**, Evan Mason** [1]**, Ananda Pascual** [1]**, Baptiste Mourre** [4]**, Emmanuel Cosme** [2] **and Julien Le Sommer** [2]

[1] Institut Mediterrani d'Estudis Avançats (IMEDEA) (CSIC-UIB), 07190 Esporles, Illes Balears, Spain; evanmason@gmail.com (E.M.); ananda.pascual@imedea.uib-csic.es (A.P.)

[2] Univ. Grenoble Alpes, CNRS, IRD, Grenoble INP, IGE, 38000 Grenoble, France; Emmanuel.Cosme@univ-grenoble-alpes.fr (E.C.); julien.lesommer@univ-grenoble-alpes.fr (J.L.S.)

[3] Institut Mines-Télécom, Telecom-Bretagne, UMR 6285 labSTICC, 29238 Brest, France; ronan.fablet@imt-atlantique.fr

[4] Balearic Islands Coastal Observing and Forecasting System (SOCIB), 07121 Palma de Mallorca, Illes Balears, Spain; bmourre@socib.es

* Correspondence: lauragomnav@gmail.com

Received: 28 March 2018; Accepted: 28 March 2018; Published: 12 April 2018

Abstract: The aim of this study is to assess the capacity of the Surface Water Ocean Topography (SWOT) satellite to resolve fine scale oceanic surface features in the western Mediterranean. Using as input the Sea Surface Height (SSH) fields from a high-resolution Ocean General Circulation Model (OGCM), the SWOT Simulator for Ocean Science generates SWOT-like outputs along a swath and the nadir following the orbit ground tracks. Given the characteristic temporal and spatial scales of fine scale features in the region, we examine temporal and spatial resolution of the SWOT outputs by comparing them with the original model data which are interpolated onto the SWOT grid. To further assess the satellite's performance, we derive the absolute geostrophic velocity and relative vorticity. We find that instrument noise and geophysical error mask the whole signal of the pseudo-SWOT derived dynamical variables. We therefore address the impact of removal of satellite noise from the pseudo-SWOT data using a Laplacian diffusion filter, and then focus on the spatial scales that are resolved within a swath after this filtering. To investigate sensitivity to different filtering parameters, we calculate spatial spectra and root mean square errors. Our numerical experiments show that noise patterns dominate the spectral content of the pseudo-SWOT fields at wavelengths below 60 km. Application of the Laplacian diffusion filter allows recovery of the spectral signature within a swath down to the 40–60 km wavelength range. Consequently, with the help of this filter, we are able to improve the observation of fine scale oceanic features in pseudo-SWOT data, and in the estimation of associated derived variables such as velocity and vorticity.

Keywords: satellite altimetry; SWOT; western Mediterranean Sea; fine scale; SWOT simulator; ROMS model; filtering

1. Introduction

The Surface Water and Ocean Topography (SWOT) satellite mission is a joint mission by the National Aeronautics and Space Administration (NASA) and the *Centre National d'Études Spatiales* (CNES), with contributions from the UK and Canadian Space Agencies [1]. Presently, the satellite's launch is planned for 2021 [2]. It will provide water elevation maps for oceanographic and hydrological purposes [3,4]. The novelty of this satellite is that it carries a wide-swath altimeter with unprecedented horizontal resolution and global coverage. On the other hand, the associated irregular temporal sampling will constitute a challenge for the exploitation of the data. SWOT will have a 21-day repeat cycle and the revisit time will vary from approximately 10 days at the equator to two days at the poles [5,6]. This implies temporal variability in spatial coverage as the number of observations per repeat cycle will increase with latitude. Moreover, there will also be a temporal variability within a cycle. During each cycle, there are periods of time with a higher temporal sampling. This is due to a longer revisit time so that SWOT also fulfills its hydrological objectives by providing coverage of the bulk of the global land surface [7]. In satellite measurements, there is always a compromise between spatial and temporal resolution. As SWOT aims for global coverage, i.e., high spatial resolution, we lose in temporal resolution (SWOT's repeat cycle will be longer than, for example, the 10-day repeat cycle of the Jason altimeter satellites [8]).

One of the primary oceanographic objectives of the SWOT mission is to characterize the ocean meso- and submesoscale circulation [9] determined from ocean surface topography at spatial resolutions of 15 km (spatial resolution is defined to be perturbation wavelength in the oceanographic context). The resolution capacity of current along-track one-dimensional altimeter data, depending on the altimeter, has been found to be between 40 and 50 km at western boundary currents and between 70 and 110 km at the eastern basins [10]. Two-dimensional gridded products based on the altimetric constellation allow for mapping wavelengths down to 200 km [11]. The SWOT mission is expected to allow to capture wavelengths down to 15 km on its two-dimensional swaths [12], therefore increasing substantially the resolution capacity of present-day altimeter data. The possibility of characterizing the submesoscale is a major breakthrough. While the mesoscale has historically received a lot of attention [13], the submesoscale has previously been out of reach. Theoretical calculations and advanced modeling suggest that submesoscale processes are key to understanding ocean fluxes [14–16]. A pertinent example is the occurrence of mid-ocean plankton blooms [17].

In the Mediterranean Sea, intense mesoscale and submesoscale variability interact across sub-basin and basin scales [18–20]. This variability has an indirect impact on the Atlantic Ocean circulation due to exchange through the Strait of Gibraltar and, subsequently, influence on the great ocean conveyor belt [21–23]. Three scales of motion are therefore overlaid, making an amalgam of intricate processes that require high resolution and can help assess the potential impact that SWOT will have on the study of processes occurring at different scales.

Understanding small scale variability in the Mediterranean Sea is important as it is a region with intrinsically smaller spatial scales than those found in other parts of the world ocean at similar latitudes. Ref. [24] showed that the grid resolution necessary to resolve the first baroclinic deformation radius in the Mediterranean is around 1/16°, whilst in the Atlantic Ocean at the same latitude it is only 1/6°. This implies that smaller structures need to be resolved in the Mediterranean Sea compared to the mid-latitudes of the Atlantic Ocean. This is further demonstrated by [25] who show that lower values of the first baroclinic Rossby radius of deformation are present in the western Mediterranean Sea. These values are approximately between 2 and 16 km, in comparison with a 20–30 km range found at mid-latitudes of the Atlantic Ocean [26]. The Mediterranean values are actually closer to the values found in the Arctic Ocean [27].

The western Mediterranean Sea is one of the areas of the global ocean that will be sampled during the SWOT fast-sampling phase [28]. This phase covers the first 60–90 days after launch, during which the satellite will provide daily high resolution Sea Surface Height (SSH) measurements over a limited repeated orbit for purposes of calibration/validation of the SWOT sensor/instrument.

The goal of this study is to assess the capacity of SWOT to resolve the fine scales in the western Mediterranean. Our first objective is to generate pseudo-SWOT data from numerical model outputs in this region in order to understand its temporal and spatial sampling pattern in this area. We then apply a noise-reduction processing technique to pseudo-SWOT data to find out the spatial scales that SWOT may ultimately be able to resolve. Given SWOT's irregular time sampling and consequent variable spatial coverage, in this paper, we focus on the spatial scales resolved within a swath.

2. Data and Methods

2.1. The SWOT Simulator

With a view to characterizing the potential of SWOT-derived SSH data, we consider a simulation-based framework using the SWOT Simulator for Ocean Science (version 1). This simulator accounts for both SWOT space-time sampling patterns and noise processes. Using as input the SSH fields from an Oceanic General Circulation Model (OGCM), the SWOT simulator generates SWOT-like outputs along a ground swath and the nadir following the orbit ground tracks [29]. Hereinafter, we refer to these outputs as SSH outputs. Note that these simulated fields correspond to Absolute Dynamic Topography (ADT) values in altimetric terminology.

A flowchart of the simulator workflow is provided in Figure 1. Two features should be pointed out. Grid files, generated in the first step of the flowchart, account for the planned orbit of the satellite and the specified domain. Instrument noise and geophysical errors are added during the last step of the flowchart, following recent technical characteristics established by the SWOT project team [29]. Instrument noise is composed of Ka-band Radar Interferometer (KaRIN) noise, roll, phase, baseline dilation and timing errors (see [30]). In this version of the simulator, the only geophysical error is associated with the wet troposphere. Therefore, it is important to keep in mind that additional noise patterns, such as sea state bias [29] or the effects of internal waves [2] are not accounted for in the generated pseudo-SWOT data.

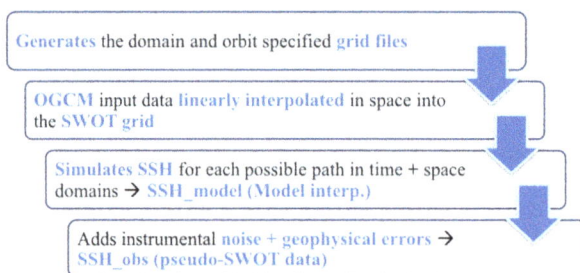

Generates the domain and orbit specified grid files

OGCM input data linearly interpolated in space into the SWOT grid

Simulates SSH for each possible path in time + space domains → SSH_model (Model interp.)

Adds instrumental noise + geophysical errors → SSH_obs (pseudo-SWOT data)

Figure 1. Flowchart of the SWOT simulator procedure.

For more details on the instrument noise and geophysical errors added by the SWOT simulator, see Appendix B.

2.2. Input Data: The Western Mediterranean OPerational (WMOP) Model

A high resolution OGCM of the western Mediterranean region provides input data for the SWOT simulator. We used the WMOP model [31] developed at SOCIB (Balearic Islands Coastal Observing and Forecasting System). More specifically, we consider a 7-year free run simulation of the model spanning the period 2009 – 2015, with spatial coverage from the Strait of Gibraltar to the Sardinia Channel (Figure 2). WMOP is a regional configuration of the Regional Oceanic Model System (ROMS) model [32] with a spatial resolution of approximately 2 km. WMOP is forced with high resolution atmospheric forcing (HIRLAM model from the Spanish Meteorological Agency AEMET), with temporal resolution of 3 h and spatial resolution of 5 km. These features make WMOP a suitable choice to evaluate the potential of SWOT-derived SSH data to resolve mesoscale processes in the western Mediterranean Sea. The presence of fine scale features of a few kilometers is illustrated in Figure 2. In Figure 2, we show snapshots of model relative vorticity (normalized by f) for days corresponding to pass 15 (Figure 2, left) and pass 168 (Figure 2, right) of cycle 2 of the SWOT orbit (see Figure 3).

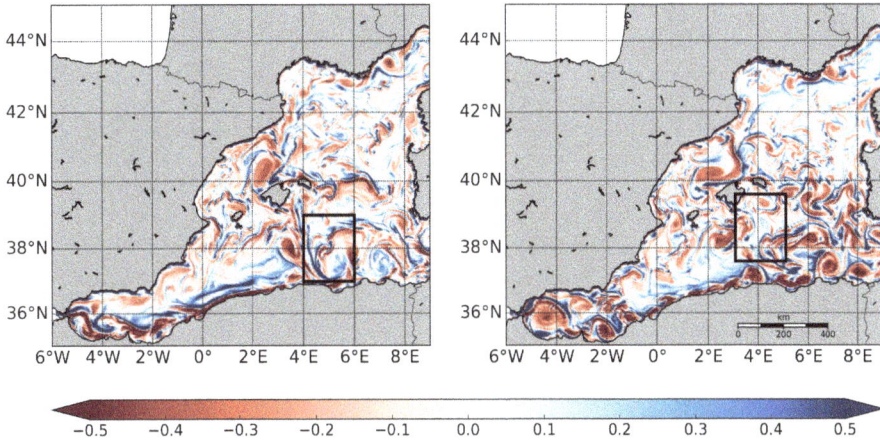

Figure 2. WMOP relative vorticity normalized by f on 23 January 2009 (**left**) and 3 February 2009 (**right**). Black boxes indicate the two regions studied in Section 3: box 1, pass 15 (**left**) and box 2, pass 168 (**right**).

2.3. Analysis and Processing of SWOT-Derived SSH Data

2.3.1. Geostrophic Velocity and Vorticity

Zonal (u_g) and meridional (v_g) (with respect to the SWOT grid) surface geostrophic velocity components are calculated as:

$$u_g = -\frac{g}{f}\frac{\partial \eta}{\partial y}, \tag{1}$$

$$v_g = \frac{g}{f}\frac{\partial \eta}{\partial x}, \tag{2}$$

where g is the gravitational acceleration, f the Coriolis parameter and η the sea level elevation. The absolute geostrophic velocity (V_g) is obtained with:

$$V_g = \sqrt{u_g + v_g{}^2}. \tag{3}$$

Geostrophic relative vorticity, ζ, is calculated from the zonal and meridional velocities:

$$\zeta = \frac{\partial v_g}{\partial x} - \frac{\partial u_g}{\partial y}. \tag{4}$$

2.3.2. Noise Filtering

As illustrated in Section 3, noise greatly affects the computation of the velocities derived from the pseudo-SWOT data. We therefore investigate filtering procedures for noise removal. The geometry of the SWOT data prevents us from using classical Fourier and convolution-based low-pass filters [33]. Fourier-based filters impose circularity constraints, which cannot be fulfilled; the masks associated with convolution-based filters should be significantly smaller than the width of the SWOT swath, which greatly limits low-pass filtering capabilities. We then considered a Partial Derivative Equation (PDE)-based formulation, such that the low-pass filtering results from an iterated Laplacian diffusion:

$$\partial_t a\,(t, y, x) - \triangle a\,(t, y, x) = 0 \iff \frac{\partial a}{\partial t} = \frac{\partial^2 a}{\partial y^2} + \frac{\partial^2 a}{\partial x^2}. \tag{5}$$

As the Green's function for the heat equation is a Gaussian kernel, the implementation of this PDE-based diffusion is equivalent to a Gaussian convolution and results in an isotropic filtering, that is to say that the filtering acts equally in all directions [34]. Using a four-neighbourhood discretization of the Laplacian operator, we can deal with missing data (e.g., nadir) or land (e.g., island) pixels. The Laplacian operator comes to compute a local mean over the four neighbours of a given pixel. Withdrawing land pixels and missing data from the computation of this local mean, we can iterate the Laplacian diffusion to reach the expected filtering level for all pixels. Each iteration of the Laplacian diffusion can be regarded as a low-pass filtering with a high cut-off frequency. The selection of the number of iterations of the Laplacian diffusion then allows us to reach lower cut-off frequencies. By contrast, the direct application of two-dimensional low-pass filters for cut-off frequencies in the range [30 km, 60 km] would result in filter supports in the range [60 km, 120 km], meaning that no filtering output could be computed for any pixel closer than 30 km (rest. 60 km) from the swath boundaries or a missing data or land pixel. Overall, the filtering level is set by the number of iterations of the Laplacian diffusion and the parameter lambda. This is shown in the following equation, which shows the implementation that we use:

$$a^{k+1} = a^k - \lambda \triangle a^k. \tag{6}$$

With this being an iterative method, in contrast to a traditional Gaussian filter, we can apply cut-off wavelengths greater than the width of a half-swath. In Appendix A, we apply the filter to white noise to show how different combinations of the filter's parameters (lambda and number of iterations) are associated with different cut-off wavelengths ($\lambda_c s$).

2.3.3. Filter Evaluation

To evaluate the performance of the filter and its different parameterizations, the following variables are calculated:

- The radial power spectral density: This variable was calculated to obtain the SWOT spatial spectra. The radially averaged power spectral density (power spectrum) of an image (in our case, the SWOT swath data) is computed.
- The Root Mean Squared Error (RMSE): The RMSE was calculated for the SSH, velocity and vorticity variables as follows:

$$RMSE = \sqrt{\frac{\sum (data - estimate)^2}{N}}, \tag{7}$$

where N is the number of points. Data is taken to be SSH_{model} (or its derived variables, i.e., velocity and vorticity) without filtering. An estimate is taken to be the simulated noisy SSH_{obs} fields (or its derived variables) without filtering, and filtered with different $\lambda_c s$. RMSE values are therefore calculated for different estimates.

3. Results

3.1. Spatial and Temporal Sampling

Pseudo-SWOT data were generated for the full WMOP time period (1 January 2009 to 11 September 2015). This was done for the Science orbit and corresponds to a total of 123 cycles. In a complete cycle, 292 passes are available over the globe, 12 of them crossing our study region (Figure 3). Data were ingested and processed by the SWOT simulator at an across and along track resolution of 2 km. As mentioned in Section 1, one of the specificities of future SWOT data will be their irregular temporal sampling. To better illustrate this, the passes of cycle 2 are plotted in Figure 3. During each cycle, there are periods of time with a higher temporal sampling. This is due to a longer revisit time so that SWOT also fulfills its hydrological objectives as described in Section 1. For instance, the temporal sampling during cycle 2 is as follows: from day 21.3 to 23.9 during which five passes are made; and from day 31.3 to 33.8 during which six passes are made. Then, from day 23.9 to 31.3 and from day 33.8 to 41.2, there are no measurements. Consequently, during each day within a cycle in this study region, there can be two, one or no passes at all. Even with this irregular sampling and without any processing of the data, the final SSH$_{obs}$ map (subplot of day 41.2) allows us to observe some features such as, for example, the signal of the Algerian Current following the north African coastline and several cyclonic and anticyclonic mesoscale eddies.

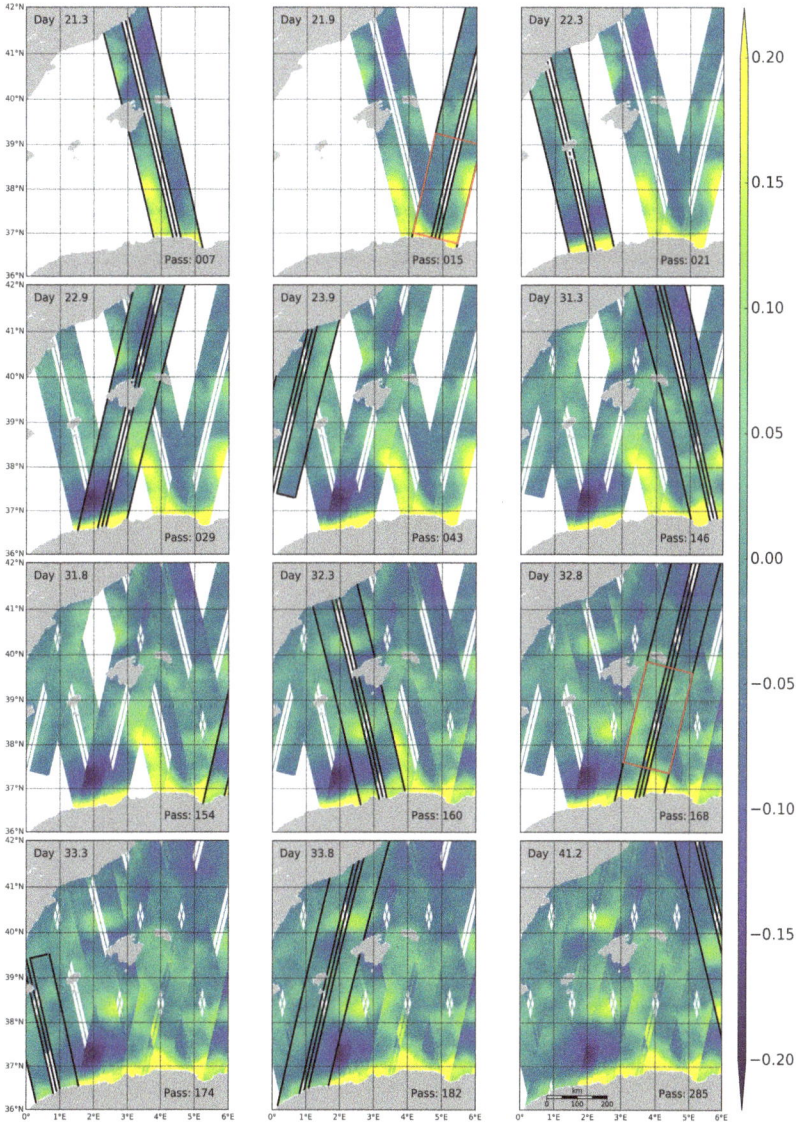

Figure 3. SSH$_{obs}$ (m) obtained for cycle 2. Time increasing from left to right, top to bottom. Days from the beginning of the simulation are shown at the top left corner and the corresponding pass number at the bottom right corner. Outline of the active pass is shown in black. The red boxes show box 1 (pass 15) and box 2 (pass 168).

3.2. Pre-Filtering Analysis of Simulator Outputs

In this study, we focus on the analysis of spatial scales of individual passes. Due to the irregular time sampling of the SWOT data, future studies will be devoted to temporal interpolation of passes. Moreover, prior swath filtering is necessary to determine the quality of the dynamical variables that can be derived from SWOT data, and how it can be improved before combining different swaths for temporal interpolation. As an illustration, we focus on the treatment of two $2° \times 2°$ boxes. Box 1 is within pass 15 and was chosen close to the north African coast as it is a region where anticyclonic eddies are shed from the Algerian Current [25,35]. For example, in the snapshot shown in Figure 4, part of an anticyclonic eddy is present on the eastern part of the domain. Box 2 is within pass 168, and this subdomain south of the Balearic island of Menorca was chosen because it contains smaller structures than in box 1 (see Figure 2). In Figure 5, filament-like structures and smaller eddies can be observed, especially at the northern part of the domain.

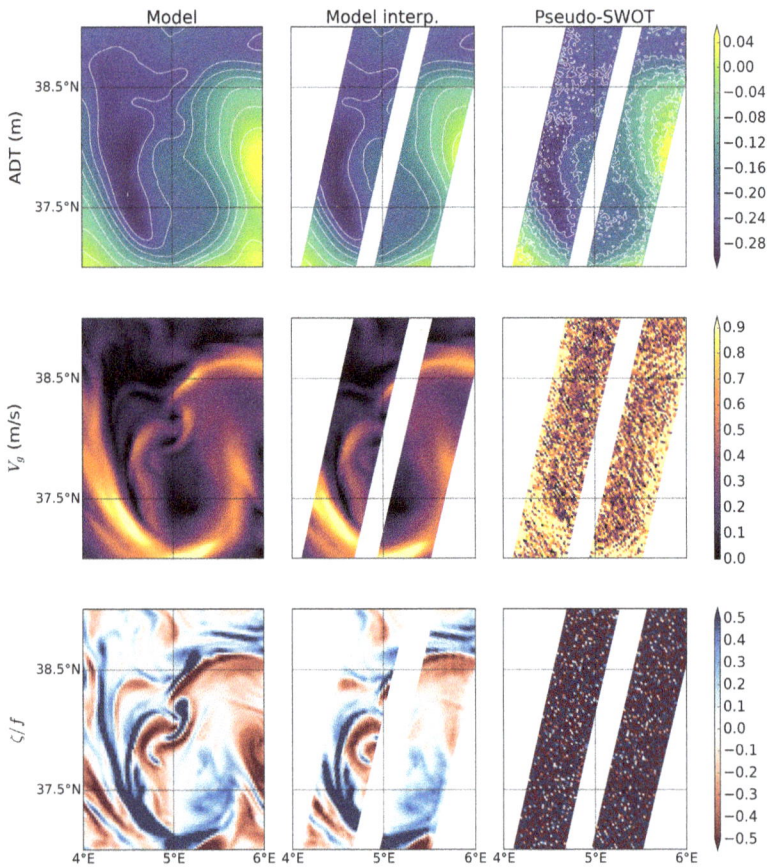

Figure 4. From top to bottom: SSH (m), geostrophic velocity (m/s) and relative vorticity (ζ) normalized by f, on 23 January 2009 corresponding to pass 15 of cycle 2 (box 1). The first, middle and last columns show the data obtained directly from the model (WMOP), from the model interpolated onto the SWOT grid (SSH$_{model}$), and with added noise (SSH$_{obs}$), respectively.

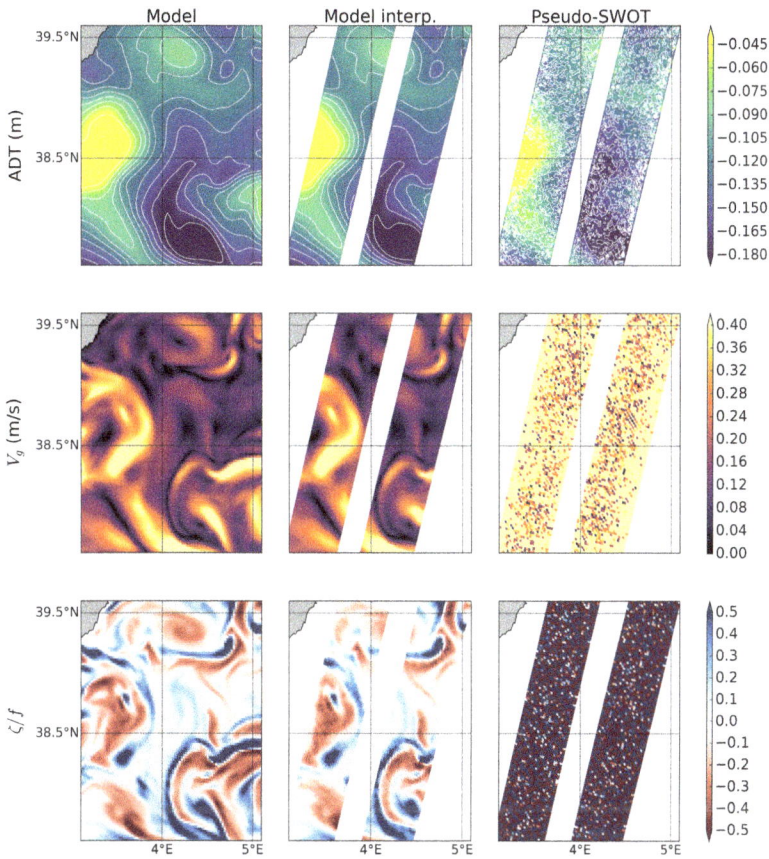

Figure 5. From top to bottom: SSH (m), geostrophic velocity (m/s) and relative vorticity (ζ) normalized by f, on 3 February 2009 corresponding to pass 168 of cycle 2 (box 2). The first, middle and last columns show the data obtained directly from the model (WMOP), from the model interpolated onto the SWOT grid (SSH$_{model}$), and with added noise (SSH$_{obs}$), respectively.

The effect of the filter is assessed for SSH and its derived dynamical variables: absolute geostrophic velocity and relative vorticity. These were calculated as explained in Section 2.3.1.

As observed in the first and middle columns (model and model interpolated onto SWOT grid data, respectively) of Figures 4 and 5, SSH and its derived variables reveal fine scale features, but the noise level masks the signal of these features when derived variables are obtained from pseudo-SWOT SSH. We can also see how the effect of the noise is lower in regions with high SSH gradients. If we compare the velocity derived from pseudo-SWOT data of box 1 and 2, for box 1, the region with high values can still be appreciated as they reach 0.9 m/s, but not for box 2 as they only reach 0.4 m/s.

To have information on the spatial scales resolved and the effect of the noise, spatial Fourier power spectra for each filter were calculated as described in Section 2.3.2. The spectra were calculated for each individual cycle, and then averaged over the 122 cycles in which both passes 15 and 168 are available (cycle 123 stops at pass 132). Figure 6 compares the spectra of model data interpolated onto the SWOT grid and the pseudo-SWOT data. The SWOT noise starts to dominate at wavelengths lower than 60 km. In the top panel of Figure 6, the red and blue curves separate at around 60 km for both boxes. If we look at the zoom inset, we see that for pass 15 the lines separate at slightly higher wavelengths than for pass 168.

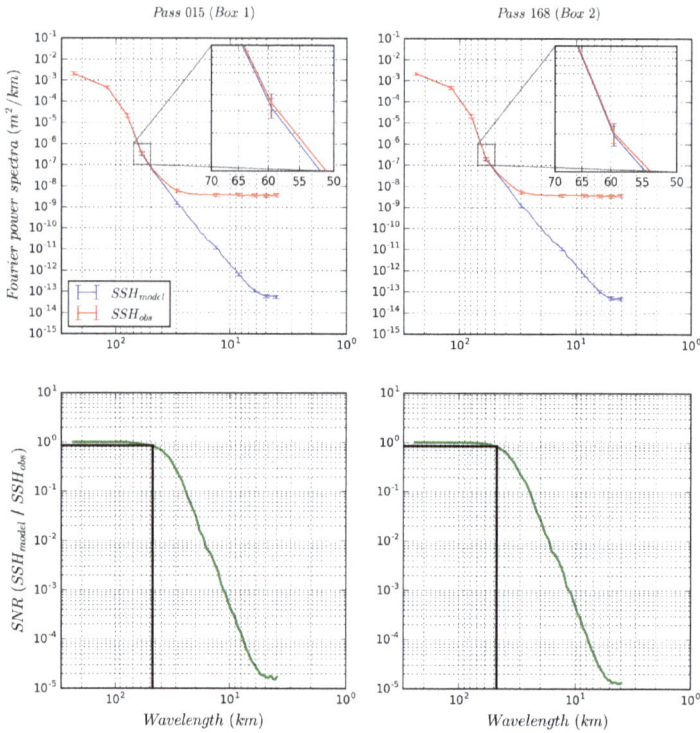

Figure 6. (**Top**) Spectra of the data before filtering, from cycle 1 to 122, corresponding to box 1 (pass 15), (**left**), and to box 2 (pass 168), (**right**). Error bars denote 95% confidence intervals. (**Bottom**) Corresponding Signal to Noise Ratio (SNR), with a horizontal black line indicating where the noise is more than 15% of the signal, and the vertical line the corresponding wavelength.

Nevertheless, there does not seem to be a significant difference between the mean spectra of both passes. Note also how the power spectral energy level of SSH_{obs} at wavelengths lower than 20 km stabilizes around 3.7×10^{-9} for both passes, whilst the energy level of SSH_{model} reduces until it reaches the grid scale. If we look at the signal to noise ratio (SNR), we find that below 50 km wavelength the energy of the noise is significant with respect to that of the signal (SNR values below 15 dB at wavelengths smaller than 47.6 km, i.e., the energy of the noise accounts for more than 15% of the energy of the signal for these scales). Such low SNR values make particularly challenging the denoising issue for scales below 50 km [36]. Consequently, we expect that the best filter parametrization will be one corresponding to λ_c between 47.6 and 60 km.

3.3. SWOT Data Filtering

The Laplacian diffusion filter was applied to remove the noise and, thus, reduce the difference between the spectrum obtained from SWOT estimates with and without noise (Figure 6). Given the results obtained from the non-filtered data spectra, λ_c is first chosen to be 60 km. We then choose smaller λ_cs (50, 40, 30 and 15 km) to see how much lower we can go with this filter. We go down to 15 km, which is the expected wavelength at which SWOT will measure SSH. For comparison, we also choose $\lambda_c = 200$ km, which is the wavelength resolved by present-day altimeter constellation fields [11]. We lastly choose $\lambda_c = 100$ km as an intermediate value between 60 and 200 km.

In Figures 7 and 8, we show the effect of the filter on SSH at different values of λ_cs. For Figure 7, the effect of the filter is mainly seen in the pseudo-SWOT data, especially in the northern part where smaller structures are present. In Figure 8, as there are more, smaller structures, we can see more differences between the filtered outputs with respect to the model interpolated and pseudo-SWOT data. These differences are not only in the shape of the structures that are present, but also in their intensity. On the top row of Figures 7 and 8, the original model is also included to show that differences do also arise from the interpolation onto the SWOT grid. In Figure 7, especially for the 200 km λ_c, we can observe that the original structure present is significantly altered. This emphasizes the importance for development of interpolation techniques to fill the gap between the two swaths with the help of the nadir altimeter data. In Figures 7 and 8, the SSH images for the different filters look very similar, but the differences are amplified when the first derivatives (Figures 9 and 10) and second derivatives (Figures 11 and 12) are calculated.

After applying the Laplacian diffusion filter, we can now retrieve the structures present in the pseudo-SWOT SSH in the absolute geostrophic velocity plots (Figures 9 and 10). With a 15 km λ_c filter, the effect of the noise can be still clearly observed, especially for box 2 where smaller structures are present. As a result, although the main structures are recovered after filtering, their shapes are not accurately retrieved. Even if spurious structures remain, with a 30 km λ_c there is a large improvement with respect to the 15 km λ_c. This improvement seems greater for box 1 than box 2, as the noise seems to have a greater effect within box 2 than box 1. For a 40 km λ_c, in box 1, we can no longer qualitatively see any remaining noise, but we can see some in box 2. For λ_cs greater than or equal to 50 km, the effect of the noise is no longer observed in either box 1 (Figure 9) or box 2 (Figure 10). On the other hand, we observe a large decrease of the magnitude of the velocities from the 15 to the 200 km λ_c. With this filtering method, the intensity of the structures present, and thus the signal, decreases with the increase of λ_c.

In the relative vorticity plots, the loss of signal with the increase of λ_c is even more evident. With no filtering, the relative vorticity of box 1 ranges from $-1.82f$ to $1.66f$ for SSH$_{model}$ and from $-15.22f$ to $18.16f$ for SSH$_{obs}$ (Figure 4). With a 200 km λ_c, this reduces to $-0.23f$ to $0.14f$ for both SSH$_{model}$ and SSH$_{obs}$ (Figure 11). For box 2, with no filtering, the relative vorticity ranges from $-0.71f$ to $1.50f$ for SSH$_{model}$ and from $-17.39f$ to $17.71f$ for SSH$_{obs}$ (Figure 5). For 200 km, it reduces to $-0.07f$ to $0.10f$ for SSH$_{model}$ and from $-0.09f$ to $0.06f$ for SSH$_{obs}$ (Figure 12). There is approximately two orders of magnitude difference between the vorticity calculated from the original data, and that filtered at $\lambda_c = 200$ km. For box 1, the velocity appears to contain no further noise with $\lambda_c = 40$ km, but this filtering is not sufficient to properly reconstruct the relative vorticity. With a 50 km λ_c, some noise is still present, and with 60 km λ_c, there appears to be no remaining noise. For box 2, the velocity appears to have no further noise with a 50 km λ_c, and, similarly to box 1, we use a 60 km λ_c to qualitatively remove remaining noise in the relative vorticity plots. The relative vorticity fields present unrealistic small-scale structures at larger λ_cs values than SSH and velocity. This is expected as the noise effects increase as higher order derivatives are reached. Nevertheless, the larger structures present in the images are recovered from the non-filtered image with a 60 km λ_c filter for both box 1 and 2. Not as much signal is lost with a 60 as with a 200 km λ_c, but some is still lost. For the mesoscale, given the relative vorticity and structures observed in Figures 11 and 12, this does not seem to have a large impact. However, there may be an impact when wanting to observe finer scales as we retrieve normalized relative vorticity much lower than 1.

Spectra were computed for λ_cs of 30, 60 and 200 km to visualize these effects. The corresponding SNR is also calculated in two different ways by using two references. One is by dividing the filtered model-interpolated data by the filtered pseudo-SWOT data, and the other by dividing the non-filtered model-interpolated data by the filtered pseudo-SWOT data. This is shown in Figures 13 and 14.

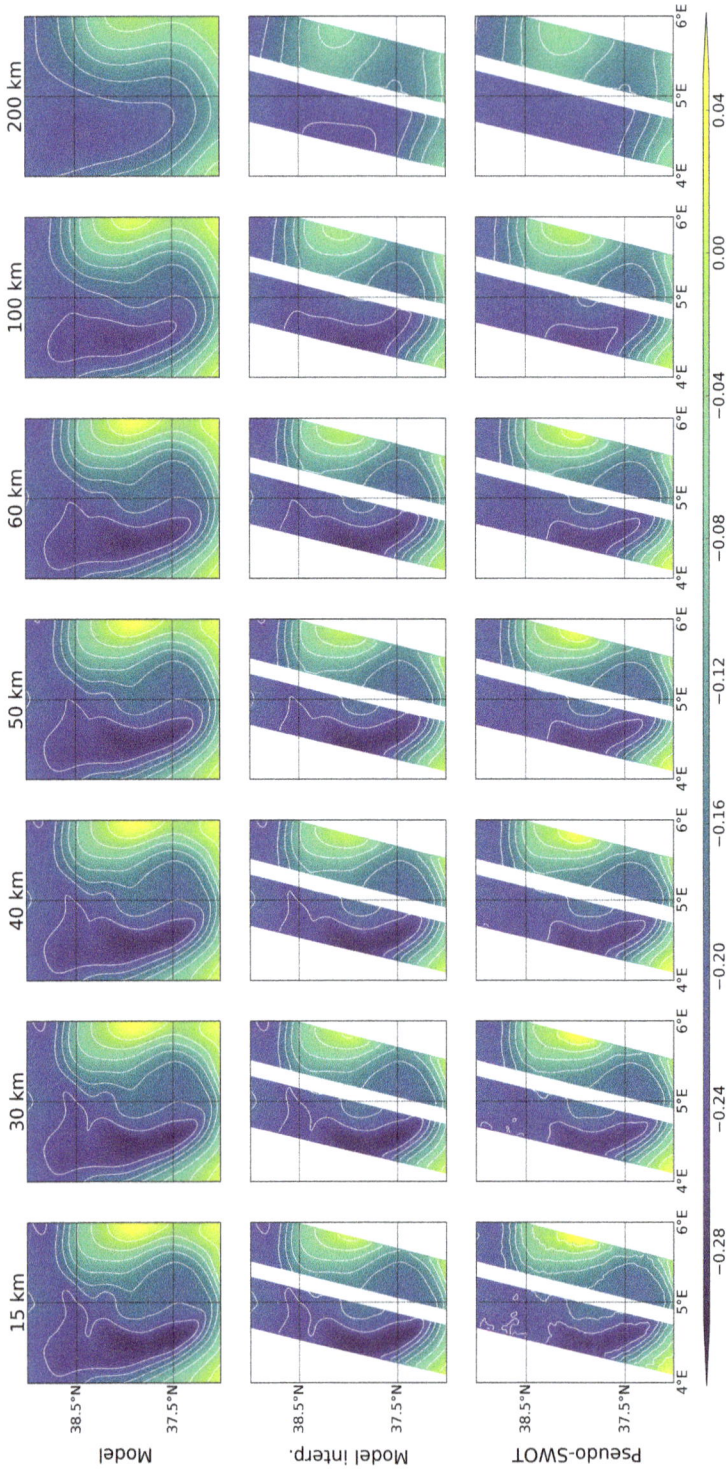

Figure 7. SSH (m) on 23 January 2009 corresponding to pass 15 of cycle 2.

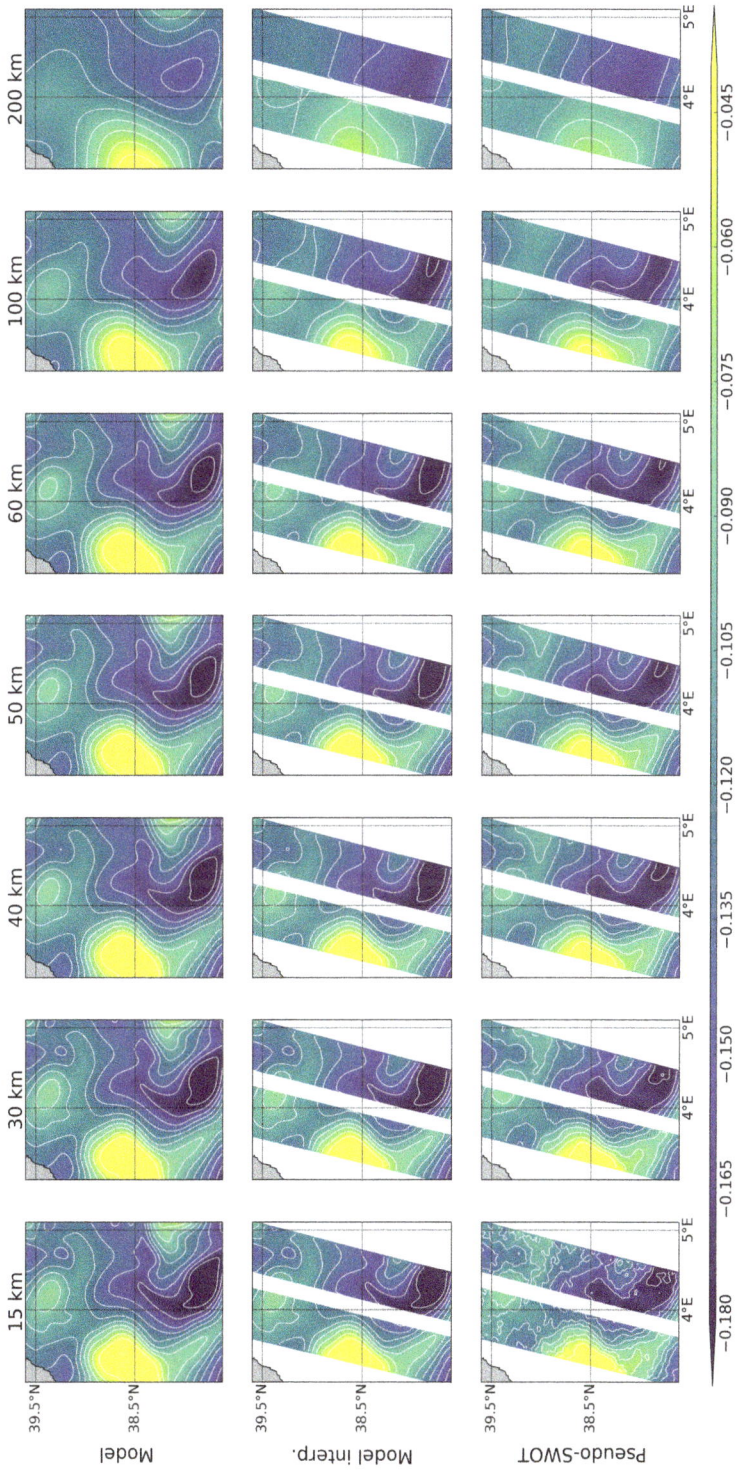

Figure 8. SSH (m) on 3 February 2009 corresponding to pass 168 of cycle 2.

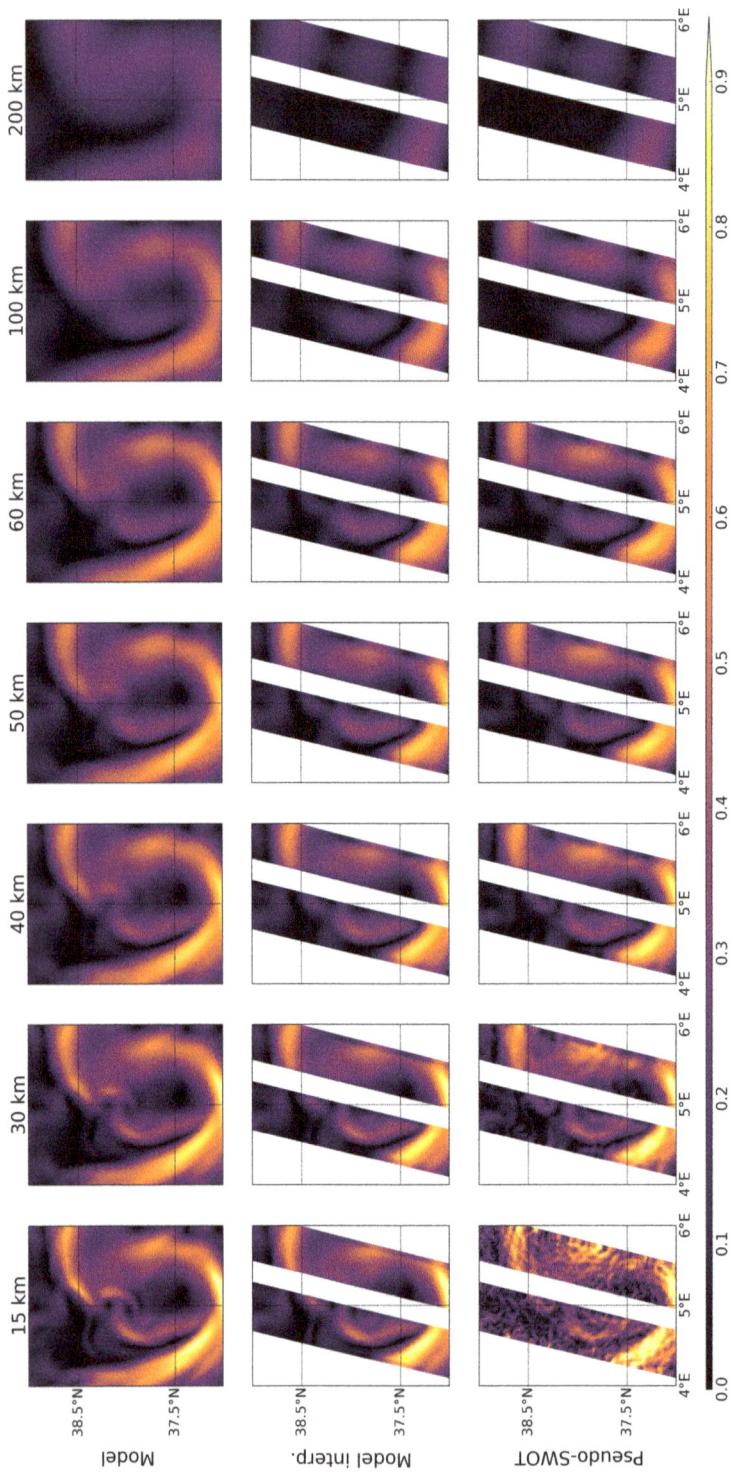

Figure 9. Absolute geostrophic velocity (m/s) on 23 January 2009 corresponding to pass 15 of cycle 2.

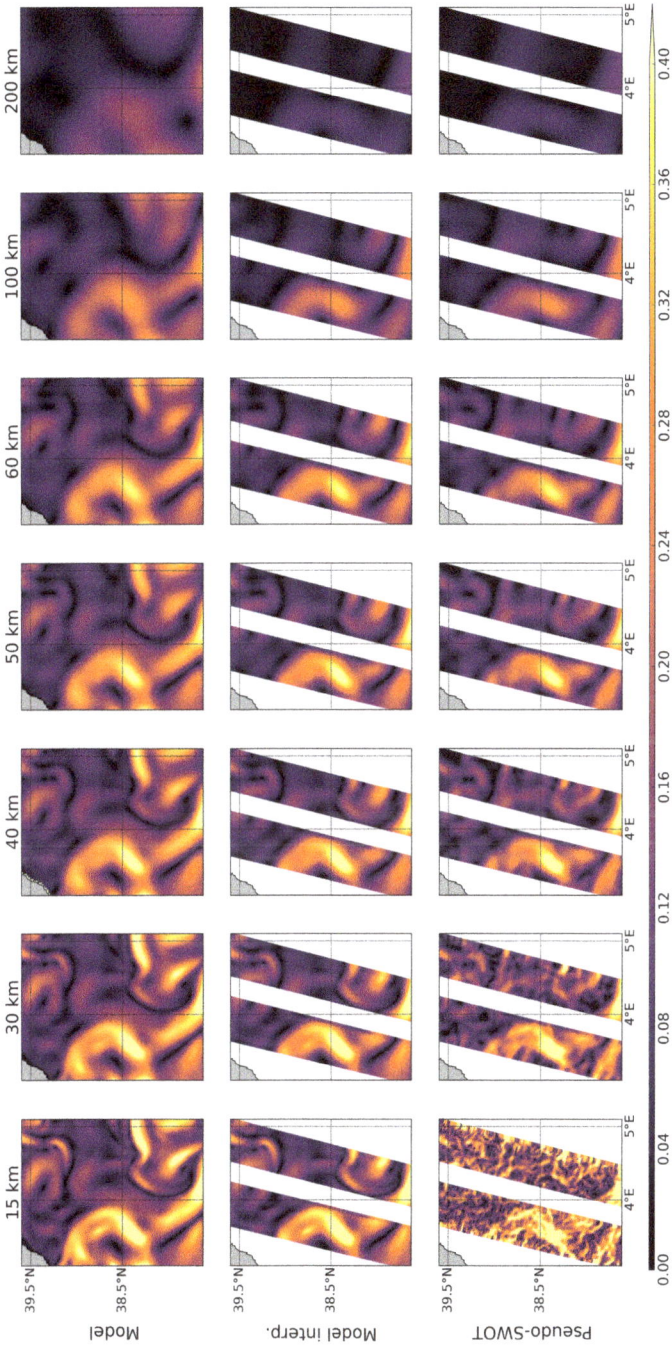

Figure 10. Absolute geostrophic velocity (m/s) on 3 February 2009 corresponding to pass 168 of cycle 2.

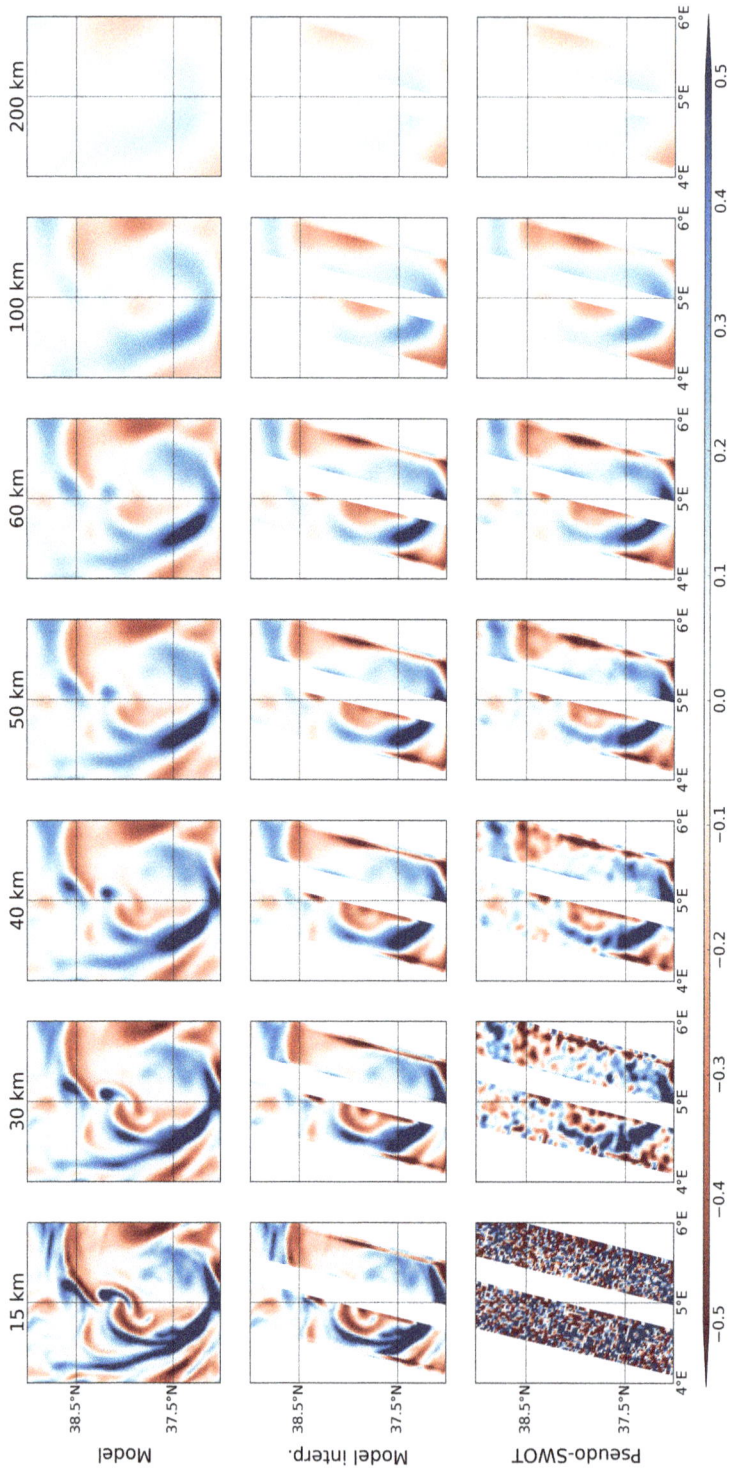

Figure 11. Relative vorticity normalized by f on 23 January 2009 corresponding to pass 15 of cycle 2.

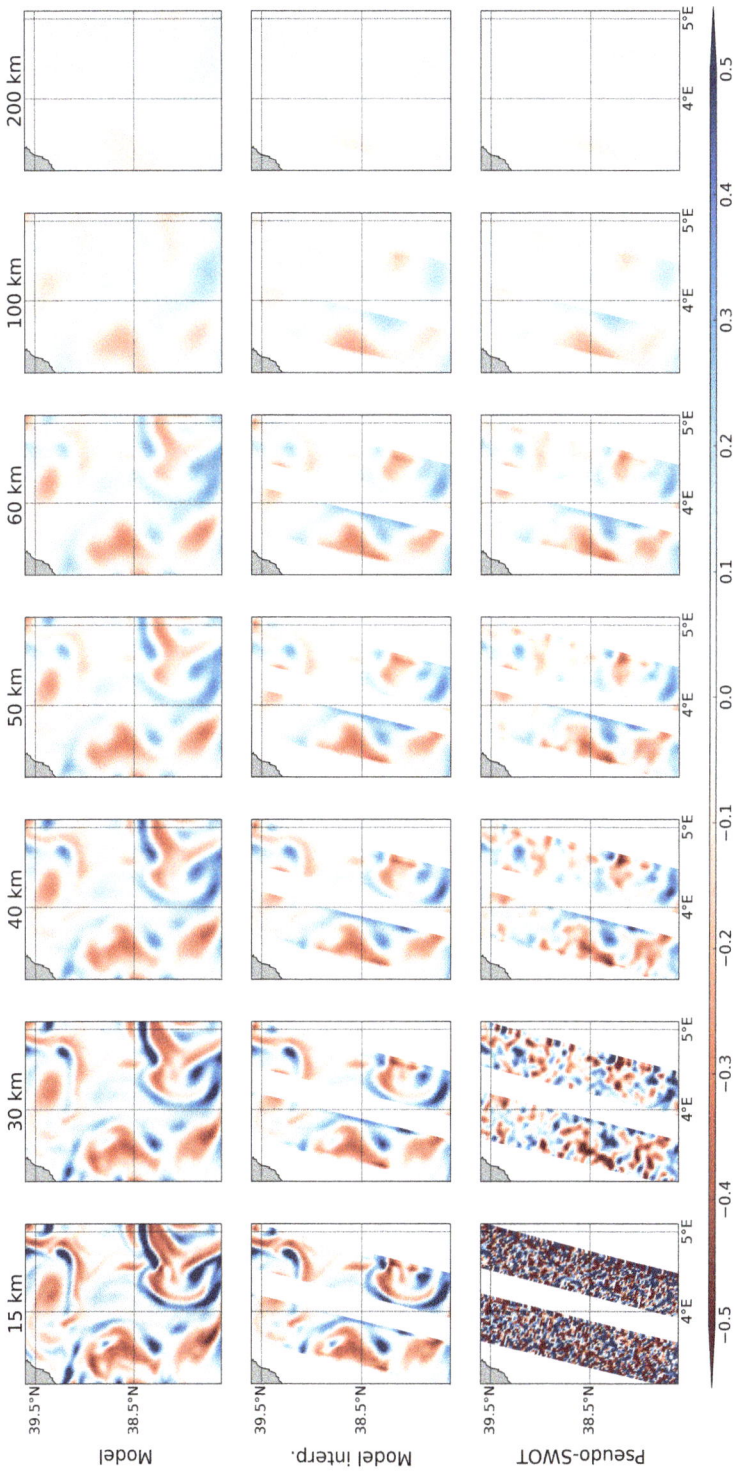

Figure 12. Relative vorticity normalized by f on 3 February 2009 corresponding to pass 168 of cycle 2.

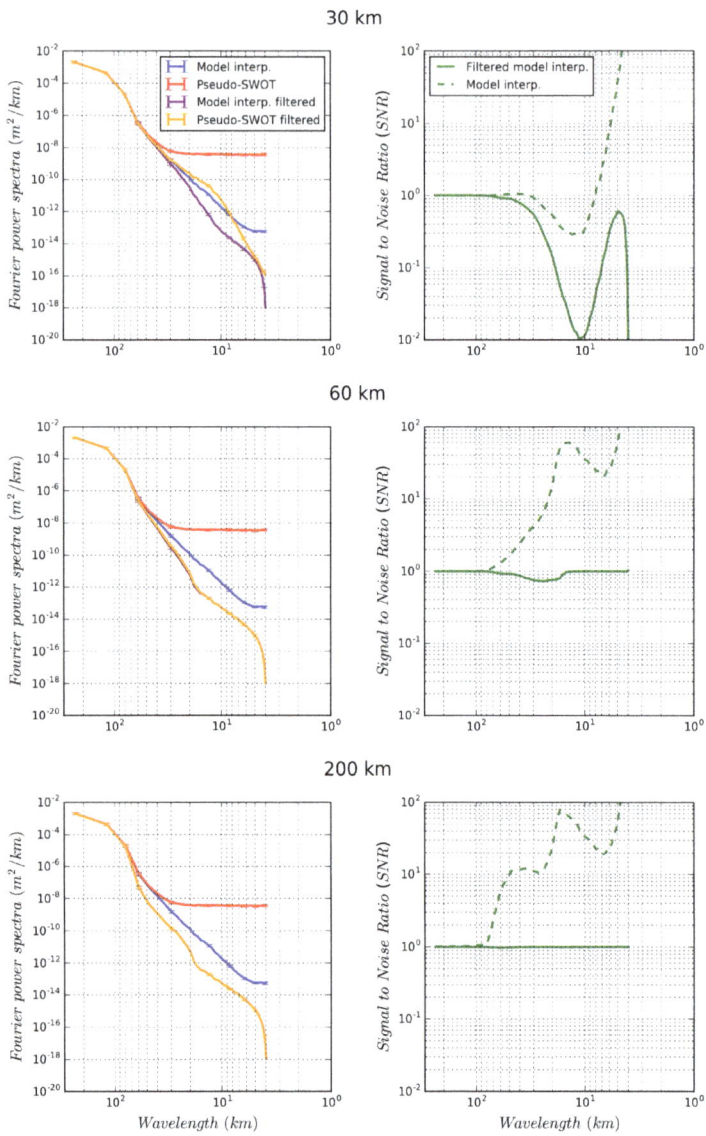

Figure 13. Box 1 region (pass 15) mean of cycles 1 to 122. (**Left**) Spectra of SSH$_{model}$ (blue) and SSH$_{obs}$ (red) before filtering and after applying the different cut-off wavelengths shown in the different rows (30, 60 and 200 km) in purple and orange, respectively. Error bars denote 95% confidence intervals. (**Right**) SNR of SSH$_{model}$ and SSH$_{obs}$, both filtered (solid line) and of SSH$_{model}$ non-filtered and filtered SSH$_{obs}$ (dashed line).

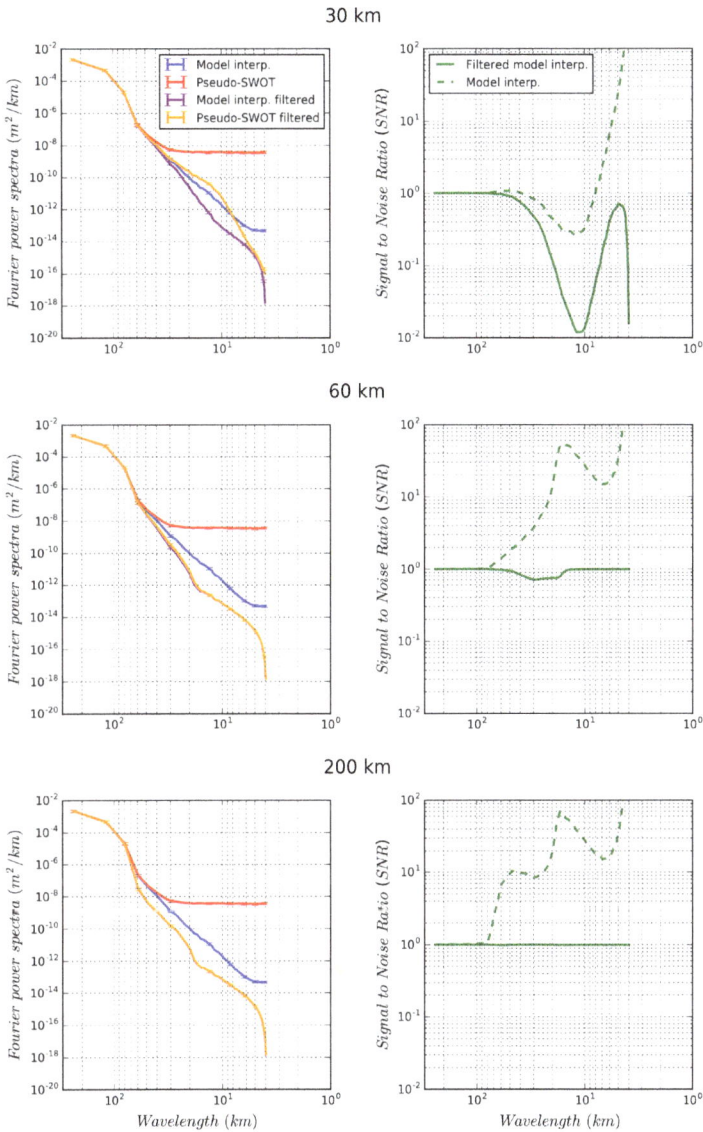

Figure 14. Box 2 region (pass 168) mean of cycles 1 to 122. (**Left**) Spectra of SSH$_{model}$ (blue) and SSH$_{obs}$ (red) before filtering and after applying the different cut-off wavelengths shown in the different rows (30, 60 and 200 km) in purple and orange, respectively. Error bars denote 95% confidence intervals. (**Right**) SNR of SSH$_{model}$ and SSH$_{obs}$, both filtered (solid line) and of SSH$_{model}$ non-filtered and filtered SSH$_{obs}$ (dashed line).

As a consequence of the application of the filter, the separation of the spectral curves of SSH$_{model}$ (model-interp.) and SSH$_{obs}$ (pseudo-SWOT) is reduced. As seen in Figure 6, with no filter, the model-interp. (blue) and pseudo-SWOT (red) curves separate at a wavelength around 60 km. With a 30 km λ_c, the noise level is still high, as observed in Figures 7–12, but the power spectra difference at wavelengths smaller than 60 km between the pseudo-SWOT filtered (yellow) and the model-interp.

(blue) curves is much smaller than it is between the model-interp (blue) and pseudo-SWOT (red) curves (top left panel of Figures 13 and 14). Moreover, the pseudo-SWOT filtered and the model-interp. curves separate at smaller wavelengths. If we look at the dashed line of the top right panel of Figures 13 and 14, we can more accurately determine this wavelength separation value by looking at where the dashed curve starts decreasing. This corresponds to 30 and 40 km wavelengths for box 1 and 2, respectively. Between wavelengths of 60 and 30 km (40 for box 2), the SNR (dashed line) is greater than 1, indicating some over-filtering/smoothing, but this value is very low (1.064 for box 1 and 1.088 for box 2). The pattern of the SNR of the filtered model-interp. over the filtered pseudo-SWOT data (solid line of top right panel of Figures 13 and 14) is similar to that of Figure 6. The noise gains importance as the wavelengths reduce from 60 km, but the SNR values of the 30 km λ_c are lower than the non-filtered ones. On the other hand, this indicates that, given that the filtered model-interp. and pseudo-SWOT spectra are still quite different, a larger λ_c is necessary. In both the left and right panels of the 30 km λ_c of Figures 13 and 14, anomalous patterns are observed below the 10 km wavelength. Not only are these spatial scales very small, but, if we look at the spectra values, they are about 10^{-12} m^2/km or lower. We consider these values to be too low for discussion.

With a 60 km λ_c, the noise is further reduced, but we lose more signal too. If we look at the continuous line of the right panel of Figures 13 and 14, we see that it remains approximately constant at 1. For box 1, it reaches a minimum SNR of 0.715 and for box 2 of 0.697. Looking at the dashed line, the SNR becomes larger than 1 at wavelengths lower than 80 km for both boxes. This means that, although we eliminate all the noise, we are also eliminating part of the signal that was initially present. At wavelengths greater than 10 km, the power spectra of the filtered pseudo-SWOT reach a maximum difference nearly two orders of magnitude smaller than the original model-interp. spectra.

Lastly, with a 200 km λ_c filtering, the model-interp. and pseudo-SWOT spectra curves are identical, and the SNR (solid line) is approximately 1 (Figures 13 and 14, bottom row): however, as observed in Figures 7–12, we lose a lot of signal. In the 20–80 km wavelength range, we can see how the SNR curves (dashed line) rapidly increase and the values are greater with a 200 km λ_c than with a 60 km λ_c. At wavelengths lower than 80 km, on average, there is about one order of magnitude difference between the filtered spectra and the original SSH$_{model}$. It is also interesting to note that the purple and yellow curves separate from the red and blue at 80 km instead of 60, showing that this cut-off exceeds that necessary to remove the noise. This also emphasizes how with SWOT a major advancement could be made as lower cut-off wavelengths will be possible, and thus the observation of smaller scale structures than with contemporary satellites. On the one hand, this result is expected thanks to the 2D swath instead of only 1D nadir data, but, on the other hand, it is important to remember that these are simulated from expected errors and that not all errors are implemented (see Appendix B).

To further quantify the differences observed between the different λ_cs in Figures 7–12, the RMSE between the interpolated model and pseudo-SWOT data shown in these Figures were calculated as described in Section 2.3.3.

In Figure 15, it is interesting to focus on the minimum points of the RMSE curves. Looking at the λ_c corresponding to the different minimum points, for both boxes, the minimum of the curve for SSH and the absolute geostrophic velocity (Vg) is found for a 30 km (29) λ_c. It is slightly higher, 40 km (41), for relative vorticity. This directly relates to the amplification of fine scale structures, and thus the effect of the noise, in the computation of second-order derivatives. It is also in accordance with what is found in the SNR in Figure 6, which shows that we cannot recover the signal at wavelengths lower than 40–50 km. With the qualitative (Figures 7–12) and the spectra (Figures 13 and 14) plots, we saw that, for box 2, as the signal is lower than in box 1, the effect of the noise is greater, and larger λ_cs are necessary. The RMSE plots show us another point of view. As the signal is not as intense in box 2 as in box 1, the over-smoothing (signal lost) due to the Laplacian diffusion filter is lower, and thus we observe lower RMSE values in Figure 15. Moreover, the improvement of the RMSE values is greater for box 2 than box 1. For SSH, the RMSE reduces by 0.05 and 0.07 m from the no filter (0 λ_c) to the minimum RMSE, for box 1 and box 2, respectively. For Vg, the RMSE reduces by 0.42 and 0.455 m/s, and for ζ/f by 3.2 and 3.35.

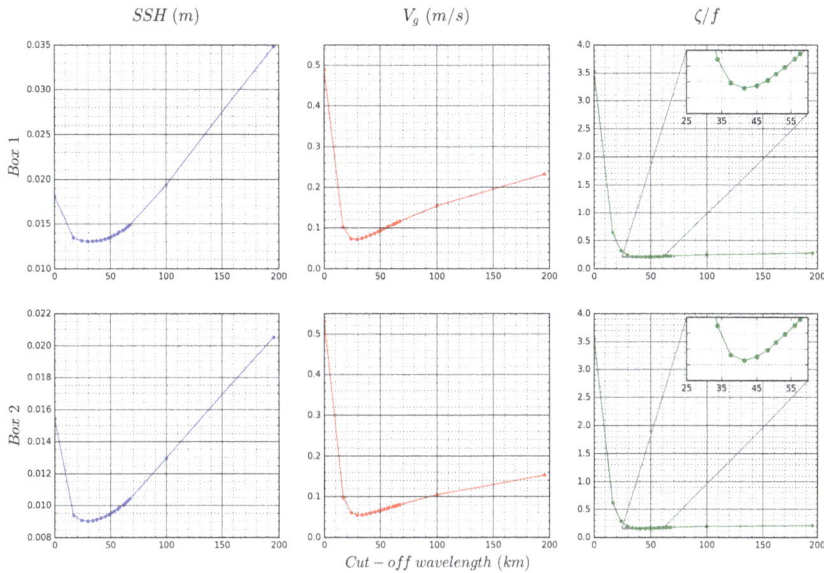

Figure 15. RMSEs of the different variables against the filter's cut-off wavelengths applied to SSH, for pass 15 (**top**) and pass 168 (**bottom**) of cycle 2. Insets show zoom of marked region for better observation of the curves' minimum points.

For SSH, we could say that with a 40 km λ_c it is sufficient, but if we do not want to see the effect of the noise in vorticity, we need a greater λ_c of 60 km. Therefore, together with what is observed in Figures 4–15, we consider that, with this filtering technique applied to this region, SWOT will be able to resolve wavelengths down to a 40–60 km wavelength range. This is the λ_c range where we found that there is a compromise between filtering out the noise of SSH and its derived variables (Vg and ζ/f), and over-smoothing the original image as little as possible. It is important to note that λ_c depends on the signal-to-noise ratio between SSH signals and instrument noise at fine scales. As such, it can be expected to change from region to region (and from season to season) depending on the energy levels at fine scales and on the noise level.

4. Discussion

We find that, by applying our filtering technique to pseudo-SWOT data in the western Mediterranean region, we cannot reach the 15 km wavelength argued for by [12]. We are however able to recover the signal within a swath down to wavelengths of 40–60 km. This wavelength range is close to the one found by [10] at the western boundary currents and nearly half the wavelength values found at the eastern basins by 1D altimeter data. In addition, with the characteristic of SWOT providing 2D SSH data, this will imply a large improvement on the 200 km wavelength resolved by present-day gridded altimetric fields [11]. The SWOT resolved wavelengths found will make it possible to detect structures of 20–30 km in diameter (following [37]) and therefore opens the possibility for observation of fine scales that are unobservable by contemporary altimetric products. This filter is a useful tool for studies comparing the capacities of pseudo-SWOT data with the present altimetric satellite constellation data. As this filter is particularly effective in removing the spatially uncorrelated

KaRIn noise, it may allow the application of already developed techniques that more effectively remove other correlated errors [38].

We find that the presence of structures of different scales and regimes governing—for example, the mesoscale (Ro. order(0.1)) or submesoscale (Ro. order(1)) affects our results on the efficiency of the Laplacian diffusion filter. This filter is therefore sensitive to the presence of different patterns, depending on the region. To reach even smaller scales, it is important to use filtering techniques that conserve and/or retrieve the gradients and, thus, the intensity of the signal present in the observed field. Nonetheless, this sensitivity of the Laplacian diffusion filter could also be due to the effects of noise, depending on the structure in question and its intensity. We also note that the differences between box 1 and 2 found in, for example, the qualitative plots (Figures 7–12) and the minimum RMSE values were not very large; this is not surprising when considering how close in time the two snapshots are.

With this improvement of individual swaths, gridded and possibly daily SWOT SSH maps could be obtained through different interpolation and/or reconstruction techniques. Current gridded altimetric products are obtained using optimal interpolation (OI) [39]. OI may not be the best interpolation method. OI exploits a covariance model of the field to be interpolated. The specification of this covariance model typically relies on a trade-off between the space-time size of the missing data areas and the space-time scales of interest. A covariance model with a large correlation length may lead to an over-smoothing of fine-scale structures, whereas shorter correlation lengths result in filling large missing areas with the background field. From a computational point of view, OI requires a matrix inversion, whose complexity evolves as the cube of the number of observation points. The image-like structure of SWOT data may then be highly computationally-demanding when considering large correlation lengths to fill in large missing areas. Multi-scale OI may be an alternative. However, we expect dynamical interpolation and other data assimilation methods [40–42] to be more adapted both in terms of computational complexity and in their ability to embed relevant dynamic priors to reconstruct horizontal scales down to a few tens of kilometres from SWOT data. On the other hand, SWOT data will greatly improve the present-day OI altimetric products [5]. In addition, SWOT gridded data could be improved in the 40–60 km wavelength range by combining it with the data of a higher temporal resolution.

Dynamic interpolation is an example of a technique that has been investigated by [40] that could help to obtain gridded, daily SSH maps from SWOT. When they apply this method to the Gulf Stream region, they recover the SSH field down to 80 km wavelength. Data-driven schemes recently introduced by [41,42] are also of interest to better reconstruct horizontal scales below 100 km. Overall, for such approaches, it is important to recover the lowest wavelengths possible as spatial resolution loss is likely when producing the gridded maps. Moreover, this spatial resolution loss might be even higher when addressing gridded maps of derived variables. Therefore, the cut-off wavelength should be adjusted to the variables that are to be studied.

Another reconstruction technique that has been investigated in the context of SWOT is a 3D multivariate reconstruction of ocean state. Ref. [43] do this by combining information from SSH and high resolution image structure observations. Once this is achieved, study of the capacity of SWOT to detect fine scale structures could be improved by, for example, better characterizing eddies. As the dataset would be of a higher spatial and temporal resolution than the L2 product, it would then be possible to apply eddy-tracking algorithms like the *py-eddy tracker* [44] or the code developed by [45], which have already been implemented in this region to characterize the western Mediterranean eddy field. A comparison could then be made with the eddies characterized in data from the WMOP model, in the presently available altimetric data and in pseudo-SWOT data.

In future work, the effect of the inter- and intra-annual (or seasonal) variability in the region on the results obtained could be studied too. Although mean spectra were obtained, we focused on two dates in winter. Refs. [46,47] found that there is a strong winter–summer difference in the upper ocean dynamics due to the change of stratification, with the mixed layer depth being deeper in winter.

For example, the reconstruction of mesoscale structures in the upper ocean from pseudo-SWOT data in the Kuroshio Extension region has been studied. They found that the simulated and reconstructed vorticity correlation coefficients varied both inter- and intra-annually [48].

The implementation of filtering techniques that take into account the first and second order SSH derivatives has been started. With this, we hope that in future studies we will be able to recover even smaller wavelengths and to conserve the intensity of the signal after having applied the filter.

New versions of the SWOT simulator will allow the simulation of pseudo-SWOT data during the fast-sampling phase. This makes it possible to start preparing for the calibration/validation phase including the comparison with high resolution in situ data collected during future intensive multi-platform experiments in the western Mediterranean Sea. On the other hand, the only source of geophysical error implemented in the SWOT simulator is still just that related to the wet troposphere. New releases of the SWOT simulator may include the effects of sea state and internal tides [49]. Internal tides and waves are important sources of geophysical errors because, at wavelengths shorter than 50 km, they can affect SWOT data [2]. Therefore, in future work, it would also be interesting to compare our results with an updated version of the SWOT simulator and other OGCMs, especially those that include tides.

5. Conclusions

We have generated simulated pseudo-SWOT data for the western Mediterranean Sea using a SWOT simulator and outputs from an ocean numerical model. To evaluate the output SWOT data, we derived absolute geostrophic velocities and relative vorticities from the pseudo-SWOT SSH data. We find that, due to the satellite's instrumental noise and geophysical errors, the features observed in the pseudo-SWOT SSH are lost in the derived dynamical variables. Looking at the spatial spectra, we find that noise dominates the signal at wavelengths smaller than 60 km. We applied a Laplacian diffusion filtering technique to attempt to remove the noise and hence observe finer scales. We estimated the appropriate cut-off wavelength for each parametrization. To filter out the noise, we applied a series of ascending cut-off wavelengths: 15, 30, 40, 50, 60, 100 and 200 km. We find that in this study region, using this technique, we cannot resolve the expected 15 km wavelength. On the other hand, we are able to recover the signal within a swath down to a 40–60 km wavelength range. This is still an improvement in comparison to wavelengths resolved by present-day 1D altimeters, especially at eastern basins. Robust swath-filtering is an important first step towards meeting our goals for reconstruction techniques that will enable us to combine SWOT and altimetric data in order to produce gridded SSH maps of significantly higher resolution than contemporary products. New versions of the SWOT simulator code include improved representation of instrumental and geophysical errors, and also give us the option to obtain pseudo-data for the SWOT fast-sampling phase. New pseudo-SWOT data will allow us to better refine the results of this study and to examine a wider range of scenarios.

Acknowledgments: The research leading these results has received funding from the Sea Level Thematic Assembly Center (SL-TAC) of the Copernicus Marine and Environment Monitoring Service (CMEMS) and from the Centre National d'Études Spatiales (CNES) through Ocean Surface Topography Science Team (OST/ST) project MANATEE. L. Gómez-Navarro acknowledges CNES and FP-7 PhD funding. This work was supported by a Short Term Scientific Mission (STSM) grant from COST Action ES140. E. Mason was supported by the Copernicus Marine Environment Monitoring Service (CMEMS) MedSUB project. E. Cosme and J. Le Sommer are supported by the CNES through the OST/ST and the SWOT Science Team. The WMOP simulation used in this study was produced in the framework of the MEDCLIC project funded by "La Caixa" Foundation. This study is a contribution to the PRE-SWOT project (CTM2016-78607-P) funded by the Spanish Research Agency and the European Regional Development Fund (AEI/FEDER, UE). Codes and output files are available online at the project repository (https://github.com/LauraGomezNavarro/paper_Gomez-Navarro_etal_2018).

Author Contributions: Ananda Pascual and Laura Gomez-Navarro designed the study; Ronan Fablet, Ananda Pascual and Laura Gomez-Navarro designed the filtering experiments; Baptiste Mourre provided the WMOP simulation data and helped on the analysis of the model data; Laura Gomez-Navarro, Ronan Fablet, Evan Mason, Ananda Pascual, Julien Le Sommer and Emmanuel Cosme contributed to the analysis of the results; Laura Gomez-Navarro wrote the manuscript and all authors contributed to the writing.

Conflicts of Interest: The authors declare no conflict of interest. The founding sponsors had no role in the design of the study; in the collection, analyses, or interpretation of data; in the writing of the manuscript, and in the decision to publish the results.

Abbreviations

The following abbreviations are used in this manuscript:

ADT	Absolute Dynamic Topography
AEMET	Spanish Meteorological Agency
CNES	Centre National d'Études Spatiales
HIRLAM	HIgh Resolution Limited Area Model
KaRIn	Ka-band Radar Interferometer
NASA	National Aeronautics and Space Administration
OGCM	Oceanic General Circulation Model
PDE	Partial Derivative Equation
RMSE	Root Mean Square Error
ROMS	Regional Oceanic Modeling System
SOCIB	the Balearic Islands Islands Coastal Observing and Forecasting System
SSH	Sea Surface Height
SNR	Signal to Noise Ratio
SWOT	Surface Water Ocean Topography
WMOP	Western Mediterranean OPerational forecasting system

Appendix A

In order to know which number of iterations and lambda to set in the filter's parametrization, the filter was applied to a set of 100 randomly generated white noise fields. Spectra were then obtained and the cut-off wavelength was found by identifying the one that corresponded to where the energy was reduced to a half. An example is shown in Figure A1.

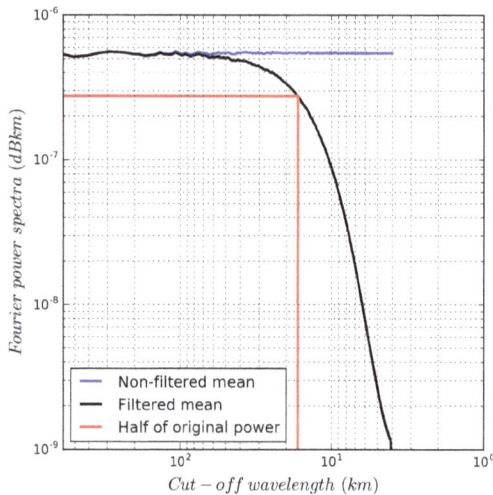

Figure A1. Illustration of how the parameterization corresponding to a 15 (16.72) km cut-off wavelength (λ_c) is estimated. The blue line represents the mean spectra of the 100 non-filtered white noise fields. The black line is the mean spectra of the 100 filtered white noise fields. The horizontal red line shows the half-power spectra of the blue line, and the vertical red line the corresponding wavelength value of the black line, and thus the cut-off wavelength.

In Table A1, we show the different λ_c obtained for a set of lambdas and number of iterations and in Figure A2 a plot of the values shown in Table A1 is presented. As can be observed in Table A1, in most cases, several combinations of lambdas and iterations can give the same cut-off wavelength. We decided to choose the combination corresponding to the smallest lambda, as the smaller the lambda, the smaller the over-smoothing.

Table A1. Cut-off wavelengths (λ_c) values and their corresponding lambda and number of iterations (iter) combinations. The cut-off wavelengths shown in Figures 7–12 are in bold.

Cut-off	Lambda	Iter	Cut-off	Lambda	Iter	Cut-off	Lambda	Iter	Cut-off	Lambda	Iter
16.72	0.05	50	71.88	0.10	450	105.09	0.20	500	141.41	0.35	500
23.95	0.05	100		0.15	300		0.25	400		0.40	450
	0.10	50		0.30	150		0.30	350		0.45	400
29.47	0.05	150		0.45	100		0.35	300		0.50	350
	0.15	50	74.5	0.10	500		0.40	250		0.55	350
33.85	0.05	200		0.20	250		0.50	200		0.60	300
	0.10	100		0.25	200		0.65	150		0.65	300
	0.20	50		0.50	100	110.78	0.25	450		0.70	250
37.58	0.05	250	77.31	0.15	350		0.45	250		0.75	250
	0.25	50		0.35	150		0.55	200	151.91	0.40	500
41.38	0.05	300		0.55	100		0.70	150		0.45	450
	0.10	150	83.63	0.15	400		0.75	150		0.50	400
	0.15	100		0.20	300	117.12	0.25	500		0.55	400
	0.30	50		0.25	250		0.30	400		0.60	350
45.02	0.05	350		0.30	200		0.35	350		0.70	300
	0.35	50		0.40	150		0.40	300		0.80	250
48.19	0.05	400		0.60	100		0.50	250	164.1	0.45	500
	0.10	200	87.19	0.15	450		0.60	200		0.50	450
	0.20	100		0.45	150		0.65	200		0.50	500
	0.40	50		0.65	100		0.80	150		0.55	450
50.58	0.05	450	91.07	0.15	500	124.24	0.30	450		0.60	400
	0.15	150		0.20	350		0.30	500		0.65	350
	0.45	50		0.25	300		0.35	400		0.65	400
53.2	0.05	500		0.30	250		0.40	350		0.70	350
	0.10	250		0.35	200		0.45	300		0.75	300
	0.25	100		0.50	150		0.50	300		0.75	350
	0.50	50		0.70	100		0.55	250		0.80	300
56.12	0.55	50		0.75	100		0.60	250	178.42	0.55	500
57.7	0.10	300	95.31	0.20	400		0.70	200		0.60	450
	0.15	200		0.40	200		0.75	200		0.60	500
	0.20	150		0.55	150	132.27	0.35	450		0.65	450
	0.30	100		0.80	100		0.40	400		0.70	400
	0.60	50	99.96	0.20	450		0.45	350		0.70	450
61.15	0.65	50		0.25	350		0.55	300		0.75	400
63.03	0.1	350		0.3	300		0.65	250		0.8	350
	0.35	100		0.35	250		0.80	200		0.80	400
	0.70	50		0.45	200				195.49	0.65	500
65.03	0.15	250		0.60	150					0.70	500
	0.25	150								0.75	450
	0.75	50								0.75	500
67.17	0.1	400								0.8	450
	0.20	200								0.8	500
	0.40	100									
	0.80	50									

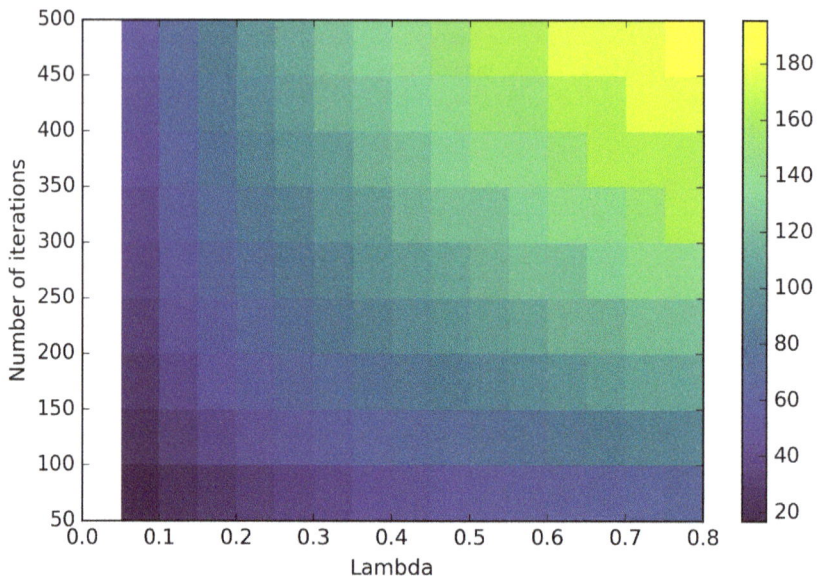

Figure A2. Laplacian diffusion cut-off wavelengths (km) for different combinations of the number of iterations and lambdas.

Appendix B

The noise added by the SWOT simulator can be divided into two types:

- Instrument errors: There are the different types of noise that can affect the signal due to the satellite itself:

  - Ka-Band Radar Interferometer (KaRIn)
  - Roll
  - Timing
  - Phase
  - Baseline dilation

Below are two example cycles of the instrument errors added by the simulator to passes 15 and 168 (Figures A3 and A4). Please note that the color-scale has been adjusted for each error type.

- Geophysical errors: In version 1 of the simulator, only the geophysical error due to the wet troposphere is implemented. Other geophysical errors include those due to the dry troposphere, the ionosphere and the sea state bias (electromagnetic bias). However, the wet troposphere is a major source of geophysical errors and it is implemented via these following two variables:

  - Path delay (pd),
  - Residual path delay (pd_err_1b).

Below, we show two example cycles of the geophysical errors added by the simulator to passes 15 and 168 (Figures A5 and A6). Please note that the color-scale has been adjusted for each error type.

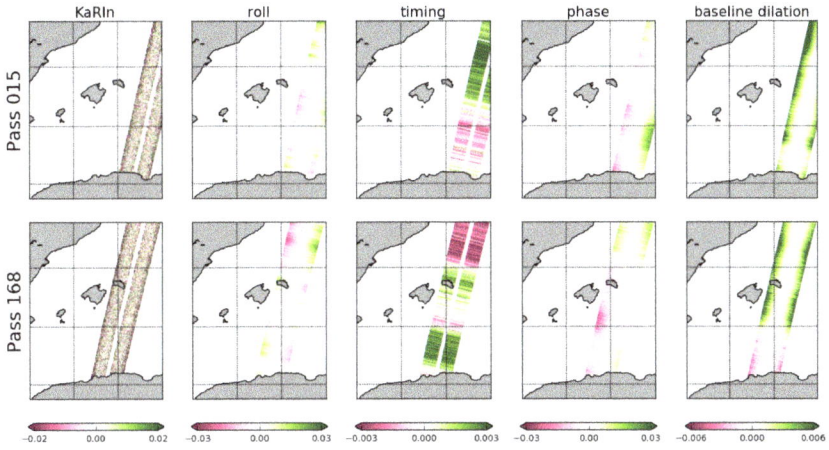

Figure A3. The different instrument noise types (m) for passes 15 pass 168 over cycle 2 are shown.

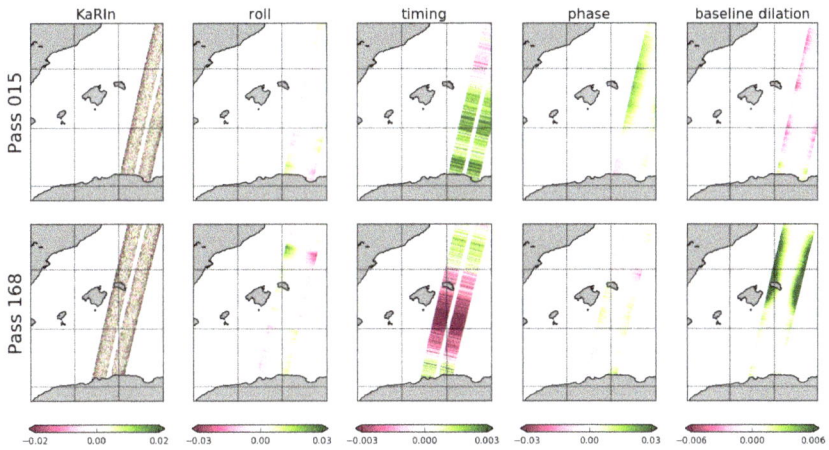

Figure A4. The different instrument noise types (m) for passes 15 and 168 over cycle 30 are shown.

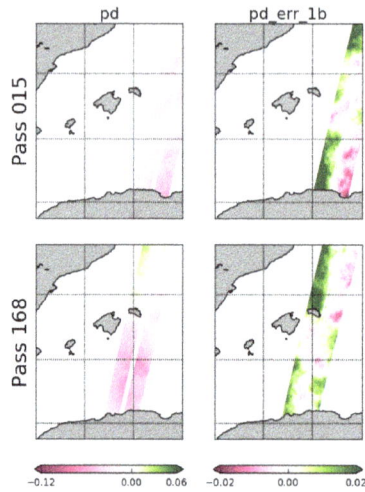

Figure A5. The different geophysical errors (m) for passes 15 and 168 over cycle 2 are shown.

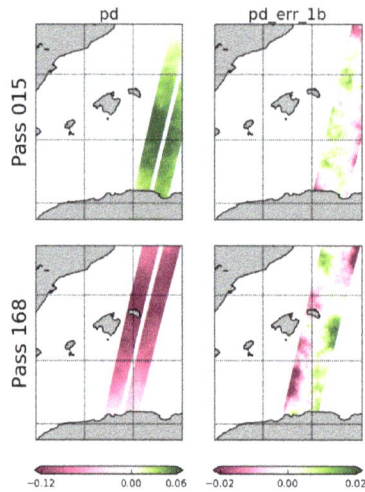

Figure A6. The different geophysical errors (m) for passes 15 and 168 over cycle 30 are shown.

For more details on instrument and geophysical errors, see [30,50].

References

1. Alsdorf, D.; Mognard, N.; Lettenmaier, D. Remote Sensing of Surface Water and Recent Developments in the SWOT Mission. In Proceedings of the Session H21J-06, AGU Fall Meeting, San Francisco, CA, USA, 5–9 December 2011.
2. Fu, L.L.; Morrow, R. A next generation altimeter for mapping the sea surface variability: Opportunities and challenges. In Proceedings of the 48th Liege Colloquium on Ocean Dynamics, Liège, Belgium, 23–27 May 2016.

3. Lee, H.; Biancamaria, S.; Alsdorf, D.; Andreadis, K.; Clark, E.; Durand, M.; Jung, H.C.; Lettenmaier, D.; Mognard, N.; Rodríguez, E.; et al. Capability of SWOT to Measure Surface Water Storage Change. In Proceedings of the Towards high-resolution of oceans dynamics and terrestrial water from space meeting, Lisbon, Portugal, 21–22 October 2010.

4. Rodríguez, E. The Surface Water and Ocean Topography (SWOT) Mission. In Proceedings of the OSTST (Ocean Surface Topography Science Team) meeting, Lisbon, Portugal, 18–20 October 2010.

5. Pujol, M.I.; Dibarboure, G.; Le Traon, P.Y.; Klein, P. Using high-resolution altimetry to observe mesoscale signals. *J. Atmos. Ocean. Technol.* **2012**, *29*, 1409–1416, doi:10.1175/JTECH-D-12-00032.1.

6. Rodríguez, E. Surface Water and Ocean Topography Mission Project Science Requirements Document. Jet Propulsion Laboratory, California Institute of Technology, JPL D-61923; March 2016; 28p. Available online: https://swot.jpl.nasa.gov/documents.htm (accessed on 12 April 2018).

7. Rogé, M.; Morrow, R.; Ubelmann, C.; Dibarboure, G. Using a dynamical advection to reconstruct a part of the SSH evolution in the context of SWOT, application to the Mediterranean Sea. *Ocean Dyn.* **2017**, *67*, 1047–1066.

8. Cipollini, P.; Calafat, F.M.; Jevrejeva, S.; Melet, A.; Prandi, P. Monitoring Sea Level in the Coastal Zone with Satellite Altimetry and Tide Gauges. *Surv. Geophys.* **2017**, *38*, 33–57.

9. Fu, L.L.; Ferrari, R. Observing Oceanic Submesoscale Processes From Space. *Eos Trans. Am. Geophys. Union* **2008**, *89*, 488, doi:10.1109/IGARSS.2011.6049757.

10. Dufau, C.; Orsztynowicz, M.; Dibarboure, G.; Morrow, R.; Le Traon, P.Y. Mesoscale resolution capability of altimetry: Present and future. *J. Geophys. Res. Ocean.* **2016**, doi:10.1002/2015JC010904.

11. Chelton, D.B.; Schlax, M.G.; Samelson, R.M.; de Szoeke, R.A. Global observations of large oceanic eddies. *Geophys. Res. Lett.* **2007**, *34*, doi:10.1029/2007GL030812.

12. Fu, L.L.; Ubelmann, C. On the transition from profile altimeter to swath altimeter for observing global ocean surface topography. *J. Atmos. Ocean. Technol.* **2014**, *31*, 560–568, doi:10.1175/JTECH-D-13-00109.1.

13. Mémery, L.; Olivier, F. Primary production and export fluxes at the Almeria-Oran front: A numerical study. In Proceedings of the EGS-AGU-EUG Joint Assembly, Nice, France, 6–11 April, 2003.

14. Lévy, M.; Klein, P.; Treguier, A.M. Impact of sub-mesoscale physics on production and subduction of phytoplankton in an oligotrophic regime. *J. Mar. Res.* **2001**, *59*, 535–565.

15. Lapeyre, G.; Klein, P. Impact of the small-scale elongated filaments on the oceanic vertical pump. *J. Mar. Res.* **2006**, *64*, 835–851.

16. Omand, M.M.; D'Asaro, E.A.; Lee, C.M.; Perry, M.J.; Briggs, N.; Cetinić, I.; Mahadevan, A. Eddy-driven subduction exports particulate organic carbon from the spring bloom. *Science* **2015**, *348*, 222–225.

17. McGillicuddy, D.J.; Anderson, L.A.; Bates, N.R.; Bibby, T.; Buesseler, K.O.; Carlson, C.A.; Davis, C.S.; Ewart, C.; Falkowski, P.G.; Goldthwait, S.A.; et al. Eddy/wind interactions stimulate extraordinary mid-ocean plankton blooms. *Science* **2007**, *316*, 1021–1026.

18. Allen, J.; Smeed, D.; Tintoré, J.; Ruiz, S. Mesoscale subduction at the almeria–oran front: Part 1: Ageostrophic flow. *J. Mar. Syst.* **2001**, *30*, 263–285.

19. Ruiz, S.; Pascual, A.; Garau, B.; Faugère, Y.; Alvarez, A.; Tintoré, J. Mesoscale dynamics of the Balearic Front, integrating glider, ship and satellite data. *J. Mar. Syst.* **2009**, *78*, S3–S16.

20. Pascual, A.; Ruiz, S.; Olita, A.; Troupin, C.; Claret, M.; Casas, B.; Mourre, B.; Poulain, P.M.; Tovar-Sanchez, A.; Capet, A.; et al. A multiplatform experiment to unravel meso-and submesoscale processes in an intense front (alborex). *Front. Mar. Sci.* **2017**, *4*, 39.

21. Bethoux, J.; Gentili, B.; Morin, P.; Nicolas, E.; Pierre, C.; Ruiz-Pino, D. The Mediterranean Sea: A miniature ocean for climatic and environmental studies and a key for the climatic functioning of the North Atlantic. *Progress Oceanogr.* **1999**, *44*, 131–146.

22. Malanotte-Rizzoli, P.; Artale, V.; Borzelli-Eusebi, G.L.; Brenner, S.; Crise, A.; Gacic, M.; Kress, N.; Marullo, S.; d'Alcalà, M.R.; Sofianos, S.; et al. Physical forcing and physical/biochemical variability of the Mediterranean Sea: A review of unresolved issues and directions for future research. *Ocean Sci.* **2014**, *10*, 281, doi:10.5194/os-10-281-2014.

23. Robinson, A.R.; Leslie, W.G.; Theocharis, A.; Lascaratos, A. Mediterranean sea circulation. In *Ocean Currents: A Derivative of the Encyclopedia of Ocean Sciences*; Elsevier: Amsterdam, The Netherlands, 2001; pp. 1689–1705.

24. Hallberg, R. Using a resolution function to regulate parameterizations of oceanic mesoscale eddy effects. *Ocean Model.* **2013**, *72*, 92–103, doi:10.1016/j.ocemod.2013.08.007.

25. Escudier, R.; Renault, L.; Pascual, A.; Brasseur, P.; Chelton, D.; Beuvier, J. Eddy properties in the Western Mediterranean Sea from satellite altimetry and a numerical simulation. *J. Geophys. Res.* **2016**, *121*, 3990–4006, doi:10.1002/2015JC011371.
26. Chelton, D.B.; Deszoeke, R.A.; Schlax, M.G.; El, K.; And, N.; Siwertz, N. Geographical Variability of the First Baroclinic Rossby Radius of Deformation. *J. Phys. Oceanogr.* **1998**, *28*, 433–460.
27. Nurser, A.J.G.; Bacon, S. Arctic Ocean Rossby radius Eddy length scales and the Rossby radius in the Arctic Ocean Arctic Ocean Rossby radius. *Ocean Sci. Discuss.* **2013**, *10*, 1807–1831, doi:10.5194/osd-10-1807-2013.
28. Wang, J.; Fu, L.L.; Qiu, B.; Menemenlis, D.; Farrar, J.T.; Chao, Y.; Thompson, A.F.; Flexas, M.M. An Observing System Simulation Experiment for the Calibration and Validation of the Surface Water Ocean Topography Sea Surface Height Measurement Using In Situ Platforms. *J. Atmos. Ocean. Technol.* **2018**, *35*, 281–297, doi:10.1175/JTECH-D-17-0076.1.
29. Gaultier, L.; Ubelmann, C.; Fu, L.L. *SWOT Simulator Documentation*; Tech. Rep. 1.0.0, Jet Propulsion Laboratory, California Institute of Technology: Pasadena, CA, USA, 2015.
30. Gaultier, L.; Ubelmann, C.; Fu, L.L. The challenge of using future SWOT data for oceanic field reconstruction. *J. Atmos. Ocean. Technol.* **2016**, *33*, 119–126, doi:10.1175/JTECH-D-15-0160.1.
31. Juza, M.; Mourre, B.; Renault, L.; Gómara, S.; Sebastián, K.; Lora, S.; Beltran, J.; Frontera, B.; Garau, B.; Troupin, C.; et al. SOCIB operational ocean forecasting system and multi-platform validation in the Western Mediterranean Sea. *J. Oper. Oceanogr.* **2016**, *9*, s155–s166.
32. Shchepetkin, A.F.; McWilliams, J.C. The regional oceanic modeling system (ROMS): A split-explicit, free-surface, topography-following-coordinate oceanic model. *Ocean Model.* **2005**, *9*, 347–404.
33. Sonka, M.; Hlavac, V.; Boyle, R. *Image Processing, Analysis, and Machine Vision*; Cengage Learning: Boston, MA, USA, 2014.
34. Aubert, G.; Kornprobst, P. *Mathematical Problems in Image Processing: Partial Differential Equations and the Calculus of Variations*; Springer Science & Business Media: Berlin, Germany, 2006; Volume 147.
35. Escudier, R.; Mourre, B.; Juza, M.; Tintoré, J. Subsurface circulation and mesoscale variability in the Algerian subbasin from altimeter-derived eddy trajectories. *J. Geophys. Res. Ocean.* **2016**, *121*, 6310–6322, doi:10.1002/2016JC011760.
36. Gunturk, B.K.; Li, X. *Image Restoration: Fundamentals and Advances*; CRC Press: Boca Raton, FL, USA, 2012.
37. Klein, P.; Morrow, R.; Samelson, R.; Chelton, D.; Lapeyre, G.; Fu, L.; Qiu, B.; Ubelmann, C.; Le Traon, P.Y.; Capet, X.; et al. Mesoscale/Sub-Mesoscale Dynamics in the Upper Ocean. 2015. NASA Surface Water and Ocean Topography (SWOT). Available online: https://www.aviso.altimetry.fr/fileadmin/documents/missions/Swot/WhitePaperSWOTSubmesoscale.pdf (accessed on 12 April 2018).
38. Ruggiero, G.; Cosme, E.; Brankart, J.; Sommer, J.L.; Ubelmann, C. An efficient way to account for observation error correlations in the assimilation of data from the future SWOT High-Resolution altimeter mission. *J. Atmos. Ocean. Technol.* **2016**, *33*, 2755–2768, doi:10.1175/JTECH-D-16-0048.1.
39. Pujol, M.I.; Faugere, Y.; Taburet, G.; Dupuy, S.; Pelloquin, C.; Ablain, M.; Picot, N. DUACS DT2014: The new multi-mission altimeter data set reprocessed over 20 years. *Ocean Sci.* **2016**, *12*, 1067–1090.
40. Ubelmann, C.; Klein, P.; Fu, L.L. Dynamic interpolation of sea surface height and potential applications for future high-resolution altimetry mapping. *J. Atmos. Ocean. Technol.* **2015**, *32*, 177–184.
41. Fablet, R.; Verron, J.; Mourre, B.; Chapron, B.; Pascual, A. Improving mesoscale altimetric data from a multitracer convolutional processing of standard satellite-derived products. *IEEE Trans. Geosci. Remote Sens.* **2018**, doi:10.1109/TGRS.2017.2750491.
42. Lguensat, R.; Viet, P.H.; Sun, M.; Chen, G.; Fenglin, T.; Chapron, B.; Fablet, R. Data-driven Interpolation of Sea Level Anomalies Using Analog Data Assimilation. 2017. Available online: https://hal.archives-ouvertes.fr/hal-01609851 (accessed on 12 April 2018).
43. Moro, M.D.; Brankart, J.M.; Brasseur, P.; Verron, J. Exploring image data assimilation in the prospect of high-resolution satellite oceanic observations. *Ocean Dyn.* **2017**, *67*, 875–895.
44. Mason, E.; Pascual, A.; McWilliams, J.C. A new sea surface height-based code for oceanic mesoscale eddy tracking. *J. Atmos. Ocean. Technol.* **2014**, *31*, 1181–1188, doi:10.1175/JTECH-D-14-00019.1.
45. Conti, D.; Orfila, A.; Mason, E.; Sayol, J.M.; Simarro, G.; Balle, S. An eddy tracking algorithm based on dynamical systems theory. *Ocean Dyn.* **2016**, *66*, 1415–1427, doi:10.1007/s10236-016-0990-7.

46. D'Ortenzio, F.; Iudicone, D.; de Boyer Montegut, C.; Testor, P.; Antoine, D.; Marullo, S.; Santoleri, R.; Madec, G. Seasonal variability of the mixed layer depth in the Mediterranean Sea as derived from in situ profiles. *Geophys. Res. Lett.* **2005**, *32*, doi:10.1029/2005GL022463.

47. Houpert, L.; Testor, P.; de Madron, X.D.; Somot, S.; D'ortenzio, F.; Estournel, C.; Lavigne, H. Seasonal cycle of the mixed layer, the seasonal thermocline and the upper-ocean heat storage rate in the Mediterranean Sea derived from observations. *Progress Oceanogr.* **2015**, *132*, 333–352.

48. Qiu, B.; Chen, S.; Klein, P.; Ubelmann, C.; Fu, L.; Sasaki, H. Reconstructability of 3-Dimensional upper ocean circulation from SWOT Sea Surface Height measurements (early online release). *J. Phys. Oceanogr. J. Phys. Ocean.* **2016**, *46*, 947–963, doi:10.1175/JPO-D-15-0188.1.

49. Lindstrom, E.; Cherchal, S.; Fu, L.L.; Morrow, R.; Pavelsky, T.; Cretaux, J.F.; Vaze, P.; Lafon, T.; Coutin-Faye, S.; Amen, L.; et al. *Summary Report of the 2nd SWOT Science Team Meeting 2017*; Techreport; Meteo-France Conference Centre: Touloue, France, 2017.

50. Esteban-Fernandez, D. SWOT Project: Mission Performance and Error Budget Document. Jet Propulsion Laboratory, California Institute of Technology, JPL D-79084. April 2017; 117p. Available online: https://swot.jpl.nasa.gov/documents.htm (accessed on 12 April 2018).

remote sensing

MDPI

Article

Evaluation of Coastal Sea Level Offshore Hong Kong from Jason-2 Altimetry

Xi-Yu Xu [1,2,3,*], Florence Birol [2] and Anny Cazenave [2,4]

[1]   The CAS Key Laboratory of Microwave Remote Sensing, National Space Science Center,
     Chinese Academy of Sciences, Beijing 100190, China
[2]   Laboratoire d'Etudes en Géophysique et Océanographie Spatiales (LEGOS), Observatoire Midi-Pyrénées,
     31400 Toulouse, France; florence.birol@legos.obs-mip.fr (F.B.); anny.cazenave@legos.obs-mip.fr (A.C.)
[3]   State Key Laboratory of Remote Sensing Science, Institute of Remote Sensing and Digital Earth,
     Chinese Academy of Sciences, Beijing 100094, China
[4]   International Space Science Institute, 3102 Bern, Switzerland
*   Correspondence: xuxiyu@mirslab.cn; Tel.: +86-10-6255-0409

Received: 23 November 2017; Accepted: 6 February 2018; Published: 12 February 2018

Abstract: As altimeter satellites approach coastal areas, the number of valid sea surface height measurements decrease dramatically because of land contamination. In recent years, different methodologies have been developed to recover data within 10–20 km from the coast. These include computation of geophysical corrections adapted to the coastal zone and retracking of raw radar echoes. In this paper, we combine for the first time coastal geophysical corrections and retracking along a Jason-2 satellite pass that crosses the coast near the Hong-Kong tide gauge. Six years and a half of data are analyzed, from July 2008 to December 2014 (orbital cycles 1–238). Different retrackers are considered, including the ALES retracker and the different retrackers of the PISTACH products. For each retracker, we evaluate the quality of the recovered sea surface height by comparing with data from the Hong Kong tide gauge (located 10 km away). We analyze the impact of the different geophysical corrections available on the result. We also compute sea surface height bias and noise over both open ocean (>10 km away from coast) and coastal zone (within 10 km or 5 km coast-ward). The study shows that, in the Hong Kong area, after outlier removal, the ALES retracker performs better in the coastal zone than the other retrackers, both in terms of noise level and trend uncertainty. It also shows that the choice of the ocean tide solution has a great impact on the results, while the wet troposphere correction has little influence. By comparing short-term trends computed over the 2008.5–2014 time span, both in the coastal zone and in the open ocean (using the Climate Change Initiative sea level data as a reference), we find that the coastal sea level trend is about twice the one observed further offshore. It suggests that in the Hong Kong region, the short-term sea level trend significantly increases when approaching the coast.

Keywords: Jason-2; Hong Kong coast; retracking; X-TRACK; ALES; PISTACH

1. Introduction

Sea level rise is one of the most threatening consequences of present-day global warming. About 10% of the world population currently lives in the world's coastal zones and this number will increase in the future. Therefore, it is crucial to monitor and understand sea level variations along coastlines [1]. Although the tide gauge network has expanded in recent years, some highly populated areas like western Africa remain devoid of any station. For 25 years, satellite altimetry routinely monitored sea level changes over the global open ocean, but was largely unexploited in the coastal areas. Indeed, satellite altimetry was originally designed to precisely measure sea level in the open ocean, where the shape of the pulse-limited radar altimeter echo (i.e., after reflection on the sea surface; called waveform), is well described by the classical mathematical Brown model [2], based on the assumption of a homogeneous rough sea surface within the radar footprint. In that case, the sea level parameters (range between satellite and sea surface, significant wave height and backscatter coefficient) are extracted from the model via a Maximum Likelihood Estimation (MLE) approach. In a coastal band of a few kilometers wide (corresponding to the footprint size of the altimeter antenna), radar echoes integrate reflections from nearby land, leading to complex waveforms that significantly depart from the standard Brown model (e.g., [3–5] and references therein). Besides, in some shallow shelf areas, the sea surface might be so calm that the waveform can display several peaks due to specular reflection. Figure 1 shows two examples of waveforms, one over the open ocean (a) and the other in a coastal zone (b).

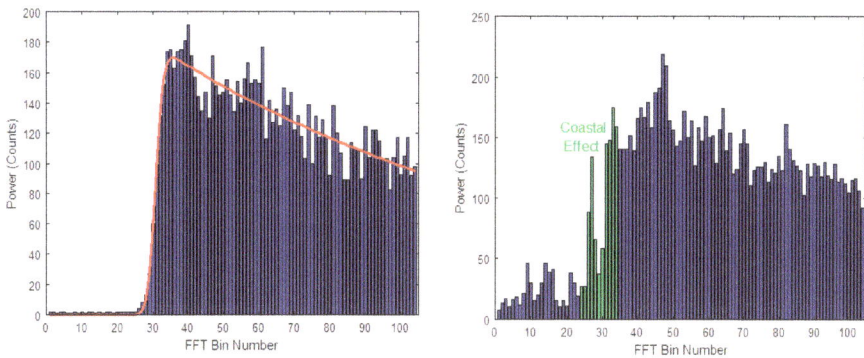

Figure 1. Examples of typical open ocean waveform (**left**; the red line corresponds to the fitted Brown model) and coastal ocean waveform (**right**).

Another difficulty arises from the geophysical corrections applied to the altimeter measurements that usually suffer large uncertainties in the coastal zone. The most limiting corrections are wet tropospheric delay, ocean tides correction, dynamic atmospheric correction (DAC), and sea state bias. Near the coast, these corrections either have large spatio-temporal variability poorly reproduced by numerical models (e.g., tides and DAC corrections), or are less precisely estimated by onboard sensors than over the open ocean (e.g., the radiometer-based wet tropospheric correction that suffers severe land contamination). As a consequence, most altimeter data in a band of 10–30 km from land are declared invalid and discarded from the standard products. For about a decade, significant efforts have been realized by the altimetry community and space agencies to overcome this difficulty and retrieve as much as possible historical altimeter measurements near the coast. Corresponding coastal altimetry products are based on improved coastal geophysical corrections [6,7] and/or dedicated analysis approaches to extract the sea level parameters (range, significant wave height and backscatter coefficient) from non-standard waveforms (a process called 'retracking') [3–5].

In this study, we investigate the relative performances of the experimental coastal altimetry products available for the Jason-2 mission. Jason-2 is chosen here because it was consistently operating for nearly one decade and, above all, because it has been reprocessed by most of the coastal altimetry groups. Satellite pass 153 is considered. We used the following coastal altimetry products: (1) X-TRACK [6] developed by LEGOS (Laboratoire d'Etudes en Géophysique et Océanographie Spatiales, France); (2) PISTACH (Prototype Innovant de Système de Traitement pour l'Altimétrie Côtière et l'Hydrologie, [8]) developed by CLS (Collecte Localisation Satellites, France); and (3) ALES (Adaptive Leading Edge Sub-waveform, [3]) developed by NOC (National Oceanography Centre, UK). The products propose either improved geophysical corrections or waveform retracking algorithms, thus have different contents. X-TRACK products provide sea level time series on a nominal mean track and are available for all altimeter missions except HY-2A and Sentinel-3A. ALES products provide products for Jason-2 and Envisat missions. The successor of the PISTACH product, PEACHI (Prototype for Expertise on Altimetry for Coastal, Hydrology, and Ice), is dedicated to the SARAL/AltiKa mission and provides sea level data along original, cycle-by-cycle tracks [9]. Because of these important differences between products, very few attempts to inter-compare coastal sea level data have been carried out so far, despite the importance of this type of exercise for defining the best processing strategy to derive coastal altimetry data.

In this paper we focus on the South China Sea (Hong-Kong area), not only because of a variety of climate processes impacting sea level (ENSO–El Niño Southern Oscillation-, eddies, storm surges, monsoon, etc.), but also for its importance in the economy and security of Southeastern Asia and the Western Pacific regions. An important consideration when selecting the study area was also the availability of external tide gauge-based sea level data for validation. A dozen tide gauges exist along the China coast, but all stations located along the Chinese mainland ceased to provide data after 1997. Fortunately, a tide gauge remained in operation during the Jason-2 mission at Quarry Bay, Hong-Kong. Corresponding hourly data are available from the University of Hawaii Sea Level Center (UHSLC) [10]. In this study, we considered the closest Jason-2 satellite track (which is ascending and numbered 153) of the HK tide gauge (10.05 km away).

The paper is organized as follows: The study area and the data set are presented in Sections 2 and 3, respectively. Section 4 describes the methodology while Section 5 shows the results. Some elements of discussion are proposed in Section 6, followed by a conclusion (Section 7).

2. Study Area

Hong-Kong (HK) is located just south of the Tropic of Cancer. The climate is predominantly subtropical and displays clear seasonal variations. The southwesterly/northeasterly monsoon give rise to warm wet summers and cool dry winters. HK is also frequently impacted by typhoons. On the western side of the HK island flows the Zhujiang River (Pearl River), which brings abundant freshwater (~3.5 × 10^{11} m^3 per year [11]), resulting in a high salinity gradient. All these factors increase the complexity of the HK environment and have an impact on the regional sea level variations at different spatio-temporal scales.

The HK coast has also an extremely complex geomorphology. As shown on Figure 2, tiny islands lie within the radar footprint of the Jason-2 track chosen for this study. As a consequence, the corresponding altimeter and radiometer measurements are expected to be severely impacted by land effects. This makes this area particularly relevant for analyzing the performances of coastal altimetry data.

The definition of the "coastal zone" in altimetry is somewhat arbitrary. In some studies the criteria of 50 km from land is used (for example, ALES data are provided only in the 50-km coastal band), while some others focus on the first 10 km or even 5 km off the coastline. In this paper we have chosen to base the definition of the area on the "rad_surf_type" parameter provided in the Geophysical Data Record products (GDRs). This surface classification flag is derived from the radiometer measurements and indicates the type of observed surface. For the Jason-2 pass #153, the coastal area based on this flag corresponds to the area where the satellite flies less than 70 km from the closest land (including small islands).

The HK coastal topography is extremely irregular. Figure 3 shows a bathymetric profile along the Jason-2 pass, from southwest to northeast (blue dashed line; from 21.8°N to 22.3°N latitude). Despite a narrow band between 21.8°N and 22°N, where the depth is steeply falling down to ~−60 m, the study area corresponds to very shallow waters. We can thus expect complex local tides and currents influencing sea level variations. Figure 3 also shows the corresponding along track distance to land (mainland or island). The "distance to land" profile is rather complicated. Coastward, it first decreases to ~8 km, slightly increases to ~10 km, and then fluctuates between 10 and 0 km. It even reaches ~0 km when the satellite flies over a small island called Wailingding Island (see Figure 2). Considering both distance to coast and water depth, we define three cases for our comparison exercise, corresponding to three different overlapping segments along the Jason-2 pass. Segments 1, 2, and 3, correspond to cases 1, 2, and 3. In Figure 2, these can be identified by the orange, green, and dashed blue colors, respectively. They cover distances of 50 km, 30 km, and 10 km, and reflect increasing coastal conditions (thus increasing difficulty to retrieve a coherent physical signal from altimetry data). For each case, each satellite cycle, and each product analyzed, we spatially averaged all available 20-Hz (~0.3 km resolution) along-track sea level data along the corresponding pass segment (up to the last valid measurement at HK coast). We finally obtained three mean sea level values for each product and each date. The corresponding altimetry-based sea level time series were then compared with the tide gauge data for validation.

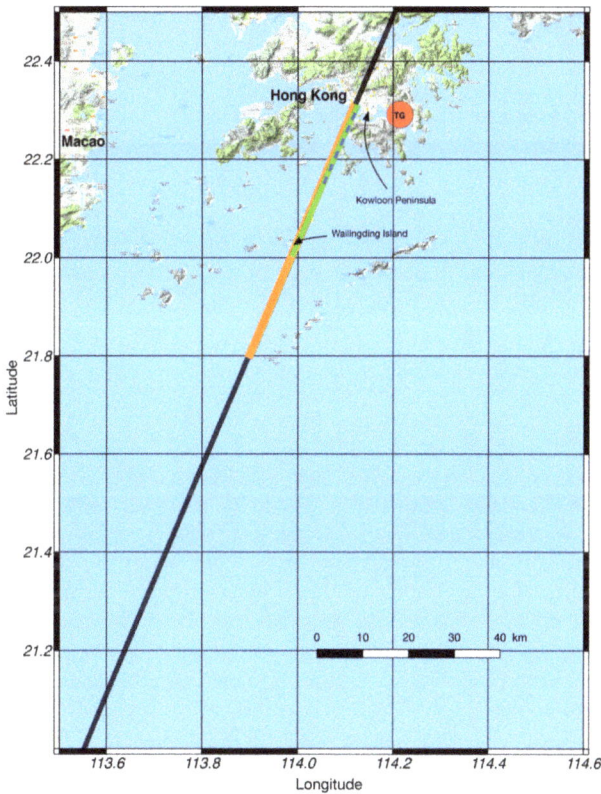

Figure 2. Map showing the study area, the selected Jason-2 pass 153 (black and colored line) and the Quarry Bay tide gauge (red circle). The latter is located ~10 km away from the Jason-2 pass. The along-track sections corresponding to study cases 1, 2, and 3 are also indicated (orange, green, and blue-green dashed lines, respectively). The background map is from Google Earth.

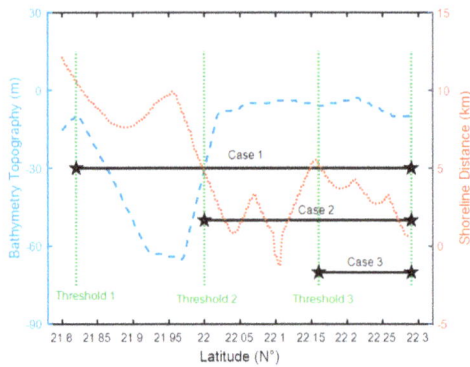

Figure 3. Bathymetry and shoreline distance at the Hong-Kong (HK) coast along 3 segments of the Jason-2 pass identified as cases 1, 2, and 3 (see location in Figure 2).

3. Data Sets

The Jason-2 mission started operating in July 2008, while the high-quality tide gauge data were accessible until 2014 only. Besides, the ALES data were available for cycle 1 to 252 (July 2008 to May 2015). These constraints define the study time period: from July 2008 to December 2014, corresponding to cycles 1-238 of Jason-2 satellite, i.e., a time span of 6.5 years.

3.1. Coastal Altimetry Data Sets

3.1.1. General Description

The experimental coastal Jason-2 products analyzed in this study are X-TRACK (CTOH/LEGOS, [6]), ALES (PODAAC, [3]) and PISTACH (CNES, [8]). These data sets are available from the following websites: ftp.legos.obs-mip.fr; ftp://podaac.jpl.nasa.gov/allData/coastal_alt/L2/ALES and ftp://ftpsedr.cls.fr/pub/oceano/pistach.

We have to keep in mind the differences between the products considered in this study. X-TRACK is a level 3 (L3) product: using the GDR data and state-of-the-art altimetry corrections, along-track sea level time series projected onto reference tracks (points at same locations for every cycle) are computed at 1-Hz (~6 km along-track resolution). Both a raw and a spatially filtered version (40 km cutoff frequency) of the product are distributed. It is simple to use, and is based on improved geophysical corrections near the coast (see [6] for details) but its current version only includes the standard MLE4 retracker adapted to open ocean conditions. ALES and PISTACH are level 2 (L2, i.e., cycle-by-cycle) products: the measurements are provided at 20-Hz (~0.3 km along-track resolution) and are not projected onto reference tracks, which means that their location varies from one cycle to another (this processing step needs to be done by the user). The ALES and PISTACH products do not include improved geophysical corrections, but provide altimeter range estimates obtained using different retracking algorithms (Section 3.1.2). PISTACH also includes a classification of the altimeter waveforms which can be used to study the type of waveform analyzed. It also indicates if the waveform shape is consistent to the retracker used.

3.1.2. Waveform Retrackers

Waveform retrackers can be classified into two categories: physically-based retrackers and empirical retrackers [12]. The model-free retracker is purely based on the statistics of the waveforms and does not require any echo model. Among the multiple model-free retrackers available, the improved threshold retracker [13] is usually considered as the best one. It combines the advantages of the OCOG (Offset Center of Gravity, [14]) and simple threshold [15] algorithms.

Over the last decade, a new approach called "sub-waveform" has been developed. It consists of fitting only the portion of the waveform that contains the leading edge but excludes the trailing edge where most artifacts appear (e.g., [3,8,16–18]).

The retrackers that are available in the different L2 products are summarized in Table 1. The standard GDRs include two solutions: MLE3 and MLE4. The rationale of these two retrackers is similar: fitting the waveform to a Brown model based on the MLE (essentially non-linear least squares) techniques. The main difference between them is the number of parameters included in the model. MLE3 estimates three parameters: epoch (i.e., altimetric range), Significant Wave Height (SWH) and amplitude (i.e., backscatter coefficient-sigma-0), while MLE4 also retrieves the square of off-nadir angle.

Table 1. Overview of different retrackers applied in different altimetry products.

Retracker	Product	Idea	Sub-Waveform	Comments
MLE4	SGDR [1]	Brown model	No	Official standard retracker.
MLE3	SGDR [1]	Brown model	No	
OCE3	PISTACH	Brown model	No	Same as MLE3
RED3	PISTACH	Brown model	Fixed: bins: $t_0 + [-10:20]$	Simplified version of ALES
ALES	ALES	Brown model	Adaptive to the SWH	Two-pass retracker
ICE1	PISTACH	Modified threshold	No	
ICE3	PISTACH	Modified threshold	Fixed: bins: $t_0 + [-10:20]$	

[1]: SGDR contains all GDR parameters, plus some supplementary items such as the retracker outputs and waveforms.

The PISTACH products provide four retrackers: OCE3, RED3, ICE1, and ICE3 [8]. OCE3 is essentially the same as the MLE3 parameters in the GDRs. ICE 1 is a modified threshold retracker. RED3 and ICE3 are the counterparts of OCE3 and ICE1 respectively, where the retracker is executed in a sub-waveform version instead of on the entire waveform.

ALES is based on an advanced retracker called sub-waveform retracker [3]. It is an improved version of RED3, and the difference between them is the estimation of the sub-waveform. In RED3, the sub-waveform has always 31 bins ($t_0 + [-10:20]$, where t_0 is the 32nd bin of the entire waveform), while in ALES, the sub-waveform length can vary from 39 bins (for SWH = 1 m) to 104 bins (i.e., the entire waveform, for SWH $\geq$17 m). In practice, the content of the ALES product is the same as the standard SGDR product (Sensor GDR) plus seven additional parameters specific to to the ALES retracker.

3.1.3. Geophysical Corrections

In PISTACH and X-TRACK, state-of-the-art geophysical corrections other than those of the official GDR are provided. For X-TRACK, only the ocean tide solution and the DAC are provided individually, while in PISTACH, two to three values are given for each correction. Different sets of correction terms obviously lead to different coastal sea level estimates.

3.2. Tide Gauge Data

The Quarry Bay tide gauge is a float-type instrument. It provides sea level data with an accuracy of 1 cm for a single measurement and is regularly calibrated every other year [19]. The tide gauge is located at 114.22°E, 22.28°N, near the northern coast of the HK Island, separated from the Kowloon Peninsula by the Victoria Harbor (see Figure 2). Note that ~95% of the Victoria Harbor shoreline is shaped by human activity [11]. Thus sea level on this area is likely influenced by anthropogenic local-scale factors, in addition to more regional and global ocean variations. Hourly tide gauge data were downloaded from the Sea Level Center of the University of Hawaii (https://uhslc.soest.hawaii.edu).

A harmonic analysis was first applied to the tide gauge data in order to compute and remove the tidal signals from the sea level time series. A time-averagedsea level value was also removed from the time series in order to be consistent with the altimetry sea level data. Finally, the hourly tide gauge data were interpolated to the time of the Jason-2 observations. Note that the dynamic atmospheric correction was not removed from the tide gauge data. This will be discussed in Section 5.5. The tide gauge-based sea level time series interpolated to the closest Jason-2 observations is shown in Figure 4. We observe a large seasonal cycle, due to the monsoon, modulated by important high-frequency variations which can reach several tens of cm. In particular, a peak is observed at cycle 228. It is caused by a storm surge associated with the violent Typhoon Kalmaegi that sideswiped the HK coast before dawn on 16 September 2014. The Jason-2 altimeter flew over the HK area at 3.45 am (local time) on 16 September 2014, and the peak in the tide gauge sea level series is coincident with the typhoon event. Therefore, this peak was eliminated in our analysis as an outlier.

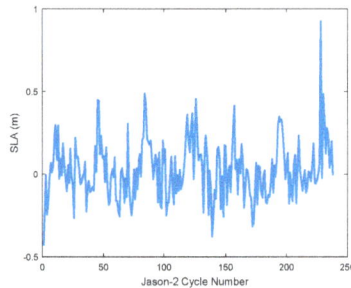

Figure 4. HK tide gauge sea level time series (in meters) interpolated to Jason-2 observations.

3.3. ESA Climate Change Initiative (CCI) Sea Level Data

Global altimetry sea level data sets are regularly produced by different groups worldwide. Five of them are solely based on the TOPEX-Poseidon/Jason-1/2/3 satellites (reference missions): AVISO (Archivage, Validation et Interprétation des données des Satellites Océanographiques) in France, NOAA (National Oceanic and Atmospheric Administration), GSFC (Goddard Space Flight Center) and CU (University of Colorado) in the United States, and CSIRO (Commonwealth Scientific and Industrial Research Organization) in Australia. The Climate Change Initiative (CCI) sea level (sl_CCI) project of the European Space Agency (ESA) has been developed in the recent years to realize the full potential of multi-mission altimetry as a significant contribution to climate research [20–22]). In the context of the CCI, altimetry missions have been reprocessed using improved geophysical corrections and improved links between missions, providing reduced errors of sea level products. In this study, we use the sl_CCI (v 2.0) data from July 2008 to December 2014 to estimate sea level trends in the open ocean away from the HK coast. The sl_CCI v 2.0 data set is a gridded product at 0.25° × 0.25° resolution. The Sl_CCI project was not dedicated to estimate sea level near the coast. However, it can be used as a good independent reference to link open ocean sea level variations to coastal altimetry sea level, thus providing a regional context to interpret the coastal results obtained in this study.

4. Methodology

4.1. Altimetry Data Processing

As indicated above, current coastal altimetry products differ in terms of content. Thus a first step consists of processing the different data sets to obtain homogeneous variables for further comparison. Because there is no waveform data in PISTACH, we used the waveforms provided in ALES, and merged PISTACH and ALES using the measurement time common to all products. In Section 5.4, we also projected all along-track, cycle-by-cycle L2 data onto the X-TRACK 1-Hz reference grids to benefit from the XTRACK improved geophysical corrections.

4.1.1. Sea Level Anomaly (SLA) Estimation

From the altimeter range, once all the propagation and geophysical corrections mentioned below are removed, we can deduce the sea surface height (SSH, i.e., sea level referred to a reference ellipsoid). If we further remove a mean sea surface in order to overcome the problem of geoid estimation, we obtain sea level anomaly (SLA) data. In this study we use SLA data, computed as follows:

$$SLA = H - R - \Delta R_{iono} - \Delta R_{dry} - \Delta R_{wet} - \Delta R_{ssb} - \Delta R_{tide} - \Delta R_{DAC} - MSS \qquad (1)$$

In Equation (1), H is the orbital height, R is the Ku-band altimeter range, ΔR_{iono} is the ionospheric correction, ΔR_{dry} and ΔR_{wet} are the dry and wet tropospheric corrections, respectively, ΔR_{ssb} is the sea state bias, ΔR_{tide} is the tide correction (composed of the ocean tide, pole tide and solid Earth tide), ΔR_{DAC} is the dynamic atmospheric correction, and MSS is mean sea surface (in practice we used the MSS_CNES_CLS-2011 model).

In some cases, R is not directly available (for instance, there is no specialized range parameter in the ALES product), so it can be computed as follows:

$$R = T + E \times (c/2) + D + M + 0.180 \qquad (2)$$

where T is the onboard tracking range in the Ku band, E is the retracked offset (with time dimension), the factor "$c/2$ (c is light velocity)" is the scaling factor from traveling time to range, D is the Doppler correction in Ku band, M is the instrument bias due to PTR and LPF features ([23,24] and 0.180 is a bias (in meters) due to wrong altimeter antenna reference point [25].

In this study, we computed SLA time series using the altimeter ranges issued from six retrackers: ALES, MLE3, MLE4, RED3, ICE1, and ICE3. For the MLE3, MLE4, and ALES retrackers, we first used Equation (2) to compute R, and then applied Equation (1) to compute the SLA (because R is often flagged in the GDR and ALES products). To validate our calculation method, we compared the MLE4 SLA obtained with the equivalent official "*ssha*" parameter provided in the GDRs, and found good consistency. For RED3, ICE1, and ICE3, we applied Equation (1) to directly compute the SLA, because there were no valid E values in the PISTACH product.

4.1.2. Choice of the Geophysical Corrections

The geophysical corrections near the coast also need specific considerations. The first error source comes from the wet tropospheric correction because the onboard radiometer suffers from land contamination in the coastal area. A simple but effective approach is to extrapolate a model-based correction (using for example atmospheric reanalyses from the European Center for Medium-Range Weather Forecasts, ECMWF) but the corresponding spatial resolution is relatively low for coastal applications. Other approaches include an improved radiometer-based correction accounting for the land contamination effect [26], or the computation of GNSS-derived Path Delay (GPD, [27]). Since the GPD has not been included in the current versions of the three coastal products analyzed here, we used the decontaminated radiometer solution provided in PISTACH.

Concerning the ionospheric correction, the imperfect coastal altimeter range measurements lead to significant errors, generating outliers in the correction values. We use the MAD (median absolute deviation) technique described in detail in Section 4.1.3 to detect and remove the outliers. The along-track profile of ionospheric corrections is further spatially low-pass filtered using a LOESS (LOcally Estimated Scatterplot Smoothing) method with a cutoff frequency at 100 km.

The coastal ocean tide corrections, provided by global models, are also far from accurate. There are five different ocean tide solutions available in the products, computed by two scientific teams: (1) the Goddard Ocean Tide (GOT) models developed by Ray et al. [28], and the Finite Element Solution (FES) models developed by Lyard et al. [29]. Two ocean tide solutions are provided in the official GDRs:

GOT4.8 and FES2004. PISTACH contains two older versions of GOT: GOT00.2 and GOT4.7 as well as an upgraded version of FES: FES 2012. In Equation (1) we adopt the GOT4.8 ocean tide solution.

Ray [30] compared different tide solutions against 196 shelf-water tide gauges and 56 coastal tide gauges. Their accuracy was characterized by the RSS (root sum square) error of the eight main tidal components (Q1, O1, P1, K1, N2, M2, S2, K2). For the shelf-water gauges, the accuracy of GOT4.8 was 7.04 cm along European coasts and 6.11 cm elsewhere, while the accuracy of FES2012 was 4.82 cm and 4.96 cm respectively for the European coasts and elsewhere. For the coastal tide gauges, the accuracy of GOT4.8 and FES2012 were 8.45 cm and 7.50 cm respectively. In comparison, the accuracy of GOT4.7 and FES2004 in shelf-water were 7.77 cm and 10.15 cm respectively. These results illustrate the significant improvement in coastal ocean tide solution during the last 10 years. In Section 5.5, we also compare three tide solutions, GOT4.8, FES 2004, and FES 2012, to assess their relative performances.

Another important altimetry correction is the sea state bias (SSB). The SSB depends on the retracking algorithm, because it contains the tracker bias. A careful analysis showed that for Jason-2 GDRs, the SLA obtained from MLE3 and MLE4 retrackers has large bias. From a statistical analysis based on cycles 1 to 238 for a couple of altimeter passes over the open ocean, we obtained: $SLA_{MLE3} - SLA_{MLE4}$ = +2.3 cm. Near the coast, this bias appeared to be even larger and even more critical as it was not constant. Figure 5 shows both MLE3 and MLE4 SSB corrections as a function of SWH for an arbitrary pass (cycle 16, pass #153). MLE3 SSB has a clear bias (~+3 cm) relative to MLE4 SSB. Moreover, MLE3 SSB seems to have many outliers, in particular near the coast. We concluded that the bias observed between MLE3 and MLE4 sea level estimates corresponds to a bias in the SSB corrections.

Deeper investigation showed that the MLE3 SSB outliers are often related to large off-nadir angle values (not shown), probably erroneous given the good attitude control of Jason-2. For that reason, we adopted the MLE4 SSB in the computation of all SLAs, resulting in a relative bias <1 cm for all retrackers.

In Equation (1), some parameters are available at 20-Hz (e.g., E), and others at 1-Hz (e.g., D). We interpolated all 1-Hz parameters at 20-Hz and finally computed 20-Hz SLA. The 20-Hz SLA is more useful for the retracker performance analysis, and can also be used for subtle feature detection.

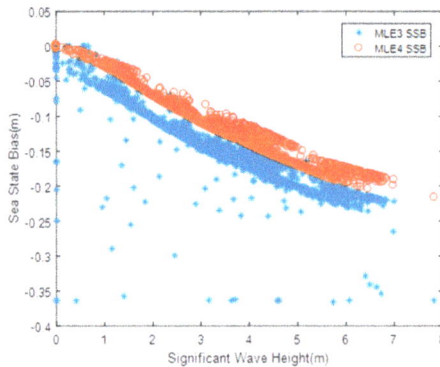

Figure 5. Sea state bias (SSB) difference with respect to Significant Wave Height.

4.1.3. Data Editing Strategy

Near the coast, the number of outliers in data sets is obviously much larger than over the open ocean. Thus the data must be edited before being used. The outliers are detected and removed using the MAD filter. It is based on the median value of the data analyzed, which is more robust than the mean value in the presence of outliers. For a Gaussian random signal, MAD is defined as:

$$MAD = \sqrt{\frac{2}{\pi}} \cdot \sigma \qquad (3)$$

where σ is the standard deviation of the data. For the Chauvenet's criterion (0.5% outliers for a Gaussian series, [31]), it will be appropriate to set a threshold of $2.81 \times \sigma$, equivalent to $3.52 \times MAD$.
Therefore, values satisfying:

$$|x - Median| > 3.52 \times MAD \qquad (4)$$

are defined as outliers. This is very similar to the criterion ($3.5 \times MAD$) adopted by Birol et al. [6] in the ionospheric correction editing process of the X-TRACK product.

Objectively, all outliers cannot be thoroughly detected by the above criterion. In the presence of distorted waveforms, retrackers such as ALES can occasionally produce unrealistic SLA values, up to tens of meters and hence increase the MAD. Some outliers may remain after editing in this case. So we have defined a threshold value before applying the MAD editing. All the SLA values beyond ± 2 m are deleted. Some high amplitude real oceanographic features (such as storm surge) may also be deleted. But for the purpose of the present study, the storm surge events can be removed from the analysis. Finally, almost all outliers are detected after this threshold is applied.

A typical along track profile of 20-Hz SLA before and after editing is shown in Figure 6 for a given cycle (cycle #5 is arbitrarily chosen here). ALES is not as robust as the MLE algorithms for the last few coastal measurements: it has many outliers in almost all cycles (in fact, the mean number of outliers beyond $\pm$ 10 m in ALES is ~9 per cycle, much larger than for MLE3 or MLE4). However, after editing, ALES performs better than MLE3 and MLE4, since it displays the lowest noise level.

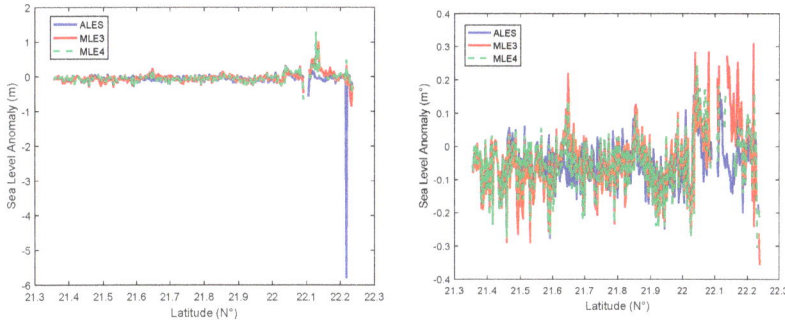

Figure 6. 20-Hz sea level anomaly (SLA) before (**left**) and after (**right**) editing. Jason-2 cycle 5 is considered here.

4.2. Sea Level Data Analysis

After computing the SLA for each cycle and each retracking, the time series can be analyzed to retrieve useful oceanography information. Because of the monsoon, the annual and semi-annual signals are both significant near the HK coast. Therefore, at a first order, SLA variations can be modeled as follows:

$$SLA(t) = a_1 \cos(2\pi t / T_{year}) + a_2 \sin(2\pi t / T_{year})$$
$$+ a_3 \cos(4\pi t / T_{year}) + a_4 \sin(4\pi t / T_{year}) + a_5 t + a_6 + \varepsilon(t) \qquad (5)$$

where T_{year} = 365.2425 days, $\varepsilon(t)$ is the residual SLA, a_1 to a_6 are the regression coefficients to be estimated: a_6 is a bias, a_5 is the sea level trend, the annual/semi-annual amplitude and phase can be deduced from a_1 to a_4:

$$A_{annual} = \sqrt{a_1^2 + a_2^2}; A_{semi-annual} = \sqrt{a_3^2 + a_4^2} \qquad (6)$$

$$\Phi_{annual} = \arctan(a_2/a_1); \Phi_{semi-annual} = \arctan(a_4/a_3) \qquad (7)$$

Coefficient uncertainty is defined as the square root of the diagonal elements in the covariance matrix of the coefficient vector.

In practice, we estimated the seasonal terms and trend separately: the seasonal coefficients were removed first from the detrended time series; then the short-term trend was estimated from the residuals (i.e., initial time series corrected for the seasonal terms) after reintroducing the initial trend. In the latter step, a nine-point moving window (corresponding to an about 3-month cutoff frequency) was applied to the residuals to reduce the intrinsic 59-day erroneous signal presented in Jason altimetry missions [32,33].

Note that the computed trends based on only 6.5 years of data essentially represent the inter-annual variability. These short-term trends are expected to be significantly different for long-term trends estimated over the whole altimetry era. In the following the term 'trend' means 'short-term trend'. In this work, it is used as a diagnostic to analyze the quality of near-coastal altimetry data and not the long term sea level trend related to climate change.

5. Results

5.1. A First Analysis Based on Jason-2 Waveforms Observed along the Track

The HK coast considered in this study has a very complex topography, so the waveforms are diverse. In order to go further in our analysis we used the waveform classification provided in the PISTACH product. It classifies the altimeter waveforms into 16 classes, including a "doubt" class. Based on considerations on all waveform shapes, we grouped these classes into five categories (see Table 2). In Table 2, for each category, we provide a general description of the surface observed and propose a specific retracker.

Table 2. Overview of the five categories defined from the 16 PISTACH classes.

Category	Waveform Characteristics	Possible Surface Observed	Retracking Strategy
Brown	The waveform is close to the Brown model.	The land contribution in the altimeter footprint is null or small.	MLE4
Distorted Brown	The waveform is similar to the Brown model, but with distortions (either with an increasing plateau, or with a sharply decreasing plateau, or with a too broad noise floor).	There is land signature at the fringe of the altimeter footprint, with different reflection than over open ocean surface.	ALES
Peak dominated	One (or a few) large peak(s) dominate(s) the waveform; there may be a Brown shape in the waveform, but its maximum power is much lower than the peak(s).	There are one or a few strong bright targets (e.g., extremely calm water surface or effective corner reflector, see [34]) within the altimeter footprint.	MLE based on Gaussian model or improved threshold
Brown + Peak	The waveform is a mixture of Brown shape and one (or a few) peak(s) with comparable power levels. The location of the peak can be in the leading edge or the plateau.	The portions of ocean and land surfaces within the altimeter footprint are equivalent.	MLE based on BAGP (Brown with Asymmetric Gaussian Peak [35])
Others	Unexplained waveform patterns (e.g., very noisy echoes or linear echoes).	Unexplainable surface features (e.g., very composite geomorphology or some extreme events).	Should be rejected.

We use these five categories to classify all Jason-2 waveforms used in our study. The resulting percentages of waveforms as a function of category and case (i.e., distance to the coast) are shown in Figure 7. In all three cases the most frequent category is "Brown", and the second is "Distorted Brown". Not surprisingly, the percentage of Brown-like waveforms decreases when approaching land. In case 2, more than 52% of the waveforms are Brown-like and in case 3 this number decreases to 37%. Concerning the distorted-Brown waveforms, logically, their percentage increases when the distance to the coast decreases. It varies from 17% in case 1 to 27% in case 3. These results suggest that within a shoreline distance of 5 km, a significant number of waveforms are Brown-like and that using a classical open-ocean retracker can still provide good quality sea level data in this very nearshore area. However, the use of ALES is expected to increase the number of retrieved accurate data, which is what we found in the previous section.

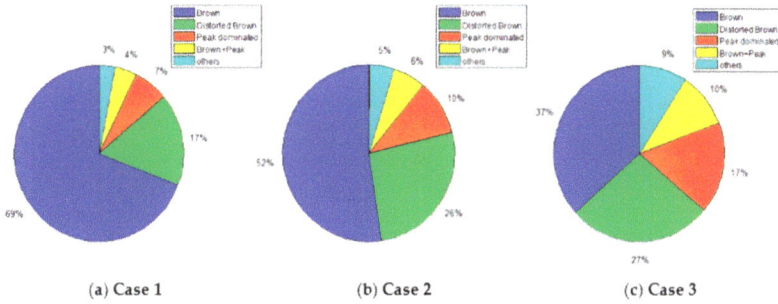

(a) Case 1   (b) Case 2   (c) Case 3

Figure 7. Percentages of waveform categories in the 3 cases.

5.2. Sea Level Trend

5.2.1. A regional View

We first used the gridded CCI sea level data to obtain a regional picture of the sea level trends from July 2008 to December 2014 (Figure 8a). To compare the gridded CCI sea level with the coastal sea level estimated in this study (see below), we interpolated the CCI trend grid along the Jason-2 pass at 1-Hz resolution (~7 km). Corresponding CCI-based along-track sea level trends are shown in Figure 8b. The uncertainty associated with the trend estimation is also provided (blue bars). These error bars are rather large because the time span of analysis is short and the inter-annual variability is large. Trend estimates using the whole CCI altimetry record (i.e., 23-years, from January 1993 to December 2015) display much smaller errors, but a behavior similar to Figure 8b is also observed, i.e., larger errors around 16°N than near the coast.

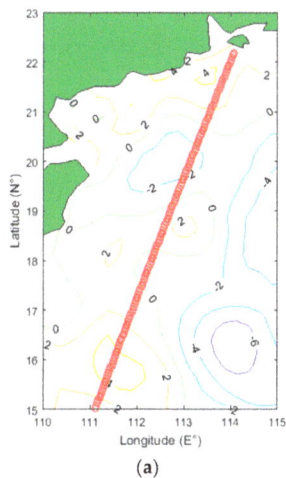

(a)

Figure 8. *Cont.*

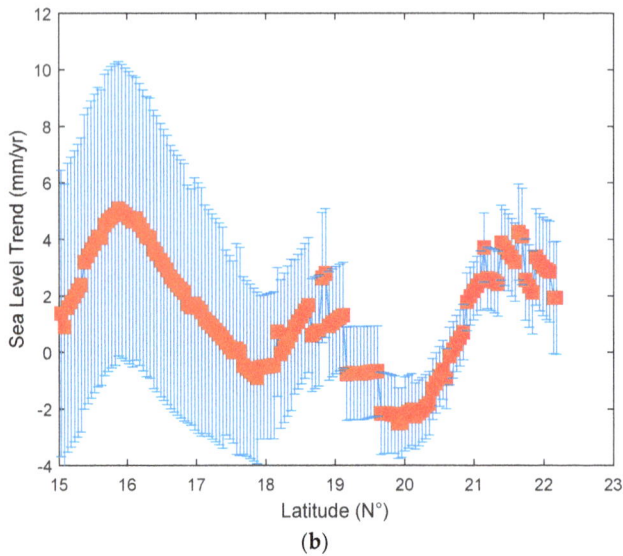

Figure 8. Regional 6.5 yr-long trends in sea level derived from the CCI sea level data. (**a**) using gridded data and (**b**) interpolated along the Jason-2 pass 153. Vertical blue bars in Figure 8b represent the uncertainty on the trend estimation.

In parallel, in order to see the impact of the different altimetry products used here (thus of the different data processing) on the results, the linear 6.5-year trends were also estimated along the Jason-2 pass from both the raw and filtered versions of the X-TRACK SLA time series. Results obtained from the filtered X-TRACK SLA are shown in Figure 9 as a function of latitude.

From Figures 8 and 9, we observe a significant spatial variability in the regional sea level trends. The CCI map (Figure 8a) displays values of −6 to −4 mm/yr over the open ocean which tend to increase to +2 to +4 mm/yr near the coast. Figure 8b illustrates how Jason 2 pass 153 captures part of this regional trend pattern, with values varying significantly from −2 mm/yr to +4 mm/yr along the track. In X-TRACK (Figure 9), around 20°N, the linear trend is also relatively small (~2–3 mm/yr), in agreement with the CCI data. A change is observed near 19°N, corresponding to the along-track influence of a narrow trench (the bathymetry rapidly decreases to ~−2500 m and then recovers to ~−100 m). As in the CCI data, the X-TRACK trends increase towards the coast but the values obtained in the vicinity of HK are much larger than for the CCI trends: +8 to +9 mm/yr against +2 mm/yr. The sea level trend at the last point of the filtered X-TRACK SLA is: +8.6 ± 2.2 mm/yr. However, gridded altimetry products, such as in the CCI data set, have too low resolution to capture near-shore sea level variations. This also explains why the trends of the gridded CCI data have larger uncertainty than the unsmoothed X-TRACK data.

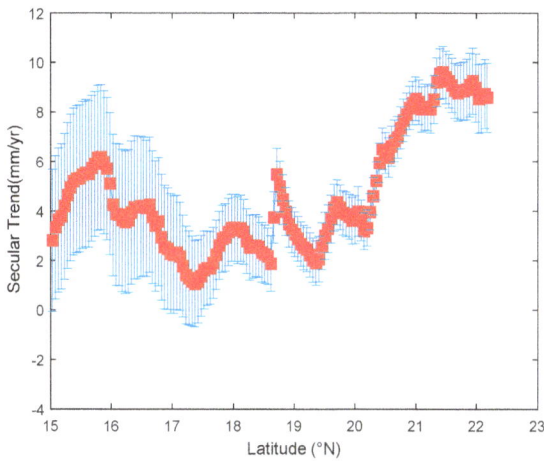

Figure 9. Regional 6.5 yr-long trends in sea level derived from filtered X-TRACK data along the Jason-2 pass 153. Vertical blue bars represent the uncertainty on the trend estimation.

5.2.2. The Tide Gauge Reference

Using Equation (5), we computed the tide gauge trend over the 6.5-yr time span considered in this study. We found a linear sea level trend value of +5.2 ± 2.0 mm/yr (after eliminating the outlier corresponding to the storm-surge event). However for further comparison with the altimetry-based sea level data, the tide gauge time series needs to be corrected for vertical land motion (VLM) [36]. This can be done using GPS precise positioning techniques. Yuan et al. (2008) [37] analyzed data from a GPS station named HKOH, which is close to the tide gauge. VLM at HKOH was estimated to +0.22 ± 0.46 mm/yr. Lau et al. (2010) [38] analyzed GPS data (2006–2009) from HKOH and obtained a VLM estimate of +0.3 mm/yr (uncertainty not provided). The time spans of these studies are different from ours, but because the VLM at HK appeared to be fairly stable over the last decade, we assume a VLM value of +0.3 mm/yr. After correcting for VLM, we find a trend of +5.5 ± 2.0 mm/yr at the tide gauge site.

5.2.3. Solutions Derived from the Different Retrackers

For each of the three cases defined in Section 2 and each retracker, we computed a spatially averaged 20-Hz SLA time series (see Section 4.1.1) as well as the associated 20-Hz noise level (estimated by the standard deviation of the 20-Hz SLA series). The results are shown in Figure 10. In all cases, ALES solution provides the lowest noise level after editing, and MLE4 is slightly less noisy than MLE3. Concerning the three experimental retrackers used in PISTACH, ICE3 has the lowest noise level, and RED3 is slightly less noisy than ICE1.

Note that a problem was detected in some of the PISTACH SLA solutions: around cycle #150, both ICE3 and RED3 have much larger SLA than ICE1, the relative bias being about 0.2 m. Comparison with the tide gauge sea level showed that ICE1 SLA roughly followed the tide gauge data, while ICE3 and RED3 data display large jumps. The latter severely influences the corresponding sea level trend estimates.

Sea level trends obtained from Equation (5) are summarized in Table 3 (except for OCE3 in PISTACH, which is the same as MLE3).

Table 3. Estimated linear trend and associated uncertainty (mm/yr) as a function of sea level data source and case.

Data Source	Case 1	Case 2	Case 3
ALES	+5.9 ± 1.5	+9.7 ± 1.6	+17.3 ± 2.3
MLE3	+5.0 ± 1.6	+8.1 ± 2.2	+2.6 ± 3.1
MLE4	+4.2 ± 1.6	+4.3 ± 2.2	+5.5 ± 3.5
ICE1	−29.1 ± 2.4	−27.5 ± 2.8	−22.9 ± 4.0
ICE3	+57.5 ± 2.3	+60.1 ± 2.5	+52.9 ± 3.0
RED3	+55.3 ± 2.1	+58.0 ± 2.3	+58.5 ± 3.0
XTRACK		+8.6 ± 2.2	
Tide Gauge (in-situ reference)		+5.2 ± 2.0	
Tide Gauge (After VLM correction)		+5.5 ± 2.0	
Regional trend from the CCI data		2.7 ± 2.0	

From Table 3 we note that, using the tide gauge as reference, case 1 provides the closest estimate to the tide gauge trend. Besides, we note a difference of less than 1 mm/yr for ALES and M LE3. This can easily be explained. As shown in Figure 2, case 2 includes a number of small islands while case 1 contains more land-free areas, thus more accurate SLA data. Finally, case 3 has very few valid altimetry measurements (usually less than five valid points), which is not enough to provide robust results.

MLE3 and ALES trends are both close to the tide gauge trend (within 0.5 mm/yr). The trends estimated from MLE4 are slightly lower than for ALES and MLE3 but the difference is within the error bar. The trends deduced from the PISTACH retrackers highly disagree with the tide gauge trend: both ICE3 and RED3 show unrealistic large values (>+5 cm/yr), while ICE1 shows a negative trend of −2 cm/yr. The ICE1 retracker may be inherently not accurate enough to derive trends, but concerning ICE3 and RED3 retrackers, since the processing procedure used in this work is homogeneous, the large errors may not come from the retracker algorithm itself, but more likely from errors in the basic products. For example, in a new PISTACH version, the 18 cm calibration bias used in Equation (2) may no longer be applied in the computation of R. Since the retracked offset E parameter is no more accessible in the current PISTACH version, this hypothesis could not be verified. Anyway, in the remaining part of the study we discard ICE1, ICE3, and RED3 solutions and concentrate on MLE3, MLE4, and ALES which, in the context of our study, appear as the best available retrackers to capture coherent coastal sea level signals from altimetry.

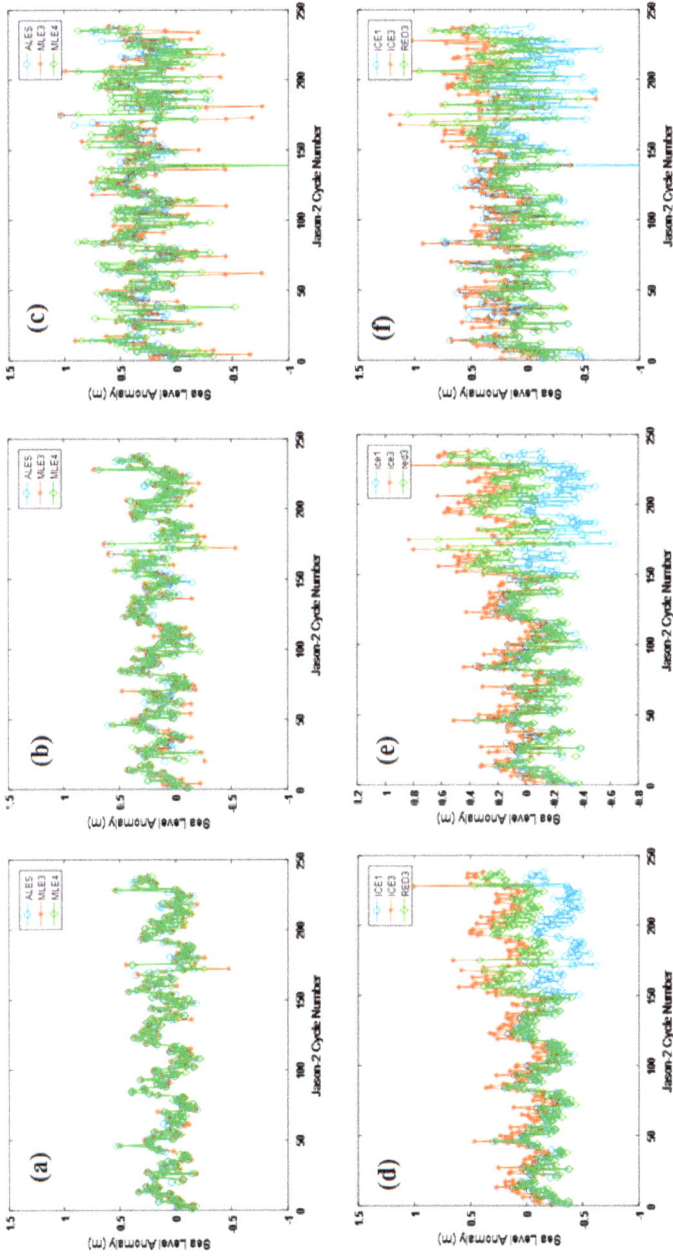

Figure 10. *Cont.*

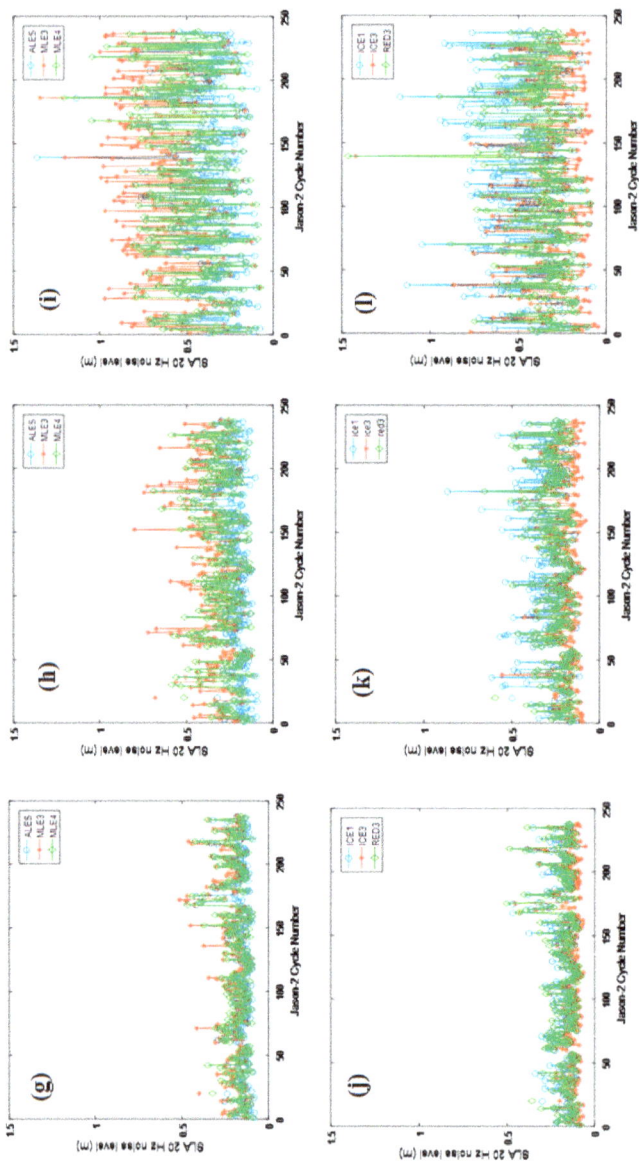

Figure 10. SLA means and noise levels. (**a**) Case 1, mean SLA for ALES, MLE3 and MLE4; (**b**) Case 2, mean SLA for ALES, MLE3 and MLE4; (**c**) Case 3, mean SLA for ALES, MLE3 and MLE4; (**d**) Case 1, mean SLA for ICE1, ICE3 and RED3; (**e**) Case 2, mean SLA for ICE1, ICE3 and RED3; (**f**) Case 3, mean SLA for ICE1, ICE3 and RED3; (**g**) Case 1, noise levels for ALES, MLE3 and MLE4; (**h**) Case 2, noise levels for ALES, MLE3 and MLE4; (**i**) Case 3, noise levels for ALES, MLE3 and MLE4; (**j**) Case 1, noise levels for ICE1, ICE3 and RED3; (**k**) Case 2, noise levels for ICE1, ICE3 and RED3; (**l**) Case 3, noise levels for ICE1, ICE3 and RED3.

5.3. Coastal Seasonal Signal along the Jason-2 Pass

Using Equation (5) we also computed the amplitude and phase of the seasonal signal for all sea level time series. The resulting values are shown in Table 4 (for the annual signal) and Table 5 (for the semi-annual signal). All three cases give rather similar results that agree well with X-TRACK. The annual phases lie between 300°–360° and are significantly larger than the tide gauge-based phase. Amplitudes are also slightly larger. The semiannual phases lie between 220°–270° and are close to the tide gauge-based phase. Amplitudes are also slightly larger. We cannot exclude that some local seasonal signal at the tide gauge site is responsible for the difference observed between altimetry and tide gauge data.

Table 4. Estimated annual amplitude (cm) and phase (degree) as a function of sea level data source and case.

Data Source	Case 1	Case 2	Case 3
ALES	13.05/344	12.49/348	11.19/2
MLE3	13.29/338	12.39/339	13.17/343
MLE4	12.96/339	11.44/340	11.12/348
XTRACK		13.23/338	
Tide Gauge		11.46/311	

Table 5. Estimated semiannual amplitude (cm) and phase (degree) as a function of sea level data source and case.

Data Source	Case 1	Case 2	Case 3
ALES	6.03/235	5.67/239	7.38/260
MLE3	6.17/241	6.76/252	9.24/270
MLE4	6.02/236	6.96/245	11.80/262
XTRACK		6.81/223	
Tide Gauge		7.62/236	

5.4. Relative Performances of MLE4, MLE3, and ALES Near Hong Kong

Here, we examine the relative performance of the MLE3, MLE4, and ALES retrackers in the H-Kcoastal zone. Because ALES and PISTACH are L2 products, we used the L3 X-TRACK product to project ALES and PISTACH MLE3 and MLE4 SLA onto regular 1-Hz reference points along the track, allowing us to obtain SLA time series over the study period. Figure 11a shows the percentage of valid measurements obtained for the different retrackers. MLE3 seems to be the most robust near the coast in the sense that it provides more valid data. Figure 11b shows the time-averaged SLA values for the three retrackers. Over the open ocean, ALES and MLE4 agree well (within 1 cm); MLE3 has a negative bias of 1–2 cm. Near the coast, ALES shows less variations, while MLE3 and MLE4 display large peaks likely due to retracking errors. Figure 11c shows the trend estimates for the different retrackers. ALES displays the smoothest pattern, which seems to indicate a better performance than MLE3 and MLE4.

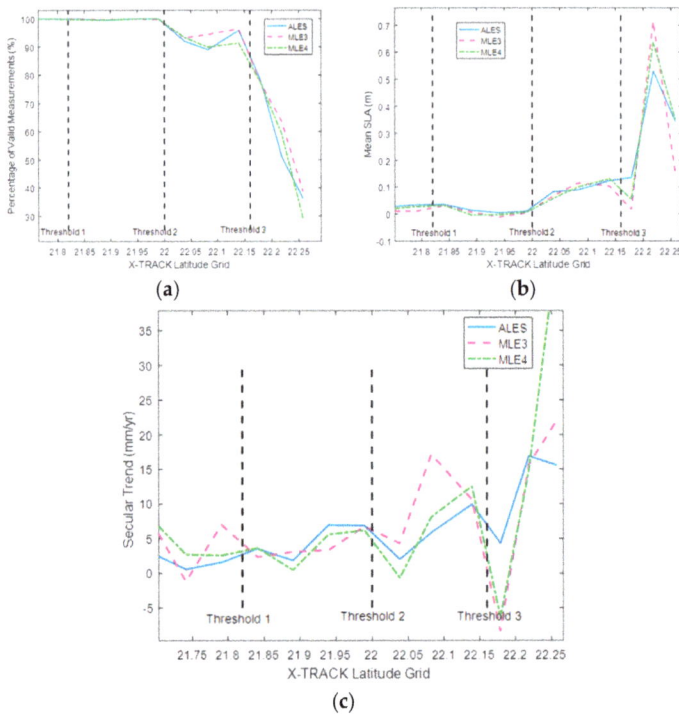

Figure 11. (**a**) Percentage of valid measurements; (**b,c**) mean values and 6.5-yr trend estimates for ALES, MLE3, and MLE4 retrackers. Results are presented as a function of latitude.

The sea level residuals obtained after removing the trend and seasonal signal are shown in Figure 12 for MLE3, MLE4, ALES and the tide gauge data. A 3-month low pass filter was applied to the different SLA time series. For most cycles, the altimetry-based SLA have variations similar to the tide gauge, but the latter occasionally shows larger anomalies. In the first few cycles, the tide gauge residuals have surprising low values, and around cycle 135–160 have a few large negative peaks. Very local small-scale tides, waves, and currents may cause these sea level signals. The standard deviations of the altimetry SLA residuals with respect to the tide gauge residuals, before and after the 3-month smoothing, are given in Table 6. The improvement due to the smoothing is significant, the standard deviations decreasing by more than 50%. The consistency between the altimetry and tide gauge residuals is about 5 cm, which is encouraging given that the study area which is quite complex. ALES SLA has slightly larger standard deviation with respect to tide gauge sea level.

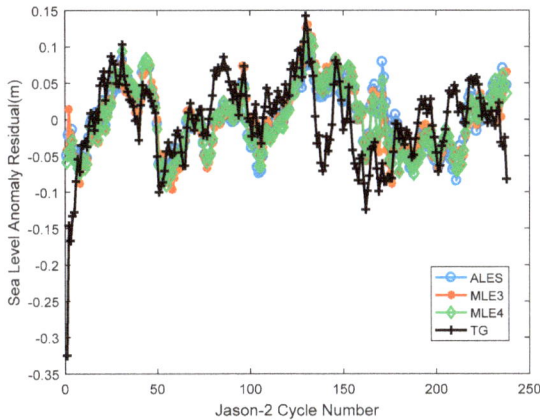

Figure 12. Detrended and deseasoned SLA time series based on ALES, MLE3, and MLE4, with 3-month smoothing (tide gauge SLA—noted TG—is shown as reference).

Table 6. Deseasoned and detrended SLA standard deviation w.r.t. tide gauge sea level (cm) for case 1.

SLA Series	ALES	MLE3	MLE4
Agreement	5.12	4.82	4.88

5.5. Impact of the Geophysical Corrections

As mentioned above, different solutions for some of the geophysical corrections are provided with the different coastal altimetry products. This is the case in particular for the wet tropospheric delay and ocean tides correction. In order to test their relative impact on the quality of the coastal altimetry sea level data, as well as to define the best choice of corrections, we use the different possibilities in Equation (1) to derive different sea level data sets (only one term varies in each case). Concerning the wet tropospheric correction, we tested both the decontaminated solution from the radiometer [26] and the composite correction derived from ECMWF and the radiometer. Concerning the tides, we considered both FES2004 and FES2012, besides GOT4.8. We also computed a version without the DAC. Given the results obtained above, we concentrated on case 1 area. Different diagnostics are considered to evaluate the impact of the correction choice:

(1) Secular trend and uncertainty (mm/yr) computed for different retrackers (results are shown in Table 7), and
(2) Standard deviation of the SLA residuals with respect to the tide gauge data (results are shown in Table 8).

Table 7. Secular trend and associated uncertainty (mm/yr) for case 1, with different geophysical corrections choices. DAC = dynamic atmospheric correction.

Choice	ALES	MLE3	MLE4
Standard choice (Table 4)	+5.9 ± 1.5	+5.0 ± 1.6	+4.2 ± 1.6
composite wet delay	+6.1 ± 1.5	+5.1 ± 1.6	+4.3 ± 1.6
FES 2004 ocean tide	+5.7 ± 1.5	+4.6 ± 1.6	+4.7 ± 1.6
FES 2012 ocean tide	+9.1 ± 1.3	+7.8 ± 1.5	+7.1 ± 1.5
No DAC applied	+6.3 ± 1.6	+5.1 ± 1.8	+4.5 ± 1.7

Table 8. Residual SLA, standard deviation w.r.t. tide gauge-based sea level (cm), for case 1, after 3-month smoothing, with different geophysical corrections choices.

Choice	ALES	MLE3	MLE4
Standard choice (Table 4)	5.12	4.82	4.88
composite wet delay	5.02	4.80	5.01
FES 2004 ocean tide	4.87	4.76	4.82
FES 2012 ocean tide	4.89	4.56	4.88
No DAC applied	4.84	4.56	4.76

Inspection of Tables 6 and 7 indicates that the impact of the wet tropospheric correction is rather low. For the trend estimates, the uncertainties are the same but the composite wet delay results are +0.1–0.2 mm/yr larger. Some values of the standard deviations of the differences w.r.t the tide gauge data increase, but some others decrease.

The impact of the ocean tide correction appears much more important. The use of FES tide solutions introduces slightly less residual errors than GOT 4.8. FES2012 decreases the trend uncertainties, but the trend estimate seems too large (large difference with respect to the tide gauge). Correlations for different tide solutions with respect to the tide gauge data are also shown in Table 9. All ocean tide solutions lead to excellent correlations between altimetry-based and tide gauge-based sea level. FES 2012 gives the highest correlation, FES2004 the lowest.

Table 9. Correlations of different tide solutions with respect to the tide gauge tide height.

Ocean Tide Solution	GOT 4.8	FES 2004	FES 2012
Correlation	0.91	0.89	0.95

Remember that we did not correct the HK tide gauge data from the DAC. Assuming that the tide gauge and altimeter roughly capture the same dynamic atmospheric effects, we should obtain better statistics if no DAC correction is applied to altimetry. It corresponds to what is observed in Table 8.

In conclusion, we note that the choice of the retracker and of the altimetry geophysical corrections, especially the tide correction, has a very significant impact on the estimated coastal sea level trend.

6. Discussion

There is a long-standing debate in the coastal altimetry community concerning the choice of the best retracking algorithms and still today, no consensus has emerged. From the results of this study, we consider that regarding the complexity of the corresponding radar echoes associated with the variety of existing coastal land surfaces, the choice of the retracker should be specific to the coastal area considered. We have however to keep in mind that this option does not suit for regional or global applications because we need homogeneous data sets, which would be difficult to obtain if one switches from one retracker to another as a function of the oceanic region considered. This would certainly produce bias in the estimated sea level.

Although the results obtained in this study with ALES in the HK coastal area are already encouraging, we believe that they could be further improved. For example, one of the most useful advances in altimeter waveform processing is the Singular Value Decomposition (SVD) approach proposed by Ollivier [39]. The idea of SVD is to eliminate the waveform components which are relative to the minor singular value, since those components probably arise from system noises rather than real ocean signals. Near the coast, the waveforms can be very different, so only the "Brown-like" waveforms were singled out (according to the "wf_class" parameter in the PISTACH product) to establish the waveform matrix. Results for cycle #100 are shown here as an arbitrary example. For cycle #100, there are 420 waveforms considered as ocean waveforms offshore HK, 342 of which are Brown-like. Therefore it is sufficient to carry out SVD processing. We tested two typical thresholds: 90% and 80%.

The 20-Hz epoch noise characteristics are provided in Table 10 (the 1-Hz noise is roughly similar to the 20-Hz one). The measurements are split into two regions: far from the coast (shoreline distance >10 km) and near the coast (shoreline distance ≤10 km). For both regions, the epoch noises are significantly improved when using the SVD. The present study is somewhat preliminary since we have not computed nor analyzed the SLA performances after the SVD processing, but this strategy looks like a serious option to consider in the future for coastal altimetry processing.

Table 10. 20-Hz epoch noise (cm) for different singular value decomposition (SVD) configurations (cycle #100).

Threshold Percentage	No SVD	90% SVD	80% SVD
Far from the coast	8.66	8.05	7.52
Near the coast	9.41	8.90	8.60
The last ocean-like waveform	39.17	36.18	34.68

7. Conclusions

In this paper, we attempted to compare for the first time the three most advanced coastal altimetry products for Jason-2 satellite altimetry: ALES, PISTACH, and XTRACK. The HK coastal zone was selected because it is the only candidate in the South China Sea with available hourly tide gauge data for comparison and validation. Besides, the HK coast presents extremely complex geomorphological and environmental conditions that lead to quite different waveforms. We considered six retrackers: MLE4, MLE3, ALES, ICE1, ICE3, and RED3, and computed Jason-2 sea level for each retracker over a 6.5-year time span (Jason-2 cycles 1–238). We generated a time series (cycle-by-cycle) for each sea level estimate in three cases, each covering different spatial scales. We found that case 1 is a good compromise between coherence and accuracy, and may represent the best sea level estimate at the HK coast. This is probably because, in the study zone, it contains more land-free areas than cases 2 and 3 much contaminated by the presence of small islands. An interesting outcome is that averaging along-track altimetry data over distances larger than 10 km provides better agreement with the tide gauge record than using only altimetry data very close to the coast.

ICE3 and RED3 retrackers show surprising large jumps (~+0.2 m) around cycle #150, that prevents further quantitative analysis. Data from other cycles need to be analyzed to determine whether these jumps result from a software bug (possibly different algorithm versions). A number of studies have shown that improved threshold retrackers sometimes outperform physically-based retrackers, so if any jump is always present, the data set needs to be corrected.

The results indicate that, in spite of the presence of large outliers in ALES (up to tens of meters), after outlier-editing, ALES performs better than MLE4 and MLE3, both in terms of noise level and uncertainty in the sea level trend estimate. We validated the coastal altimetry-based sea level by comparing with data from the HK tide gauge (located ~10 km away). Applying a 3-month smoothing, the standard deviation between the altimetry and tide gauge sea level time series is about 5 cm, which is quite encouraging given the complexity of the study area.

Another interesting result is that the computed sea level trend within 5 km from the coast is about twice the trend estimated at larger distances from the coast. This result, based on merging the CCI open ocean-based sea level with the retracked coastal data, suggests that offshore of the HK region, the short-term trend significantly increases when approaching the coast. This result cannot be generalized for several reasons, in particular because the estimated short-term trends essentially represent the inter-annual variability that may be different from one region to another.

Although the results presented here are encouraging, as shown in Section 6, the HK coastal altimetry sea level data could be further improved by applying an SVD processing before the waveform analysis. There is also certainly room to develop better altimeter waveform retrackers. Geophysical corrections, in particular tide corrections, also appear as an important limiting factor for coastal altimetry applications. In parallel, even if they provide much shorter time series of data than the TOPEX/Poseidon and -Jason missions, we could consider additional altimetry missions (such as HY-2A as well as SARAL/AltiKa and Sentinel-3 based respectively on Ka-band and delay-Doppler technology). In the near future, it will be also highly beneficial to systematically combine the best retrackers and geophysical corrections, to provide a coastal sea level data set with global coverage, usable for climate studies and coastal impact investigations.

Acknowledgments: This study is supported by the National Key R & D Program of China under contract #2017YFB0502800 and #2017YFB0502802, and by the Open Fund of State Key Laboratory of Remote Sensing Science (Grant No. OFSLRSS201705). We are very grateful to Fernando Nino for providing us with Figure 2. Xi-Yu Xu was funded by the China Scholarship Council (Grant No. 201604910329) during a one year visit in LEGOS. Jason-2 SGDR data are archived and maintained by AVISO. X-TRACK, ALES, and PISTACH data are generated and distributed by LEGOS/CTOH, NOC and CLS respectively. Special gratitude is also given to the three peer-reviewers, whose insightful comments have improved the manuscript significantly.

Author Contributions: Xi-Yu Xu collected and processed the coastal altimetry and CCI data and computed all results presented in the paper. Florence Birol processed the tide gauge data and checked the results. All three authors contributed to discuss the results and wrote the manuscript.

Conflicts of Interest: The authors declare no conflict of interest.

References

1. Nicholls, R.J. Planning for the impacts of sea level rise. *Oceanography* **2011**, *24*, 144–157. [CrossRef]
2. Brown, G. The average impulse response of a rough surface and its applications. *IEEE Trans. Antennas Propag.* **1977**, *25*, 67–74. [CrossRef]
3. Passaro, M.; Cipollini, P.; Vignudelli, S.; Quartly, G.D.; Snaith, H.M. ALES: A multimission adaptive subwaveform retracker for coastal and open ocean altimetry. *Remote Sens. Environ.* **2014**, *145*, 173–189. [CrossRef]
4. Cipollini, P.; Calafat, F.M.; Jevrejeva, S.; Melet, A.; Prandi, P. Monitoring sea level in the coastal zone with coastal altimetry and tide gauges. *Surv. Geophys.* **2017**, *38*, 33–57. [CrossRef]
5. Cipollini, P.; Birol, F.; Fernandes, M.J.; Obligis, E.; Passaro, M.; Strub, P.T.; Valladeau, G.; Vignudelli, S.; Wilkin, J. Satellite altimetry in coastal regions. In *Satellite Altimetry over Oceans and Land Surfaces*; Stammer, D., Cazenave, A., Eds.; CRC Press: Boca Raton, FL, USA, 2017; ISBN 978-1-4987-4345.
6. Birol, F.; Fuller, N.; Lyard, F.; Cancet, M.; Niño, F.; Delebecque, C.; Fleury, S.; Toublanc, F.; Melet, A.; Saraceno, M. Coastal applications from nadir altimetry: Example of the X-TRACK regional products. *Adv. Space Res.* **2017**, *59*, 936–953. [CrossRef]
7. Vignudelli, S.; Kostianoy, A.G.; Cipollini, P.; Benveniste, J. (Eds.) *Coastal Altimetry*; Springer: Berlin/Heidelberg, Germany, 2011; 565p. [CrossRef]
8. Mercier, F.; Rosmorduc, V.; Carrere, L.; Thibaut, P. *Coastal and Hydrology Altimetry Product (PISTACH) Handbook*; CLS-DOS-NT-10-246; CNES: Paris, France, 2010.
9. Valladeau, G.; Thibaut, P.; Picard, B.; Poisson, J.C.; Tran, N.; Picot, N.; Guillot, A. Using SARAL/AltiKa to improve Ka-band altimeter measurements for coastal zones, hydrology and ice: The PEACHI prototype. *Mar. Geodesy* **2015**, *38* (Suppl. 1), 124–142. [CrossRef]
10. Caldwell, P.C.; Merrifield, M.A. *Joint Archive for Sea Level Data Report*; JIMAR Contribution No. 15-392, Data Report No. 24; University of Hawaii at Manoa, Joint Institute for Marine and Atmospheric Research: Manoa, HI, USA, 2015.
11. Lai, W.S.; Matthew, J.P.; Ho, K.Y.; Astudillo, J.C.; Yung, M.N.; Russell, B.D.; Williams, G.A.; Leung, M.Y. Hong Kong's marine environments: History, challenges and opportunities. *Reg. Stud. Mar. Sci.* **2016**, *8*, 259–273. [CrossRef]
12. Gommenginger, C.; Thibaut, P.; Fenoglio-Marc, L.; Quartly, G.; Deng, X.; Gómez-Enri, J.; Challenor, P.; Gao, Y. Retracking Altimeter Waveforms Near the Coasts. In *Coastal Altimetry*; Vignudelli, S., Kostianoy, A., Cipollini, P., Benveniste, J., Eds.; Springer: Berlin/Heidelberg, Germany, 2011; pp. 61–101.
13. Hwang, C.; Guo, J.; Deng, X.; Hsu, H.-Y.; Liu, Y. Coastal gravity anomalies from retracked Geosat/GM altimetry: Improvement, limitation and the role of airborne gravity data. *J. Geodesy* **2006**, *80*, 204–216. [CrossRef]

14. Wingham, D.J.; Rapley, C.G.; Griffiths, H. New techniques in satellite tracking systems. In Proceedings of the IGARSS'86 Symposium, Zürich, Switzerland, 8–11 September 1986; pp. 1339–1344.
15. Davis, C.H. Growth of the Greenland ice sheet: A performance assessment of altimeter retracking algorithms. *IEEE Trans. Geosci. Remote Sens.* **1995**, *33*, 1108–1116. [CrossRef]
16. Mercier, F.; Picot, N.; Thibaut, P.; Cazenave, A.; Seyler, F.; Kosuth, P.; Bronner, E. CNES/PISTACH Project: An Innovative Approach to Get Better Measurements over in-Land Water Bodies from Satellite Altimetry: Early Results. In *EGU General Assembly Conference Abstracts, Proceedings of the European Geosciences Union General Assembly 2009, Vienna, Austria, 19–24 April 2009;* Copernicus Publications: Gettingen, Germany, 2009; Volume 11, p. 11674.
17. Bao, L.; Lu, Y.; Wang, Y. Improved retracking algorithm for oceanic altimeter waveforms. *Prog. Nat. Sci.* **2009**, *19*, 195–203. [CrossRef]
18. Yang, L.; Lin, M.; Liu, Q.; Pan, D. A coastal altimetry retracking strategy based on waveform classification and subwaveform extraction. *Int. J. Remote Sens.* **2012**, *33*, 7806–7819. [CrossRef]
19. Chan, Y.W. Tide Reporting and Applications in Hong Kong, China. 2006. Available online: www.gloss-sealevel.org/publications/documents/hong_kong2006.pdf (accessed on 7 February 2018).
20. Ablain, M.; Cazenave, A.; Larnicol, G.; Balmaseda, M.; Cipollini, P.; Faugère, Y.; Fernandes, M.J.; Henry, O.; Johannessen, J.A.; Knudsen, P.; et al. Improved sea level record over the satellite altimetry era (1993–2010) from the Climate Change Initiative project. *Ocean Sci.* **2015**, *11*, 67–82. [CrossRef]
21. Ablain, M.; Legeais, J.F.; Prandi, P.; Marcos, M.; Fenoglio-Marc, L.; Dieng, H.B.; Benveniste, J.; Cazenave, A. Satellite altimetry-based sea level at global and regional scales. *Surv. Geophys.* **2017**, *38*, 7–31. [CrossRef]
22. Legeais, J.F.; Ablain, M.; Zawadzki, L.; Zuo, H.; Johannessen, J.A.; Scharffenberg, M.G.; Fenoglio-Marc, L.; Fernandes, J.; Andersen, O.B.; Rudenko, S.; et al. An Accurate and Homogeneous Altimeter Sea Level Record from the ESA Climate Change Initiative, submitted. *Earth Syst. Sci. Data* **2017**. [CrossRef]
23. Thibaut, P.; Amarouche, L.; Zanife, L.O.Z.; Stunou, N.; Vincent, P.; Raizonville, P. Jason-1 altimeter ground processing look-up correction tables. *Mar. Geodesy* **2004**, *27*, 409–431. [CrossRef]
24. Xu, X.Y.; Xu, K.; Wang, Z.Z.; Liu, H.G.; Wang, L. Compensating the PTR and LPF Features of the HY-2A Satellite Altimeter Utilizing Look-Up Tables. *IEEE J. Sel. Top. Appl. Earth Obs. Remote Sens.* **2015**, *8*, 149–159. [CrossRef]
25. Dumont, J.P.; Rosmorduc, V.; Carrere, L.; Picot, N.; Bronner, E.; Couhert, A.; Desai, S.; Bpnekamp, H.; Scharroo, R.; Lillibridge, J. OSTM/Jason-2 Products Handbook (Issue: 1 rev 11). SALP-MU-M-OP-15815-CN. 2017. Available online: https://www.aviso.altimetry.fr/fileadmin/documents/data/tools/hdbk_j2.pdf (accessed on 7 February 2018).
26. Brown, S. A Novel Near-Land Radiometer Wet Path-Delay Retrieval Algorithm: Application to the Jason-2/OSTM Advanced Microwave Radiometer. *IEEE Trans. Geosci. Remote Sens.* **2010**, *48*, 1986–1992. [CrossRef]
27. Fernandes, M.J.; Lázaro, C.; Ablain, M.; Pires, N. Improved wet path delays for all ESA and reference altimetric missions. *Remote Sens. Environ.* **2015**, *169*, 50–74. [CrossRef]
28. Ray, R.D.; Egbert, G.D.; Erofeeva, S.Y. Tide predictions in shelf and coastal waters: Status and prospects. In *Coastal Altimetry*; Vignudelli, S., Kostianoy, A.G., Cipollini, P., Benveniste, J., Eds.; Springer: Berlin/Heidelberg, Germany, 2011; Chapter 7.
29. Lyard, F.; Lefevre, F.; Letellier, T.; Francis, O. Modelling the global ocean tides: Modern insights from FES2004. *Ocean Dyn.* **2006**, *56*, 394–415. [CrossRef]
30. Ray, R. Status of Modeling Shallow-water Ocean Tides: Report from Stammer international model comparison project. In Proceedings of the NASA/CNES Surface Water and Ocean Topography (SWOT) Science Definition Team (SDT) Meeting, Toulouse, France, 26–28 June 2014.
31. Glover, D.M.; Jenkins, W.J.; Doney, S.C. *Modeling Methods for Marine Science*; Cambridge University Press: Cambridge, UK, 2011; 571p.
32. Masters, D.; Nerem, R.S.; Choe, C.; Leuliette, E.; Beckley, B.; White, N.; Ablain, M. Comparison of global mean sea level time series from TOPEX/Poseidon, Jason-1, and Jason-2. *Mar. Geodesy* **2012**, *35*, 20–41. [CrossRef]
33. Chambers, D.P.; Cazenave, A.; Champollion, N.; Dieng, H.; Llovel, W.; Forsberg, R.; Schuckmann, K.; Wada, Y. Evaluation of the Global Mean Sea Level Budget between 1993 and 2014. *Surv. Geophys.* **2017**, *38*, 309–327. [CrossRef]

34. Xu, X.Y.; Liu, H.G.; Yang, S.B. Echo phase characteristic of interferometric altimeter for case of random surface plus one strong point scatter. In Proceedings of the 2016 IEEE International Geoscience and Remote Sensing Symposium (IGARSS), Beijing, China, 10–15 July 2016; pp. 6456–6459.
35. Halimi, A.; Mailhes, C.; Tourneret, J.-Y.; Thibaut, P.; Boy, F. Parameter estimation for peaky altimetric waveforms. *IEEE Trans. Geosci. Remote Sens.* **2013**, *51*, 1568–1577. [CrossRef]
36. Cazenave, A.; Dominh, K.; Ponchaut, F.; Soudarin, L.; Crétaux, J.F.; Le Provost, C. Sea level changes from Topex-Poseidon altimetry and tide gauges, and vertical crustal motions from DORIS. *Geophys. Res. Lett.* **1999**, *26*, 2077–2080. [CrossRef]
37. Yuan, L.G.; Ding, X.L.; Chen, W.; Guo, Z.H.; Chen, S.B.; Hong, B.S.; Zhou, J.T. Characteristics of daily position time series from the Hong Kong GPS fiducial network. *Chin. J. Geophys.* **2008**, *51*, 1372–1384. [CrossRef]
38. Lau, D.S.; Wong, W.T. *Monitoring Crustal Movement in Hong Kong Using GPS: Preliminary Results*; Hong Kong Observatory: Hong Kong, China, 2010.
39. Ollivier, A. Nouvelle Approche Pour L'extraction de Paramtres Gophysiques Partir Des Mesures en Altimtrie Radar. Ph.D. Thesis, Institut National Polytechnique de Grenoble, Grenoble, France, 2006.

![remote sensing logo] *remote sensing*

MDPI

Article

Monitoring Sea Level and Topography of Coastal Lagoons Using Satellite Radar Altimetry: The Example of the Arcachon Bay in the Bay of Biscay

Edward Salameh [1,2,*], **Frédéric Frappart** [1,3], **Vincent Marieu** [4], **Alexandra Spodar** [4,5], **Jean-Paul Parisot** [4], **Vincent Hanquiez** [4], **Imen Turki** [2] and **Benoit Laignel** [2]

[1] Laboratoire d'Etudes en Géophysique et Océanographie Spatiales (LEGOS), Université de Toulouse, IRD, CNES, CNRS, UPS, 31400 Toulouse, France; frederic.frappart@legos.obs-mip.fr

[2] Normandie University, UNIROUEN, UNICAEN, CNRS, M2C, Morphodynamique Continentale et Côtière, 76000 Rouen, France; imen.turki@univ-rouen.fr (I.T.); benoit.laignel@univ-rouen.fr (B.L.)

[3] Géosciences Environnement Toulouse (GET), Université de Toulouse, IRD, CNES, CNRS, UPS, 31400 Toulouse, France

[4] Environnements et Paléoenvironnements Océaniques et Continentaux (EPOC), UMR 5805, allée Geoffroy St Hilaire, 33615 Pessac CEDEX, France; vincent.marieu@u-bordeaux.fr (V.M.); alexandra.spodar@univ-littoral.fr (A.S); Parisotjp@wanadoo.fr (J.-P.P.); vincent.hanquiez@u-bordeaux.fr (V.H.)

[5] Laboratoire d'Océanologie et de Géosciences (LOG), UMR 8187, 59140 Dunkerque, France

[*] Correspondence: edward.salameh@legos.obs-mip.fr; Tel.: +33-6-30-21-26-72

Received: 21 December 2017; Accepted: 10 February 2018; Published: 14 February 2018

Abstract: Radar altimetry was initially designed to measure the marine geoid. Thanks to the improvement in the orbit determination from the meter to the centimeter level, this technique has been providing accurate measurements of the sea surface topography over the open ocean since the launch of Topex/Poseidon in 1992. In spite of a decrease in the performance over land and coastal areas, it is now commonly used over these surfaces. This study presents a semi-automatic method that allows us to discriminate between acquisitions performed at high tides and low tides. The performances of four radar altimetry missions (ERS-2, ENVISAT, SARAL, and CryoSat-2) were analyzed for the retrieval of sea surface height and, for the very first time, of the intertidal zone topography in a coastal lagoon. The study area is the Arcachon Bay located in the Bay of Biscay. The sea level variability of the Arcachon Bay is characterized by a standard deviation of 1.05 m for the records used in this study (2001–2017). Sea surface heights are very well retrieved for SARAL (R~0.99 and RMSE < 0.23 m) and CryoSat-2 (R > 0.93 and RMSE < 0.42 m) missions but also for ENVISAT (R > 0.82 but with a higher RMSE >0.92 m). For the topography of the intertidal zone, very good estimates were also obtained using SARAL (R~0.71) and CryoSat-2 (R~0.79) with RMSE lower than 0.44 m for both missions.

Keywords: radar altimetry; coastal altimetry; sea surface height; topography of the intertidal zone; ERS-2; ENVISAT; SARAL; CryoSat-2

---

1. Introduction

Coastal regions represent only 5% of Earth's land area, yet their societal and economical importance are larger than their surface area suggests [1]. The land area within 100 km from the coast accommodates about 39% of the global population according to the CIESIN (Center for International Earth Science Information Network) [2]. Coastal systems are experiencing high pressures due to population growth and the overexploitation of their resources. Anthropogenic pressures exacerbated by sea level rise and the increase of global temperature lead to a rapid and threatening environmental change of these systems, which requires effective long-term coastal management initiatives.

Common features of coastal systems are coastal lagoons, occupying 13% of coastal areas worldwide [3]. Coastal lagoons are defined as "inland water bodies, separated from the ocean by a barrier, connected to the ocean by one or more restricted inlets which remain open at least intermittently, and have water depth which seldom exceed a few meters" [4]. They are subject to forcings from rivers, wind stress, tides, precipitation to evaporation balance, and surface heat balance [4]. These ecosystems provide important services and societal benefits (e.g., food provision, recreational, water regulation, etc.); however their subsistence is threatened by global climate change [5]. Understanding the physical dynamics of these systems is of great importance in order to direct the planning and implementation of coastal management strategies in coastal lagoons.

In the need for a better understanding of lagoons' dynamics, satellite radar altimetry measuring the variation of the surface elevation could be a very useful tool providing key information, especially for non-monitored areas. However, using altimetry in coastal regions remains a great challenge due to numerous issues including land contamination in the footprint that impacts the radar echo (or waveform), but also the lower quality of the corrections applied to the distance between the satellite and the surface (or altimeter range) than over open ocean [6]. Despite these shortcomings, recent improvements in processing techniques (e.g., careful recovering of flagged data, applying specialized retracking, improving the correction terms) extended the capabilities of altimeters in coastal areas [7].

This study analyzes the performance of radar altimetry to monitor sea level and to provide, for the very first time, topography of the intertidal zone along the altimeter tracks in the Arcachon Bay, a coastal lagoon situated in the south-west of France. The choice of this lagoon as a case study was motivated by the coverage provided by ERS-2, ENVISAT, SARAL, and CryoSat-2 altimetry missions. Our goal is to assess the evolution of measurement accuracy at Ku-band (ERS-2, ENVISAT) and the benefits of the Ka-band (SARAL) in Low Resolution Mode (LRM). An assessment of instrumental performance was undertaken as well for observations made by the Ku-band satellite mission CryoSat-2, the first altimeter to operate in Synthetic Aperture Radar (SAR) mode.

The aim of this study is threefold: (i) to investigate the limitations and potential improvements of altimetry to monitor lagoons Sea Surface Height (SSH), (ii) to test the capability of altimetry to retrieve topographic variations and (iii) to optimize (time wise and accuracy wise) the processing of altimetry data in coastal lagoons environment. Tide gauge measurements and lidar topography datasets were used to evaluate SSHs and topography estimation made by satellite radar altimetry.

2. Study Area

The Arcachon Bay (44°40′N, 1°10′W) is a mesotidal shallow semi-confined lagoon, located in the southeast of the Bay of Biscay (Figure 1). The total lagoon surface (174 km^2) is composed of channels (57 km^2) that drain the intertidal area (117 km^2). The main channels have a maximum depth around 20 m and are extended by a complex network of secondary channels [8]. The tidal cycle is semi-diurnal with a weak diurnal inequality. The tide amplitudes vary from 0.8 to 4.6 m for neap and spring tides respectively. The Arcachon Bay connects to the Atlantic Ocean through two narrow passes of 1–1.5 km width and around 12 km long. The two passes are separated by the Arguin Bank. Important seawater exchanges, reaching up to 384.106 m^3 occur during each tidal cycle [9]. Freshwater inputs from small rivers and groundwater are coming mostly from the Eyre River and the Porges Canal, located south-east and north of the Bay respectively (see Figure 1). They represent more than 95% (73% and 24% respectively) of the total annual freshwater inflows [10]. The intertidal area is composed of a mix of muddy and sandy material [8]. A large zone of 70 km^2 of the mudflats in the inner lagoon is covered with Zostera noltii seagrass [11].

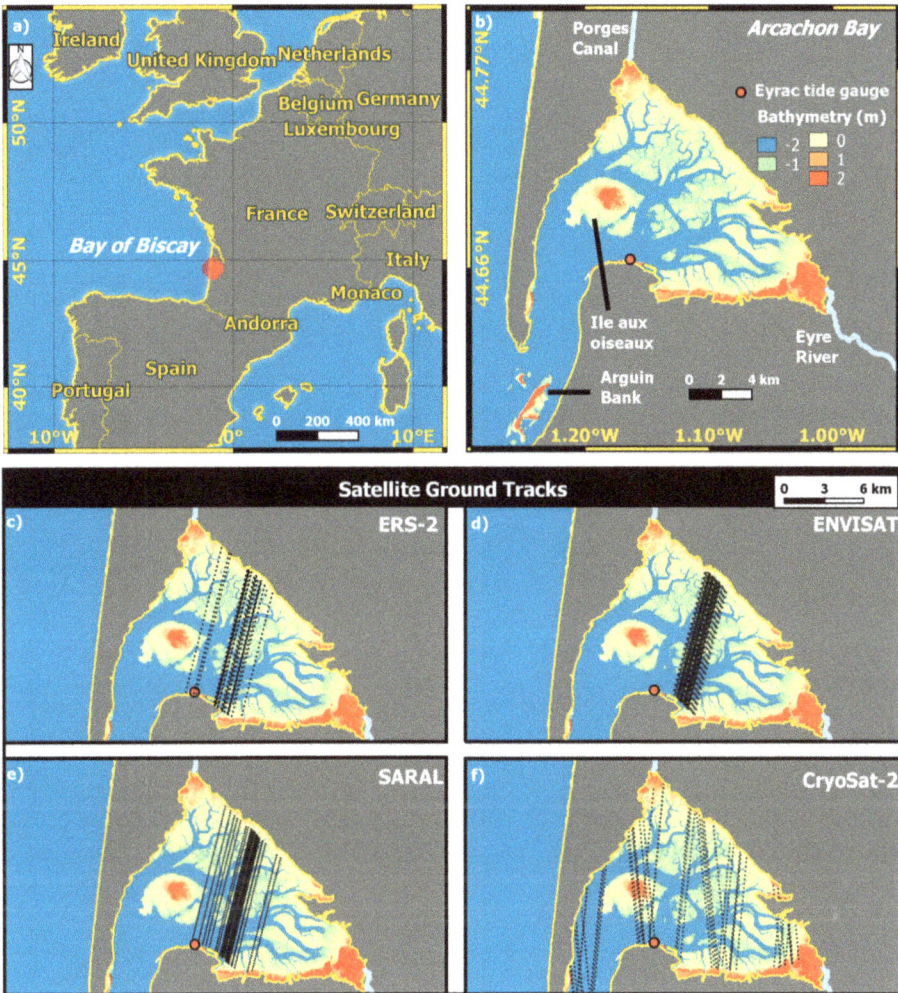

Figure 1. (**a**) The Arcachon lagoon is located in the Bay of Biscay along the south part of the French Atlantic coast. (**b**) The Arcachon lagoon is a mesotidal shallow semi-confined lagoon. Several altimetry missions' ground-tracks cover the lagoon: (**c**) ERS-2 (1993–2003, since 2003 ERS-2 has experienced a number of failures), (**d**) ENVISAT (2002–2010 on the nominal orbit), (**e**) SARAL (2013–2016 on the nominal orbit), and (**f**) CryoSat-2 (since 2010).

3. Datasets

3.1. Altimetry Data

Table 1 presents a summary of the main characteristics of the altimetry missions used in this study, which are described in more detail below.

Table 1. Main characteristics of the altimetry missions used in this study.

Mission	ERS-2	ENVISAT	SARAL	CryoSat-2
Agency	ESA	ESA	CNES/ISRO	ESA
Launch on	21/04/1995	01/03/2002	25/02/2013	08/04/2010
End date	06/07/2011	08/06/2012	Present	Present
Altimeter name	RA	RA-2	AltiKa	SIRAL
Radar frequency	Ku-band	Ku and S-bands	Ka-band	Ku-band
Altitude	785 km	790 km	790 km	717 km
Orbit inclination	98.52°	98.54°	98.54°	92°
Repetitivity	35 days	35 days	35 days	369 days
Ground-track spacing at the equator	85 km	85 km	85 km	7.5 km
Along track sampling	20 Hz (350 m)	18 Hz (~400 m)	40 Hz (175 m)	20 Hz (350 m)

3.1.1. ERS-2

The ERS-2 satellite (European Remote Sensing-2) was launched in 1995 by ESA (European Space Agency). Its payload is composed of several sensors, including a radar altimeter (RA), operating at Ku-band (13.8 GHz). It was sun-synchronously orbiting at an altitude of 785 km with an inclination of 98.52° with a 35-day repeat cycle. This orbit has a ground-track spacing about 85 km at the equator. ERS-2 provided observations of the topography of the Earth from 82.4° latitude north to 82.4° latitude south. ERS-2 data are available from 17 May 1995 to 9 August 2010 but with a limited coverage after 22 June 2003.

3.1.2. ENVISAT

ENVISAT (ENVIronmental SATellite) mission was launched on 1st March 2002 by ESA. It carries 10 instruments including the advanced radar altimeter (RA-2). RA-2 is a nadir-looking pulse-limited radar altimeter operating at two frequencies at Ku- (13.575 GHz) and S- (3.2 GHz) bands. ENVISAT orbits at an altitude of 790 km, with an inclination of 98.54°, on a sun-synchronous orbit with a 35-day repeat cycle, providing observations of the Earth surface (ocean and land) from 82.4° latitude North to 82.4° latitude South. This orbit was formerly used by ERS-1 and 2, with an equatorial ground-track spacing of about 85 km. ENVISAT remains on its nominal orbit until October 2010 [12]. From November 2010 to April 2012, ENVISAT was put into the extending phase consisting of a drifting on a 30-day orbit lowered by 17 km.

3.1.3. SARAL

SARAL (Satellite for Argos and ALtika) is a CNES-ISRO (Centre National d'Etudes Spatiales—Indian Space Research Organization) joint-mission that was launched on 25 February 2013. Its payload is composed of the AltiKa radar altimeter and bi-frequency radiometer, and a triple system for precise orbit determination: the real-time tracking system DIODE of DORIS instrument, a Laser Retroflector Array (LRA), and the Advanced Research and Global Observation Satellite (ARGOS-3). Its orbital characteristics are the same as ENVISAT (see above). The first four cycles of SARAL do not follow precisely the ENVISAT orbit. AltiKa radar altimeter is a solid-state mono-frequency altimeter that provides accurate range measurements. It is the first altimeter to operate at Ka-band (35.7 GHz). Its accuracy is expected to be about 1 cm over ocean. Over the coastal regions, it is expected to provide measurements significantly better than those from the previous Ku band missions. Improvements come from the reduced footprint of the Ka-band (about ten times smaller in surface than it is in Ku-band) and from the higher along-track sampling rate of 40 Hz (~175 km), twice that of ENVISAT [13].

3.1.4. CryoSat-2

CryoSat-2 mission was launched on 8 April 2010 by ESA. This mission is dedicated mainly to polar observations. However, its acquisitions can be useful for ocean and inland monitoring as it provides a global monitoring of the Earth's surface [14]. The mission's main payload consists of a radar altimeter, SIRAL (Synthetic Aperture Interferometric Radar Altimeter), operating at Ku-band

(13.575 GHz) in three different modes: Low Resolution Mode (LRM), Synthetic Aperture Radar mode (SAR), and Synthetic Aperture Interferometric mode (SARIn). CryoSat-2 orbits at an altitude of 717 km, with an inclination of 92°, on a non-sun-synchronous orbit with a 369-day repeat cycle. The equatorial ground-track spacing is about 7.5 km shifting every 30 days. The short inter-track distance increases the sampling over the study area during one cycle [15]. For this study region and period, CryoSat-2 operated in SAR mode.

All altimetry data used in this study come from the Geophysical Data Records (GDR) made available by the Centre of Topography of the Oceans and the Hydrosphere (CTOH—http://ctoh.legos. obs-mip.fr/). They are sampled along the altimeter track at 18 Hz for ENVISAT, 20 Hz for ERS-2 and CryoSat-2, and 40 Hz for SARAL (high-frequency mode commonly used over land and coastal areas where the surface properties are changing more rapidly than over the open ocean).

3.2. Ancillary Data

3.2.1. Arcachon-Eyrac Tide Gauge

The Arcachon-Eyrac tide gauge is managed by the French hydrographic service (Service Hydrographique et Océanographique de la Marine—SHOM) and the Gironde sea and land state office (Direction Départementale des Territoires et de la Mer—DDTM). It is operating since November 1967. The tide gauge (Figure 1) (1.163550021°W and 44.66500092°N) is a non-contact radar sensor providing sea level measurements at 1-min time intervals since June 2000. These data are made available by REFMAR (available online: http://refmar.shom.fr/, accessed on 10 February 2018). Altimetry and tide gauge data timeline is shown in Figure 2.

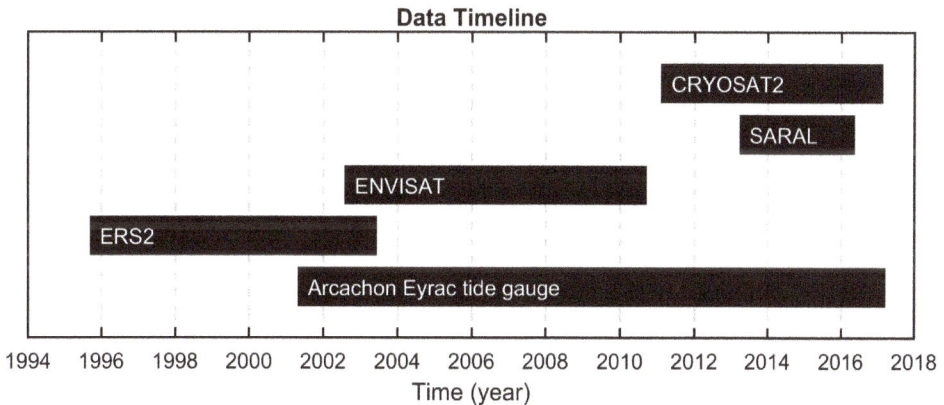

Figure 2. Temporal coverage of the in situ Arcachon Eyrac tide gauge and altimetry datasets used in this study.

3.2.2. Lidar-Derived Topography of the Intertidal Zone

The LIDAR data is extracted from RGE ALTI® product provided by the French national institute for geography and forest information (IGN). Arcachon Bay raw data have been acquired by airborne topographic LIDAR at low tide, on 25 June 2013, and interpolated on a regular 1 × 1 m grid for RGE ALTI® product. The controlled altimetric precision for this data is 0.2 m. For the convenience of the study, these data have been subsampled on a 10 × 10 m grid.

4. Methods

The following flowchart exhibits the different steps of the methodology (Figure 3). All steps are described in the corresponding sections.

Altimetry data processing
section 4.1

Made using MAPS software to obtain sea surface height (SSH)

Leveling all data to a common datum
section 4.2

All data were referenced to the Mean Sea Level (MSL)

Extraction of the topography along the ground tracks
section 4.3

Filling the topography with water using tide-gauge data
section 4.4

To discriminate between submerged and emerged points of measurement

Altimetry data selections for SSH (or water level) estimation
section 4.5

Topography comparisons using emerged points
Altimetry-lidar comparisons

MAPS manual selections

Automatic selections

MAPS software is used to process altimetry data and to manually select valid measurements

Cycles submerged/emerged classification
section 4.5.1

Grouping of each cycle measurements into quartiles
4.5.2

Data selections
section 4.5.3

Selected-measurements:
-The least dispersive quartile for submerged cycles
-The lowest quartile for emerged cycles

Passing-Bablok for methods comparison
section 4.6

Topography: altimetry/lidar
SSH: altimetry/tide-gauge

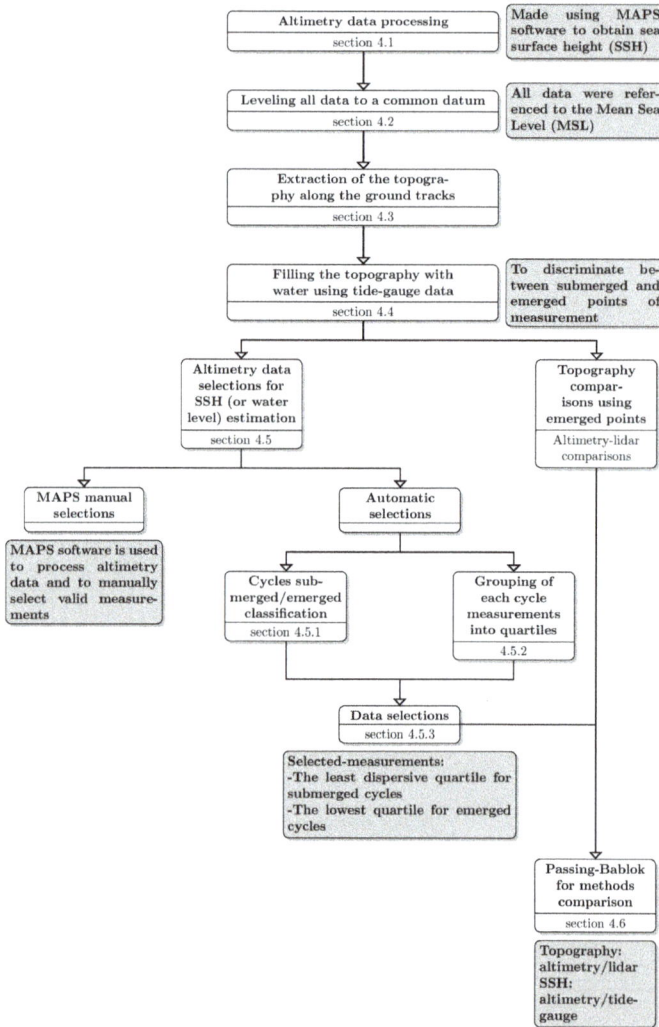

Figure 3. The different steps of the method presented in the form of a flow chart.

4.1. Altimetry Data Processing

The principle of radar altimetry is the following: the altimeter emits a radar pulse and measures the two-way travel-time from the satellite to the surface. The distance between the satellite and the Earth surface—the altimeter range (R)—is thus derived with a precision of a few centimeters. The satellite altitude (H) referred to an ellipsoid is also accurately known from orbitography modeling. Taking into account propagation delays due to interactions of electromagnetic wave in the atmosphere and geophysical corrections, the height of the reflecting surface (h) with reference to an ellipsoid or a geoid can be estimated as [16,17]:

$$h = H - \left(R + \sum \Delta R_{propagation} + \sum \Delta R_{geophysical} \right) \qquad (1)$$

where H is the height of the center of mass of the satellite above the ellipsoid, R is the nadir altimeter range from the center of mass of the satellite to the sea surface taking into account instrumental corrections, $\Delta R_{propagation}$ and $\Delta R_{geophysical}$ are the sums of the environmental and geophysical corrections to apply to the range and respectively given by Equations (2) and (3).

$$\sum \Delta R_{propagation} = \Delta R_{ion} + \Delta R_{dry} + \Delta R_{wet} + \Delta R_{SSB} \tag{2}$$

ΔR_{ion} is the atmospheric refraction range delay due to the free electron content associated with the dielectric properties of the ionosphere, ΔR_{dry} is the atmospheric refraction range delay due to the dry gas component of the troposphere, ΔR_{wet} is the atmospheric refraction range delay due to the water vapor and the cloud liquid water content of the troposphere, ΔR_{SSB} is the range correction due to the interaction of the electromagnetic pulse emitted by the altimeter with the scatterometers within the footprint. It is known as Sea State Bias (SSB) and is the sum of the electromagnetic (EM), skewness, and tracker biases.

$$\sum \Delta R_{geophysical} = \Delta R_{ocean} + \Delta R_{solid\ Earth} + \Delta R_{pole} + \Delta R_{atm} \tag{3}$$

where ΔR_{ocean} is the value of the ocean tide, $\Delta R_{solid\ Earth}$ and ΔR_{pole} are the corrections respectively accounting for crustal vertical motions due to the solid Earth and polar tides, ΔR_{atm} is the dynamic atmosphere correction.

Altimetry-derived heights are automatically obtained from the GDR data using the Multi-mission Altimetry Processing Software (MAPS) that is commonly used for the selection of valid altimetry data and their processing over land and ocean [18–20]. More details on MAPS can be found in Frappart et al. [21]. As the ΔR_{SSB} values were most of the times flagged in the GDR, they were not applied for the consistency of the series of observations. ΔR_{SSB} corresponds to cm-level correction applied to the range as it varies typically from -1% to 4% of the Significant Wave Height (SWH) (e.g., Gaspar et al. [22]). In the Arcachon Bay, SWH is generally lower than 0.2 m but can exceptionally reach 0.5 to 0.7 m for wind intensities greater than 20 m·s^{-1} [23]. ΔR_{ocean} and ΔR_{atm} were not applied as they also affect the measurements from the Arcachon-Eyrac tide-gauge. Due to the width of the altimeters and radiometers footprints of several kilometers and several tenths of kilometers, radar echoes are a mix of returns from ocean, intertidal zone, and from the surrounding environment (forests, cities, roads, etc.) and microwave brightness temperatures are the combination of emissions from a very inhomogeneous environment. The same set of corrections used for land hydrology will be applied in this study.

The geophysical corrections applied to the range are derived from the Global Ionospheric Maps (GIM) and Era Interim model outputs from the European Centre Medium-Range Weather Forecasts (ECMWF) for the ionosphere and the dry and wet troposphere range delays respectively. The environmental corrections are obtained from the solid Earth tide [24] and polar tide [25] tables for SARAL, and using the IERS (International Earth Rotation and Reference Systems) convention for ENVISAT. Ranges used to derive altimeter heights were those processed with the Ice-1 retracking algorithm [26,27] (formulation in Appendix A) because they are present in the ERS-2, ENVISAT and SARAL GDRs and were shown to be more suitable for hydrological studies in terms of accuracy of water levels and availability of the data (e.g., Frappart et al. [28,29]). For CryoSat-2, altimeter heights were those processed by the Sea-Ice retracker [30].

4.2. Leveling to a Common Datum

The datasets used in this study are referenced to various datums. ENVISAT and CryoSat-2 altimetry data are referenced to WGS84 ellipsoid, SARAL altimetry data to the Topex/Poseidon (T/P) ellipsoid, the lidar-based topography of the intertidal zone to the French reference system NGF/IGN69, and the Arcachon-Eyrac tide-gauge to the French chart datum. A datum conversion from T/P ellipsoid to WGS84 is automatically performed for SARAL measurements using Equation (4) adapted from [31] in the new version of MAPS implemented for this study:

$$\Delta h = \frac{a'\left(1 - e'^2\right)}{\sqrt{1 - e'^2 sin^2\varphi}} - \frac{a\left(1 - e^2\right)}{\sqrt{1 - e^2 sin^2\varphi}} \tag{4}$$

where Δh is the variation of height at latitude φ due to the change of ellipsoid from T/P to WGS84 datum, $a = 6{,}378{,}137$ m and $e = 0.081819190842621$ are the semi-major axis and the eccentricity of the WGS84 datum, $a' = 6{,}378{,}136.3$ m and $e' = 0.081819221456$ are the semi-major axis and the eccentricity of the T/P datum.

For comparison purposes, they were all referenced to the Mean Sea Level (MSL). Datum conversion for all data (altimetry, topography, and tide gauge data) were made using the French maritime altimetry references [32] giving the vertical differences between the different datum (−44.8 m between WGS84 and the French chart, −1.98 m between the French chart and the French reference system, and −2.48 m between the French chart and the MSL).

4.3. Extraction of the Topography of the Intertidal Zone under the Altimeter Tracks

For comparison purposes, the topography of the intertidal zone was extracted along the altimeter ground tracks. For each altimetry measurement, the closest topography point was selected. The maximum distance obtained between an altimetry measurement and its corresponding topography point is 10 m, which is consistent with the spatial resolution of the lidar-derived topography (10 m). This extraction was performed for every ERS-2, ENVISAT, SARAL, and CryoSat-2 cycle as the orbit of the satellites is not exactly repetitive and generally varies within 1 km around the nominal track (Figure 1).

4.4. Manual Classifications of Altimetry Measurements and Cycles

Two types of classification based on ancillary data were made. The first one consists of separating altimetry measurements between submerged and emerged measurement points. This classification was made in order to use submerged points for SSH estimation, and emerged points for topography measurements. The resulting altimetry levels were compared to Arcachon Eyrac in-situ gauge measurements and lidar topography data respectively. To discriminate between land and water, the topography along the altimeter tracks is filled with the water levels measured at the Arcachon-Eyrac tide gauge. For each gridpoint of the Lidar-based topography, the gridpoint is considered submerged if the water level is greater than the topography, emerged if not (see Figures 6 and 7 in Section 5).

The second type is a classification by cycle and it was made to separate cycles with emerged land from completely submerged cycles. The cycle is considered as an emerged cycle if more than 20% of the altimeter ground track flew over land. This classification was performed to assess the performance of the automatic classification presented in Section 4.5.1.

4.5. Automatic Selections of Valid Altimetry Measurements

A method to automatically select the altimetry measurements (used for SSH estimation) over emerged and submerged areas was tested. It is composed of the following steps.

4.5.1. Classification of Cycles between Submerged and Emerged Cycles

This first step discriminates submerged from emerged cycles using intrinsic altimetry parameters in order to be completely independent of in-situ data. The rationale behind this step is the use of different selection criteria for submerged and emerged cycles. The classification was made using the unsupervised k-means clustering algorithm [33]. The number of clusters (k) was chosen to be two for submerged and emerged classes and the distance measure technique used is the cityblock (Manhattan) technique [34]. In contrast to the Euclidean distance (straight line distance between two points in Euclidean space), the cityblock distance is calculated as the distance in x plus the distance in y (in 2D). The backscattering coefficient and the peakiness (the average of these parameters per cycle) are the two parameters used to perform the clustering technique (peakiness formulation can be found in Appendix A). Before data clustering, the latter parameters were normalized and centered in order to unify their influence on distance. The use of these two parameters is based on a priori knowledge that they present significant descrepancies between water-dominated footprints and water-land-mixed footprints.

4.5.2. Grouping of Cycle's Measurements into Four Equal Parts

In this step, the altimetry measurements made during a given cycle were separated into four equally-sized groups. The four groups were separated by the following scores: Quartile 0 (the minimum), Quartile 1 (larger than 25% of the data points), Quartile 2 (the median), Quartile 3 (bigger than 75% of the data points), and Quartile 4 (the maximum). For each group, the standard deviation is computed to assess its dispersion.

4.5.3. Data Automatic Selections

For cycles classified as submerged cycles, the groups with the least standard deviation (least dispersion) are preserved. For cycles classified as emerged cycles, the first groups with the lowest 25% of values (values between Quartile 0 and Quartile 1) are preserved because these groups are the only groups able to reach the water when land emerges.

The results of the two types of selections are compared in Section 5.2.

4.6. Passing-Bablok Regression for Method Comparisons

Passing-Bablok is the regression method used to account for agreement and systematic bias between two methods (altimetry/tide-gauges or altimetry/lidar). We opted for Passing-Bablok method instead of ordinary linear regression because it is not sensitive to the outliers or the distribution of errors and because the independent variable (water level from tide gauge records or bathymetry from lidar) is not free of error. This robust, non-parametric method consists of fitting a line describing the relationship between the two variables (X and Y) and testing whether the slope is 1 and the intercept is 0. In-depth details on the method can be found in [35]. The results are presented as a scatter plot between X (method 1) and Y (method 2), a regression line, and a linear regression equation where the slope and the intercept represent proportional and constant systematic bias respectively. A statistical test of the assumption of linearity is performed using cumulative sum linearity test. Non-linear samples are not suitable for concluding on method agreement. Furthermore, a 95% Confidence Interval (CI) is also computed for the slope and the intercept to test the hypothesis that the slope is equal to 1 or the intercept is equal to 0. These hypotheses are accepted if 1 corresponds to the slope's 95% CI (if not, there is a proportional difference between the two methods) and if 0 corresponds to the intercept's 95% CI (if not, there is a constant difference or bias between the two methods).

4.7. Absolute Calibration of Altimetry Missions over the Intertidal Zone

Comparisons between altimetry-based and in situ SSH from tide gauge were performed. They require simultaneous in situ and altimetry measurements in the same terrestrial reference frame at the exact same location or comparison point (e.g., Cancet et al. [36]). The absolute altimeter bias ($Bias_{altimeter}$) is estimated as follows [37]:

$$Bias_{altimeter} = < h_{altimeter} - h_{in\ situ} > \tag{5}$$

where $h_{altimeter}$ and $h_{in\ situ}$ are the height of the reflecting surface estimated from altimeter and in situ measurements respectively. In this study, an absolute calibration is performed for estimating the bias of ERS-2, ENVISAT, SARAL and CryoSat-2 for sea level measurements and topography of the intertidal zone in the Arcachon Bay.

5. Results

Depending on their overflight time, radar altimetry missions acquired observations of the Arcachon Bay all over the tidal cycle. During low tide, they provide observations of the surface topography of the intertidal zone whereas, at high tides, they monitor the sea surface height. These changes in the nature of the reflecting surface, from wet sand and mud to sea water, modify the radar echo acquired by the altimeter. Lidar-based topography profiles were extracted along the altimetry ground tracks in the intertidal zone of the Arcachon Bay. They were filled with water using the record from Arcachon-Eyrac tide-gauge corresponding to the altimeter overflights. They were compared to the along track profiles of altimeter height, backscattering coefficients and waveform peakiness (not available for ERS-2 in the CTOH GDR and only available at 1 Hz for ENVISAT) estimated using the Ice-1 retracking algorithm for ERS-2, ENVISAT and SARAL and the SeaIce retracking algorithm for CryoSat-2 (see [27]) for details about the computation of these two latter parameters that are available in the altimeter GDRs). Examples for high and low tides using SARAL and CryoSat-2 data are respectively presented in Figures 4 and 5 (same type of figures is available in the Supplementary Material for ERS-2 (Figure S1) and ENVISAT (Figure S2)). SSH estimated using altimetry data was given by the median of submerged points' altimeter heights. For high and low tides, SARAL and CryoSat-2 missions show relatively good correspondence with in situ measurements while ERS-2 and ENVISAT show low accuracy, especially at low tides. ERS-2 is unable to accurately retrieve topography variations due to both, the coarse resolution of its footprint and changes in the range resolution between ocean (bandwidth of 300 MHz) and ice modes (bandwidth of 20 MHz). As for ENVISAT, despite the important bias obtained, altimetry measurements follow the topography variations and give an acceptable correlation coefficient as a result (see Sections 5.2 and 5.3). We observe a significant increase in the backscattering coefficients and waveform peakiness values between high and low tide (Figure 4c,d, and Figure 5c,d). This is due to the contribution of the emerging land to the received signal. A slight variation of these parameters is observed as well at the extreme sides for high tides where land and water fall together in the altimeter footprint.

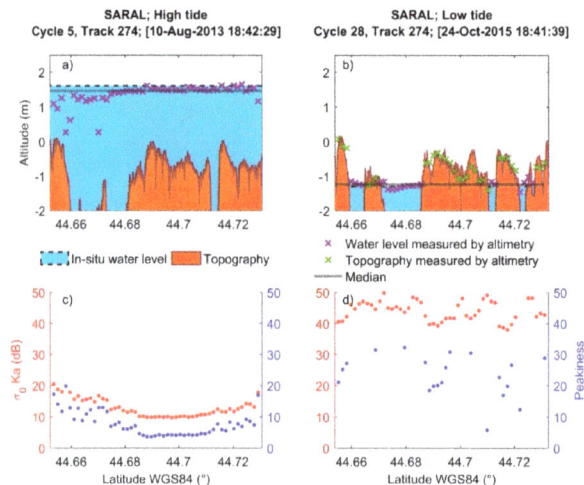

Figure 4. (**a,b**) Examples of SARAL along-track profiles of altimetry height over water (purple crosses) and land (green crosses) at high (**a**) and low (**b**) tides, the topography under the altimeter ground track is represented in brown and it is filled with water (in blue) using leveled tide-gauge records; (**c,d**) Variation of Ice-1 backscattering coefficients of Ka-band (red dots) and Peakiness (blue dots) at high (**c**) and low (**d**) tides.

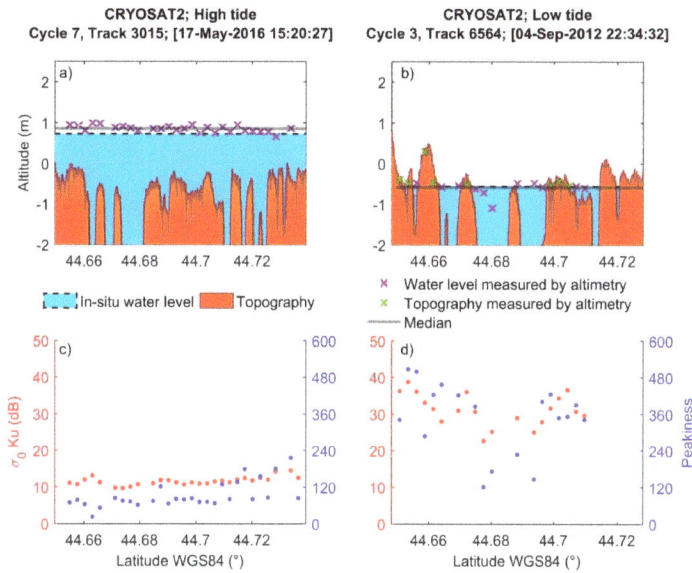

Figure 5. (**a**,**b**) Examples of CryoSat-2 along-track profiles of altimetry height over water (purple crosses) and land (green crosses) at high (**a**) and low (**b**) tides, the topography under the altimeter ground track is represented in brown and it is filled with water (in blue) using leveled tide-gauge records; (**c**,**d**) Variation of Retracker 1 backscattering coefficients of Ku-band (red dots) and Peakiness (blue dots) at high (**c**) and low (**d**) tides.

5.1. Auto-Classification Using the k-Means Algorithm

Using the k-means algorithm, the altimetry cycles were classified between emerged and submerged cycles. The evaluation of the auto-classification was made using the classification based on in-situ data as explained in Section 4.4. It should be noted that the clustering technique was made using two parameters (the backscattering coefficient and peakiness) for SARAL and CryoSat-2 (Figure 6), and using only the backscattering coefficient for ERS-2 and ENVISAT. SARAL showed the best classification results with 100% accordance with manual classification. A good consistency was shown also by CryoSat-2 and ENVISAT with ~80% accordance and ERS-2 with 82%.

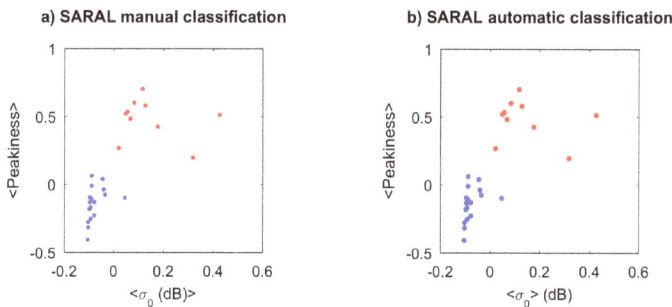

Figure 6. *Cont.*

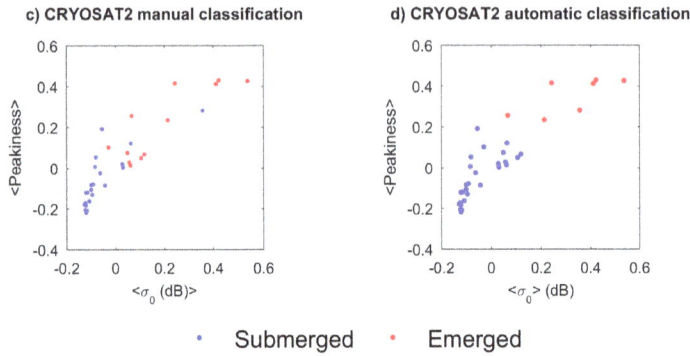

Figure 6. Comparisons between manual (left) and automatic (right) classifications (made using an unsupervised clustering technique) of cycles for SARAL (**a,b**) and CryoSat-2 (**c,d**).

5.2. Water Levels Comparison

For the selections made using MAPS, SSH estimation was given by the median of submerged altimetry measurements made during one crossing (cycle). An automatic selection of altimetry data was performed in addition to the manually refined data selections in MAPS. The automatic technique uses the less dispersive quartile of measurements in each cycle for cycles classified as submerged by the clustering technique, and the lowest quartiles for cycles classified as emerged. The SSH was given by the median of the chosen quartile without eliminating measurement points classified as emerged by the classification based on ancillary data. Figure 7 presents the comparisons made between SSHs acquired by altimetry and tide gauge measurements for MAPS manual selections (left) and for automatic selections (right) for ERS-2 (a and b), ENVISAT (c and d), SARAL (e and f), and CryoSat-2 (g and h).

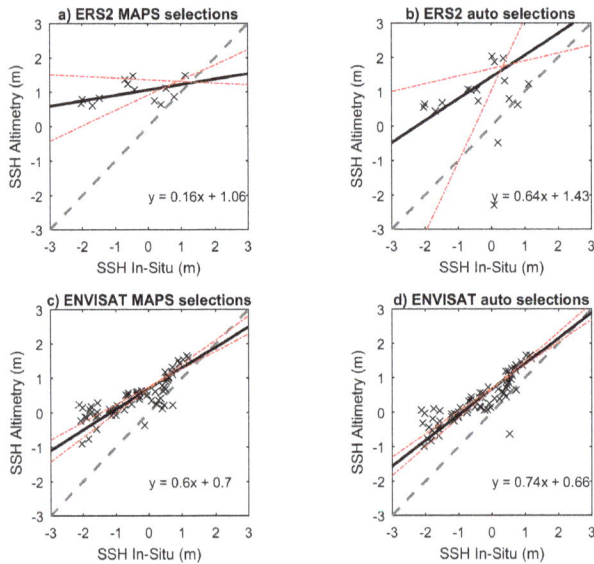

Figure 7. *Cont.*

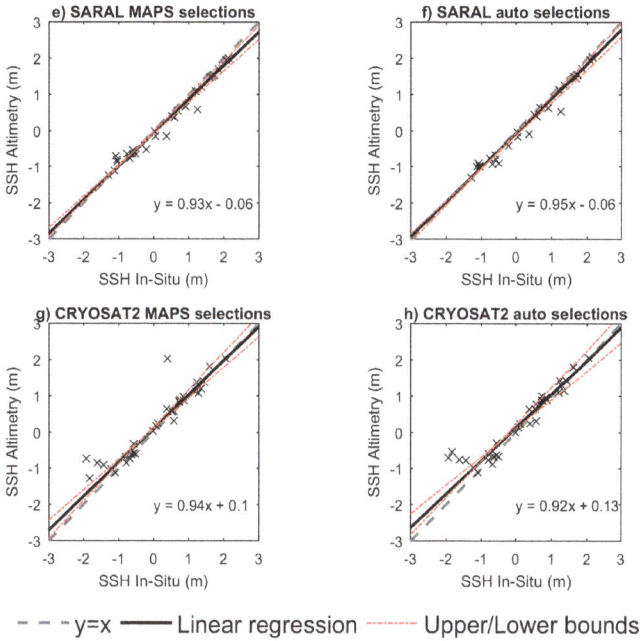

Figure 7. Comparisons between in-situ and altimetry-based SSH for ERS-2 (**a,b**), ENVISAT (**c,d**), SARAL (**e,f**), and CryoSat-2 (**g,h**) using two different approaches for selecting the valid measurements: MAPS manual selections (left) and automatic selections (right).

Linearity exists between altimetry and the tide gauge measurements, for ERS-2, SARAL, and CryoSat-2 data for the two types of selections and for the automatic selections made for ENVISAT (Table 2).

Table 2. Passing-Bablok regression model and statistical results for the comparisons made between in-situ and altimetry-based SSH (ERS-2, ENVISAT, SARAL, and CryoSat-2) for manual (MAPS) and automatic selections.

Mission	MAPS				Auto			
	ERS-2	ENVISAT	SARAL	CryoSat-2	ERS-2	ENVISAT	SARAL	CryoSat-2
Linearity [3]	1	0	1	1	1	1	1	1
Slope	0.16	0.60	0.93	0.94	0.64	0.74	0.95	0.92
Slope LB [1]	−0.05	0.51	0.87	0.85	0.23	0.67	0.91	0.79
Slope UB [2]	0.44	0.71	0.97	1.02	2.07	0.83	1.01	1.01
Intercept	1.06	0.70	−0.06	0.10	1.43	0.66	−0.06	0.13
Intercept LB	1.36	0.74	−0.09	0.10	1.68	0.69	−0.13	0.11
Intercept UB	0.91	0.69	−0.02	0.15	1.02	0.65	−0.05	0.22
R	0.39	0.82	0.99	0.93	0.11	0.86	0.99	0.95
RMSE (m)	1.75	1.04	0.23	0.42	1.70	0.92	0.22	0.39
Mean Bias (m)	1.47	0.88	0.18	0.23	1.50	0.79	0.17	0.24

[1] Lower Bound. [2] Upper Bound. [3] 1 for accepted linearity and 0 for refused linearity.

ERS-2 showed the most unsatisfactory results with slopes lower than 0.64 and intercepts higher than 1 (Table 2). The results obtained by the Passing-Bablok method could not be interpreted for ENVISAT MAPS selections since no linear relationship exists with the tide gauge measurements (a break is observed in the slope at 0 m (Figure 7c)). Low accuracy for MAPS selections made for ENVISAT could be inferred from the RMSE and the mean bias obtained (1.04 m and 0.88 m respectively). The automatic selections significantly improved the estimation of SSH obtained using ENVISAT data and rendered the linearity test positive. The slopes given by ENVISAT increased from 0.60 to 0.74 and were accompanied by an improvement of R (from 0.82 to 0.86) and RMSE that remains important (RMSE~0.92 m) (Table 2). As explained in Section 4.6, the method does not show proportional or constant biases if 1 belongs to the slope CI (slope(LB) < 1 < slope(UB)) and 0 belongs to the intercept CI (intercept (LB) < 0 < intercept (UB)) respectively. For SARAL, a better result is obtained using the automatic selection showing only constant bias with no proportional bias (the confidence interval of the slope contains the value 1). Slight proportional and constant biases are obtained for MAPS selections. MAPS selections present better results for CryoSat-2 that shows slight constant biases with no proportional biases for the two types of selections.

5.3. Topography Comparison

Figure 8 presents topography comparisons between lidar and altimetry measurements for ERS-2 (a), ENVISAT (b), SARAL (c), and CryoSat-2 (d). For topography comparisons all measurements from every cycle were used along with the corresponding lidar topography measurements. Table 3 lists all regression models obtained. Linearity was rejected only for ENVISAT data. Therefore, the slope and the intercept obtained for ENVISAT could not be interpreted. Good comparison results are obtained for SARAL giving a slope of 1.06 and an intercept at 0.01 with no proportional or constant bias observed for topography monitoring. CryoSat-2 provides as well acceptable results with slight systematic and proportional biases. Showing a positive linearity test, ERS-2 slope and intercept showed large differences from 1 and 0.

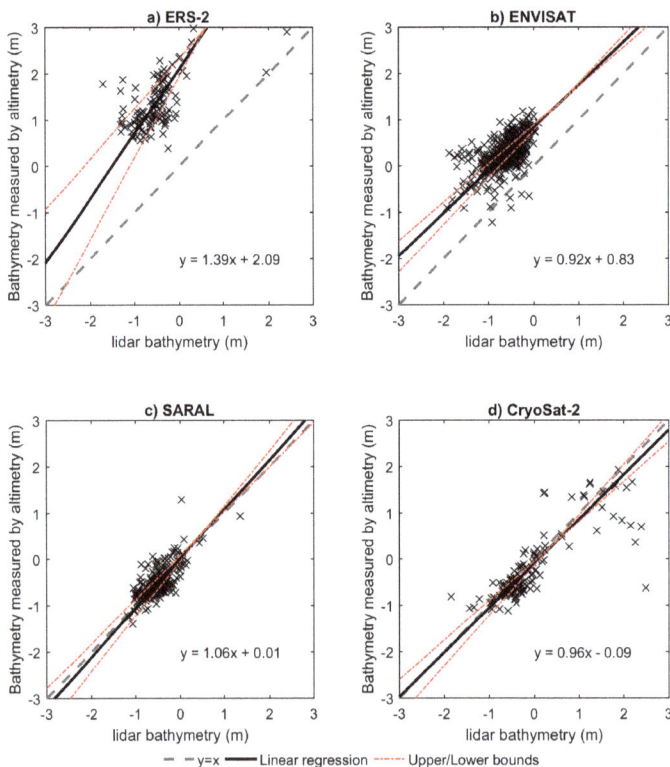

Figure 8. Comparisons between lidar and Altimetry-based topography measurements for: (**a**) ERS-2, (**b**) ENVISAT, (**c**) SARAL, and (**d**) CryoSat-2.

Table 3. Passing-Bablok regression model and statistical results for the comparisons made between lidar and altimetry-based topography estimation for ERS-2, ENVISAT, SARAL, and CryoSat-2.

Mission	ERS-2	ENVISAT	SARAL	CryoSat-2
Linearity Test [3]	1	0	1	1
Slope	1.39	0.92	1.06	0.96
Slope LB [1]	1.08	0.84	0.96	0.86
Slope UB [2]	1.75	1.02	1.19	1.09
Intercept	2.09	0.83	0.01	−0.09
Intercept LB	2.31	0.89	0.08	−0.04
Intercept UB	1.92	0.77	−0.04	−0.13
R	0.17	0.54	0.71	0.79
RMSE (m)	2.01	0.95	0.23	0.44
Mean Bias (m)	1.93	0.90	0.17	0.25

[1] Lower Bound. [2] Upper Bound. [3] 1 for accepted linearity and 0 for refused linearity.

6. Discussion

For each altimetry cycle of ERS2, ENVISAT, SARAL, and CryoSat-2, the water level of the Arcachon Bay was estimated by computing the median of submerged points, and compared to tide gauge measurements carried out at the same time (in a 5 min interval). Altimetry measurements above emerged points were considered as topography measurements and were compared to the nearest lidar bathymetry points.

Over coastal areas, the altimeter footprint is contaminated by signals from different surface types which reduce the accuracy of the measurements [6]. For water levels estimation, SARAL operating at Ka-band frequency shows significant improvement in the measurement accuracy (vertical resolution) comparing to the ERS-2 and ENVISAT Ku-band altimeters, allowing better observations in coastal areas like lagoons [38,39]. This improvement comes from the reduced footprint area (~6 km) of SARAL which reduces the impact of land on SARAL waveforms. It should be noted that the use of Ka-band is not the only factor that governs the size of the footprint, the larger bandwidth (~480 MHz) used by SARAL has an impact on the footprint size as well. According to [6], SARAL AltiKa's footprint is ~0.8 times smaller than a Ku-band altimeter for a given bandwidth. CryoSat-2 shows as well major improvements in the accuracy for water levels estimation comparing to ERS-2 and ENVISAT. Operating in the same band (Ku) as the latter missions, the better performance of CryoSat-2 comes from the use of the SAR mode of acquisition. The processing of CryoSat-2 SAR data is based on the delay Doppler technique that reduces the footprint size in the along-track direction which makes CryoSat-2 more suitable for coastal monitoring. The lower consistency with in-situ measurements was obtained using data from LRM missions operating at Ku-band (ERS-2 and ENVISAT) especially at low tides where its spatial resolution could not resolve the complex water-land mixture of the bay. Figure 7 shows that above mean sea level ENVISAT was capable of estimating the water level of the bay with a relatively low bias.

The results also show a great potential for radar altimetry to retrieve the topographic variation of the intertidal zone. SARAL and CryoSat-2 show a large consistency with the lidar measurements. ENVISAT was able to follow the variation of the intertidal zone bathymetry producing a correlation coefficient (R) of 0.57 but with an important bias (due to the large footprint of ENVISAT).

CryoSat-2 was the only mission that flew over the "ile aux oiseaux" sandbank (Figure 1b). Some altimetry measurements made by CryoSat-2 underestimated the elevation of topography in this area. It is very likely that the cause of the high discrepancy is due to the penetration of the electromagnetic wave in the dry soil (long time after high water). We calculated the maximum penetration depth that corresponds to the electromagnetic wave frequency and the dielectric properties of the sand (Appendix B). The differences obtained between the underestimated measurements made by altimetry and the corresponding lidar measurements are lower than the maximum penetration depth of 1.35 m which is consistent with our hypothesis. We eliminated the underestimated points from CryoSat-2 measurements and we observed an improvement in terms of slope (from 0.96 to 0.98), intercept (from −0.09 to −0.08), and R (from 0.79 to 0.89).

In this study, an automatic classification of cycles was performed to discriminate between submerged and emerged cycles. An automatic process was also envisaged to classify each measurement point, using the backscattering coefficient that tends to increase for land-dominated areas. However, an important limitation was encountered during the automating process. During low tides, some altimetry points above water showed as well high backscattering coefficients, which complicated the separation of emerged and submerged points using the backscattering coefficient parameter. The high values of the backscattering coefficient obtained for water at low tides is most likely due to smoother water surfaces in the channels than in the bay, caused by the reduced wave activity. This reduced roughness increases the specular reflection and thus the power received by the sensor (the backscattering coefficient). An example is shown in Figure 9 (left) for the cycle 20 of SARAL. The backscattering coefficient of water in large channels (zone 5 and zone 3) is lower than the backscattering coefficient of water in narrow channels (zone 4 and zone 2) showing values that match land-dominated areas (zone 1).

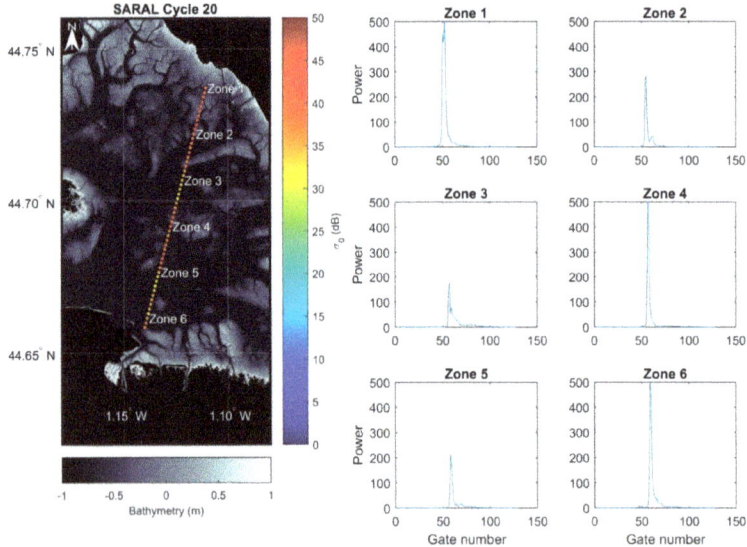

Figure 9. Spatial variation of the backscattering coefficient for SARAL (cycle 20) at low tide along with the corresponding waveforms of the indicated zones.

Figures 9–12 show the spatial variation of the backscattering coefficient for a given cycle and the waveforms corresponding to different points in the bay. Figures 10 and 12 correspond to cycles crossing the bay at high tides for SARAL and CryoSat-2 respectively (same type of figures is available in the supplementary material for ENVISAT (Figures S3 and S4)). For SARAL (Figure 10) and CryoSat-2 (Figure 12), we observe at the middle of the bay (zone 3 and zone 4) that the waveforms look like typical ocean waveforms. However, the waveforms at the edges of the bay (zones 1, 2, 5, and 6 for SARAL and zones 1 and 6 for CryoSat-2) show a sharp rise. It is not the case for ENVISAT at high tides (Figure S3). The waveforms of ENVISAT in the middle of the bay (zone 3 and zone 4) differ from the typical ocean waveforms due to the larger footprint of ENVISAT. It should be noted that the waveforms of CryoSat-2 at high tides maintain typical ocean waveforms closer to the coastline than SARAL due to the SAR mode of CryoSat-2 that increases the resolution in the along-track direction of the satellite. Another consequence of the SAR mode is the highly energetic waveforms.

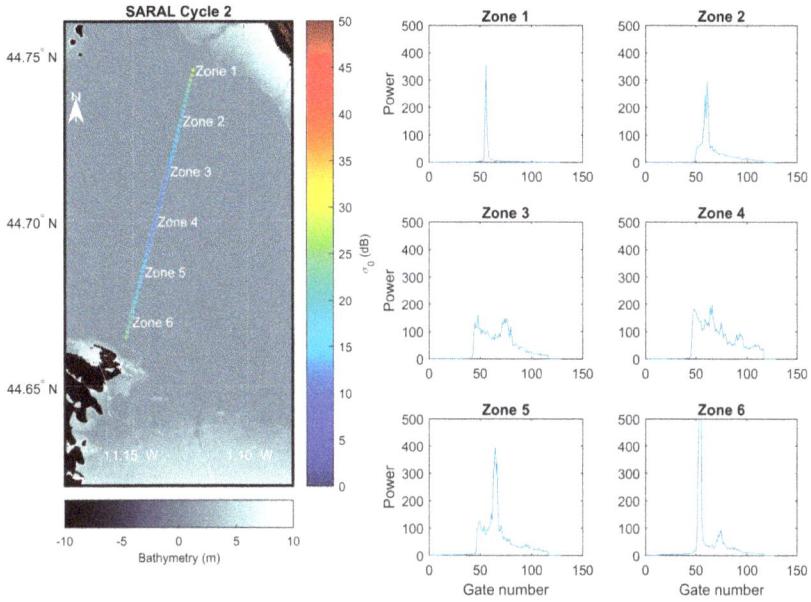

Figure 10. Spatial variation of the backscattering coefficient for SARAL (cycle 2) at high tide along with the corresponding waveforms of the indicated zones.

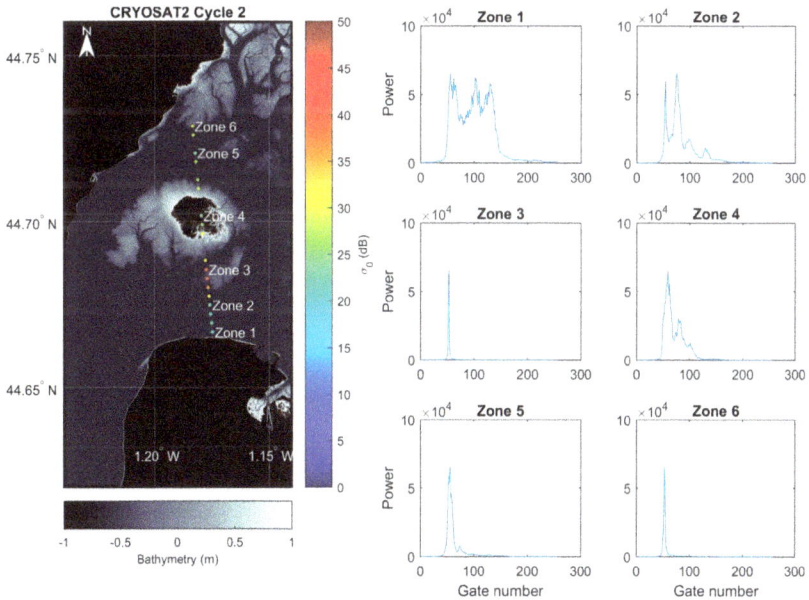

Figure 11. Spatial variation of the backscattering coefficient for CryoSat-2 (cycle 2) at low tide along with the corresponding waveforms of the indicated zones.

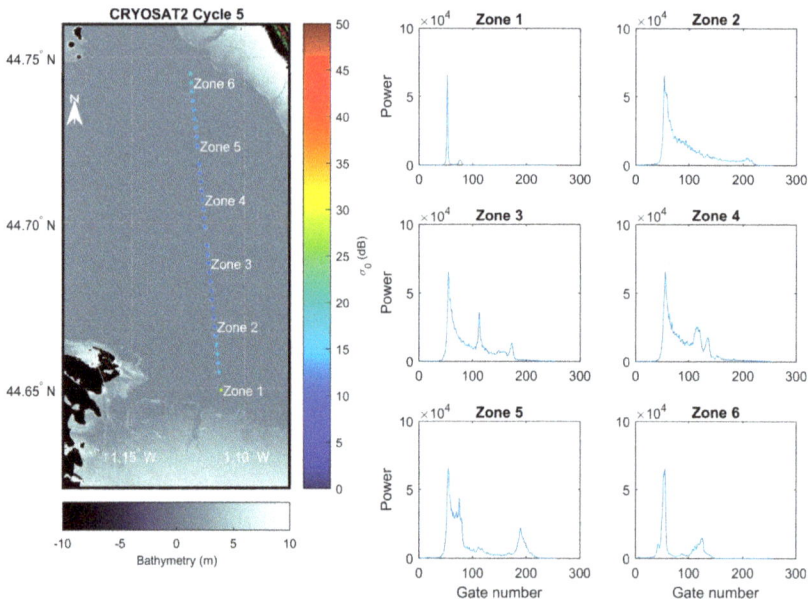

Figure 12. Spatial variation of the backscattering coefficient for CryoSat-2 (cycle 5) at high tide along with the corresponding waveforms of the indicated zones.

Figures 9 and 11 correspond to cycles crossing the bay at low tides for SARAL and CryoSat-2 respectively. At low tides, SARAL (Figure 9) shows peak shaped waveforms (relatively similar to ENVISAT waveforms at low tides (Figure S4)) while CryoSat-2 (Figure 11) shows mixed shaped waveforms. Typical ocean waveforms were not observed for low tides even in areas dominated by water.

The use of satellite radar altimetry to monitor SSH and topography has the potential to be applied in any tidal zone especially in remote areas where no tide gauge record or topography data are available. Radar altimetry parameters (SSH, σ_0 and peakiness) can be used to discriminate between high (parameter values almost constant for a given cycle) and low (larger variability) tides. Until now, only large submerged and emerged areas can be discriminated during low tides. Analyses of the radar echoes (waveforms) can provide valuable information to improve the characterization of the nature of the surface (emerged or submerged).

The new-era coastal altimetry satellites contribute to the advancement of coastal ocean observing systems [40]. The launch of the SWOT mission in 2021 will offer new opportunities for monitoring of coastal lagoons topography and changes in time thanks to (i) the acquisitions of elevation along two swaths of 60 km of width that will provide 2D maps, (ii) its better vertical accuracy, (iii) its shorter repeat period, (iv) the use of the Ka-band that has a smaller penetration depth.

7. Conclusions

This study is the first to demonstrate the capabilities of altimetry to retrieve the Sea Surface Height and the topography of the intertidal zone in a coastal lagoon. Data from four altimetry missions were analyzed over Arcachon's Bay coastal lagoon located in the South-East of the Bay of Biscay: (i) ERS-2, (ii) ENVISAT, (iii) SARAL, and (iv) CryoSat-2. The two first missions were operating at Ku-band in low resolution mode whereas SARAL is the first mission to operate in Ka-band in LRM and CryoSat-2 is the first mission to operate at Ku-band in SAR mode. Altimetry measurements were validated using tide gauge and lidar bathymetry datasets.

Considering data from the classical LRM missions operating at Ku-band (ERS-2 and ENVISAT), SSH and topography retrievals from ERS-2 are inaccurate whereas ENVISAT measurements provide better results for SSH (R > 0.82 and RMSE < 1.04 m) than for topography (R~0.54 and RMSE~0.95 m). However, SARAL and CryoSat-2 showed very satisfying correspondence with ancillary data with correlation coefficient (R) higher than 0.93 and RMSE lower than 0.42 m for SSH estimations and R higher than 0.71 for topography with RMSE lower than 0.44 m. It should be noted that the standard deviation of the whole sea level records is 1.05 m.

Furthermore, a successful and simple statistical method (based on a combination of a non-supervised clustering technique and grouping altimetry data into quartiles) was introduced in order to select valid altimetry data. The purpose of this method is to accelerate the processing phase of altimetry data and to optimize the accuracy of Sea Surface Height estimation. This selection method showed very satisfying results with similar accuracy performance as the manual selections made using MAPS software.

The most important limitation lies in the large footprint of altimeters. This coarse resolution prevented us from automatically discriminating land from water using intrinsic altimetry parameters like the backscattering coefficient or waveforms peakiness. However, we were able to discriminate acquisitions made during low tides from the ones made during high tides, because at the scale of the bay good contrast between low and high tides was obtained by the backscattering coefficient and the peakiness.

While altimetry was able to monitor the topography of the bay, caution must be taken when using altimetry data to retrieve topographic variations in similar environments. Dry sandy features are most likely problematic due to the important penetration depth of electromagnetic waves in such mediums. It is recommended to use this approach in areas exposed frequently to water.

Supplementary Materials: The following are available online at www.mdpi.com/2072-4292/10/2/297/s1, **Figure S1**: (a,b) Examples of ERS-2 along-track profiles of altimetry height over water (purple crosses) and land (green crosses) at high (a) and low (b) tides, the topography under the altimeter ground track is represented in brown and it is filled with water (in blue) using leveled tide-gauge records; (c,d) Variation of Ice-1 backscattering coefficients of Ku-band (black dots) at high (c) and low (d) tides., **Figure S2**: (a,b) Examples of ENVISAT along-track profiles of altimetry height over water (purple crosses) and land (green crosses) at high (a) and low (b) tides, the topography under the altimeter ground track is represented in brown and it is filled with water (in blue) using leveled tide-gauge records; (c,d) Variation of Ice-1 backscattering coefficients of Ku-band (black dots) at high (c) and low (d) tides. **Figure S3**: Spatial variation of the backscattering coefficient for ENVISAT (cycle 21) at high tide along with the corresponding waveforms of the indicated zones. **Figure S4**: Spatial variation of the backscattering coefficient for ENVISAT (cycle 9) at low tide along with the corresponding waveforms of the indicated zones.

Acknowledgments: This work was supported by the CNES. It is based on SWOT space mission and the SWOT COTEST project and was also supported by the Normandy Region.

Author Contributions: Edward Salameh and Frédéric Frappart conceived and designed the experiments; Edward Salameh, Frédéric Frappart, Alexandra Spodar, Vincent Marieu, and Jean-Paul Parisot performed the experiments and analyzed the data; Edward Salameh and Frédéric Frappart wrote the paper. All authors were involved in the analysis of the results and contributed to the discussion of results and the preparation of the manuscript.

Conflicts of Interest: The authors declare no conflict of interest.

Appendix A.

The Ice-1 (OCOG—Offset Centre of Gravity) retracking algorithm approach consists of replacing the waveform with a box that has the same center of gravity as the waveform. The box is defined using three measurements:

$$\text{centre of gravity} = \frac{\sum_{n=1+aln}^{n=N-aln} n y^2(n)}{\sum_{n=1+aln}^{n=N-aln} y^2(n)} \tag{A1}$$

$$\text{amplitude} = \sqrt{\frac{\sum_{n=1+aln}^{n=N-aln} y^4(n)}{\sum_{n=1+aln}^{n=N-aln} y^2(n)}} \tag{A2}$$

$$width = \frac{\left(\sum_{n=1+aln}^{n=N-aln} y^2(n)\right)^2}{\sum_{n=1+aln}^{n=N-aln} y^4(n)} \tag{A3}$$

where N is the total gate number, aln is the number of eliminated gate in the starting and ending of waveform and y(n) is the power of the nth gate. The middle of the leading edge point (LEP) is given by:

$$LEP = centre\ of\ gravity - 0.5 \times width \tag{A4}$$

Peakiness is a measure of how sharply peaked an echo is. It is the ratio of the maximum power (highest bin value) to the accumulated echo power above the retracking point. The peakiness in AltiKa GDR is given by:

$$Peakiness_{AltiKa} = \frac{P_{max} \times N_{right}}{\sum_{i=1}^{N_{WF}} P_i} \tag{A5}$$

where N_{WF} is the total number of the registering gates, and N_{right} is a number of points to the right from the tracking point (i.e., above the retracking point). For SIRAL:

$$Peakiness_{SIRAL} = \frac{P_{max} \times 30}{\sum_{i=1}^{N_{WF}} P_i} \tag{A6}$$

Appendix B.

The penetration depth (δp) of an electromagnetic wave in an homogeneous medium can be expressed as a function of its dielectric permittivity. According to [41], it can be written as follows:

$$\delta_p = \frac{\lambda \sqrt{\varepsilon\prime}}{2\pi\varepsilon\prime\prime} \tag{A7}$$

where λ is the wavelength of the electromagnetic wave, $\varepsilon\prime$ and $\varepsilon\prime\prime$ are the real and complex parts of the dielectric permittivity respectively. Empirical relationships between dielectric permittivity (ε) and soil nature and its water content were established. In this study, we used the empirical relationships from [42] that have the following form:

$$\varepsilon = a_0 + a_1 S + a_2 C + (b_0 + b_1 S + b_2 C) m_v + (c_0 + c_1 S + c_2 C) m_v \tag{A8}$$

where S and C are the sand and clay fraction of soil given in percent, m_v is the volumetric water content given in percent, $a_0, a_1, a_2, b_0, b_1, b_2, c_0, c_1, c_2$ are the constants of the empirical relationship given in Table 2 of (Hallikainen et al., 1985).

Maximum penetration depth are obtained for dry soils (i.e., $m_v = 0$). (A1) becomes:

$$max(\delta_p) = \frac{\lambda \sqrt{a_0\prime + a_1\prime S + a_2\prime C}}{2\pi(a_0\prime\prime + a_1\prime\prime S + a_2\prime\prime C)} \tag{A9}$$

Over the sandbank known as "île aux oiseaux", we consider $S = 100$ and $C = 0$. This latter equation becomes:

$$max(\delta_p) = \frac{\lambda \sqrt{a_0\prime + 100a_1\prime}}{2\pi(a_0\prime\prime + 100a_1\prime\prime)} \tag{A10}$$

References

1. Agardy, T.; Alder, J. Coastal Systems. In *Ecosystems and Human Well-Being: Current State and Trends*; Hassan, R., Scholes, R., Ash, N., Eds.; Island Press: Washington, DC, USA, 2005; pp. 513–550.
2. CIESIN—Center for International Earth Science Information Network. Gridded Population of the World (GPW) Version 3 Beta. Available online: http://sedac.ciesin.columbia.edu (accessed on 20 September 2017).

3.   Barnes, R.S.K. *Coastal Lagoons (Cambridge Studies in Modern Biology 1)*; Cambridge University Press: Cambridge, UK, 1980.

4.   Kjerfve, B.; Magill, K.E.; Baruch, B.W. Geographic and hydrodynamic characteristics of shallow coastal lagoons. *Mar. Geol.* **1989**, *88*, 187–199. [CrossRef]

5.   Chapman, P.M. Management of coastal lagoons under climate change. *Estuar. Coast. Shelf Sci.* **2012**, *110*, 32–35. [CrossRef]

6.   Vignudelli, S.; Kostianoy, A.G.; Cipollini, P.; Benveniste, J. (Eds.) *Coastal Altimetry*; Springer: Berlin/Heidelberg, Germany, 2011.

7.   Cipollini, P.; Calafat, F.M.; Jevrejeva, S.; Melet, A.; Prandi, P. Monitoring Sea Level in the Coastal Zone with Satellite Altimetry and Tide Gauges. *Surv. Geophys.* **2017**, *38*, 33–57. [CrossRef]

8.   Deborde, J.; Anschutz, P.; Auby, I.; Glé, C.; Commarieu, M.-V.; Maurer, D.; Lecroart, P.; Abril, G. Role of tidal pumping on nutrient cycling in a temperate lagoon (Arcachon Bay, France). *Mar. Chem.* **2008**, *109*, 98–114. [CrossRef]

9.   Plus, M.; Dumas, F.; Stanisiè Re, J.-Y.; Maurer, D. Hydrodynamic characterization of the Arcachon Bay, using model-derived descriptors. *Cont. Shelf Res.* **2009**, *29*, 1008–1013. [CrossRef]

10. Rimmelin, P.; Dumon, J.-C.; Maneux, E.; Gonçalves, A. Study of Annual and Seasonal Dissolved Inorganic Nitrogen Inputs into the Arcachon Lagoon, Atlantic Coast (France). *Estuar. Coast. Shelf Sci.* **1998**, *47*, 649–659. [CrossRef]

11. Blanchet, H.; De Montaudouin, X.; Chardy, P.; Bachelet, G. Structuring factors and recent changes in subtidal macrozoobenthic communities of a coastal lagoon, Arcachon Bay (France). *Estuar. Coast. Shelf Sci.* **2005**, *64*, 561–576. [CrossRef]

12. Envisat, C.Z. RA-2 Advanced Radar Altimeter: Instrument Design and Pre-Launch Performance Assessment Review. *Acta Astronaut.* **1999**, *44*, 323–333.

13. Verron, J.; Sengenes, P.; Lambin, J.; Noubel, J.; Steunou, N.; Guillot, A.; Picot, N.; Coutin-Faye, S.; Sharma, R.; Gairola, R.M.; et al. The SARAL/AltiKa Altimetry Satellite Mission. *Mar. Geodesy* **2015**, *38*, 2–21. [CrossRef]

14. Nielsen, K.; Stenseng, L.; Andersen, O.; Knudsen, P. The Performance and Potentials of the CryoSat-2 SAR and SARIn Modes for Lake Level Estimation. *Water* **2017**, *9*, 374. [CrossRef]

15. Wingham, D.J.; Francis, C.R.; Baker, S.; Bouzinac, C.; Brockley, D.; Cullen, R.; de Chateau-Thierry, P.; Laxon, S.W.; Mallow, U.; Mavrocordatos, C.; et al. CryoSat: A mission to determine the fluctuations in Earth's land and marine ice fields. *Adv. Space Res.* **2006**, *37*, 841–871. [CrossRef]

16. Frappart, F.; Blumstein, D.; Cazenave, A.; Ramillien, G.; Birol, F.; Morrow, R.; Rémy, F. Satellite Altimetry: Principles and Applications in Earth Sciences. In *Wiley Encyclopedia of Electrical and Electronics Engineering*; John Wiley & Sons, Inc.: Hoboken, NJ, USA, 2017; pp. 1–25.

17. Chelton, D.B.; Ries, J.C.; Haines, B.J.; Fu, L.L.; Callahan, P.S. Satellite Altimetry. In *Satellite Altimetry and the Earth Sciences: A Handbook of Techniques and Applications*; Fu, L.L., Cazenave, A., Eds.; Academic Press: San Diego, CA, USA, 2001; pp. 1–131.

18. Frappart, F.; Fatras, C.; Mougin, E.; Marieu, V.; Diepkilé, A.T.; Blarel, F.; Borderies, P. Radar altimetry backscattering signatures at Ka, Ku, C, and S bands over West Africa. *Phys. Chem. Earth* **2015**, *83–84*, 96–110. [CrossRef]

19. Frappart, F.; Roussel, N.; Biancale, R.; Martinez Benjamin, J.J.; Mercier, F.; Perosanz, F.; Garate Pasquin, J.; Martin Davila, J.; Perez Gomez, B.; Gracia Gomez, C.; et al. The 2013 Ibiza Calibration Campaign of Jason-2 and SARAL Altimeters. *Mar. Geodesy* **2015**, *38*, 219–232. [CrossRef]

20. Biancamaria, S.; Frappart, F.; Leleu, A.-S.; Marieu, V.; Blumstein, D.; Desjonquères, J.-D.; Boy, F.; Sottolichio, A.; Valle-Levinson, A. Satellite radar altimetry water elevations performance over a 200 m wide river: Evaluation over the Garonne River. *Adv. Space Res.* **2017**, *59*, 128–146. [CrossRef]

21. Frappart, F.; Papa, F.; Marieu, V.; Malbeteau, Y.; Jordy, F.; Calmant, S.; Durand, F.; Bala, S. Preliminary Assessment of SARAL/AltiKa Observations over the Ganges-Brahmaputra and Irrawaddy Rivers. *Mar. Geodesy* **2015**, *38*, 568–580. [CrossRef]

22. Gaspar, P.; Ogor, F.; Le Traon, P.-Y.; Zanife, O.-Z. Estimating the sea state bias of the TOPEX and POSEIDON altimeters from crossover differences. *J. Geophys. Res.* **1994**, *99*, 24981–24994. [CrossRef]

23. Parisot, J.-P.; Diet-Davancens, J.; Sottolichio, A.; Crosland, E.; Drillon, C.; Verney, R. Modélisation des agitations dans le bassin d'Arcachon. *Xèmes Journées Sophia Antip.* **2008**, 435–444. [CrossRef]

24. Cartwright, D.E.; Edden, A.C. Corrected Tables of Tidal Harmonics. *Geophys. J. R. Astron. Soc.* **1973**, *33*, 253–264. [CrossRef]

25. Wahr, J.M. Deformation induced by polar motion. *J. Geophys. Res. Solid Earth* **1985**, *90*, 9363–9368. [CrossRef]

26. Wingham, D.J.; Rapley, C.G.; Griffiths, H. New Techniques in Satellite Altimeter Tracking Systems. In Proceedings of the IGARSS 86 Symposium, Zurich, Switzerland, 8–11 September 1986; pp. 1339–1344.

27. Bamber, J.L. Ice sheet altimeter processing sheme. *Int. J. Remote Sens.* **1994**, *15*, 925–938. [CrossRef]

28. Frappart, F.; Minh, K.D.; L'hermitte, J.; Cazenave, A.; Ramillien, G.; Le Toan, T.; Mognard-Campbell, N. Water volume change in the lower Mekong from satellite altimetry and imagery data. *Geophys. J. Int.* **2006**, *167*, 570–584. [CrossRef]

29. Frappart, F.; Calmant, S.; Cauhopé, M.; Seyler, F.; Cazenave, A. Preliminary results of ENVISAT RA-2-derived water levels validation over the Amazon basin. *Remote Sens. Environ.* **2006**, *100*, 252–264. [CrossRef]

30. Laxon, S. Sea ice altimeter processing scheme at the EODC. *Int. J. Remote Sens.* **1994**, *15*, 915–924. [CrossRef]

31. Jekeli, C. *Geometric Reference System in Geodesy*; Division of Geodesy and Geospatial Science School of Earth Sciences, Ohio State University: Columbus, OH, USA, 2006.

32. SHOM. Références Altimétriques Maritimes (RAM). Available online: http://diffusion.shom.fr/ (accessed on 10 February 2018).

33. Hartigan, J.A.; Wong, M.A. A K-Means Clustering Algorithm. *J. R. Stat. Soc.* **1979**, *28*, 100–108.

34. Craw, S. Manhattan Distance. In *Encyclopedia of Machine Learning*; Sammut, C., Webb, G.I., Eds.; Springer Science & Business Media: Boston, MA, USA, 2011.

35. Passing, H.; Bablok, W. A new biometrical procedure for testing the equality of measurements from two different analytical methods. *J. Clin. Chem. Clin. Biochem.* **1983**, *21*, 709–720. [PubMed]

36. Cancet, M.; Bijac, S.; Chimot, J.; Bonnefond, P.; Jeansou, E.; Laurain, O.; Lyard, F.; Bronner, E.; Féménias, P. Regional in situ validation of satellite altimeters: Calibration and cross-calibration results at the Corsican sites. *Adv. Space Res.* **2013**, *51*, 1400–1417. [CrossRef]

37. Ménard, Y.; Jeansou, E.; Vincent, P. Calibration of the TOPEX/POSEIDON altimeters at Lampedusa: Additional results at Harvest. *J. Geophys. Res.* **1994**, *99*, 24487. [CrossRef]

38. Vu, P.; Frappart, F.; Darrozes, J.; Marieu, V.; Blarel, F.; Ramillien, G.; Bonnefond, P.; Birol, F.; Blarel, F. Multi-Satellite Altimeter Validation along the French Atlantic Coast in the Southern Bay of Biscay from ERS-2 to SARAL. *Remote Sens.* **2018**, *10*, 93. [CrossRef]

39. Bonnefond, P.; Verron, J.; Aublanc, J.; Babu, K.; Bergé-Nguyen, M.; Cancet, M.; Chaudhary, A.; Crétaux, J.-F.; Frappart, F.; Haines, B.; et al. The Benefits of the Ka-Band as Evidenced from the SARAL/AltiKa Altimetric Mission: Quality Assessment and Unique Characteristics of AltiKa Data. *Remote Sens.* **2018**, *10*, 83. [CrossRef]

40. Liu, Y.; Kerkering, H.; Weisberg, R.H. *Coastal Ocean Observing Systems*; Elsevier (Academic Press): London, UK, 2015.

41. Ulaby, F.; Moore, R.; Fung, A. *Microwave Remote Sensing: Active and Passive. Volume 2, Radar Remote Sensing and Surface Scattering and Emission Theory*; Addison-Wesley: Boston, MA, USA, 1982.

42. Hallikainen, M.; Ulaby, F.; Dobson, M.; El-rayes, M.; Wu, L. Microwave Dielectric Behavior of Wet Soil-Part 1: Empirical Models and Experimental Observations. *IEEE Trans. Geosci. Remote Sens.* **1985**, *GE-23*, 25–34. [CrossRef]

remote sensing

MDPI

Article

Evolution of the Performances of Radar Altimetry Missions from ERS-2 to Sentinel-3A over the Inner Niger Delta

Cassandra Normandin [1,*], Frédéric Frappart [2,3], Adama Telly Diepkilé [4], Vincent Marieu [1], Eric Mougin [2], Fabien Blarel [3], Bertrand Lubac [1], Nadine Braquet [5] and Abdramane Ba [6]

[1] Oceanic and Continental Environments and Paleoenvironments (EPOC), Mixed Research Unit (UMR) 5805, University of Bordeaux, Allée Geoffroy Saint-Hilaire, 33615 Pessac, France; vincent.marieu@u-bordeaux.fr (V.M); bertrand.lubac@u-bordeaux.fr (B.L)

[2] Geosciences Environment Toulouse (GET), University of Toulouse, National Center for Scientific Reaseach (CNRS), Institute for Research and Development (IRD), UPS. Observatory Midi-Pyrénées (OMP), 14 Av. E. Belin, 31400 Toulouse, France; frederic.frappart@legos.obs-mip.fr (F.F.); eric.mougin@get.omp.eu (E.M.)

[3] Laboratory of Studies on Spatial Geophysics and Space Oceanography (LEGOS), University of Toulouse, National Center for Space Studies (CNES), CNRS, IRD, UPS. OMP, 14 Av. E. Belin, 31400 Toulouse, France; fabien.blarel@legos.obs-mip.fr

[4] Department of Education and Research (DER) Math-Informatics, Faculty of Sciences and Technology (FST)/University of Sciences, Techniques and Technologies of Bomako (USTTB), Bamako 3206, Mali; Adama.Diepkile@USherbrooke.ca

[5] National Research Institute of Science and Technology for Environment and Agriculture (IRSTEA), IRD, 361 rue Jean-François Breton, 34196 Montpellier, France; nadine.braquet@ird.fr

[6] Laboratory of Optics, Spectroscopy and Atmospheric Sciences (LOSSA), Department of Education and Research (DER) Physics, Faculty of Sciences and Technology (FST)/University of Sciences, Techniques and Technologies of Bomako (USTTB), Bamako 3206, Mali; abdramaneba55@yahoo.fr

* Correspondence: cassandra.normandin@u-bordeaux.fr

Received: 14 March 2018; Accepted: 23 May 2018; Published: 25 May 2018

Abstract: Radar altimetry provides unique information on water stages of inland hydro-systems. In this study, the performance of seven altimetry missions, among the most commonly used in land hydrology (i.e., European Remote-Sensing Satellite-2 (ERS-2), ENVIronment SATellite (ENVISAT), Satellite with Argos and ALtika (SARAL), Jason-1, Jason-2, Jason-3 and Sentinel-3A), are assessed using records from a dense in situ network composed of 19 gauge stations in the Inner Niger Delta (IND) from 1995 to 2017. Results show an overall very good agreement between altimetry-based and in situ water levels with correlation coefficient (R) greater than 0.8 in 80% of the cases and Root Mean Square Error (RMSE) lower than 0.4 m in 48% of cases. Better agreement is found for the recently launched missions such as SARAL, Jason-3 and Sentinel-3A than for former missions, indicating the advance of the use of the Ka-band for SARAL and of the Synthetic-aperture Radar (SAR) mode for Sentinel-3A. Cross-correlation analysis performed between water levels from the same altimetry mission leads to time-lags between the upstream and the downstream part of the Inner Niger Delta of around two months that can be related to the time residence of water in the drainage area.

Keywords: altimetry; water levels; validation; Inner Niger Delta

1. Introduction

Surface waters, which are part of the continental branch of the terrestrial water cycle, play an essential role in supplying fresh water for basic human and economic needs. They are strongly impacted by climate changes and anthropogenic pressures caused by population growth and changes in agricultural practices [1–3]. Despite the importance of their monitoring for addressing integrated water resource management, use in operational flood forecasting or disaster mitigation, reliable in situ measurements of water stage and discharge has become increasingly scarce information due to either the disappearance of the gauge networks or the difficulty to get access to data [4,5].

Satellite radar altimetry, initially developed for the measurement of the ocean surface topography through the measurement of the distance between the Earth's surface and the spaceborne radar altimeter [6], has demonstrated its efficiency for deriving water levels of inland water bodies (see Crétaux et al. [7] for a recent review). Radar altimetry, was initially used over land to retrieve water levels over homogeneous surfaces such as large lakes and enclosed seas [8,9], but also at cross-sections between rivers and altimetry ground-tracks of several kilometers of width in large river basins [10,11]. These early results were obtained using Geosat and Topex/Poseidon (T/P) ranges (i.e., the distance between the satellite and the surface) derived from the Ocean retracking algorithm. Root Mean Square Errors (RMSE) lower than 0.05 m and 1.1 m through comparisons with in situ water stages were obtained over lakes and rivers respectively. The comparisons are performed at the so-called Virtual Stations. Virtual stations (VS) are defined as the cross-sections of an altimetry ground-track and a water body (i.e., lake, reservoir river channel, floodplain, or wetland) where the temporal variations of the height from one cycle to the next can be associated with changes in water level [7]. With the launch of ENVISAT in 2002, ranges processed using other retracking algorithms were included in the Geophysical Data Records (GDR) made available by space agencies. Among them, the Offset Center Of Gravity (OCOG, also known as Ice-1) was found to provide, most of the time, the most accurate estimate of river water levels (with RMSE generally lower than 0.3 m and correlation coefficient R greater than 0.9) [12]. Combined with the availability of land-dedicated corrections of the ionosphere, wet troposphere delays and improvements in the data processing, this allowed the generalization of the use of radar altimetry for the monitoring of inland waters [7,13].

Thanks to these different improvements and the use of high-frequency data (10, 18, 20 or 40 Hz depending on the altimetry mission) instead of 1 Hz data (~7 km of sampling along the track) as over the open ocean water bodies of a few or below one hundred meters of width can now be monitored with very good accuracy (e.g., [14–16]). As radar altimetry data have global coverage and are freely available, they are now commonly used in a wide range of hydrological applications (see Crétaux et al. [7] for a recent review), and, even in support for the management of in situ networks [17]. Until now, no study provided a systematic assessment of the performance over rivers of different altimetry missions that were operating since the beginning of the high-precision altimetry era, which started with the launch of Topex/Poseidon (T/P) in 1992, contrary to that done over lakes [18,19].

The goal of this study is to evaluate the quality of altimetry-based water levels for all missions in repetitive orbits whose data contained in the GDR were processed using the OCOG retracking algorithm. For this purpose, comparisons between altimetry-based water stages and in situ measurements from a dense gauges network were performed in the Inner Niger Delta (IND).

Several studies already used altimetry-based water levels to better understand spatio-temporal dynamics of the flood in this region [20–22] and to estimate river discharges [23,24]. The IND was chosen as study area as it is densely covered with (i) in situ gauge stations whose records are available over the whole high-precision altimetry era and (ii) cross-sections between altimetry ground-tracks from different missions and rivers of various widths.

A dense network of VS was built in the IND, composed of 52, 63, 623 VS for European Remote-Sensing Satellite-2 (ERS-2), ENVIronment SATellite (ENVISAT), Satellite with Argos and ALtika (SARAL) respectively, 31 for Sentinel-3A and 8, 8, 9 for Jason-1, Jason-2 and Jason-3 respectively. At each virtual station, time variations of river levels from radar altimetry are constructed.

In this study, comparisons between altimetry-based water stages derived from acquisitions of Jason-1, Jason-2 and Jason-3, ERS-2, ENVISAT, SARAL and Sentinel-3A and in situ water levels from 19 gauge stations located in the IND are presented in terms of RMSE and R. Intra-mission results consistency were also assessed through cross-correlations between virtual stations along the river.

2. Method

2.1. Principle of Radar Altimetry and Data Processing

2.1.1. Principle of Altimetry Measurement

The principle of radar altimetry is the following: a radar altimeter emits an electromagnetic wave in the nadir direction and measures its round-trip time. The distance between the satellite and the Earth surface—the altimeter range (R_0)—is derived with a precision of a few centimeters. The satellite altitude (H) referred to an ellipsoid is determined from precise orbitography technique with accuracy better than 2 cm. Taking into account propagation corrections caused by delays resulting from interactions of electromagnetic wave with the atmosphere, and geophysical corrections, the height of the reflecting surface (h) with reference to an ellipsoid can be estimated as [25,26]:

$$h = H - \left(R_0 + \sum \left(\Delta R_{propagation} + \Delta R_{geophysical} \right) \right) \tag{1}$$

where H is the height of the center of mass of the satellite above the ellipsoid estimated using precise orbit determination (POD) technique, R_0 is the nadir altimeter range from the center of mass of the satellite to the sea surface taking into account instrumental corrections.

$$\sum \Delta R_{propagation} = \Delta R_{ion} + \Delta R_{dry} + \Delta R_{wet} \tag{2}$$

where ΔR_{ion} is the atmospheric refraction range correction due to the free electron content associated with the dielectric properties of the ionosphere, ΔR_{dry} is the atmospheric refraction range correction due to the dry gas component of the troposphere, ΔR_{wet} is the atmospheric refraction range correction due to the water vapor and the cloud liquid water content of the troposphere.

$$\sum \Delta R_{geophysical} = \Delta R_{solid\ Earth} + \Delta R_{pole} \tag{3}$$

where $\Delta R_{solid\ Earth}$ and ΔR_{pole} are the corrections respectively accounting for crustal vertical motions due to the solid Earth and pole tides.

2.1.2. Time Variations of River Levels from Radar Altimetry Measurements

In this study, the Multi-mission Altimetry Processing Software (MAPS), developed by Frappart et al. [27] was used to visualize and process the altimetry data over land [22,28–30] and ocean [31,32] to build the VS in the IND. Data processing is composed of three main steps: (i) a coarse delineation of the VS using Google Earth; (ii) a refined selection of the valid altimetry data based on visual inspection; and (iii) the computation of the time series of water level. The altimetry-based water level is computed for each cycle using the median of the selected altimetry heights, along with their respective deviation (i.e., mean absolute deviation). This process is repeated each cycle to construct the water level time series at the virtual stations and illustrated in Figure 1.

Altimetry datasets are referenced either to WGS84 ellipsoid or to Topex/Poseidon ellipsoid. A datum conversion from T/P ellipsoid to WGS84 is automatically performed using Equation (4) adapted from Jekeli et al. [33] and implemented in the version of MAPS used in Salameh et al. [34]:

$$\Delta h = \frac{a' \left(1 - e'^2 \right)}{\sqrt{1 - e'^2 \sin^2 \varphi}} - \frac{a \left(1 - e^2 \right)}{\sqrt{1 - e^2 \sin^2 \varphi}} \tag{4}$$

where Δh is the variation of height at latitude φ due to the change of ellipsoid from T/P to WGS84 datum, $a = 6,378,137$ m and $e = 0.081819190842621$ are the semi-major axis and the eccentricity of the WGS84 datum, $a' = 6,378,136.3$ m and $e' = 0.081819221456$ are the semi-major axis and the eccentricity of the T/P datum.

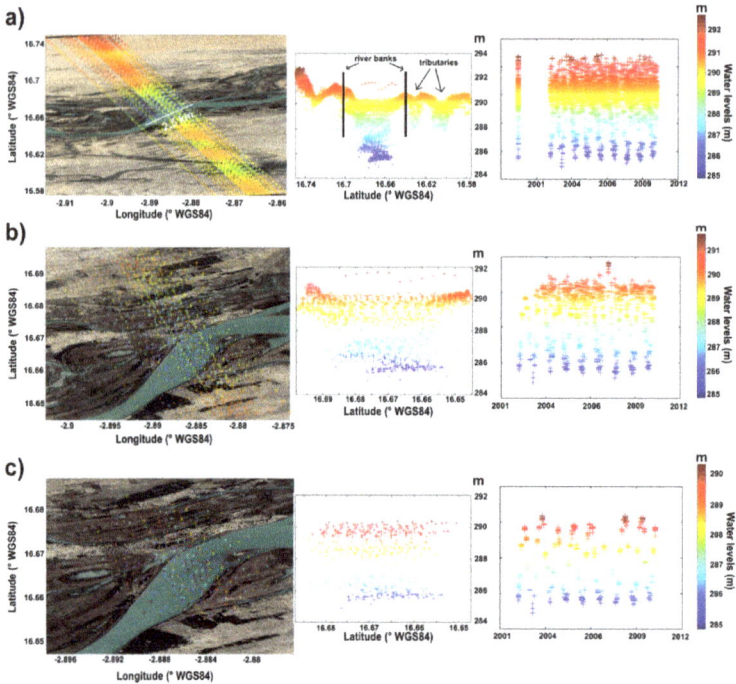

Figure 1. The different steps of the altimetry data using Multi-mission Altimetry Processing Software (MAPS). First, a rough selection of the altimetry data (represented with crosses of color) is performed: (**a**) all altimetry data located a few kilometers away from the center of the river are selected (left panel), the major topographic features, such hills, river banks, tributaries, etc. can be identified (central panel), temporal variations of the altimetry signal cannot be related to any hydrological signal (right panel). Then, a more accurate selection is made; (**b**) the number of data is decreasing and outliers are removed (left panel), the shape of the river and its temporal variations in width (central panel) and height (left panel) becomes clearer. This process is repeated until final selection is achieved; (**c**) all subfigures are derived from the MAPS Graphical User Interface (GUI).

Some along-track altimetry profiles exhibit a parabolic shape caused by non-nadir reflections known as hooking effect (see Figure 2). Hooking effect is corrected as follows:

$$h(s_0) = h(s_i) + \frac{1}{2R_{corr}(s_0)}\left(1 + \left(\frac{\partial H}{\partial s}(s_i)\right)^2\right)ds^2 \tag{5}$$

where s is the along-track coordinate, $h(s_0)$ is the altimeter height at nadir, $R_{corr}(s_0)$ the altimeter range at nadir corrected from the geophysical and environmental effects, s_0 the location of the nadir along the altimeter track, s_i the coordinates of the slant measurements, $\partial H/\partial s$ the rate of altitude variation of the satellite along the orbital segment, and ds the along track difference between s_0 and s_i. Then, the altimeter height at nadir is computed using the summit of the parabola representing the actual water level:

$$h(s_0) = as_0^2 + bs_0 + c \tag{6}$$

where a, b and c are parabola coefficients calculating using a least-square fitting of the altimeter data affected by hooking.

Finally, s_0 and $h(s_0)$ are defined as follow:

$$s_0 = -\frac{b}{2a} \text{ and } h(s_0) = c - \frac{b}{4a} \tag{7}$$

$$h(s_0) = as_0^2 + bs_0 + c \tag{8}$$

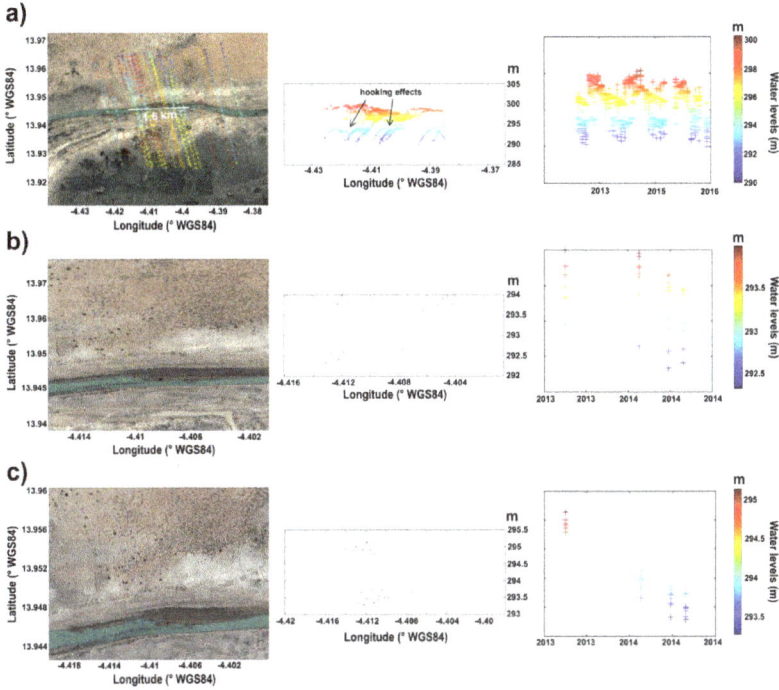

Figure 2. Example of the hooking correction. (**a**) On a rough selection, parabolic profiles in different cycles were identified. (**b**) Zooming on them, it appears that they are responsible for deviation of several tenths of centimeters of the river levels. (**c**)Once the correction of the hooking effect is applied, the deviation is reduced to a maximum of a couple of tenths cm).

2.2. Validation of the Altimetry-Based Water Levels

Validation of the altimetry-based water levels is performed against records from the closest in situ gauge stations. The along-stream distance between a VS and the closest in situ gauge stations is generally lower than 100 km (see Tables S1–S5). Root Mean Square Error (RMSE), R and R^2 values were estimated between altimetry-based water levels and in situ ones measured the same day using the classical formulas:

$$\text{RMSE} = \left(\frac{1}{n} \sum_{i=1}^{n} (h_{alti}(t_i) - h_{in\ situ}(t_i))^2 \right)^{1/2} \tag{9}$$

$$R = \frac{\sum_{i=1}^{n} (h_{alti}(t_i) - \langle h_{alti}(t_i) \rangle)(h_{in\ situ}(t_i) - \langle h_{in\ situ}(t_i) \rangle)}{\left(\sum_{i=1}^{n} (h_{alti}(t_i) - \langle h_{alti}(t_i) \rangle)^2 \right)^{1/2} \left(\sum_{i=1}^{n} (h_{in\ situ}(t_i) - \langle h_{in\ situ}(t_i) \rangle)^2 \right)^{1/2}} \tag{10}$$

$$R^2 = \frac{\sum_{i=1}^{n}(h_{alti}(t_i) - \langle h_{in\ situ}(t_i)\rangle)^2}{\sum_{i=1}^{n}(h_{in\ situ}(t_i) - \langle h_{in\ situ}(t_i)\rangle)^2} \qquad (11)$$

where h_{alti} and $h_{in\ situ}$ are the altimetry-based and the in situ water stages respectively, t_i is the measurement time and n the number of common observations. The average of a variable x is written <x>.

As the in situ gauge stations are leveled against a reference unavailable to us, no bias estimates were computed between the in situ and the altimetry-based water levels, but they were between the different missions in the same orbit as follows:

$$\text{Bias} = \frac{1}{n}\sum_{i=1}^{n}(h_{alti1}(t_i) - h_{alti2}(t_i)) \qquad (12)$$

while h_{alti1} is the more recent mission in the orbit and h_{alti2} is the older one.

The consistency of the intra-mission altimetry-based water levels as well as likely time-lag between water stages in the IND were estimated using the maximum of the cross-correlation function R_{hh} and the argument of the maximum:

$$R_{hh}(\tau) = \frac{\sum_{i=1}^{n}(h_{alti}(t_i) - \langle h_{alti}(t_i)\rangle)(h_{in\ situ}(t_i - \tau) - \langle h_{in\ situ}(t_i)\rangle)}{\left(\sum_{i=1}^{n}(h_{alti}(t_i) - \langle h_{alti}(t_i)\rangle)^2\right)^{1/2}\left(\sum_{i=1}^{n}(h_{in\ situ}(t_i - \tau) - \langle h_{in\ situ}(t_i)\rangle)^2\right)^{1/2}} \qquad (13)$$

where τ is the time displacement.

3. Study Area and Datasets

3.1. Study Area

The IND is an extensive Sahelian floodplain located between longitudes 3–5° W and latitudes 13–17° N in Central Mali (Figure 3a). It is encompassed between the in situ gauge stations of Macina (−5.37° W, 13.95° N), on the Niger River, and Douna (−5.9° W, 13.22 °N), on the Bani River, upstream, and Diré (−3.38° W, 16.27° N), downstream. Its drainage area represents a surface of 73,000 km² [35]. The flooded area extent depends on the intensity of the West African Monsoon and can reach 35,000 km² during the wettest rainy seasons [36–40]. The flooding period ranges from August to December and during the dry season, from March to May, the area dries out with the exception of the rivers mainstem and the permanent lakes3.2. Radar Altimetry Data

The data used in this study come from the acquisitions of the following radar altimetry missions in their nominal orbit: Jason-1 (2002–2008), Jason-2 (2008–2016), Jason-3 (since 01/2016), ERS-2 (05/1995–06/2003), ENVISAT (03/2002–10/2010), SARAL (02/2013–2016), Sentinel-3A (since 02/2016). The main characteristics of these missions are presented below.

3.1.1. Missions with a 35-Day Repeat Period (European Remote-Sensing Satellite-2 (ERS-2), ENVIronment SATellite (ENVISAT), Satellite with Argos and ALtika (SARAL))

ERS-2, ENVISAT and SARAL orbited at an average altitude of 790 km, with an inclination of 98.54°, in a sun-synchronous orbit with a 35-day repeat cycle. They provided observations of the Earth surface (ocean, land, and ice caps) from 82.4° latitude north to 82.4° latitude south. This orbit was formerly used by ERS-1 mission, with an equatorial ground-track spacing of about 85 km.

ERS-2 was launched in 1995 by the European Space Agency (ESA) as ERS-1 follow-on mission. The satellite carries, among other instruments, a radar altimeter (RA) operating at Ku-band (13.8 GHz) developed for measuring height over ocean, land and ice caps. ERS-2 data are available from 17 May 1995 to 9 August 2010. After 22 June 2003, the dataset coverage is limited to ground station visibility.

ENVISAT mission was launched on 1 March 2002 by ESA. It carried 10 instruments including the advanced radar altimeter (RA-2). It was based on the heritage of the sensor on-board the ERS-1 and

2 satellites. RA-2 was a nadir-looking pulse-limited radar altimeter operating at two frequencies at Ku-(13.575 GHz), as ERS-1 and 2, and S-(3.2 GHz) bands [41]. ENVISAT remained in its nominal orbit until October 2010 and its mission ended on 8 April 2012. RA-2 stopped operating correctly at S-band in January 2008.

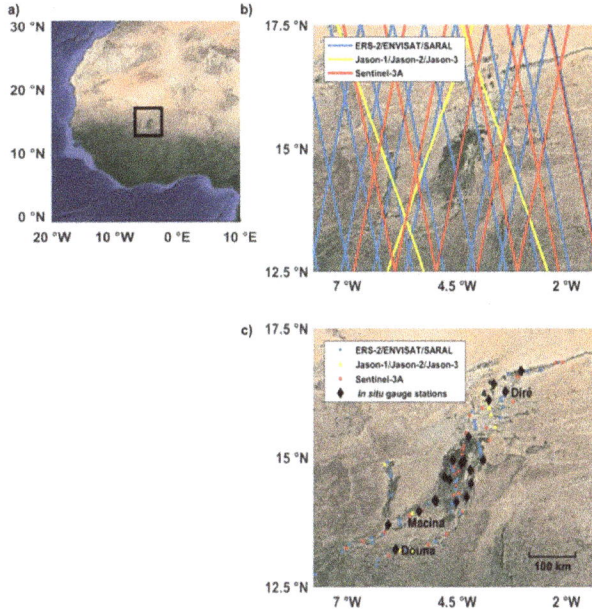

Figure 3. (**a**) Location of the IND in Africa; (**b**) Altimetry tracks over the IND from European Remote-Sensing Satellite-2 (ERS-2), ENVIronment SATellite (ENVISAT), Satellite with Argos and ALtika (SARAL) (blue dots), Jasons-1/Jason-2/Jason-3 (yellow dots) and Sentinel-3A (red dots); and (**c**) Location of virtual stations to calculate water levels (colored circles) using altimetry data and in situ gauge stations (black diamonds) in IND.

SARAL mission was launched on 25 February 2013. SARAL is a new collaboration between Centre National d'Etudes Spatiales (CNES) and Indian Space Research Organization (ISRO). Its payload comprises the AltiKa radar altimeter and bi-frequency radiometer, and a triple system for precise orbit determination: the real-time tracking system Détermination Immédiate d'Orbite par Doris embarqué (DIODE) of the Doppler Orbitography and Radio-positioning Integrated by Satellite (DORIS) instrument, a Laser Retroflector Array (LRA), and the Advance Research and Global Observation Satellite (ARGOS-3). AltiKa radar altimeter is a solid-state mono-frequency altimeter that provides accurate range measurements. It is the first altimeter to operate in the Ka-band (35.75 GHz) [42]. It has been put in a drifting orbit since July 2016.

3.1.2. Missions with a 10-Day Repeat Period (Jason-1, Jason-2 and Jason-3)

Jason-1, Jason-2 and Jason-3 orbit at an altitude of 1336 km, with an inclination of 66°, on a 10-day repeat cycle, providing observations of the Earth surface (ocean and land) from 66° latitude North to 66° latitude South, with an equatorial ground-track spacing of about 315 km. This orbit was formerly used by Topex/Poseidon mission.

Jason-1 mission was launched on 7 December 2001 by a cooperation between CNES and National Aeronautics and Space Administration (NASA). Jason-1 sensors are based on the former Topex/Poseidon missions, composed of the Poseidon-2 altimeter which is a two-frequency altimeter with C (5.3 GHz) and Ku (13.575 GHz) -bands. Its payload is also composed of the Jason Microwave Radiometer from NASA and a triple system for precise orbit determination: DORIS instrument from the CNES, Black Jack Global Positioning System receiver from NASA and a LRA from NASA/Jet Propulsion Laboratory (JPL) [43]. Jason-1 remained in its nominal orbit until 26 January 2009 and was decommissioned on 21 June 2013.

Jason-2 mission was launched on 20 June 2008 as a cooperation between CNES, the European Organization for the Exploitation of Meteorological Satellites (EUMETSAT), NASA and the National Oceanic and Atmospheric Administration (NOAA). Its payload is mostly composed of the Poseidon-3 radar altimeter from CNES, the Advanced Microwave Radiometer (AMR) from JPL/NASA, and a triple system for precise orbit determination: the real-time tracking system DIODE of DORIS instrument from CNES, a Global Navigation Satellite System (GNSS) receiver and a LRA from NASA/JPL. Poseidon-3 radar altimeter is a two-frequency solid-state altimeter that measures accurately the distance between the satellite and the surface (range) and provides ionospheric corrections over the ocean [44]. It operates at Ku and C bands. Raw data are processed by SSALTO (Segment Sol multimissions d'ALTimétrie, d'Orbitographie). Jason-2 remained in its nominal orbit until 3 July 2016.

Jason-3 mission was launched on 17 January 2016 as cooperation between CNES, EUMETSAT, NASA and NOAA. This satellite is composed of Poseidon-3B radar altimeter with a Precise Orbit Determination (POD) package with a Global Positioning System (GPS) receiver, DORIS and a LRA from NASA/JPL.

3.1.3. Mission with a 27-Day Repeat Period (Sentinel-3A)

Sentinel-3A mission was launched on 16 February 2016 by ESA to an orbit of altitude 814 km. The satellite caries one altimeter radar called SRAL (SAR Radar ALtimeter), a dual-frequency SAR altimeter (Ku-band at 13.575 GHz and C-band at 5.41 GHz). Its payload comprises also a Microwave Radiometer (MWR) instrument for wet path delay measurements and a triple system for precise orbit determination: a POD including a GPS receiver, a LRA and a DORIS instrument [45].

All this information is summarized in Table 1.

Table 1. Major characteristics of the high-precision radar altimetry missions used in this study.

Mission	Jason-1/2/3	ERS-2 ENVISAT	SARAL	Sentinel-3A
Instrument	Poseidon-2 Poseidon-3 Poseidon-3B	Radar Altimeter (RA) Radar Altimeter (RA-2)	AltiKa	Sar Radar Altimeter (SRAL)
Space agency	Centre National d'Etudes Spatiales (CNES), National Aeronautics and Space Administraion (NASA)	European Space Agency (ESA)	CNES, Indian Space Research Organization (ISRO)	European Space Agency (ESA)
Operation	2001–2013 Since 2008 Since 2016	1995–2003 2002–2012	Since 2013	Since 2016
Acquisition mode	Low Resolution Mode (LRM)	LRM	LRM	Pseudo Low Resolution Mode (PLRM), SAR
Acquisition	Along-track	Along-track	Along-track	Along-track
Frequency (GHz)	13.575 (Ku) 5.3 (C)	13.8 (Ku) 13.575 (Ku) 3.2 (S)	35.75 (Ka)	13.575 (Ku) 5.41 (C)
Altitude (km)	1315	800	800	814.5
Orbit inclination (°)	66	98.55	98.55	98.65
Repetitively (days)	9.9156	35	35	27
Equatorial cross-track separation (km)	315	75	75	104

The data used in this study are summarized in Table 2. Ranges used to derive altimeter heights and backscattering coefficients are those processed with OCOG/Ice-1/Ice retracking algorithm [46]. Previous studies showed that Ice-1-derived altimetry heights are the more suitable for hydrological studies in terms of accuracy of water levels and availability of the data (e.g., [12,47,48]) among the commonly available retracked data present in the GDRs.

Table 2. Major characteristics of the high-precision radar altimetry missions used in this study.

Altimetry Mission	Jason-1	Jason-2	Jason-3	ERS-2	ENVISAT	SARAL	Sentinel-3A
GDR	E	D	D	Centre de Topographie des Océans et de l'Hydrosphère (CTOH) [13]	V2.1	T	ESA IPF 06.07 land
Along-track sampling	20 Hz	20 Hz	20 Hz	20 Hz	18 Hz	40 Hz	20 Hz
Retracker	ICE	ICE	ICE	ICE-1	ICE-1	ICE-1	Offset Centre of gravity (OCOG)
ΔR_{iono}	GIM-based						
ΔR_{dry}	European Centre for Medium-Range Weather Forecasts (ECMWF)-based using Digital Elevation Model (DEM)			ECMWF-based using h from altimeter	ECMWF-based using DEM		
ΔR_{wet}	ECMWF-based using DEM						
$\Delta R_{solid\ Earth}$	Based on Catwright et al. [49]						
ΔR_{pole}	Based on Wahr et al. [50]						

3.2. In Situ Water Levels

Daily stage records from 19 in situ gauge stations located in the IND were used in this study to validate altimetry-based water levels (see Table 3 for their names, locations and periods of data availability and Figure 3c for their locations). Measurements were acquired at 12:00 a.m. local time. They are made available by the Malian water agency (Direction Nationale de l'Hydraulique—DNH).

Table 3. List of in situ gauge stations in the IND used this study.

In Situ Gauge Station	Longitude (°)	Latitude (°)	Validation Period
Akka	−4.23	15.39	1992–2017
Diondiori	−4.78	14.61	2008–2010
Diré	−3.38	16.27	1991–2017
Douna	−5.90	13.22	1991–2004
Goundam	−3.65	16.42	2009–2017
Kakagnan	−4.33	14.93	2008–2010
Kara	−5.01	14.16	1992–2011
Kirango	−6.07	13.7	2015–2017
Konna	−3.9	14.95	1992–1999
Koryoumé	−3.03	16.67	1992–2017
Macina	−5.29	14.14	1991–2017
Mopti	−4.18	14.48	1991–2017
Sévéri	−4.19	14.75	2008–2010
Sormé	−4.4	14.87	2008–2010
Sossobé	−4.67	14.56	2008–2010
Tilembeya	−4.98	14.15	1991–2006
Toguéré Kou	−4.59	14.93	2008–2010
Tonka	−3.76	16.11	1991–2017
Tou	−4.52	14.13	2008–2010

4. Results

4.1. Direct Validation of the Altimetry-Based Water Stages

The nominal altimetry ground-tracks from ERS-2, ENVISAT, SARAL (35-day repeat orbit), Sentinel-3A (27-day repeat orbit), Jason-1, Jason-2 and Jason-3 (10-day repeat orbit) missions present a large number of cross-sections with river streams and floodplains in the IND (see Figure 3b). A dense network of virtual stations from different missions was defined in the IND (see Table 4). Virtual station (VS) locations in the IND are presented in Figure 3c.

Table 4. Number of virtual stations defined in the IND for each mission.

Mission	ERS-2	ENVISAT	SARAL	Sentinel-3A	Jason-1	Jason-2	Jason-3
Number of virtual stations (VS)	52	63	62	31	8	8	9

Altimetry-based water levels were compared to water stage records from close in situ gauge. These comparisons were performed for VS located on the rivers and not on the floodplains for distances between the in situ gauge and the VS lower than 100 km. In situ gauge records from 19 stations were used to perform the 89 following comparisons:

- 19 against ERS-2-based water stages;
- 32 against ENVISAT-based water stages;
- 14 against SARAL-based water stages;
- 3 against Jason-1 and Jason-2-based water stages;
- 2 against Jason-3-based water stages;
- 16 against Sentinel-3A-based water stages.

The complete results of these comparisons (distance between the in situ gauge and the VS, number of data used for comparisons (N), RMSE, R and R^2) are presented in Tables S1 to S5 for ERS-2, ENVISAT, SARAL, Jason-1/2/3 and Sentinel-3A missions (in supplementary information). The results of these comparisons are also presented as maps in Figure 4 for the altimetry missions that were launched before 2010 (ERS-2, ENVISAT, Jason-1 and 2) and in Figure 5 for the most recent missions. The number of data used for the comparison is, most of the time, statistically significant, except for 15 comparisons against ENVISAT (less than 20 common observations), 4 against SARAL (less than 15 common observations) as well as the whole comparisons against Sentinel-3A as only 16 cycles were used:

- between 28 and 70 for the 19 ERS-2-based time series of water level (out of 85 available cycles);
- between 7 and 81 for 32 ENVISAT-based time series of water level (out of 89 available cycles);
- between 6 and 28 for the 14 SARAL-based time series of water level (out of 35 available cycles);
- between 46 and 147 for the 3 Jason-1-based time series of water level (out of 262 available cycles);
- between 37 and 72 for the 3 Jason-2-based time series of water level (out of 303 available cycles);
- between 45 and 50 for the 2 Jason-3-based time series of water level (out of 55 available cycles);
- between 3 and 15 for the 16 Sentinel-3A-based time series of water level (out of 16 available cycles).

Very good agreements were generally found between altimetry-based and in situ water stages for all the missions over a total of 89 comparisons performed. Values of R greater than 0.95 were obtained 41 times (45%), between 0.95 and 0.9, 18 times (20%), and between 0.8 and 0.9, 18 times (20%). Correlation coefficients R lower than 0.7 were obtained only 4 times (4%) (Figure 6a). The minimum R value is 0.57. RMSE lower than 0.3 m were obtained 12 times (13%), between 0.3 and 0.5 m, 17 times (19%), between 0.5 and 0.75 m, 29 times (32%) and above 1 m, 19 times (21%) (Figure 6b).

Better agreement was found for the recent missions such as SARAL, Jason-3 and Sentinel-3A than for the older ones (ERS-2, ENVISAT, Jason-1 and Jason-2). Focusing on the correlations, for instance R was greater than 0.9 in 10 out of 19 comparisons for ERS-2 (52%), in 19 out of 32 (59%) for ENVISAT, in 10 out of 14 (71%) for SARAL, in 14 out of 16 (88%) for Sentinel-3A, 1 (R = 0.89) out of 3 (33%) for Jason-1, 3 out of 3 for Jason-2 (100%) and 3 out of 3 (100%) for Jason-3 (Tables S1–S5). The first results of the use of Sentinel-3A are very encouraging as only very few cycles were available, keeping in mind that among the three correlation coefficients lower than 0.9, three were already higher than 0.85.

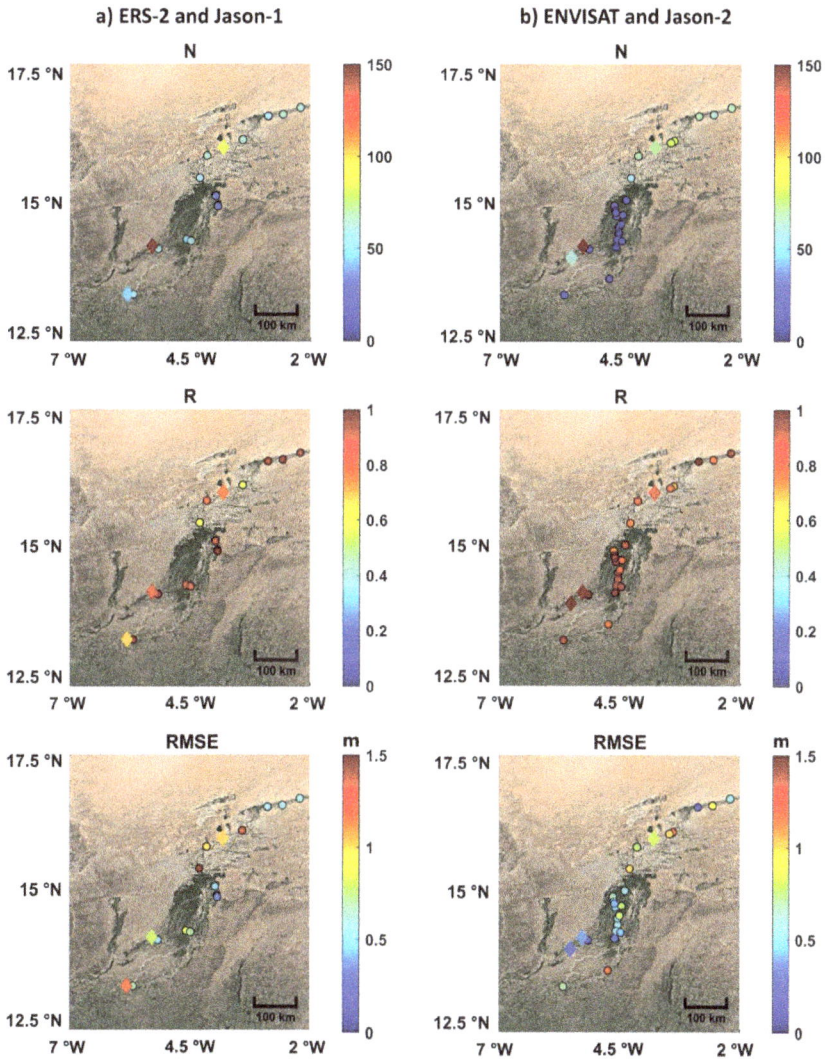

Figure 4. Comparisons between in situ and altimetry water levels for several missions (**a**) ERS-2 and Jason-1 data and (**b**) ENVISAT and Jason-2. For each comparison, the number of samples (N), correlation (R), and RMSE is presented. Diamonds points correspond to Jason data and circles with black contours to ERS-2/ENVISAT.

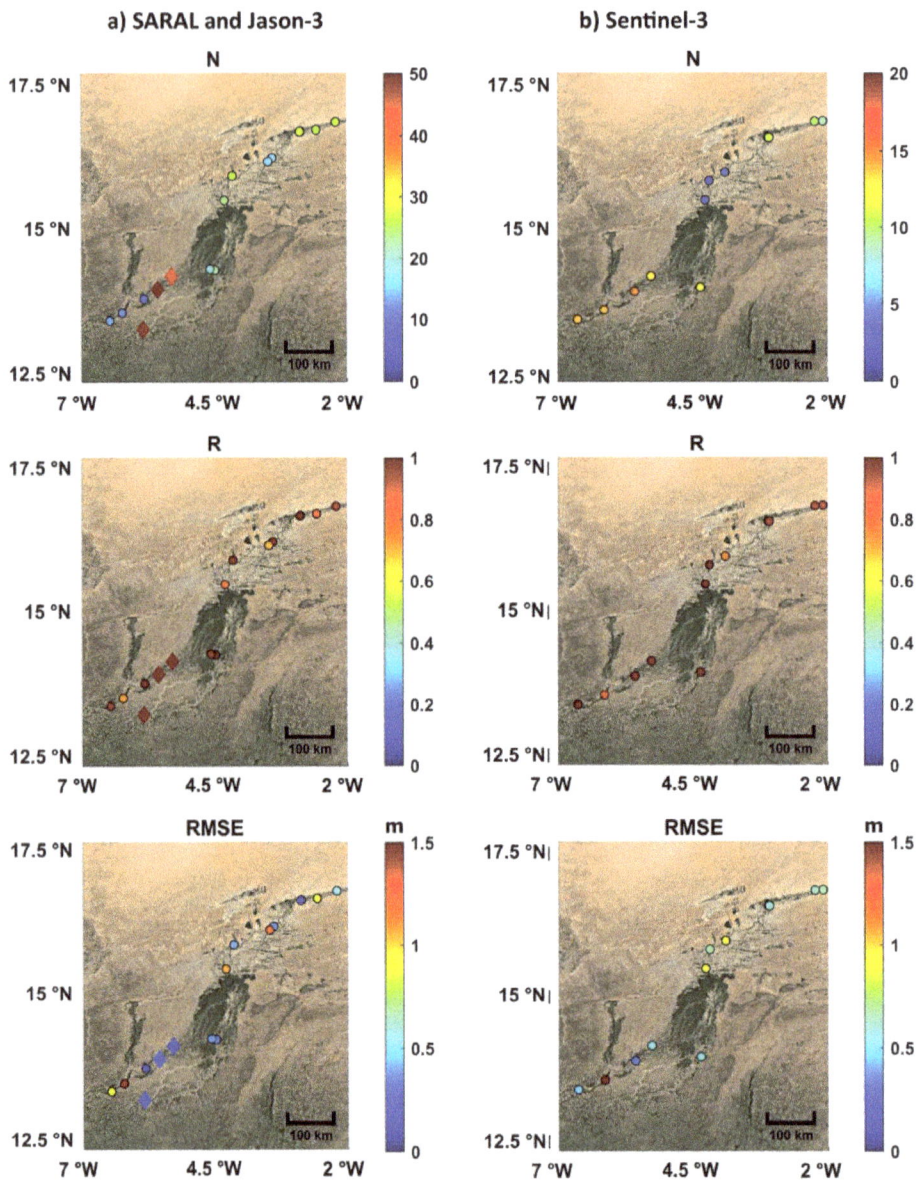

Figure 5. Comparisons between in situ and altimetry water levels for several missions (**a**) SARAL and Jason-3 data and (**b**) Sentinel-3A. For each comparison, the number of samples (N), correlation (R), and RMSE is presented. Diamonds points correspond to Jason data and circles with black contours to SARAL/Sentinel-3A.

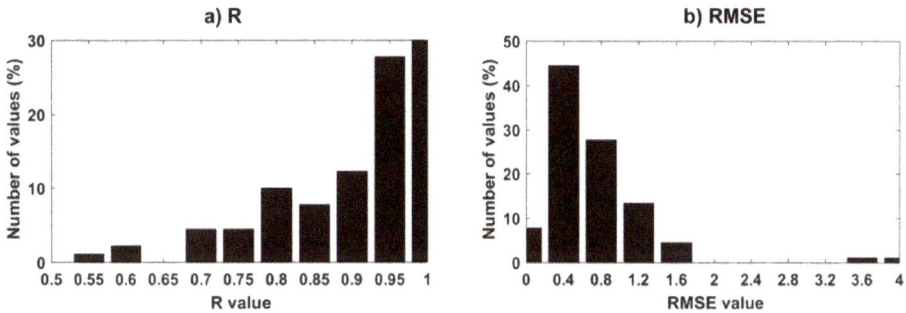

Figure 6. Histograms of (**a**) correlations and (**b**) RMSE between altimetry-based and in situ water stages for all missions.

Combining altimetry-based time series of water levels from missions in the same nominal orbits (i.e., ERS-2, ENVISAT and SARAL, and Jason-1, Jason-2 and Jason-3) or at inter-mission cross-overs (ERS-2/ENVISAT/SARAL with Sentinel-3A ground-tracks), multi-mission time series of water levels were obtained. In this latter case, the maximum difference in time between SARAL and Sentinel-3A acquisitions is half the length of the shortest repeat period of the two satellites (i.e., 13 days considering the 27 days of Sentinel-3A repeat period). Inter-mission biases were removed:

- using the acquisitions made during tandem phases when two missions were in the same orbit a few seconds or minutes apart from each other (e.g., Jason-1 and Jason-2, Jason-2 and Jason-3, ERS-2 and ENVISAT);
- averaging the acquisitions made during the common period of observations at low water stages (April–May–June) for Sentinel-3A and SARAL;
- averaging the acquisitions made during low water periods (April–May–June) on different years for ENVISAT (2003–2010) and SARAL (2013–2016).

Low water periods were chosen, rather than high water periods or the complete hydrological cycle, to minimize the effect of the difference in temporal sampling, assuming that water levels are more stable during low water stages.

Examples of multi-mission time series of water levels are presented in Figures 7 and 8 for ERS-2/ENVISAT/SARAL and Sentinel-3A (when there is a cross-over) and for Jason-1, Jason-2 and Jason-3 respectively along with in situ gauge station gauge records of Diré upstream part of the IND, Mopti, central part and Macina, downstream part, over 1995–2017 (Figure 7a–c respectively), and of Macina, over 2002–2017 (Figure 8).

In examples presented in Figure 7, a very good agreement is found between altimetry-based and in situ water stages. There is a gap in the time series between November 2010 and January 2013 as no altimetry mission was in the 35-day repeat orbit during this period. In spite of the distance between the VS and the two first in situ gauge stations (77 and 40 km, with Diré and Mopti stations respectively, but only 1 km from the Macina station under the ERS-2/ENVISAT and SARAL ground-tracks, and 8 km from Macina station under Sentinel-3A ground-tracks), better results were found in the first examples than in the latter one, with higher R^2 and lower RMSE for ERS-2 and ENVISAT and similar ones for SARAL and Sentinel-3A. An underestimation of the annual amplitude of the water levels is observed during the ERS-2 observation period. In the example presented in Figure 8, the VS and the Macina in situ gauge station are separated by only 1 km. The quality of the water stage retrieval is increasingly better from Jason-1 to Jason-3. It is important to mention that Jason-1 data contained in the GDR E, released in May 2016, allow the accurate and continuous estimation of water stages over the IND contrary to the previous GDR versions that contained few useful data over land (except over large lakes, see [18,51,52] for instance).

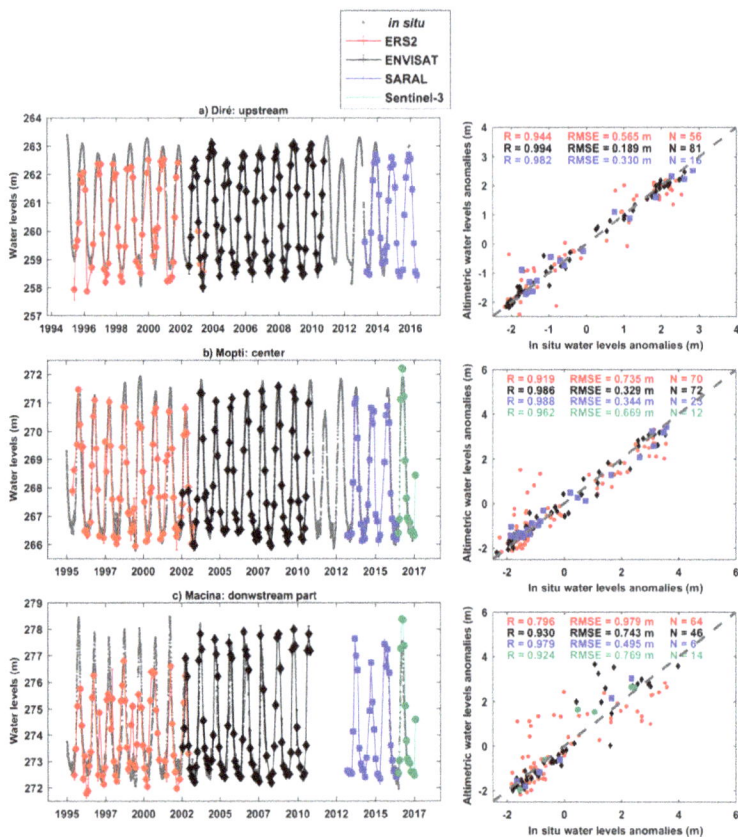

Figure 7. Altimetry-based water levels from 1995 to 2017 from ERS-2/ENVISAT/SARAL/Sentinel-3A (red/black/blue/green respectively) and in situ (grey) data (left). Scatter plots of water levels anomalies from radar altimetry and in situ gauge stations at (**a**) Diré (upstream IND), (**b**) Mopti (center IND) and (**c**) Macina (downstream IND).

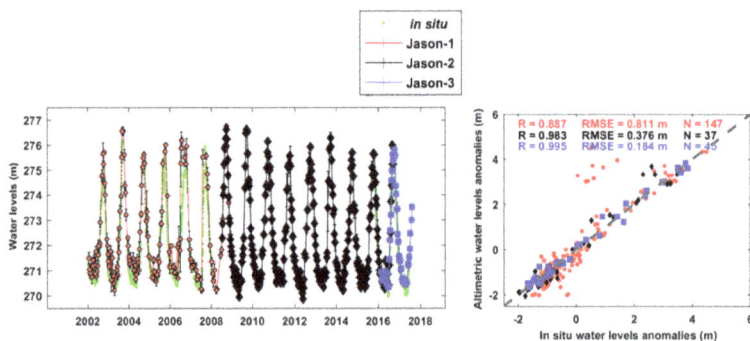

Figure 8. Altimetry-based water levels from 2002 to 2017 from Jason-1/Jason-2/Jason-3 (red/black/blue respectively) and in situ water levels (light green) data (left). Scatter plots of water levels anomalies from radar altimetry and in situ gauge stations at Macina (right figure, downstream IND).

4.2. Intermission Water Stage Comparison

Comparisons between water levels estimated by altimetry missions during their tandem phase (i.e., few cycles during which two missions were orbiting a few minutes apart from one another) were performed between ERS-2 and ENVISAT from June 2002 to July 2003 (11 cycles in common), Jason-1 and Jason-2 from July 2008 to January 2009 (21 cycles in common), Jason-2 and Jason-3 from January to September 2016 (23 cycles in common). They allow the increase of the number of comparisons in the IND, not only on the rivers but also on the wetlands that are monitored using in situ gauges.

A total of 48 comparisons between ERS-2 and ENVISAT-based water levels were performed in the IND with several samples (N) varying from 3 to 11 (Figure 9). On the total number of comparisons between ERS-2 and ENVISAT missions, 22 were performed on more than 8 samples (45%) (Figure 9a). Very good agreement was generally obtained between altimetry-based water stages from the two missions (Figure 9a,b). The value R greater than 0.95 was obtained 24 times (50%), between 0.95 and 0.9, 7 times (15%), and between 0.8 and 0.9, 7 times (15%) (Figure 10a). Correlation coefficients (R) lower than 0.5 were obtained 5 times (10%) (Figure 9b). Values of RMSE lower than 0.3 m were obtained 13 times (27%), between 0.3 and 0.5 m, 10 times (21%), between 0.5 and 0.75 m, 13 times (27%) and above 1 m, 8 times (17%) (Figure 10b). Large biases are observed between ERS-2 and ENVISAT (-1.16 ± 0.38 m on average) (Figure 9d).

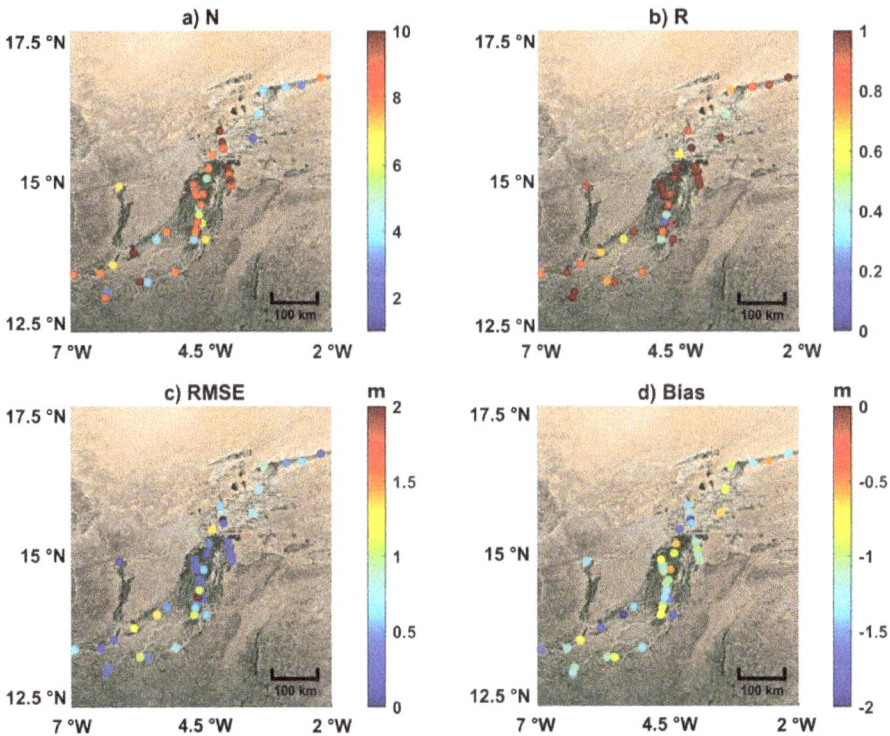

Figure 9. Comparisons between ERS-2 and ENVISAT water levels in terms of (**a**) number of samples (N), (**b**) correlation (R), (**c**) RMSE, and (**d**) bias.

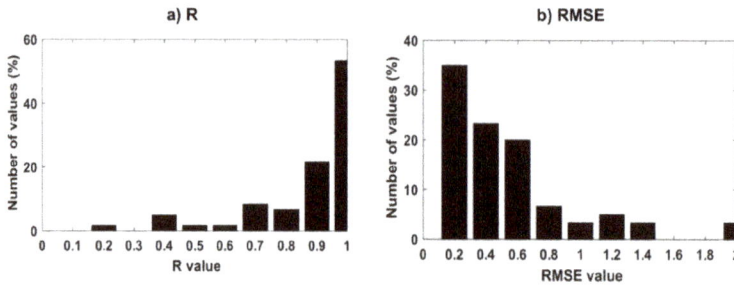

Figure 10. Histograms of (**a**) correlation coefficients and (**b**) RMSE for ERS-2/ENVISAT intermission water stages comparisons.

A total of 5 and 7 comparisons between Jason-1 and Jason-2, and Jason-2 and Jason-3-based water levels were performed in the IND with several samples (N) varying from 5 to 15 for Jason-1/Jason-2, and from 13 to 22 for Jason-2/Jason-3. Three of these comparisons were performed on more than 10 samples (60%) for Jason-1 (Figure 11a). Very good agreement was generally found between altimetry-based water stages from Jason-1 and 2. Values of R greater than 0.95 were obtained 3 times (60%). In the two other cases, R equal to 0.80 and 0.69 were found (Figure 11b). Values of RMSE lower than 0.3 m were obtained 3 times (60%). In the two other cases, RMSE lower than 0.35 m were found (Figure 11c). Biases between Jason-1 and 2 ranged between 0.34 and 1.06 m (0.75 ± 0.28 m) (Figure 11d).

Better agreement is found between altimetry-based water stages from Jason-2 and 3. Correlation coefficients R greater than 0.95 were obtained in all the cases except one (86%) for which R equals to 0.84 (Figure 12b). Values of RMSE lower than 0.30 m were obtained in all the cases except one (86%) for which RMSE equals to 0.40 m (Figure 12c). Biases between Jason-1 and 2 are very similar, ranging from -0.36 to -0.14 m (-0.27 ± 0.08 m) (Figure 12d).

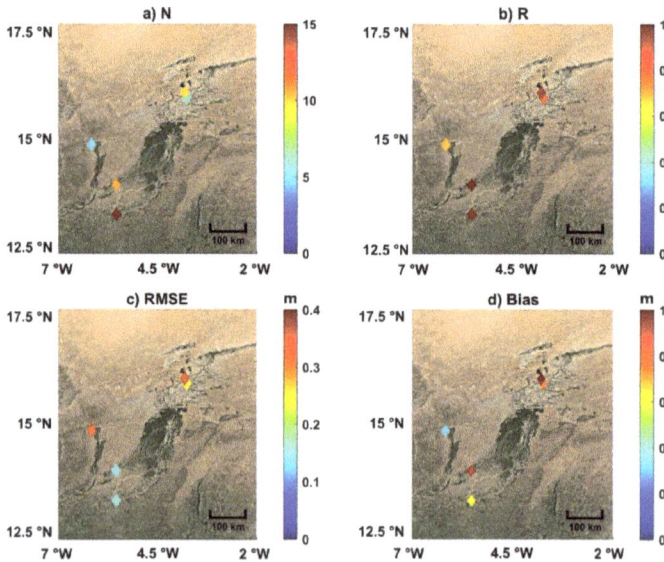

Figure 11. Comparisons between Jason-1 and Jason-2 water levels in terms of (**a**) number of samples (N); (**b**) correlation (R); (**c**) RMSE; and (**d**) bias.

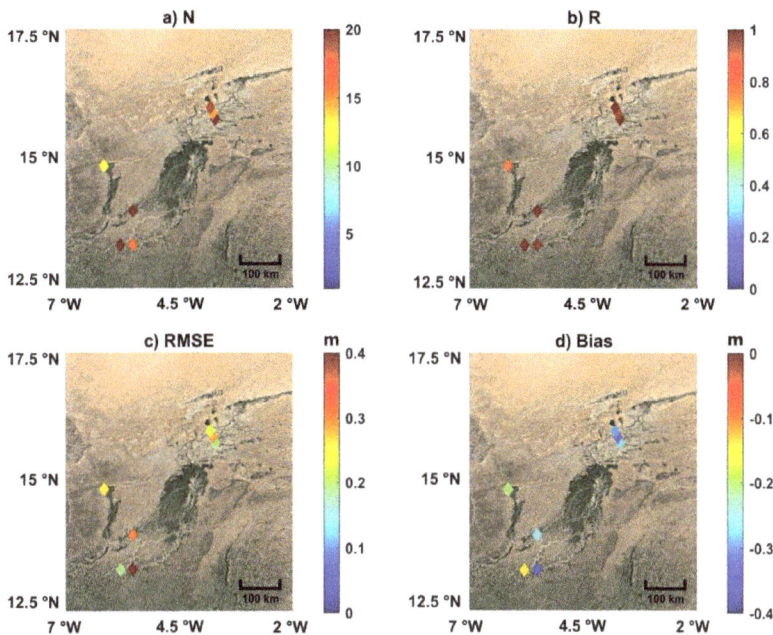

Figure 12. Comparisons between Jason-2 and Jason-3 water levels in terms of (**a**) number of samples (N); (**b**) correlation (R); (**c**) RMSE and (**d**) bias.

Examples of time series combining ERS-2 and ENVISAT, Jason-1 and 2, and Jason-2 and 3 are presented on Figure 13a–c. Virtual stations from close locations were chosen: ERS-2/ENVISAT VS 0545-a and Jason-1/Jason-2/Jason-3 VS 046-d. They are located in the upstream part of the IND at an approximate along-stream distance of 45 km. Seasonal amplitudes of between 4 and 5 m are observed during the common period of availability of the different altimetry-based water levels. The agreement is better for the recent missions than for the older ones: R increases from 0.92 to 0.99 whereas RMSE decreases from 0.62 to 0.22 m as well as the bias from −1.35 to 0.36 m. Lower deviations are generally observed on the time series from the more recent missions than other the older ones: $0.13 \pm 0.11, 0.13 \pm 0.15, 0.19 \pm 0.22, 0.21 \pm 0.16, 0.15 \pm 0.11$ m were obtained averaging the mean absolute deviation from individual cycles for ERS-2, ENVISAT, Jason-1, 2 and 3 respectively. Please note that, for readability purpose, biases between the time series were removed in Figure 13.

4.3. Multi-Mission Time Series on Floodplains

Floodplains and wetlands are generally not monitored using in situ gauges. Radar altimetry is a unique tool for the long-term observations of the changes in water levels over inundated areas [53–57]. Among the VS defined in the IND, 16 were built in floodplains under ERS-2/ENVISAT/SARAL ground-tracks and 9 under Sentinel-3A ground-tracks, but none under Jason-1, Jason-2 and Jason-3 ground-tracks. Their consistency was checked during the tandem phase between ERS-2 and ENVISAT (Figure 14). Values of R greater than 0.95 were obtained 10 times (63%), between 0.95 and 0.9, twice (12%), and between 0.8 and 0.9, 3 times (19%). However, R lower than 0.5 was obtained once (6%) (Figure 14a). Values of RMSE lower than 0.3 m were obtained 7 times (44%), between 0.3 and 0.5 m, 5 times (31%), between 0.5 and 0.75 m, 3 times (19%) and above 1 m, once (6%) (Figure 14b).

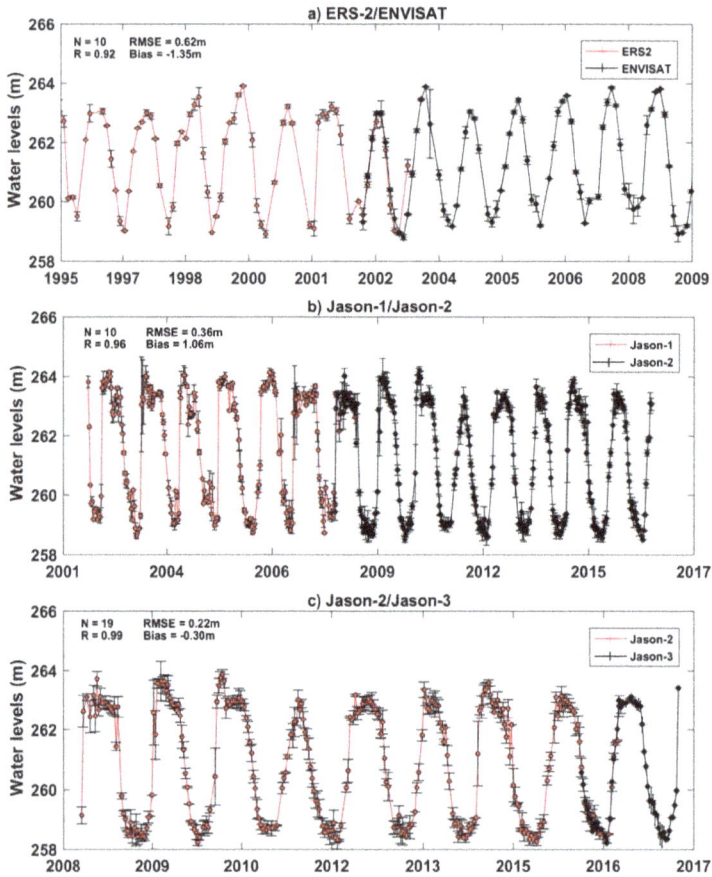

Figure 13. Intermission quality time series: (**a**) for the VS 0545-a located in the upstream part in the IND between ERS-2/ENVISAT; (**b**) for the VS 046-d (upstream part) between Jason-1/Jason-2; and (**c**) for the VS 046-d between Jason-2/Jason-3. The two VS are separated by 45 km.

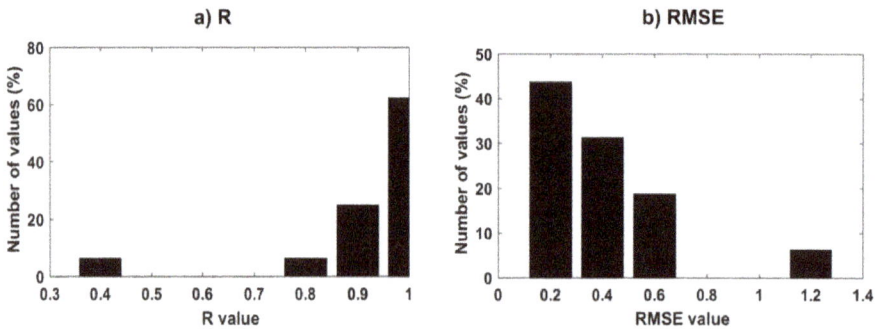

Figure 14. Histograms of (**a**) R and (**b**) RMSE for ERS-2/ENVISAT intermission water stages comparisons on floodplains virtual stations.

Three examples of multi-mission time series of water levels (corrected for inter-mission bias) over the IND floodplains are presented for VS located in the center, in the north east and in the south (Figure 15a–c respectively). They provide time-variations of water stages between 1995–2016 with the exception of a gap between November 2010 and January 2013.

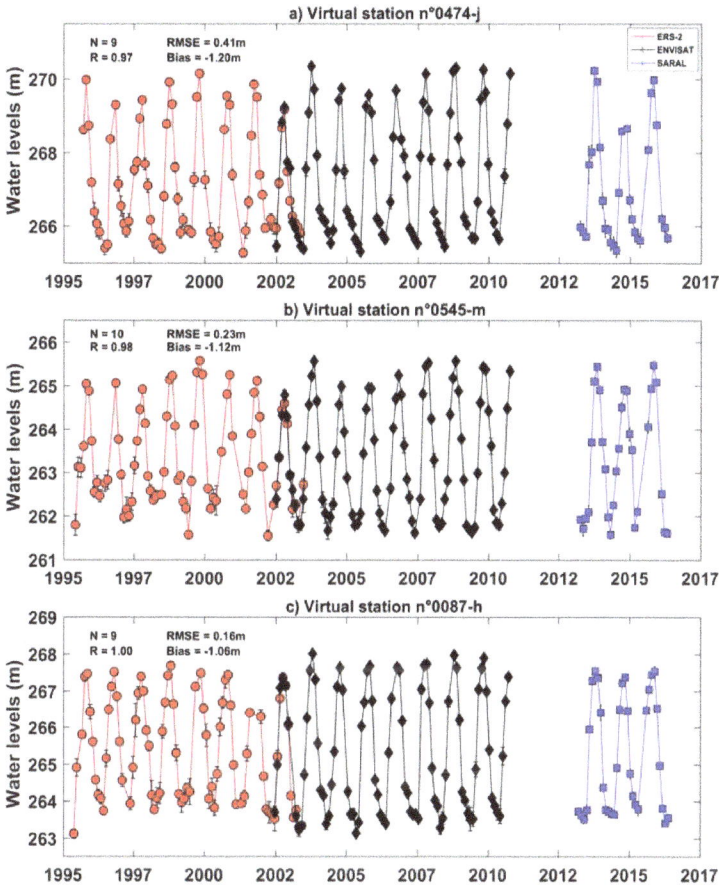

Figure 15. Water levels time series on floodplains derived from radar altimetry multi-mission (ERS-2/ENVISAT and SARAL) from 1995 to 2016 at 3 different locations in the Inner Niger Delta: the center of the delta (**a**); the north eastern (**b**) and the southern (**c**) parts.

4.4. Consistency of the Altimetry-Based Water Levels in the Inner Niger Delta (IND)

A consistency check was performed between a reference VS located either in Niger River mainstem or in the Bani major tributary and the other VS from the same altimetry mission located upstream and downstream on the same river course estimating the maximum of correlation using the cross-correlation function and the associated time-lag, similarly as in Bogning et al. [29]. Cross-correlation function maxima and associated time-lags are presented for ERS-2 and ENVISAT in Figure 16 and Figure S1 respectively and for SARAL and Sentinel-3A in Figure 17 and Figure S2 respectively for three different VS locations in the IND: near Douna, over the Bani tributary, near Mopti, in the downstream part of the IND, and near Diré in the upstream part of the IND. As the repeat period of these missions is either 35 or 27 days in their nominal orbit, only time-lags of plus or minus one repeat period and a half (i.e., 53 and 41 days) were considered due to the relatively small scale of the IND. Due to the changes of the river features (slope, depth, width, etc.), biases and RMSEs were not computed between the time series of altimetry-based water levels.

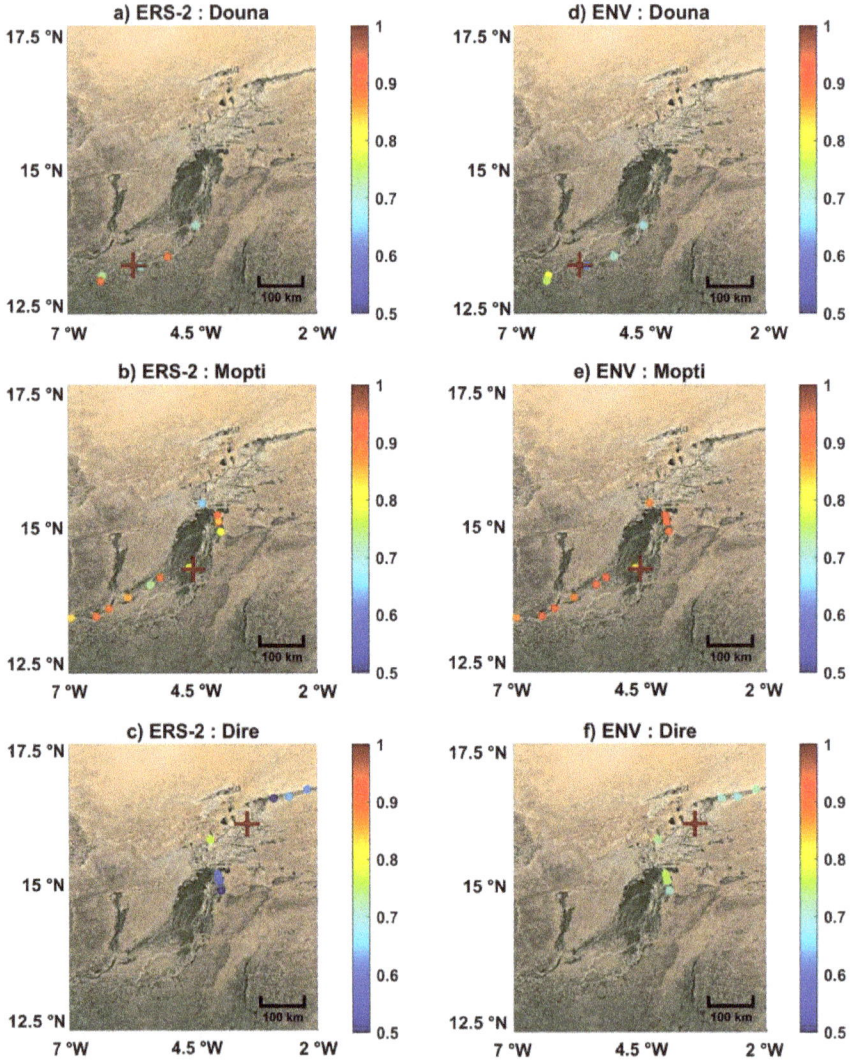

Figure 16. Maxima of cross-correlation between time series of water levels derived from ERS-2 (**left**) and ENVISAT (**right**) data in the IND for three different VS near Douna (**a,d**); Mopti (**b,e**) and Diré (**c,f**). Red crosses symbolize the VS chosen as reference (auto-correlation of 1).

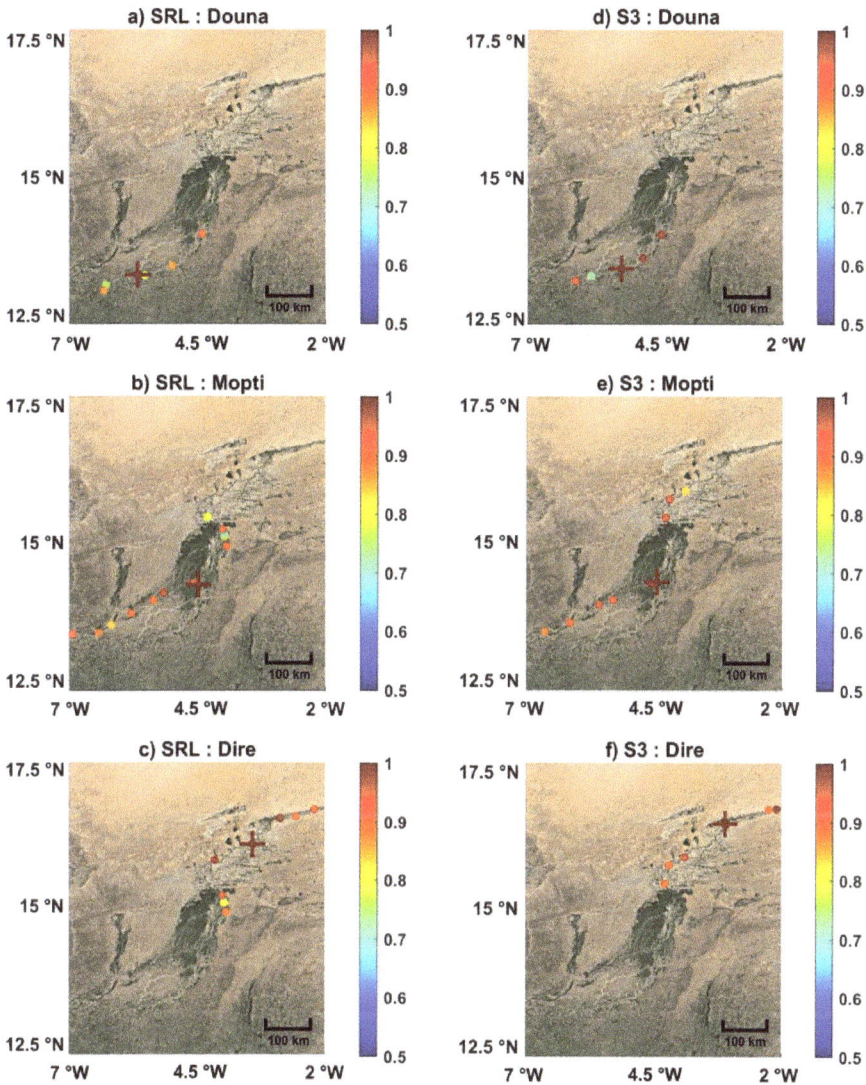

Figure 17. Maxima of cross-correlation between time series of water levels derived from SARAL (left) and Sentinel-3A (right) data in the IND for three different VS near Douna (**a,d**); Mopti (**b,e**) and Diré (**c,f**). Red crosses symbolize the VS chosen as reference (auto-correlation of 1).

Correlation coefficients (R) higher than 0.7 are generally observed for ERS-2 and ENVISAT, except in the upstream part of the delta for ERS-2, with correlation ranging from 0.5 to 0.7 (Figure 16c). Correlations higher than 0.8 are observed in the IND downstream and central parts for ENVISAT (Figure 16). Very high correlations, generally higher than 0.85, especially in the downstream part, except on a few locations, were observed for SARAL and, especially, for Sentinel-3A (Figure 17). No time-lag was observed over the Bani River for none of the altimetry missions (Figures S1a, d and S2a,d). Time lags of plus or minus one cycle were respectively found for the upper and lower stations along the Niger mainstem (Figures S1 and S2).

5. Discussion

The coverage of altimetry tracks from missions of the 35-day (ERS-2/ENVISAT/SARAL) and the 27-day (Sentinel-3A) repeat orbits allows the construction of a dense network of VS on both rivers and floodplains in the IND, completed by a few VS from missions in the 10-day repeat orbit (Figure 3c). Yet, a larger number (more than 15%) of ENVISAT (6) and SARAL-based (62) VS than ERS-2-based (52) ones were built on the ground-tracks of the 35-day repeat orbit. These VS with no valid ERS-2-based water levels are located on the upstream part of the IND for 5 of them, central part for 3 others and downstream part for the last 3 ones.

Due to the relative flatness of the IND, the no-construction of VS cannot be attributed to data losses caused by tracking issues, but to incorrect range estimates. They were likely to be caused by the narrow width of the river streams, especially in the upstream part, where they are generally lower than 300 m. They can also be accounted for the possible complexity of the waveforms (e.g., multi-peaked over areas with several bright targets encompassed in the altimeter footprint such as several river streams that can reach 1 km of width as in the downstream part of the IND, or several floodplain lakes as in the central part).

Comparisons with a high number of in situ gauge records show an overall very good agreement between in situ and altimetry-based water levels, increasingly better for the most recent missions than the older ones. Contrary to what was found in other study areas with variations of topography (e.g., [28,29]), the data used to build the VS were mostly acquired at 330 and 320 MHz Ku chirp bandwidth operation mode by both ERS-2 and ENVISAT (i.e., with the better range resolution in more than 95% of the cases for any VS—Figure 18). This is the reason quite similar results when comparing ERS-2 and ENVISAT-based water stages with in situ water levels were found in terms of R, with generally higher RMSE for ERS-2 (see Figure 4, Tables S1 and S2), and also a good agreement when comparing ERS-2 and ENVISAT-based water levels during their tandem phase, except in terms of bias (Figure 10). In the same orbit, better results were found using SARAL data (Figure 5 and Table S3), as already observed in other river basins, resulting from the use of the Ka-band with its smaller footprint and higher chirp frequency of 500 MHz [29,58].

Very similar numbers of VS were built using Jason-1, Jason-2 and Jason-3 data. The release of the Jason-1 GDR E, which contained valid data over land and not mostly only over large lakes as in the previous versions, allowed the extension of the duration of the time series of water levels over rivers for the missions in the 10-day repeat orbit down to 2001, at the expense of lower accuracy than using Jason-2 and Jason-3 data (Figures 4 and 5 and Table S4). This lower accuracy can be accounted for the small river widths, most of the time lower than 500 m, except in some locations in the downstream part of the IND.

On the contrary, Jason-2 and Jason-3 confirm their great capability for detecting and accurately estimating water levels over narrow rivers, especially over flat areas [28,30]. Comparisons performed during the tandem phase logically confirm the lower agreement between Jason-1 and Jason-2 than between Jason-2 and Jason-3 (Figures 11 and 12). In terms of bias, if high and variable biases were found between Jason-1 and Jason-2-based time series of water levels, much lower ones with low variability were found between Jason-2 and Jason-3 ones during their tandem period. Poseidon-3, on-board Jason-3, is nominally operating in open loop or DIODE/DEM tracking mode over land surfaces, meaning that the reception window is controlled by an a-priori elevation from an along-track DEM loaded in the altimeter [30].

Very good results were also found using one year and a half of measurements from Sentinel-3A, the first altimeter to operate in SAR over all types of surfaces. During this short time period, the results obtained are almost as good as the ones obtained using SARAL, confirming the strong potential of this technique for monitoring inland water stages. This is likely due to the sharper waveform obtained in SAR mode compared with Low Resolution Mode (LRM) [59] as OCOG is particularly well adapted for very specular echoes from a single reflector (e.g., a river stream). On the contrary, less accurate estimates can be expected from multi-peak waveforms.

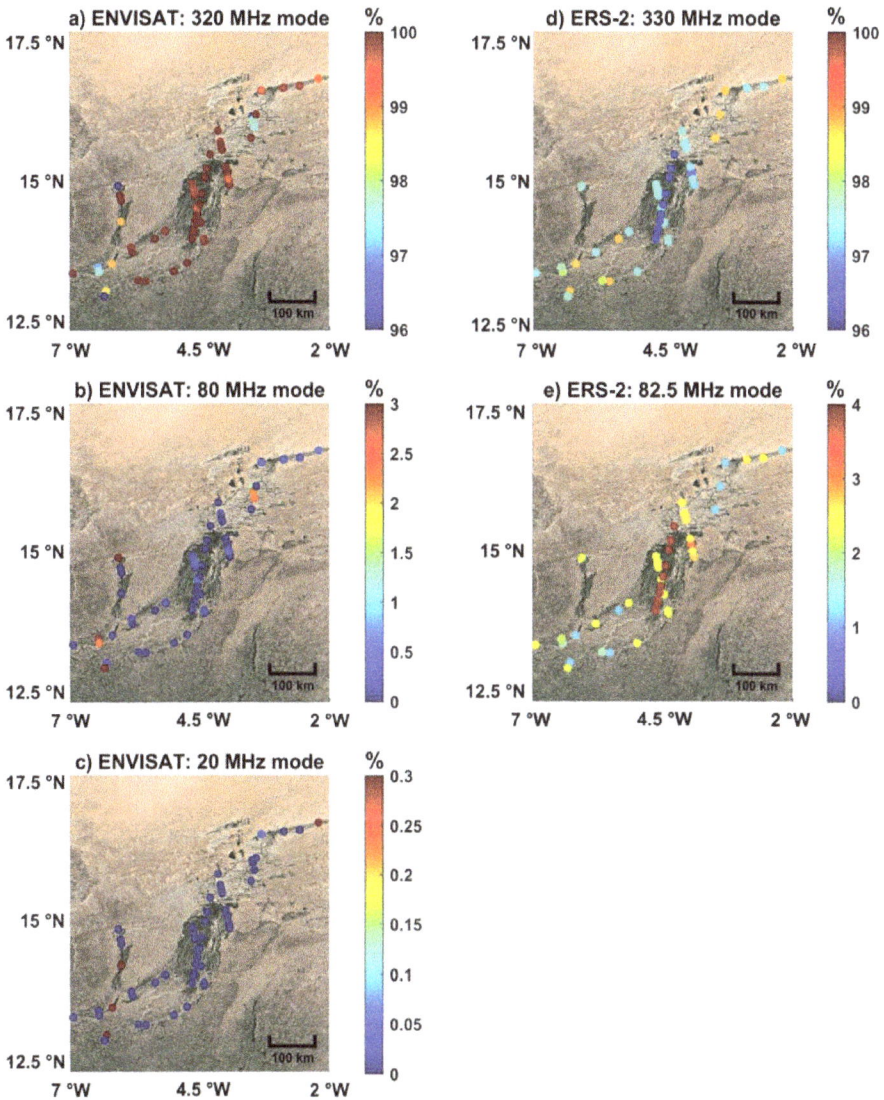

Figure 18. Percentage of acquisition in the different Ku chirp bandwidth operation mode for each VS in the IND for ENVISAT at 320 MHz (**a**); at 80 MHz; (**b**) and at 20 MHz; (**c**) ERS-2 at 330 MHz (**d**) and at 82.5 MHz (**e**).

The consistency of intra-mission water levels derived from radar altimetry measurements was analyzed using cross-correlations for ERS-2, ENVISAT, SARAL and Sentinel-3A. In this case, too, higher correlations were found for the most recent missions operating in (LRM) at Ka band (SARAL) and in SAR at Ku band (Sentinel-3A). Time-lags corresponding to the maximum of cross-correlation (Figures S1 and S2) also showed that there is a time-lag of around two months between the upstream and the downstream part of the IND, in accordance with the water residence time estimated using satellite images [39].

6. Conclusions

An extensive assessment of the performance of almost all the missions put in orbit from ERS-2 to Sentinel-3A was performed over the IND. Around 90 comparisons between in situ and altimetry-based were performed as well as more than 75 intermission comparisons between ERS-2 and ENVISAT, Jason-1 and Jason-2, and Jason-2 and Jason-3. Results of these comparisons show (i) a better agreement between altimetry-derived and in situ water stages over rivers of width varying from a few hundreds of meters to ~1.5 km and floodplains, (ii) an increase in accuracy in water level estimates for the most recent missions such as SARAL, Jason-3 and Sentinel-3A benefiting from the use of the Ka-band for SARAL and of the SAR acquisition mode for Sentinel-3A.

Due to the small number of VS defined on Jason-3 ground-tracks, the DIODE/DEM was not evaluated. Intra-mission consistency check performed on ERS-2, ENVISAT, SARAL and Sentinel-3A missions exhibits a time-lag for the maximum of around 2 months that can be related to the residence time of the water in the IND.

Very good performance of the recently launched altimeter SARAL onboard Sentinel-3A ensures the continuity of the monitoring of the IND whose data density should increase with the launch, in 2018, of Sentinel-3B, in the same orbit as Sentinel-3A but Sentinel-3B flies at ±140° out of phase with Sentinel-3A.

This network of VS will present a strong interest for (i) assessing the impacts of climate variability and human effects (e.g., dam operation and rice production), and (ii) validating the measurement of the future Surface Water and Ocean Topography (SWOT) mission, the first mission to operate in close nadir interferometry SAR for providing elevations in two swaths, to be launched in 2021, on both rivers and floodplains.

Supplementary Materials: The following are available online at http://www.mdpi.com/2072-4292/10/6/833/s1, Figure S1: Time-lag corresponding to maxima of cross-correlation between time series of water levels derived from SARAL (left) and SENTINEL-3A (right) data in the IND for three different VS near Douna (a and d), Mopti (b and e) and Diré (c and f). Light blue crosses symbolize the VS chosen as reference (auto-correlation of 1 and time-lag null), Figure S2: Time-lag corresponding to maxima of cross-correlation between time series of water levels derived from SARAL (left) and SENTINEL-3A (right) data in the IND for three different VS near Douna (a and d), Mopti (b and e) and Diré (c and f). Light blue crosses symbolize the VS chosen as reference (auto-correlation of 1 and time-lag null), Table S1: Results of the comparisons between in situ and altimetry water levels derived from ERS-2 mission: number of sample (N), correlation coefficient (R), determination coefficient (R^2) and Root Mean Square Error (RMSE), Table S2: Results of the comparisons between in situ and altimetry water levels derived from ENVISAT mission: number of sample (N), correlation coefficient (R), determination coefficient (R^2) and Root Mean Square Error (RMSE), Table S3: Results of the comparisons between in situ and altimetry water levels derived from SARAL mission: number of sample (N), correlation coefficient (R), determination coefficient (R^2) and Root Mean Square Error (RMSE), Table S4: Results of the comparisons between in situ and altimetry water levels derived from Jason-1, Jason-2 and Jason-3 missions: number of sample (N), correlation coefficient (R), determination coefficient (R^2) and Root Mean Square Error (RMSE), Table S5: Results of the comparisons between in situ and altimetry water levels derived from Sentinel-3A mission: number of sample (N correlation coefficient (R), determination coefficient (R^2) and Root Mean Square Error (RMSE).

Author Contributions: C.N., F.F., A.T.D. and E.M. designed the study. C.N. processed the altimetry data to produce the time series of water levels. A.T.D. and N.B. provided the in situ data and gave important information on their quality. All of the authors of the present work contributed to the discussion of the results, as well as the writing of the manuscript.

Funding: This study was supported by Centre National d'Etudes Spatiales (CNES) through "Centre de Topographie des Océans et de l'Hydrosphère—CTOH" grant. Cassandra Normandin is supported by Ministère de l'Enseignement Supérieur et de la Recherche through a PhD grant.

Acknowledgments: We thank four anonymous reviewers who helped us improving our manuscript.

Conflicts of Interest: The authors declare no conflict of interest. The founding sponsors had no role in the design of the study; in the collection, analyses, or interpretation of data; in the writing of the manuscript, and in the decision to publish the results.

References

1. Vörösmarty, C.J.; Green, P.; Salisbury, J.; Lammers, R.B. Global water resources: vulnerability from climate change and population growth. *Science* **2000**, *289*, 284–288. [CrossRef] [PubMed]
2. Oki, T.; Kanae, S. Global hydrological cycles and world water resources. *Science* **2006**, *313*, 1068–1072. [CrossRef] [PubMed]
3. Haddeland, I.; Heinke, J.; Biemans, H.; Eisner, S.; Flörke, M.; Hanasaki, N.; Konzmann, M.; Ludwig, F.; Masaki, Y.; Schewe, J.; et al. Global water resources affected by human interventions and climate change. *Proc. Natl. Acad. Sci. USA* **2014**, *111*, 3251–3256. [CrossRef] [PubMed]
4. Gleick, P.H. Global freshwater resources: soft-path solutions for the 21st century. *Science* **2003**, *302*, 1524–1528. [CrossRef] [PubMed]
5. Alsdorf, D.E.; Rodríguez, E.; Lettenmaier, D.P. Measuring surface water from space. *Rev. Geophys.* **2007**, *45*, RG2002. [CrossRef]
6. Stammer, D.; Cazenave, A. *Satellite Altimetry over Oceans and Land Surfaces*; Taylor & Francis: Boca Raton, FL, USA, 2017; ISBN 978-1-4987-4345-7.
7. Crétaux, J.-F.; Nielsen, K.; Frappart, F.; Papa, F.; Calmant, S.; Benveniste, J. Hydrological applications of satellite altimetry: rivers, lakes, man-made reservoirs, inundated areas. In *Satellite Altimetry Over Oceans and Land Surfaces*; Earth Observation of Global Changes; Stammer, D., Cazenave, A., Eds.; CRC Press: Boca Raton, FL, USA, 2017; pp. 459–504.
8. Morris, C.S.; Gill, S.K. Variation of Great Lakes water levels derived from Geosat altimetry. *Water Resour. Res.* **1994**, *30*, 1009–1017. [CrossRef]
9. Birkett, C.M. The contribution of TOPEX/POSEIDON to the global monitoring of climatically sensitive lakes. *J. Geophys. Res.* **1995**, *100204*, 25179–25204. [CrossRef]
10. Koblinsky, C.J.; Clarke, R.T.; Brenner, A.C.; Frey, H. Measurement of river level variations with satellite altimetry. *Water Resour. Res.* **1993**, *29*, 1839–1848. [CrossRef]
11. Birkett, C.M. Contribution of the TOPEX NASA Radar Altimeter to the global monitoring of large rivers and wetlands. *Water Resour. Res.* **1998**, *34*, 1223. [CrossRef]
12. Frappart, F.; Calmant, S.; Cauhopé, M.; Seyler, F.; Cazenave, A. Preliminary results of ENVISAT RA-2-derived water levels validation over the Amazon basin. *Remote Sens. Environ.* **2006**, *100*. [CrossRef]
13. Frappart, F.; Legrésy, B.; Niño, F.; Blarel, F.; Fuller, N.; Fleury, S.; Birol, F.; Calmant, S. An ERS-2 altimetry reprocessing compatible with ENVISAT for long-term land and ice sheets studies. *Remote Sens. Environ.* **2016**, *184*. [CrossRef]
14. Baup, F.; Frappart, F.; Maubant, J. Use of satellite altimetry and imagery for monitoring the volume of small lakes. In *International Geoscience and Remote Sensing Symposium (IGARSS)*; 2014.
15. Sulistioadi, Y.B.; Tseng, K.-H.; Shum, C.K.; Hidayat, H.; Sumaryono, M.; Suhardiman, A.; Setiawan, F.; Sunarso, S. Satellite radar altimetry for monitoring small rivers and lakes in Indonesia. *Hydrol. Earth Syst. Sci.* **2015**, *19*, 341–359. [CrossRef]
16. Frappart, F.; Papa, F.; Malbeteau, Y.; León, J.G.J.G.; Ramillien, G.; Prigent, C.; Seoane, L.; Seyler, F.; Calmant, S. Surface freshwater storage variations in the orinoco floodplains using multi-satellite observations. *Remote Sens.* **2015**, *7*, 89–110. [CrossRef]
17. da Silva, J.S.; Calmant, S.; Seyler, F.; Moreira, D.M.; Oliveira, D.; Monteiro, A. Radar Altimetry Aids Managing Gauge Networks. *Water Resour. Manag.* **2014**, *28*, 587–603. [CrossRef]
18. Birkett, C.; Reynolds, C.; Beckley, B.; Doorn, B. From Research to Operations: The USDA Global Reservoir and Lake Monitor. In *Coastal Altimetry*; Springer Berlin Heidelberg: Berlin/Heidelberg, Germany, 2011; pp. 19–50.
19. Ričko, M.; Birkett, C.M.; Carton, J.A.; Crétaux, J.-F. Intercomparison and validation of continental water level products derived from satellite radar altimetry. *J. Appl. Remote Sens.* **2012**, *6*, 61710. [CrossRef]
20. Cretaux, J.-F.; Berge-Nguyen, M.; Leblanc, M.; Rio, R.A.D.; Delclaux, F.; Mognard, N.; Lion, C.; Pandey, R.-K.; Tweed, S.; Calmant, S.; et al. Flood mapping inferred from remote sensing data. *Int. Water Technol. J.* **2011**, *1*, 48–62.
21. Goita, K.; Diepkile, A.T. Radar altimetry of water level variability in the Inner Delta of Niger River. In Proceedings of the IEEE International Geoscience and Remote Sensing Symposium, Munich, Germany, 22–27 July 2012; pp. 5262–5265.

22. Frappart, F.; Fatras, C.; Mougin, E.; Marieu, V.; Diepkilé, A.T.; Blarel, F.; Borderies, P. Radar altimetry backscattering signatures at Ka, Ku, C, and S bands over West Africa. *Phys. Chem. Earth* **2015**, *83–84*, 96–110. [CrossRef]

23. Tarpanelli, A.; Amarnath, G.; Brocca, L.; Massari, C.; Moramarco, T. Discharge estimation and forecasting by MODIS and altimetry data in Niger-Benue River. *Remote Sens. Environ.* **2017**, *195*, 96–106. [CrossRef]

24. Tourian, M.J.; Schwatke, C.; Sneeuw, N. River discharge estimation at daily resolution from satellite altimetry over an entire river basin. *J. Hydrol.* **2017**, *546*, 230–247. [CrossRef]

25. Chelton, D.B.; Ries, J.C.; Haines, B.J.; Fu, L.-L.; Callahan, P.S. Chapter 1 Satellite Altimetry. In *Satellite Altimetry and Earth Sciences A Handbook of Techniques and Applications*; Elsevier, 2001; Volume 69, pp. 1–131. ISBN 0074-6142.

26. Frappart, F.; Blumstein, D.; Cazenave, A.; Ramillien, G.; Birol, F.; Morrow, R.; Rémy, F. Satellite Altimetry: Principles and Applications in Earth Sciences. In *Wiley Encyclopedia of Electrical and Electronics Engineering*; John Wiley & Sons, Inc.: Hoboken, NJ, USA, 2017; pp. 1–25. ISBN 047134608X.

27. Frappart, F.; Papa, F.; Marieu, V.; Malbeteau, Y.; Jordy, F.; Calmant, S.; Durand, F.; Bala, S. Preliminary Assessment of SARAL/AltiKa Observations over the Ganges-Brahmaputra and Irrawaddy Rivers. *Mar. Geod.* **2015**, *38*. [CrossRef]

28. Biancamaria, S.; Frappart, F.; Leleu, A.S.; Marieu, V.; Blumstein, D.; Desjonquères, J.D.; Boy, F.; Sottolichio, A.; Valle-Levinson, A. Satellite radar altimetry water elevations performance over a 200 m wide river: Evaluation over the Garonne River. *Adv. Sp. Res.* **2017**, *59*, 128–146. [CrossRef]

29. Bogning, S.; Frappart, F.; Blarel, F.; Niño, F.; Mahé, G.; Bricquet, J.P.; Seyler, F.; Onguéné, R.; Etamé, J.; Paiz, M.C.; Braun, J.J. Monitoring water levels and discharges using radar altimetry in an ungauged river basin: The case of the Ogooué. *Remote Sens.* **2018**, *10*, 350. [CrossRef]

30. Biancamaria, S.; Schaedele, T.; Blumstein, D.; Frappart, F.; Boy, F.; Desjonquères, J.D.; Pottier, C.; Blarel, F.; Niño, F. Validation of Jason-3 tracking modes over French rivers. *Remote Sens. Environ.* **2018**, *209*, 77–89. [CrossRef]

31. Frappart, F.; Roussel, N.; Biancale, R.; Martinez Benjamin, J.J.; Mercier, F.; Perosanz, F.; Garate Pasquin, J.; Martin Davila, J.; Perez Gomez, B.; Gracia Gomez, C.; et al. The 2013 Ibiza Calibration Campaign of Jason-2 and SARAL Altimeters. *Mar. Geod.* **2015**, *38*. [CrossRef]

32. Vu, P.; Frappart, F.; Darrozes, J.; Marieu, V.; Blarel, F.; Ramillien, G.; Bonnefond, P.; Birol, F. Multi-Satellite Altimeter Validation along the French Atlantic Coast in the Southern Bay of Biscay from ERS-2 to SARAL. *Remote Sens.* **2018**, *10*, 93. [CrossRef]

33. Jekeli, C. *Geometric Reference System in Geodesy*; Division of Geodesy and Geospatial Science School of Earth Sciences, Ohio State University: Columbus, OH, USA, 2006.

34. Salameh, E.; Frappart, F.; Marieu, V.; Spodar, A.; Parisot, J.P.; Hanquiez, V.; Turki, I.; Laignel, B. Monitoring sea level and topography of coastal lagoons using satellite radar altimetry: The example of the Arcachon Bay in the Bay of Biscay. *Remote Sens.* **2018**, *10*, 297. [CrossRef]

35. Mahé, G.; Bamba, F.; Soumaguel, A.; Orange, D.; Olivry, J.C. Water losses in the inner delta of the River Niger: water balance and flooded area. *Hydrol. Process.* **2009**, *23*, 3157–3160. [CrossRef]

36. De Noray, M.-L. Delta intérieur du fleuve Niger au Mali–quand la crue fait la loi: l'organisation humaine et le partage des ressources dans une zone inondable à fort contraste. *VertigO-la Rev. électronique en Sci. l' Environ.* **2003**, *4*, 1–9. [CrossRef]

37. Zwarts, L. *The Niger, A Lifeline: Effective Water Management in the Upper Niger Basin*; RIZA: Lelystad, 2005; ISBN 978-90-807150-6-6.

38. Jones, K.; Lanthier, Y.; van der Voet, P.; van Valkengoed, E.; Taylor, D.; Fernández-Prieto, D. Monitoring and assessment of wetlands using Earth Observation: The GlobWetland project. *J. Environ. Manag.* **2009**, *90*, 2154–2169. [CrossRef] [PubMed]

39. Bergé-Nguyen, M.; Crétaux, J.-F. Inundations in the Inner Niger Delta: Monitoring and Analysis Using MODIS and Global Precipitation Datasets. *Remote Sens.* **2015**, *7*, 2127–2151. [CrossRef]

40. Ogilvie, A.; Belaud, G.; Delenne, C.; Bailly, J.-S.; Bader, J.-C.; Oleksiak, A.; Ferry, L.; Martin, D. Decadal monitoring of the Niger Inner Delta flood dynamics using MODIS optical data. *J. Hydrol.* **2015**, *523*, 368–383. [CrossRef]

41. Benveniste, J.; Roca, M.; Levrini, G.; Vincent, P.; Baker, S.; Zanife, O.; Zelli, C.; Bombaci, O. The radar altimetry mission: RA-2, MWR, DORIS and LRR. *ESA Bull.* **2001**, *106*, 25101–25108.

42. Steunou, N.; Desjonquères, J.D.; Picot, N.; Sengenes, P.; Noubel, J.; Poisson, J.C. AltiKa Altimeter: Instrument Description and In Flight Performance. *Mar. Geod.* **2015**, *38*, 22–42. [CrossRef]

43. Taylor, P.; Perbos, J.; Escudier, P.; Parisot, F.; Zaouche, G.; Vincent, P.; Menard, Y.; Manon, F.; Kunstmann, G.; Royer, D.; et al. Jason-1: Assessment of the System Performances Special Issue: Jason-1 Calibration/Validation. *Mar. Geod.* **2003**, *26*, 37–41. [CrossRef]

44. Desjonquères, J.D.; Carayon, G.; Steunou, N.; Lambin, J. Poseidon-3 Radar Altimeter: New Modes and In-Flight Performances. *Mar. Geod.* **2010**, *33*, 53–79. [CrossRef]

45. Donlon, C.; Berruti, B.; Buongiorno, A.; Ferreira, M.H.; Féménias, P.; Frerick, J.; Goryl, P.; Klein, U.; Laur, H.; Mavrocordatos, C.; et al. The Global Monitoring for Environment and Security (GMES) Sentinel-3 mission. *Remote Sens. Environ.* **2012**, *120*, 37–57. [CrossRef]

46. Wingham, D.J.; Rapley, C.G.; Griffiths, H. New Techniques in Satellite Altimeter Tracking Systems. *Proc. IGARSS Symp. Zurich* **1986**, 1339–1344.

47. Frappart, F.; Do Minh, K.; L'Hermitte, J.; Cazenave, A.; Ramillien, G.; Le Toan, T.; Mognard-Campbell, N. Water volume change in the lower Mekong from satellite altimetry and imagery data. *Geophys. J. Int.* **2006**, *167*. [CrossRef]

48. Santos da Silva, J.; Calmant, S.; Seyler, F.; Rotunno Filho, O.C.; Cochonneau, G.; Mansur, W.J. Water levels in the Amazon basin derived from the ERS 2 and ENVISAT radar altimetry missions. *Remote Sens. Environ.* **2010**, *114*, 2160–2181. [CrossRef]

49. Cartwright, D.E.; Edden, A.C. Corrected Tables of Tidal Harmonics. *Geophys. J. R. Astron. Soc.* **1973**, *33*, 253–264. [CrossRef]

50. Wahr, J.M. Deformation induced by polar motion. *J. Geophys. Res.* **1985**, *90*, 9363. [CrossRef]

51. Crétaux, J.F.; Jelinski, W.; Calmant, S.; Kouraev, A.; Vuglinski, V.; Bergé-Nguyen, M.; Gennero, M.C.; Nino, F.; Abarca Del Rio, R.; Cazenave, A.; Maisongrande, P. SOLS: A lake database to monitor in the Near Real Time water level and storage variations from remote sensing data. *Adv. Sp. Res.* **2011**, *47*, 1497–1507. [CrossRef]

52. Frappart, F.; Biancamaria, S.; Normandin, C.; Blarel, F.; Bourrel, L.; Aumont, M.; Azemar, P.; Vu, P.-L.; Le Toan, T.; Lubac, B.; et al. Influence of recent climatic events on the surface water storage of the Tonle Sap Lake. *Sci. Total Environ.* **2018**, *636*, 1520–1533. [CrossRef]

53. Frappart, F.; Papa, F.; Famiglietti, J.S.; Prigent, C.; Rossow, W.B.; Seyler, F. Interannual variations of river water storage from a multiple satellite approach: A case study for the Rio Negro River basin. *J. Geophys. Res. Atmos.* **2008**, *113*. [CrossRef]

54. Lee, H.; Shum, C.K.; Yi, Y.; Ibaraki, M.; Kim, J.W.; Braun, A.; Kuo, C.Y.; Lu, Z. Louisiana wetland water level monitoring using retracked TOPEX/POSEIDON altimetry. *Mar. Geod.* **2009**, *32*, 284–302. [CrossRef]

55. Frappart, F.; Papa, F.; Güntner, A.; Werth, S.; Santos da Silva, J.; Tomasella, J.; Seyler, F.; Prigent, C.; Rossow, W.B.; Calmant, S.; Bonnet, M.-P. Satellite-based estimates of groundwater storage variations in large drainage basins with extensive floodplains. *Remote Sens. Environ.* **2011**, *115*. [CrossRef]

56. da Silva, J.S.; Seyler, F.; Calmant, S.; Filho, O.C.R.; Roux, E.; Araújo, A.A.M.; Guyot, J.L. Water level dynamics of Amazon wetlands at the watershed scale by satellite altimetry. *Int. J. Remote Sens.* **2012**, *33*, 3323–3353. [CrossRef]

57. Zakharova, E.A.; Kouraev, A.V.; Rémy, F.; Zemtsov, V.A.; Kirpotin, S.N. Seasonal variability of the Western Siberia wetlands from satellite radar altimetry. *J. Hydrol.* **2014**, *512*, 366–378. [CrossRef]

58. Bonnefond, P.; Verron, J.; Aublanc, J.; Babu, K.; Bergé-Nguyen, M.; Cancet, M.; Chaudhary, A.; Crétaux, J.-F.; Frappart, F.; Haines, B.; et al. The Benefits of the Ka-Band as Evidenced from the SARAL/AltiKa Altimetric Mission: Quality Assessment and Unique Characteristics of AltiKa Data. *Remote Sens.* **2018**, *10*, 83. [CrossRef]

59. Keith Raney, R. The delay/doppler radar altimeter. *IEEE Trans. Geosci. Remote Sens.* **1998**, *36*, 1578–1588. [CrossRef]

remote sensing

MDPI

Article

River Levels Derived with CryoSat-2 SAR Data Classification—A Case Study in the Mekong River Basin

Eva Boergens [1,*], Karina Nielsen [2], Ole Baltazar Andersen [2], Denise Dettmering [1] and Florian Seitz [1]

[1] Deutsches Geodätisches Forschungsinstitut der Technischen Universität München (DGFI-TUM), Arcisstraße 21, 80333 Munich, Germany; denise.dettmering@tum.de (D.D.); florian.seitz@tum.de (F.S.)
[2] Division of Geodesy, DTU Space, National Space Institute, DK-2800 Kongens Lyngby, Denmark; karni@space.dtu.dk (K.N.); oa@space.dtu.dk (O.B.A.)
* Correspondence: eva.boergens@tum.de; Tel.: +49-89-23031-1263

Received: 13 October 2017; Accepted: 27 November 2017; Published: 30 November 2017

Abstract: In this study we use CryoSat-2 SAR (delay-Doppler synthetic-aperture radar) data in the Mekong River Basin to estimate water levels. Compared to classical pulse limited radar altimetry, medium- and small-sized inland waters can be observed with CryoSat-2 SAR data with a higher accuracy due to the smaller along track footprint. However, even with this SAR data the estimation of water levels over a medium-sized river (width less than 500 m) is still challenging with only very few consecutive observations over the water. The target identification with land–water masks tends to fail as the river becomes smaller. Therefore, we developed a classification approach to divide the observations into water and land returns based solely on the data. The classification is done with an unsupervised classification algorithm, and it is based on features derived from the SAR and range-integrated power (RIP) waveforms. After the classification, classes representing water and land are identified. Better results are obtained when the Mekong River Basin is divided into different geographical regions: upstream, middle stream, and downstream. The measurements classified as water are used in a next step to estimate water levels for each crossing over a river in the Mekong River network. The resulting water levels are validated and compared to gauge data, Envisat data, and CryoSat-2 water levels derived with a land–water mask. The CryoSat-2 water levels derived with the classification lead to more valid observations with fewer outliers in the upstream region than with a land–water mask (1700 with 2% outliers vs. 1500 with 7% outliers). The median of the annual differences that is used in the validation is in all test regions smaller for the CryoSat-2 classification results than for Envisat or CryoSat-2 land–water mask results (for the entire study area: 0.76 m vs. 0.96 m vs. 0.83 m, respectively). Overall, in the upstream region with small- and medium-sized rivers the classification approach is more effective for deriving reliable water level observations than in the middle stream region with wider rivers.

Keywords: satellite altimetry; inland water; CryosSat-2 SAR; Mekong Basin; water level time series; classification; stack data

---

1. Introduction

River waters are crucial as suppliers of water for irrigation and fresh water for drinking. However, river floods can destroy crops, settlements, and infrastructure. For this reason, it is essential to monitor the water level of river systems. However, an increasing number of in situ gauges have been derelicted since the 1980s [1], or the data is not publicly available. Therefore, it is more and more important to close this data gap with satellite altimetry.

In recent years many studies have been published that apply satellite pulse-limited altimetry over rivers of various sizes (e.g., [2–7]). CryoSat-2, launched in 2010, is the first satellite carrying a delay-Doppler altimeter [8] that operates in three measuring modes: the classical pulse-limited low resolution (LR) mode, the delay-Doppler synthetic-aperture radar (SAR) mode, and the SAR interferometric (SARin) mode.

Compared to conventional radar altimeters, delay-Doppler SAR measurements have a higher along-track resolution and a smaller footprint. This improves the observation of water levels of inland water bodies like lakes (e.g., [9–11]) or rivers (e.g., [12,13]). The advantage of SAR altimetry observations is that they are especially effective for measuring smaller inland waters like rivers. However, CryoSat-2 has a long repeat time of 369 days compared to 35 days for Envisat and SARAL, and 10 days for Topex/Poseidon, Jason-1, Jason-2, and Jason-3. This restricts the estimation of water level time series over lakes and even more over rivers. The advantage of the long repeat time is the very dense spatial distribution of observations. This spatial distribution enables better river monitoring in the continuous progression. Unlike lakes, rivers can change their water levels rapidly over their course which makes a denser spatial distribution of observations desirable.

To derive water levels from lakes or rivers it is necessary to identify the water returns of the altimeter. A land–water mask can be used for this purpose, like the mask provided by the World Wildlife Fund (https://www.worldwildlife.org/pages/global-lakes-and-wetlands-database). Such a mask is constant over time; therefore, it neither accounts for the seasonal variations of the water extent nor inter-annually shifting river and lake banks. Extracting dynamic land–water masks from optical remote sensing images is difficult in the tropics, which includes the study area of the Mekong Basin, since cloud-free optical data is only available during the dry season with low water level. Moreover, SAR images with sufficient spatial resolution are only available from the launch of Sentinel-1 in 2014. Although a high accuracy land–water mask is provided by the Mekong River Commission (http://portal.mrcmekong.org/map_service) for the study area with an accuracy of 30 m, this accuracy might not be sufficient for medium- and small-sized rivers. Additionally, the mask has no seasonal variations included.

Within this study, an innovative classification approach for SAR altimetry data is developed and validated, which is independent of the accuracy and availability of land–water masks. With this approach, reliable river levels, even for small and medium-sized rivers, are derived. Compared to classical approaches based on land–water masks, the number of valid measurements is increased and the water level precision is improved.

For the classical pulse-limited altimeter data, classification has been done successfully for the last decade worldwide (e.g., [14,15]). Even very small water areas in wetlands have been classified successfully with Envisat data by [16].

In these classifications, the shape of the waveform is used to discriminate between different reflecting surfaces. Also, CryoSat-2 SAR data has been classified based on the SAR waveform for lakes [11], lakes and rivers [17], or ice [18]. This study takes a step further and uses not only the waveform but also the range-integrated power (RIP) for a classification of the altimeter measurements in water and non-water returns over the Mekong River Basin in Southeast Asia. The RIP is only available for delay-Doppler SAR altimetry and gives additional insight to the reflective surface that the waveform alone could not provide [19].

The unsupervised *k-means* algorithm is employed for the classification [20] as not enough reliable training data is available for a supervised classification. The *k-means* algorithm is a widely used unsupervised clustering algorithm and has been used for altimetry classification before (e.g., [11]).

This paper is structured as follows: First, an introduction is given about the study area of the Mekong River Basin in Section 2, afterwards more information of the CryoSat-2 SAR data is given in Section 3. The classification and the used features are described in Section 4.1 followed by an explanation of the water level estimation in Section 4.2. The results and validations are presented in Section 5. These results are discussed in Section 6 and an outlook is given. The paper ends with the conclusions in Section 7. An overview over all relevant processing steps of this study is given in Figure 1.

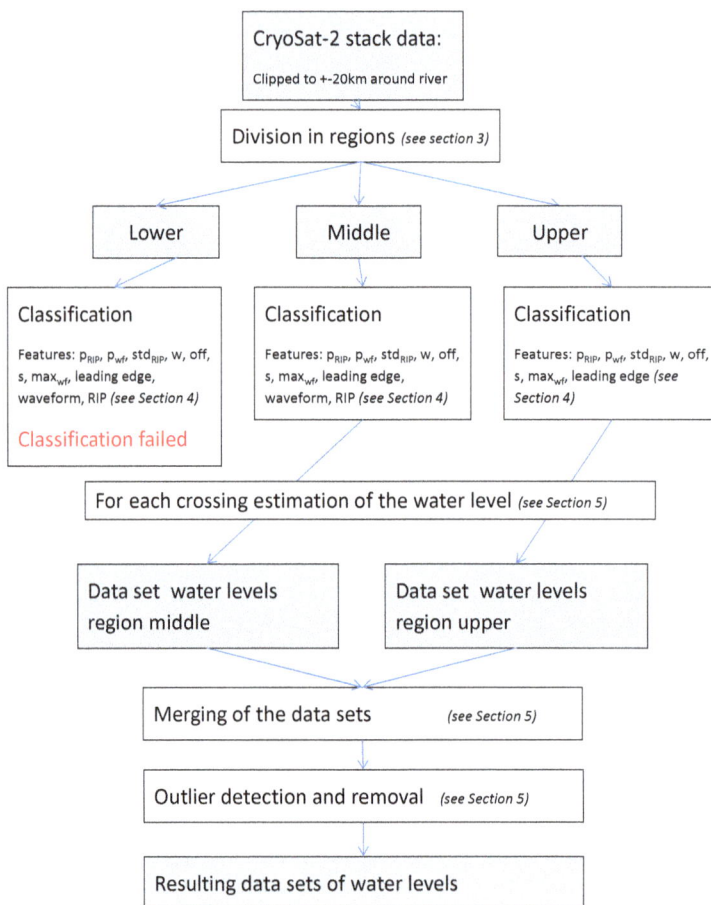

Figure 1. Processing steps used in this study for extracting water levels from the CryoSat-2 synthetic-aperture radar (SAR) data. RIP: range-integrated power.

2. Study Area

The Mekong River Basin in Southeast Asia (China, Myanmar, Thailand, Laos, Cambodia, and Vietnam) is investigated in this study, with a focus on the part of the basin south of the Chinese border. Further upstream the river gorge is too narrow for reliable altimeter observations. Downstream, the study area is limited by by the confluence with the Tonle Sap River from where the river is under a tidal influence. The tributaries, namely the large left bank side tributaries in Laos, are investigated as well. The hydrology of the Mekong Basin is primarily influenced by the precipitation on the Tibetan Plateau and the south-eastern monsoon [21].

In the Mekong River Basin the river width varies between 20 m to more than 2 km. Most of the tributaries and the upstream reach of the main river stream are less than 100- m-wide small rivers. The medium rivers, which are less than 500- m but more than 100- m-wide, are the main tributaries for the upstream main river. In the downstream reach of the river, before it splits into the delta, the river has a width of over 2 km (see also Figure 2 for a map of the basin).

The Mekong River and its tributaries flow through different topographic regions (Figure 2). The main river upstream from Vientiane and the left bank tributaries in Laos are surrounded by mountainous areas with steep banks where the rivers have a greater slope and have a width smaller than 500 m or even less than 100 m. Downstream of Vientiane and up to the Mekong Falls the river widens and flows with less slope over the Khorat plateau. Below the Mekong Falls the river is surrounded by seasonal wetlands and widens to more than 1 km. For further processing we defined three overlapping data masks according to these regions (Figure 2). The regions are determined by the roughness of a topography model and the absolute height. Afterwards a margin around each subregion allows for an overlap.

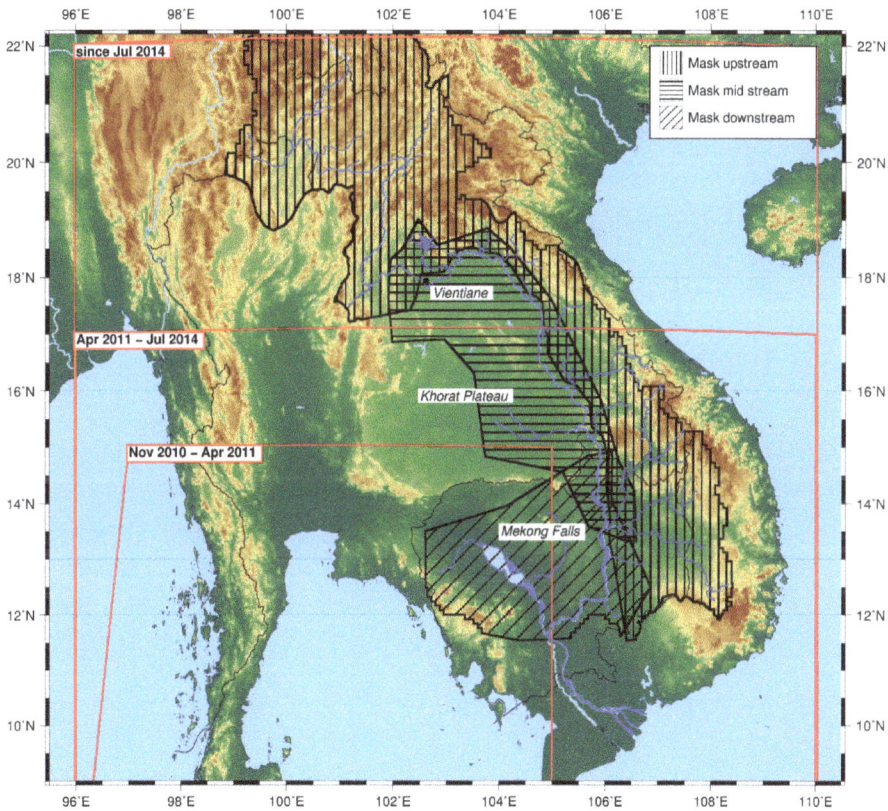

Figure 2. Map of the study area with the regional masks (black areas with different hachures) and the SAR mode mask with their validity period (red boxes).

3. Data

3.1. CryoSat-2 SAR Data

In this study we use delay-Doppler SAR altimeter data measured by CryoSat-2 between 2010 and 2016. CryoSat-2 measures in three different modes (LR, SAR, and SARIn mode), which are set in a geographical mask (https://earth.esa.int/web/guest/-/geographical-mode-mask-7107): The LRM is active mostly over the oceans and the interior of the ice sheets of Antarctica and Greenland, whereas the SAR mode measures over sea ice and other selected regions, and SARin focuses mostly on glaciated regions [22]. This mask has changed over the life time of the satellite. The entire study area of the Mekong River Basin has only been measured in SAR mode since July 2014 (see Figure 2 for the extent of the SAR mode mask). In SAR mode the along-track foot print size is reduced to 300 m while it remains at 14 km in the across-track direction [23].

The delay-Doppler SAR altimeter measures a point on the surface several times from different looking angles [24]. Unlike pulse limited altimetry, which is only able to measure the returning signal from the whole foot print, SAR altimetry discretizes the returning signal corresponding to 300- m-wide stripes of the footprint in along track direction. The signals from each of these stripes are recorded in the so-called single-look waveform. All single-look waveforms for the same surface area are collected in the stack matrix.

In Figure 3, two exemplary stack matrices are presented. The first (a) is measured over the Tonle Sap lake and the second (b) over a medium-sized river in the upstream region of the Mekong River. Each row is a single-look waveform. The integration of this matrix over all single-looks results in the multi-look SAR waveform (in Figure 3 with integration over each row of the stack) hereafter referred to as the waveform. The integration over the range bins results in the range-integrated power (RIP). In Figure 3 this corresponds to the integration over the columns. Detailed information on the delay-Doppler measurements is described in [8].

Here, we use the CryoSat-2 baseline C SAR Level 1b data provided by ESA GPOD SARvatore (https://gpod.eo.esa.int/) for the period 2010–2016. SARvatore is the official ESA platform for the dissemination of the CryoSat-2 level 1b SAR data. The data is processed on user request with different processing parameters. The most important data processing parameters of the data in this study are: the applied Hamming window, fft zero-padding, 20-Hz data, and stack data in the output file.

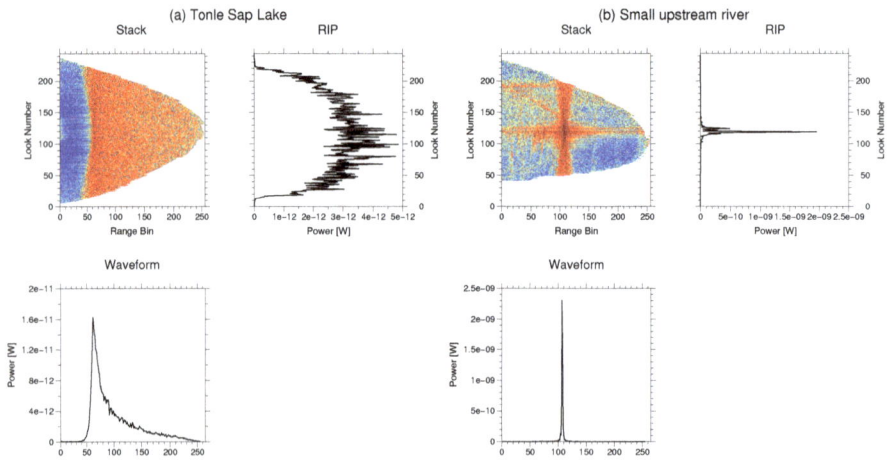

Figure 3. Two exemplary stack matrices with their RIP and waveform. The color of the stack plot indicates the power of the signal. The example on the left-hand side is measured over the Tonle Sap lake, the one on the right-hand side over a medium upstream river.

3.2. Envisat Data

Envisat water level observations are used for the validation in this study. The water levels are taken from the DAHITI database [4]. DAHITI uses a throughout outlier detection in combination with a Kalman filter approach to derive reliable water level time series for lakes and rivers. More information on the processing strategy is available in [4]. The Envisat water level time series in the Mekong Basin has a temporal resolution of 35 days, even though data gaps may occur for some cycles. For small tributaries, a correction for the Hooking effect is applied additionally in [5].

3.3. In Situ Data

For validation, in situ gauge data is used which is provided by the Mekong River Commission (http://ffw.mrcmekong.org/). The gauge data has a daily temporal resolution but the time series ends with 2012. No absolute height of the gauge is provided.

Additionally, we use a river polygon which is also provided by the Mekong River Commission (http://portal.mrcmekong.org/map_service). The polygon was derived from aerial images and topographic maps. The accuracy of the river mask is ~30 m, but no information about the seasonality of the polygon is given.

4. Methods

4.1. Classification Approach

For the medium and small rivers in our study area of the Mekong basin no reliable land–water mask is available. Thus, a classification by means of the *k-means* algorithm is performed to extract the water measurements.

The *k-means* algorithm [20] is an unsupervised method to cluster the data on the basis of different features. For the land–water classification a set of features derived from the CryoSat-2 stack data over the intermediate step of the waveform and the RIP is used. The features are summarized in Table 1. The features derived from the waveform are the *maximum power*, the *peakiness*, and the *position of the leading edge*. It is well known that waveforms of water reflections have a higher power than those of land reflections. Medium, and to a greater degree small, water bodies have smooth mirror-like surfaces which can only be measured by signals emitted close to nadir. This leads to a very peaky waveform and RIP with a high power. Following [25] the peakiness p_{wf} is calculated with

$$p_{wf} = \frac{\max(wf_i)}{\sum_i wf_i},$$ (1)

where wf is the waveform and wf_i the power of the *i*th bin.

Table 1. Features used for the classification.

RIP Features	Waveform Features
Peakiness: p_{RIP}	p_{wf}
Standard deviation: std_{RIP}	Maximum power: max_{wf}
Width: w	Relative position of leading edge
Off-center: off	
Symmetry: s	

To estimate the relative *position of the leading edge* in the waveform, the waveform is retracked using an improved threshold retracker with a threshold of 50% on the best sub-waveform [26]. The on-board tracking system always tries to hold the leading edge of the main reflection at the nominal tracking point. This is not always possible and leads to a deviation of the leading edge from the nominal tracking point. Over wider rivers the tracking system can manage to keep the leading edge close to the tracking point. In Figure 4 (left panel), one exemplary waveform with its features *maximum power* and *position of the leading edge* is shown (the *peakiness* cannot be displayed).

Features based on the RIP are the *peakiness* p_{RIP}, the *standard deviation* std_{RIP}, the *width*, the *off-center*, and the *symmetry*. Water reflections over large water bodies result in a overall smoother RIP than water reflections over small water bodies that in turn have a smoother RIP than land reflections(see Figure 3). This smoothness is measured with a standard deviation feature std_{RIP}. The std_{RIP} is then

$$std_{\text{RIP}} = \sqrt{\frac{1}{N} \sum_i (\text{RIP}_i - mean(\text{RIP}))^2}, \tag{2}$$

where RIP_i is the ith entry of the RIP and N the number of looks in the RIP, usually 246.

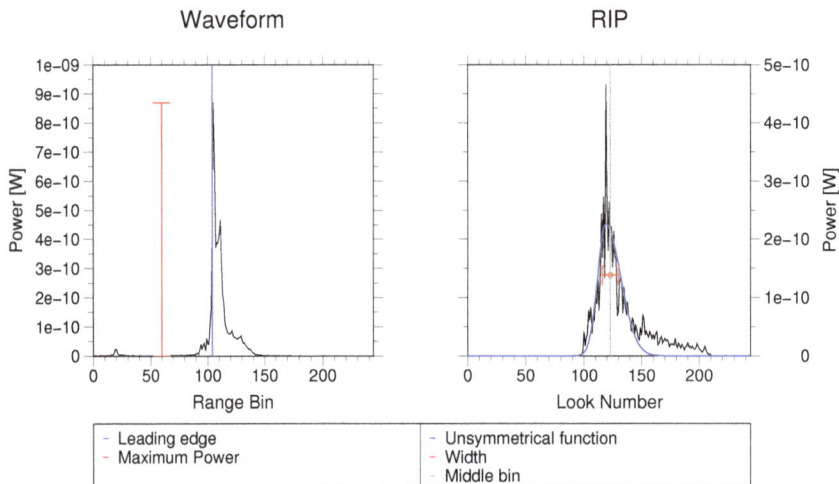

Figure 4. One example of a waveform and the corresponding RIP with some of their derived features.

As mentioned before, small and medium inland waters with a smooth surface only reflect the signal back to the satellite at near nadir. Therefore, the RIP is both very peaky and narrow. The width w is derived with:

$$w = \frac{\left(\sum_i \text{RIP}_i^2 \right)^2}{\sum_i \text{RIP}_i^4}. \tag{3}$$

The off-center feature off describes the deviation of the main reflection from the nadir point. It should be close to zero for measurements of water, whereas land measurements are more disturbed and often show the maximum return in the side lobes. We measure the off-center feature off as the difference between the middle look of the RIP and the mean point of the RIP which is calculated with:

$$off = \frac{246}{2} - \frac{\sum_i i \text{RIP}_i^2}{\sum_i \text{RIP}_i^2}. \tag{4}$$

A positive off value indicates that the majority of the returning power was detected before the satellite passed the nadir position, a negative value vice versa.

The last feature is a measure of the symmetry of the RIP s. For an ideal smooth water reflection, like a small lake, the RIP should be perfectly symmetrical. However, for a sloped target, such as a river, the reflection depends on the relative orientation between the satellite and the water surface. The reflection is stronger when the satellite looks on a water surface that is sloped towards it. A positive s indicates a water surface sloped towards the approaching satellite. This effect leads to an unsymmetrical RIP. To quantify this, an unsymmetrical exponential function $\overline{\text{RIP}}$ is fitted to the RIP with

$$\overline{\text{RIP}}_i = \begin{cases} a \exp\left(\frac{(i-b)^2}{2c_1^2}\right), & \text{if } i < b \\ a \exp\left(-\frac{(i-b)^2}{2c_2^2}\right), & \text{if } i \geq b. \end{cases} \qquad (5)$$

Here, a is the amplitude of the exponential function, b the look where the function reaches its maximum, and c_1 and c_2 are the two decay parameters. The *symmetry* feature is then

$$s = c_1 - c_2. \qquad (6)$$

Figure 4 (right-hand side), displays an RIP with the feature w marked. The off-center feature off is too small to be visible in this example, but the symmetry, or the lack thereof, is clearly showing.

Additional to these eight features, both the whole waveform and the whole RIP are used as features. Each bin is then considered as a single feature. The waveform needs to be shifted so that the leading edge is positioned on the nominal tracking point. We found that including waveform and RIP as features only improves the results for the middle region and does not change them in the upstream regions. Therefore, the upstream region is classified without waveform and RIP as features. Since the features span different orders of magnitude, it is necessary to normalize the feature set. All of these features were chosen according to their sensitivity for the posed problem of water classification in the Mekong River Basin and independently from each other. More features were tested but discarded because they were either not sensitive for the classification or highly correlated to one of the used features.

The *k-means* algorithm is used to cluster the data on the basis of the above features in 20 classes. An unsupervised clustering algorithm is applied because no reliable training data is available. The *k-means* algorithm assumes normally distributed features with equal variance, which we ensured and tested by the normalization of the features.

The number of classes depends on the application and variation in the input features. An estimate for the number of classes can be done with knowledge of the classified data. In our study case, a look at the spatial distribution of the features tells us that only two classes, land and water, are not sufficient as altimeter measurements of land can be very diverse (this holds also for water measurements, but they are less diverse than land). The diversity of the returning waveform and RIP can be explained by the reflective properties of e.g., land, water, vegetation. With this it can be concluded that at least 10 classes are needed. We tested the classification and validated resulting water levels for several numbers of classes (10, 15, 20, 30) and found similar results for all with the results of 20 classes slightly superior.

Each of the clusters is defined by their centroid which are the mean features of all points in this cluster. New data is classified by grouping it to the closest centroid. Here, the clustering is done on one randomly drawn third of the data. The residual two-thirds of the data are then classified into the cluster classes. The clustering is not done on the whole data set due to computational efficiency. The repeatability of the clustering and classification will be validated in Section 5. After the classification, it is determined which classes represent water and land returns. This was done by visual inspection of the mean waveform and RIP for each class and the locations of the observations in each class related to the approximate location of the river known from the land–water mask (see Section 3).

As described in Section 2 the Mekong Basin is divided into different regions: upstream, middle, and downstream. We classify each of the regions separately as they are too diverse in the reflectivity properties of the water bodies to be classified together. Additionally, the classification is done only on altimeter data not further away than 20 km from the river polygon due to computational efficiency (the polygon can be seen in Figure 2).

4.2. Water Level Estimation Approach

4.2.1. Altimetric Water Levels

The classification results in a set of measurements considered as water returns. From these measurements the water level at each crossing of the satellite track with a river in the Mekong River Basin is determined. A river polygon is employed to locate all crossings in the river basin (see Section 3). We used all measurements less than 5 km away from the river crossing that were classified as water and retracked the SAR waveforms with an improved threshold retracker with 50% threshold [26]. The retracked range is corrected for dry and wet tropospheres, ionospheres, solid Earth and pole tides, and geoid undulation. To ensure consistency for both the Envisat and CryoSat-2 data, the same corrections are used. Instead of using a median or mean over all classified measurements, we search for a horizontal line in the heights, which is assumed to represent the water surface. It is still possible that some of the water-classified measurements do not represent the river surface and need to be excluded from the water level computation (across-track of nadir effects or water bodies surrounding the river). These outliers do not necessarily have to be at the margin of the river but can also be located in the middle due to islands or sandbanks in the river. This can restrict the use of an along-track standard deviation of the heights for outlier detection.

To find the line of equal water height, a histogram of the water levels with Doane bins [27] is used. Doane bins are more suitable for small (less than 30) non-normally distributed data sets than the classical Sturge bins [28]. The Doane bin method not only uses the number of data points to estimate the number of bins and thus bin size, but the shape of the distribution is incorporated with a skewness parameter as well. The range of the data in the data set is indirectly considered in the Doane bins. If a horizontal line is present in the heights, one of the bins is distinctively larger, e.g., contains more observations, than the others and collects the heights of nearly equal water level. The median of the heights in this bin is then taken as water level. If less than five height points are classified as water, the median of the heights is taken as the water level. The advantage of this approach is that it is better suited for rivers wider than 1 km with islands and sandbanks that cause outliers in the heights. However, in many cases this histogram approach or only taking the median of all observations delivers similar results.

4.2.2. Outlier Detection

In spite of careful data selection through the classification and in the height retrieval, some of the water levels have to be considered as outliers. The height retrieval with a horizontal line only prevents outliers if more than five observations are labeled as water observations. In particular, along the small and medium-sized rivers many crossings have less than five observations. A considerable number of crossings have only one observation. The long repeat time of the CryoSat-2 satellite of 369 days complicates the detection of outliers. Unlike altimetric water levels of short repeat time missions, such as Envisat and the Jasons, we do not have a water level time series at a location which can be investigated for outliers by the estimation of and the comparison to an annual signal.

To find the CryoSat-2 outliers we make use of the said repeat time of 369 days. With the knowledge of the very stable annual signal of the Mekong River one can assume that two measurements of the same CryoSat-2 track 369 days apart should measure a similar height [21]. Based on this, a water level is considered as an outlier if the mean difference to all other heights of the same pass is larger than 7 m; the choice of threshold will be discussed below. This is only applicable if other water level measurements of the same track exist. Due to the changing mode mask (see Section 3) some regions are only measured in the last two years. To overcome this, a second outlier detection is applied which compares the water level with water levels of other tracks that are close in space and season (day of the year). To this end, we used all measurements that were less than 10 km away along the river and less than 30 days of the year apart. If the water level was different by more than 10 m from the distance weighted mean water level of all these points, it was considered as an outlier.

The thresholds for the outlier detection were chosen as a conservative upper bound. It is expected that there is an average water level difference of 40 to 60 cm in the five days during the rising water season, but it could be as high as 4 or 5 m [29]. Additionally, some inter-annual changes in the flood season can be expected, and the slope of the river has to be considered which is has a median value of 30 cm/km for the Mekong River. Of the three thresholds used for the outlier detection the difference of 7 m w.r.t. the year is the most sensitive for the later result. The time and distance weighted mean in the second part of the outlier detection limits the sensitivity of the other threshold.

4.2.3. Merging of the Overlap Regions

From the classification we derive a set of heights for each of the different geographical regions which have a certain overlap (see Figure 2 and Section 2). In this overlap, for the same crossing two water levels were computed. Therefore, it has to be decided which height shall be used. To resolve this, we use the distance-weighted mean water level as in the outlier detection (see Section 4.2.2). The water level that is closest to this mean water level is applied.

5. Results and Validation

We applied the described methodology for the classification and water level determination on CryoSat-2 SAR data in the Mekong River Basin. In this section, both the results of the classification and the water level determination are presented and validated.

5.1. Results of the Classification

After the clustering and classification of the CryoSat-2 measurements in the Mekong Basin we select the classes of water returns. In the upstream region we identify three and in the middle region six out of 20 classes as water classes. In the downstream region the rivers are surrounded by seasonal wetland, which makes it difficult to distinguish water return related to the river from the surroundings. Therefore, the downstream region will not be investigated further.

In Figure 5 the mean waveform and mean RIP of some classes are shown (note the different power axes). The classes displayed are selected to best represent all 20 classes for the upstream and middle region. As can be seen, the shape of the mean waveform and mean RIP of water classes in the upstream region reappear in the middle region, but not as water classes. In the middle region small lakes or paddy fields have the same signature as the river upstream. This observation leads to the decision to classify the two regions separately. Some classes have very similar mean waveforms and RIPs (e.g., second water class and first land class in the middle region) but are either classified as water or as land. In this case the maximum power is the important discriminant between the classes. In the upstream region not all stacks over land are 'full', i.e., not every single-look recorded the returning power. This leads to strongly distorted RIPs as shown for the third land class. All mean waveforms and RIPs are displayed in Appendix A for the interested reader.

In Figure 6, a section of the river network in the upstream region with the results of the classification is shown. The course of the river is well depicted, however, not every crossing of the satellite track with the river water is identified. At some crossings no water reflection of the river is measured since the river is too narrow. On the other hand, some points classified as water are not close to the given polygon (blue line). However, the topography model (HydroSHEDs, as presented in Lehner et al. [30]) shown in the background indicates river valleys in the three circled areas. Therefore, one can assume that the classification is able to find rivers that are so small (down to 20 m width) that they are not present in the high-resolution river polygon provided by the MRC. Figure 7 shows the classification for one exemplary track in the upstream region. The measurements classified as water (red dots) line up to a nearly constant water level at all crossings of the satellite track with the river.

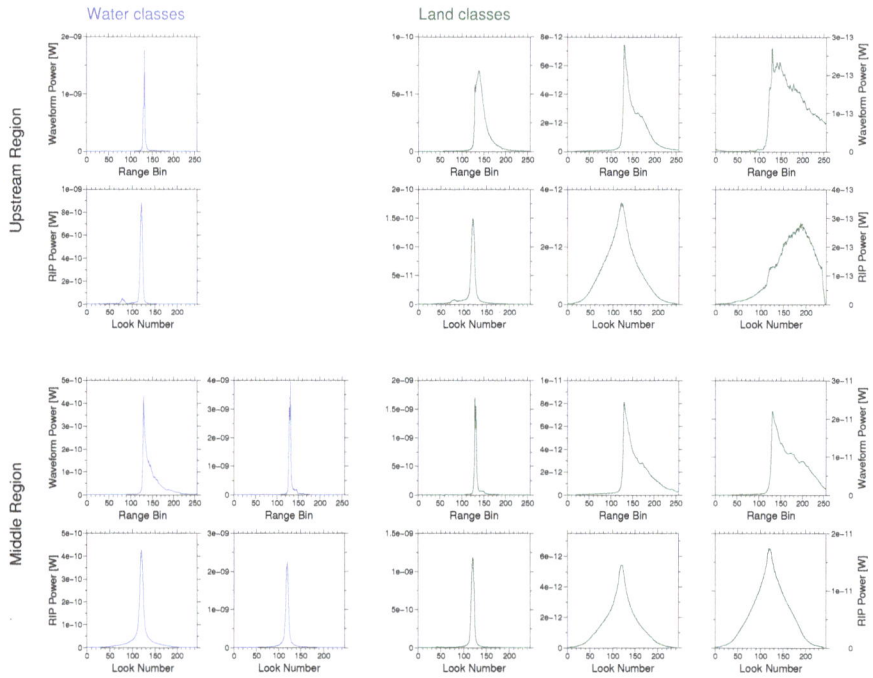

Figure 5. The mean waveforms and RIP of some selected classes; the waveforms and RIP above each other belong to the same class.

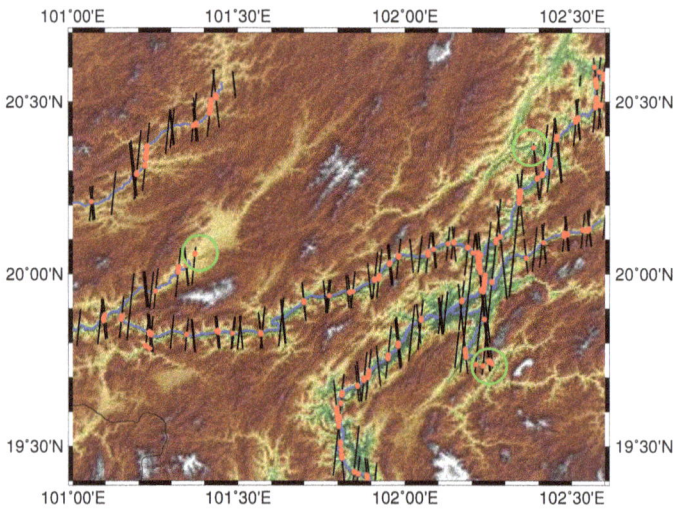

Figure 6. An example of the classification. The red dots are classified as water measurements, and black lines are the land-classified measurements. The three circles indicate areas where water was detected in river valleys which are not included in the river polygon. The background shows the HydroSHEDs topography model.

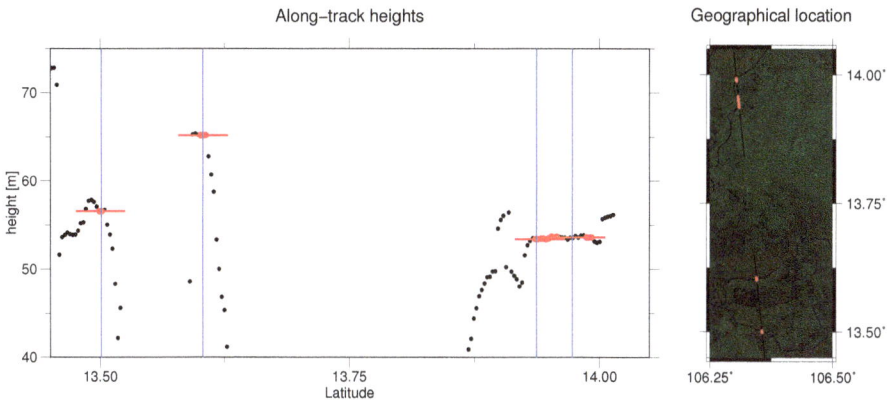

Figure 7. Heights along one track which crosses a river at four locations. The map on the right side shows the geographical surroundings with the background as a Landsat-8 scene where the rivers are well visible. The black dots are all retracked heights with the red dots indicating which measurements were classified as water. The blue vertical lines show the location of the crossing of the track with the river polygon and the horizontal lines are the estimated water level at each crossing.

5.2. Resulting Water Level

In the entire Mekong Basin we estimate water levels at more than 2000 crossings, which means approximately one measurement every 4 km along the main river (compared to 50 km for Envisat). The water level was not measured at every crossing. As mentioned before, at some crossings the river was so small that a reliable measurement could not be made in every pass, and some other water levels were discarded during outlier detection. Furthermore, at some crossings the classification failed to identify the water. However, we are still able to retrieve at least some measurements from rivers as narrow as 20 m. In Figure 8 all measured heights at all dates are presented in a map, which shows the overall topography of the river network well but cannot show details like seasonal variations.

For one track the heights and the classification are displayed in Figure 7 with a Landsat map of the river network and the surrounding area. In this track four water crossings are found where the two most northern ones are very close together with a difference of the water level of 20 cm. There the river meanders under the track which causes two crossings close together. The two southern crossings belong to two different rivers which explains the large height difference. It is visible that only few measurements are used to estimate the water level at each crossing. Approximately 180 water levels (or 8%) are estimated by just one measurement, with the majority in the upstream region. Hence, the outlier detection is more important in this area. In the upstream region, the outlier detection removes 113 of 1740 water level observations; all of them are detected by the comparison with surrounding water level measurements. The comparison with water levels of the same pass reveals only 37 of the 113 outliers. In the middle region only 13 of 529 observations are rejected as outliers: 10 by the comparison with surrounding observations, 9 by the comparison with the same pass, and 6 by both outlier detections.

For crossings with more than one water measurement we can calculate the standard deviation of the measurements used for water level estimation. More than 85% of the water levels have a standard deviation of less than 0.5 m.

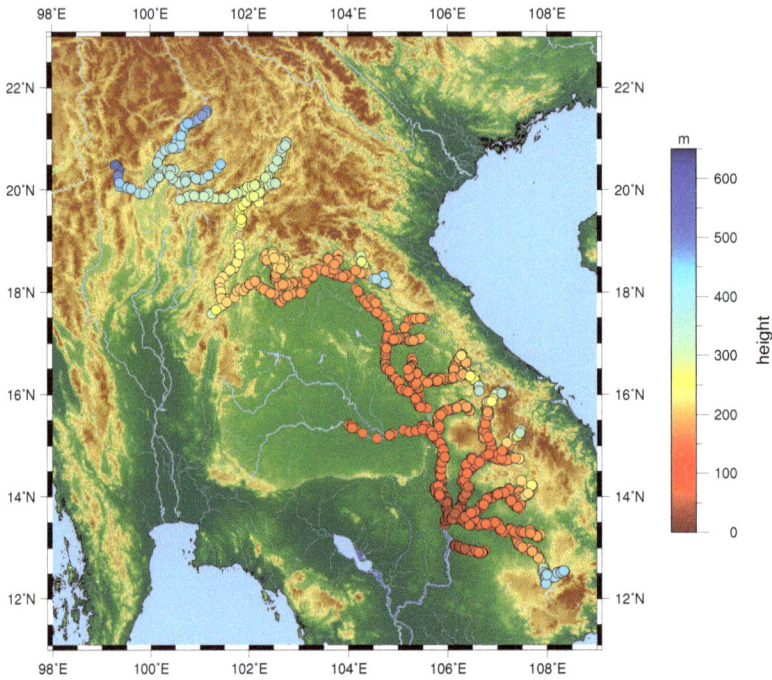

Figure 8. Resulting water levels in the Mekong River Basin.

5.3. Validation

In this section both the classification and the water levels are validated. We test and validate the repeatability of the classification with cross-validation. The altimetric water levels cannot be validated against in situ gauge data without an absolute height reference as is common for short repeat time missions because no time series can be built from CryoSat-2 data without reducing the unknown topography of the river. Therefore, we validate the resulting water levels with respect to the stable seasonal signal and compare these results with the performance of Envisat water levels and CryoSat-2 data extracted with a land–water mask in the same validation.

5.3.1. Validation of the Classification

The classification is validated with a cross-validation. To this end, we cluster the data in a first step. The resulting classes are considered as the 'true' classes. In the second step only parts of the data are clustered with the other parts classified in this second clustering. The cross-validation compares the resulting classifications of the two steps.

Table 2 summarizes the results of the cross-validation; water and non-water classes are distinguished. The overall accuracy is 97.9%. This cross-validation shows that the classification is stable and does not change with the data subset used for the clustering.

Table 2. Results of the cross-validation. The first classification is a clustering on the whole data set, and the second classification is based on clustering on a subset of data with classification of the residual data set.

		Second Classification	
		Water	No Water
First Classification	Water	7321	205
	No water	423	22,660

5.3.2. Validation of Water Levels

Unlike water level time series measured by short repeat time missions, CryoSat-2 measurements cannot be validated against the time series of in situ gauges without reducing the topography as done by [12]. The Mekong River and its tributaries have topography that is too complex to allow for reliable reduction. Besides this, the temporal overlap between the CryoSat-2 data and the gauge data is only about 1.5 years or even less (April 2011 until December 2012).

To validate the water levels we use the nearly one-year repeat time of CryoSat-2. We investigate the differences between two subsequent tracks at the same river crossing. A histogram of the differences is shown in Figure 9a. Table 3 displays the median, mean, and standard deviation of these differences for the merged results as well as for the two regions (upstream and middle) separately. The results of the validation are compared to a validation with in situ gauge data, Envisat data, and CryoSat-2 data with a land–water mask.

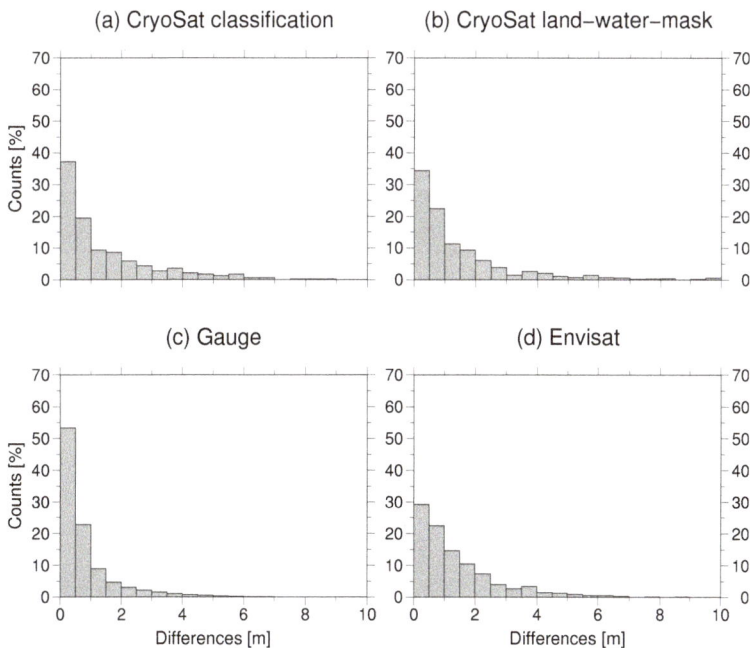

Figure 9. Histogram of the differences of height measurements 369 days apart for CryoSat-2 water levels with classification, CryoSat-2 water levels inside the land–water mask, gauge water level, and Envisat water level.

For this validation, we take the differences between gauge measurements that are 369 days apart and Envisat measurements where the day of the year has a less than 5-day difference. The gauges give a measure of how stable the annual signal is in the Mekong Basin, i.e., a benchmark on how small the differences between the water levels could be. The Envisat observations are the most commonly used data for inland waters with a pulse-limited altimeter. The comparison to the water levels which are derived from CryoSat-2 by simply averaging measurements inside the land–water mask allows for evaluation of the benefits of the classification approach. The land–water mask water levels underwent the same outlier detection as used on the results of the CryoSat-2 classification for better comparability. The median of the differences in Table 3 for the CryoSat-2 classification results are in both the upstream and the middle stream region better than the ones from the Envisat results. In the upstream region the improvements of the CryoSat-2 classification results compared to the Envisat results are more apparent than in the middle region where the validation results are more similar. However, the CryoSat-2 classification results have a higher standard deviation in the upstream regions, and thus larger differences than the Envisat results. The land–water mask approach for CryoSat-2 in the upstream region has even larger differences, which can also be seen in Figure 9. In the middle region no large differences between the validation of the two CryoSat-2 approaches are to be seen. The land–water mask approach yields to more valid water level observations in the middle region but at the same time also a higher number of outliers in both regions (see Table 4).

Table 3. Analysis of the differences of height measurements 369 days apart for the whole study area, only the upstream region, and only the middle stream region.

	Median (m)	Mean (m)	Standard Deviation (m)
Whole study area			
CryoSat-2 classification	0.76	1.43	1.59
CryoSat-2 land–water mask	0.83	1.86	4.55
Gauge	0.45	0.82	1.09
Envisat	0.96	1.42	1.44
Upstream region			
CryoSat-2 classification	0.79	1.54	1.70
CryoSat-2 land–water mask	0.85	2.00	5.44
Gauge	0.42	0.72	1.05
Envisat	1.01	1.46	1.49
Middle region			
CryoSat-2 classification	0.76	1.15	1.10
CryoSat-2 land–water mask	0.84	1.55	1.87
Gauge	0.54	1.00	1.14
Envisat	0.81	1.26	1.26

Table 4. Comparison of the two approaches for CryoSat-2 water levels in terms of absolute number of water levels and number of outliers.

	Upstream Region		Middle Region	
	No. of Water Levels	% of Outliers	No. of Water Levels	% of Outliers
Classification approach	1703	2	520	2
Land–water mask approach	1534	7	1364	4

6. Discussion

In the results and validation section above we demonstrated the advantages and limitations of the classification approach for CryoSat-2 SAR data over the Mekong Basin. Compared to CryoSat-2 water levels extracted with a fixed land–water mask, the classification approach yields water levels with a higher quality according to the used measures with fewer outliers. The difference between the two approaches is most pronounced in the upstream region with rivers as small as 20 m in width. However, the land–water mask approach yields more water levels in the middle region. This reveals the opportunity that SAR altimetry provides for rivers in the Mekong Basin which are too small to be reliably identified in optical (e.g., Landsat, resolution: 30 m) or SAR (e.g., Sentinel-1, resolution: 10 m) images. As shown in Section 5.1 and Figure 6 the classification of SAR altimetry is able to identify rivers which are not visible in the land–water mask derived from satellite images. On the other hand, the classification fails to identify all crossings with a river, which leads to data gaps in the water level measurements. Some of the undetected rivers are very small and shadowed by surrounding river gorges. At other crossings the returning signal is too noisy and disturbed to be identified correctly even by visual inspection.

The better performance of the classification approach in the upstream part of the river is probably caused by the feature selection which is more focused on the reflective properties of small water bodies. The main drawback of the classification approach is its failure in the downstream area. The classification is not able to discern between river water and water in seasonal wetland regions. This problem can only be solved by using a high-resolution seasonal land–water mask that could be derived from Sentinel-1 or 2 data for example.

The existence of outliers after the classification process indicates a certain amount of misclassification. Some of these might be related to reflections from river banks, concrete areas, or other flat regions causing river-like waveforms. Moreover, water surfaces close to the river (such as paddy fields) can provoke outliers.

The inevitable outlier detection is challenging due to the long-repeat orbit of CryoSat-2 preventing the application of the usual virtual station concept. In this study, the outlier detection approach exploits the stable annual water level cycle within the Mekong Basin. Hence, this method is not transferable to other river systems with less stable annual signals. The second outlier detection approach of this study relies on a certain density of water level observations along the river. Without any observation close by a water level cannot be checked to be an outlier.

The validation inherits the problems of the outlier detection as no direct comparison with in situ gauge data is possible. As the gauge data has no absolute reference, not even the absolute height of CryoSat-2 water levels in the vicinity of a gauging station can be validated.

Using CryoSat-2 over rivers improves the knowledge for the river topography, as with this mission the spatial distribution of water level observations is much denser than with any prior satellite altimetry mission (as can be seen in Figure 8). Even small tributaries are observed successfully with CryoSat-2 SAR, which allows monitoring of the entire river network. However, this high spatial resolution entails a sparse repeat time and thus, precludes the application of the classical virtual station concept for data screening, interpretation, and validation.

Nevertheless, using the derived CryoSat-2 data set—in combination with short repeat altimetry mission results—for building a mathematical water level model (as done in [31]) or for assimilation in hydrological models [32] will certainly improve our knowledge of the water regime in the Mekong region.

7. Conclusions

We demonstrate in this study the advantages of CryoSat-2 SAR altimetry data for measuring rivers in the Mekong Basin identified by a classification which is independent of an accurate land–water mask. The classification uses features derived not only from the waveform but also from the RIP. The RIP contains more information about the reflecting surface than the waveform on its own can provide. This improves the classification and allows us to identify even very small rivers with a width as small as 20 m. In fact, the classification works better on medium and small rivers than on large rivers. The cross-validation of the classification shows that it is stable and repeatable. However, we were not able to use this classification to isolate the river in the downstream region where the Mekong River is surrounded by seasonal wetlands.

The classification in water and land measurements is used to derive water levels at the crossings of the CryoSat-2 track with a river in the whole Mekong Basin. Overall, more than 2000 water levels are measured, after outlier detection. However, it is not possible to derive a water level at every crossing. The altimeter is not able to measure a water return at every possible river crossing due to too-small rivers or overly-disturbed returns. Additionally, some measured water levels are discarded in the outlier detection.

The water levels are validated using the 369-day return time of CryoSat-2 and the very stable annual signal in the basin. The same validation is performed on Envisat water levels, gauge measurements, and by using a precise land–water mask on CryoSat-2 data. In particular, for small rivers in the upstream region the classification improves the water level determination compared to the use of a land–water mask: The median differences between consecutive passes in the upstream region are 0.79 m compared to 0.85 m for the land–water mask approach. The reduction of the standard deviation by a factor of four is even more significant. Moreover, the number of outliers is significantly smaller for the classification approach than for the land–water mask approach, with only 2% for the classification but 5.6% for the CryoSat-2 land–water mask.

Compared to Envisat water levels the CryoSat-2 water levels are of higher quality in the whole river basin due to the smaller footprint of the SAR compared to pulse-limited altimeter on Envisat. The median differences for Envisat are as high as 1.01 m in the upstream region, and 0.96 m for the whole basin (CryoSat-2 classification 0.76 m).

The classification of CryoSat-2 data allows to exploit the full along-track resolution of the data set. Therefore, it is possible to observe the water levels of rivers as small as 20 m that are usually not correctly mapped in land–water masks. Moreover, due to the dense spatial resolution of CryoSat-2 most of the smaller tributaries are observed, which is not possible with the short repeat missions Envisat, SARAL, or Jason-2 and Jason-3. Information on the water level variations for all rivers within the basin helps to further understand the hydrology of the river network.

The resulting water levels of this study will be used in combination with other altimetric water levels following the ideas of [31] to build basin-wide multi-mission water level time series. With CryoSat-2 data we will be able to significantly improve the spatial resolution of the water level observations and better close the data gap between the end of the Envisat mission and the launch of the SARAL mission. Since the launch of Sentinel-3A in February 2016, SAR altimetry data with a short repeat time of 27 days has been available. This enables testing of the transferability of the application and validation of the proposed classification approach based on a new data set.

Acknowledgments: We want to thank the ESA GPOD team who provided us with the full data set through the SARvatore database and who were very helpful with questions regarding the data. This work was supported by the German Research Foundation (DFG) through the TUM International Graduate School of Science and Engineering (IGSSE) and the Technical University of Munich (TUM) in the framework of the Open Access Publishing Program.

Author Contributions: Eva Boergens developed the method, conducted the data analysis, and wrote the majority of the paper. Karina Nielsen helped with the development of the method and validation. Ole B. Andersen and Denise Dettmering contributed to the discussion of the method and results and the writing of the manuscript. Florian Seitz supervised the research and contributed to manuscript writing and organization.

Conflicts of Interest: The authors declare that they have no conflict of interest.

Appendix A. Mean Waveforms and RIPs

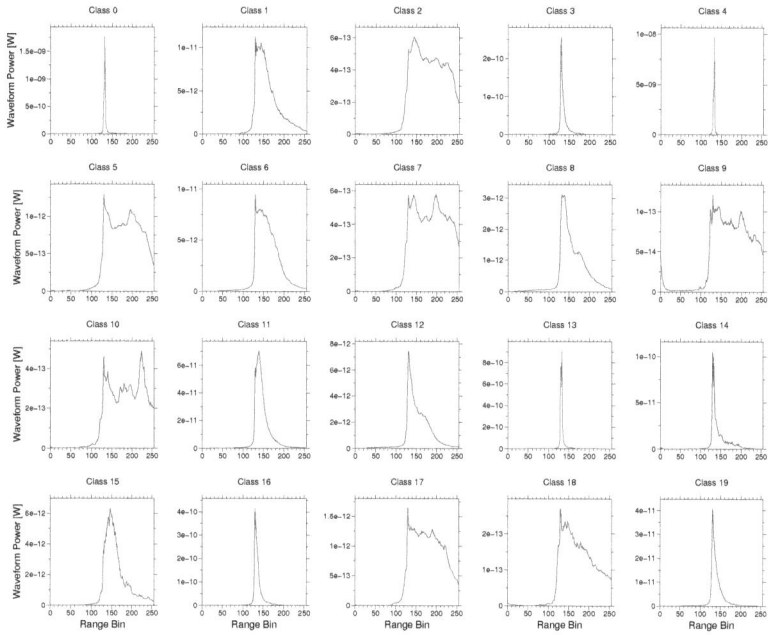

Figure A1. Upstream region mean waveforms. Water classes: 0, 4, 13.

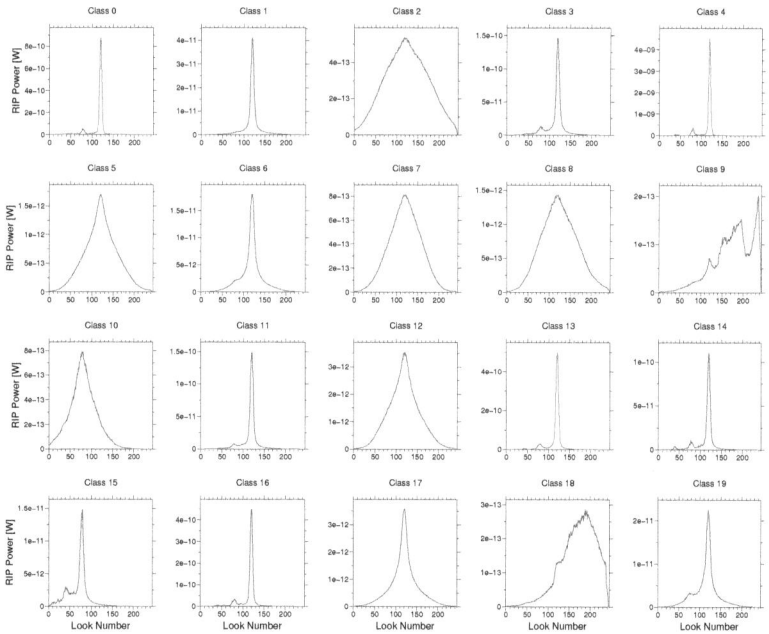

Figure A2. Upstream region mean RIPs. Water classes: 0, 4, 13.

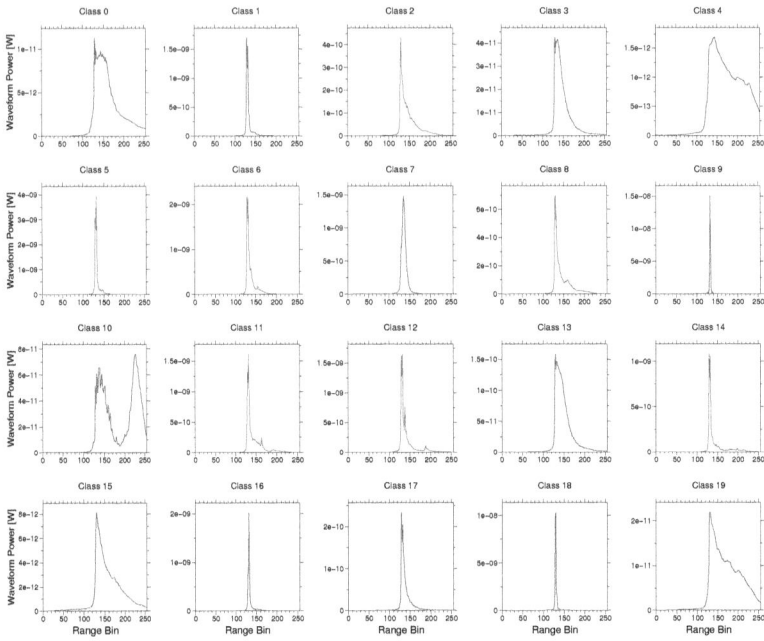

Figure A3. Middle region mean waveforms. Water classes: 2, 5, 6, 8, 11, 12, 14.

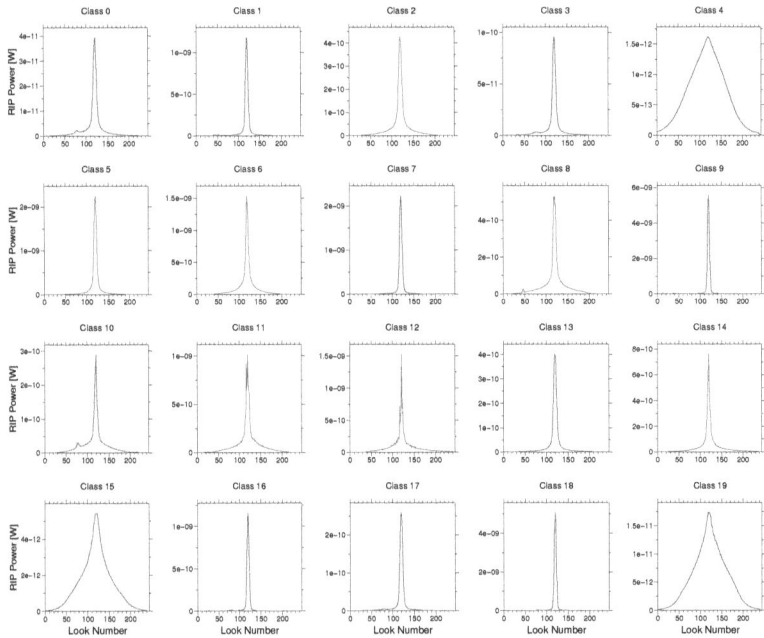

Figure A4. Middle region mean RIPs. Water classes: 2, 5, 6, 8, 11, 12, 14.

References

1. Global Runoff Data Center. *Long-Term Mean Monthly Discharges and Annual Characteristics of GRDC Stations*; Technical Report; Federal Institute of Hydrology: Koblenz, Germany, 2013.
2. Birkett, C.M. Contribution of the TOPEX NASA radar altimeter to the global monitoring of large rivers and wetlands. *Water Resour. Res.* **1998**, *34*, 1223–1239.
3. Santos da Silva, J.; Calmant, S.; Seyler, F.; Rotunno Filho, O.C.; Cochonneau, G.; Mansur, W.J. Water levels in the Amazon basin derived from the ERS-2 and ENVISAT radar altimetry missions. *Remote Sens. Environ.* **2010**, *114*, 2160–2181.
4. Schwatke, C.; Dettmering, D.; Bosch, W.; Seitz, F. DAHITI—An innovative approach for estimating water level time series over inland waters using multi-mission satellite altimetry. *Hydrol. Earth Syst. Sci.* **2015**, *19*, 4345–4364.
5. Boergens, E.; Dettmering, D.; Schwatke, C.; Seitz, F. Treating the Hooking Effect in satellite altimetry data: A case study along the Mekong River and its tributaries. *Remote Sens.* **2016**, *8*, 91.
6. Frappart, F.; Do Minh, K.; L'Hermitte, J.; Cazenave, A.; Ramillien, G.; Le Toan, T.; Mognard-Campbell, N. Water volume change in the lower Mekong from satellite altimetry and imagery data. *Geophys. J. Int.* **2006**, *167*, 570–584.
7. Biancamaria, S.; Frappart, F.; Leleu, A.S.; Marieu, V.; Blumstein, D.; Desjonquères, J.D.; Boy, F.; Sottolichio, A.; Valle-Levinson, A. Satellite radar altimetry water elevations performance over a 200 m wide river: Evaluation over the Garonne River. *Adv. Space Res.* **2016**, *59*, 1–19.
8. Raney, R.K. The delay/doppler radar altimeter. *IEEE Trans. Geosci. Remote Sens.* **1998**, *36*, 1578–1588.
9. Nielsen, K.; Stenseng, L.; Andersen, O.B.; Villadsen, H. Validation of CryoSat-2 SAR mode based lake levels. *Remote Sens. Environ.* **2015**, *171*, 162–170.
10. Kleinherenbrink, M.; Lindenbergh, R.C.; Ditmar, P.G. Monitoring of lake level changes on the Tibetan Plateau and Tian Shan by retracking Cryosat SARIn waveforms. *J. Hydrol.* **2015**, *521*, 119–131.
11. Göttl, F.; Dettmering, D.; Müller, F.L.; Schwatke, C. Lake level estimation based on CryoSat-2 SAR altimetry and multi-looked waveform classification. *Remote Sens.* **2016**, *8*, 1–16.
12. Villadsen, H.; Andersen, O.B.; Stenseng, L.; Nielsen, K.; Knudsen, P. CryoSat-2 altimetry for river level monitoring—Evaluation in the Ganges-Brahmaputra River basin. *Remote Sens. Environ.* **2015**, *168*, 80–89.
13. Bercher, N.; Dinardo, S.; Lucas, B.M.; Fleury, S.; Picot, N. A review of cryosat-2/siral applications for the monitoring of river water levels. *ESA Living Planet* **2013**, *2013*, 1–7.
14. Berry, P.A.M.; Garlick, J.D.; Freeman, J.A.; Mathers, E.L. Global inland water monitoring from multi-mission altimetry. *Geophys. Res. Lett.* **2005**, *32*, 1–4.
15. Desai, S.; Chander, S.; Ganguly, D.; Chauhan, P.; Lele, P.; James, M. Waveform Classification and Water-Land Transition over the Brahmaputra River using SARAL/AltiKa & Jason-2 Altimeter. *Indian Soc. Remote Sens.* **2015**, *43*, 475–485.
16. Dettmering, D.; Schwatke, C.; Boergens, E.; Seitz, F. Potential of ENVISAT Radar Altimetry for Water Level Monitoring in the Pantanal Wetland. *Remote Sens.* **2016**, *8*, 596.
17. Villadsen, H.; Deng, X.; Andersen, O.B.; Stenseng, L.; Nielsen, K.; Knudsen, P. Improved inland water levels from SAR altimetry using novel empirical and physical retrackers. *J. Hydrol.* **2016**, *537*, 234–247.
18. Armitage, T.W.K.; Davidson, M.W.J. Using the interferometric capabilities of the ESA CryoSat-2 mission to improve the accuracy of sea ice freeboard retrievals. *IEEE Trans. Geosci. Remote Sens.* **2014**, *52*, 529–536.
19. Wingham, D.J.; Francis, C.R.; Baker, S.; Bouzinac, C.; Brockley, D.; Cullen, R.; de Chateau-Thierry, P.; Laxon, S.W.; Mallow, U.; Mavrocordatos, C.; et al. CryoSat: A mission to determine the fluctuations in Earth's land and marine ice fields. *Adv. Space Res.* **2006**, *37*, 841–871.
20. MacQueen, J. Some methods for classification and analysis of multivariate observations. *Proc. Fifth Berkeley Symp. Math. Stat. Probab.* **1967**, *1*, 281–297.
21. Mekong River Commission. (Ed.) *Overview of the Hydrology of the Mekong Basin*; Mekong River Commission: Phnom Penh, Cambodia, 2005.
22. ESA. *CryoSat-2 Geographical Mode Mask*; ESA: Paris, France, 2016.
23. Scagliola, M. *CryoSat Footprints*. 2013. Available online: https://earth.esa.int/documents/10174/125271/CryoSat_Footprints_TN_v1.1.pdf/2a5d996b-8b77-4d1c-ae7b-fbf93848c35d;jsessionid=B1FF8C50A1B0F2A0879F6FA028844644.eodisp-prod4040?version=1.0 (accessed on 23 November 2017).

24. Cullen, R.A.; Wingham, D.J. CryoSat Level 1b Processing Algorithms and Simulation Results. In Proceedings of the 2002 IEEE International Geoscience and Remote Sensing Symposium, Toronto, ON, Canada, 24–28 June 2002; pp. 1762–1764.
25. Laxon, S.W. Sea-Ice Altimeter Processing Scheme at the EODC. *Int. J. Remote Sens.* **1994**, *15*, 915–924.
26. Gommenginger, C.; Thibaut, P.; Fenoglio-Marc, L.; Quartly, G.; Deng, X.; Gómez-Enri, J.; Challenor, P.; Gao, Y. Retracking altimeter waveforms near the coasts. In *Coastal Altimetry*; Benveniste, J., Cipollini, P., Kostianoy, A.G., Vignudelli, S., Eds.; Springer: Berlin, Germany, 2011; pp. 61–101.
27. Doane, D.P. Aesthetic frequency classifications. *Am. Stat.* **1976**, *30*, 181–183.
28. Sturges, H. The choice of a class interval. *J. Am. Stat. Assoc.* **1926**, *21*, 65–66.
29. Mekong River Commision. *Annual Mekong Flood Report 2008*; Mekong River Commission: Phnom Penh, Cambodia, 2009.
30. Lehner, B.; Verdin, K.; Jarvis, A. New global hydrography derived from spaceborne elevation data. *Eos* **2008**, *89*, 93–94.
31. Boergens, E.; Buhl, S.; Dettmering, D.; Klüppelberg, C.; Seitz, F. Combination of multi-mission altimetry data along the Mekong River with spatio-temporal kriging. *J. Geodesy* **2017**, *91*, 519–534.
32. Schneider, R.; Nygaard Godiksen, P.; Villadsen, H.; Madsen, H.; Bauer-Gottwein, P. Application of CryoSat-2 altimetry data for river analysis and modelling. *Hydrol. Earth Syst. Sci.* **2017**, *21*, 751–764.

remote sensing

MDPI

Article

Impact of Surface Soil Moisture Variations on Radar Altimetry Echoes at Ku and Ka Bands in Semi-Arid Areas

Christophe Fatras [1,2,*], **Pierre Borderies** [1], **Frédéric Frappart** [2,3], **Eric Mougin** [2],
Denis Blumstein [3,4] **and Fernando Niño** [3]

[1]   Office National d'Etudes et de Recherches Aerospatiales (ONERA), 2 Avenue Edouard Belin,
      31055 Toulouse CEDEX 4, France; pierre.borderies@onera.fr
[2]   Geosciences Environnement Toulouse (GET), UMR 5563, CNRS/IRD/UPS, Observatoire Midi-Pyrenees
      8 (OMP), 14 Avenue Edouard Belin, 31400 Toulouse, France; frederic.frappart@legos.obs-mip.fr (F.F.);
      eric.mougin@get.omp.eu (E.M.)
[3]   Laboratoire d'Etudes en Geophysique et Oceanographie Spatiales (LEGOS), UMR 5566, CNRS/IRD/UPS,
      Observatoire Midi-Pyrenees 8 (OMP), 14 Avenue Edouard Belin, 31400 Toulouse, France;
      denis.blumstein@cnes.fr (D.B.); fernando.nino@ird.fr (F.N.)
[4]   Centre National d'Etudes Spatiales (CNES), 18 Avenue Edouard Belin, 31400 Toulouse, France
[*]   Correspondence: christophe.fatras@gmail.com; Tel.: +33-561-332-970

Received: 28 February 2018; Accepted: 3 April 2018; Published: 9 April 2018

Abstract: Radar altimetry provides information on the topography of the Earth surface. It is commonly used for the monitoring not only sea surface height but also ice sheets topography and inland water levels. The radar altimetry backscattering coefficient, which depends on surface roughness and water content, can be related to surface properties such as surface soil moisture content. In this study, the influence of surface soil moisture on the radar altimetry echo and backscattering coefficient is analyzed over semi-arid areas. A semi-empirical model of the soil's complex dielectric permittivity that takes into account that small-scale roughness and large-scale topography was developed to simulate the radar echoes. It was validated using waveforms acquired at Ku and Ka-bands by ENVISAT RA-2 and SARAL AltiKa respectively over several sites in Mali. Correlation coefficients ranging from 0.66 to 0.94 at Ku-band and from 0.27 to 0.96 at Ka-band were found. The increase in surface soil moisture from 0.02 to 0.4 (i.e., the typical range of variations in semi-arid areas) increase the backscattering from 10 to 15 dB between the core of the dry and the maximum of the rainy seasons.

Keywords: radar altimetry; waveform; dielectric permittivity; soil moisture

1. Introduction

The semi-arid region of West-Africa has been identified by [1] as a hot-spot for surface-atmosphere coupling, and where the routine monitoring of soil moisture would improve boreal summer seasonal forecasting. Moreover, the monitoring of the temporal variations of the surface soil moisture (SSM), i.e., the water content in the upper soil profile, is of primary importance in semi-arid regions, since SSM drives the evapotranspiration flux to the atmosphere via the partition between infiltration and run-off, and between soil evaporation and plant transpiration [2]. The surface soil moisture thus modulates the latent and sensible energy fluxes at the surface [3]. In semi-arid conditions, SSM also influences the main ecohydrological surface processes such as the plant germination, growth and mortality [4], the degradation and mineralization of the organic matter and the emission of gaseous compounds [5]. Accordingly, in semi-arid regions (i.e., in water limited systems), SSM is probably the main relevant Essential Climate Variable (ECV) that needs to be monitored at different temporal and spatial scales.

Spaceborne active microwave remote sensing offers different techniques and instruments to monitor SSM in semi-arid regions. Low spatial resolution systems between approximately 25 and 50 km, but with a high temporal resolution of a few days, such as the wind scatterometers onboard the European Remote Sensing (ERS) Satellites, QuickSCAT, and the Advanced SCATterometer, (ASCAT), on board METOP, and radiometers such as the advanced microwave scanning radiometer, AMSR-E and the salinity moisture and ocean salinity satellite, SMOS, have shown considerable potential for monitoring SSM over semi-arid regions [6–10].

Specifically, it was demonstrated with the wind-scatterometer on-board ERS-1&2 [6,11,12] that spaceborne scatterometers can be used for estimating SSM. Wind scatterometers derived-SSM products are now widely used in combination with microwave radiometers products [13–15]. At higher spatial resolutions, synthetic aperture radar (SAR) such as ERS-1 and -2 SARs and ASAR onboard ENVISAT, have also shown great potentialities for deriving SSM [16–19].

Following the study conducted in Australia by the authors of [20], who demonstrated the effects of soil moisture and soil roughness on the nadir C- and Ku-band backscattering coefficients, recent studies demonstrated the capabilities of spaceborne altimeters for estimating SSM in semi-arid regions such as in west-Africa [21–24] and in Australia [25]. Indeed, thanks to their nadir-looking capability, altimeters minimize the attenuation by the vegetation layer and a relationship can be directly found with the moisture content of the underlying soil. Over the Sahel region where the vegetation density is generally low, a very good linear correlation was found between the backscattering coefficient measured at Ku-band by the ENVISAT RA2 radar altimeter and SSM, with correlation values higher than those found with SAR data over the same sites [19,21]. However, the effects of the altimeter frequency and of soil roughness were not investigated. Moreover, the dynamic range of the backscattered coefficient must be known in order to be taken into consideration in the inversion procedure of SSM.

The present study aims to investigate how the spatio-temporal variations of SSM and soil roughness impact the radar altimetry echoes or waveforms at Ku- and Ka- bands over a semi-arid region. To this end, a physically-based model which takes into account the interaction between the incident wave and the surface was developed. This manuscript is organized as follows: the altimetry waveform model, named Continental waveform ALtimetry Model (CALM), is presented in Section 2. Then the dataset used for simulating the scenes and used for validating the model, as well as the study sites are described in Sections 3 and 4, respectively. Modeling results are confronted to real data acquired by the Ku-band ENVISAT and Ka-band AltiKa altimeters, over five contrasted sites in terms of landscape, relief, roughness and heterogeneity of the illuminated surface, including the presence of open waters. The performance of the model in terms of waveform generation is analyzed in Section 5 and the impact of SSM on the altimetry signal is eventually studied through a sensitivity analysis (Section 6).

2. Radar Altimetry Backscattering Modeling

To simulate the electromagnetic interaction between the wave emitted by the altimeter antenna and the illuminated soil, a two scale approach is used. The antenna footprint on the soil is sampled using triangular facets in order to be able to take into account any topography. Each facet is characterized by small scale roughness parameters and soil dielectric permittivity. The approach consists in vectorially summing the complex contributions of all the facets at the receiver. The vectorial contribution of each facet is computed assuming that the field distribution over the facet is the same as if it were infinite. Note that the large scale sampling may be adapted to a multi-scale one all the more easily as the computations are very fast.

The Kirchoff model was used to take into account the small-scale roughness. It is used to simulate the electromagnetic (EM) response of a rough surface taking into account rms height (hrms) and the correlation length of the local surface. The macroscopic scale integrates the effect of the surface topography in the altimeter footprint (several to several tenths of km^2 depending on the wavelength of the EM wave) through the use of a Digital Elevation Model (DEM). Due to the roughness parameters used in this study, the Kirchoff model was preferred to other models at our wavelengths [26].

2.1. Simulation of the Surface Backscattering Using the Kirchoff Model

2.1.1. Kirchoff Model of the Stationary Phase

The Kirchhoff model for the stationary phase, also known as Geometric Optics (GO) backscattering model, is based upon the tangent plane approximation. The surface is represented as a mosaic of randomly orientated planes, each one of them locally tangent to the surface. This model can be applied under the following assumptions:

$$kl > 6 \tag{1}$$

$$l^2 > 2.76s\lambda \tag{2}$$

$$ks\cos\theta > 1.5 \tag{3}$$

where $k = 2\pi/\lambda$ is the wavenumber and λ the wavelength, s the root mean square height (m) representing the surface roughness, l the surface correlation length (m), θ the wave incidence angle. s and l are the rms height and the correlation length.

In the GO model, each elementary surface has a proper backscattering coefficient (which is valid for VV and HH polarization) as expressed in the following equation:

$$\sigma_{HH}(\theta) = \sigma_{VV}(\theta) = \frac{\Gamma_0}{2m^2(\cos\theta)^4} \exp\left(-\frac{(\tan\theta)^2}{2m^2}\right) \tag{4}$$

where Γ_0 is the Fresnel reflection coefficient at nadir incidence and m the autocorrelation function that depends on the nature of the surface.

For a Gaussian surface, the autocorrelation function is:

$$m = \frac{\sqrt{2}s}{l} \tag{5}$$

For an exponential surface, the autocorrelation function is:

$$m = \frac{s}{l} \tag{6}$$

As Low Resolution Mode (LRM) altimeters are nadir-looking sensors, if the local slope is small, (3) becomes:

$$ks > 1.5 \tag{7}$$

The Fresnel coefficients at nadir incidence can be written as:

$$\Gamma_0 = \left|\frac{\sqrt{\varepsilon} - 1}{\sqrt{\varepsilon} + 1}\right|^2 \tag{8}$$

where ε is the soil dielectric permittivity.

2.1.2. Soil Dielectric Permittivity Estimates at Ku and Ka Bands

Soil dielectric permittivity estimates are determined using the semi-empirical from [27,28]. According to this model that extends earlier works from [29,30], the complex dielectric permittivity of a medium (ε_m) is defined as:

$$\varepsilon_m = \left(1 + \frac{\rho_b}{\rho_s}(\varepsilon_s{}^\alpha - 1) + m_v{}^\beta \varepsilon_{fw}{}^\alpha - m_v\right)^{\frac{1}{\alpha}} \tag{9}$$

where ε_s and ε_{fw} are the relative dry soil and effective free water permittivities respectively, ρ_b and ρ_s are the soil bulk and specific soil densities respectively, m_v is the volumetric soil moisture given in percent, α and β are the shape and empirical constants respectively. All details about the complex dielectric permittivity from [27,28] are given in Appendix A.

To our knowledge, no dielectric permittivity model has been specifically designed for the Ka-band. In this study, the semi-empirical model for soil complex dielectric permittivity, validated for the Ku-band was extended to the Ka-band as in previous studies [30,31].

2.2. Altimeter Waveform Generation

2.2.1. Vectorial Polarimetric Backscattering

The Forward Scattering Alignment convention (FSA) was chosen for all the computation at each elementary surface (see Appendix B for the description of the surface discretization), following the conventions detailed in [32]. To take into account the polarization, the emitted wave is considered to be linearly polarized with the electric field parallel to the X-axis (Figure A1).

Over a given triangular sample of barycenter O_i, the incident electromagnetic wave emitted by the altimeter (S for source in Figure 1) is directed along the wave vector $\vec{k_i}$, defined as:

$$\vec{k_i} = \frac{\vec{SO_i}}{\left|\vec{SO_i}\right|} \tag{10}$$

The components of the incident electric field $\vec{E_i}$ following the three axes of the global vector set are defined by:

$$\vec{E_i} = \vec{k_i} \wedge \vec{y} = \begin{bmatrix} E_{Xi} \\ E_{Yi} \\ E_{Zi} \end{bmatrix} \tag{11}$$

where y is the norm vector in the y direction.

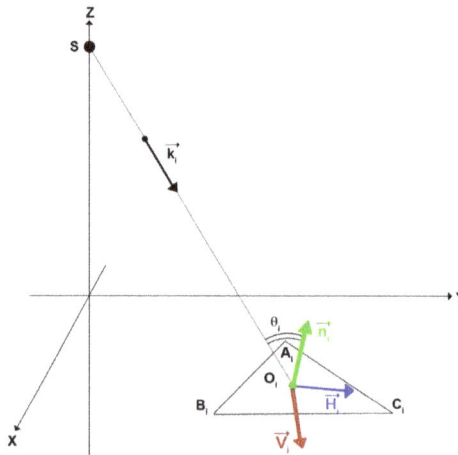

Figure 1. Forward Scattering Alignment (FSA) convention applied to an elementary triangle surface of the meshgrid used in the study.

The incident electric field is then projected on the unitary vectors $\overrightarrow{H_i}$ and $\overrightarrow{V_i}$ which form with $\overrightarrow{k_i}$ a direct base. The $\overrightarrow{H_i}$ vector is defined by:

$$\overrightarrow{H_i} = \frac{\overrightarrow{n_i} \wedge \overrightarrow{k_i}}{|\overrightarrow{n_i} \wedge \overrightarrow{k_i}|} = \begin{bmatrix} H_{Xi} \\ H_{Yi} \\ H_{Zi} \end{bmatrix} \tag{12}$$

where $\overrightarrow{n_i}$ is the perpendicular vector to the elementary surface considered. The $\overrightarrow{V_i}$ vector is defined as:

$$\overrightarrow{V_i} = \overrightarrow{H_i} \wedge \overrightarrow{k_i} = \begin{bmatrix} V_{Xi} \\ V_{Yi} \\ V_{Zi} \end{bmatrix} \tag{13}$$

The projection of $\overrightarrow{E_i}$ on the $(\overrightarrow{H_i}, \overrightarrow{V_i})$ base is done through a scalar product:

$$E_{Hi} = \begin{bmatrix} E_{Xi} \\ E_{Yi} \\ E_{Zi} \end{bmatrix} \cdot \begin{bmatrix} H_{Xi} \\ H_{Yi} \\ H_{Zi} \end{bmatrix} \tag{14}$$

$$E_{Vi} = \begin{bmatrix} E_{Xi} \\ E_{Yi} \\ E_{Zi} \end{bmatrix} \cdot \begin{bmatrix} V_{Xi} \\ V_{Yi} \\ V_{Zi} \end{bmatrix} \tag{15}$$

Now the components of the incident electric field in terms of horizontal polarization H and vertical polarization V are known in the global set (O,X,Y,Z). A local set is defined at the sample i with center O_i, vertical axis O_i Z directed along the normal to the facet plane and horizontal axes in the sample plane. The vectors $\overrightarrow{H_i}$ and $\overrightarrow{V_i}$ are projected on the local polarization vectors $\overrightarrow{H_{il}}$ and $\overrightarrow{V_{il}}$ through a rotation matrix [R].

The scattering matrix [S] of the facet in the local set is defined as follows, with the backscattering coefficient σ_{HH} and σ_{VV} corresponding to the HH and VV polarizations respectively [33]. The ground cross-polarization is neglected in this study.

$$S = \begin{bmatrix} \sqrt{\sigma_{HH}} & 0 \\ 0 & \sqrt{\sigma_{VV}} \end{bmatrix} \tag{16}$$

The scattered electric field components towards the backscattering direction (same directions as $\overrightarrow{H_i}$ and $\overrightarrow{V_i}$ are obtained as follows:

$$\begin{bmatrix} E_{Hsl} \\ E_{Vsl} \end{bmatrix} = S \cdot \begin{bmatrix} E_{Hil} \\ E_{Vil} \end{bmatrix} \tag{17}$$

The Equations (11)–(17) define the aspects relative to the electric field scattering. The local scattered field is afterward expressed in the global set, and then projected on the incident polarization. In order to take into account all factors impacting the attenuation amplitude of the EM field, a weighting coefficient is applied to the electric field for each elementary surface, corresponding to the attenuation factor present in the radar equation, and defined as follows:

$$C_i = \frac{A_i G^2 P \lambda^2}{R_i^4 (4\pi)^3} W(\theta_i) \tag{18}$$

where A_i is the elementary triangle area (see Appendix B), G the antenna gain, P the emitted power, λ the wavelength, R_i the distance between the antenna and the ith surface element (SO$_i$ in Figure 1), $W(\theta_i)$ the antenna diagram used in the radar system (see Appendix C for the definition of the radar parameters).

The final electric field backscattered by the elementary surface following the X-, Y- and Z-axis is then determined by:

$$
\begin{bmatrix} E_{XF_i} \\ E_{YF_i} \\ E_{ZF_i} \end{bmatrix} = E_{HS}C_i \begin{bmatrix} H_{Xi} \\ H_{Yi} \\ H_{Zi} \end{bmatrix} + E_{VS}C_i \begin{bmatrix} V_{Xi} \\ V_{Yi} \\ V_{Zi} \end{bmatrix}
\tag{19}
$$

A final phase component is added, taking into account the propagation term and a random phase Φ_i for removing the speckle component:

$$
E_{Yi} = e^{-jk_iR_i + j\Phi_i}
\tag{20}
$$

The final electric field backscattered by the elementary surface is the complex sum of the fields scattered by all the facets.

2.2.2. Monopulse Waveform

When reaching the surface, the electric field is reflected and partly backscattered to the satellite following a power distribution, function of time. This power distribution of the radar echo is known as an altimeter waveform. The received power is sampled in time gates, each of them integrating the backscattered power during its aperture time. The contribution of each portion of the illuminated surface is received at a different time related to the topography of the surface (Figure 2). Based on the distance between the satellite and the surface, the recording of any individual waveforms starts when the first backscattered signal reaches the satellite position. The backscattered signal is sampled in a collection of time gates. The time gates have a duration (d_{gate}) of length τ, function of the radar system bandwidth B_w, as defined in [34]:

$$
d_{gate} = \tau = \frac{1}{B_w}
\tag{21}
$$

Each elementary surface backscatters an electric field towards the satellite, which arrives at time t_i after the emission time t_0. The three components of the waveform (WF) for any gate n can be expressed as:

$$
WF_X(n) = \sum_{t_i=T_{n-1}}^{T_n} |E_{XFi}(ti)|^2, WF_Y(n) = \sum_{t_i=T_{n-1}}^{T_n} |E_{YFi}(ti)|^2, WF_Z(n) = \sum_{t_i=T_{n-1}}^{T_n} |E_{ZFi}(ti)|^2
\tag{22}
$$

where T_{n-1} and T_n are the starting and ending times of the gate n.

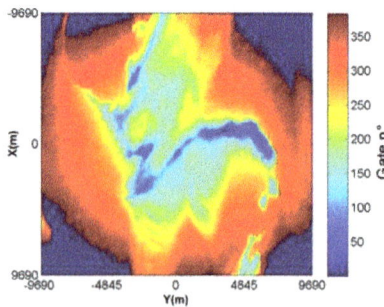

Figure 2. Example of modeled surface. The colors correspond to the corresponding radar time gate (here for 384 time gates).

Each waveform corresponds to a duration of a few milliseconds (a typical time gate lasts a few nanoseconds). The number of time gates generally varies from 64 to 128 on the current satellite altimeters and corresponds to the duration of the reception mode of the radar altimeter sensor. The opening of the reception window of the radar altimeter sensor is based either on the series of estimated ranges from a series of previous echoes (closed-loop mode) or, on the most recent altimetry missions, is forced by an estimate of the range based on a DEM to limit the problem of non-detection of open water targets as rivers flowing at the bottom of narrow valleys (open-loop mode, see [35] for more details). The altimeter waveform model developed in this study does not take into account these modes of operation. It is necessary to select n time gates over all the possible responses to obtain the individual altimeter waveform (red square on Figure 3). In order to calculate the backscattered power at a particular time of interest, a time lag variable is introduced corresponding to the switch of the sensor to the reception mode (see Figure 3). This time lag is determined manually by seeking the maximum cross-correlation between modeled and measured waveforms using 512 time-gate samples. The retained time-lag value for each site is then used for modeling the proper 128 time-gates waveforms.

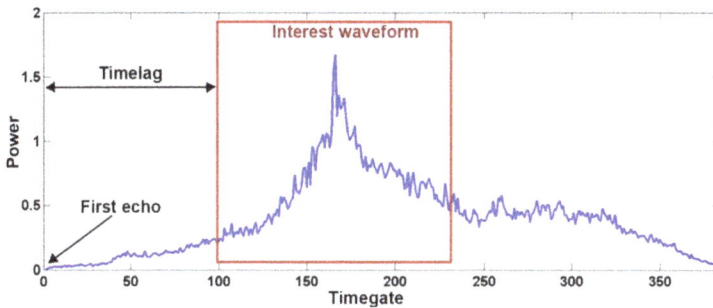

Figure 3. Example of a "raw" waveform simulated using the Continental waveform ALtimetry Model (CALM). We show the position of the first return signal, the time lag duration, and the interest waveform window (here composed of 128 points).

For each one of these time gates, the 3 components of the electric fields corresponding to the elementary surfaces whose surface-altimeter travel time corresponds to the gate's time interval are added. Finally, for each gate, the total complex electric field is multiplied by its conjugate to obtain the power received by the altimeter for an individual waveform.

2.2.3. Averaged Waveform

In order to model as precisely as possible the satellite's orbit impact on the radar measurement, the satellite movement on its orbit was taken into account. From the satellite's orbit speed, which can be approximated as a radial speed, the satellite's movement is simulated. In the model, Earth surfaces are represented using a planar approximation that is acceptable considering the dimension of the surfaces used in the simulation (a couple of tenths kilometers by a couple of tenths kilometers typically). The satellite's movement can be assumed as a one dimension translation, following the Y-axis in CALM (see Appendix B).

The radar altimetry systems on orbit operate continuously along their tracks. The emission of the EM wave towards the surface is regularly performed according to the pulse repetition frequency. The pulse duration itself depends on the radar system parameters, and is different from one satellite to another. The repetition frequency, along with the satellite's orbit speed, not only provides the apparent satellite ground speed, but also the ground space between two consecutive altimetry measurements.

During the backscattering process of the electromagnetic field, some of the coherent reflections vary much from one measurement to another during the satellite's movement. Thus it is preferable to make an average over a similar surface of many consecutive waveforms, which from a statistical point of view measure very similar surfaces. That is the procedure made onboard of both satellites, except for the "burst mode". Examples of simulated waveforms taken from a series of consecutive 100 modeled radar echoes are presented in Figure 4a. The corresponding average waveform is presented in Figure 4b.

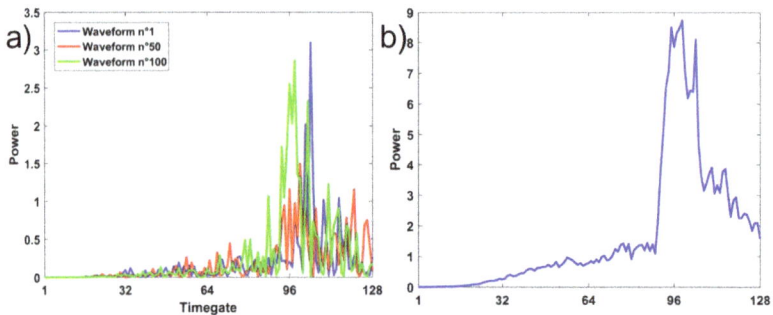

Figure 4. (**a**) Three examples of simulated waveforms (numbers 1, 50 and 100) taken from a series of consecutive radar echoes; (**b**) Mean waveform obtained averaging 100 simulated burst echoes.

3. Datasets

3.1. Radar Altimetry Data

Radar altimetry data used in this study comes from the acquisitions on the nominal orbit of the following missions: ENVISAT (2002–2010) and SARAL (2013–2016). These two satellites orbited a sun-synchronous polar orbit at an altitude of about 800 km, with a 35-day repeat cycle and an inclination of 98.6°, providing observations of the Earth from 81.5° latitude North to 81.5° latitude South, with a ground-track spacing of around 80 km at the Equator. RA-2, onboard the European Space Agency (ESA) ENVISAT mission, is a nadir-looking pulse-limited radar altimeter operating at Ku- (13.575 GHz) and S- (3.2 GHz) bands. The latter is available until January 2008, and is not used in the present study. SARAL is a CNES-ISRO joint-mission that was launched on 25 February 2013. Its payload is composed of the AltiKa radar altimeter and bi-frequency radiometer, and a triple system for precise orbit determination: the real-time tracking system DIODE (Détermination Immédiate d'Orbite par Doris Embarqué—Immediate Onboard Orbit Determination by Doris) of the DORIS (Doppler Orbitography and Radio-positioning Integrated by Satellite) instrument, a Laser Retroflector Array (LRA), and the Advanced Research and Global Observation Satellite (ARGOS-3). It is the first altimeter to operate at Ka-band (35.75 GHz). The main characteristics of RA-2 and AltiKa are summarized in Table 1.

The Sensor Geophysical Data Records (SGDRs) delivered by CNES and ESA include accurate satellite positions (longitude, latitude and altitude of the satellite on its orbit) and timing, altimeter ranges, instrumental, propagation and geophysical corrections applied to the range, measured waveforms, and several other parameters such as the backscattering coefficients. These parameters are available at high frequency (i.e., 18 and 40 Hz for ENVISAT and SARAL data respectively). The waveforms from ENVISAT RA-2 at Ku band are sampled on 128 timegates, whereas waveforms from SARAL/AltiKa are sampled on 116 timegates. In this study, data from cycles 75 to 85 of ENVISAT RA-2 (January–December 2009) and cycles 1 to 35 (from February 2013 to June 2016) for SARAL AltiKa. These data were made available by the Centre de Topographie des Océans et de l'Hydrosphère (CTOH) [36].

Table 1. Main characteristics of ENVISAT and SARAL altimetry missions and of RA-2 and AltiKa sensors from [37,38].

	ENVISAT RA-2	SARAL AltiKa
Mean satellite altitude (km)	800	800
Velocity on orbit (km s^{-1})	7.45	7.47
Apparent ground velocity (km s^{-1})	6.62	6.64
Frequency bands	Ku (13.575 GHz), S (3.2 GHz)	Ka (35.75 GHz)
Pulse duration (chirp sweep time) (μs)	20	110
Effective pulse duration (ns)	3.125	1
Bandwidth (MHz)	320, 80, 20 (Ku) 160 (S)	500
Pulse repetition frequency (Hz)	1795 (Ku) 449 (S)	~3800
Time between two pulses (T, en μs)	557 (Ku) 2230 (S)	~263

3.2. ASTER DEM

ASTER is a Japanese sensor which is one of the five instruments onboard the Terra satellite, launched in 1999 by the NASA in low Earth orbit. It has collected ground data since February 2000. ASTER provides high-resolution images of the Earth in 14 bands of the electromagnetic spectrum, from the visible domain to the thermic infrared. The resolution of these images varies between 15 m and 90 m. Data from ASTER are used to create detailed maps of the surface temperature, emissivity, reflectance, and what is of interest to us, maps of elevation.

The first version of the ASTER Global Data Elevation Model (GDEM), which was released in 2009, has been generated using stereoscopic images collected by the instrument. The global coverture was extending from 83°S to 83°N, covering 99% of the global continental surface. The second version [37] added 260,000 stereoscopic images to the previous version, improving the global coverage and reducing the number of artefacts. The data production algorithm provides a better spatial resolution, an improved horizontal and vertical precision, along with an improved water surfaces coverture. The distributed data format remains GeoTIFF, and the same sampling grid is used, with a resolution of 30m and cells of 1° × 1°. ASTER GDEM data are freely available worldwide. They can be downloaded from the internet on the Land Processes Distributed Active Archive Center (LP DAAC) (https://lpdaac.usgs.gov/).

4. Study Area

The selected sites are distributed along a north-south rainfall gradient in the Gourma region, Mali in the west-African Sahel. The Gourma region (14.5–17.5°N and 1–2°W), is a vast peneplain at between 250 and 330 m altitude and has been previously used by numerous remote sensing studies, based on active and passive microwave sensors (e.g., [7,9–11,18,19,21]). Indeed, the presence of large homogeneous and flat surfaces characterized by a high seasonal and inter-annual variability makes this region particularly well suited both for methodological development and for multi-product validation exercises [39]. Predominantly sandy (58% of the region) or shallow (23%) soils are distributed in large alternant swaths of contrasted land cover (fixed sand dunes, eroded surfaces locally capped by iron pan). Besides these two major landforms, remnants of alluvial systems and lacustrine depressions form a web of narrow bands often slotted in between sandy and shallow soils [39]. In this region, the mean annual rainfall varies between 150 and 400 mm from north to south with interannual variations ranging between 15 and 30% of the annual rainfall. Rainfall occurs during the northern hemisphere summer, starting between May and July until September or October with a maximum in August [40].

Five study sites are selected on the 302, 373 and 846 ground tracks of ENVISAT and SARAL altimeters (see Figure 5). The site 1 is located on SARAL/ENVISAT track 846. The satellite goes from north to south over the reference point at latitude 16.75°N and longitude 1.842°W, at the north of the In Zaket area. This site presents mostly sand dunes along with some rock formation in the center of the area. At the east of the site, the beginning of a great dune can be seen. No open water surface is present here.

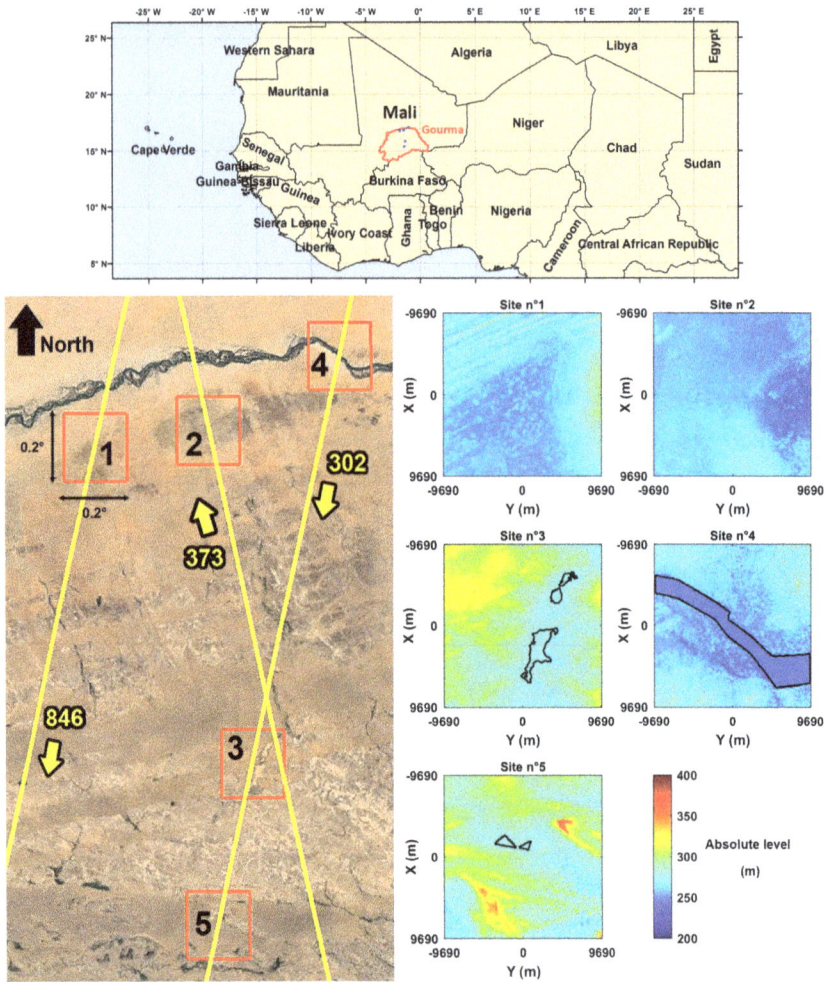

Figure 5. Location of the five study sites distributed on satellite tracks 302, 373 and 846 of ENVISAT/SARAL over the Gourma region, Mali © Google Earth. The azimuthal direction of the satellite on orbit, the site numbers and the corresponding DEM extraction area are indicated. The water areas limits over each site are marked with black lines.

Site 2 is located on SARAL/ENVISAT track 373, with the satellite going from south to north. The reference point is at latitude 16.8°N and longitude 1.478°W. The site is a sandy plane with nearly no dune pattern. No open water surface is present in this area.

Site 3 is located on SARAL/ENVISAT track 302, close to the Gossi village. The satellite goes from north to south over the latitude 15.82°N and longitude 1.339°W reference point. The site presents mainly sand dunes, which in the center presents a structure of three ponds along with a flood area in the southern part. The DEM has been reworked to simulate these water ponds by quadrangular areas, with heights of 278 m, 277 m and 275 m from north to south respectively.

Site 4 is located at coordinates 17.02°N and 1.059°W on the SARAL/ENVISAT track 302. The site is composed of a sandy soil, through which flows the Niger River from northwest to southeast of the extraction area (see Figure 5). The Niger River is simulated by six quadrangular areas at the same altitude of 245 m, which is considered to be the height of the high river bed at the end of the monsoon.

Site 5, located on SARAL/ENVISAT track 302, is close to the Agoufou pond. With coordinates of 15.345°N and 1.45°W, this sandy plain with dune formations presents a pond composed of two triangular areas (see). These two ponds are simulated by two triangular areas with the same altitude of 275 m. A rocky peak at the northeast of this area (the red "V" on Figure 5) implies adding a consequent time-lag on the waveform formation.

5. Results

5.1. Comparison between Simulated and Real Waveforms

Radar altimetry waveform simulations were performed over the five sites presented in Section 4. Study area for both ENVISAT RA-2 and SARAL AltiKa missions. The representation of the scene is described in Appendix D. Parameters used in this study are presented in Appendix E. As no information on the power amplification applied to the received signal in the electronics onboard the satellites is available, an empirical gain determination was applied (see Appendix F).

Radar altimetry waveforms were simulated for the dry season (i.e., SSM = 0.02) over the five test sites for both ENVISAT RA-2 and SARAL AltiKa missions. They were compared to radar altimetry waveforms present in the SGDR. The real waveforms were selected at the closest acquisition location to the reference location of the test sites. As the orbit slightly change with time, the footprint center is not exactly the same from one cycle to another, the scene encompassed is different.

In our study, ENVISAT RA-2 measurements for the thirty-five chosen cycles, corresponding to the dry season from January to May, are close to the reference point and, are encompassed in a circle of 500 m of radius. SARAL/AltiKa are more scattered, within a 2500 m buffer along the nominal orbit track. This deviation is similar for the five study sites, and introduces variations on each waveform observed. These variations are limited for ENVISAT data, the footprints being similar between measurement times. The impact on SARAL/AltiKa data is stronger, seeing that a distance of 2500 m between two different footprints implies that only 60% of the same scene is observed. Finally, the selected waveforms were averaged over the whole period of observation for each satellite. The resulting averaged waveforms are presented in Figure 6.

An overall good agreement is found between the averaged observed waveforms and the simulated ones with correlation ranging from 0.66 to 0.94 at Ku-band for ENVISAT RA-2 and from 0.27 to 0.96 at Ka-band for SARAL/AltiKa. Over site 1 at Ku-band, both simulated and real waveforms have a similar shape with a correlation coefficient of 0.94. On the same site at Ka-band, the modeled waveform presents a slower increase of the leading edge contrary to the average of real waveforms, but with a similar trailing edge. The correlation coefficient between simulated and waveforms is of 0.96. Over site 2, both simulated and averaged observed waveforms have close profiles at Ku and Ka-bands with correlation coefficients of 0.91 for both bands. However, the simulated waveform has a much lower amplitude (150 against 400) at Ka-band compared to Ku-band. Over site 3, both simulated and real waveforms are very similar until gates 80 at Ka-band and 90 at Ku-band. After those time gates, the modeled waveforms present a huge peak at both Ku and Ka-bands which is not present in the average measured waveforms. This peak is due to the presence of temporary ponds in the scene. The issue of the pond levels, considered for the simulation as high and which is oppositely low for the dry season measured waveforms considered, leads to these differences and low correlation coefficient. Besides, due to the drifts of the satellite ground-track from one cycle to another, larger for SARAL than for ENVISAT, the location of the signature of the ponds also changes. Correlation coefficients of 0.66 and 0.27 were found at Ku and Ka-bands respectively. Over site 4, corresponding to the cross-section with the Niger River, the average of the observed radar echoes are very specular at both Ku- and Ka- bands. Very huge power is reflected by the surface of the river acting as a mirror at microwave frequencies. The modeled waveforms exhibit more important trailing edge slopes resulting in larger radar echoes. This may result from issues in the modeling of water surfaces considered as flat surfaces, and hence neglecting the surface roughness responsible for the decrease of the signal aside from the nadir-looking angle. The correlation coefficients of 0.72 and 0.88 were found at Ku and Ka-bands respectively. Over site 5, the modeled and the average of the waveforms are characterized by one large (at Ka-band) or two peaks (at Ku-band) corresponding to the two small ponds present in the scene. The correlation coefficients are of 0.88 and 0.79 at Ku and Ka-bands respectively. As it can be seen in Figure 6, important changes in amplitude can be observed from one cycle to another, that cannot be attributed neither to the filling of the temporary ponds nor to the drift in the altimeter orbits.

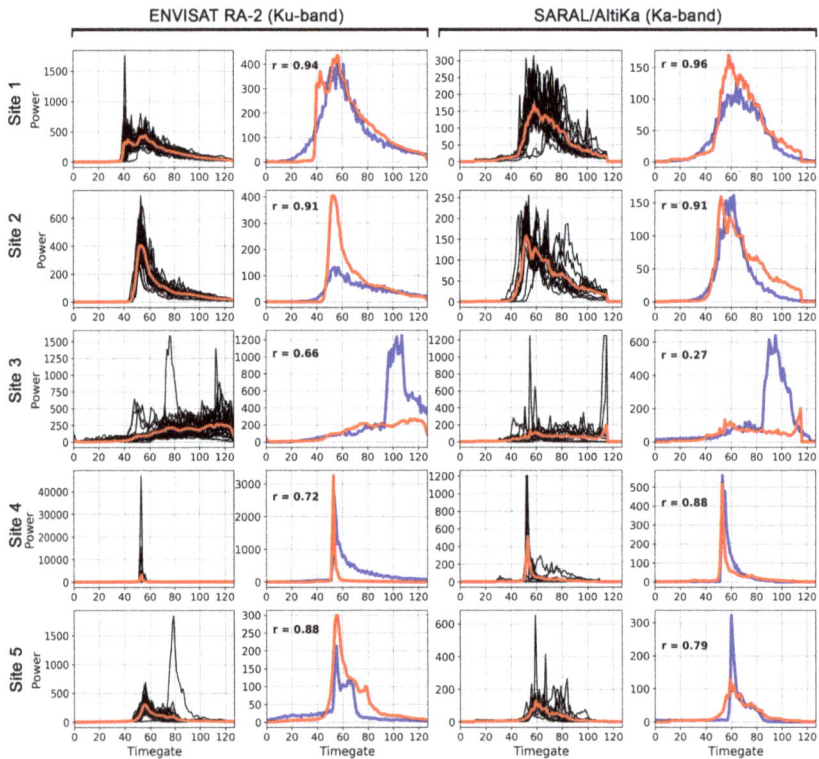

Figure 6. Individual (black) and mean (red) waveforms on the five study sites over the observation periods from ENVISAT RA-2 at Ku-band (left of the left panel) and from SARAL/AltiKa at Ka-band (left of the right panel). Comparison over the five study sites between the modeled waveform (in blue) and the mean observed waveform (in red), with the correlation coefficient r indicated from ENVISAT RA-2 at Ku-band (right of the left panel) and from at SARAL/AltiKa at Ka-band (right of the right panel).

5.2. Impact of Surface Soil Moisture on Altimetry Signal

Previous studies showed that the temporal variations in SSM in semi arid areas have a strong impact on the radar altimetry backscattering coefficients reaching up to 25–30 dB between the dry and the wet seasons [20–25]. Over the five sites considered previously, new simulations were run at Ku and Ka-bands considering the following values of SSM: 0.05, 0.1, 0.2, 0.3 and 0.4 that encompass the range of SSM variations in the Gourma region of Mali. The backscattering coefficients derived from the modeled waveforms were also estimated. Following [20–25], backscattering coefficients were derived using the Offset Center of Gravity (OCOG) or Ice-1 retracking algorithm [41] that is commonly used for the application of satellite altimetry to land hydrology [42]. The results of these simulations are presented in Figure 7 along with the variations of the backscattering coefficient against SSM.

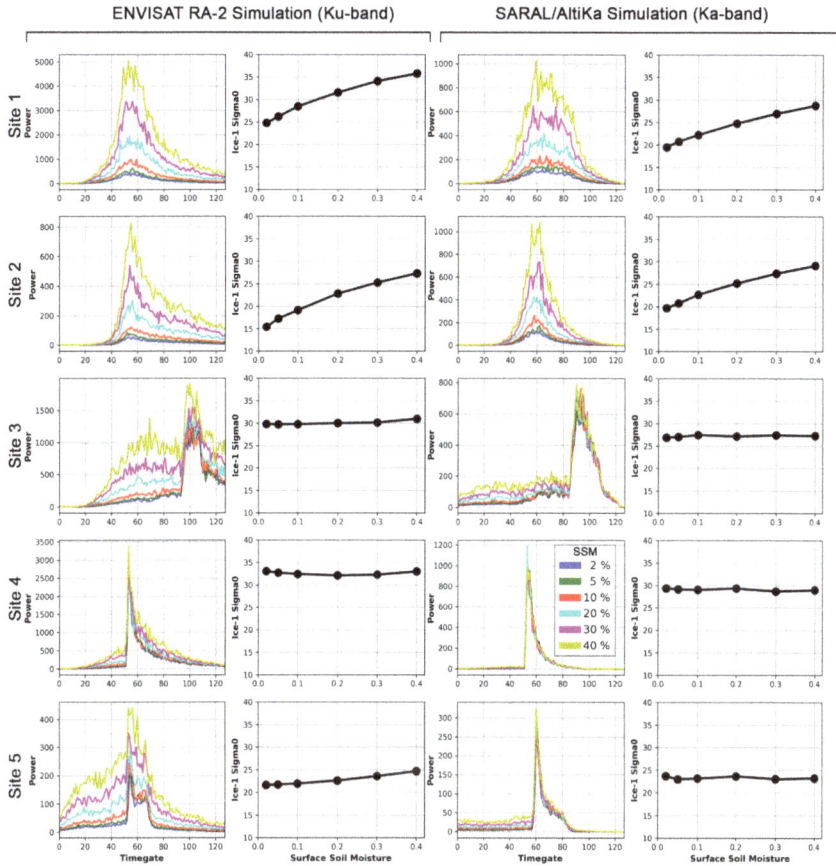

Figure 7. Simulated waveforms for increasing values of surface soil moisture (SSM) (2%—blue, 5%—green, 10%—red, 20%—light blue, 30%—purple, 40%—yellow) on the five study sites at Ku-band (left of the left panel) and at Ka-band (left of the right panel). Corresponding backscattering coefficients (dB) as a function of SSM (%) at Ku-band (right of the left panel) and from at Ka-band (right of the right panel).

The amplitude of the simulated waveforms rises with increasing SSM over sandy sites without water (sites 1 and 2), at both Ku and Ka-bands. The amplitude of the maximum of the waveform is multiplied by a factor between 5 and 8 at Ku-band and between 4 and 6 at Ka-band for variations in SSM between 0.1 and 0.4. Backscattering values range from ~25 to ~35 dB over site 1 and from ~15 to ~30 dB over site 2 at Ku-band, and from ~20 to ~30 dB over the two sites at Ka-band. These results are in good agreement with what was observed in earlier studies [20–25]. Nevertheless, lower values of backscattering during the dry season and higher amplitude would have been expected at Ka rather than at Ku band. This discrepancy is likely to be due to (i) the configuration of the scenes that are not homogeneous; (ii) the extension of the semi-empirical model of the soil complex permittivity at

Ka-band that is not completely valid; and (iii) the values of the correlation length and of the RMS of the roughness height that are not completely adequate, especially at Ka-band (see Appendix E).

Over site 4, corresponding to the presence of the Niger River in the scene, the radar echo is dominated the peaky waveform corresponding to the signature of the river. Increases in returned power are observed on the sides of the peak corresponding to changes in SSM on the river banks. They are negligible compared with the value of the maximum observed at the peak. As a consequence, the backscattering coefficient is constant for any value of the SSM and equals to ~35 dB at Ku-band and ~30 dB at Ka-band. Please note that the width of the Niger River remains constant for all the simulations, i.e., effects of the flood were not taken into account. An increase of the width of the river would have caused an increase in the backscattering coefficient at both bands, as observed in [23].

The configuration of sites 3 and 5 leads to a mix between the two former situations. The ponds, considered permanent, and hence present in all the simulations, occupy a smaller surface than the Niger River in site 4. As a consequence, the amplitude of the simulated waveforms rises over all the gates, even the ones corresponding to the ponds (i.e., from gates 95 to 100 and from gates 80 to 110 over site 3, from gates 50 to 75 and from gates 60 to 85 at Ku and Ka-bands respectively) with increasing SSM. However, the backscattering coefficient remains almost constant and equals to ~30 dB (with slight increase at ~31 dB for SSM equals 0.4) and ~25 dB at Ku and Ka-bands respectively over site 3, almost constant and equals to ~24 dB at Ka-band over site 5, with a slight increase from ~21 to ~24 dB at Ku-band over site 5. The reason for these lacks of change in the backscattering coefficient with SSM when the radar altimetry waveforms are impacted is due to the retracking algorithm. In the case of OCOG/Ice-1, the waveform is simulated by a rectangular box encompassing the maximum of the power returned. Over the two sites, the box width corresponds to the extent of the signature of the small ponds. Most of the increase of the SSM on the backscattered power is filtered out. As a consequence, the backscattering coefficient is almost constant. This is the reason why, the measurements contaminated by temporary floods were excluded from SSM inversion by [21].

6. Conclusions

This study analyzed the impact of SSM on the radar altimetry waveform over semi-arid areas, and as a consequence, on radar altimetry backscattering coefficient through a modeling approach. A two-scale model, coupled with a semi-empirical model for the complex dielectric permittivity of the soil, was applied to simulate the altimetry radar echoes at Ku and Ka-bands through an approach taking into account the surface roughness (microscopic scale) and the surface topography (macroscopic scale). Comparisons between simulated and real waveforms from ENVISAT RA-2 and SARAL AltiKa acquired at Ku and Ka-bands show a good agreement with correlation ranging from 0.66 to 0.94 at Ku-band and from 0.27 to 0.96 at Ka-band over the five study sites representative of the Sahelian zone of Gourma in Mali. Increase in SSM over typical range of variations in semi-arid conditions (from 0.02 to 0.4), show a rise of the power received by the altimeter over sandy areas that causes a raise of 10 to 15 dB of the backscattering coefficient. Such large changes can be observed when the radar altimeter made its acquisitions a short time after an intense rain event. The presence of open water in the scene observed by the altimeter (river and even small ponds) cancelled the effect of the increase in SSM. The waveform is dominated by the peak corresponding to the presence of open water, especially in the case of a large river such as the Niger. As the retracking algorithm derives the waveform parameters from this peak, including the backscattering coefficient, changes in SSM are negligible compared with the presence of open water. These results are in good agreement with what was previously observed analyzing radar altimetry backscattering coefficients acquired over semi-arid areas.

The modeling approach developed in this study (CALM) could benefit from the following improvements:

(1) a more realistic description of the open water areas taking into account the small undulations of the surface as in [43],

(2) accurate values of the roughness parameters at Ku and Ka-bands and their spatio-temporal variations over the study sites,

(3) DEM at higher spatial resolution and with a better accuracy, to obtain more accurate simulation results.

The next steps would be the analysis of the impact of the vegetation on the simulation results and the application of this model to the SAR acquisition mode operating at low incidence angles and the comparisons to the Sentinel-3A data.

Acknowledgments: The authors were supported by CNES and ONERA through C.F PhD grant. They were also supported by by CNES/EUMETSAT Ocean Surface Topography Science Team (OSTST) through the funding of the "new Perspectives for higher ResolutIon Altimetry—a Multi-disciplinary approach (PRIAM)" project and by CNES TOSCA through the funding of the "CAractérisation des Surfaces Continentales par Altimétrie et Diffusiométrie (CASCAD)" project.

Author Contributions: C.F., P.B., F.F. and E.M. conceived and designed the study. C.F. performed the experiments. All the authors analyzed the data and contributed to the redaction of the paper.

Conflicts of Interest: The authors declare no conflict of interest. The founding sponsors had no role in the design of the study; in the collection, analyses, or interpretation of data; in the writing of the manuscript, and in the decision to publish the results.

Appendix A

The parameters of the soil complex dielectric permittivity model from [27,28] are detailed below. The effective free water permittivity ε_{fw} of the soil is defined as:

$$\varepsilon_{fw} = \varepsilon_{w\infty} + \frac{\varepsilon_0 - \varepsilon_{w\infty}}{1 + j2\pi f\tau} - j\frac{\sigma_{eff}}{2\pi f\varepsilon_0} * \frac{\rho_s - \rho_b}{\rho_s * m_v} \tag{A1}$$

where f is the frequency of the electromagnetic wave emitted by the sensor, m_v is the volumetric soil moisture given in percent, The water dielectric permittivity for very high frequencies converge towards an "infinite" value $\varepsilon_{w\infty}$ taken at $\varepsilon_{w\infty} = 4.9$, ε_{w0} is the non-salted water permittivity for the null frequency estimated using the following equation [29]:

$$\varepsilon_{w0} = 87.74 - 0.4008 * T + \left(9.398 \times 10^{-4}\right) * T^2 + \left(1.410 \times 10^{-6}\right) * T^3 \tag{A2}$$

T is the ground temperature in Celsius degrees, τ is the water relaxation time, which is also function of the temperature, and defined as follows:

$$2\pi\tau = \left(1.1109 \times 10^{-10}\right) - \left(3.824 \times 10^{-12}\right) * T + \left(6.938 \times 10^{-14}\right) * T^2 - \left(5.096 \times 10^{-16}\right) * T^3 \tag{A3}$$

σ_{eff} is the effective water conductivity in the ground, and depends on the soil bulk density ρ_b and on the sand and clay composition of the ground S and C:

$$\sigma_{eff} = -1.645 + 1.939\rho_b - 0.02013 * S + 0.01594 * C \tag{A4}$$

The relative dry soil permittivity ε_s is defined as:

$$\varepsilon_s = (1.01 + 0.44\rho_s)^2 - 0.062 \tag{A5}$$

where ρ_S is the specific soil density defined as:

$$\rho_s = \frac{\rho_b}{1 - v_a} \tag{A6}$$

v_a is the void fraction of the ground (taken further on at 0.65), and ρ_b is the soil bulk density, in g/cm^3. The shape constant (α) is further on taken at 0.65 and the empirical constant (β) is defined as:

$$\beta = \beta_r + j * \beta_{im} \tag{A7}$$

where

$$\beta_r = (127.48 - 0.519 * S - 0.152 * C)/100 \tag{A8}$$

$$\beta_{im} = (1.33797 - 0.603 * S - 0.166 * C)/100 \tag{A9}$$

Appendix B

In the Continental waveform ALtimetry Model (CALM), the surface illuminated by the EM wave emitted by the sensor (i.e., the radar altimeter footprint) is represented using a meshgrid of triangles. This meshgrid is considered uniform in both directions (X and Y-axis, see Figure A1). The altimetry backscattering is assumed to occur in a rectangular surface, which size is determined by the number of mesh Nx in the X-axis (a mesh corresponds to a basic elementary triangle) and Ny in the Y axis encompassing the theoretical altimeter footprint. The surface resolution R is then the distance between two consecutive meshes, taken in this study as constant in both axes. The surface parameters are defined in a matrix that has a 2Nx + 1 by 2Ny + 1 size representing each point position on the meshgrid (position in X and Y axis, point height in the Z axis and the surface of each triangle). The center of the matrix is the origin point of the coordinate system defined as the instant of emission of the EM wave for each measurement, and is located on the satellite groundtrack.

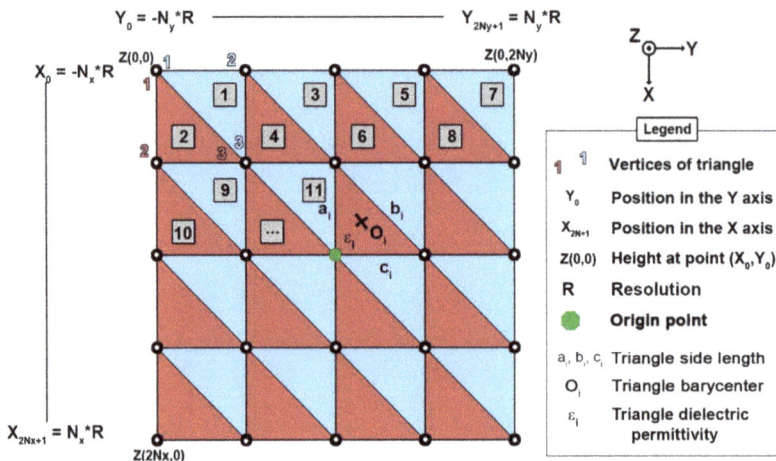

Figure A1. Meshgrid representing the coordinate system and norms used in CALM. In this example, Nx = Ny = 2.

Each elementary surface composing the meshgrid is characterized by the lengths a_i, b_i and c_i of the triangle sides and its barycenter O_i, as represented in Figure A1. By numbering each elementary surface, we create a matrix with—among other information—the coordinates of the triangles vertices and their surface, calculated from the Heron formula:

$$A_i = \sqrt{p_i(p_i - a_i)(p_i - b_i)(p_i - c_i)} \tag{A10}$$

where

$$p_i = \frac{1}{2}(a_i + b_i + c_i) \tag{A11}$$

The perpendicular vectors to each triangle are determined using a normalized vectorial product. Each elementary surface is also attributed to a proper dielectric permittivity ε_{ri}, computed as explained in Section 2.1.2. In the different cases considered in this study, dielectric permittivity and roughness parameters (root mean square height, correlation function type and correlation length) have the same value for all the non-inundated grid elements at each time step.

Finally, each elementary surface is given an 11-variables vector:

- the mean X position (noted as X_i further on),
- the mean Y position (noted as Y_i further on),
- the mean Z position (noted as Z_i further on),
- the X axis component of the perpendicular vector (noted as N_x),
- the Y axis component of the perpendicular vector (noted as N_y),
- the Z axis component of the perpendicular vector (noted as N_z),
- the nature of the elementary surface (-1 for water, 0 for ground is the default value),
- the root mean square height hrms (noted as h_i),
- the surface correlation length (noted as l_i),
- the triangle area (noted as A_i),
- the surface complex dielectric permittivity (noted as ε_{ri}),

All these parameters characterize each elementary surface, and can be further on changed to attribute new properties to selected areas.

Appendix C

The antenna field of view depends on the wavelength λ and the antenna diameter D, and is expressed in radians by the following formula:

$$\theta_{3dB} = \frac{70\lambda\pi}{D180} \tag{A12}$$

Knowing this antenna field of view and the distance between the satellite orbit and the ground, the observed ground area can be limited to a disc of radius corresponding to the observable ground, given by the 3 dB power attenuation.

The antenna gain is expressed by:

$$G = \frac{4\log(2)}{\theta_{3dB}^2} \tag{A13}$$

Taking into account the antenna diagram, the ponderation is then defined as:

$$W(\theta) = \exp(-2G\theta) \tag{A14}$$

Appendix D

The definition of the simulation scene is composed of three main steps: the resizing of the DEM data according the altimeter footprint, the delineation of open water areas present in the scene and the orientation of the scene depending whether the altimetry tracks are ascending or descending.

- Resizing of the DEM

    This first step allows to take into account the different altimeter footprint sizes (see Figure A2), the location of the open water surfaces, and to improve the computation time. As the extracted gridded ASTER DEM data are provided by tiles of $1° \times 1°$, each tile largely encompasses the size of an individual altimeter footprint. For each individual footprint, the DEM is cropped to encompass the altimeter footprint in the square of minimum area containing it. For instance, using the 30 m resolution GDEM ASTER data, and for an ENVISAT RA-2 simulation at Ku-band, a square of $N_x = N_y = 646$ grid elements is selected. This implies corresponds to a scene of 19,380 m, to be compared with the approximated 18,000 m footprint diameter of ENVISAT RA-2 at Ku-band (see the example in Figure A3).

- Delineation of surface water extent

    Water surface extent in the study areas were delineated using Google Earth and added to the scene. As the elevation of this surface is often non-uniform in the DEM, the lowest elevation in this area and its surroundings is attributed to the whole open water area (see Figure A4). As many information on these surfaces is missing, a default backscattering coefficient of 17 and 20 dB at Ku and Ka bands respectively was attributed. Those values correspond to the mean backscattering value over the Niger River at both bands.

- Orientation of the scene

    In CALM, the orbit trajectory is considered linear following the Y-axis of the modeled surface. The original DEM extraction is orientated with the North as the top of the extraction area. To take into account the satellite movement on orbit it is then necessary to apply a surface rotation of 90°, clockwise for a movement from south to north, and anticlockwise from north to south (green arrow in Figure A2). The real and simulated orbits are not exactly in the same direction, but rather present an angle with the meridian. Rigorously, the rotation of the observed area should respect the same angle to be in the same measurement conditions. Nevertheless, over a hundred successive measurements, the satellite ground movement does not exceed 500 m, for a footprint diameter going from 8 km to 18 km for the considered altimetry satellites. The error on orbit simulation is considered as negligible.

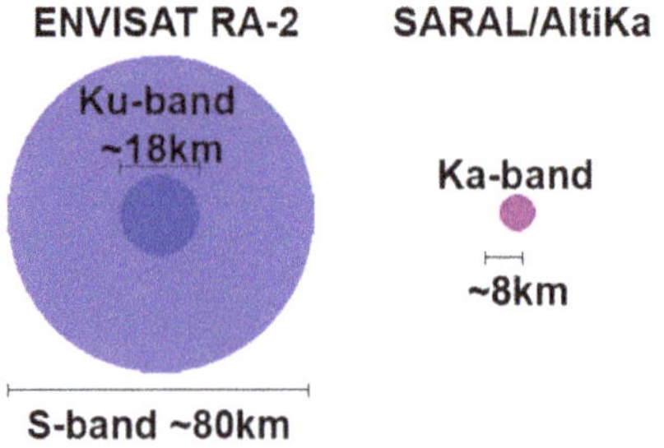

Figure A2. Size of the theoretical footprints for a totally plane surface as a function of the frequency used for ENVISAT RA-2 and SARAL AltiKa.

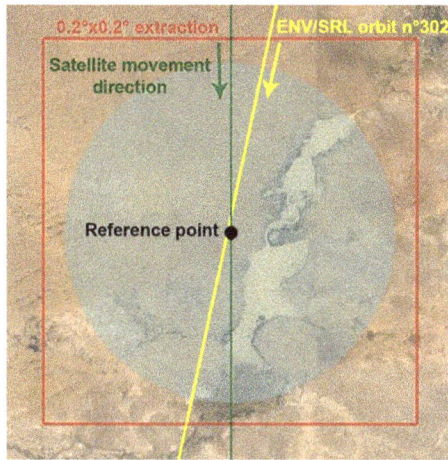

Figure A3. Example of a scene using ASTER GDEM data centered close to the Gossi pond in Mali. The real altimeter satellite path 0302 of ENVISAT RA-2 is represented in yellow, and the simulated orbit (green), along with the reference point of the simulation. The shadowed area around the reference point corresponds to the Ku-band footprint of ENVISAT RA-2.

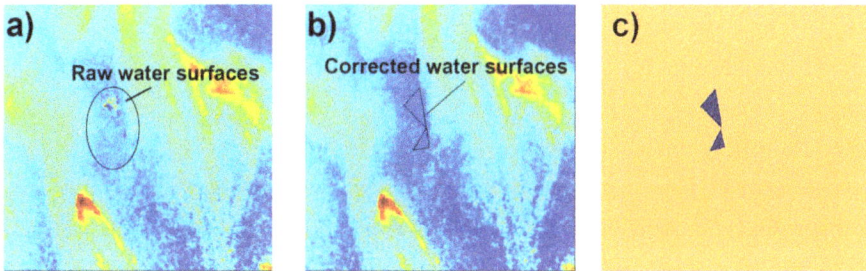

Figure A4. Example of processing applied to surfaces with water areas. (**a**) The raw DEM without any water surface treatment; (**b**) DEM with corrected water surfaces; (**c**) Surface nature as entered in CALM (water in blue, ground in orange).

Appendix E

All the parameters used in the CALM simulation for ENVISAT RA-2 at Ku-band and SARAL/AltiKa at Ka-band are listed in Table A1.

Table A1. Main characteristics of ENVISAT and SARAL altimetry missions and of RA-2 and AltiKa sensors from [37,38].

	ENVISAT RA-2	SARAL AltiKa
Mean satellite altitude (km)	800	800
Velocity on orbit (km s^{-1})	7.45	7.47
Apparent ground velocity (km s^{-1})	6.62	6.64
Frequency bands	Ku (13.5 GHz), S (3.2 GHz)	Ka (35.5 GHz)
Pulse duration (chirp sweep time) (µs)	20	110
Effective pulse duration (ns)	3.125	1
Bandwidth (MHz)	320, 80, 20 (Ku) 160 (S)	500
Pulse repetition frequency (Hz)	1795 (Ku) 449 (S)	~3800
Time between two pulses (*T*, µs)	557 (Ku) 2230 (S)	~263
Antenna parabola diameter (m)	1.2	1
Footprint diameter (km)	~18	~8
Antenna gain	37 dB	44 dB
Emitted power (W)	161	100
Distance between two consecutive measurements	3.69 m	1.66 m
Void fraction (v_a)	0.36	
Sand ground component S	60	
Clay ground component C	20	
Soil bulk density (rb)	1.69	
Alpha parameter	0.65	
Temperature (°C)	30	
Correlation length	4.5 cm	
Height rms (roughness)	0.35 cm	

Very few in situ data are available over this area for the roughness parameters (hrms, correlation length). The roughness data were measured at the Agoufou site in August 2005 (60 profiles) and February 2006 (16 profiles) in the heart of the wet and dry seasons respectively [19]. These data were obtained using a 2 m-long profilometer linear needle roughness device, which needles are separated from one another by 1 cm, allowing the estimation of standard deviation of roughness heights and the autocorrelation length. The mean values determined over the Agoufou site allowed to estimate a mean hrms height of 0.7/0.5 cm, for a mean correlation length of 12.6/13.5 cm during the dry and the rainy seasons. The 1 cm separation length between two needles is too long for Ku (2.3 cm of wavelength) and Ka (0.8 cm wavelength) bands. As these in situ roughness values cannot be used in this study, roughness parameters determined in [26] for dry and smooth sandy soils similar to the Gourma soil types (hrms of 0.35 cm and correlation length of 4.5 cm). Due to the limited anthropogenic impact in this region, the ground is considered Gaussian. Using this values, the validity of the assumptions allowing the use of this GO model were checked (Table A2). All the criteria are validated at Ka band, and 2 out of 3 at Ku-band, except ks > 1.5.

Table A2. Variables used for the validity of the GO model.

	Ku Band	Ka Band
kl (rad)	12.72	33.46
l^2 (10^{-3} m^2)	2.0	2.0
2.67sλ (10^{-3} m^2)	0.21	0.08
ks (rad)	0.99	2.60

Appendix F

The electronic systems onboard the satellites contain amplificators that increases the power of received electromagnetic signal, strongly attenuated after 1600 km of propagation. No information concerning the amplificators onboard the satellites is avaialble. In order to take into account the amplificatory gains, the Automatic Gain Control (AGC) and the system calibration on the waveforms modeled with CALM to compare them with real satellite waveforms, a gain factor is applied. This gain is determined empirically to make for the simulated waveforms to correspond to the mean measured satellite waveform over the observation periods (11 cycles for ENVISAT, 18 for AltiKa). Along with the added time lag, this leads to a decent waveform modeling with the proper timeframe.

Table A3 shows that the empirical gain to apply is important. It remains though quite constant for each satellite (only a maximum 10 dB difference). Of course, this gain is insufficient to provide precise modeling, but this method has proven to give satisfactory results. The timelag is inherent to each observed surface, and thus vary strongly from one site to another. In order for the modeled waveform to correspond to the mean measured waveform, the time-lag has been determined using cross-correlation maxima.

Table A3. Empirical gains and time lags applied to each study site for both ENVISAT RA-2 and SARAL/AltiKa configurations.

Simulation Site	ENVISAT RA-2		SARAL/AltiKa	
	Empirical Gain (dB)	Timelag (ns)	Empirical Gain (dB)	Timelag (ns)
Site 1	343.42	0	339.03	33
Site 2	337.78	−30	336.53	18
Site 3	341.76	35	336.02	158
Site 4	339.03	100	327.78	160
Site 5	335.44	425	330	164
Mean	340.37	n/a	335.59	n/a
STD	2.28	n/a	2.62	n/a

References

1. Koster, R.D.; Dirmeyer, P.A.; Guo, Z.; Bonan, G.; Chan, E.; Cox, P.; Gordon, C.T.; Kanae, S.; Kowalczyk, E.; Lawrence, D.; et al. GLACE Team Regions of strong coupling between soil moisture and precipitation. *Science* **2004**, *305*, 1138–1140. [CrossRef] [PubMed]
2. Entekhabi, D.; Rodriguez-Iturbe, I. Analytical framework for the characterization of the space-time variability of soil moisture. *Adv. Water Resour.* **1994**, *17*, 35–45. [CrossRef]
3. Small, E.E.; Kurc, S.A. Tight coupling between soil moisture and the surface radiation budget in semiarid environments: Implications for land-atmosphere interactions. *Water Resour. Res.* **2003**, *39*. [CrossRef]
4. Mott, J.J. Factors Affecting Seed Germination in Three Annual Species from an Arid Region of Western Australia. *J. Ecol.* **1974**, *62*, 699. [CrossRef]
5. Delon, C.; Mougin, E.; Serça, D.; Grippa, M.; Hiernaux, P.; Diawara, M.; Galy-Lacaux, C.; Kergoat, L. Modelling the effect of soil moisture and organic matter degradation on biogenic NO emissions from soils in Sahel rangeland (Mali). *Biogeosciences* **2015**, *12*, 3253–3272. [CrossRef]
6. Wagner, W.; Lemoine, G.; Rott, H. A Method for Estimating Soil Moisture from ERS Scatterometer and Soil Data. *Remote Sens. Environ.* **1999**, *70*, 191–207. [CrossRef]
7. Jarlan, L.; Mougin, E.; Frison, P.L.; Mazzega, P.; Hiernaux, P. Analysis of ERS wind scatterometer time series over Sahel (Mali). *Remote Sens. Environ.* **2002**, *81*, 404–415. [CrossRef]
8. Gruhier, C.; de Rosnay, P.; Kerr, Y.; Mougin, E.; Ceschia, E.; Calvet, J.C.; Richaume, P. Evaluation of AMSR-E soil moisture product based on ground measurements over temperate and semi-arid regions. *Geophys. Res. Lett.* **2008**, *35*. [CrossRef]
9. Gruhier, C.; De Rosnay, P.; Hasenauer, S.; Holmes, T.; De Jeu, R.; Kerr, Y.; Mougin, E.; Njoku, E.; Timouk, F.; Wagner, W.; et al. Soil moisture active and passive microwave products: Intercomparison and evaluation over a Sahelian site. *Hydrol. Earth Syst. Sci.* **2010**, *14*, 141–156. [CrossRef]

10. Louvet, S.; Pellarin, T.; al Bitar, A.; Cappelaere, B.; Galle, S.; Grippa, M.; Gruhier, C.; Kerr, Y.; Lebel, T.; Mialon, A.; et al. SMOS soil moisture product evaluation over West-Africa from local to regional scale. *Remote Sens. Environ.* **2015**, *156*, 383–394. [CrossRef]
11. Frison, P.L.; Mougin, E.; Hiernaux, P. Observations and Interpretation of Seasonal ERS-1 Wind Scatterometer Data over Northern Sahel (Mali). *Remote Sens. Environ.* **1998**, *63*, 233–242. [CrossRef]
12. Frison, P.-L.; Jarlan, L.; Mougin, E. Using Satellite Scatterometers to Monitor Continental Surfaces. In *Land Surface Remote Sensing in Continental Hydrology*; Elsevier: New York, NY, USA, 2016; pp. 79–113, ISBN 9781785481048.
13. De Jeu, R.A.M.; Wagner, W.; Holmes, T.R.H.; Dolman, A.J.; van de Giesen, N.C.; Friesen, J. Global Soil Moisture Patterns Observed by Space Borne Microwave Radiometers and Scatterometers. *Surv. Geophys.* **2008**, *29*, 399–420. [CrossRef]
14. Naeimi, V.; Scipal, K.; Bartalis, Z.; Hasenauer, S.; Wagner, W. An Improved Soil Moisture Retrieval Algorithm for ERS and METOP Scatterometer Observations. *IEEE Trans. Geosci. Remote Sens.* **2009**, *47*, 1999–2013. [CrossRef]
15. Holgate, C.M.; De Jeu, R.A.M.; van Dijk, A.I.J.M.; Liu, Y.Y.; Renzullo, L.J.; Vinodkumar; Dharssi, I.; Parinussa, R.M.; Van Der Schalie, R.; Gevaert, A.; et al. Comparison of remotely sensed and modelled soil moisture data sets across Australia. *Remote Sens. Environ.* **2016**, *186*, 479–500. [CrossRef]
16. Tansey, K.J.; Millington, A.C.; Battikhi, A.M.; White, K.H. Monitoring soil moisture dynamics using satellite imaging radar in northeastern Jordan. *Appl. Geogr.* **1999**, *19*, 325–344. [CrossRef]
17. Moran, M.S.; Hymer, D.C.; Qi, J.; Sano, E.E. Soil moisture evaluation using multi-temporal synthetic aperture radar (SAR) in semiarid rangeland. *Agric. For. Meteorol.* **2000**, *105*, 69–80. [CrossRef]
18. Baup, F.; Mougin, E.; Hiernaux, P.; Lopes, A.; De Rosnay, P.; Chenerie, I. Radar Signatures of Sahelian Surfaces in Mali Using ENVISAT-ASAR Data. *IEEE Trans. Geosci. Remote Sens.* **2007**, *45*, 2354–2363. [CrossRef]
19. Baup, F.; Mougin, E.; De Rosnay, P.; Hiernaux, P.; Frappart, F.; Frison, P.L.; Zribi, M.; Viarre, J. Mapping surface soil moisture over the Gourma mesoscale site (Mali) by using ENVISAT ASAR data. *Hydrol. Earth Syst. Sci.* **2011**, *15*. [CrossRef]
20. Ridley, J.; Strawbridge, F.; Card, R.; Phillips, H. Radar backscatter characteristics of a desert surface. *Remote Sens. Environ.* **1996**, *57*, 63–78. [CrossRef]
21. Fatras, C.; Frappart, F.; Mougin, E.; Grippa, M.; Hiernaux, P. Estimating surface soil moisture over Sahel using ENVISAT radar altimetry. *Remote Sens. Environ.* **2012**, *123*. [CrossRef]
22. Fatras, C.; Frappart, F.; Mougin, E.; Frison, P.-L.; Faye, G.; Borderies, P.; Jarlan, L. Spaceborne altimetry and scatterometry backscattering signatures at C- and Ku-bands over West Africa. *Remote Sens. Environ.* **2015**, *159*. [CrossRef]
23. Frappart, F.; Fatras, C.; Mougin, E.; Marieu, V.; Diepkilé, A.T.; Blarel, F.; Borderies, P. Radar altimetry backscattering signatures at Ka, Ku, C, and S bands over West Africa. *Phys. Chem. Earth* **2015**, *83–84*. [CrossRef]
24. Bonnefond, P.; Verron, J.; Aublanc, J.; Babu, K.; Bergé-Nguyen, M.; Cancet, M.; Chaudhary, A.; Crétaux, J.-F.; Frappart, F.; Haines, B.; et al. The Benefits of the Ka-Band as Evidenced from the SARAL/AltiKa Altimetric Mission: Quality Assessment and Unique Characteristics of AltiKa Data. *Remote Sens.* **2018**, *10*, 83. [CrossRef]
25. Uebbing, B.; Forootan, E.; Braakmann-Folgmann, A.; Kusche, J. Inverting surface soil moisture information from satellite altimetry over arid and semi-arid regions. *Remote Sens. Environ.* **2017**, *196*, 205–223. [CrossRef]
26. Fatras, C.; Borderies, P.; Baghdadi, N.; Zribi, M.; El Hajj, M.; Frappart, F.; Mougin, E. Radar Backscattering Coefficient over Bare Soils at Ka-Band Close to Nadir Angle. *IEEE Geosci. Remote Sens. Lett.* **2016**, *13*, 1290–1294. [CrossRef]
27. Hallikainen, M.; Ulaby, F.; Dobson, M.; El-Rayes, M.; Wu, L. Microwave Dielectric Behavior of Wet Soil-Part 1: Empirical Models and Experimental Observations. *IEEE Trans. Geosci. Remote Sens.* **1985**, *23*, 25–34. [CrossRef]
28. Dobson, M.; Ulaby, F.; Hallikainen, M.; El-rayes, M. Microwave Dielectric Behavior of Wet Soil-Part II: Dielectric Mixing Models. *IEEE Trans. Geosci. Remote Sens.* **1985**, *23*, 35–46. [CrossRef]
29. Malmberg, C.G.; Maryott, A.A. Dielectric constant of water from 0 to 1000 °C. *J. Res. Natl. Bur. Stand.* **1956**, *56*, 1–8. [CrossRef]
30. Nashashibi, A.; Ulaby, F.T.; Sarabandi, K. Measurement and modeling of the millimeter-wave backscatter response of soil surfaces. *IEEE Trans. Geosci. Remote Sens.* **1996**, *34*, 561–572. [CrossRef]
31. Ulaby, F.T. *Radar Polarimetry for Geoscience Applications*; Fawwaz, T., Elachi, C., Eds.; Artech House: Norwood, MA, USA, 1990; ISBN 0890064067.

32. Ruck, G.T.; Barrick, D.E.; Stuart, W.D.; Krichbaum, C.K. *Radar Cross Section Handbook*; Plenum Press: New York, NY, USA, 1970; Volume 1.

33. Chelton, D.B.; Ries, J.C.; Haines, B.J.; Fu, L.-L.; Callahan, P.S. Chapter 1 Satellite Altimetry. In *Satellite Altimetry and Earth Sciences: A Handbook of Techniques and Applications*; Elsevier: New York, NY, USA, 2001; Volume 69, 131p, ISBN 0074-6142.

34. Biancamaria, S.; Frappart, F.; Leleu, A.-S.; Marieu, V.; Blumstein, D.; Desjonquères, J.-D.; Boy, F.; Sottolichio, A.; Valle-Levinson, A. Satellite radar altimetry water elevations performance over a 200 m wide river: Evaluation over the Garonne River. *Adv. Space Res.* **2017**, *59*, 128–146. [CrossRef]

35. CTOH (Center for Topographic studies of the Ocean and Hydrosphere). Available online: http://ctoh.legos.obs-mip.fr (accessed on 9 April 2018).

36. Zelli, C. ENVISAT RA-2 advanced radar altimeter: Instrument design and pre-launch performance assessment review. *Acta Astronaut.* **1999**, *44*, 323–333. [CrossRef]

37. Tachikawa, T.; Hato, M.; Kaku, M.; Iwasaki, A. The characteristics of ASTER GDEM version 2. In Proceedings of the IEEE International Geoscience and Remote Sensing Symposium, Vancouver, BC, Canada, 24–29 July 2011. [CrossRef]

38. Steunou, N.; Desjonquères, J.D.; Picot, N.; Sengenes, P.; Noubel, J.; Poisson, J.C. AltiKa Altimeter: Instrument Description and In Flight Performance. *Mar. Geod.* **2015**, *38*, 22–42. [CrossRef]

39. Mougin, E.; Hiernaux, P.; Kergoat, L.; Grippa, M.; de Rosnay, P.; Timouk, F.; Le Dantec, V.; Demarez, V.; Lavenu, F.; Arjounin, M.; et al. The AMMA-CATCH Gourma observatory site in Mali: Relating climatic variations to changes in vegetation, surface hydrology, fluxes and natural resources. *J. Hydrol.* **2009**, *375*. [CrossRef]

40. Frappart, F.; Hiernaux, P.; Guichard, F.; Mougin, E.; Kergoat, L.; Arjounin, M.; Lavenu, F.; Koité, M.; Paturel, J.-E.; Lebel, T. Rainfall regime across the Sahel band in the Gourma region, Mali. *J. Hydrol.* **2009**, *375*, 128–142. [CrossRef]

41. Wingham, D.J.; Rapley, C.G.; Griffiths, H. New Techniques in Satellite Altimeter Tracking Systems. In Proceedings of the 1986 International Geoscience and Remote Sensing Symposium on Remote Sensing, Zurich, Switzerland, 8–11 September 1986; pp. 1339–1344.

42. Frappart, F.; Calmant, S.; Cauhopé, M.; Seyler, F.; Cazenave, A. Preliminary results of ENVISAT RA-2-derived water levels validation over the Amazon basin. *Remote Sens. Environ.* **2006**, *100*. [CrossRef]

43. Boisot, O.; Pioch, S.; Fatras, C.; Caulliez, G.; Bringer, A.; Borderies, P.; Lalaurie, J.-C.; Guérin, C.-A. Ka-band backscattering from water surface at small incidence: A wind-wave tank study. *J. Geophys. Res. Oceans* **2015**, *120*, 3261–3285. [CrossRef]

remote sensing

MDPI

Article

On the Desiccation of the South Aral Sea Observed from Spaceborne Missions

Alka Singh [1],*, Ali Behrangi [1,2], Joshua B. Fisher [1] and John T. Reager [1]

[1]   Jet Propulsion Laboratory, California Institute of Technology, Pasadena, CA 91109, USA;
      ali.behrangi@jpl.nasa.gov (A.B.); Joshua.B.Fisher@jpl.nasa.gov (J.B.F.); John.Reager@jpl.nasa.gov (J.T.R.);
      alka.singh@jpl.nasa.gov (A.S.)
[2]   Department of hydrology and atmospheric sciences, The University of Arizona, Tucson, AZ 85721, USA;
      behrangi@email.arizona.edu
*   Correspondence: alka.singh@jpl.nasa.gov; Tel.: +1-818-354-4179

Received: 3 April 2018; Accepted: 17 May 2018; Published: 19 May 2018

Abstract: The South Aral Sea has been massively affected by the implementation of a mega-irrigation project in the region, but ground-based observations have monitored the Sea poorly. This study is a comprehensive analysis of the mass balance of the South Aral Sea and its basin, using multiple instruments from ground and space. We estimate lake volume, evaporation from the lake, and the Amu Darya streamflow into the lake using strengths offered by various remote-sensing data. We also diagnose the attribution behind the shrinking of the lake and its possible future fate. Terrestrial water storage (TWS) variations observed by the Gravity Recovery and Climate Experiment (GRACE) mission from the Aral Sea region can approximate water level of the East Aral Sea with good accuracy (1.8% normalized root mean square error (RMSE), and 0.9 correlation) against altimetry observations. Evaporation from the lake is back-calculated by integrating altimetry-based lake volume, in situ streamflow, and Global Precipitation Climatology Project (GPCP) precipitation. Different evapotranspiration (ET) products (Global Land Data Assimilation System (GLDAS), the Water Gap Hydrological Model (WGHM)), and Moderate-Resolution Imaging Spectroradiometer (MODIS) Global Evapotranspiration Project (MOD16) significantly underestimate the evaporation from the lake. However, another MODIS based Priestley-Taylor Jet Propulsion Laboratory (PT-JPL) ET estimate shows remarkably high consistency (0.76 correlation) with our estimate (based on the water-budget equation). Further, streamflow is approximated by integrating lake volume variation, PT-JPL ET, and GPCP datasets. In another approach, the deseasonalized GRACE signal from the Amu Darya basin was also found to approximate streamflow and predict extreme flow into the lake by one or two months. They can be used for water resource management in the Amu Darya delta. The spatiotemporal pattern in the Amu Darya basin shows that terrestrial water storage (TWS) in the central region (predominantly in the primary irrigation belt other than delta) has increased. This increase can be attributed to enhanced infiltration, as ET and vegetation index (i.e., normalized difference vegetation index (NDVI)) from the area has decreased. The additional infiltration might be an indication of worsening of the canal structures and leakage in the area. The study shows how altimetry, optical images, gravimetric and other ancillary observations can collectively help to study the desiccating Aral Sea and its basin. A similar method can be used to explore other desiccating lakes.

Keywords: lake level; lake volume; evaporation; streamflow; Gravity Recovery and Climate Experiment (GRACE); altimetry; Landsat; Aral Sea

1. Introduction

Lakes and reservoirs store 87% of the Earth's total fresh open water [1]. Unfortunately, many of them are gradually receding over the years due to climatic or/and anthropogenic forcings [2].

There are several feedbacks between the lake and its environment. Thus, quantifying the changes in the lake is critical to understanding the interactions among various components of the region better. For example, fluctuations in a lake/reservoir can be linked to different climatic changes at a regional scale, including desertification, dust storms, melting of glaciers, and changes in the vegetation and land types. For instance, lake volume loss reduces its heat capacity, and thus it can warm up and cool off faster than before.

One of the examples of massive lake volume reduction is the Aral Sea. In the 1960s, Soviet Russia started the world's second largest irrigation program, under which the Amu Darya and the Syr Darya rivers were diverted across the Karakum desert. The average annual combined pre-irrigation streamflow of the Amu Darya and the Syr Darya into the Aral Sea was 56 km^3 [3], which was reduced to less than 10 km^3 by 2002 [4]. The world's fourth-largest freshwater lake (until the 1960s) eventually separated into two parts, the vast South Aral Sea located in the south and the small North Aral Sea situated in the north. After the construction of Dike Kokaral dam between the north and south part of the Aral Sea, the Syr Darya streamflow has stabilized the North Aral Sea. However, in the past few decades, the perennial Amu Darya has transformed to an intermittent river as it runs through the desert and Khorezm oasis before merging into the South Aral Sea [5]. Consequently, the South Aral Sea continued its journey of desiccation and became a hypersaline and almost non-habitable lake. The Aral Sea is in the lowland climate zone [6], but studies have suggested that it may move towards a monsoon climate [4], which is characterized by seasonal climate change due to warming and cooling of the Aral Sea.

Hydrologic analysis of the Aral Sea and surrounding regions has been an active area of research in the last two decades [7–11]. The present study tries to advance previous studies by quantifying different hydrological parameters of the lake using various remote-sensing datasets that can complement each other. This is an essential step as in situ observations are limited and often not available. A comprehensive analysis of the dynamics of the South Aral Sea is performed using a mass balance equation as an example of how integrated multisensor data can be used to monitor different hydrological variables in an endorheic basin. This study demonstrates a framework of how a lake volume calculated from different remote-sensing data can act as a thermometer of the hydrological state of the basin. Lake volume variations are used to estimate evaporation loss from the lake, and to evaluate existing evaporation products, which are hard to validate otherwise. Furthermore, we estimated runoff from two different methods. Runoff is a product of hydrological processes acting in the watershed. To evaluate the cause of change in runoff, the saptio-temporal changes of the entire basin are analyzed.

2. Study Area

Here we study the South Aral Sea (Figure 1, blue polygon) and the Amu Darya basin (Figure 1, green polygon). The South Aral Sea is a remnant of the vast Aral Sea where the Amu Darya terminates. It has a shallow, broad east lobe and a deep, elongated west lobe, and a narrow channel connects them towards the north. The Amu Darya primarily runs through Turkmenistan and Uzbekistan covering an area of nearly 617,000 km^2. It receives water almost entirely from glaciers in the Pamir Mountains and the Hindu Kush and mainly originates from Tajikistan and northern Afghanistan, forming the border between the two countries. The high Pamir and Tian Shan ranges are significantly wet, creating massive glaciers and snowfields, and the Amu Darya streamflow brings the water to the severely arid southeast of the Aral Sea. The streamflow from these glaciers is heaviest during the spring thaw.

The Amu Darya basin mask is derived from the global river basin database obtained from http://www.wsag.unh.edu/Stn-30/stn-30.html [12,13]. The South Aral Sea land-water mask time series are generated from the multi-temporal Landsat dataset [14].

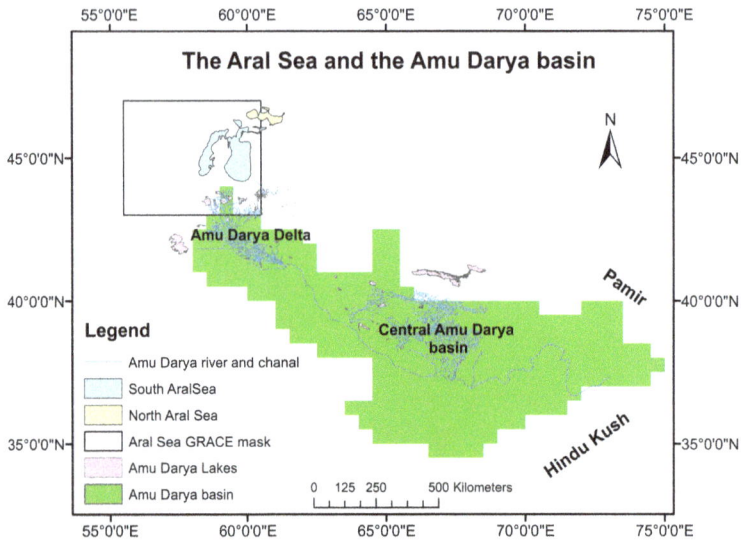

Figure 1. Study area: the Amu Darya originates from the glaciers of Hindukush and Pamir and terminates into the South Aral Sea (blue polygon).

3. Observation and Model Data

3.1. Lake Water Storage (LWS)

Remote-sensing-based lake-water-storage estimation can be done using altimetry and optical images. Since 1992, satellite altimetry is used successfully for monitoring the water levels of large rivers, lakes, and floodplains [15–17]. Satellite altimetry measures the water surface height along its pass based on the reflection from the ground of the emitted radiations from the satellite. Different missions have different pass location, repeat cycle, and footprints. However, there are significant inter-pass gaps, which lead to limited coverage of the terrestrial water bodies. For instance, the East Aral Sea could not have reliable altimetry observations for more than four years (2012–2016) as none of the altimetry mission's pass was over the dry lake. Singh et al. (2015) demonstrated the use of Landsat and bathymetry to obtain the water level, as an alternative to altimetry. In this study, a linear regression between altimetry water level and the Gravity Recovery and Climate Experiment (GRACE) terrestrial water storage (TWS) is examined as another alternative to estimate water level (discussed in Section 4.1).

Many studies intersected the digital elevation model (DEM) with the water level obtained from satellite altimetry to compute water storage changes of a lake/reservoir [18–20]. In this study also, the South Aral Sea volumetric variations are estimated by intersecting the Aral Sea bathymetry with the satellite altimetry water level time-series obtained from the Database for Hydrological Time Series of Inland Waters (DAHITI), developed by the Technische Universität München (http://dahiti.dgfi.tum.de/en/) [21]. Jason-1, Jason-2, and Jason-3, (10 days repeat pass) observe the East Aral Sea, while the West Aral Sea has additional observations from Envisat and Saral/Altika (35 days repeat pass).

Multiple Landsat missions' (Thematic Mapper (TM), Enhanced Thematic Mapper Plus (ETM+), and Operational Land Imager (OLI)) near-infrared (NIR) images are mosaiced and classified to generate land-water mask time series. Landsat bathymetry intersection is used to create water level time-series (Singh et al. 2015).

3.2. Terrestrial Water Storage (TWS)

Since 2002, variations in the terrestrial water storage within the Earth's system is routinely observed by GRACE [22–24]. The availability of monthly observations from the time variable GRACE mission has revolutionized the hydrological estimations by providing a possibility for the closure of water budget over a study area. In this study, 0.5°-gridded, JPL mascon solutions are downloaded from https://podaac.jpl.nasa.gov [25–27]. The mascon cell size varies with latitude, and for this region, it is $4° \times 5°$. Thus the GRACE signal includes mass changes occurring in the large extended area ($4° \times 5°$) as shown in Figure 1 (black box). The study uses GRACE signals from the lake region (Figure 1, black box) to estimate its water level. However, the lake area is less than one-tenth of the area observed by GRACE, but lake volume is a major driver in the mass variation of the region. Additionally, the GRACE signal from the Amu Darya basin is also analyzed to examine the spatiotemporal pattern of the basin and to estimate long-term streamflow into the East Aral Sea. For the analysis of the Amu Darya basin, the GRACE mascon solution is scaled to sub mascon 0.5° resolution by multiplying it with 0.5° gain factor obtained from https://podaac.jpl.nasa.gov. These gain factors derived from the Community Land Model do not have a lake component. Therefore, we can not use the scaled mascon product for the lake analysis, and the mass conservation within a mascon unit is more reliable than in 0.5° sub-mascon level.

3.3. Vegetation Index

The global Moderate-Resolution Imaging Spectroradiometer (MODIS) based normalized difference vegetation index (NDVI) product (MYD13C2) is retrieved from the National Aeronautics and Space Administration (NASA) Land Processes Distributed Active Archive Center (LP DAAC) https://lpdaac.usgs.gov/ [28].

3.4. Evapotranspiration (ET)

Since 2002, the MODIS satellite has been providing a wide range of information about global dynamics at 250 m to 1 km spatial resolution. The spectral signatures captured by its 36 spectral bands are used for many applications such as optical imaging of the Earth, estimation of radiation budget, and calculation of vegetation indices. The study investigated two MODIS derived ET products: MOD16 and Priestley-Taylor Jet Propulsion Laboratory (PT-JPL).

Global monthly 0.5° MOD16 ET product is downloaded from http://ntsg.umt.edu/project/modis/mod16.php [29,30]. The MOD16 algorithm [31] is based on the Penman–Monteith (1965) equation for ET.

The PT-JPL actual evapotranspiration (AET) product [32] is based on the Priestley-Taylor potential evapotranspiration (PET) formulation. To reduce PET to AET, Fisher et al. introduced ecophysiological constraint functions based on atmospheric moisture (vapor pressure deficit and relative humidity) and vegetation indices (NDVI and soil-adjusted vegetation index). The driving equation in the model is the following:

$$AET = ET_s + ET_c + ET_i \qquad (1)$$

where ET_s, ET_c, and ET_i are evaporation from the soil, canopy and intercepted water, respectively. Each is calculated explicitly based on relative surface wetness, green canopy fraction, plant temperature constraint, plant moisture constraint and soil moisture constraint. No calibration or site-specific parameters are required for this approach.

In addition to MODIS-derived ET, we have also evaluated ET from global hydrological models (GHMs). Monthly ET estimates from the Water Gap Hydrological Model (WGHM) obtained by personal contact and the National Oceanic, and Atmospheric Administration's (NOAA) Global Land Data Assimilation System (GLDAS) is retrieved from the http://disc.sci.gsfc.nasa.gov/ [33,34].

3.5. In Situ Data

The Aral Sea was one of the well-sampled inland water bodies on the planet until the 1980s. However, following the collapse of the USSR, field research into its advanced stages of desiccation has reduced significantly [4]. The Amu Darya monthly streamflow (2003–2010) and the historical annual hydrological data of the lake (1780–2009) are obtained from the http://www.cawater-info. net/ database. The Aral Sea bathymetry (from the 1960s) at 1 m contour spacing is provided by Prof. Renard [35] by personal communication. The East Aral Sea section of the bathymetry is updated to 30 m spatial resolution by Singh et al. [36] by combining Landsat and altimetry, and it is publicly available. Therefore, in this study, the West Aral Sea analysis is done using 1960s bathymetry while the East Aral Sea analysis used the updated bathymetry of Singh et al.

3.6. Precipitation

Monthly precipitation estimates are obtained from the latest Global Precipitation Climatology Center (GPCC V6) [37], and Global Precipitation Climatology Project (GPCP V2.3), and 1° daily data are retrieved from https://www.esrl.noaa.gov/ [38,39].

All datasets (except streamflow, Landsat, and altimetry) are harmonized to monthly, mm level water height on 0.5° (180° meridian) grid. Landsat-based 30 m resolution masks of the East and West Aral Sea are resampled to a 0.5° grid. The land-water mask time series is interested in the respective monthly evaporation and precipitation data to estimate the volumetric variations in evaporation and precipitation only from the water body.

4. Methods and Results

In this section, we demonstrate how remote-sensing observations are used to quantify the water budget components of the Aral Sea and Amu Darya basin. The retrieval method, challenges, solutions, and results are discussed as follows:

4.1. South Aral Sea Volume Dynamics

For hundreds of years (historical records began in the 1780s) until the 1960s, the Aral Sea level fluctuation was less than 5 m (Figure 2a). However, predominantly due to the massive irrigation project in the 1960s, it receded in shape and size dramatically. By 1985, the small North Aral Sea separated from the vast South Aral Sea. The North Aral Sea went through some fluctuations as seen in Figure 2a (red plot) due to many failed dam-building attempts between the two parts of the Aral Sea. It eventually stabilized after the construction of dike Kokaral dam in 2005 (Figure 2a, magenta plot). However, the South Aral Sea continued to shrink and further separated into the shallow East Aral Sea and the deep West Aral Sea by 2003 [9].

The annual water level of the South Aral Sea was estimated by a set of altimetry sensors (i.e., Jason1, Jason2, Envisat and Saral/altika) and compared with the available in situ water level. The results show a good agreement (Figure 2a) with more than 0.99 correlation and ~60 cm RMSE (root mean square error) or 1.8% normalized RMSE (by the mean of the observed data). Figure 2b (red and blue plots) shows that East and West Aral Sea had similar water level until 2009 and they were disconnected for the first time in late 2009. However, the 2010 flood resumed the equipotential status between the two parts of the South Aral Sea for nearly a year. Afterward, they started their independent paths of desiccation. The reason behind their different progression since 2010 has been that the East Aral Sea reached approximately its lake bed later in 2009, so it needs to rise for about 2 m to get reconnected with the West Aral Sea. The West Aral gets water only from the East Aral overflow; there was a 3–4 months lag between the increase in water level in 2010 among the two parts. Meanwhile, the West Aral continuously undergoes evaporation loss, as its lakebed is 13 m below mean sea level (MSL), while the flat East Aral dries up at 27 m.

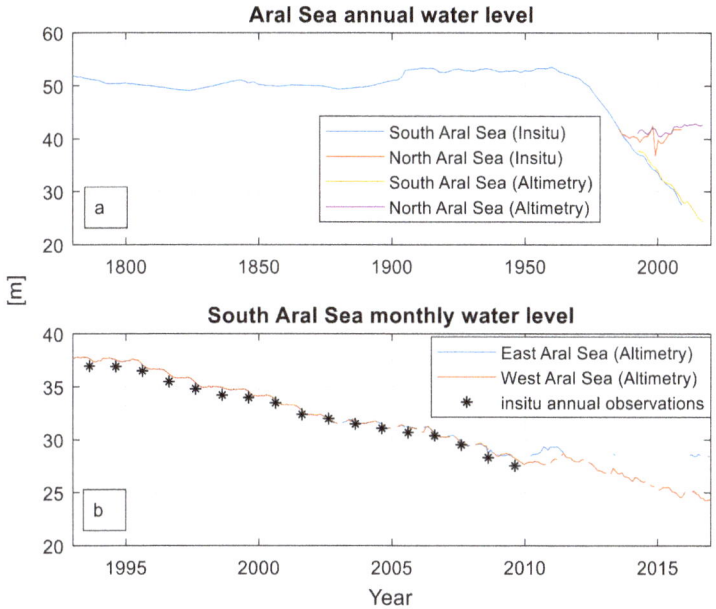

Figure 2. Water level above mean sea level (**a**) historical annual Aral Sea data (**b**) altimetry-based monthly South Aral Sea observations.

Satellite altimetry missions have continuously observed the West Aral Sea. However, the East Aral Sea lacks reliable observations after its water level went below 28.5 m (Figure 2b, red plot) because none of the altimetry missions has passed through the dry lake. Therefore, other remote-sensing observations have to be explored to explain the missing water level for nearly four years from 2011. Singh et al. (2015) demonstrated that Landsat together with bathymetry can provide 0.97 correlation with the altimetry water level and can act as a good alternative. However, optical images fail in bad weather (dust storm, and cloud). Consequently, observations from Landsat are not available for many months, especially during winters over the Aral Sea (Figure 3b). Therefore, we examined another potential way to fill the gaps by linear regression between GRACE-based terrestrial water storage (TWS) from the Aral Sea region (Figure 1). Study area black box) and the East Aral Sea water level from altimetry. First, seasonal components from the TWS and water level time series are removed. Then deseasonalized TWS and water level (Figure 3a) are linearly regressed using nine years (2003–2011) of monthly data. The best fit model ($r^2 = 0.82$, RMSE = 49 cm) is used for the estimation of the deseasonalized water level for the 2003–2017 time frame, and then the altimetry-based seasonal component is added. The derived GRACE-based water level and altimetry observations showed 0.9 correlation and ~54 cm RMSE or 1.8% normalized RMSE. Figure 3b shows that GRACE-based estimates agreed well with the altimetry observations except for the 2010 flood. The GRACE-based estimate observed an early peak in spring 2010, while altimetry saw it in summer 2010. It can be attributed to the integral nature of the GRACE signal, which observes mass changes not only from the surface water but also from the soil moisture and groundwater.

Although the TWS variability in this mascon also includes signals from the West Aral Sea, it obtains water only from the East Aral Sea overflow. Therefore, the empirical relation between the increase in the TWS and water level of the East Aral Sea remains valid. The East Aral acquires a similar water surface area as the West Aral as soon as it floods due to its flat topography and eventually has similar evaporation loss progression. However, the impact of consecutive evaporation loss from the West Aral on TWS may lead to the overestimation of the water-level drop, but it soon reaches the cutoff as the East Aral dries up.

Therefore, the GRACE-derived sea level for the East Aral Sea stops at 27 m, as in late 2014, and it resumes with an increase in the TWS of the lake. Whenever the two parts of the South Aral are connected, then they have mostly similar progression but different water levels since 2010.

However, due to the limited spatial resolution of GRACE, the GRACE mask (Figure 1, black box) extends beyond the lake. Spring 2010 was relatively wet, (which increased the soil moisture in the entire $4° \times 5°$ box and the lake size was at its diminished level. It seems that whenever snow accumulation or soil moisture variation is more than lake mass change, TWS has less ability to estimate the water level. Furthermore, the GRACE mask also includes the Amu Darya delta, which absorbs a significant amount of the floodwater before it reaches the South Aral Sea (discussed in Section 4.2). Figure 3b also shows that in spring 2009, GRACE-based estimates were nearly 29.5 m, which cannot be correct because there is no altimetry observation for that period, suggesting that water level had to be below 28.5 m. In this case, the time series is interpolated to fill the data gaps.

Nevertheless, for rest of the time-series, TWS in the GRACE box is driven by the lake mass variation, and thus the GRACE-based water level shows good agreement with altimetry observations.

Figure 3b shows the complete disappearance of the East Aral Sea in 2014, consistent with the analysis of the Landsat data (Singh et al. 2016). The East Aral Sea bed is approximately 27 m above MSL, and when its water level reaches 28.5 m from MSL, the lake expands enough to be observed by altimetry missions. Therefore, to obtain a complete time series, the water level is estimated from a combination of altimetry (when the water level is above 28.5 m) and Landsat (when the water level is below 28.5 m). GRACE-based estimates are used when neither altimetry nor Landsat could is available.

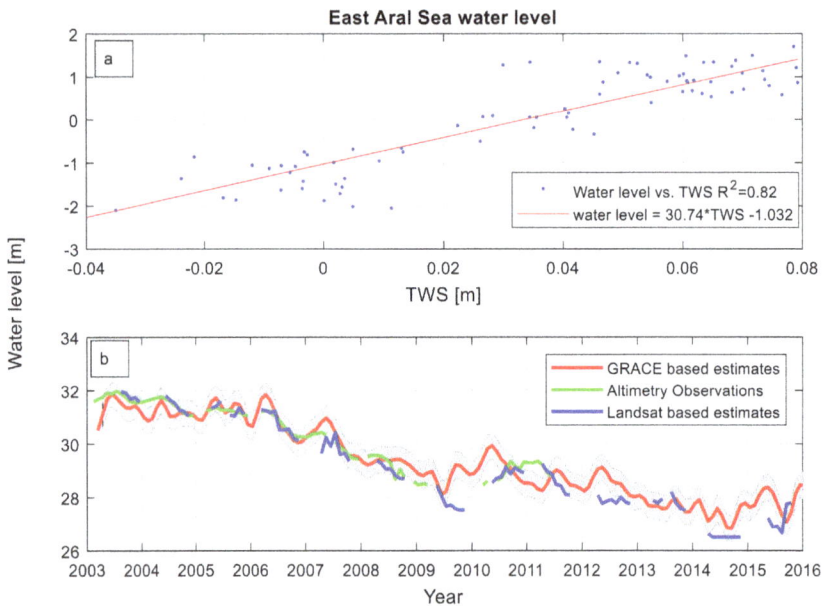

Figure 3. The East Aral Sea water level. (**a**) Best fit between de-seasonalized terrestrial water storage (TWS) and altimetry water level; and (**b**) altimetry water level observations compared with the derived water height from Landsat and Gravity Recovery and Climate Experiment (GRACE). The gray area shows the uncertainty range of the GRACE-based estimate, calculated by the ± root mean square error (RMSE).

In this study, we have three independently generated datasets of lake structure. Lake level is estimated by satellite altimetry, lake water-surface-area or contours are estimated by the Landsat imageries and bathymetry of the lake derived from 1960s field observations. A combination of any of the two can calculate the third, for example, the altimetry and Landsat/MODIS combination can generate the bathymetry of a lake [36,40] or estimate lake volume from [18,41]. In this work, the bathymetry of the East and West Aral Sea are intersected with their respective water levels to obtain water surface area time-series due to data gaps in the Landsat observations (Figure 3b). Volumetric variations of the lakes are calculated using a truncated pyramid model (Equation (2)) by integrating the change in water level and surface area [18].

$$VV(t) = \sum_{t=1}^{n} \frac{1}{3} \times (H_t - H_{t-1}) \times \left(A_t + A_{t-1} + \sqrt{(A_t \times A_{t-1})} \right) \tag{2}$$

where,

$VV(t)$ = Volumetric variations with respect to the initial state (t_0) at the nth month

A_t = Area of the water extent at month t

A_{t-1} = Area of the water extent at the previous month

H_t = Level of the water body at month t

H_{t-1} = Level of the water body at the previous month.

n = Number of months.

Between 1993 and 2017, the South Aral Sea lost approximately 195 km^3 water (Figure 4a). Since 2009, the East Aral Sea has been fluctuating on its almost flat lakebed. Eventually, most of the streamflow evaporates within a year. The West Aral Sea seasonally receives some water when the East Aral Sea water level exceeds 28.5 m above MSL. However, due to evaporation loss, the West Aral Sea was still losing water at approx. 2.7 km^3/annum rate and reduced below 25 m by late 2016 (Figure 2b). Considering the geometry of the West Aral Sea (Figure 4b), nearly 43 km^3 of its volume might still exist below the 25 m water level. If the current trend of water loss (almost 2.7 km^3/year) continues, then the West Aral Sea might disappear by nearly 2032. The seasonal expansion and shrinking of the East Aral Sea are neither enough to revive the east part nor maintain the West Aral Sea due to massive evaporation loss. The South Aral Sea dynamics are driven by the Amu Darya streamflow and evaporation loss from the lake. Therefore, we further explored these two parameters.

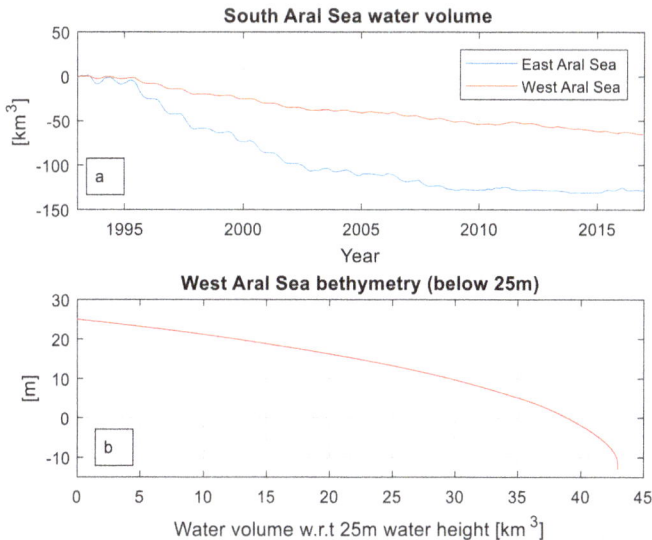

Figure 4. (a) the South Aral Sea volume variations and (b) the West Aral Sea bathymetry below 25 m MSL.

4.2. Evaporation from the South Aral Sea

The Aral Sea is a terminal inland water body, and evaporation is the only major water outlet. The annual evaporation rate of the lake was nearly 66 km^3 [9] before the 1960s. The historical vertical flux data from the Aral Sea shows a substantial difference between the total annual precipitation and evaporation to/from the waterbody (Figure 5a). However, the recent annual fluxes from the lake shown in Figure 5b demonstrate that MOD16 ET is almost equal to GPCP precipitation. It cannot be realistic, knowing the fact that the lake is shrinking and it has an additional Amu Darya streamflow into it.

Figure 5b shows that the difference between annual AET and PET has decreased due to accelerated AET with the decreasing lake size. The positive feedback between sea surface temperature (SST) and evaporation can explain the increase in the rate of AET from the Aral Sea. As the lake loses water, it not only reduces water surface area but also becomes shallower. Shallow water like the shrunken East Aral Sea (especially since 2009) heats up faster than deep water as it has less volume per square area. With decreasing lake area, specific humidity near the lake surface is decreased, and thus the rate of evaporation is increased. Another distinguishing factor of the East Aral Sea is its increasing salinity, which is above 130 ppm [10]. The salinization of the lake leads to vertical stratification, which is characterized by a rapid change in water temperature and salinity level at a given horizontal or vertical region. Consequently, the surface of the lake heats up faster as the salt concentration is not distributed evenly but stratified from lower salt concentration at the surface to the highest at the bottom of the lake [42].

Considering the significance of the evaporation for the water balance of the Aral Sea, we compared potential ET from PT-JPL, and actual ET estimates from WGHM, GLDAS, MOD16, and PT-JPL for the South Aral Sea.

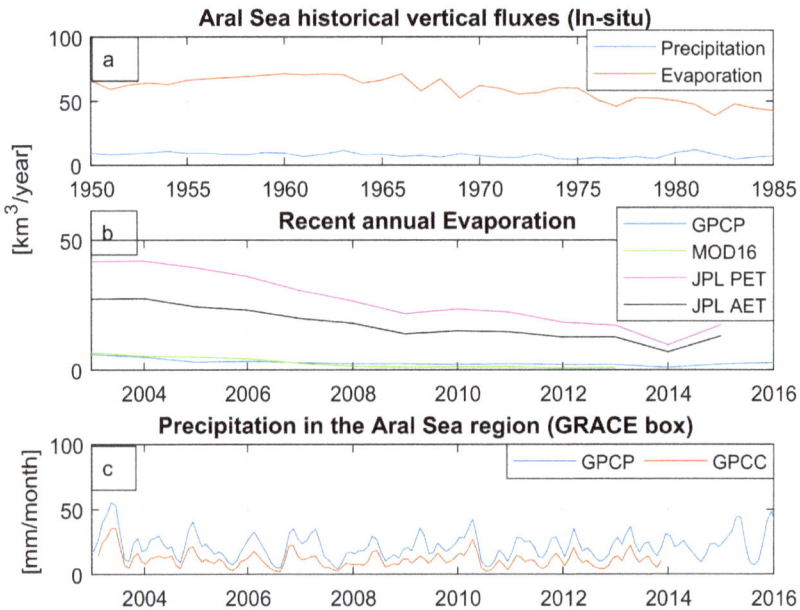

Figure 5. Time series of the South Aral Sea precipitation and evaporation; (**a**) historical annual in-situ observations; (**b**) recent annual evapotranspiration (ET) and Global Precipitation Climatology Project (GPCP) observations; and (**c**) monthly precipitation observations over the lake.

Evaporation from open water is considerably different from land. Terrestrial open water evaporation dynamics needs to be added in the current evaporation estimates taking into account the land-water conversion. The presence of a lake and its size have a significant impact on its surrounding climate. Total evaporation decreases with the reducing size of the lake, but most of the ET models do not use a dynamic water mask. Consequently, the impact of available water in the system can be miscalculated. In this study, the time-variable Landsat-based water-surface masks are intersected with all the ET estimates to calculate volumetric evaporation loss from the waterbody.

As an independent estimate of evaporation, we back-calculated evaporation from the lake using lake volume dynamics (Figure 4a) and based on the water balance equation (Equation (3)):

$$BCE = \Delta(VV(t)) + R - P \tag{3}$$

where,

BCE = Back-calculated evaporation from the lake (magenta plot in Figure 6)
$\Delta(VV(t))$ = Diffrential of the lake volume (calculated by Equation (2)) with respect to its previous month
R = Amu Darya streamflow into the South Aral Sea
P = Precipitation

This alternative and independent approach to calculate the lake evaporation provides an additional tool for evaluating the evaporation products. However, groundwater infiltration into the lake has been ignored in this estimation due to non-availability of the data. The lake volume estimates have limited uncertainty compared to other hydrological parameters because of the well-established water level estimation methods from satellite altimetry with cm-level accuracy [21]. However, the bathymetry of the lake is more than half a century old, and it has experienced some restructuring due to floods [18]. This may introduce an unknown uncertainty (likely to be small) in the estimation of the lake volume. Nevertheless, the estimated lake volume variations can be considered as a strong constraint in the water balance equation.

Additional uncertainty in back-calculated evaporation (BCE) comes from the lack of the exact coordinate of the streamflow gauge, which measures Amu Darya streamflow into the Aral Sea. The Amu Darya terminates in a massive delta with many small lakes. During the low/normal flow, water reaches directly into the East Aral Sea. However, during heavy streamflow (2005 and 2010) a significant amount of water is consumed in the delta region. Therefore, except during the flood months, the BCE can be considered as the nearest approximation of the evaporation loss from the South Aral Sea. The BCE estimates (Figure 6, magenta plot) are limited by the availability of the streamflow data (until 2010).

Precipitation is relatively a smaller direct input in the region with 80–200 mm annual range [43]. GPCP observations are closer to this range compared to GPCC (Figure 5c). Therefore, for the back-calculation of evaporation from the lake (Equation (3)), GPCP is used. Furthermore, GPCP utilizes the nearest rain-gauge stations to reduce potential biases. Figure 6 shows that most of the products severely underestimate ET compared to the BCE estimation. The PT-JPL ET, however, indicates notably different and reasonable agreement with the BCE.

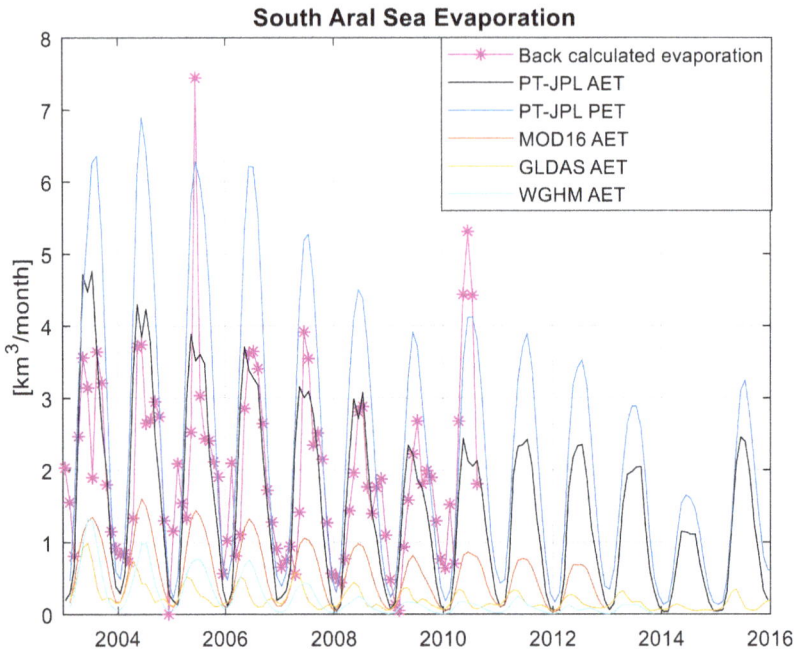

Figure 6. Evaporation from the South Aral Sea back-calculated evaporation (BCE) compared with the other ET products.

The BCE shows an even better agreement with the PT-JPL AET given a month's time lag because the BCE is estimated based on changes in the lake volume compared to its previous month. Except for the flooding months, the PT-JPL AET shows 0.83 correlation with the BCE and 0.76 cubic km RMSE (42% NRMSE). Theoretically, AET should be similar to PET over open water, but we found that they differ because a lot of the radiation/energy to drive ET gets absorbed deep into the water without actually ever evaporating any water. Therefore, the actual ET ends up being less than the potential. The MOD16 and GHMs ET showed very low seasonality in the lake volume estimation. Thus, our analysis indicates that PT-JPL ET is more consistent with the rest of the water cycle component for the evaluation of evaporation over the open water. PT-JPL has been widely and independently assessed as the top-performing global ET remote-sensing algorithm for evaporation [44–51]. However, this paper is the first to compare it with other ET products over open water. This shows that PT-JPL probably has a good representation of surface-atmosphere moisture coupling.

4.3. Amu Darya Streamflow into the Lake

As global lakes are not always well instrumented, this analysis provides insights on how well streamflow might be estimated using other water-cycle components. The study attempted to determine the Amu Darya streamflow into the South Aral Sea by two independent methods.

- A water balance-based streamflow estimate (R1, Figure 7b, green plot) is generated by combining PT-JPL ET (assuming it as actual evaporation from the lake), GPCP and South Aral Sea volumetric variations (Equation (4)). The average annual Amu Darya streamflow into the lake (except 2005 and 2010 flow) ranges between 0–1 km^3/month while the accumulated error from different datasets in Equation (4) is more than one km^3/month. Consequently, accurate estimation of the streamflow is not possible with this method. Therefore, three-monthly weighted-average (3MWA) by 0.25, 0.5, 0.25 weights, is calculated to obtain a long-term trend of the streamflow into the lake. The derived estimate (R1, Figure 7b) showed 0.71 correlation with the in situ 3MWA streamflow.

$$R1 = 3MWA \ (\Delta(VV) - P + ET) \tag{4}$$

where,

R1 = Streamflow estimated from lake water budget (green plot in Figure 7b)
3MWA = three-monthly weighted-average
ET = Evaporation from the lake (PT-JPL ET) and P = Precipitation (GPCP)

- Second streamflow (R2, Figure 7b, red plot) is calculated from the deseasonalized GRACE signal obtained from the Amu Darya basin (DGADB) (Figure 1, green polygon). An empirical relation between 3MWA of the in-situ Amu Darya streamflow and 3MWA of the DGADB is used to generate GRACE-based streamflow (R2). The Least-absolute-residuals method based two-degree polynomial curve showed a good agreement (r^2 = 0.94 and RMSE = 0.2 km^3) between the two. The derived curve (R2, Figure 7b) showed 0.68 correlation with the in situ 3MWA streamflow.

Furthermore, the dashed vertical lines in Figure 7a shows that the DGADB observes flood peaks into the lake about two months earlier than the in situ data (compare red and blue dashed lines). Reager et al. [52] also demonstrated that basin-scale TWS could be used to characterize regional flood potential with longer lead times in flood warnings. This two months' early warning of heavy streamflow into the Amu Darya delta could be useful for the water resource management of the region.

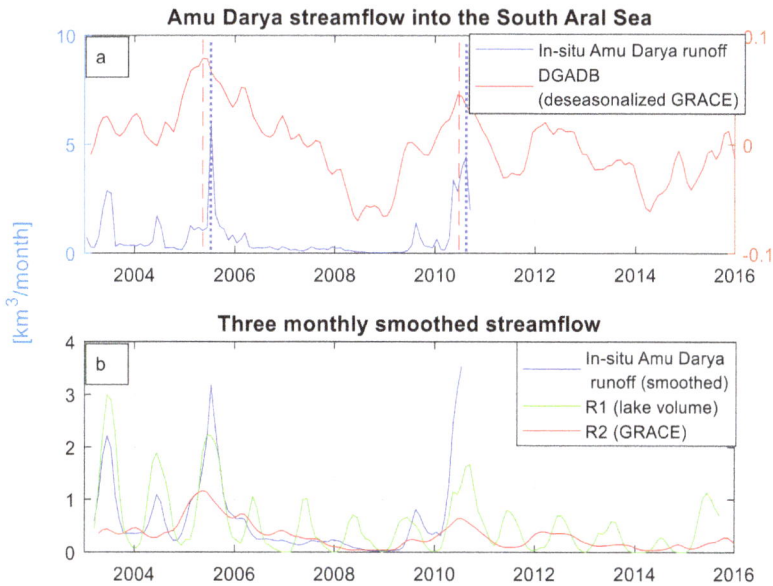

Figure 7. The Amu Darya streamflow. (**a**) Deseasonalized GRACE signal from the Amu Darya basin (DGADB) compared with the in-situ streamflow. The red vertical lines are peaks of DGADB, and the blue vertical lines are peaks of in situ streamflow; (**b**) three monthly-smoothed Amu Darya streamflow observed by in situ data compared with the two derived estimates (R1 and R2).

As discussed above, the streamflow during the flood events (2005 and 2010) was significantly absorbed in the massive Amu Darya delta. Therefore, the estimated R1 and R2 have a relatively smaller peak in 2005 and 2010 than in situ streamflow. Figure 7b suggests that R1 and R2 provide only a general idea of the long-term trend and are likely to be far from true streamflow. However, they have captured well the extreme weather events in the long-term patterns of the streamflow into the lake, i.e., 2005 and 2010 floods and 2009 and 2014 droughts. R2 produces low flow reasonably well (except in 2004), but underestimates peak flow, yet the pattern of the peak flow is captured. R1 is relatively skillful in capturing peak flows but overall provides many false signals not observed by either in situ observation or GRACE. The GRACE signal (DGADB, Figure 7a red plot) indicates that 2015 and 2016 had high seasonality, but they do not exceed 2005/2010 peak flows or 2009/2014 low flows.

4.4. The Amu Darya Basin

Figure 8 shows the spatiotemporal linear trend in the Amu Darya basin between 2003 and 2017 of (a) TWS from GRACE, (b) precipitation from GPCP, (c) MODIS NDVI and (d) ET from MOD16. The figure shows that during this period TWS went down in the Amu Darya delta and the southwestern Hindukush region. The decrease in the delta TWS can be attributed to the increasing absorption of water in the central and northeastern part of the Amu Darya basin. This area is a major irrigation belt (other than a delta region) of the basin. The figure shows that in the irrigation area water mass increased (Figure 8a), ET decreased (Figure 8d), and the precipitation trend remained relatively flat (Figure 8b), which indicates a possible increase in infiltration. Furthermore, a decrease in total annual Amu Darya streamflow into the Aral Sea from more than 10 km^3 in 2002 to less than 4 km^3 by 2009 shows that either precipitation has gone down during those years (which is not seen in GPCP dataset) or abstraction of water has increased upstream. On the other hand, the reduction of NDVI in the central part of the basin (Figure 8c) supports the assumption of increased infiltration loss. However, vegetation loss can also be attributed to increasing salt deposition in the soil profiles and desertification [8]. Forkutsa et al. [53] discussed the fact that more than 40% of the river water is lost to evaporation and infiltration from the canal. Water escapes the route to form lakes and ponds along the way. Consequently, the rise in groundwater level has brought soil salt to the surface, leading to widespread salinization [9]. It is probable that during this time frame the canal conditions are deteriorated, leading to more infiltration loss.

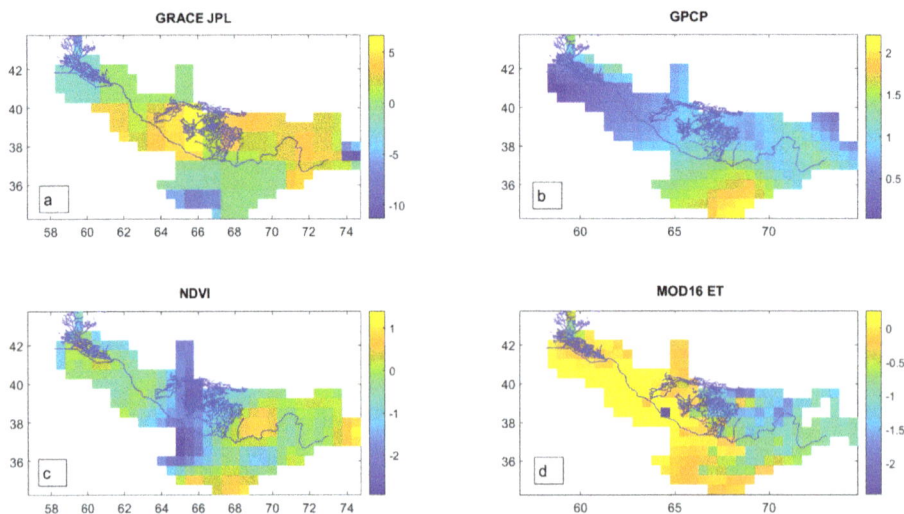

Figure 8. The Amu Darya basin: (**a**) GRACE TWS; (**b**) GPCP; (**c**) normalized difference vegetation index (NDVI); and (**d**) MODIS based ET (MOD16).

5. Discussion

Accurate monitoring of the world's lakes in a changing climate is increasingly essential, as lakes often significantly contribute to regional climatology and ecosystems. Several lakes are drying (e.g., Aral Sea, Urmia, Lake Chad) due to climate change and anthropogenic activities (e.g., through water abstractions for irrigation). These not only lead to loss of Lake Habitat and biodiversity but also can trigger several adverse impacts on the environment and life in the surrounding region, like desertification, dust storms, and the salinization of land. All these processes can eventually offset economic growth near the lakes. Similar processes can be seen in various global lakes like Lake Urmia [54,55], Lake Chad [56,57], the Dead Sea [58–60], and Lake Poopo [61,62]. Therefore, careful monitoring of the changes in lake/inland sea properties is becoming critical more than ever. Unfortunately, many of these lakes are not well instrumented. Thus exploiting the potential of integrating multi-dimensional remote-sensing data with other complementary data is essential. The following are the highlights of this research:

1. Lake level estimate: this paper suggests methods for filling gaps in the altimetry observations. These data gaps may occur due to intermission time lag or loss of altimetry ground track due to changes in the shape of the water bodies. Landsat images together with bathymetry can provide an alternative water level estimate. However, sometimes, optical images have limitations during lousy weather. In that case, GRACE signals from lakes like the Aral Sea have a potential to estimate water level. The linear regression between the TWS and water level has been explored to generate the water level from GRACE.

2. The rate of evaporation loss: most of the models/data products do not estimate evapotranspiration (ET) from inland waterbodies well, except for one. We have back-calculated the lake evaporation (BCE) by integrating altimetry-based lake volume variations, with the in situ runoff and GPCP precipitation. This study found the PT-JPL ET estimate to have the closest approximation to the BCE compared to the other existing ET products MODIS (MOD16) and hydrological models (WGHM and GLDAS). While PT-JPL has never been tested over open water bodies, our findings are consistent with multiple studies that have consistently found PT-JPL to be the top-performing ET remote-sensing algorithm over terrestrial vegetation [44,45,48,49,51,63].

3. Estimating river streamflow to the lake: the study also suggests that the GRACE signal from the Amu Darya basin can provide a long-term trend of streamflow into the lake and may predict flood events one or two months in advance. Another streamflow is estimated based on the lake water budget, which showed a good long-term progression but has some false highs. The back-calculated streamflow (R1) indicated strikingly high seasonality, which demonstrates possible seasonal groundwater infiltration into the lake, assuming error in other datasets are not seasonally biased. Nevertheless, in the absence of any in-situ streamflow, these methods can be explored.

4. Assessing the spatiotemporal variations in the water cycle of the Amu Darya basin: finally, we monitored the spatial changes of the Amu Darya basin to examine the cause of reducing streamflow. Various insights could be gained through analyzing the maps of a temporal trend in ET, TWS, NDVI, and Precipitation. The decrease in TWS in the Amu Darya delta region is mainly due to the increase in water mass in the central part of the Amu Darya basin, which is probably due to rising infiltration with the worsening of the canal system. This assumption cannot be validated due to lack of ground-based observations but is supported by the decrease in ET and NDVI in the region with the increase in TWS.

5. Future of the Aral Sea: the low Amu Darya streamflow and huge evaporation loss from the vast open body have endangered the existence of the South Aral Sea. If the present trend continues, the remnant West Aral Sea will also disappear by nearly 2032 or reach the level of its base flow. One possible solution is to drain the Amu Darya streamflow directly into the West Aral Sea to avoid evaporation loss from the vast shallow East Aral Sea. Assuming 4 km^3/year water flows

into the West Aral Sea based on the current the annual Amu Darya streamflow (without any flood), the West Aral Sea will start increasing at a rate of more than 1 km^3/year. Additionally, a dam is also required to be built between the East and West Aral Sea to stop flooding from the west when it reaches more than 28 m above MSL.

6. Concluding Remarks

In this paper, we show how the combination of various remote-sensing products can help gain more information on the fate of the Aral Sea lake. A framework that can be extended to study other endorheic lakes. The Aral Sea was well instrumented in the Soviet era, after which it gradually lost ground-based stations. However, some in situ records for streamflow to the lake (until 2010), and lake bathymetry (from the 1960s) are available. These enable us to assess current remote-sensing potential and determine challenging areas for further improvement. Our study over the Aral Sea mainly consists of using remote-sensing data to quantify changes in the lake level and volume, the rate of evaporation loss, estimating river streamflow to the lake, and assessing the spatiotemporal variations in the water cycle of both the lake and the Amu Darya basin. The study demonstrated that in addition to altimetry, optical and GRACE data can help to estimate the lake level. Also, GRACE may predict flood events one or two months in advance. Furthermore, lake evaporation was calculated using runoff, altimetry, and precipitation. In comparison to the other global ET products (for example WGHM, GLDAS, MOD16), PT-JPL ET showed the best approximation to the estimated ET for the open water. The Aral Sea basin analysis by the GRACE and other datasets suggests a rise in infiltration, potentially due to worsening of the canals.

Efforts are ongoing to monitor other world's inland lake. Given the lack of quality in situ observations, remote sensing has shown great potential. However, several areas require further improvement by the community, some of which are highlighted below:

- Higher spatial resolution GRACE signals can improve its application tremendously by reducing the impact of contributions from other hydrological compartments.
- Evaporation estimates from the waterbodies need to be better estimated. The lake's volume variations and its salinity need to be incorporated in the models.
- With the recent operation of Global Precipitation Measurements (GPM) and Soil Moisture Active Passive (SMAP) missions, precipitation, soil moisture is expected to be monitored better than before. The role of new observations in studies like that presented here needs to be further investigated.
- The upcoming Surface Water and Ocean Topography (SWOT) mission is expected to provide volumetric variations of most of the inland water bodies because of its wide swath altimetry. This can potentially advance water balance studies such as that investigated in this work.
- By increasing confidence in the quality of surface/sub-surface estimates (surface water and soil moisture), the role of groundwater dynamics can be better explored from GRACE.

The study demonstrated the potential of deriving some proxy estimates of the missing or inadequate hydrological variables by combining multi-dimensional, multi-sensor and multi-mission earth observation datasets to undertake a comprehensive analysis of the hydrological state of a basin.

Author Contributions: The study was conceptualized and investigated by A.S. She prepared the original draft and took care of modifications. A.B. is the principal investigator of the project. He consistently improved the work by important guidance, ideas, and effective discussions. J.B.F. contributed to the paper by providing the PT-JPL data and its analysis. A.B. and J.B.F. helped in the improvement of the original draft by thorough revisions. J.T.R. and A.B. provided funding for the project. All the co-authors assisted in the finalization of the manuscript.

Acknowledgments: Authors acknowledge different agencies for proving various data, including NASA Physical Oceanography Distributed Active Archive Center (PO.DAAC) for GRACE TWS, NASA Land Process Distributed Active Archive Center (LP DAAC) for NDVI, Numerical Terradynamic Simulation Group for MOD16, Cawater group for the Aral Sea in-situ data, DAHITI for altimetry data, NASA Goddard Earth Sciences Data and Information Services Center (GES DISC) for GLDAS ET, NOAA's Earth System Research Laboratory for GPCP

Remote Sens. **2018**, *10*, 793

and GPCC data. The authors are thankful to Prof. Andreas Guentner (GFZ, Potsdam) for proving WGHM data. Gregory Halverson processed the PT-JPL ET data. The research described in this paper was carried out at the Jet Propulsion Laboratory, California Institute of Technology, under a contract with the National Aeronautics and Space Administration. Financial support was also made available from NASA GRACE, and GRACE-FO (NNH15ZDA001NGRACE) and NASA Energy and Water Cycle Study, (NH13ZDA001N-NEWS) awards; J.B.F. was also supported by NASA SUSMAP. Government sponsorship is acknowledged.

Conflicts of Interest: The authors declare no conflict of interest. The founding sponsors had no role in the design of the study; in the collection, analyses, or interpretation of data; in the writing of the manuscript, and in the decision to publish the results.

References

1. Gleick, P.H.; Pacific Institute for Studies in Development, Environment; Security, Stockholm Environment Institute. *Water in Crisis: A Guide to the World's Fresh Water Resources*; Oxford University Press: New York, NY, USA, 1993; ISBN 978-0-19-507627-1.
2. Nicholson, S.E. Historical Fluctuations of Lake Victoria and Other Lakes in the Northern Rift Valley of East Africa. In *Environmental Change and Response in East African Lakes*; Lehman, J.T., Ed.; Springer Netherlands: Dordrecht, The Netherlands, 1998; Volume 79, pp. 7–35. ISBN 978-90-481-5043-4.
3. Bortnik, V.N.; Chistyaeva, S.P. Hydrometeorology and Hydrochemistry of the USSR Seas. *Aral Sea Leningr. Gidrometeoizdat* **1990**, *VII*, 196.
4. Zavialov, P.O. *Physical Oceanography of the Dying Aral Sea*; Springer Science & Business Media: Berlin, Germany, 2005; ISBN 978-3-540-22891-2.
5. Dodson, J.; Betts, A.V.G.; Amirov, S.S.; Yagodin, V.N. The nature of fluctuating lakes in the southern Amu-dar'ya delta. *Palaeogeogr. Palaeoclimatol. Palaeoecol.* **2015**, *437*, 63–73. [CrossRef]
6. Micklin, P.P. The Water Management Crisis in Soviet Central Asia. *Carl Beck Pap. Russ. East Eur. Stud.* **1991**, 131. [CrossRef]
7. Asokan, S.M.; Rogberg, P.; Bring, A.; Jarsjö, J.; Destouni, G. Climate model performance and change projection for freshwater fluxes: Comparison for irrigated areas in Central and South Asia. *J. Hydrol. Reg. Stud.* **2016**, *5*, 48–65. [CrossRef]
8. Bosch, K.; Erdinger, L.; Ingel, F.; Khussainova, S.; Utegenova, E.; Bresgen, N.; Eckl, P.M. Evaluation of the toxicological properties of ground- and surface-water samples from the Aral Sea Basin. *Sci. Total Environ.* **2007**, *374*, 43–50. [CrossRef] [PubMed]
9. Micklin, P. Introduction to the Aral Sea and Its Region. In *The Aral Sea*; Springer Earth System Sciences; Springer: Berlin/Heidelberg, Germany, 2014; pp. 15–40. ISBN 978-3-642-02355-2.
10. Roget, E.; Khimchenko, E.; Forcat, F.; Zavialov, P. The internal seiche field in the changing South Aral Sea (2006–2013). *Hydrol. Earth Syst. Sci.* **2017**, *21*, 1093–1105. [CrossRef]
11. Singh, A.; Seitz, F.; Eicker, A.; Güntner, A. Water Budget Analysis within the Surrounding of Prominent Lakes and Reservoirs from Multi-Sensor Earth Observation Data and Hydrological Models: Case Studies of the Aral Sea and Lake Mead. *Remote Sens.* **2016**, *8*, 953. [CrossRef]
12. Vörösmarty, C.J.; Fekete, B.M.; Meybeck, M.; Lammers, R.B. Geomorphometric attributes of the global system of rivers at 30-minute spatial resolution. *J. Hydrol.* **2000**, *237*, 17–39. [CrossRef]
13. Fekete, B.M.; Vörösmarty, C.J.; Lammers, R.B. Scaling gridded river networks for macroscale hydrology: Development, analysis, and control of error. *Water Resour. Res.* **2001**, *37*, 1955–1967. [CrossRef]
14. Singh, A.; Seitz, F.; Schwatke, C. Application of Multi-Sensor Satellite Data to Observe Water Storage Variations. *IEEE J. Sel. Top. Appl. Earth Obs. Remote Sens.* **2013**, *6*, 1502–1508. [CrossRef]
15. Birkett, C.; Murtugudde, R.; Allan, T. Indian Ocean Climate event brings floods to East Africa's lakes and the Sudd Marsh. *Geophys. Res. Lett.* **1999**, *26*, 1031–1034. [CrossRef]
16. Crétaux, J.-F.; Biancamaria, S.; Arsen, A.; Bergé-Nguyen, M.; Becker, M. Global surveys of reservoirs and lakes from satellites and regional application to the Syrdarya river basin. *Environ. Res. Lett.* **2015**, *10*, 15002. [CrossRef]
17. Kleinherenbrink, M.; Ditmar, P.G.; Lindenbergh, R.C. Retracking Cryosat data in the SARIn mode and robust lake level extraction. *Remote Sens. Environ.* **2014**, *152*, 38–50. [CrossRef]
18. Singh, A.; Kumar, U.; Seitz, F. Remote Sensing of Storage Fluctuations of Poorly Gauged Reservoirs and State Space Model (SSM)-Based Estimation. *Remote Sens.* **2015**, *7*, 17113–17134. [CrossRef]

19. Medina, C.E.; Gomez-Enri, J.; Alonso, J.J.; Villares, P. Water level fluctuations derived from ENVISAT Radar Altimeter (RA-2) and in-situ measurements in a subtropical waterbody: Lake Izabal (Guatemala). *Remote Sens. Environ.* **2008**, *112*, 3604–3617. [CrossRef]

20. Crétaux, J.-F.; Abarca-del-Río, R.; Bergé-Nguyen, M.; Arsen, A.; Drolon, V.; Clos, G.; Maisongrande, P. Lake Volume Monitoring from Space. *Surv. Geophys.* **2016**, *37*, 269–305. [CrossRef]

21. Schwatke, C.; Dettmering, D.; Bosch, W.; Seitz, F. DAHITI—An innovative approach for estimating water level time series over inland waters using multi-mission satellite altimetry. *Hydrol. Earth Syst. Sci.* **2015**, *19*, 4345–4364. [CrossRef]

22. Wahr, J.; Swenson, S.; Zlotnicki, V.; Velicogna, I. Time-variable gravity from GRACE: First results. *Geophys. Res. Lett.* **2004**, *31*, L11501. [CrossRef]

23. Rodell, M.; Famiglietti, J.S.; Chen, J.; Seneviratne, S.I.; Viterbo, P.; Holl, S.; Wilson, C.R. Basin scale estimates of evapotranspiration using GRACE and other observations. *Geophys. Res. Lett.* **2004**, *31*, L20504. [CrossRef]

24. Richey, A.S.; Thomas, B.F.; Lo, M.-H.; Reager, J.T.; Famiglietti, J.S.; Voss, K.; Swenson, S.; Rodell, M. Quantifying renewable groundwater stress with GRACE. *Water Resour. Res.* **2015**, *51*, 5217–5238. [CrossRef] [PubMed]

25. Watkins, M.M.; Wiese, D.N.; Yuan, D.-N.; Boening, C.; Landerer, F.W. Improved methods for observing Earth's time variable mass distribution with GRACE using spherical cap mascons. *J. Geophys. Res. Solid Earth* **2015**, *120*, 2648–2671. [CrossRef]

26. Wiese, D.N.; Landerer, F.W.; Watkins, M.M. Quantifying and reducing leakage errors in the JPL RL05M GRACE mascon solution. *Water Resour. Res.* **2016**, *52*, 7490–7502. [CrossRef]

27. Wiese, D.N.; Yuan, D.-N.; Boening, C.; Landerer, F.W.; Watkins, M.M. *JPL GRACE Mascon Ocean, Ice, and Hydrology Equivalent Water Height RL05M.1 CRI Filtered Version 2*, PO.DAAC. 2016. [CrossRef]

28. Didan, K. *MYD13C2 MODIS/Aqua Vegetation Indices Monthly L3 Global 0.05Deg CMG V006*; NASA: Washington, DC, USA, 2015.

29. Running, S.W.; Kimball, J.S. Satellite-Based Analysis of Ecological Controls for Land-Surface Evaporation Resistance. In *Encyclopedia of Hydrological Sciences*; Anderson, M.G., McDonnell, J.J., Eds.; John Wiley & Sons, Ltd.: Hoboken, NJ, USA, 2006; ISBN 978-0-470-84894-4. [CrossRef]

30. Mu, Q.; Heinsch, F.; Zhao, M.; Running, S. Development of a global evapotranspiration algorithm based on MODIS and global meteorology data. *Remote Sens. Environ.* **2007**, *111*, 519–536. [CrossRef]

31. Mu, Q.; Zhao, M.; Running, S.W. Improvements to a MODIS global terrestrial evapotranspiration algorithm. *Remote Sens. Environ.* **2011**, *115*, 1781–1800. [CrossRef]

32. Fisher, J.B.; Tu, K.P.; Baldocchi, D.D. Global estimates of the land–atmosphere water flux based on monthly AVHRR and ISLSCP-II data, validated at 16 FLUXNET sites. *Remote Sens. Environ.* **2008**, *112*, 901–919. [CrossRef]

33. Rodell, M. *GLDAS CLM Land Surface Model L4 Monthly 1.0 × 1.0 degree, Version 1*; Goddard Earth Sciences Data and Information Services Center (GES DISC): Greenbelt, MD, USA. [CrossRef]

34. Derber, J.C.; Parrish, D.F.; Lord, S.J. The New Global Operational Analysis System at the National Meteorological Center. *Weather Forecast.* **1991**, *6*, 538–547. [CrossRef]

35. Benduhn, F.; Renard, P. A dynamic model of the Aral Sea water and salt balance. *J. Mar. Syst.* **2004**, *47*, 35–50. [CrossRef]

36. Singh, A.; Seitz, F. Updated bathymetric chart of the East Aral Sea, supplement to: Singh, Alka; Kumar, Ujjwal; Seitz, Florian (2015): Remote sensing of storage fluctuations of poorly gauged reservoirs and State Space Model (SSM)-based estimation. *Remote Sens.* **2015**, *7*, 17113–17134. [CrossRef]

37. Schneider, U.; Becker, A.; Finger, P.; Meyer-Christoffer, A.; Rudolf, B.; Ziese, M. *GPCC Full Data Reanalysis Version 7.0 at 0.5°: Monthly Land-Surface Precipitation from Rain-Gauges built on GTS-based and Historic Data*; NCAR: Boulder, CO, USA, 2015. [CrossRef]

38. Adler, R.F.; Sapiano, M.; Huffman, G.J.; Bolvin, D.; Wang, J.-J.; Nelkin, E.; Xie, P.; Chiu, L.; Ferraro, R.; Schneider, U.; et al. New Global Precipitation Climatology Project monthly analysis product corrects satellite data drifts. *GEWEX News* **2016**, *26*, 7–9.

39. Adler, R.F.; Huffman, G.J.; Chang, A.; Ferraro, R.; Xie, P.-P.; Janowiak, J.; Rudolf, B.; Schneider, U.; Curtis, S.; Bolvin, D.; et al. The Version-2 Global Precipitation Climatology Project (GPCP) Monthly Precipitation Analysis (1979–Present). *J. Hydrometeorol.* **2003**, *4*, 1147–1167. [CrossRef]

40. Arsen, A.; Crétaux, J.-F.; Berge-Nguyen, M.; del Rio, R. Remote Sensing-Derived Bathymetry of Lake Poopó. *Remote Sens.* **2013**, *6*, 407–420. [CrossRef]

41. Gao, H.; Birkett, C.; Lettenmaier, D.P. Global monitoring of large reservoir storage from satellite remote sensing: Global monitoring of large reservoir storage from space. *Water Resour. Res.* **2012**, *48*, W09504. [CrossRef]

42. The Aral Sea Crisis. Available online: http://www.columbia.edu/~tmt2120/introduction.htm (accessed on 10 November 2017).

43. Issanova, G.; Abuduwaili, J.; Galayeva, O.; Semenov, O.; Bazarbayeva, T. Aeolian transportation of sand and dust in the Aral Sea region. *Int. J. Environ. Sci. Technol.* **2015**, *12*, 3213–3224. [CrossRef]

44. Chen, Y.; Xia, J.; Liang, S.; Feng, J.; Fisher, J.B.; Li, X.; Li, X.; Liu, S.; Ma, Z.; Miyata, A.; et al. Comparison of satellite-based evapotranspiration models over terrestrial ecosystems in China. *Remote Sens. Environ.* **2014**, *140*, 279–293. [CrossRef]

45. Ershadi, A.; McCabe, M.F.; Evans, J.P.; Chaney, N.W.; Wood, E.F. Multi-site evaluation of terrestrial evaporation models using FLUXNET data. *Agric. For. Meteorol.* **2014**, *187*, 46–61. [CrossRef]

46. Fisher, J.B.; Melton, F.; Middleton, E.; Hain, C.; Anderson, M.; Allen, R.; McCabe, M.F.; Hook, S.; Baldocchi, D.; Townsend, P.A.; et al. The future of evapotranspiration: Global requirements for ecosystem functioning, carbon and climate feedbacks, agricultural management, and water resources: The future of evapotranspiration. *Water Resour. Res.* **2017**, *53*, 2618–2626. [CrossRef]

47. Fisher, J.B.; Malhi, Y.; Bonal, D.; Da Rocha, H.R.; De Araãšjo, A.C.; Gamo, M.; Goulden, M.L.; Hirano, T.; Huete, A.R.; Kondo, H.; et al. The land–atmosphere water flux in the tropics. *Glob. Chang. Biol.* **2009**, *15*, 2694–2714. [CrossRef]

48. McCabe, M.F.; Ershadi, A.; Jimenez, C.; Miralles, D.G.; Michel, D.; Wood, E.F. The GEWEX LandFlux project: Evaluation of model evaporation using tower-based and globally gridded forcing data. *Geosci. Model Dev.* **2016**, *9*, 283–305. [CrossRef]

49. Miralles, D.G.; Jiménez, C.; Jung, M.; Michel, D.; Ershadi, A.; McCabe, M.F.; Hirschi, M.; Martens, B.; Dolman, A.J.; Fisher, J.B.; et al. The WACMOS-ET project – Part 2: Evaluation of global terrestrial evaporation data sets. *Hydrol. Earth Syst. Sci.* **2016**, *20*, 823–842. [CrossRef]

50. Polhamus, A.; Fisher, J.B.; Tu, K.P. What controls the error structure in evapotranspiration models? *Agric. For. Meteorol.* **2013**, *169*, 12–24. [CrossRef]

51. Vinukollu, R.K.; Wood, E.F.; Ferguson, C.R.; Fisher, J.B. Global estimates of evapotranspiration for climate studies using multi-sensor remote sensing data: Evaluation of three process-based approaches. *Remote Sens. Environ.* **2011**, *115*, 801–823. [CrossRef]

52. Reager, J.T.; Thomas, B.F.; Famiglietti, J.S. River basin flood potential inferred using GRACE gravity observations at several months lead time. *Nat. Geosci.* **2014**, *7*, 588–592. [CrossRef]

53. Forkutsa, I.; Sommer, R.; Shirokova, Y.I.; Lamers, J.P.A.; Kienzler, K.; Tischbein, B.; Martius, C.; Vlek, P.L.G. Modeling irrigated cotton with shallow groundwater in the Aral Sea Basin of Uzbekistan: I. Water dynamics. *Irrig. Sci.* **2009**, *27*, 331–346. [CrossRef]

54. Hassanzadeh, E.; Zarghami, M.; Hassanzadeh, Y. Determining the Main Factors in Declining the Urmia Lake Level by Using System Dynamics Modeling. *Water Resour. Manag.* **2012**, *26*, 129–145. [CrossRef]

55. Jeihouni, M.; Toomanian, A.; Alavipanah, S.K.; Hamzeh, S. Quantitative assessment of Urmia Lake water using spaceborne multisensor data and 3D modeling. *Environ. Monit. Assess.* **2017**, *189*, 572. [CrossRef] [PubMed]

56. Lemoalle, J.; Bader, J.-C.; Leblanc, M.; Sedick, A. Recent changes in Lake Chad: Observations, simulations and management options (1973–2011). *Glob. Planet. Chang.* **2012**, *80*, 247–254. [CrossRef]

57. Okpara, U.T.; Stringer, L.C.; Dougill, A.J. Lake drying and livelihood dynamics in Lake Chad: Unravelling the mechanisms, contexts and responses. *Ambio* **2016**, *45*, 781–795. [CrossRef] [PubMed]

58. Ali, W. Environment and Water Resources in the Jordan Valley and Its Impact on the Dead Sea Situation. In *Water Security in the Mediterranean Region*; Scozzari, A., El Mansouri, B., Eds.; Springer Netherlands: Dordrecht, The Netherlands, 2011; pp. 229–238. ISBN 978-94-007-1622-3.

59. Rawashdeh, S.A.; Ruzouq, R.; Al-Fugara, A.; Pradhan, B.; Ziad, S.H.A.-H.; Ghayda, A.R. Monitoring of Dead Sea water surface variation using multi-temporal satellite data and GIS. *Arab. J. Geosci.* **2013**, *6*, 3241–3248. [CrossRef]

60. Shafir, H.; Alpert, P. Regional and local climatic effects on the Dead-Sea evaporation. *Clim. Chang.* **2011**, *105*, 455–468. [CrossRef]

61. Satgé, F.; Espinoza, R.; Zolá, R.P.; Roig, H.; Timouk, F.; Molina, J.; Garnier, J.; Calmant, S.; Seyler, F.; Bonnet, M.-P. Role of Climate Variability and Human Activity on Poopó Lake Droughts between 1990 and 2015 Assessed Using Remote Sensing Data. *Remote Sens.* **2017**, *9*, 218. [CrossRef]

62. Zola, R.P.; Bengtsson, L. Long-term and extreme water level variations of the shallow Lake Poopó, Bolivia. *Hydrol. Sci. J.* **2006**, *51*, 98–114. [CrossRef]

63. Michel, D.; Jiménez, C.; Miralles, D.G.; Jung, M.; Hirschi, M.; Ershadi, A.; Martens, B.; McCabe, M.F.; Fisher, J.B.; Mu, Q.; et al. The WACMOS-ET project – Part 1: Tower-scale evaluation of four remote-sensing-based evapotranspiration algorithms. *Hydrol. Earth Syst. Sci.* **2016**, *20*, 803–822. [CrossRef]

remote sensing

MDPI

Article

Monitoring Water Levels and Discharges Using Radar Altimetry in an Ungauged River Basin: The Case of the Ogooué

Sakaros Bogning [1,2,3,*], Frédéric Frappart [3,4], Fabien Blarel [3], Fernando Niño [3], Gil Mahé [5], Jean-Pierre Bricquet [5], Frédérique Seyler [6], Raphaël Onguéné [2], Jacques Etamé [1], Marie-Claire Paiz [7] and Jean-Jacques Braun [4]

[1] Département de Sciences de la Terre, Université de Douala, BP 24 157 Douala, Cameroun; etame.jacques@yahoo.fr

[2] Jeune Equipe Associée à l'IRD—Réponse du Littoral Camerounais aux Forçages Océaniques Multi-Échelles (JEAI-RELIFOME), Université de Douala, BP 24 157 Douala, Cameroun; ziongra@yahoo.fr

[3] LEGOS, Université de Toulouse, CNES, CNRS, IRD, UPS OMP, 14 Av. E. Belin, 31400 Toulouse, France; frederic.frappart@legos.obs-mip.fr (F.F.); fabien.blarel@legos.obs-mip.fr (F.B.); fernando.nino@ird.fr (F.N.)

[4] GET, Université de Toulouse, CNRS, IRD, UPS OMP, 14 Av. E. Belin, 31400 Toulouse, France; jjbraun1@gmail.com

[5] HydroSciences Montpellier, Université de Montpellier, CNRS, IRD, 300 Av. Pr E. Jeanbrau, 34090 Montpellier, France; gil.mahe@ird.fr (G.M.); jean-pierre.bricquet@ird.fr (J.-P.B.)

[6] ESPACE-DEV, Université de Montpellier, IRD, Université des Antilles, Université de Guyane, Université de La Réunion, Maison de la Télédétection, 500 Rue J-F. Breton, 34093 Montpellier, France; frederique.seyler@ird.fr

[7] The Nature Conservancy Gabon Program Office, Lot 114 Haut de Gué-Gué, 13553 Libreville, Gabon; mcpaiz@tnc.org

*   Correspondence: sakaros.bogning@legos.obs-mip.fr; Tel.: +33-561-332-970

Received: 30 January 2018; Accepted: 22 February 2018; Published: 24 February 2018

Abstract: Radar altimetry is now commonly used for the monitoring of water levels in large river basins. In this study, an altimetry-based network of virtual stations was defined in the quasi ungauged Ogooué river basin, located in Gabon, Central Africa, using data from seven altimetry missions (Jason-2 and 3, ERS-2, ENVISAT, Cryosat-2, SARAL, Sentinel-3A) from 1995 to 2017. The performance of the five latter altimetry missions to retrieve water stages and discharges was assessed through comparisons against gauge station records. All missions exhibited a good agreement with gauge records, but the most recent missions showed an increase of data availability (only 6 virtual stations (VS) with ERS-2 compared to 16 VS for ENVISAT and SARAL) and accuracy (RMSE lower than 1.05, 0.48 and 0.33 and R^2 higher than 0.55, 0.83 and 0.91 for ERS-2, ENVISAT and SARAL respectively). The concept of VS is extended to the case of drifting orbits using the data from Cryosat-2 in several close locations. Good agreement was also found with the gauge station in Lambaréné (RMSE = 0.25 m and R^2 = 0.96). Very good results were obtained using only one year and a half of Sentinel-3 data (RMSE < 0.41 m and R^2 > 0.89). The combination of data from all the radar altimetry missions near Lamabréné resulted in a long-term (May 1995 to August 2017) and significantly improved water-level time series (R^2 = 0.96 and RMSE = 0.38 m). The increase in data sampling in the river basin leads to a better water level peak to peak characterization and hence to a more accurate annual discharge over the common observation period with only a 1.4 $m^3 \cdot s^{-1}$ difference (i.e., 0.03%) between the altimetry-based and the in situ mean annual discharge.

Keywords: altimetry; water level; discharge

1. Introduction

Inland waters have a crucial role in the Earth's water cycle through complex processes at interfaces with the atmosphere and oceans. They also strongly influence socio-economic practices through their impacts on primary needs, agricultural and industrial activities [1]. Recent and future global changes and increase of population will intensify the stress on water resources [2,3]. However, in many parts of the world, reliable field measurements of water level and water discharge are either completely unavailable or difficult to access for addressing integrated water resource management, use in operational flood forecasting or disaster mitigation [4,5]. In the large rainforest of Central Africa, hosting the Congo river basin and associated small neighboring river basins, the number of hydrological stations has dramatically dropped off because of the irregular maintenance of stations; furthermore, the spatial distribution of the stations often hinder the effectiveness of the network [6].

Spaceborne radar altimetry, although originally designed for the study of ocean topography by continuously measuring the distance between the Earth's surface and the sensor onboard the satellite [7], has proved to be very useful for continental hydrology [8]. In spite of limitations over land, radar altimetry has demonstrated a strong capability to accurately retrieve water levels of large lakes and enclosed seas where the observed surfaces are sufficiently homogeneous [9,10] but also in large river basins where the cross-sections between river and altimetry ground-tracks can reach several kilometers [11,12]. These early results were obtained using Geosat and Topex/Poseidon data processed with the Ocean retracking algorithm. With the launch of ENVISAT in 2002, other retracking algorithms started to be commonly used for processing radar altimetry data. Among them, the Offset Center Of Gravity (OCOG, also known as Ice-1) was found to provide, most of the times, the best results for the determination of river water stages [13]. Combined with availability of land dedicated corrections of the ionosphere and the wet troposphere delays and improvements in the data processing, this allowed the generalization of the use of radar altimetry for the monitoring of inland waters. Currently, all these improvements allow detection of water bodies of a few or below one hundred meters of width (e.g., [14–16]).

With the decrease in number and availability of river discharges around the world, altimetry-based water stages are used to estimate river discharges among several other techniques based on remote sensing [17]. Among them, some commonly used are (i) the application of either power law or a polynomial relationship between stage and discharge (rating curve) to altimetry-based water stages [18–21], (ii) the use of flood routing and hydrodynamics models to derive rating curves under altimetry ground-tracks [22–24], (iii) the calibration of hydrodynamics and hydrological models using altimetry-based water stages [25–29].

This study presents the multi-mission altimetry-based hydrological network setup in the Ogooué River Basin (ORB) to provide a continuous water stage and discharge monitoring in this almost ungauged basin. It aims at answering the following questions:

- what are the performances of the different altimetry missions, from ERS-2 to Sentinel-3A, to retrieve water levels?
- how does the combination of data from several altimetry virtual stations improve the retrieval of the annual discharge in the ORB?

2. Study Area

The Ogooué River is the largest Gabonese river. Its length is about 900 km from its source in the Mounts Ntalé, in Congo, at an altitude close to 840 m.a.s.l. to its mouth in the southern part of the Atlantic coast of Gabon. The Ogooué river flows northwestward in upstream until the confluence with the Ivindo and southwestward from the confluence with the Ivindo river to a 100 km long and 100 km width delta it forms in the south of Port Gentil where it discharges into the Atlantic ocean [30]. The Ogooué river basin (ORB) is located between 9° and 15°E, and 3°S and 2.5°N (Figure 1) stretching on about 80% of the total area of Gabon and It is bounded on the east by the Congo basin, on the south by the Niari and Nyanga basins, on the west and north-west by the coastal river basins [31].

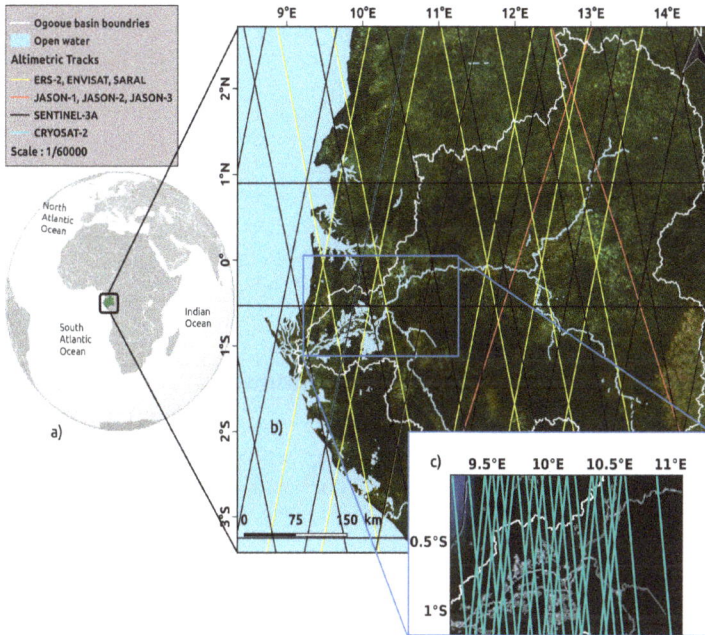

Figure 1. (**a**) Location of the Ogooué River Basin in Gabon in Equatorial Africa. (**b**) In this basin, delineated with a white line, the Ogooué and its major tributaries appear in light blue. Altimetry tracks are represented in red for the missions on a 10-day repeat cycle on their nominal track (Jason-1/2/3), in black for Sentinel-3A on its nominal track (27-day repeat cycle), in yellow for the missions on a 35-day repeat cycle on their nominal track (ERS-2/ENVISAT/SARAL), (**c**) zoom of the downstream of the Ogooué River Basin with altimetric tracks of Cryosat-2 on its nominal track (369-day repeat cycle) in cyan lines.

Due to its location crossing the Equator, the ORB receives the largest annual precipitation in Africa (1600–2200 mm yr^{-1}), making the annual discharge of the river of 4750 m$^3 \cdot$s^{-1}, the third along the African West Coast after the Congo River (40,000 m$^3 \cdot$s^{-1}) and the Niger River (5590 m$^3 \cdot$s^{-1}) [30,32]. The annual variation of the discharge of the Ogooué river passes by two maximum : in spring and autumn corresponding to the rainy seasons [31]. On the ORB at Lambaréné, before the 1970's, river discharge of spring floods were equivalent to those of the autumn floods. After the 1970's, spring floods differ significantly from autumn floods with differences between 2000 and 3000 m$^3 \cdot$s^{-1} [33].

3. Datasets

3.1. Radar Altimetry Data

Radar altimetry data used in this study comes from the measurements on the nominal orbit of the following missions: Jason-2 (06/2008–10/2016), Jason-3 (since 01/2016), ERS-2 (06/1995–07/2003), ENVISAT (06/2002–10/2010), SARAL (02/2013–07/2016), Sentinel-3A (since 02/2016) and Cryosat-2 (since 04/2010—operating in low resolution mode—LRM), but also from the second (drifting) orbit of ENVISAT (10/2010–06/2012). The Jason missions have a 10-day, ERS-2, ENVISAT and SARAL a 35-day, Cryosat-2 and 369-day and Sentinel-3A a 27-day repeat-periods. Jason-2 and Jason-3 data come from Geophysical Data Records (GDRs) D, GDR v2.1 for ENVISAT, GDR T for SARL, GDR C for Cryosat-2 and GDR ESA IPF 06.07 land products for Sentinel-3A delivered by CNES/ESA/NASA processing centers. These data were made available by Centre de Topographie des Océans et de l'Hydrosphère (CTOH) [34]. ERS-2 data were reprocessed by CTOH to ensure the continuity with ENVISAT for land studies [35].

3.2. In Situ Water Levels and Discharges

Long-term datasets of field measured water level and discharge are not available in the ORB since it is was completely ungauged between the 1980s and 2001 [31]. Only the Lambaréné gauge station gathered data from July 2001 to September 2017, sometimes with non operating periods up to a month. These in situ data of water level (collected at Lambaréné by the Société de l'Energie et de l'Eau du Gabon, SEEG) were used to calculate the river discharge, using a historical calibration formula provided by the HydroSciences Montpellier (HSM) laboratory of the University of Montpellier (France). Both are used for validating altimetry-based water level and discharge.

4. Methods

4.1. Altimetry-Based Water Levels

Initially developed to provide accurate measurements of the sea surface topography, radar altimetry is now commonly used for the monitoring of inland water levels (see [8] for a recent review). The variations of the altimeter height from one cycle to the other can be associated to changes in water level.

In this study, we used the Multi-mission altimetry Processing Software (MAPS), frequently used for processing altimetry data over land and ocean (e.g., [36–40]), that allows a refined selection of the valid altimeter data to build time-series of water levels at a so-called virtual station. Data processing is composed of four main steps:

(i)   the rough delineation of the cross-section between the altimeter tracks and the rivers using Google Earth,
(ii) the loading of the altimetry over the study area and the computation of the altimeter heights from the raw data contained in the GDRs,
(iii) a refined selection of the valid altimetry data through visual inspection,
(iv) the computation of the water level time-series as the median of the selected water levels every cycle.

A detailed description of the processing of altimetry data using MAPS can be found in [41]. MAPS is made available by CTOH. Previous studies showed that Ice-1-derived altimetry heights are the more suitable for hydrological studies in terms of accuracy of water levels and availability of the data (e.g., [13,37]) among the commonly available retracked data present in the GDRs. In this study, the data used were processed using the Offset Center of Gravity (OCOG) [42] also named Ice-1 or Ice retracking algorithm depending on the mission for all the missions.

Time series of water levels are generally obtained processing data from altimetry missions with a repeat period between 10 and 35 days [8]. It is what was done in this study to build the network of altimetry virtual stations under Jason-1/2/3, ERS-2/ENVISAT/SARAL and Sentinel-3A groundtracks. Considering the life cycles of these different missions and excluding the Jason missions whose crosstrack is too coarse (315 km at the Equator) to allow the definition of a dense network of virtual stations in, most of the river basins, there is a lack of data between the end of the ENVISAT mission in October 2010 on the nominal orbit, or in April 2012 considering its extended orbit, and the launch of the SARAL mission in February 2013. The only option to fill this gap is to consider the data acquired by the Cryosat-2 mission. Due to its long revisit time (369 days), its data are used to retrieve time series of water levels over large lakes [43], but are generally considered useless to define virtual stations over rivers [44,45]. The long repeat cycle of Cryosat-2 is compensated by a small crosstrack of 7.5 km at the Equator leading to a large number of cross-sections between the river and the altimetry tracks in a close vicinity. Under the assumption that the temporality of water stages is not changing on a few tenths of kilometers of distance, it is possible to build VS gathering Cryosat-2 from tracks in a short distance thanks to the high spatial coverage and the sub-cycle period of 30 days of this mission. Changes in river characteristics (width, depth) over distances of a few kilometers are likely to impact the amplitude of the water stage but not its dynamics, a linear regression between altimetry-based water levels and in-situ data was applied to altimetry data to retrieve time series of water levels (see (2) in sub-section 4.2 Conversion into river discharges).

4.2. Discharge Estimates

River discharge is classically estimated from water level measurements through a functional relationship between the two quantities known as stage-discharge rating or rating curve (Rantz et al., 1982). It has the following form:

$$Q(t) = \alpha(h(t) - h_0)^\beta \qquad (1)$$

where Q is the river discharge, h the water level, h_0 the null-discharge elevation, and α and β are related to the geometry of the channel cross-section and to the friction coefficient modulating the discharge.

This technique have been successfully applied to estimate river discharge using altimetry-based water levels when it was possible to derive the rating curve a common period of observation. It permits to derive river discharge with accuracy better than 20% (e.g., [18–20,22,46–48]).

As the altimetry crossing over the river does not generally occur at the location of the in-situ station but several tenths of kilometers upstream or downstream, the flow cross-sectional area is likely to vary over these short distances. To avoid errors caused by these changes, previous studies used a linear regression between altimetry and in situ water levels (h_{alti} and h_{insitu} respectively) when the dynamic of the flow is quite similar (e.g., [20,46]):

$$h_{insitu} = ah_{alti} + b \qquad (2)$$

where a and b are the coefficients of the linear regression that allows the radar altimetry data to exactly fit the variations from the in situ gauge used to estimate the river discharge.

5. Results

5.1. Altimetry-Based Network of Gauging Stations

An altimetry-based network of 34 virtual stations (VS) was defined across the Ogooué river and its major tributaries (Ivindo and Ngounie rivers in northeast and southwest of the ORB respectively). Water level time series were mostly derived from Envisat, Saral and Sentinel-3A observations. As ENVISAT and SARAL missions were orbiting on the same nominal orbit, 15 of them provide a pluri-annual record from June 2002 to October 2010 and from February 2013 to June 2016, one of them from June 2002 to October 2010 and one of them from February 2013 to June 2016. As this orbit was formerly used by ERS-2, this record was extended from May 1995 to July 2003 for the 6 VS located closer to the mouth. Due to the small width of the rivers in the upstream part of the ORB and the presence of topography, VS could not be defined 125 km upstream Lambaréné because of the narrow width of the Ogooué river and its surrounding rugged topography. This network is completed with 2 VS created using observations from Jason-2 (from July 2008 to August 2016) and Jason-3 (since January 2016). Due to the large cross-track of these missions (315 km at the Equator), very few cross-sections between the rivers and the altimeter ground tracks occur. The network was also completed by 11 VS defined under Sentinel-3A ground tracks, which was launched in February 2016. The locations of the VS from the different altimetry missions are presented in Figure 2 and Table 1.

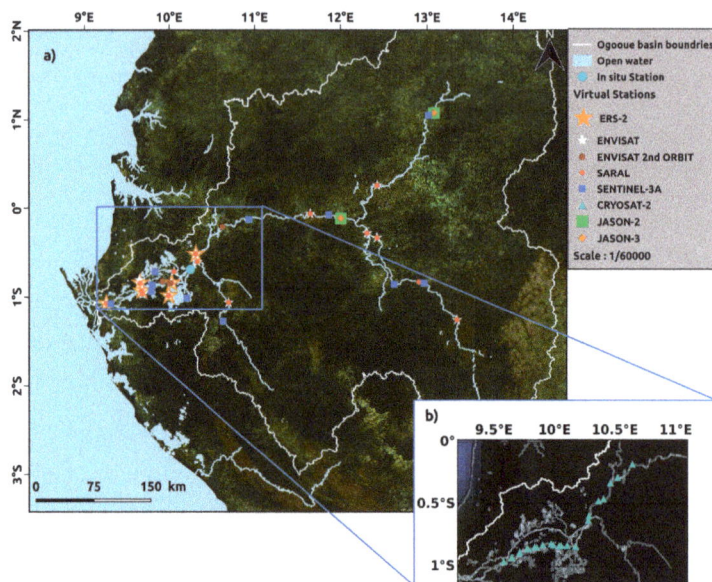

Figure 2. Locations of the altimetry virtual stations in the Ogooué River Basin. VS from ERS-2, ENVISAT, ENVISAT 2nd orbit, SARAL, Sentinel-3A, Cryosat-2, Jason-1, Jason-2, Jason-3 are represented using orange stars, white stars, brown dots, red dots, blue squares, cyan triangles, green squares and orange diamonds respectively. For readability purpose, virtual stations from missions with repeat period shorter than 35 days are presented in (**a**) and virtual stations from Cryosat-2 are presented in (**b**).

Table 1. List of VS where water stages are derived from altimetry measurements.

Virtual Stations	Missions	Longitude (°)	Latitude (°)	River Width (km)	Distance to the River Mouth (km)
SV_229_Ogooué	ENVISAT, SARAL	13.3533	−1.3082	0.22	693.09
SV_272_Ivindo	ENVISAT, SARAL	12.4228	0.2542	0.20	524.91
SV_272_Ogooué	ENVISAT, SARAL	12.3051	−0.2816	0.36	477.59
SV_315_Ogooué	ENVISAT, SARAL	11.6422	−0.0618	0.37	386.57
SV_358_Ngounié	ENVISAT, SARAL	10.6490	−1.2761	0.20	274.948
SV_401_Ngounié	ENVISAT, SARAL	10.3227	−0.5958	0.62	193.531
SV_401_Ogooué	ERS-2, ENVISAT, SARAL	10.3045	−0.5129	1.30	198.222
SV_444_Ogooué	ERS-2, ENVISAT, SARAL	9.2569	−1.0722	0.59	42.582
SV_730_Ogooué	SARAL	12.9154	−0.8316	0.19	607.233
SV_773_Ogooué	ENVISAT, SARAL	12.4700	−0.5595	0.23	494.696
SV_902_lake_Onangué	ERS-2, ENVISAT, SARAL	9.9912	−1.0001		160.97
SV_902_Ogooué	ERS-2, ENVISAT, SARAL	10.0280	−0.8323	1.25	141.66
SV_902_Ogooué_2	ENVISAT, SARAL	10.0551	−0.7091	0.32	152.215
SV_945_lake_Louandé	ENVISAT	9.6497	−0.8047		107.174
SV_945_lake_Ogognié	ENVISAT, SARAL	9.6844	−0.9624		101.671
SV_945_Ogooué	ERS-2, ENVISAT, SARAL	9.6755	−0.9220	1.19	97.413
SV_945_Ogooué_2	ERS-2, ENVISAT, SARAL	9.6571	−0.8382	0.47	103.125
Station 1	ENVISAT 2nd orbit	10.6414	−0.1864	0.49	127.826
Station 2	ENVISAT 2nd orbit	10.0208	−0.8328	0.88	138.166
Station 3	ENVISAT 2nd orbit	9.9445	−0.8082	1.17	252.246
SV_128_Ogooué	SENTINEL-3A	9.2788	−1.0638	1.23	45.271
SV_378_Ogooué	SENTINEL-3A	9.8069	−0.8454	1.24	112.654
SV_185_lake_Onangué	SENTINEL-3A	10.1962	−1.0009		169.463
SV_050_Ogooué	SENTINEL-3A	10.9045	−0.1177	0.30	298.329
SV_107_Ogooué	SENTINEL-3A	11.8457	−0.0803	0.37	412.048
SV_164_Ogooué	SENTINEL-3A	12.6126	−0.8438	0.31	564.01
SV_356_Ogooué	SENTINEL-3A	12.9676	−0.8423	0.30	617.561
SV_164_Ivindo	SENTINEL-3A	13.0280	1.0330	0.19	677.105
SV_050_Ngounié	SENTINEL-3A	10.6477	−1.2728	0.35	301.24

Table 1. *Cont.*

Virtual Stations	Missions	Longitude (°)	Latitude (°)	River Width (km)	Distance to the River Mouth (km)
SV_378_lake_Avanga	SENTINEL-3A	9.7878	−0.9345		112.299
SV_050_Ngounié	SENTINEL-3A	9.8386	−0.7021	0.23	301.24
Lambaréné	CRYOSAT-2	10.2220	−0.7139		
SV_185_Ogooué	JASON-2, JASON-3	12.0035	−0.1148	0.36	430.318
SV_096_Ivindo	JASON-2, JASON-3	13.0790	1.0758	0.18	687.576

5.2. Altimetry-Based Water Levels Validation Using the Lambaréné Gauge Record

The network of gauge stations was not maintained after the 1980s in the ORB. Only the gauge station from Lambaréné provided measurements of water stages and discharge estimates during the period of acquisition of altimetry data. Four ERS-2/ENVISAT/SARAL tracks (401, 444, 902 and 945) are crossing the Ogooué River close to Lambaréné, at 22 and 135.5 km upstream and at 38 and 89 downstream respectively. Sentinel-3 tracks 050, 128 and 378 are crossing the Ogooué River at 121 km upstream and 66 and 133 downstream of Lambaréné respectively, as well as 32 Cryosat-2 tracks. Comparisons between altimetry-based and in situ water stages were performed for measurements acquired the same day. The results are presented in Figures 3–7 for ERS-2/ENVISAT/SARAL, Sentinel-3A and ENVISAT 2nd orbit and Cryosat-2 respectively. For the different missions except Cryosat-2, the water levels are presented from upstream (a) to downstream (c or d), (b) corresponding to the closest distance between the VS and the in situ station of Lambaréné.

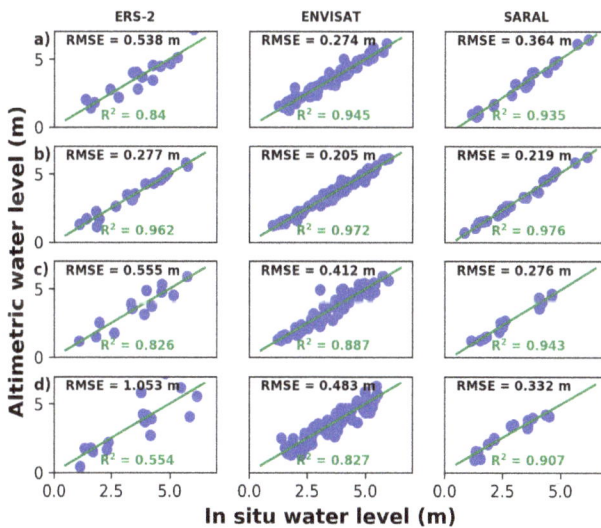

Figure 3. Results of the comparison between the altimetry-based water stages from ERS-2/ENVISAT/SARAL for tracks 401 (**a**), 902 (**b**), 945 (**c**), 444 (**d**) and the in situ ones from Lambaréné gauge station.

Overall very good results are obtained for all these stations. As expected, the quality of the water stage retrieval decreases as the distance to the in situ station increases. Better results were generally obtained using SARAL data, the first mission to operate in Ka-band, than using ERS-2 and ENVISAT ones (Ku-band, Figures 3 and 4). Results obtained using Sentinel-3A data, the first mission to operate in SAR mode on a frequent repetitive orbit, are very encouraging (Figure 5). In spite of the few cycles available (15), altimetry-based water levels obtained using data from this mission already exhibit a similar quality as the ones obtained using data from SARAL when considering the closer distance to the Lambaréné in situ station (Figure 5b). A very good agreement is also found with Cryosat-2 at the

Lambaréné (R^2 = 0.96 and RMSE = 0.25 m) demonstrating the potential of this mission for the retrieval of water stages (Figure 6).

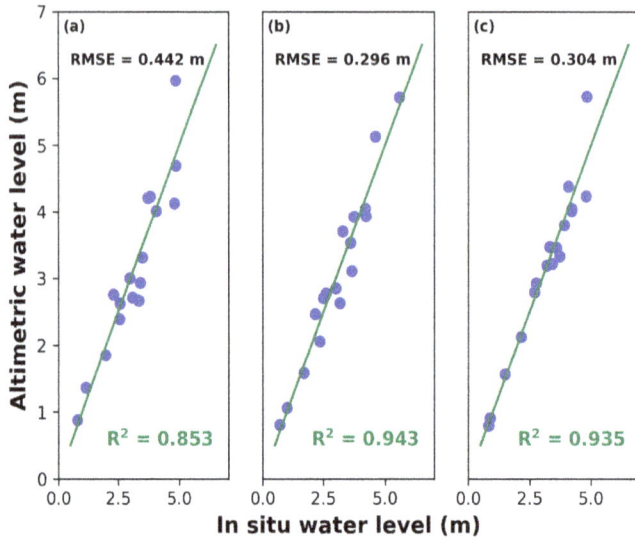

Figure 4. Results of the comparison between the altimetry-based water stages from ENVISAT on its second orbit for (**a**) Station 1, (**b**) Station 2 and (**c**) Station 3 and the in-situ ones from Lambaréné gauge station.

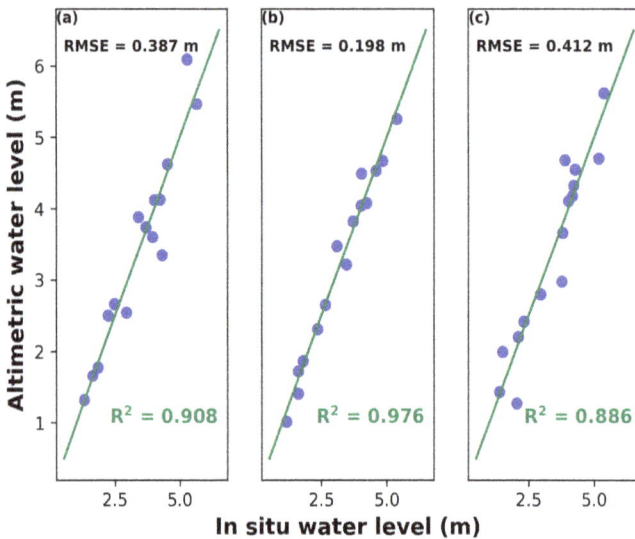

Figure 5. Results of the comparison between the altimetry-based water stages from Sentinel-3A for tracks (**a**) 050, (**b**) 378 and (**c**) 128 and the in-situ ones from Lambaréné gauge station.

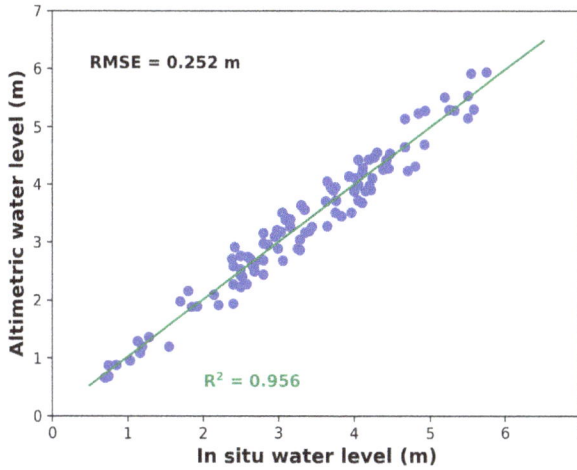

Figure 6. Results of the comparison between the altimetry-based water stages from Cryosat-2 and the in-situ ones from Lambaréné gauge station.

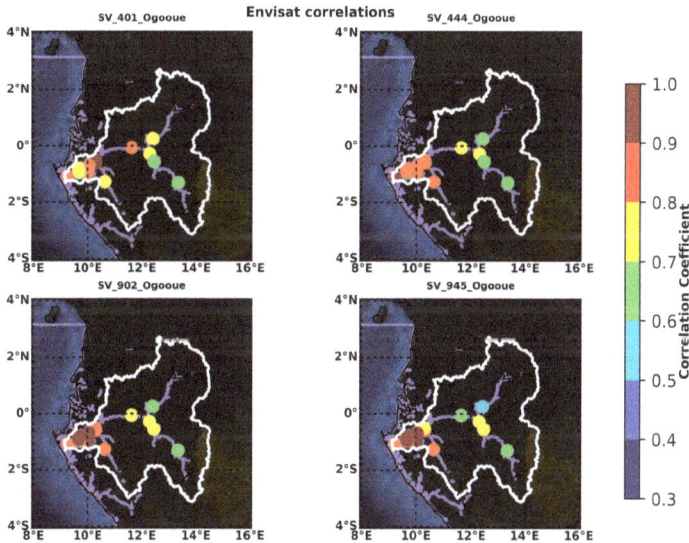

Figure 7. Maps of maximum of cross-correlation between time series from ENVISAT data in the ORB for the four VS around Lambaréné.

5.3. Consistency of the Altimetry-Based Water Levels in the Ogooué River Basin

Only one in-situ gauge station record had water stage measurements during the altimetry period in the ORB. Even if this record can be used to validate 4 SV from ERS-2/ENVISAT/SARAL, 3 from Sentinel-3A, and one combining Cryosat-2 water level estimates at different cross-sections, a consistency check was performed for the other stations in the ORB. It consists in estimating the cross-correlation between the different time-series of water levels from the same mission. Maxima of correlation were reported on Figures 6 and 7 for ENVISAT and SARAL respectively. As the repeat period of these mission is 35 days on their nominal orbit, only time-lags of plus or minus 53 days (one repeat period and a half) were considered due to the relatively small scale of the ORB. Due to the changes of the river features (slope, depth, width, ...), bias and RMSE were not computed between the time-series of altimetry-based water levels. This consistency-check was applied neither to the ERS-2 VS that are not sufficiently numerous (only 6 located on the downstream part of the river) nor to the ENVISAT VS from the second orbit because of the short period of operation of ENVISAT (17 cycles available).

It can be seen that maximum values of cross-correlation between the time series are greater downstream than upstream of the ORB whatever the VS considered to make the comparisons (Figures 7 and 8). Larger maxima of cross-correlation are obtained with the ENVISAT mission than with the SARAL mission. They are above 0.7 between ENVISAT VS in the downstream part of the basin up to the confluence between its two major tributaries, the Ivindo and the Ngounié Rivers. On these two tributaries, the maximum of cross-correlations generally range from 0.6 to 0.8 (Figure 7). Much lower agreement was found between the SARAL VS. If maxima of cross-correlation are generally above 0.6 on the downstream part of the ORB but they rapidly decrease down to 0.4 upstream. Higher consistency is found using ENVISAT VS on a larger part of the ORB.

No time-lag is present in most of the cases. Nevertheless, a few maxima cross-correlation coefficients between time series occurred with a time-lag of one month, for both ENVISAT and SARAL missions. This situation happened in the case of two stations located upstream of the Ogooué River (ENVISAT SV_229_Ogooué using SV_401_Ogooué, SV_902_Ogooué and SV_945_Ogooué validated against records from Lambaréné in situ gauge as reference and SARAL SV_773_Ogooué using SV_401_Ogooué and SV_945_Ogooué as reference) and in the case of the station SV_945_Lake_Ogognié located on the Lake Ogognié downstream of the ORB in cross-correlation with the validated time series from SV_401_Ogooué.

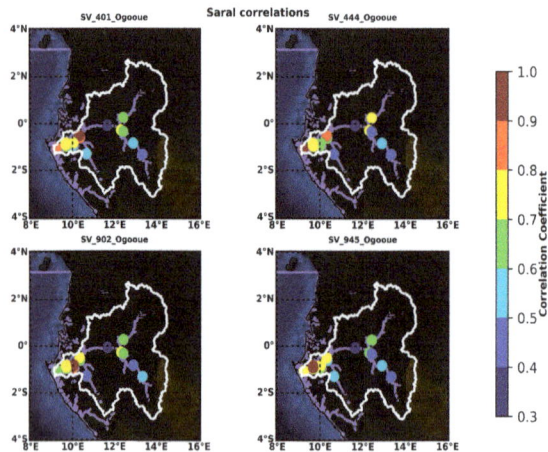

Figure 8. Maps of maximum of cross-correlation between time series from SARAL data in the ORB for the four VS around Lambaréné.

Due to the lack of in situ gauge station in the ORB with the exception of the one from Lambaréné, no validation was performed for the water stages derived from Jason-2 and Jason-3 data. As the two cross-sections between Jason-2 and 3 ground-tracks (185 on the Ogooué and 096 on the Ivindo) are close from cross-sections between Sentinel-3A ground-tracks (107 on the Ogooué and 164 on the Ivindo) (see Table 1 and Figure 2), cross-comparisons between the water levels derived from Jason-2, Jason-3 and Sentinel-3A were performed during the common period of availability of the different datasets. For this purpose, data from the 10-day repeat orbit missions (Jason-2 and Jason-3 that were orbiting 2 minutes appart) were interpolated at the date of acquisition of Sentinel-3A. In both cases, Jason-2 time-series of water levels exhibit a clear seasonal cycle, especially for track 096 on the Ivindo, with high stages observed during the primary peak from October to December and the secondary peak in April-May and low stages observed in January-February (small dry season) and from June to August for the large dry season, which is consistent our knowledge of the hydrological cycle in the ORB (Figure 9). This also the case for Jason-3 time-series that agree well with the ones from Jason-2 (R = 0.88 and 0.87 and RMSE = 0.51 and 0.68 m for tracks 185 on the Ogooué and 096 on the Ivindo respectively). If similar but smoother water levels variations were observed in the Ivindo at the SV built under Sentinel-3A track 164 (R = 0.92 and 0.95, RMSE = 0.71 and 0.56 m, Figure 9a), no realistic water level variations were observed at the SV built under Sentinel-3A track 107 (Figure 9b). This SV was discarded.

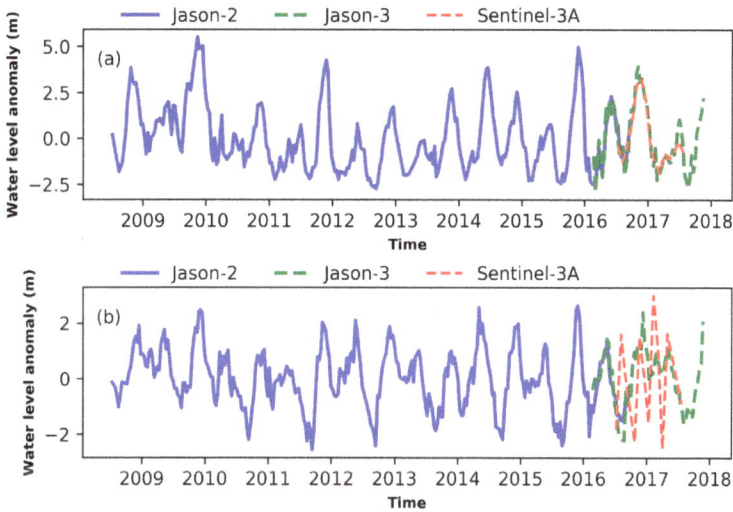

Figure 9. Time series of water level from Jason-2 (blue), Jason-3 (dashed green) and Sentinel-3A (dashed red) on the Ivindo (**a**) and upstream Ogooué (**b**) rivers.

5.4. Multi-Mission Discharge Estimates

The time-series of water level from the VS, that were validated against the stage record from the Lambaréné in situ gauge, were combined using Equation (2) to form a unique time series over May 1995-July 2017 time period. It is presented in Figure 10 along with the time-series of water levels from Lambaréné gauge station. The combined altimetry-based time series of water levels is in good agreement with the in-situ one (R^2 = 0.96 and RMSE = 0.38 m). Contrary to the individual altimetry-based time-series of water levels used in this study that are constrained by the revisit time of the satellite (27 days for ENVISAT on its second orbit, 35 days for ERS-2/ENVISAT and SARAL on their nominal orbit, 369 days for Cryosat-2), the combined time-series present the advantage to have between 3 to 4 (during the ERS-2 and the ENVISAT periods between 1995 and 2010) and 9 (during the SARAL and the Sentinel-3A periods since 2013 when combining with the Cryosat-2 measurements available since 2010) measurements each month.

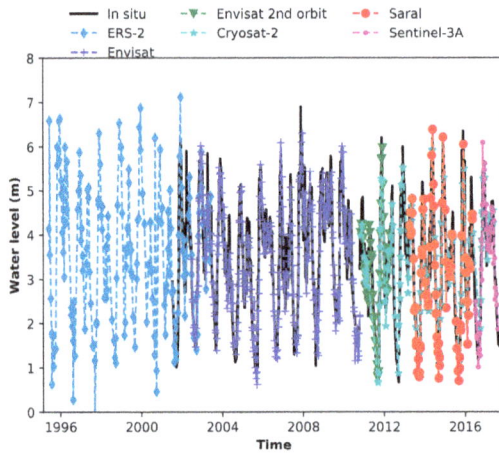

Figure 10. Time series of water level at Lambaréné from the in-situ gauge record (black continuous line), the multi-mission altimetry-based record (ERS-2 data are represented with diamonds, ENVISAT with blue crosses on its nominal orbit and with green triangles on its second orbit, Cryosat-2 with green-blue stars, SARAL with red circles, Sentinel-3 with purple dots).

Applying the rating curve relating water stages and discharge, established for the Lambaréné gauge using in situ measurements of water levels and discharge, an altimetry-based time series of discharge was obtained. Figure 11 shows discharges estimated from altimetry-based and in situ water levels at Lambaréné. A very good agreement between both sources (R = 0.94 and RMSE = 701.6 $m^3 \cdot s^{-1}$ for a mean annual discharge of 4253.4 $m^3 \cdot s^{-1}$ or 16.5% of the mean annual discharge over the observation period). Considering the total annual discharge over the common observation period, altimetry-based and in situ estimates only differs by 1.4 $m^3 \cdot s^{-1}$ (or 0.03% of the mean annual discharge) in average (Table 2).

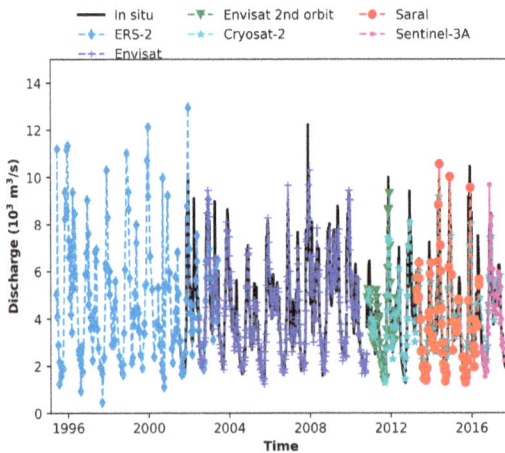

Figure 11. Time series of river discharge at Lambaréné from the in-situ gauge record (black continuous line), the multi-mission altimetry-based record (ERS-2 data are represented with diamonds, ENVISAT with blue crosses on its nominal orbit and with green triangles on its second orbit, Cryosat-2 with green-blue stars, SARAL with red circles, Sentinel-3 with purple dots).

Table 2. Average annual flow derived for the different VS of every altimetry mission and also combining their different records.

Missions	Stations	Estimated Discharge (m$^3 \cdot$s^{-1})	Relative Error (%)	R(-)	RMSE (m$^3 \cdot$s^{-1})
In situ	Lambaréné	4253.427	0	1	0
ERS-2	SV_401_Ogooué	4951.475	16.41	0.850	1107.647
	SV_444_0gooué	5465.756	21.45	0.141	2466.262
	SV_902_Ogooué	4498.633	5.76	0.987	332.152
	SV_945_Ogooué	5104.026	19.99	0.886	920.226
	Combined	4848.378	13.99	0.755	1440.860
ENVISAT	SV_401_Ogooué	4311.470	1.36	0.969	431.757
	SV_444_Ogooué	4281.369	0.66	0.884	804.946
	SV_902_Ogooué	4316.212	1.47	0.981	352.028
	SV_945_Ogooué	4342.597	2.09	0.922	722.334
	Combined	4340.594	2.049	0.942	604.321
SARAL	SV_401_Ogooué	3927.343	−7.67	0.975	470.446
	SV_444_Ogooué	3103.856	−27.023	0.947	409.759
	SV_902_Ogooué	3851.170	−9.46	0.986	346.980
	SV_945_Ogooué	2898.672	−31.85	0.973	335.693
	Combined	4060.427	−4.537	0.978	396.820
ENVISAT 2nd Orbit	Station 1	3850.274	−9.47	0.816	868.273
	Station 2	3740.297	−12.06	0.963	450.838
	Station 3	3841.379	−9.69	0.895	654.027
	Combined	3839.296	−9.736	0.898	679496
SENTINEL-3A	SV_050_Ogooué	4454.109	4.72	0.915	757.845
	SV_378_Ogooué	4210.519	−1.01	0.983	316.740
	SV_128_Ogooué	4373.213	2.82	0.931	629.888
	Combined	4262.683	0.22	0.942	597.610
CRYOSAT-2	Lambaréné	4188.220	−1.53	0.971	408.739
SENTINEL-3A + CRYOSAT-2	Combined	4118.9275	−3.162	0.967	462.141
CRYOSAT-2 + SARAL	Combined	4136.3508	−2.752	0.977	405.978
CRYOSAT-2 + ENVISAT 2nd orbit	Combined	4243.2487	−0.239	0.956	515.057
All missions	Combined	4252.052	−0.03	0.936	701.645

6. Discussion

Despite the relatively small size of the ORB and the small length of the Ogooué River (~900 km), the distance between Equatorial cross-tracks of the missions (80 km for ERS-2/ENVISAT/SARAL, and 315 km for Topex-Poseidon/Jason-1/Jason-2/Jason-3 on their nominal orbits), make possible the construction of a dense network of altimetric VS. It is composed of 16 SV using ENVISAT and SARAL, 6 SV using ERS-2 data, 10 SV using Sentinel-3A data, and only 2 SV using Jason-2 and Jason-3 data. In spite of the improvements made in the processing of data from early high precision missions, too few valid data were found in the GDR E data for Jason-1 to provide a continuous monitoring of water stages at two locations where the SV were constructed. These two cross-sections are located in the upstream parts of the basin where the river widths are between 180 and 360 m. Similarly, even if a very good agreement was found between ERS-2 based water levels in the downstream part of the basin, up to several tenths of kilometers upstream Lambaréné in situ station, no SV was defined either on the upstream Ogooué River or on its tributaries. Due to the small width of the river and the presence of higher topography, ERS-2 was affected more by tracking loss than ENVISAT. Furthermore, the accuracy of the data was degraded due to more frequent changes of acquisition modes than ENVISAT (Table 3). ENVISAT was, most of the time, operating in the 320 MHz Ku chirp bandwidth acquisition mode than in the 80 MHz or the 20 MHz modes, that is to say with the better range resolution as the size of the range detection window is 64, 256 and 1024 m respectively. For ERS-2, different cases were observed.

When no valid water level time series were obtained from ERS-2 acquisitions, it can be attributed to either rapid changes in the topography along the track causing changes in Ku chirp bandwidth acquisition modes (with a large number of acquisitions at 82.5 MHz causing a loss of accuracy of the data) or altimeter lock on the hills on top of the river (with acquisitions mostly at 330 MHz but a few tenths of meters above the river, see the example presented in [37] for ENVISAT). When time-series of water levels were derived, the percentage of acquisition at high frequency is high but lower than the one for ENVISAT causing a loss of accuracy compared with ENVISAT (Table 1).

Compared to earlier missions on the same nominal orbit operating at Ku-band (ERS-2 and ENVISAT), SARAL, the first mission to operate at Ka-band, agrees better with gauge records. Higher R and lower RMSE were found between the time-series of water stages from in-situ and from altimetry-based measurements for the four SV close to Lambaréné (Figure 3). In spite of the short period of observation of SARAL on its nominal orbit (35 cycles from February 2013 to June 2016 and even shorter considering that SARAL started to drift since July 2016), the benefits of the Ka-band smaller footprint compared to the Ku-one can be clearly observed as it was reported in other locations [49]. The use of other cycles would have lead to large errors in the water stage retrievals because of (i) slope effects causing large changes of water base levels, (ii) changes in the characteristics of the river (width, depth) responsible changes in amplitude of the water levels, (iii) detection of other water bodies (e.g., lakes) than the Ogooué River.

Only considering ERS-2, ENVISAT and SARAL on their nominal orbit would have resulted in gaps in the monitoring of water levels in the ORB between October 2010 and February 2013 and after July 2016, and especially in Lambaréné where the river discharge is estimated. To complete the time-series, data from the ENVISAT 2nd orbit were considered from October 2010 to April 2012. Three SV were build giving water estimates of equivalent quality as the ones obtained on the nominal orbit (Figure 4). Applying an innovative processing, time-series of water levels were derived from Cryosat-2 measurements, the first altimetry mission to operate in SAR mode. The results show, that the data from this mission, that are of very good quality (R^2 and RMSE of very similar quality as ENVISAT or SARAL, see Figure 6). The monitoring of the water levels in the Ogooué can continue using acquisitions from the Sentinel-3A mission and soon, from the Sentinel-3B. Comparisons performed against the in-situ gauge records from Lambaréné of one year and a half of Sentinel-3A-based water levels show the high quality of the data acquired in SAR mode (R^2 = 0.98 and RMSE < 20 cm for the SV located at the closer distance to Lambaréné station, see Figure 5).

Table 3. Acquisition modes of the ERS-2 and ENVISAT measurements for the VS in the ORB.

Virtual Stations	Rivers or Lakes	Missions	ENVISAT Data Modes			ERS-2 Data Modes	
			320 Hz (%)	80 Hz (%)	20 Hz (%)	330 Hz (%)	82.5 Hz (%)
SV_229_Ogooué	Ogooué	ENVISAT, SARAL	99.786	0.213	0	100	0
SV_272_Ivindo	Ivindo	ENVISAT, SARAL	96.893	1.804	1.302	58.0	42.0
SV_272_Ogooué	Ogooué	ENVISAT, SARAL	98.267	0.533	1.2	6.25	93.75
SV_315_Ogooué	Ogooué	ENVISAT, SARAL	91.896	6.212	1.890	100	0
SV_358_Ngounié	Ngounié	ENVISAT, SARAL	88.76	10.3	0.939	70.270	29.729
SV_401_Ngounié	Ngounié	ENVISAT, SARAL	98.776	1.146	0.077	81.132	18.867
SV_401_Ogooué	Ogooué	ERS-2, ENVISAT, SARAL	98.776	1.146	0.077	81.132	18.867
SV_444_Ogooué	Ogooué	ERS-2, ENVISAT, SARAL	99.887	0.113	0	85.0	15.0
SV_730_Ogooué	Ogooué	SARAL	92.347	6.152	1.5005	34.426	65.573
SV_773_Ogooué	Ogooué	ENVISAT, SARAL	99.178	0.821	0	70.588	29.411
SV_902_lake_Onangué	lake Onangué	ERS-2, ENVISAT, SARAL	99.669	0.259	0.070	87.804	12.195
SV_902_Ogooué	Ogooué	ERS-2, ENVISAT, SARAL	99.669	0.259	0.070	87.804	12.195
SV_902_Ogooué_2	Ogooué	ENVISAT, SARAL	99.669	0.259	0.070	87.804	12.195
SV_945_lake_Louandé	lake Louandé	ENVISAT	99.513	0.487	0	100	0
SV_945_lac_Ogognié	lake Ogognié	ENVISAT, SARAL	99.513	0.487	0	100	0
SV_945_Ogooué	Ogooué	ERS-2, ENVISAT, SARAL	99.513	0.487	0	100	0
SV_945_Ogooué_2	Ogooué	ENVISAT	99.513	0.487	0	100	0

Due to the presence of only one in-situ station with records during the altimetry acquisition period, the consistency of the altimetry-based time-series of water levels was checked for ENVISAT and SARAL through cross-correlation estimates. They were performed choosing as reference the SV close to Lambaréné station whose records were validated. The difference observed for the cross-correlations between ENVISAT and SARAL occur in the upstream parts of the basin (Figures 7 and 8 respectively). The lower scores for SARAL can be attributed to the shorter observation period of SARAL compared with ENVISAT (35 cycles for SARAL against 85 for ENVISAT, that is to say, more or less a ratio of one third). This difference in number of observations has a smaller impact on the downstream part of the ORB as the water stages are mostly driven by the flow from the upstream parts. On the upstream part, changes in rainfall conditions between two VS are likely to strongly modify the water stage at these two locations. Even a few of this kind of changes can reduce the correlation between the two time-series of river levels between these two stations, especially for shorter records. Due to the relatively small length of the basin and the quite long repeat period of ENVISAT and SARAL (35 days), maxima of correlation are generally observed with no time-lag, except for a limited number of stations in the upstream parts of the ORB where differences lower than one month were obtained.

Based on the good agreement found between in situ and altimetry-based water stages for the VS located around Lambaréné gauge station, the stage discharge rating curve was applied to any individual VS but also for the combination of VS for the same mission. The results are presented in Table 2. This combination represents a good compromise between quality of the resulting time series of discharge and number of observations. For all the missions, the results of the combination are very close to the best result obtained with a single one. It also decreases the risk to miss a rapid flood event that will modify the annual discharge. The combination of data from several missions (ENVISAT 2nd orbit + Cryosat-2, SARAL + Cryosat-2, Sentinel-3A + Cryosat-2) also increase the quality of the discharge estimate. When combining the data from the missions (Cryosat-2 period of availability of the data is overlapping the ones from ENVISAT, ENVISAT 2nd orbit, SARAL and Sentinel-3A), very good performances are obtained (see Table 2).

7. Conclusions

This study provides one of the first assessments of the performances of multiple satellite altimetry-based water levels in a river basin, from ERS-2 to Sentinel-3A. Comparisons between altimetry water stages and in situ gauge records from Lambaréné, the unique station still in operation in the ORB showed the improvement in performances of the missions in operation since ERS-2. In spite of good performances in the downstream part of the ORB ($R^2 > 0.82$ and RMSE < 0.6 m for 3 of the four comparisons), no VS using ERS-2 data were created in the upstream part of the basin due to the small width of the rivers and the presence of topography. On the contrary, the whole basin was sampled using ENVISAT ($R^2 > 0.82$ and RMSE < 0.5 m) and SARAL ($R^2 > 0.90$ and RMSE < 0.4 m) on the same orbit. A very good consistency was also found for these two missions when computing cross-correlations between altimetry-based water stages all over the basin. As SARAL orbit control was not as strict as is generally is for high precision altimetry missions, the drifts of several kilometers from the nominal paths lead to lower agreement in the upstream parts of the basin than using ENVISAT. Missions operating in SAR mode also exhibit very good accuracy ($R^2 > 0.88$ and RMSE < 0.4 m for only one year and a half of Sentinel-3A data and $R^2 = 0.96$ and RMSE = 0.25 m for Cryosat-2). For Cryosat-2, the concept of VS was extended: data acquired on close cross-sections were combined to build the time series of water level. The use of Cryosat-2 (and ENVISAT 2nd orbit) data allowed a continuous monitoring of the ORB for all the altimetry period, showing the potential of the data acquired during geodetic (drifting) orbit of the altimetry missions for land hydrology.

The altimetry-based time-series of water levels from the different altimetry missions close to the Lambaréné in situ stations were converted to discharges using the rating curves from the in situ station. Very good agreement was also found, the quality of which is directly linked to the accuracy of the water levels at the VS for the different missions. This study also shows the interest to combine the data from different missions to improve quality of the monitoring of level and discharge in terms of sampling frequency and annual discharge estimates.

The network of altimetry VS built in this study and associated discharge estimates has a strong interest for the scientific community. It will allow (i) the continuous monitoring of water stages in an almost ungauged basin, (ii) the analysis of possible effects of climate variability and anthropogenic effects (i.e., deforestation) on the hydrological cycle of the ORB. It will be very useful for the calibration and validation of the NASA/CNES Surface Water and Ocean Topography (SWOT) mission, that will use the SAR interferometry technique to map surface water elevation at a spatial resolution of 100 m as its one day calibration orbit encompass the downstream part of the Ogooué Basin.

Acknowledgments: This work is dedicated to Gaston Liénou who sadly passed away in 2015. He enormously contributed to the continuity of study operations in the Ogooué river basin. The authors are grateful to the Center for Topographic studies of the Ocean and Hydrosphere (CTOH) at LEGOS (Toulouse, France) for providing the altimetry dataset. We also thank the Société de l'Énergie et de l'Eau du Gabon (SEEG) branch of Lambaréné (Lambaréné, Gabon) for supplying of the Ogooué water level from the Lambaréné hydrological station. Aurélie Flore KOUMBA PAMBO, Scientific Cell coordinator, Agence Nationale des Parcs Nationaux (ANPN) and Centre National de la Recherche Scientifique et Technologique (CENAREST), Libreville, Gabon, is warmly thanked for her assistance with respect to the reasearch authorizations. Other thanks go to the "Service Coopération et de l'Action Culturelle (SCAC)" of the French Embassy in Cameroon and TNC-Gabon, and the project "Centre de Topographie des Océans et de l'Hydrosphère" funded by CNES. This work was supported by the French national programme EC2CO-Biohefect "Régime d'Altération/Érosion en Afrique Centrale (RALTERAC)". Final acknowledgements are for the project "Jeune Equipe Associée Internationale—Réponse du Littoral Camerounais aux Forçages Océaniques Multi-Échelles (JEAI-RELIFOME)" of the University of Douala and the Laboratoire Mixte International (LMI) Dynamique des écosystèmes continentaux d'Afrique Centrale en contexte de changements globaux (DYCOFAC) for its support in the accomplishment of this work. The authors want to thank two anonymous Reviewers for their helpful comments.

Author Contributions: Sakaros Bogning and Frédéric Frappart conceived and designed the study; Sakaros Bogning performed the altimetry data processing with the help of Fabien Blarel, Frédéric Frappart, Fernando Niño and Frédérique Seyler; Jean-Pierre Briquet and Gil Mahé processed the in situ data; all the authors analyzed the results and contributed to the writing of the paper.

Conflicts of Interest: The authors declare no conflict of interest.

References

1. Younger, P.L. *Water*; Hodder & Stoughton: London, UK, 2012.
2. Vörösmarty, C.J.; Green, P.; Salisbury, J.; Lammers, R.B. Global water resources: Vulnerability from climate change and population growth. *Science* **2000**, *289*, 284–288. [CrossRef] [PubMed]
3. Oki, T.; Kanae, S. Global hydrological cycles and world water resources. *Science* **2006**, *313*, 1068–1072. [CrossRef] [PubMed]
4. Gleick, P.H. Global freshwater resources: Soft-path solutions for the 21st century. *Science* **2003**, *302*, 1524–1528. [CrossRef] [PubMed]
5. Alsdorf, D.E.; Rodríguez, E.; Lettenmaier, D.P. Measuring surface water from space. *Rev. Geophys.* **2007**, *45*, RG2002. [CrossRef]
6. Alsdorf, D.; Beighley, E.; Laraque, A.; Lee, H.; Tshimanga, R.; O'Loughlin, F.; Mahé, G.; Dinga, B.; Moukandi, G.; Spencer, R.G.M. Opportunities for hydrologic research in the Congo Basin. *Rev. Geophys.* **2016**, *54*, 378–409. [CrossRef]
7. Stammer, D.; Cazenave, A. *Satellite Altimetry over Oceans and Land Surfaces*; Taylor & Francis: Boca Raton, FL, USA, 2017.
8. Crétaux, J.-F.; Nielsen, K.; Frappart, F.; Papa, F.; Calmant, S.; Benveniste, J. Hydrological applications of satellite altimetry: Rivers, lakes, man-made reservoirs, inundated areas. In *Satellite Altimetry over Oceans and Land Surfaces*; Stammer, D., Cazenave, A., Eds.; Earth Observation of Global Changes; CRC Press: Boca Raon, FL, USA, 2017; pp. 459–504.
9. Morris, C.S.; Gill, S.K. Variation of Great Lakes water levels derived from Geosat altimetry. *Water Resour. Res.* **1994**, *30*, 1009–1017. [CrossRef]
10. Birkett, C.M. The contribution of TOPEX/POSEIDON to the global monitoring of climatically sensitive lakes. *J. Geophys. Res.* **1995**, *100204*, 179–225. [CrossRef]

11. Koblinsky, C.J.; Clarke, R.T.; Brenner, A.C.; Frey, H. Measurement of river level variations with satellite altimetry. *Water Resour. Res.* **1993**, *29*, 1839–1848. [CrossRef]

12. Birkett, C.M. Contribution of the TOPEX NASA Radar Altimeter to the global monitoring of large rivers and wetlands. *Water Resour. Res.* **1998**, *34*, 1223. [CrossRef]

13. Frappart, F.; Calmant, S.; Cauhopé, M.; Seyler, F.; Cazenave, A. Preliminary results of ENVISAT RA-2-derived water levels validation over the Amazon basin. *Remote Sens. Environ.* **2006**, *100*, 252–264. [CrossRef]

14. Baup, F.; Frappart, F.; Maubant, J. Use of satellite altimetry and imagery for monitoring the volume of small lakes. In Proceedings of the International Geoscience and Remote Sensing Symposium (IGARSS), Quebec City, QC, Canada, 13–18 July 2014.

15. Sulistioadi, Y.B.; Tseng, K.-H.; Shum, C.K.; Hidayat, H.; Sumaryono, M.; Suhardiman, A.; Setiawan, F.; Sunarso, S. Satellite radar altimetry for monitoring small rivers and lakes in Indonesia. *Hydrol. Earth Syst. Sci.* **2015**, *19*, 341–359. [CrossRef]

16. Frappart, F.; Papa, F.; Malbeteau, Y.; León, J.G.; Ramillien, G.; Prigent, C.; Seoane, L.; Seyler, F.; Calmant, S. Surface freshwater storage variations in the orinoco floodplains using multi-satellite observations. *Remote Sens.* **2015**, *7*, 89–110. [CrossRef]

17. Bjerklie, D.M.; Lawrence Dingman, S.; Vorosmarty, C.J.; Bolster, C.H.; Congalton, R.G. Evaluating the potential for measuring river discharge from space. *J. Hydrol.* **2003**, *278*, 17–38. [CrossRef]

18. Kouraev, A.V.; Zakharova, E.A.; Samain, O.; Mognard, N.M.; Cazenave, A. Ob' river discharge from TOPEX/Poseidon satellite altimetry (1992–2002). *Remote Sens. Environ.* **2004**, *93*, 238–245. [CrossRef]

19. Zakharova, E.A.; Kouraev, A.V.; Cazenave, A.; Seyler, F. Amazon River discharge estimated from TOPEX/Poseidon altimetry. *Comptes Rendus Geosci.* **2006**, *338*, 188–196. [CrossRef]

20. Papa, F.; Durand, F.; Rossow, W.B.; Rahman, A.; Bala, S.K. Satellite altimeter-derived monthly discharge of the Ganga-Brahmaputra River and its seasonal to interannual variations from 1993 to 2008. *J. Geophys. Res. Ocean.* **2010**, *115*, C12013. [CrossRef]

21. Birkinshaw, S.J.; Moore, P.; Kilsby, C.G.; O'Donnell, G.M.; Hardy, A.J.; Berry, P.A.M. Daily discharge estimation at ungauged river sites using remote sensing. *Hydrol. Process.* **2014**, *28*, 1043–1054. [CrossRef]

22. Leon, J.G.; Calmant, S.; Seyler, F.; Bonnet, M.-P.; Cauhopé, M.; Frappart, F.; Filizola, N.; Fraizy, P. Rating curves and estimation of average water depth at the upper Negro River based on satellite altimeter data and modeled discharges. *J. Hydrol.* **2006**, *328*, 481–496. [CrossRef]

23. Tarpanelli, A.; Barbetta, S.; Brocca, L.; Moramarco, T. River Discharge Estimation by Using Altimetry Data and Simplified Flood Routing Modeling. *Remote Sens.* **2013**, *5*, 4145–4162. [CrossRef]

24. Paris, A.; Dias de Paiva, R.; Santos da Silva, J.; Medeiros Moreira, D.; Calmant, S.; Garambois, P.-A.; Collischonn, W.; Bonnet, M.-P.; Seyler, F. Stage-discharge rating curves based on satellite altimetry and modeled discharge in the Amazon basin. *Water Resour. Res.* **2016**, *52*, 3787–3814. [CrossRef]

25. Wilson, M.D.; Bates, P.; Alsdorf, D.; Forsberg, B.; Horritt, M.; Melack, J.; Frappart, F.; Famiglietti, J. Modeling large-scale inundation of Amazonian seasonally flooded wetlands. *Geophys. Res. Lett.* **2007**, *34*. [CrossRef]

26. Getirana, A.; Bonnet, M.; Calmant, S.; Roux, E.; Rotunno Filho, O.C.; Mansur, W.J. Hydrological monitoring of poorly gauged basins based on rainfall-runoff modeling and spatial altimetry. *J. Hydrol.* **2009**, *379*, 205–219. [CrossRef]

27. Milzow, C.; Krogh, P.E.; Bauer-Gottwein, P. Combining satellite radar altimetry, SAR surface soil moisture and GRACE total storage changes for model calibration and validation in a large ungauged catchment. *Hydrol. Earth Syst. Sci. Discuss.* **2010**, *7*, 9123–9154. [CrossRef]

28. Pereira-Cardenal, S.; Riegels, N.D.; Berry, P.A.M.; Smith, R.G.; Yakovlev, A.; Siegfried, T.U.; Bauer-Gottwein, P. Real-time remote sensing driven river basin modeling using radar altimetry. *Hydrol. Earth Syst. Sci.* **2011**, *15*, 241–254. [CrossRef]

29. De Paiva, R.C.D.; Buarque, D.C.; Collischonn, W.; Bonnet, M.P.; Frappart, F.; Calmant, S.; Bulhões Mendes, C.A. Large-scale hydrologic and hydrodynamic modeling of the Amazon River basin. *Water Resour. Res.* **2013**, *49*, 1226–1243. [CrossRef]

30. Mignard, S.L.A.; Mulder, T.; Martinez, P.; Charlier, K.; Rossignol, L.; Garlan, T. Deep-sea terrigenous organic carbon transfer and accumulation: Impact of sea-level variations and sedimentation processes off the Ogooue River (Gabon). *Mar. Pet. Geol.* **2017**, *85*, 35–53. [CrossRef]

31. Mahe, G.; Lerique, J.; Olivry, J.-C. Le fleuve Ogooué au Gabon: Reconstitution des débits manquants et mise en évidence de variations climatiques à l'équateur. *Hydrol. Cont.* **1990**, *5*, 105–124.

32. Lambert, T.; Darchambeau, F.; Bouillon, S.; Alhou, B.; Mbega, J.D.; Teodoru, C.R.; Nyoni, F.C.; Massicotte, P.; Borges, A.V. Landscape Control on the Spatial and Temporal Variability of Chromophoric Dissolved Organic Matter and Dissolved Organic Carbon in Large African Rivers. *Ecosystems* **2015**, *18*, 1224–1239. [CrossRef]

33. Lienou, G.; Mahe, G.; Paturel, J.E.; Servat, E.; Sighomnou, D.; Ekodeck, G.E.; Dezetter, A.; Dieulin, C. Evolution des régimes hydrologiques en région équatoriale camerounaise: Un impact de la variabilité climatique en Afrique équatoriale? *Hydrol. Sci. J.* **2008**, *53*, 789–801. [CrossRef]

34. Home—CTOH. Available online: http://ctoh.legos.obs-mip.fr/ (accessed on 24 October 2017).

35. Frappart, F.; Legrésy, B.; Niño, F.; Blarel, F.; Fuller, N.; Fleury, S.; Birol, F.; Calmant, S. An ERS-2 altimetry reprocessing compatible with ENVISAT for long-term land and ice sheets studies. *Remote Sens. Environ.* **2016**, *184*, 558–581. [CrossRef]

36. Blarel, F.; Frappart, F.; Legrésy, B.; Blumstein, D.; Fatras, C.; Mougin, E.; Papa, F.; Prigent, C.; Rémy, F.; Niño, F.; et al. Radar altimetry backscattering signatures at Ka, Ku, C and S bands over land. In Proceedings of the Living Planet Symposium, Prague, Chech Republic, 9–13 May 2016.

37. Biancamaria, S.; Frappart, F.; Leleu, A.-S.; Marieu, V.; Blumstein, D.; Desjonquères, J.-D.; Boy, F.; Sottolichio, A.; Valle-Levinson, A. Satellite radar altimetry water elevations performance over a 200 m wide river: Evaluation over the Garonne River. *Adv. Space Res.* **2017**, *59*, 128–146. [CrossRef]

38. Frappart, F.; Roussel, N.; Biancale, R.; Martinez Benjamin, J.J.; Mercier, F.; Perosanz, F.; Garate Pasquin, J.; Martin Davila, J.; Perez Gomez, B.; Gracia Gomez, C.; et al. The 2013 Ibiza Calibration Campaign of Jason-2 and SARAL Altimeters. *Mar. Geodesy* **2015**, *38*, 219–232. [CrossRef]

39. Vu, P.; Frappart, F.; Darrozes, J.; Marieu, V.; Blarel, F.; Ramillien, G.; Bonnefond, P.; Birol, F. Multi-Satellite Altimeter Validation along the French Atlantic Coast in the Southern Bay of Biscay from ERS-2 to SARAL. *Remote Sens.* **2018**, *10*, 93. [CrossRef]

40. Salameh, E.; Frappart, F.; Marieu, V.; Spodar, A.; Parisot, J.-P.; Hanquiez, V.; Turki, I.; Laignel, B. Monitoring Sea Level and Topography of Coastal Lagoons Using Satellite Radar Altimetry: The Example of the Arcachon Bay in the Bay of Biscay. *Remote Sens.* **2018**, *10*, 297. [CrossRef]

41. Frappart, F.; Papa, F.; Marieu, V.; Malbeteau, Y.; Jordy, F.; Calmant, S.; Durand, F.; Bala, S. Preliminary Assessment of SARAL/AltiKa Observations over the Ganges-Brahmaputra and Irrawaddy Rivers. *Mar. Geodesy* **2015**, *38*, 568–580. [CrossRef]

42. Wingham, D.J.; Rapley, C.G.; Griffiths, H. New Techniques in Satellite Altimeter Tracking Systems. In Proceedings of the International Geoscience and Remote Sensing Symposium (IGARSS), Zurich, Switzerland, 1986; pp. 1339–1344.

43. Nielsen, K.; Stenseng, L.; Andersen, O.; Knudsen, P. The Performance and Potentials of the CryoSat-2 SAR and SARIn Modes for Lake Level Estimation. *Water* **2017**, *9*, 374. [CrossRef]

44. Schneider, R.; Godiksen, P.N.; Villadsen, H.; Madsen, H.; Bauer-Gottwein, P. Application of CryoSat-2 altimetry data for river analysis and modelling. *Hydrol. Earth Syst. Sci.* **2017**, *21*, 751–764. [CrossRef]

45. Jiang, L.; Schneider, R.; Andersen, O.; Bauer-Gottwein, P. CryoSat-2 Altimetry Applications over Rivers and Lakes. *Water* **2017**, *9*, 211. [CrossRef]

46. Birkinshaw, S.J.; O'Donnell, G.M.; Moore, P.; Kilsby, C.G.; Fowler, H.J.; Berry, P.A.M.M. Using satellite altimetry data to augment flow estimation techniques on the Mekong River. *Hydrol. Process.* **2010**, *24*, 3811–3825. [CrossRef]

47. Papa, F.; Bala, S.K.; Pandey, R.K.; Durand, F.; Gopalakrishna, V.V.; Rahman, A.; Rossow, W.B. Ganga-Brahmaputra river discharge from Jason-2 radar altimetry: An update to the long-term satellite-derived estimates of continental freshwater forcing flux into the Bay of Bengal. *J. Geophys. Res. Ocean.* **2012**, *117*, C11021. [CrossRef]

48. Getirana, A.C.V.; Peters-Lidard, C. Estimating water discharge from large radar altimetry datasets. *Hydrol. Earth Syst. Sci.* **2013**, *17*, 923–933. [CrossRef]

49. Bonnefond, P.; Verron, J.; Aublanc, J.; Babu, K.; Bergé-Nguyen, M.; Cancet, M.; Chaudhary, A.; Crétaux, J.-F.; Frappart, F.; Haines, B.; et al. The Benefits of the Ka-Band as Evidenced from the SARAL/AltiKa Altimetric Mission: Quality Assessment and Unique Characteristics of AltiKa Data. *Remote Sens.* **2018**, *10*, 83. [CrossRef]

remote sensing

MDPI

Article

Lead Detection in Polar Oceans—A Comparison of Different Classification Methods for Cryosat-2 SAR Data

Denise Dettmering *, Alan Wynne, Felix L. Müller, Marcello Passaro and Florian Seitz

Deutsches Geodätisches Forschungsinstitut, Technische Universität München, Arcisstraße 21, 80333 Munich, Germany; alan.wynne@tum.de (A.W.); felix-lucian.mueller@tum.de (F.L.M.); marcello.passaro@tum.de (M.P.); florian.seitz@tum.de (F.S.)
* Correspondence: denise.dettmering@tum.de; Tel.: +49-89-23031-1198

Received: 6 July 2018; Accepted: 26 July 2018; Published: 28 July 2018

Abstract: In polar regions, sea-ice hinders the precise observation of Sea Surface Heights (SSH) by satellite altimetry. In order to derive reliable heights for the openings within the ice, two steps have to be fulfilled: (1) the correct identification of water (e.g., in leads or polynias), a process known as lead classification; and (2) dedicated retracking algorithms to extract the ranges from the radar echoes. This study focuses on the first point and aims at identifying the best available lead classification method for Cryosat-2 SAR data. Four different altimeter lead classification methods are compared and assessed with respect to very high resolution airborne imagery. These methods are the maximum power classifier; multi-parameter classification method primarily based on pulse peakiness; multi-observation analysis of stack peakiness; and an unsupervised classification method. The unsupervised classification method with 25 clusters consistently performs best with an overall accuracy of 97%. Furthermore, this method does not require any knowledge of specific ice characteristics within the study area and is therefore the recommended lead detection algorithm for Cryosat-2 SAR in polar oceans.

Keywords: leads; satellite altimetry; CryoSat-2; classification; peakiness; polar ocean

---

1. Introduction

Satellite altimeter missions have been providing precise measurements of the Sea Surface Height (SSH) on a nearly global scale for more than 25 years. This data set has significantly improved our understanding of the sea level dynamics. However, in the polar regions, the altimetry measurements are contaminated by the sea-ice resulting in a degradation in the precision or even usefulness of the results. Radar altimeter observations in the vicinity of sea-ice typically result in an over-estimation of SSH [1]. Knowledge of the locations of openings within the sea-ice, known as leads, can help improve the SSH estimation and the understanding of the global climate. For instance, the SSH derived through leads is assumed to be the instantaneous ocean elevation which can be used to estimate the sea-ice freeboard [2,3]. Therefore, it is clear to see that a reliable SSH is essential for the estimation of the freeboard thickness. Freeboard estimates can be negatively biased when an unreliable lead detection is imposed and the SSH precision is degraded by ice samples detected as leads [1]. Accordingly, the reliable identification of leads is an essential criteria for precisely monitoring the Arctic ice volume decrease, as well as for Arctic sea level change studies.

Indications to the usefulness of satellite altimetry for finding open water within sea-ice were provided as early as 1980 by Dwyer and Godin [4]. They were able to determine the sea-ice boundary using GEOS-3 satellite altimetry data with the knowledge that within sea-ice regions, the altimetry backscattering properties are in stark contrast to that of open water. The altimeter returned signal, known as a waveform, provides this backscattering information. Leads are known to produce high power and single peak specular returns in altimeter waveforms. This is because leads typically have a smooth (relative to the radar wavelength) surface as they generally are absent of any wave field, meaning a large percentage of the radiated energy is reflected back towards the antenna of the satellite [4,5].

To date, many methods to find leads based on altimeter observations have been implemented. Laxon [5] developed a parameter based on the maximum power of ERS satellite altimeter waveforms, pulse peakiness, which could be used for charting the open water areas within the Arctic regions. Connor et al. [6] also used the pulse peakiness parameter with Envisat data to find leads in order to make estimates of the sea-ice freeboard. Röhrs and Kaleschke [7] used a maximum power threshold on Cryosat-2 (CS-2) SAR altimeter waveforms to identify leads. Zygmuntowska et al. [8] developed a supervised classification method based on Bayesian classification [9] applied to ASIRAS (airborne version of the SIRAL instrument on-board CS-2) waveform shape parameters. Passaro et al. [10] developed a parameter based on the stack waveforms of CS-2 known as the stack peakiness. This parameter also considers multiple consecutive observations in an attempt to isolate a lead at the nadir position. Müller et al. [11] developed an unsupervised classification method based on K-medoids partitional clustering [12] of model waveform features and classification of observation data relative to the model using K-nearest neighbor (KNN) classification [13]. Shen et al. [14] aimed to find an optimal classifier–feature assembly for sea-ice classification. They tested a number of waveform feature statistical methods with different combinations of waveform features. Just recently, Lee et al. [15] proposed a promising new method for detecting leads using CS-2 waveforms. Their method is based on a waveform mixture analysis, which assumes that each footprint of the waveform is a linear combination of leads and various types of sea-ice.

Most of the above mentioned studies include a validation of the presented results. However, since they are based on different input data, different ground truth data sets and different accuracy metrics, they are not directly comparable. To overcome this deficiency and to recommend the best classification method for further applications is the motivation of this investigation.

Within this study, four lead classification methods are tested for CS-2 SAR data and evaluated using high resolution imagery acquired during CS-2 dedicated underflights. The aim is to find the best existing CS-2 SAR altimeter lead classification algorithm that provides low false classification. The four classification methods under investigation are (1) the waveform maximum power as introduced, e.g., by Röhrs and Kaleschke [7]; (2) the multi-parameter classification method primarily based on pulse peakiness developed by Ricker et al. [16]; (3) a method based on the stack peakiness presented by Passaro et al. [10]; and (4) the unsupervised classification method published by Müller et al. [11].

This is the first study that compares different classification approaches using exactly the same input data and the same very high resolution (<1 m) ground truth. Previous studies have used space-based evaluation techniques that have suffered from a relatively low resolution, meaning that small scale leads are not observed by the ground truth and, therefore, they bias the classification results. Furthermore, observations with time differences of multiple hours have been utilized, also distorting the results. In contrast, here, almost time coincident ground truth data is used.

This paper is structured as follows. Section 2 describes the CS-2 altimeter SAR data sets and some of their processing parameters as well as the ground truth image data sets. A brief description of each employed lead classification method together with an overview of the threshold optimization technique is given in Section 3. A quantitative comparison of all lead classification methods relative to the ground truth combined with a discussion of the results is provided in Section 4. Finally, the conclusion summarizes the results and provides the reader with a recommendation for the "best" lead classification method based on the available data.

2. Data Sets and Study Areas

This section will discuss the CS-2 data sets, the ground truth used for evaluating the classification methods, and the corresponding study areas.

2.1. CryoSat-2 Data

The CS-2 mission is the first satellite altimeter utilizing the so-called Delay-Doppler or SAR concept in some geographical regions, among them large parts of the Polar Oceans (depending on a changing mode mask [17]). The basic idea behind the SAR principle is that the antenna size is synthetically increased (in the along-track direction) allowing to create a coherence of the emitted pulses [18]. SAR altimeters are described as being beam-limited in the along-track direction while being pulse-limited in the across-track direction. Along-track processing improves the respective resolution and furthermore, the multi-looked processing greatly improves the signal-to-noise ratio [19]. While the footprint of CS-2 yields about 300 m in the along-track direction, in the cross-track direction up to 15 km are covered by the signal [20].

For this study, in addition to the averaged multi-looked waveforms, full stack data is needed. ESA grants access to an online service called Grid Processing On-Demand (GPOD), which provides the full stack (L1B-S) for the CS-2 mission. The CS-2 data sets used within this study (mainly multi-looked waveforms and derived parameters) are all taken from the SAR Versatile Altimetric Toolkit for Ocean Research & Exploitation (SARvatore) GPOD service (see acknowledgments). The GPOD service allows for free and easy access to two main data products, namely ESA Baseline C L1B-S and L2 data [21]. The online service allows for tailored processing of each product. Dinardo [22] has described in detail the processing steps for L1B data. For the present study, a Hamming window is applied before the along-track Fast Fourier Transform in an attempt to stop signals coming from leads that are outside the synthetic along-track 3 dB beam-width (main-lobe), as recommended by Passaro et al. [10].

2.2. Ground Truth Data

For evaluating lead classification results, the contemporaneity between altimetry measurements and ground truth observations is of great importance in order to exclude any differences due to sea-ice drifts. The NASA Operation IceBridge (OIB) mission [23] includes several CS-2 underflights, following the CS-2 groundtracks in Arctic and Antarctica with very short time discrepancy in overpassing [24]. Moreover, airborne sensors acquire cloud-free images with a very high resolution.

The ground truth validation data used within this study is the Operation IceBridge Digital Mapping System (DMS) L1B geolocated and orthorectified images [25]. These images are panchromatic and provide a very high resolution, making them an ideal choice for lead detection. The pixel resolution for a flight height of 450 m is about 10 cm × 10 cm, resulting in an image size of approximately 775 m × 690 m. Within this study, the images are resampled into 1 m resolution in order to reduce the computational costs. This resolution is deemed sufficient for detecting even small leads, especially since the CS-2 resolution is significant lower (see Section 2.1).

2.3. Flight Lines

In order to ensure a reliable validation, CS-2 observations are selected on tracks that are simultaneous and almost time-coincident with the airborne underflights from the NASA Operation IceBridge mission. In total, 12 data sets were found defining the validation data set from polar regions within this study (eight on the northern hemisphere, four on the southern hemisphere). Figure 1 illustrates the locations of these data sets for the Arctic and Antarctic regions, respectively. The majority of the data sets are taken from the spring season.

Airborne images that were acquired later than 90 min after the satellite overpass and earlier than 90 min before the satellite overpass are discarded from the analysis. This threshold is chosen as an optimum balance between having sufficient data to perform this study and preventing lead positional errors due to the sea-ice motion which may bias the statistical results. Additionally, adapting the methodology from Ricker et al. [16] to coarsely discriminate between open ocean and sea-ice, only those data observed with a sea-ice concentration greater than 70% are used, since the theoretical basis for lead detection may no longer hold in low sea-ice concentration areas. The sea-ice concentration values used in this study are the interpolated sea-ice concentration values available from the Ocean and Sea-Ice Satellite Application Facility (OSI SAF) (see acknowledgments).

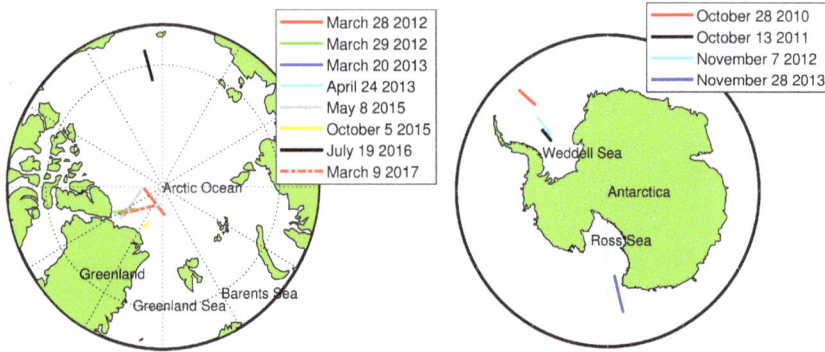

Figure 1. Illustration of Operational IceBridge (OIB) flight lines taken for validation in the Arctic Region (**left**) and the Antarctic Region (**right**).

3. Methodology

This section gives an overview of the employed four lead classification methods, whose performances will be compared within this paper. Moreover, the methods used for ground truth image classification and for optimal threshold definition will be described.

3.1. Maximum Power Classifier (MAX)

The maximum power classifier [7] has the simplest implementation of all the classification methods used within this study. It relies on the application of a threshold to the maximum power of multi-looked altimetry waveforms. If the maximum power is above the threshold, then the corresponding observation is classified as a lead. In a similar previous study, Wernecke and Kaleschke [1] found the maximum power classifier to be the best lead classification method with their ground truth validation data—MODIS satellite imagery. Although the maximum power classifier has an easy implementation, it does require knowledge of the study area in order to correctly define the threshold, which is usually not assured. Moreover, the maximum waveform power—and thus the correct threshold—is known to vary substantially across different regions and seasons [10].

3.2. Multi-Parameter Classification Method (MULTI)

The second lead classification method used in this study is the one proposed by Ricker et al. [16], which uses six parameters in order to discriminate a lead observation from one over sea-ice. The parameters are:

- Pulse Peakiness (PP): Defined as the ratio of the waveform maximum power to the accumulated power and also in this case scaled by the number of range bins. Larger values are indicative of the presence of a lead within the altimeter footprint.
- Stack Kurtosis (K): An additional measure of the peakiness of the range integrated stack power [26]. A high value suggests that the distribution is prone to an outlier, e.g., caused by the presence of a lead within the altimeter footprint.
- Stack Standard Deviation (SSD): This value provides information to the variation of surface backscattering power with incidence angle [26]. Therefore, small values of SSD are used as an indicator for the presence of leads.

- Modified PP (two parameters PP_l and PP_r to consider the bins "left" and "right" of the maximum power bin): This is meant to disregard lead observations which are not at nadir. The assumption here is that off-nadir lead reflections do no show as specular a reflection as an observation at nadir.
- Sea-ice concentration: This is used as a coarse discrimination between ocean and ice areas. Only observations in areas with significant sea-ice concentration (>70%) are allowed as leads in case all other conditions are met.

The MULTI method also requires the definition of thresholds for all parameters, which are used to discriminate lead observations from sea-ice observations. Furthermore, the correct setup of the thresholds requires knowledge of the sea-ice characteristics and are usually not transferable between different regions.

3.3. Stack Peakiness Classifier (STACK)

The stack kurtosis and SSD can be used to classify a waveform that might contain a lead, but they are independently insufficient in isolating a nadir return from a stack of waveforms which has observed a lead backscatter. This was the motivation of Passaro et al. [10] to develop the Stack Peakiness (SP) parameter that is defined using the Range Integrated Power (RIP) normalized against its maximum value. For CS-2 Baseline C, the SP parameter must be computed using the stack data available from the GPOD website, as it is not available in the official Baseline C L2 data product. However, it is included in the Baseline D L2 data.

The assumption behind the algorithm is that larger local maxima in SP are likely to be evoked by a lead at nadir. Therefore, a multi-observation analysis occurs before a lead classification is made, which considers the current, forward and aft observations. The forward and aft observations must be above the median SP (SP_{median}) plus Median Absolute Deviation MAD (SP_{MAD}) of all CS-2 tracks within this study for the current observation to be considered a lead. This is done in an attempt to reduce false lead observations as a lead within the altimeter footprint influences the SP value even when not at nadir. The current observation must also be above a predefined minimum threshold ($SP_{minlead}$) in order for it to be classified as a lead.

The SP strategy developed in Passaro et al. [10] had the objective of isolating one single return in each lead, without distinguishing other reflecting surfaces. In contrast, within this study all observations above a lead are considered as part of the statistical analysis. This therefore increases the percentage of points considered as erroneous ice detections in this study (see Section 4).

The performance of this classification method depends on the correct threshold definition as much as the two previous methods. Section 3.5 will handle the threshold optimization.

3.4. Unsupervised Classifier (UNSU)

The unsupervised classification method developed by Müller et al. [11] is believed to be the first attempt using this method applied to satellite altimetry over sea-ice areas. There are many different approaches to unsupervised learning including clustering (e.g., K-means, K-medoids [27]), latent variable models (e.g., Hidden Markov Models [28]) and others. Müller et al. [11] employed the clustering K-medoids algorithm for their unsupervised classification on pulse-limited altimetry waveforms (Envisat and SARAL). In this study, the unsupervised classification method is tested for the first time with SAR altimeter data.

The first step in the unsupervised classification process is to cluster or group model data applying the K-medoids algorithm. The six waveform features defined by Müller et al. [11] are also used for the clustering in this study, with minor changes in some computation parameters:

- Waveform maximum (Wm): As described in Section 3.1, the maximum power can be used to characterize the surface below the satellite altimeter.

- **Trailing edge decline (Ted):** The Ted is a characterization of the trailing edge of the waveform, i.e., from the maximum power range bin to the last range bin, by means of a fitting to a power series model. A rapid decay (low value) would be associated to the typical waveform shape of a lead.
- Waveform noise (Wn): In the context presented here, the Wn represents the MAD of the residuals to the fitting of the power series model. Very small values for specular lead type returns are expected.
- Waveform width (Ww): The amount of range bins with their power greater than 1% of the waveform maximum is used to determine the waveform width. A small waveform width is expected in the presence of a lead.
- **Leading edge slope (Les):** The first waveform bin containing more than 12.5% of the waveform maximum power subtracted from the maximum power bin provides relative information regarding the Les. Again, low values are indicative of a lead surface.
- Trailing edge slope (Tes): Conversely, the Tes is obtained by a subtraction of the maximum waveform power bin from the last bin position containing more than 12.5% of the waveform maximum power. The characteristics of the Tes is similar to the Les for single peak waveforms, but it can also be used to identify strong multiple peaks.

The features produce varying orders of magnitudes, which causes biased weighting within the K-medoids algorithm. Therefore, before used for clustering, all features are reduced using the standard score (sometimes referred to as z-score) to give more reliable clustering results [29].

In order to have a model that is representative of as many waveform types as possible, a large collection of waveforms is required from the CS-2 mission. This has to include waveforms from sea-ice, leads and ocean surfaces. The maximal extent of the Arctic sea-ice is in March [30] and therefore, the beginning of the melt-season (April/May) is assumed as the most likely period to cover as many sea-ice types as possible. Figure 2 shows the location of the model data taken for this study from the Greenland Sea. The data used corresponds to when CS-2 was operating in SAR mode during the time period of April through to May 2015, corresponding to 361,977 model waveforms.

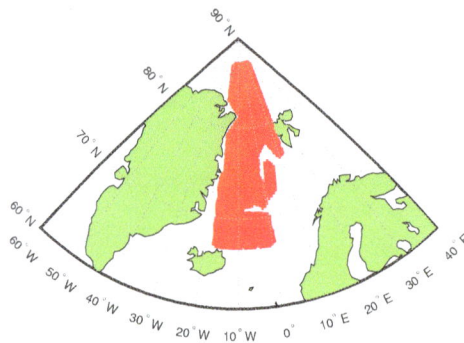

Figure 2. Map of the locations (red areas) of CS-2 waveforms used for the unsupervised classification model taken within the time period of April–May 2015.

To initialize the K-medoids algorithm, the number of clusters k needs to be decided. The value for k should be larger than the final amount of surface classifications, i.e., $k > 3$ (lead, sea-ice and ocean). Müller et al. [11] used $k = 30$ which was empirically derived by means of visual analysis of the resulting clusters and sums of distances. Within this study, a test of seven cluster numbers is performed ranging from $k = 5$ to $k = 35$ in steps of 5.

After the K-medoids algorithm has finished, the clusters have to be classified manually to one of the surface types. The final step is then to classify observation data relative to the classified cluster model. This is done using K-Nearest Neighbor (KNN) classification. The KNN size must also be chosen and this is done using a 10-fold cross-validation method as described in detail in Müller et al. [11].

3.5. Threshold Optimization

Three of the four classification methods under investigation (namely MAX, MULTI, and STACK) depend on one or more predefined thresholds that can be determined using a reliable ground truth. Wernecke and Kaleschke [1] developed an optimization technique, which is also applied here. It minimizes a cost function to derive the threshold value Θ

$$\text{cost}(\Theta) = w \times \text{False Ice}(\Theta) + \text{False Lead}(\Theta) \tag{1}$$

where w is a weighting defining how the false classifications are minimized. "False Ice" are observations classified as ice but are actually lead samples and "False Leads" are observations classified as leads but are actually ice samples. For small values of w (i.e., <1) False Lead observations are primarily reduced and for $w = 1$, the total false classifications are minimized. The ground truth is used to randomly separate the data into a training (50%) and testing (50%) subset with an equal number of sea-ice and lead samples within each subset. Θ is derived from the training subset by testing all potential threshold values as initial guesses to the Nelder and Mead [31] minimization algorithm. The random sub-sampling is repeated 200 times in order to find the global minimum of the cost function and to have an estimation of the spread of performance on each testing subset. This is done for different values of w.

For MULTI, only the PP and modified PP parameters thresholds are derived using this optimization. This is done in an attempt to reduce computation time. The sea-ice concentration value does not depend on the satellite altimeter so the threshold of 70% remains. The thresholds of 40 and 4 from Ricker et al. [16] for K and SSD respectively also remain as it was found that the vast majority of lead observations fell inside these thresholds.

For STACK, the SP_{median} and SP_{MAD} found through the ground tracks used in this study resulted in $SP_{median} = 9.76$ and $SP_{MAD} = 6.33$, meaning a slightly lower sum compared to Passaro et al. [10]. For this reason, it was decided to use the values from the literature. When developing the $SP_{minlead}$, due to the random sampling of the data, the multi-observation analysis was not conducted for the threshold optimization.

The optimized threshold values and their corresponding weight value are given in Table 1. All of them are used to generate different classification results that are shown and discussed in Section 4.

Table 1. Optimized thresholds and their corresponding weight.

w	Max. Power	PP	PP_l	PP_r	SP
0.5	4.356×10^{-11} W	120.75	175.08	30.84	345.02
1	2.714×10^{-11} W	116.18	152.58	158.11	300.43
2	1.892×10^{-11} W	99.93	63.25	145.29	156.61
3	1.051×10^{-11} W	99.67	63.38	124.28	102.86
4	4.513×10^{-12} W	96.71	66.05	69.28	91.28
5	4.958×10^{-12} W	97.01	63.92	51.59	78.61
6	3.609×10^{-12} W	97.03	64.15	50.89	73.47
7	2.219×10^{-12} W	97.00	62.25	56.19	71.14
8	1.654×10^{-12} W	97.15	53.80	51.43	70.41
9	1.399×10^{-12} W	90.18	63.20	53.98	70.41
10	1.359×10^{-12} W	80.47	57.32	30.40	54.03

3.6. Ground Truth Image Classification

In order to be able to compare and validate the different altimeter lead classification methods, a reliable ground truth image classification must be used. In optical images, such as the ones from IceBridge used here, leads are known to have typically dark optical features that translates to low image pixel intensities. Lead attributes can be dynamic due to different types of thin ice within the lead itself, formed by rapid ice growth due to the cold polar temperatures. Additionally, leads with thin ice may even be covered by snow and frost flowers [32].

The ground truth data for this study is set up applying the automatic Sea-Ice Lead Detection Algorithm using Minimum Signal (SILDAMS) developed by Onana et al. [33], which has been shown to give favorable results using NASA Operation IceBridge image data. The SILDAMS algorithm is based on an affine time-frequency distribution analysis which uses the minimum signal transform in order to perform a localization around low frequencies. The minimum signal is given by

$$S^z_{min}(f) = f^{2\pi z}e^{-2\pi zf} \tag{2}$$

where f represents the frequency or image pixel intensities and z is a positive parameter for adjusting the spectral width (bandwidth). With a careful selection of the z parameter, it is possible to highlight features of interest from the image, such as leads. The bandwidth parameter is defined as $z = 0.001$, since this value has been shown by Onana et al. [33] to provide good results for images acquired in clear conditions with daylight. As one of the OIB data sets (taken on 28 October 2010) was acquired in ever decreasing lighting conditions. From a visual analysis of the images, it was decided to change z according to the lighting conditions from $z = 0.003$ for the first 450 images, to $z = 0.005$ (from 451 to 1035), $z = 0.006$ (from 1036 to 1103), and $z = 0.015$ (for the remainder of the images).

The last step in the image classification is to apply a threshold to the minimum signal transformation of the image. Within this study, only a binary classification is used between sea-ice and lead/open water. In accordance with Onana et al. [33], a threshold of 0.3 is used for water detection (smaller values = sea ice; larger values = lead). This results in a binary map for each image with a one indicating a lead pixel and a zero indicating a sea-ice pixel. Nearest-neighbor interpolation is then used to associate the binary classification to the CS-2 surface sample ground track.

All images are assumed to be cloud-free since a manual inspection of completeness of laser measurements from the laser-based Airborne Topographic Mapper (ATM) revealed almost no missing values.

Onana et al. [33] specify the lead detection capability of SILDAMS to be 99%. The manual inspection of the ground truth data set established in the present study shows that all open water areas are correctly flagged without any false detections. However, the distinction between open water and thin ice was only correct at about 90%. For this reason, the open water class also contains thin ice—which is uncritical for altimetry waveform classification which can also not differentiate very thin ice from leads [6].

4. Results and Discussion

In order to be able to assess each altimeter classification method, they must be compared with the same ground truth, which is derived from the image classification described earlier. This section provides and discusses the results of the ground truth evaluation and comparison between each altimeter classification method. In total, there are 14,231 satellite observations from all the data sets, and of these, 365 observations (i.e., about 2.5%) are defined as leads (including leads with thin ice) according to the ground truth.

4.1. Quantitative Comparison between Altimeter Classification and Ground Truth

A Receiver Operating Characteristics (ROC) graph [34] is used to visualize the performance of the different classifiers (see Figure 3). It shows the change in performance of the different altimeter classification methods relative to the ground truth depending on different thresholds. Two criteria are computed for the evaluation: the True Lead Rate (TLR), which is the percentage of leads classified by the altimeter classification and confirmed by the ground truth classification (see Equation (3)) and the False Lead Rate (FLR), which is the percentage of altimeter lead classifications that are classified as sea-ice by the ground truth to the total sea-ice ground truth samples (see Equation (4)).

$$\text{TLR} = \frac{\text{True Lead}}{\text{True Lead} + \text{False Ice}} \tag{3}$$

$$\text{FLR} = \frac{\text{False Lead}}{\text{False Lead} + \text{True Ice}} \tag{4}$$

True Lead is the number of observations classified as lead confirmed by the ground truth. False Ice is classified as ice observations but are lead samples according to the ground truth. False Lead observations are classified as a lead but are ice samples according to the ground truth. True Ice are observations classified as ice samples and are confirmed by the ground truth.

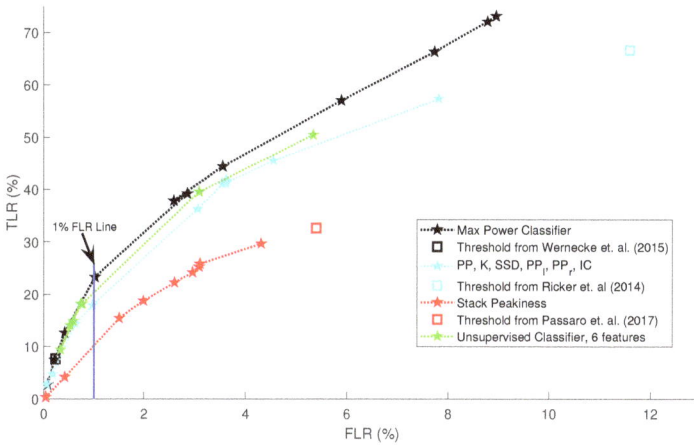

Figure 3. ROC graph depicting the change in performance for different thresholds and different cluster numbers. Each line and marker represents a different altimeter classification method and the different weight for optimized threshold respectively. The square markers represent thresholds that were determined in previous studies. The 1% False Lead Rate (FLR) line is also shown.

The aim of the ROC graphs in Figure 3 is threefold: It helps to find the best threshold for each classification method, it can be used for an inter-comparison of the methods, and it provides a quantitative performance measure for all classifiers. In general, with an increasing weight w (as defined in Equation (1)) there is an increase in the TLR and also the FLR for all classifiers with a higher rate of increase in the TLR. However, this behavior is not linear, and there seems to be a maximum TLR reachable for each method. When further increasing the weight (i.e., lowering the threshold) this only results in a larger FLR. This effect was already shown by Wernecke and Kaleschke [1].

Simply from looking at Figure 3, it is not immediately clear what is the best threshold/number of clusters to use. This strongly depends on the application, i.e., on the acceptable FLR. In case it is important to reach a very low FLR, a conservative threshold (small weight) should be used. This comes at the cost of identified leads. However, if the application requires as many leads as possible, one should use a higher w. All three classifiers reach TLR between 30% and 55% when FLR of 5% are acceptable. From Figure 3, it seems that the MAX classification method performs best for the majority of weights compared to all other classification methods, followed closely by UNSU and MULTI. The STACK classification method appears to be the poorest performing classifier, showing a narrow ROC curve with a small TLR/FLR ratio that is about half of that from MAX. The highest overall TLR of about 75% is achieved by MAX (with FLR of about 10%).

In order to quantify and compare the performance of the different classifiers, a fixed FLR is defined and the corresponding TLR are compared. A FLR of 1% is chosen as an optimum balance between a relatively high TLR and low number of false open water observations. For the MAX classifier, this results in a TLR at 23.05% (closest threshold $w = 3$; 1.05×10^{-11} W). The MULTI method results in a 18.09% TLR (closest threshold $w = 3$; $PP = 99.67$, $PP_l = 63.38$, $PP_r = 124.28$). The STACK gives a performance of 10.38% TLR (closest threshold $w = 3$; $SP_{minlead} = 102.86$). The above TLR values were derived by a polynomial curve fitting to the points of each threshold classifier. From all threshold based classifiers, this appears to show that the maximum power classifier is performing the best relative to the ground truth.

The UNSU classification method is independent from any threshold definition. Instead, the classification performance clearly depends on the number of clusters used within the process (see Section 3.4). It can be noted that for smaller k, there is typically a larger TLR and FLR. However, in contrast to the threshold-based methods, a continuous change in performance is not observed for a change in cluster number. This is illustrated in Figure 4; the figure shows the ROC graph for UNSU in dependence of k. The optimal cluster number depends on the size and variability of the input waveform data set. If only a small number of clusters is allowed, the number of observations per cluster (i.e., the cluster size) increases and the probability for correct classification decreases. In other words: the clustering is too coarse [35]. This is the case for $k \leq 10$. If k is large enough, the results only change marginally, depending on the manual preprocessing step assigning each cluster to one surface type (lead, ice, or ocean). The cluster number with FLR closest to 1% will be chosen as the optimum for this study. This corresponds to $k = 25$ and results in a 18.08% TLR and 0.73% FLR. The interpolated value at 1% FLR is 22.37% TLR.

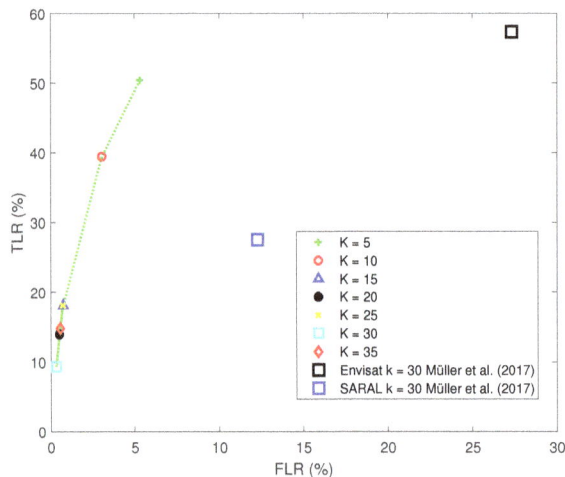

Figure 4. ROC graph depicting the change in performance for different cluster sizes. The performances calculated from the study from Müller et al. [11] are also shown.

ROC graphs are useful for analyzing the performance of classifiers relative to a ground truth, and to identify the best parameters (i.e., thresholds or number of clusters) for the respective application. However, in order to understand the overall performance of the different altimeter classifiers, different accuracy metrics must be analyzed. Producer and user accuracies are used for this purpose for both sea-ice and lead samples. While producer accuracies give an indication of the probability that a ground truth sample will be correctly classified, the user accuracy is an indication to the reliability of the classification and is defined as the ratio of the sum of correct classifications (leads and ice) to the total number of classifications [36].

Table 2 lists the overall lead and sea-ice accuracies for each altimeter classification with respect to the ground truth. The largest overall accuracy is achieved by the unsupervised classification method at 97.2%. This method also yields the best lead user accuracy and the largest ice producer accuracy.

However, the lead producer accuracy scores only second behind the maximum power classifier. Even if the numbers in some of the metrics only slightly differ, it is obvious that the two methods utilizing the maximum power for the classification (UNSU and MAX) score better than the other two (MULTI and STACK). While interpreting the numbers, it should be kept in mind,that all values have been optimized to score a FLR of about 1%. Since only integer weighting factors w are used, slightly different FLR are achieved and the numbers are not fully comparable. The good performance of UNSU is even emphasized, taking the lower FLR into account.

Table 2. Accuracies for each classifier in (%). Values provided for FLR of about 1% ($w = 3$, $k = 25$). The best performance for each of the accuracy metrics is marked in bold.

	MAX	MULTI	STACK	UNSU
FLR	1.02	0.79	1.51	0.73
Overall Accuracy	97.03	96.94	96.36	**97.18**
Lead User Accuracy	37.44	32.50	21.13	**39.29**
Lead Producer Accuracy (TLR)	**23.29**	17.81	15.34	18.08
Ice User Accuracy	**98.00**	97.86	97.79	97.87
Ice Producer Accuracy	98.98	99.03	98.49	**99.26**

4.2. Comparison to Other Studies

The statistics presented above are not directly comparable to results from previous studies. This is mainly due to different input data sets (e.g., different missions [11] or different Cryosat-2 Baselines [1]), different thresholds used within the studies [1,10,16], and a different ground truth used for validation. Most of the older studies used ground truth data sets whose resolutions were limited to many meters (40 m to 250 m; instead of 1 m used here) and therefore, potentially, many leads have been missed. Furthermore, the time between image acquisition and satellite overpass has been limited here to 90 min, meaning there is less of a possibility of induced sea-ice drift related biases in the ground truth results.

In comparison to Passaro et al. [10], the validation approach differs significantly. Their study was focused on the detection of leads with different sizes, whereas here, the detection of altimetry returns over leads is the aim. Thus, instead of counting each lead observation separately, they considered consecutive lead observations to be the same lead. If a single observation from a lead with multiple observations was correctly classified, the overall lead was assumed to be correctly classified. This directly influences the corresponding statistics and for that reason, the results presented in this study appear poorer. However, for lead prospection, they found that the MULTI classification performed on par with the STACK when compared using the ratio of correctly classified leads to false detections.

Müller et al. [11] used exactly the same metrics to validate their results, however, a different ground truth was applied. From Figure 4 it seems that in this study a better performance can be achieved. This is presumably related to the fact that CS-2 has an improved along-track resolution with respect to Envisat and SARAL (i.e., SAR stack data compared to classical pulse-limited altimetry performs better). Though, in order to prove this, the validation of Müller et al. [11] should be repeated based on a high-resolution ground truth.

4.3. Discussion of Altimeter Classification Methods

Threshold Definition: From the previous sections it became clear that those classification methods based on the maximum waveform power perform better than the rest. However, the success of MAX strongly depends on the defined threshold. Many factors affect the backscattering power received at the altimeter antenna. This is clearly evident in Figure 5, which shows the mean and standard deviation of the maximum power at lead surfaces (at nadir) according to the ground truth for each data set used within this study. The mean power varies greatly across each data set which can be a completely different region or season. The backscattering power is not just a function of the lead within the altimeter footprint but also the sea-ice, lead width, lead orientation, refrozen areas of the lead, melt-ponds and more within the illuminated area. These all contribute to varying statistics across different study areas and also make it very difficult to establish an absolute threshold that can be used for all regions and seasons. Figure 5 also displays the threshold developed for the MAX classifier. Since it is optimized based on all data, it matches relatively well. However, some leads are completely missed. This can be even worse for other regions, seasons, or conditions not included in this study.

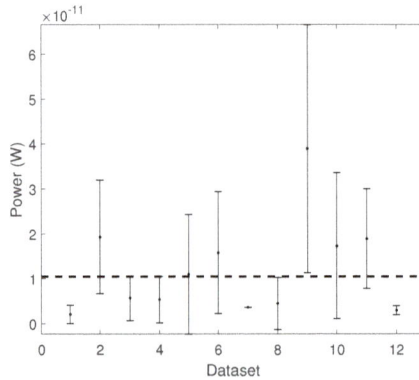

Figure 5. Mean and standard deviation of maximum power for leads across all data sets with the optimized threshold (1.05×10^{-11} W) for FLR = 1% developed here shown as a dotted line.

Nevertheless, it is clear that the maximum power is a good discriminator between leads and sea-ice compared to other index-based classification methods. The definition of a relative power threshold as proposed by Passaro et al. [10] could account for the problem of finding an absolute value and still exploit the returned power as classifier. In that study, the relative power threshold was calculated as the ratio between the maximum power of CS-2 returns classified as leads to the median of the maximum power observations for a given region/segment. A fixed power ratio threshold of 10 was derived a posteriori based on a comparison with Sentinel-1 SAR images across different regions and time of the year. The subsequent classification based on this threshold gave the best ratio of correctly classified leads to false detections among the methods trialled in their study.

Lead Location: Another important effect influencing the performance of all classifiers are off-nadir leads. When located within the satellites footprint but not directly in nadir, they strongly impact the returned radar echo even if not detected by the ground truth. This effect becomes more evident when high-resolution ground truth is used. MULTI and STACK both define dedicated parameters to disregard off-nadir leads. These methods rely on the assumption that the multi-looked waveform (in case of MULTI) or the RIP (in case of STACK) will be less peaky for a lead that is not at the nadir position. However, this is not always true. From Figure 6, it can clearly be seen that even for sea-ice observations (waveform #1159), the multi-looked waveform still exhibits a very peaky shape and therefore, the PP_l and PP_r will unlikely disregard these waveforms. This is confirmed by the classification result of MULTI, which identifies all measurements except the first one as lead observation (waveforms #1156 to #1159). Once a lead falls inside the altimeter footprint, the backscatter return is highly influenced by the presence of this lead and therefore, doubt is cast in the ability of the modified PP in disregarding off-nadir returns as the multi-looked waveform can still be influenced by leads in off-nadir positions.

The STACK method also claims to handle off-nadir leads. However, it relies on the assumption that a lead within the vicinity of the altimeter will be crossed by the satellite track (e.g., will be the local maximum), which is not always the case. Cross-track off-nadir leads cannot be detected by this method.

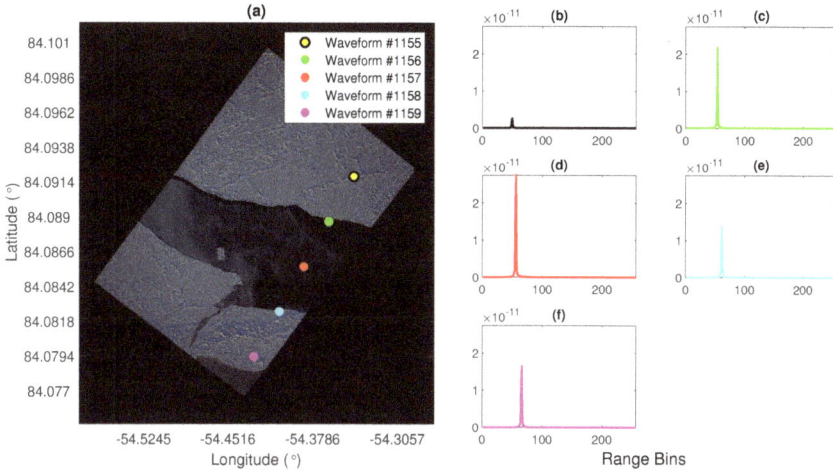

Figure 6. CS-2 descending track #15632 (taken on 20 March 2013) showing five waveform surface sample locations (**a**). Image is acquired at 17:03:15 GPS time, approximately 30 min before satellite overpass. Waveforms associated with the locations are also shown. (**b**) is waveform #1155; (**c**) #1156; (**d**) #1157; (**e**) #1158 and (**f**) #1159.

This is evident in Figure 7: local maxima in SP (i.e., measurements identified as leads; blue triangles) do not always correspond to a lead at the nadir position (red squares). An example of when the SP works well is visible between 84.05° and 84.1° latitude. Here, the two SP peaks correspond to lead surfaces at nadir according to the ground truth. These particular lead observations are orientated along-track relative to the satellite and belong to the same lead.

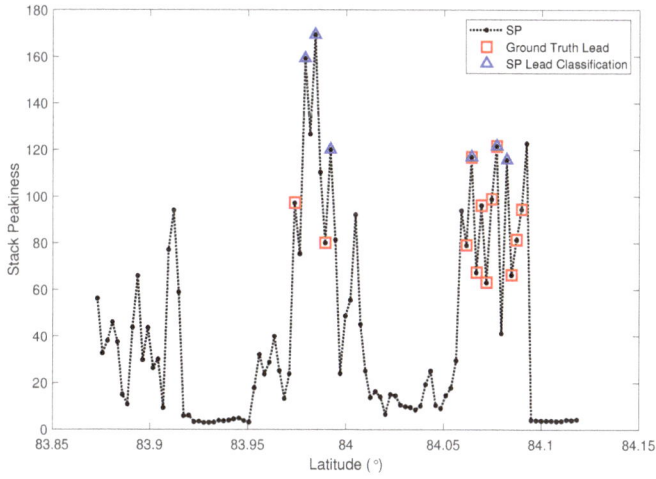

Figure 7. Evolution of CS-2 stack peakiness in data set taken on 20 March 2013. Observations classified as leads by the STACK classifier are marked in blue. Actual nadir leads identified by the ground truth are highlighted in red.

All results presented here are also influenced by a potential CS-2 mispointing angle in along-track direction (pitch angle) related to biases of the on-board star trackers [37]. Such an effect may result in the location of the CS-2 surface sample also being biased. This is especially relevant for narrow leads where, for example, the SP would peak in the observation before the lead (assuming satellite pitched up) resulting in a false classification.

In order to unambiguously isolate nadir leads from off-nadir leads, the usage of the CS-2 InSAR mode will be beneficial. The exploitation of signals from both antennas will allow for an exact location of the reflection point of the signal (in cross track direction). However, depending on the CS-2 geographically mode mask, currently, these data sets are only available in a few areas, mostly not in sea-ice ocean regions.

Stack Information: From the four methods under investigation, STACK and MULTI use the full CS-2 stack information, with respectively derived parameters (SSD and SP). Theoretically, one may expect that this additional information helps to improve the lead classification. However, it seems that using only stack data without taking the maximum power into account is not an optimal method. This might also be related to residual side-lobe effects in the RIP even after the application of the Hamming window function. Passaro et al. [10] have shown that without the use of the Hamming window, leads that have influenced the signal through the side-lobes can influence the statistics of the stack data. The SP value is being influenced by high power returns in the non-nadir look angles. This has the effect of making the RIP return for these observations wider than what might be expected for a lead at nadir and therefore, it contributes to a lower SP value. An analysis of the RIP for the locations shown in Figure 6 found that the sea-ice (#1155) SP was much larger than that of the central lead location (#1157) and this was attributed to residual side-lobe effects as much larger RIP was observed in the non-nadir look angles. In spite of this, it will be useful to combine stack information with maximum power information to develop an improved classification approach.

Unsupervised Classification: From all the lead classification methods presented in this study, the UNSU method has performed the best when aiming at a low FLR of better than 1%. In general, the UNSU method has a similar performance as the MAX classifier, and the difference between the two classification methods is minimal. The most prominent feature among the six features used in the unsupervised classification method is the maximum power for distinguishing lead surfaces. This may explain the similar performances when using the MAX approach.

The most distinct advantage of the unsupervised classification method is that no training data is necessary and that the approach does not rely on the development of a threshold. Therefore, no prior knowledge of the study area is required. The only requirement is that the general characteristics of altimeter waveforms are understood and meaningful features are defined. The clustering can be done in one region (e.g., the Greenland Sea) and transferred to other study areas (such as the Arctic Ocean or the Weddell Sea) in case all relevant waveform types are included in the clustering process. This will not affect the classification results.

The used cluster number k has only marginal influence on the classification results in case the order of magnitude is correct and matches to the number of waveform features and variability of waveforms used for the clustering process. Moreover, the approach is applicable to different altimetry missions (e.g., Envisat and SARAL [11]) and different measurement techniques (i.e., traditional pulse-limited and SAR altimetry) with only slight modifications to the waveform features used for the clustering. In contrast, all threshold-based methods must be tuned for each mission separately. Last but not least, the UNSU method can be adopted for the marginal sea-ice zones to include ocean returns in the classification whereas the other methods are not easily adaptable to ocean retrievals.

5. Conclusions and Outlook

The aim of this paper is to find a CS-2 altimeter lead classification algorithm that provides low false classification. Four altimeter classification methods are trialled and compared to high resolution automatically classified airborne images. Twelve dedicated airborne underflight image data sets acquired within 90 min before and after satellite overpass are used for the validation of all altimeter classification methods.

The unsupervised classification method (UNSU) is found to be the best classification method within this study with an overall accuracy at 97%. The maximum power classification method (MAX) closely followed with basically the same overall accuracy. The achieved lead user accuracies are 39% and 37%, respectively. Both methods perform very similar and in the case that larger false detection rates (larger than about 1%) are acceptable, the maximum power classifier is superior. The performance is expected to be further improved when a relative power threshold is applied instead a fixed one for all data sets. It is clear from this study that the maximum power index remains the most favorable index for reducing false lead observations. The two methods that do not use this parameter perform worse within this study.

In contrast to all other methods under investigation, the unsupervised classification method does not rely on any knowledge of the study area, especially on ground truth for threshold optimization. Moreover, it is applicable to all altimetry missions independent of the used measurement technique (pulse-limited or SAR). For these reasons, the unsupervised classification method (with 20 to 30 clusters) is the recommended lead classification method for Cryosat-2 SAR data.

The tested altimeter classifications within this paper have high misclassification compared to previous studies. This can be primarily associated with a much more robust and high resolution ground truth validation, which is much higher than the CS-2 along-track resolution.

The CS-2 waveforms are found to be very sensitive to the presence of leads within the altimeter footprint even when the lead does not cross the nadir point. This influences each waveform feature in the vicinity of a lead and thus, can result in many false classifications. The modified pulse peakiness (MULTI) and the stack peakiness (STACK) were designed to disregard off-nadir leads. However, these parameters are still strongly influenced by the presence of leads within the altimeter footprint, especially if these are located cross-track (in parallel to the satellite's track).

The utilization of CS-2 stack data in modified pulse peakiness and the stack peakiness methods cannot outperform the approaches using the maximum power parameter (alone or in connection with other parameters). Thus, a combination of stack data with a relative maximum power is expected to help to improve the existing classifiers.

The CS-2 mission also provides the opportunity to operate in SAR-interferometric (SARIn) mode which may help to reliably separate all types of off-nadir leads from nadir leads. Armitage and Davidson [38] have shown that the localization of nadir leads can be improved compared to the use of SAR mode. However, currently, the availability of SARIn mode is limited to certain regions.

Author Contributions: A.W. implemented the classification and validation methods, conducted the data analysis, and wrote the first version of the paper. D.D. supervised the study, wrote the final manuscript and made major contributions to the discussion and interpretation of the classification methods and results. F.L.M. and M.P. both helped with the discussion of the classification methods and results and also contributed to the manuscript writing. F.S. supervised the research and also contributed to the discussion of the applied methods and results.

Funding: This research received no external funding.

Acknowledgments: This work was supported by the German Research Foundation (DFG) and the Technical University of Munich (TUM) in the framework of the Open Access Publishing Program. The authors would like to acknowledge the European Space Agency for allowing access to the CryoSat-2 data from their G-POD website (https://gpod.eo.esa.int/). Also, thanks go to NASA and the NSIDC for the distribution of the airborne image data key to this study (https://nsidc.org/icebridge/portal/map). Furthermore, thanks are afforded to the OSI SAF for the sea-ice concentration products (http://osisaf.met.no/p/ice/index.html).

Conflicts of Interest: The authors declare no conflict of interest.

References

1. Wernecke, A.; Kaleschke, L. Lead Detection in Arctic Sea Ice from Cryosat-2: Quality Assessment, Lead Area Fraction and Width Distribution. *Cryosphere* **2015**, *9*, 2167–2200. [CrossRef]
2. Laxon, S.; Peacock, N.; Smith, D. High Interannual Variability of Sea Ice Thickness in the Arctic Region. *Lett. Nat.* **2003**, *425*, 947–950. [CrossRef] [PubMed]
3. Farrell, S.L.; Laxon, S.W.; McAdoo, D.C.; Yi, D.; Zwally, H.J. Five Years of Arctic Sea Ice Freeboard Measurements from the Ice, Cloud and land Elevation Satellite. *J. Geophys. Res.* **2009**, *114*. [CrossRef]
4. Dwyer, R.E.; Godin, R.H. *Determining Sea-Ice Boundaries and Ice Roughness using GEOS-3 Altimeter Data*; Technical Report; NASA Wallops Flight Center: Wallops Island, VA, USA, 1980.
5. Laxon, S. Sea Ice Altimeter Processing Scheme at the EODC. *Int. J. Remote Sens.* **1994**, *15*, 915–924, doi:10.1080/01431169408954124. [CrossRef]
6. Connor, L.N.; Laxon, S.W.; Ridout, A.L.; Krabill, W.B.; McAdoo, D.C. Comparison of Envisat Radar and Airborne Laser Altimeter Measurements Over Arctic Sea Ice. *Remote Sens. Environ.* **2009**, *113*, 563–570, doi:10.1016/j.rse.2008.10.015. [CrossRef]
7. Röhrs, J.; Kaleschke, L. An Algorithm to Detect Sea Ice Leads by Using AMSR-E Passive Microwave Imagery. *Cryosphere* **2012**, *6*, 343–352, doi:10.5194/tc-6-343-2012. [CrossRef]
8. Zygmuntowska, M.; Khvorostovsky, K.; Helm, V.; Sandven, S. Waveform Classification of Airborne Synthetic Aperture Radar Altimeter over Arctic Sea Ice. *Cryosphere* **2013**, *7*, 1315–1324, doi:10.5194/tc-7-1315-2013. [CrossRef]
9. Friedman, N.; Kohavi, R. Bayesian Classification. In *Handbook of Data Mining and Knowledge Discovery*; Oxford University Press: New York, USA, 2002; pp. 282–288
10. Passaro, M.; Müller, F.L.; Dettmering, D. Lead Detection using CryoSat-2 Delay Doppler Processing and Sentinel-1 SAR images. *Adv. Space Res.* **2017**, doi:10.1016/j.asr.2017.07.011. [CrossRef]
11. Müller, F.L.; Dettmering, D.; Bosch, W.; Seitz, F. Monitoring the Arctic Seas: How Satellite Altimetry can be used to Detect Open Water in Sea-Ice Regions. *Remote Sens.* **2017**, *9*, 551. [CrossRef]
12. Park, H.S.; Jun, C.H. A Simple and Fast Algorithm for K-medoids Clustering. *Expert Syst. Appl.* **2009**, *36*, 3336–3341, doi:10.1016/j.eswa.2008.01.039. [CrossRef]
13. Cover, T.M.; Hart, P.E. Nearest Neighbor Pattern Classification. *IEEE Trans. Inf. Theory* **1967**, *13*. [CrossRef]
14. Shen, X.; Zhang, J.; Zhang, X.; Meng, J.; Ke, C. Sea Ice Classification Using Cryosat-2 Altimeter Data by Optimal Classifier–Feature Assembly. *IEEE Geosci. Remote Sens. Lett.* **2017**, *14*, 1948–1952. [CrossRef]
15. Lee, S.; Kim, H.C.; Im, J. Arctic Lead Detection using a Waveform Mixture Algorithm from CryoSat-2 Data. *Cryosphere* **2018**, *12*, 1665–1679, doi:10.5194/tc-12-1665-2018. [CrossRef]
16. Ricker, R.; Hendricks, S.; Helm, V.; Skourup, H.; Davidson, M. Sensitivity of CryoSat-2 Arctic Sea-ice Freeboard and Thickness on Radar-Waveform Interpretation. *Cryosphere* **2014**, *8*, 1607–1622, doi:10.5194/tc-8-1607-2014. [CrossRef]
17. Jiang, L.; Schneider, R.; Andersen, O.B.; Bauer-Gottwein, P. CryoSat-2 Altimetry Applications over Rivers and Lakes. *Water* **2017**, *9*, 211. [CrossRef]
18. Raney, R.K. The Delay/Doppler Radar Altimeter. *IEEE Trans. Geosci. Remote Sens.* **1998**, *36*, 1578–1588. [CrossRef]
19. Martin-Puig, C.; Ruffini, G. SAR Altimeter Retracker Performance Bound over Water Surfaces. In Proceedings of the 2009 IEEE International Geoscience and Remote Sensing Symposium, Cape Town, South Africa, 12–17 July 2009; pp. 449–452.
20. European Space Agency. Cryosat Product Handbook. 2012. Available online: http://emits.sso.esa.int/emits-doc/ESRIN/7158/CryoSat-PHB-17apr2012.pdf (accessed on 21 September 2017).
21. Scagliola, M.; Fornari, M. *Main Evolutions and Expected Quality Improvements in Baseline C Level 1b Products*; Aresys Technical Note, C2-TN-ARS-GS-5154, Issue 1.3; 2015. Available online: https://earth.esa.int/documents/10174/1773005/C2-BaselineC_L1b_improvements_1.3 (accessed on 27 July 2018).
22. Dinardo, S. *Guidelines for the SAR (Delay-Doppler) L1b Processing*; Technical Note, XCRY-GSEG-EOPS-TN-14-0042 Version 2; European Space Agency, ESRIN: Frascati, Italy, 2013. Available online: https://wiki.services.eoportal.org/tiki-download_wiki_attachment.php?attId=2540 (accessed on 30 September 2017).

23. Studinger, M.; Koenig, L.; Martin, S.; Sonntag, J. Operation icebridge: Using instrumented aircraft to bridge the observational gap between icesat and icesat-2. In Proceedings of the 2010 IEEE International Geoscience and Remote Sensing Symposium, Honolulu, HI, USA, 25–30 July 2010; pp. 1918–1919.

24. Xia, W.; Xie, H. Assessing three waveform retrackers on sea ice freeboard retrieval from Cryosat-2 using Operation IceBridge Airborne altimetry datasets. *Remote Sens. Environ.* **2018**, *204*, 456–471. [CrossRef]

25. Dominguez, R. *IceBridge DMS L1B Geolocated and Orthorectified Images, Version 1*; Updated 2017; NASA National Snow and Ice Data Center Distributed Active Archive Center: Boulder, CO, USA, 2010; doi:10.5067/OZ6VNOPMPRJ0.

26. Wingham, D.J.; Francis, C.R.; Baker, S.; Bouzinac, C.; Brockley, D.; Cullen, R.; de Chateau-Thierry, P.; Laxon, S.; Mallow, U.; Mavrocordatos, C.; et al. CryoSat: A Mission to Determine the Fluctuations in Earth's Land and Marine Ice Fields. *Adv. Space Res.* **2006**, *37*, 841–871. [CrossRef]

27. Hastie, T.; Tibshirani, R.; Friedman, J. *The Elements of Statistical Learning*; Springer-Verlag: New York, NY, USA, 2009; Volume 2.

28. Durbin, R.; Eddy, S.R.; Krogh, A.; Mitchison, G. *Biological Sequence Analysis*; Cambridge University Press: Cambridge, UK, 1998.

29. Banks, D.; House, L.; McMorris, F.; Arabie, P.; Gaul, W. (Eds.) *Classification, Clustering, and Data Mining Applications: Proceedings of the Meeting of the International Federation of Classification Societies (IFCS), Illinois Institute of Technology, Chicago, 15–18 July 2004*; Studies in Classification, Data Analysis, and Knowledge Organization; Springer: Berlin/Heidelberg, Germany, 2004.

30. Kvingedal, B. On Sea Ice Variability in the Nordic Seas. Ph.D. Thesis, University of Bergen, Bergen, Norway, 2005. Available online: http://web.gfi.uib.no/publikasjoner/pdf/Kvingedal.pdf (accessed on 3 November 2017).

31. Nelder, J.A.; Mead, R. A Simplex Method for Function Minimization. *Comput. J.* **1965**, *7*, 308–313, doi:10.1093/comjnl/7.4.308. [CrossRef]

32. Perovich, D.K.; Richter-Menge, J.A. Surface Characteristics of Lead Ice. *J. Geophyis. Res.* **1994**, *99*, 16341–16350. [CrossRef]

33. Onana, V.D.P.; Kurtz, N.T.; Farrell, S.L.; Koenig, L.S.; Studinger, M.; Harbeck, J.P. A Sea-Ice Lead Detection Algorithm for Use With High-Resolution Airborne Visible Imagery. *IEEE Trans. Geosci. Remote Sens.* **2013**, *51*, doi:10.1109/TGRS.2012.2202666. [CrossRef]

34. Fawcett, T. An Introduction to ROC analysis. *Pattern Recogn. Lett.* **2006**, *27*, 861–874, doi:10.1016/j.patrec.2005.10.010. [CrossRef]

35. Xue, R.; Wunsch, D.C.; IEEE Computational Intelligence Society. *Clustering*; IEEE Press Series on Computational Intelligence; IEEE Press: Hoboken, NJ, USA; Wiley: Piscataway, NJ, USA, 2009.

36. Story, M.; Congalton, R.G. Accuracy Assessment: A User's Perspective. *Am. Soc. Photogramm. Remote Sens. Remote Sens. Brief* **1986**, *52*, 397–399.

37. Galin, N.; Wingham, D.J.; Cullen, R.; Francis, R.; Lawrence, I. Measuring the Pitch of CryoSat-2 Using the SAR Mode of the SIRAL Altimeter. *IEEE Geosci. Remote Sens. Lett.* **2014**, *11*, 1399–1403, doi:10.1109/LGRS.2013.2293960. [CrossRef]

38. Armitage, T.W.K.; Davidson, M.W.J. Using the Interferometric Capabilities of the ESA CryoSat-2 Mission to Improve the Accuracy of Sea Ice Freeboard Retrievals. *IEEE Trans. Geosci. Remote Sens.* **2014**, *52*, doi:10.1109/TGRS.2013.2242082. [CrossRef]

remote sensing

MDPI

Article

Monitoring the Arctic Seas: How Satellite Altimetry Can Be Used to Detect Open Water in Sea-Ice Regions

Felix L. Müller *, Denise Dettmering, Wolfgang Bosch and Florian Seitz

Deutsches Geodätisches Forschungsinstitut, Technische Universität München, Arcisstraße 21, 80333 Munich, Germany; denise.dettmering@tum.de (D.D.); wolfgang.bosch@tum.de (W.B.); florian.seitz@tum.de (F.S.)
* Correspondence: felix-lucian.mueller@tum.de; Tel.: +49-089-23031-1115

Academic Editors: Frédéric Frappart, Ole Andersen, Sergey Lebedev, Guillaume Ramillien, Deepak R. Mishra and Prasad S. Thenkabail
Received: 24 February 2017; Accepted: 29 May 2017; Published: 1 June 2017

Abstract: Open water areas surrounded by sea ice significantly influence the ocean-ice-atmosphere interaction and contribute to Arctic climate change. Satellite altimetry can detect these ice openings and enables one to estimate sea surface heights and further altimetry data derived products. This study introduces an innovative, unsupervised classification approach for detecting open water areas in the Greenland Sea based on high-frequency data from Envisat and SARAL. Altimetry radar echoes, also called waveforms, are analyzed regarding different surface conditions. Six waveform features are defined to cluster radar echoes into different groups indicating open water and sea ice waveforms. Therefore, the partitional clustering algorithm K-medoids and the memory-based classification method K-nearest neighbor are employed, yielding an internal misclassification error of about 2%. A quantitative comparison with several SAR images reveals a consistency rate of 76.9% for SARAL and 70.7% for Envisat. These numbers strongly depend on the quality of the SAR images and the time lag between the measurements of both techniques. For a few examples, a consistency rate of more than 90% and a true water detection rate of 94% can be demonstrated. The innovative classification procedure can be used to detect water areas with different spatial extents and can be applied to all available pulse-limited altimetry datasets.

Keywords: satellite altimetry; Envisat; SARAL; unsupervised classification; K-medoids; Greenland Sea; Fram Strait

---

1. Introduction

The Arctic Ocean, including its peripheral seas, e.g., the Greenland Sea, is considered one of the most important components of the Earth's climate system [1]. In particular, these areas show strong responses to global warming and may affect climate conditions globally, for example, by changing the oceanic thermohaline circulation. The north polar regions are crucial contributors to the global ocean current system by carrying cold and fresh water southwards. Most of the Arctic Ocean is covered by varying extents of sea ice with open water areas and floes with different spatial extents as well as fully closed ice surfaces. The seasonal fluctuations of ice covers significantly impacts the atmosphere-ocean interaction (e.g., ice-albedo). While a closed sea ice cover prevents the ocean from heat emission, openings in the ice lead to a warming of the first atmospheric layers.

The evolution of sea ice is strongly influenced by sea surface temperature, wind, waves and ocean currents [1]. During recent decades, increasing sea surface temperatures and an enhanced warm water inflow in the Arctic Ocean resulted in decreased sea ice extent and volume [2,3]. Additionally, the Greenland ice sheet experienced strong environmental changes due to an increasing mass loss enhancing melt water influx into the Arctic Ocean [4].

The monitoring of the changing north polar ocean conditions, especially in the Greenland Sea, allows investigating interconnections between land, ocean, and atmospheric processes as well as their climate forcing. Today, remote sensing systems provide a large set of different sensors for monitoring the polar regions. Radar satellite altimetry is able to provide quantitative information about sea surface heights, significant wave heights, and dynamic ocean topography [5,6]. However, in order to derive reliable altimetry products, a careful selection of measurements from open water areas is necessary. By analyzing the radar return signal of the altimeter, the so-called waveform, information about the reflecting surface can be derived. This allows the classification of waveforms in water- and sea ice-returns and the detection of open water areas in sea ice regions. For example, calm open water areas within the footprint cause a very single-peak shape. With an along-track resolution of less than a kilometer using high-frequency data, conventional satellite altimetry missions (such as Envisat and SARAL) are able to detect small open water areas that might be missed by imaging Synthetic Aperture Radar (SAR) satellite missions (Sentinel-1A/B, Radarsat-1/2, etc.) in case a high-resolution acquisition mode is not available. Additionally, small water areas have insufficient backscatter properties to be mapped by passive microwave satellite missions (e.g., Special Sensor Microwave Imager (SSM/I) and SSM/I Sounder (SSM/IS)). However, altimeter radar echoes reflected from non-uniform scatterers, like sea ice regions, are challenging to interpret because the large surface footprint of several kilometers usually covers several ice types.

The first studies dealing with satellite altimetry in sea ice regions were published in 1980 by Dwyer and Godin [7] and in 1992 by Fetterer et al. [8]. After the launch of the ESA satellites ERS-1 and ERS-2, covering high latitudes in a repeat orbit, further studies were conducted by Laxon [5] and Laxon et al. [9]. They analyzed the potential of ERS-1 sea ice monitoring and the interannual variability of sea ice thickness by employing ERS-1 and ERS-2 altimetry data. Furthermore, Peacock [10] provides a first sea surface height determination in the Arctic ocean. In recent years, several sea ice applications have been explored e.g., the detection of openings in the ice. Connor et al. [11] applied a peakiness parameter, defined by Peacock [10] in order to detect small open water bodies in the sea ice cover using high-frequency data of Envisat. Zakharova et al. [12] continues with the development of a lead detection algorithm by using the Centre National d' Études Spatiales (CNES) and Indian Space Research Organisation (ISRO) satellite SARAL and maximum power threshold. Currently, all pulse-limited altimetry-based approaches for detecting water returns in sea ice regions use thresholds for different parameters. This has the disadvantage that the thresholds have to be set manually and individually for every altimetry mission. Furthermore, a deep knowledge about the different scatter characteristics in sea ice regions is required. Besides Zygmuntowska et al. [13] developed another approach using the waveforms shape for classifying airborne SAR altimeter echoes over the Arctic sea ice in a supervised way.

The present study proposes a new strategy to detect open water areas based on an unsupervised classification of high-frequency altimetry radar echoes. The approach is able to detect water domains with different spatial extents and can be easily applied without any deeper knowledge about surface-dependent backscatter characteristics. The method is applicable to all available pulse-limited altimeter data and is independent of mission-specific radar frequencies and characteristics. Furthermore, the results are compared to processed SAR images using the method described in [14] to obtain quantitative information about the classification performance.

The present paper is structured into three main parts. First, the study area and the applied datasets are introduced. Section 3 presents the method and processing procedure as well as the comparison process of the obtained results. Section 4 presents the classification results and provides evidence of the classification performance. At first, quantitative information considering the entire available validation dataset (Section 4.1) is derived before some visual comparisons between the SAR images and the altimetry overflights are provided. The paper finishes with a conclusion and an outlook to future research.

2. Study Area and Data Sets

This section provides an introduction to the study area and the different remote sensing datasets used for classification and validation.

2.1. Greenland Sea and Fram Strait

The study area ranges from 70.0°N to 81.5°N in latitude and from the north-east coast of Greenland to 28.0°E in longitude and covers the Greenland Sea and the Fram Strait (see Figures 1 and 2). The Greenland Sea belongs to the peripheral seas of the Arctic ocean. It connects the Fram Strait in the north, a narrow passage between north-east Greenland and Svalbard, with the Norwegian Sea as well as the Iceland Sea in the south. This area is affected by the East Greenland Current (EGC), which transports more than 90% of fragmented sea ice from the Arctic Ocean through the Fram Strait southwards [15]. Therefore, the EGC represents the main and most important freshwater outlet of the Arctic Ocean. According to Serreze and Barry [16] the Greenland Sea and the region of the Fram Strait is strongly influenced by rapid atmospheric and changing sea ice conditions as well as comparatively fast ocean currents with a mean velocity of 20–30 cm/s [17] and maxima up to 80 cm/s [18]. The sea ice state reaches from a nearly closed sea ice cover, showing straight lined and circular shaped open water bodies, leads and polynyas, up to individual ice floes ranging from a few meters to kilometers in diameter [16]. Applying open water detection to the Greenland Sea and the Fram Strait offers one the chance to sensitize the unsupervised classification method for a various number of different sea ice and ocean conditions.

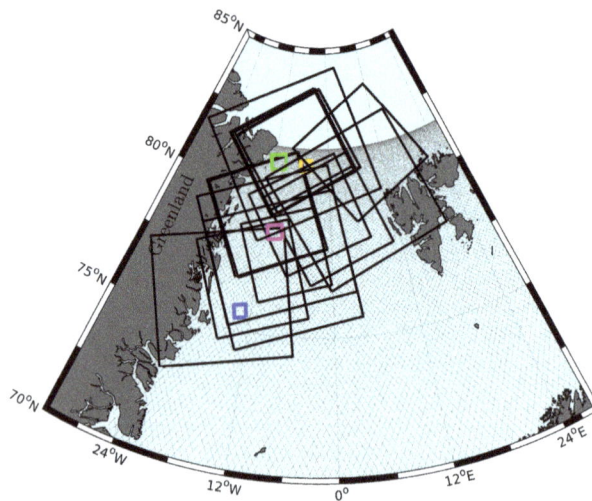

Figure 1. Black rectangles indicate locations of the SAR images from ALOS and Radarsat-2 used for comparison with Envisat classification results against the background of nominal sun-synchronous ground tracks of one Envisat cycle. The four subsets discussed in Section 4.2 are highlighted by different colors.

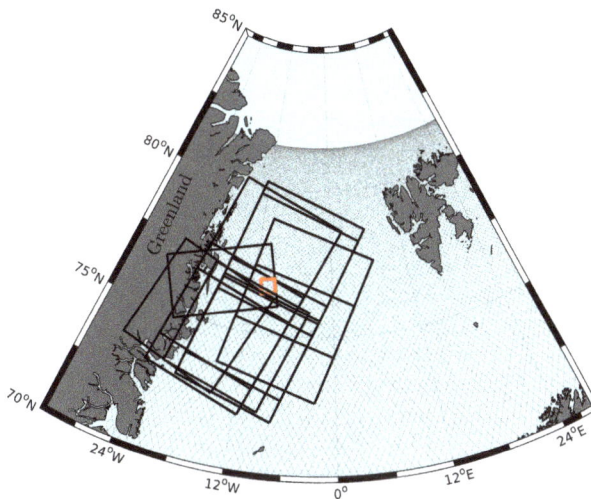

Figure 2. Black rectangles indicate locations of the SAR images from Sentinel-1A used for comparison with SARAL classification results against the background of nominal sun-synchronous ground tracks of one SARAL cycle. One subset discussed in Section 4.2 is highlighted in orange.

2.2. Radar Altimetry Data

In the present investigation, the high-frequency radar altimetry data of the ESA satellite Envisat and the CNES/ISRO altimetry satellite SARAL are used. Data of the missions Jason-1, Jason-2 and Jason-3 are disregarded due to their low orbit inclination (about 66°) not covering the Greenland Sea and Fram Strait.

Envisat and SARAL carry pulse-limited radar altimeters and are placed on the same 35 day repeat-orbit covering polar areas up to ±81.5° geographical latitude. Envisat was launched in March 2002 and orbits the Earth at an altitude of nearly 800 km. In October 2010, Envisat left the repeat-orbit and started to drift until in May 2012, the ESA mission was decommissioned after an unexpected signal loss. SARAL was placed in orbit in February 2013 and is still active even though in July 2016, the satellite started its drifting orbit phase without fix repeat period.

All computations and methodologies used in this study are based on official high-frequency Sensor Geophysical Data Record (SGDR) v2.1 dataset of Envisat's radar altimeter (RA-2) and the SGDR-T dataset of the AltiKa radar altimeter mounted on SARAL. In case of SARAL, data until July 2016 and in case of Envisat, data until the end of the mission are used. In this study, waveforms observed in the Greenland Sea and Fram Strait (see Section 2.1) are employed in the classification process. In order to calculate the altimeter backscatter values, different features stored in the SGDR dataset, for example, atmospheric attenuation and instrumental corrections (e.g., sigma naught calibration factor) are additionally used.

The two satellite missions differ mainly in the emitted radar bandwidth, the pulse repetition frequency and the footprint size of the illuminating area onto the surface. RA-2 emits *Ku*-band signals with an repetition frequency of 1800 pulses per second, covering an nominal elliptic area of approximately up to 10 km diameter [11] depending on the surface conditions. Before transmitting to earth, the waveforms are sampled to 18 Hz by the on-board processing. AltiKa works in the *Ka*-band, with a repetition frequency of 4 kHz, generates 40 Hz averaged waveforms and has half the antenna aperture of Envisat. This leads to a smaller footprint size up to 8 km diameter and an improved spatial resolution [19]. Beside instrumental influences, the waveform's shape is mainly affected by various surface characteristics. Detailed explanations referring to the representation of the varying waveform's shape can be found in Section 3.1.

2.3. Imaging Synthetic Aperture Radar (SAR) Data

A possible source for validating the classification results is the usage of imaging synthetic aperture radar (SAR) data. Beside the altimeter satellites, several multispectral and SAR imaging satellite missions regularly provide snapshots of periodically changing ocean conditions. In contrast to multispectral sensors, working mostly in the visible and infrared spectrum, SAR sensors are unaffected by cloudiness and lighting conditions, which makes it easier to identify appropriate scenes. However, SAR sensors are side-looking instruments, which can cause a shadowing of very flat and smooth surface structures (e.g., leads or polynyas) due to interjacent higher topography (e.g., ice floes, ridges). Additionally, the recorded backscatter values do not only depend on the surface characteristics (e.g., roughness) but also on the incidence angle of the reflected radar waves, which makes it more complex to provide information about different surface types. Furthermore, it has to be mentioned that most SAR satellites are placed on sun-synchronous orbits, which allows for a uniform capture of ice state but limits the minimum time lag between the acquisition dates of the SAR images and the altimetry measurements of Envisat and SARAL also using sun-synchronous orbits.

Aiming at a small time lag between SAR images and satellite altimetry, wide swath data are qualified best since these images cover a spatially extended area with medium pixel spacing. For this investigation, SAR images of the JAXA Advanced Land Observing Satellite (ALOS) [20], MDA's Radarsat-2 [21] and ESA's Sentinel-1A (S-1A) [22] are used. The Envisat classification results are compared with ALOS PALSAR Level 1.5 Wide Beam (WB) images offered by the Alaska Satellite Facility (ASF) DAAC and with Radarsat-2 (R-2) Scan SAR Mode data provided by ESA. The SARAL classification outcomes are compared with Level-1, S-1A extra wide swath mode data. S-1A images are made available through the ESA/Copernicus Sentinel Data Hub. Specifications, temporal availability in the target region, and information about the used imaging SAR products are listed in Table 1. To distinguish between open water pixels, appearing in near black, and sea ice pixels, appearing in bright gray, HH-polarized images are used. For more information regarding SAR polarization and the influence of different surface scattering see Dierking W. [23] and Jackson et al. [24].

Table 1. Synthetic Aperture Radar (SAR) image specifications [20–22] and altimeter satellites covering same time periods.

SAR Satellite	Band	Mode	Swath Width (km)	Pixel Size (m)	Period (mm/yyyy)	Altimeter Satellite
ALOS	L-Band	Wide Beam	250–350	100 × 100	June 2007–May 2008	Envisat
Radarsat-2	C-Band	Scan SAR Wide	500	50 × 50	June 2008–present	Envisat/SARAL
Sentinel-1A	C-Band	Extra Wide	400	40 × 40	October 2014–present	SARAL

In order to ensure similar sea ice conditions and allow for an unbiased comparison between SAR and altimetry, only images with a time lag less than about 3.5 h, with respect to the altimetry crossings, are used. The comparison is based on 16 grayscaled SAR images during the lifetime of Envisat and 19 images for SARAL. The SAR data are selected from different epochs considering a varying sea surface state with a focus on periods with various sea ice coverage. Figures 1 and 2 display the locations of all used SAR images. The scenes are mainly located in the Fram Strait and near the north-east coast of Greenland. Tables 2 and 3 list sensor and temporal information for all conducted comparison pairs. Two of the R-2 images are used for multiple satellite overflights. In the case of SARAL classification, it has to be mentioned that, due to sun-synchronous orbits and fixed revisit times of Sentinel-1A and SARAL, it is not possible to find suitable pairs for comparison that show good spatio-temporal coverage with a time gap smaller than 2 h 40 min during the study period.

Table 2. Acquisition date of the SAR images and time gap between altimetry observations and imaging data used for comparison with Envisat classification results.

SAR-Satellite	Acquisition Date	Time Gap hh-mm
ALOS	14 June 2007	02-30
ALOS	1 October 2007	02-57
ALOS	7 October 2007	01-55
ALOS	10 November 2007	03-07
ALOS	10 December 2007	02-50
ALOS	26 December 2007	02-13
ALOS	5 January 2008	02-40
ALOS	7 January 2008	01-46
ALOS	12 January 2008	02-49
ALOS	4 May 2008	01-25
R-2	4 November 2008	01-47
R-2	20 April 2009	02-09 00-29
R-2	21 April 2009	02-07
R-2	10 February 2010	03-04
R-2	14 March 2010	01-52 00-13 01-27
R-2	16 October 2010	02-04

Table 3. Sentinel-1A acquisition date of the SAR images and time gap between altimetry observations and imaging data used for comparison with SARAL classification results.

Acquisition Date	Time Gap hh-mm
23 October 2014	02-41
16 November 2014	03-34
14 November 2014	02-49
18 November 2014	02-42
3 December 2014	02-40
6 December 2014	02-58
27 December 2014	03-33
1 January 2015	02-59
15 January 2015	03-20
18 January 2015	03-27
16 March 2015	03-08
6 February 2015	03-29
6 February 2015	03-28
10 February 2015	03-22
22 February 2015	02-59
2 March 2015	02-44
9 March 2015	02-56
19 April 2015	02-53
15 May 2015	02-54

2.4. Sea Ice Data

Polar sea areas are affected by moving sea ice due to the influences of wind and ocean currents [16]. This results in a rapid change and high diversity of the sea surface conditions. To reach a realistic comparison of altimetry results and SAR images, the compensation for sea ice motion within the time interval between the two observation sets is required. For this purpose, daily ice vector velocity fields are exploited within the validation process. Therefore, the "Polar Pathfinder Daily 25 km EASE-Grid Sea Ice Motion Vectors, Version 3" of the National Snow and Ice Data Center (NSIDC) are employed [25]. This dataset contains zonal and meridional sea ice velocity observations of active and passive sensors as well as in situ measurements interpolated to a 25 km spacing grid referring to an azimuthal equal area map projection. This dataset covers the entire altimetry era until the end of May 2015.

The sea ice velocity data are used to shift the SAR image, respectively, the image pixel coordinates, assuming an averaged ice motion (direction and velocity) over the time interval between the altimetry measurement and the SAR image. For this purpose, only homogeneous data represented by small standard deviations in direction and velocity inside a predefined box (±35 km) around the altimetry track are selected to compute a mean displacement vector. Sea ice velocity vectors located close to the coastlines (within 25 km) are eliminated due to erroneous ice observations [25].

The comparison is performed only in areas affected by sea ice to suppress the influence of falsely detected SAR ice pixels caused by diffuse scattering behavior due to rough swell in the open ocean. Therefore, daily "Sea Ice Concentrations from Nimbus-7 SMMR and DMSP SSM/I-SSM/IS Passive Microwave Data, Version 1" of the NSIDC [26] with a spatial resolution of 25 km × 25 km are interpolated to the altimetry high-frequency data. Observations outside the ice edge without sea ice are excluded from the comparison process.

3. Methods

This study is based on an unsupervised classification process of altimetry waveforms. Unsupervised classification algorithms group unassigned data into a predefined number of classes without any background information about the data and their sources using only "natural" and hidden intra-cluster similarities [27]. The classification is performed based on a set of features characterizing the input data. In contrast, supervised classification is based on a-priori information of a well known or labeled dataset to classify and assign the observations [28]. Examples for unsupervised classification methods are artificial neural networks (e.g., Self-Organizing Maps [29]) or partitional clustering algorithms (e.g., K-means and K-medoids [30,31]). In the present investigation, a partitional cluster algorithm, K-medoids, is used for separating a set of unlabeled waveform data into clusters indicating different waveform properties. Therefore, features have to be defined describing various waveform characteristics. Based on the clustering results, K-nearest-neighbor is applied to assign unclassified waveform data.

In this section, at first, features for describing the various waveform shapes and their characteristics are specified and explained. This is followed by the description of the methodical background of the clustering and classification process. The last part of Section 3 presents the validation approach for the classification procedure. The presented methods are applied independently to Envisat and SARAL.

3.1. Waveform Features

The shape of altimetry waveforms strongly depends on the surface characteristics within the altimeter footprint. Figure 3 shows Envisat/RA-2 and SARAL/AltiKa radar pulses reflected by ocean, leads, and sea ice. Major differences can be detected in the power magnitude and the number and shape of the signal peaks. Leads produce very narrow and peaky waveforms due to the specular scattering of calm and flat water. In contrast, radar pulses originating from ocean or sea ice surfaces are influenced by waves or interlaced and piled ice floes, respectively, leading to multi peaks and wider, noisier shapes.

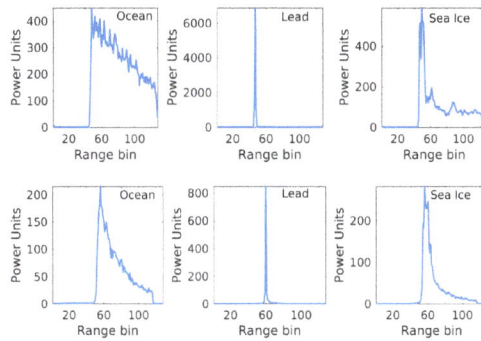

Figure 3. Waveform examples for Envisat (**top row**) and SARAL (**bottom row**) for three different surface scatterers: Ocean (**left**), Lead (**middle**), and Sea Ice (**right**).

In order to characterize a waveform and identify the main evocative surface scatterer, a number of waveform features are defined. The computed values constitute a waveform feature space that provides the input for the clustering and classification process. To increase the efficiency of the algorithm and to get a reliable open water detection, the selected features should fulfill the following conditions:

1. The features should characterize different waveform types.
2. The selected features should be stand alone and without linear dependence and major correlations among each other.
3. The feature space should be adaptable to any altimetry waveform.
4. All features should exhibit the same order of magnitude for equal weighting among each other.

In the present investigation, six features are defined to describe the waveforms mainly focusing on the reflected radar pulse shape (width) and the recorded power intensity. These features are applicable for each pulse-limited altimeter waveform, i.e., for Envisat as well as for SARAL.

- Waveform maximum (Wm)
  The waveform height is described by the maximal power of the returning radar pulse σ_{max}. It provides information about the backscatter of calm or rough surface conditions. To compute σ_{max} for Envisat as well as SARAL, the maximum waveform power and mission specific rectifications are applied by using instrumental and atmospheric corrections from the provided datasets (see Section 2.2).

- Trailing edge decline (Ted)
  The trailing edge decline is computed by fitting an exponential function, considering an exponential decay of AltiKa waveforms, from the waveform power maximum to the last bin. The estimated decay rate is used to characterize the decline of the trailing edge after the maximum.

- Waveform noise (Wn)
  This feature quantifies the trailing edge scattering. It is computed as median absolute deviations of the trailing edge fitting (see Ted) residuals. This parameter is very small for single peak waveforms (leads) and moderate for oceans.

- Waveform width (Ww)
  The number of bins where the power is equal to zero provides information about the waveform's width.

- Leading edge slope (Les)
  The leading edge slope is obtained by subtracting the first bin position containing more than 30%

of the power maximum from the bin position of the maximum power. The difference provides relative information about the width and steepness of the leading edge independent of the absolute position of the leading edge, i.e., the range.

- Trailing edge slope (Tes)
  In contrast to the leading edge slope, the trailing edge slope is obtained by subtracting the last bin position containing more than 30% of the maximum from the bin position of the maximum power. This difference provides similar information to Ted in the case of single-peak waveforms but supports the identification of strong specular peaks in front of an ocean-like trailing edge.

The selected features in the present investigation show a varying order of magnitudes, which results in an irregular weighting in the clustering algorithm. In order to comply with condition 4 (see above) a standardization has to be processed. Before conducting the unsupervised classification procedure, the features are reduced by subtracting their average and divided by their standard deviation (standard-score).

The features are calculated for RA-2 and AltiKa waveforms in the same way. In the case of SARAL, the maximum power is limited to 1250 counts. Power counts above this limit are not recorded due to too high backscatter values that cannot be resolved by the tracking window [12]. The waveforms are cut without a clear maximum peak in the radar echo, which makes it impossible to compute all features (e.g., leading edge slope) and to constitute the complete feature space. These waveforms, which are not flagged in the SGDR dataset, are skipped from the further classification process. Furthermore, all waveforms are neglected, for which no reliable computation of the defined features is possible (e.g., if trailing edge fitting is impossible with 95% confidence).

3.2. Clustering

Within the clustering process a representative subset of all waveforms from a single mission will be used to define waveform groups, so-called clusters, that will later be used to also classify all remaining observations. In a first step, this reference model has to be created. For this purpose, a set of several waveforms, containing a majority of all possible scatter types, has to be selected. To this end, waveform data covering an area in the central Greenland Sea within bounds of 15°W/10°E longitude and 68°N/80°N latitude are used. To cover as many sea ice types as possible, the epoch is selected at the beginning of the melting period in early summer from April to May [32]. For Envisat, Cycle 57 (2007, containing about 307,000 waveforms), and for SARAL, Cycle 12 (2014, ca. 670,000 waveforms) are selected.

To group the reference data, a K-medoids cluster algorithm is implemented that clusters unsupervised data into K clusters. K-medoids performs a distance minimization between the features and the most centrally located feature (medoids) based on the feature space itself. Thereby, K-medoids is more robust to outliers and noise in contrast to K-means, which tries iteratively to estimate an optimal partition of unlabeled data by minimizing the distances between the coordinates of a mean cluster center (centeroids) and the features. However, in contrast to K-medoids, K-means integrates every value of the feature space into the arithmetic average [27].

At first, K-medoids randomly chooses K medoids of the feature space and computes the distances to every feature. In the next steps, the algorithm rearranges every single feature until there is no motion within the K clusters and the minimal distances to the medoids are found. However, the clustering result depends on the initial randomly chosen medoids. This is why the algorithm is repeated several times and the best solution is selected by analyzing the final sum of all distances within the clusters. This leads to high computational efforts by employing large input datasets, but it is considerable that the clustering has to be performed just once per altimetry mission. To reduce the computation times, the algorithm examines only a random sample of cluster members during each medoids updating step. The size of the sample is set by default to 0.1% of the total number of data points. The iteration terminates if the medoids are stabilized.

Partition clustering algorithms require an initialization of the number of clusters K. In the present investigation, K is chosen empirically after several test runs by evaluating the best segmentation results [31]. Indicators for defining an appropriate K are, for example, the analysis of the sum of all distances within the clusters and, additionally, a visual analysis of all clusters. In order to obtain a clear partitioning of waveforms, it is useful to set K larger than the desired number of the three surface types, the present investigation is looking for, namely, calm open water, ocean or sea ice conditions. Figure 4 shows the clustering for 30 classes based on derived waveform features of about 307,000 Envisat waveforms (the clustering results for SARAL waveforms can be found in the Supplementary section, Figure S1). The displacements between the points and the medoids are computed using Euclidean distances.

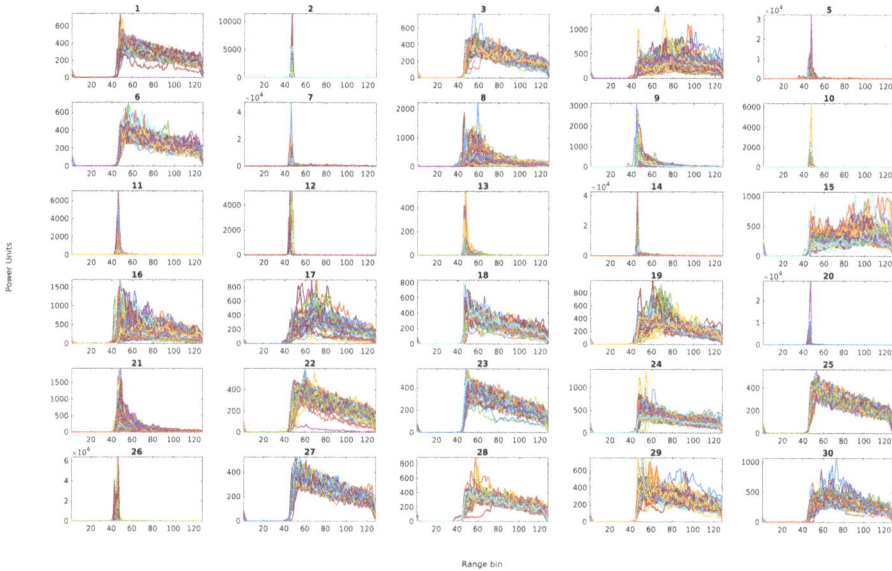

Figure 4. Envisat waveform clusters (K = 30) after K-medoids clustering showing segmented waveforms (every twenty fifth per cluster).

After running K-medoids, each cluster has to be assigned to one surface condition. The 30 clusters need to be manually condensed to three main classes indicating ocean, sea ice, and lead/polynya returns. This is done based on the feature statistic per cluster (see Figure 5) and knowledge on the physical backscattering behavior of different surfaces. It is well known that radar returns from dominant scatterers (i.e., a lead with a calm, mirror-like surface) cause single-peak waveforms with high power and narrow shape. Radar echoes nearly entirely reflected by sea-ice show a more diffuse scattering, weak power and no clear peaks. Using these relationships and transferring them to the cluster statistics enable a nearly unambiguous assignment. However, questionable clusters with ambiguous feature properties remain and are labeled as "undefined".

Figure 5 indicates the cluster assignment by different colors. Lead and polynya returns (clusters 2, 10–12, 20, and 26) are characterized by a very narrow and peaky shape and high maximum power values. In contrast, ocean returns (clusters 1, 3, 6, 18, 22, 23, 25, 27–29) are wider and show a greater trailing edge decay. Waveforms belonging to the ice class (clusters 5, 7–9, 13, 14, 16, 21, 24, and 30) are between these two groups. They are defined by a smaller trailing edge decay and slope as well as bigger power values than ocean returns. However, there are clusters (4, 15, 17, and 19), that cannot clearly be assigned to one surface type. As an example, cluster 19 shows an ocean like behavior, but is characterized by an indistinct leading edge as well as a steeper ice-like trailing edge. Undefined waveform classes show, apart from more noise in the cluster itself, no clear signature or trend to the underlying feature space or to the three main surface classes.

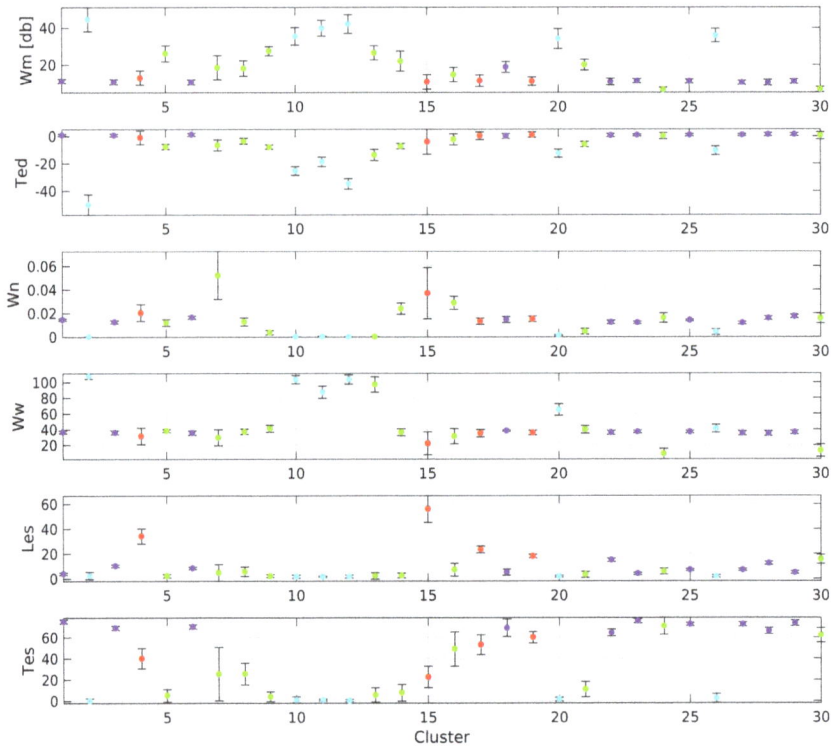

Figure 5. Means and standard deviations of waveform features (see Section 3.1) per cluster. Four classes are illustrated using different colors: "lead/polynya" (cyan), "ocean" (purple), "ice" (green), and "undefined" (red).

3.3. Classification

The waveform model created by the clustering (Section 3.2) can now be used to classify all waveforms. For this purpose the K-Nearest Neighbor (K-NN) classifier is employed. In general, K-NN belongs to the memory-based classifiers and does not require a stochastic model [28]. Basically, K-NN searches for the closest distance between a query point and a given input model. Similar to the K-medoids algorithm, the K-NN uses the euclidean distance. However, K has a completely different meaning than in the K-medoids algorithm. The K is now defined as the number of neighbors used for the classification. The cluster assignment of a specific waveform is done based on the majority of clusters of these K nearest neighbors. K must be set before the classification process starts.

In the present study, K is estimated by performing a 10-fold cross-validation. Therefore the reference model used for the clustering and already assigned to the clusters is divided into 10 randomly sorted, but equally sized subsets and validated against each other. This means, that every subset is used as a test sample and the remaining subsets as training sets. In order to find an appropriate K for K-NN, the cross-validation is performed based on different numbers of neighbors. Figure 6 shows the mean misclassification error as a percentage of the 10-fold cross-validation in the case of SARAL and Envisat. Similar errors can be expected for the classification of the remaining unlabeled waveforms. The minimum error defines the optimal number of neighbors. SARAL displays less variability and a smaller misclassification rate than Envisat. The K-NN method seems to be more stable with clustered SARAL than with Envisat waveforms, which can be explained due to less variability in the AltiKa waveforms and a more robust waveform clustering. For SARAL a minimum error rate is obtained with K = 20 (1.93 ± 0.05%). In the case of Envisat, nearest-neighbor number K = 44 (2.3 ± 0.08%) is used, providing a good balance between low error and variance.

The misclassification rate in connection with the defined number of neighbors gives information about the K-NN prediction error based on the reference model and class labels. This parameter can be used to estimate the internal precision of the classification approach. In this study, a minimal error of about 2% has to be expected from the methodology itself.

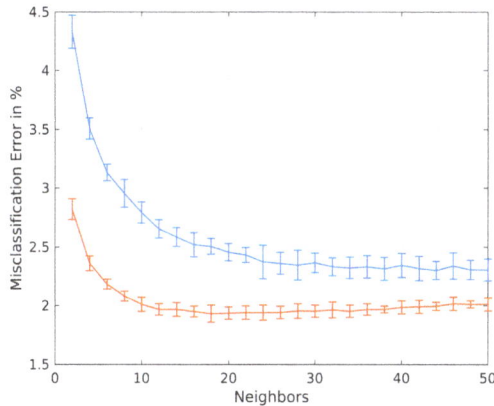

Figure 6. Misclassification error and its standard deviation for SARAL (red) and Envisat (blue) with varying number of K neighbors as computed by 10-fold cross-validation.

After defining an appropriate K, the remaining waveforms are applied to K-NN. In the end of the classification process every waveform is labeled by a certain cluster and, consequently, assigned to a specific surface type.

3.4. Validation Approach

In order to conduct an external validation for the waveform classification, a comparison with independent SAR images is performed. For this purpose, the defined waveform classes of ocean, lead/polynya, sea ice are assigned to water (ones) and non-water (zeros) observations. Undefined waveforms classes are also labeled with zeros. In order to provide quantitative information about the classification performance, it is necessary to compare the results to an external dataset. For this purpose, imaging SAR data are used, as they regularly provide snapshots of different sea surface states in the study area.

Before performing an automatic comparison between SAR and the classification results, the SAR images are pre-processed by using the ESA toolbox SNAP, version 4.0.0 for Sentinel-1A as well as Radarsat-2 and the MapReady toolbox, version 3.1.22 for ALOS image data offered by ASF. Basically, the following standard routines are applied to the imaging SAR data:

1. Thermal noise removal (only S-1A)
2. Radiometric calibration
3. Speckle filter (only Radarsat-2)
4. Delay Doppler terrain correction
5. Reprojection to Lambert Azimuthal Equal Area map projection
6. Converting backscatter values to db
7. Datatype conversion in *uint8*

On the pre-processed SAR images, linear and circular shaped black and near black areas indicate openings inside the sea ice cover generated by a smooth surface and specular reflection of the radar waves. To automatically extract these areas, the SAR images have to be converted into binary pixel values by applying several image processing tools. The applied approach is described in detail by Passaro et al. [14]. Briefly summarized, the images undergo a noise and minimum filtering in order to emphasize dark pixel regions, followed by an adaptive thresholding that considers local illumination changes. Finally, a mathematic morphological closing operation is applied to the black and white coded images to link fragmented open water regions. To control the effect of the morphological closing operation a structure element (kernel) or convolution matrix is needed. Regarding linear and circular shapes of open water areas, an octagon with various size, considering the nominal pixel spacing of the SAR images, is employed. In the case of ALOS, the octagon size is six pixels around the center pixel, and in the case of Sentinel-1A and Radarsat-2, a kernel size of 12 pixels is used. Moreover, the image coordinates are shifted to compensate for sea ice-motion, for the acquisition time difference between altimetry and SAR (see Section 2.4). In a last step, the locations of the altimetry returns are interpolated to the SAR pixel locations by using nearest neighbor method.

4. Results and Discussion

In this study, 15,025 Envisat and 19,919 SARAL observations are investigated for which SAR image classification results are available for validation. 31.2% of the Envisat waveforms and 15.0% of the SARAL returns are assigned to water classes. Furthermore, 4.7% of Envisat and 14.2% of SARAL waveforms are set to undefined and defined as non-water returns. For a quantitative rating, 19 comparison pairs for SARAL and Envisat, respectively (see Section 2.3), are used. The results of this comparison are presented in the following section. Afterwards, examples are displayed to illustrate and discuss the functionality of the validation approach.

4.1. Automatic Comparison to SAR Images

As mentioned above, the automatic comparison process only relies on observations in areas with a semi-closed sea ice layer. This allows one to reduce false SAR classifications outside the ice edge due to an unreliable SAR image processing. Table 4 provides the numbers of measurements assigned to water and ice by the two observation techniques and, therefore, allows for an assessment of the altimetry classification performance. The absolute number of water and ice detections are listed column-wise for the altimetry classification results and row-wise for the SAR open water detection. The table shows that 1124 of the 15025 Envisat observations are identified as water by both, altimetry and SAR, whereas 837 locations are assigned to non-water by altimetry and to water by SAR. Assuming the SAR to be the ground truth validating the altimetry water detection, four dependencies are derived to rate the classification results. The total consistency rate, $P(CR)$ is computed by summing up the bold values and dividing them by the total number of comparison points. In addition, three conditional frequencies are derived: The true water detection rate ($P(Alt|SAR)$) is computed by dividing the "correct" altimetry water detections by the total number of SAR water observations, whereas the false water detection rate ($P(Alt|\overline{SAR})$) is defined as the relation between the water altimetry detections not confirmed by SAR and the total number of SAR ice detections. Moreover, the percentage of correctly classified water returns $P(SAR|Alt)$ represents the "correct" water altimetry detections in relation to the total number of open water detections by altimetry.

In the case of Envisat, a consistency rate of 70.7% is reached. In detail, nearly 60% of SAR water detections are truly classified by Envisat ($P(Alt|SAR) = 0.57$) in contrast to below 30% of SAR ice observations that are falsely assigned to water areas by Envisat ($P(Alt|\overline{SAR}) = 0.27$). However, only about a quarter of all Envisat open water detections are also classified by SAR ($P(SAR|Alt) = 0.24$).

The comparison between SARAL and Sentinel-1A water detection yields a higher consistency rate of 76.9% but a smaller true water detection rate of less than 30% ($P(Alt|SAR) = 0.28$). At the same time, the false water detection rate $P(Alt|\overline{SAR})$ is very small and yields only 12.3%. Moreover, the correctly classified water return rate $P(SAR|Alt) = 33\%$ is better as for Envisat.

It has to be noticed that for the interpretation of these numbers it is important to consider that inconsistencies are not only due to altimetry classification but that the SAR open water detection as well as the sea ice-motion correction also contribute to the error budget. For example, most of the SARAL comparisons take place during the sea ice maximum between January and mid March, when the pack ice is very close and exhibits only small openings in the ice, which makes it challenging to be detected by the SAR image processing.

Analyzing the absolute water detection numbers of Envisat versus SAR images, it is remarkable that the number of open water points differs by 2732 between the SAR detection and the Envisat classification. The Envisat classification identifies significantly more open water areas than the SAR processing (factor of nearly 2.4). In the case of SARAL, a transposed situation can be found. This can be explained by different SAR sensor characteristics and an insufficient pixel resolution as well as an imprecise SAR image processing, including an unreliable sea ice-motion correction. Additionally, the altimeters are affected by off-nadir returns, which can cause an enhanced number of open water detections. In the case of Envisat, a larger footprint size than SARAL intensifies off-nadir effects.

Overall, it is important to understand, that the classification performance numbers of SARAL and Envisat are not directly comparable with each other. The underlying different instrumental, sensor, and spatio-temporal conditions differ too strongly to provide qualitative information that would allow for a comparative assessment of the two altimetry satellites. More details related to the impacts of SAR and altimetry processing on the quantitative comparison process can be found in Section 4.2.

Analyzing, for example, $P(CR)$ of Envisat and SARAL, the quantitative comparison confirms the reliability of the altimetry-based classification method and a good performance of their results. However, it has to be kept in mind that a data comparison of two totally different Earth observation techniques for open water detection in a very dynamic study area is not possible without a variety of uncertainties and inaccuracies. In order to provide a better impression of the difficulties of a quantitative comparison approach, the next section shows a couple of examples in a visual comparison.

Table 4. 2D contingency tables based on Envisat—ALOS/R-2 (top) and SARAL—S-1A (bottom) comparisons. The table shows the number of points classified as water/ice from altimetry (Alt) with the corresponding classification from SAR.

	Alt (Water)	Alt (Ice)	Σ
SAR (water)	1124	837	1961
SAR (ice)	3569	9495	13064
Σ	4693	10332	15025
	Alt (Water)	Alt (Ice)	Σ
SAR (water)	987	2600	3587
SAR (ice)	2007	14325	16332
Σ	2994	16925	19919

4.2. Visual Comparison

Using different SAR image subsets, this section will provide some visual comparisons between open water classification by altimetry and SAR images. The images were selected in order to indicate possible difficulties due to uncertainties in the SAR image or altimetry processing as well as sea ice motion correction. Figures 7 and 8 show five visual examples before (left) and after (right) image processing. The altimetry measurement locations are superimposed on the SAR image. Cyan colored altimetry observations identify open water classifications. These regions are plotted in white in the binary coded SAR images (right column). Figure 7a–f display Envisat-ALOS and Figure 7g,h Envisat—Radarsat-2 comparisons. Figure 8 shows a visual example of one SARAL—Sentinel-1A comparison. Metadata information, i.e., acquisition date and applied sea ice motion correction on each comparison is provided next to the images (visualized classification results without class assignment can be found in supplementary Figures S2–S6). Moreover, Table 5 displays quantitative comparison results.

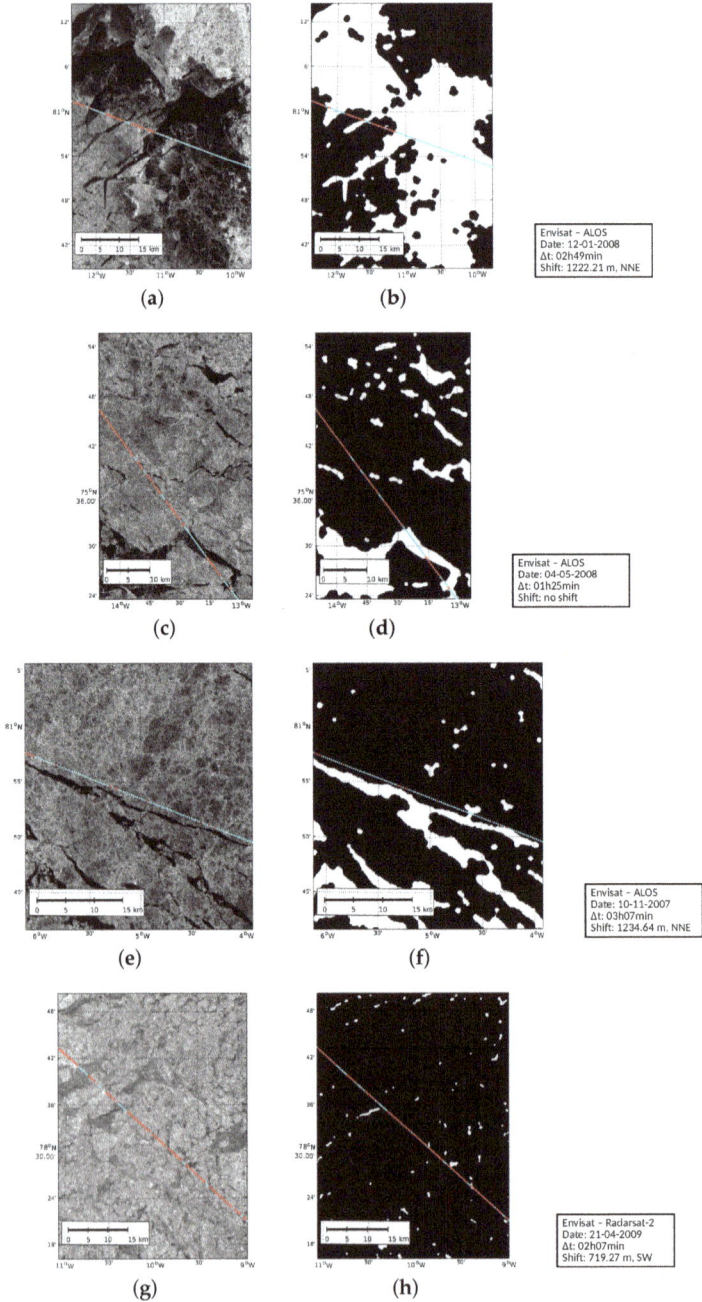

Figure 7. Examples of open water detection from Envisat against ALOS (**a–f**) and Radarsat-2 (**g,h**) before (**left**) and after SAR image processing (**right**) with open water indicated in white. Boxes provide additional image and processing information. Red: ice detection, cyan: open water detection. The geographical locations of the image subsets are displayed in Figure 1; from top to bottom in green, blue, yellow and magenta.

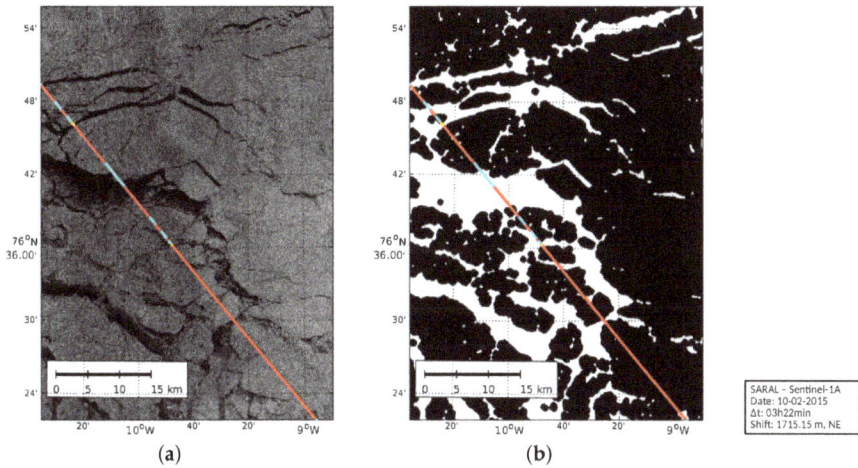

Figure 8. Example (orange highlighted in Figure 2) of open water detection from SARAL against Sentinel-1A before (**a**) and after SAR image processing (**b**) with open water indicated in white. Box provides additional image and processing information. Red: ice detection, cyan: open water detection, yellow: saturated AltiKa observations.

Table 5. Table providing percentage statistical information about conditional $P(X|Y)$ and consistency $P(CR)$ rates of visual examples discussed in Section 4.2. "SAR" and "Alt" indicate imaging SAR and altimetry. Ice detections are indexed by overline character marked shortcuts.

Subset	$P(CR)$	$P(Alt|\overline{SAR})$	$P(SAR|Alt)$	$P(Alt|SAR)$
Figure 7b	90.63%	15.91%	91.86%	94.05%
Figure 7d	85.10%	16.04%	64.58%	88.57%
Figure 7f	12.61%	88.99%	2.02%	100.00%
Figure 7h	76.22%	22.36%	0.00%	0.00%
Figure 8b	72.07%	12.41%	40.68%	25.53%

The first example (Figure 7a) demonstrates very good accordance between altimetry and SAR classification. The L-band image displays different sea ice and open water conditions. From West to East, various sized open water areas ranging from 200 m up to 3.5 km are visible. A large region appearing in dark reaches from the image center at 11°W to the eastern edge. It indicates a mixture of differently sized ice floes interrupted by open water sections. Analyzing the colored altimetry observations, the open water detection is in good accordance with the grayscaled as well as with the binary coded SAR image (Figure 7b). The quantitative comparison yields a consistency rate $P(CR)$ of 90.63%. Moreover, the altimetry classification approach provides a true classification rate close to 100% ($P(Alt|SAR) = 0.94$). This comprises small leads as well as larger areas of open water.

An almost perfect accordance between altimetry classification and SAR images can also be observed in Figure 7c, detecting an expanded lead in the southern image part and some small leads in the central part of the image. However, comparing Figure 7c and d to the altimetry classification results at 75°36.00′N, it is clearly visible that SAR image processing is not always able to segment very narrow lead fragments. This might happen because of a poor spatial pixel resolution of the SAR sensor (100 m), an insufficient identification of the ice-water transition, or a too restrictive threshold level in the SAR image processing. This deficiency in the automatic SAR image processing results in reduced performance in the quantitative comparison with a total consistency rate of about 85% and a true water detection rate of 88.6%.

However, there are also problems related to the altimetry observation technique. The example displayed in Figure 7e,f is characterized by a very long (ca. 47 km) and in most parts narrow lead located parallel to the satellite track. The SAR image is shifted about 1.2 km in the northeastern direction, assuming a steady sea ice motion. Even if the altimeter track is still located northwards at the off-nadir position of the lead, almost all measurements are classified as open water. As a consequence, in the quantitative comparison, just 2% of all Envisat open water detections are confirmed by SAR classification although they can be visually connected with the dominant lead in the image center. The overall consistency rate yields only 12.6%—probably due to the fact that the altimetry classification approach is not able to separate off-nadir water returns from nadir water returns or the mean sea ice motion correction is not enough to consider the total sea ice drift.

Additional discrepancies between altimetry and SAR classification can occur in areas with new, very thin ice coverage. Figure 7g shows those areas, appearing light gray in C-band, only a little darker than the surrounding older ice. These areas are correctly set to ice by the SAR image processing (see Figure 7h) because of the small brightness differences between the thin and surrounding ice types. In contrast, the altimetry returns within these areas are falsely classified and interpreted as calm open water since they show a very narrow and single-peaked lead/polynya-like shape. One explanation for this mis-interpretation is the dominant scattering of all flat and specular surfaces. Connor et al. [11] found that strong reflective surfaces, for example, leads/polynyas, can also affect the waveform shape if covered by very thin ice. A distinction from open water is not possible based on the altimetry waveform's shape. Since the ice is very thin, the retracked ranges should represent the water level well enough, even if the classification is wrong.

Related to the comparison process itself, uncertainties in sea ice motion correction can reduce the quantitative consistency rate. Figure 8a,b are corrected by ice-motion considering a time difference of more than 3 h. Analyzing Figure 8b, it can be shown that only 25.53% of the SAR detected ice openings are well identified by the altimetry data. A visual image inspection suggests that the applied ice motion correction is too small to completely compensate for the effect of the time lag.

Further challenging issues using SARAL SGDR-T data are so-called saturated waveforms. Zakharova et al. [12] pointed out that leads or strongly reflecting surfaces can exceed the maximum permissible power count value of 1250. The waveforms are cut and feature no clear peak due to a saturated power tracking window. Figure 8 highlights saturated SARAL observations in yellow. They are mainly located near small and calm open water areas, producing very high backscatter returns. In the classification process, they are omitted because of an unknown maximum peak position. In general, saturated SARAL waveforms are mainly traceable within the sea ice edge, but can provide evidence about the location of further open water areas. However, just 0.14% (i.e., 288 waveforms) of the comparison data are affected by a saturated power tracking window.

The present section shows a number of challenging and unavoidable impacts on the validation of the waveform classification process. Considerable parts of the inconsistencies do not originate from the altimetry classification but from the SAR classification or the ice-motion correction. In order to adequately rate the quantitative comparison results, it is necessary to keep these effects in mind.

5. Conclusions and Outlook

The present paper introduces an unsupervised classification approach based on pulse-limited multi-mission altimetry data to detect open water areas in a largely sea ice covered region. The study demonstrates the successful application of the clustering of pulse-limited altimeter waveforms for the automatic identification of open ocean, sea ice, lead and polynya observations. The approach is based on known partition cluster strategies (i.e., K-medoids) and memory-based classification methods (i.e., K-nearest-neighbor). A 10-fold cross-validation for the assessment of the precision of the classification method is performed. It indicates an internal misclassification error of about 2% for Envisat and SARAL. The algorithm is applicable to every pulse-limited altimetry satellite mission without requiring any deeper knowledge about mission specific details. Moreover, it can be assumed that the developed approach also works for SAR altimetry waveforms if the waveform feature space is adapted adequately. Additionally, the presented method can be adapted to a number of open water detection or waveform classification tasks, e.g., for the identification of lake returns [33] or in inundation areas.

In order to evaluate the classification results, a comparison with SAR images is performed. In contrast to previous studies, the present validation relies not only on visual and manually selected examples, but also on a larger set of images and an automated comparison procedure. The comparison procedure allows for a quantitative assessment of the classification performance by assigning the altimetry observations to open water and sea ice returns and checking them against processed SAR images that indicate sea ice and open water areas. We reach consistency rates of 70.7% for Envisat and 76.9% for SARAL. However, it has to be underlined that the quantitative comparison results of Envisat and SARAL are not directly comparable because of significant differences in the underlying sensor and instrument characteristics of the available SAR missions.

When interpreting the comparison results, different sources of inconsistencies have to be considered, e.g., effects from the altimetry data and their classification procedure and uncertainties in SAR image processing as well as in the ice-motion correction. The Fram Strait and the Greenland Sea are one of the most dynamic areas on Earth. Fast changing sea ice conditions due to short, periodic melting and refreezing as well as rapid climate change make it hard to provide a high reliability in the comparison as well as in the altimetry classification results. Local phenomena, such as melt ponds (i.e., open water pools on the sea ice surface) and their impacts on the open-water detection, have to be investigated. Over specific sea ice types, altimetry waveforms show ambiguities, which prevents a clear attribution to sea ice or open water returns. In particular, specular thin and flat ice produces very specular returns resembling open water returns. In contrast, big ice floes or landfast ice can imitate ocean-like returns due to similarities in ocean surface roughness and reflectivity.

Further improvements of the classification method are possible. In particular, saturated SARAL waveforms have to be included in the classification process. In addition, the application of more recent sea ice motion data in combination with Sentinel-1B data could lead to a better spatio-temporal ratio within the validation process.

A reliable classification is an indispensable requirement for a meaningful estimation and an efficient computation of sea surface heights in the Arctic by retracking only open water waveforms. In addition to Envisat and SARAL, more pulse-limited (e.g., ERS-1/2) as well as delay-doppler altimetry data (e.g., CryoSat-2, Sentinel-3A) may be employed in the classification process and, thus, contribute to the generation of a long-term sea level record for the Arctic ocean.

Supplementary Materials: The following folder and figures are available online at www.mdpi.com/2072-4292/9/6/551/s1, Figure S1 displays SARAL waveform clusters, Figures S2–S6 and folder S2 contain Figures 7 and 8 showing the classification results without class assignment.

Acknowledgments: The authors thank the following institutions and agencies for providing their data and software products under the terms of the GNU General Public License. ESA for operating and managing Envisat, Sentinel-1A and the SNAP Toolbox. CNES for providing SARAL data. JAXA/METI and ASF DAAC for making ALOS PALSAR L1.5 data and MapReady Toolbox available as well ESA for maintaining MDA Radarsat-2 data in framework of Third Party Mission program. This work was supported by the German Research Foundation (DFG) through grants, BO1228/13-1 and DE2174/3-1. The publication is funded by the Technical University of Munich (TUM) within the framework of the Open Access Publishing Program. We thank five anonymous reviewers for their valuable comments that helped to improve the manuscript.

Author Contributions: Felix L. Müller developed the classification and validation methods, conducted the data analysis and wrote the majority of the paper. Denise Dettmering supervised the present study, contributed to the manuscript writing and helped with the discussions of the applied methods and results. Wolfgang Bosch initiated the study. Florian Seitz supervised the research. Both were involved in the writing process and discussed the methods presented in the manuscript as well.

Conflicts of Interest: The authors declare no conflict of interest.

References

1. Comiso, J. *Polar Oceans from Space*; Atmospheric and Oceanographic Sciences Library, Springer: New York, NY, USA, 2010.
2. Stroeve, J.C.; Markus, T.; Boisvert, L.; Miller, J.; Barrett, A. Changes in Arctic melt season and implications for sea ice loss. *Geophys. Res. Lett.* **2014**, *41*, 1216–1225.
3. Polyakov, I.V.; Beszczynska, A.; Carmack, E.C.; Dmitrenko, I.A.; Fahrbach, E.; Frolov, I.E.; Gerdes, R.; Hansen, E.; Holfort, J.; Ivanov, V.V.; et al. One more step toward a warmer Arctic. *Geophys. Res. Lett.* **2005**, *32*, L17605.
4. Sasgen, I.; van den Broeke, M.; Bamber, J.L.; Rignot, E.; Sørensen, L.S.; Wouters, B.; Martinec, Z.; Velicogna, I.; Simonsen, S.B. Timing and origin of recent regional ice-mass loss in Greenland. *Earth Planet. Sci. Lett.* **2012**, *333–334*, 293–303.
5. Laxon, S.W. Sea-Ice Altimeter Processing Scheme at the EODC. *Int. J. Remote Sens.* **1994**, *15*, 915–924.
6. Rio, M.H.; Hernandez, F. A mean dynamic topography computed over the world ocean from altimetry, in situ measurements, and a geoid model. *J. Geophys. Res. Oceans* **2004**, *109*, doi:10.1029/2003JC002226.
7. Dwyer, R.; Godin, R. *Determining Sea-Ice Boundaries and Ice Roughness Using GEOS-3 Altimeter Data*; Technical Report; NASA Wallops Flight Center: Wallops Island, VA, USA, 1980.
8. Fetterer, F.M.; Drinkwater, M.R.; Jezek, K.C.; Laxon, S.W.C.; Onstott, R.G.; Ulander, L.M.H. Sea Ice Altimetry. In *Microwave Remote Sensing of Sea Ice*; American Geophysical Union: Washington, DC, USA, 2013; pp. 111–135.
9. Laxon, S.; Peacock, N.; Smith, D. High interannual variability of sea ice thickness in the Arctic region. *Nature* **2003**, *425*, 947–950.
10. Peacock, N.R. Sea surface height determination in the Arctic Ocean from ERS altimetry. *J. Geophys. Res.* **2004**, *109*, C07001.
11. Connor, L.N.; Laxon, S.W.; Ridout, A.L.; Krabill, W.B.; McAdoo, D.C. Comparison of Envisat radar and airborne laser altimeter measurements over Arctic sea ice. *Remote Sens. Environ.* **2009**, *113*, 563–570.
12. Zakharova, E.A.; Fleury, S.; Guerreiro, K.; Willmes, S.; Rémy, F.; Kouraev, A.V.; Heinemann, G. Sea Ice Leads Detection Using SARAL/AltiKa Altimeter. *Mar. Geod.* **2015**, *38*, 522–533.
13. Zygmuntowska, M.; Khvorostovsky, K.; Helm, V.; Sandven, S. Waveform classification of airborne synthetic aperture radar altimeter over Arctic sea ice. *Cryosphere* **2013**, *7*, 1315–1324.
14. Passaro, M.; Müller, F.L.; Dettmering, D. Lead Detection using Cryosat-2 Delay-Doppler Processing and Sentinel-1 SAR images. *Adv. Space Res.* **2017**, under review.
15. Rudels, B.; Friedrich, H.J.; Quadfasel, D. The Arctic Circumpolar Boundary Current. *Deep-Sea Res. Part II* **1999**, *46*, 1023–1062.
16. Serreze, M.; Barry, R. *The Arctic Climate System*; Cambridge Atmospheric and Space Science Series; Cambridge University Press: Cambridge, UK, 2014.
17. Bersch, M. On the circulation of the northeastern North Atlantic. *Deep-Sea Res. Part II* **1995**, *42*, 1583–1607.
18. Woodgate, R.A.; Fahrbach, E.; Rohardt, G. Structure and transports of the East Greenland Current at 75°N from moored current meters. *J. Geophys. Res. Oceans* **1999**, *104*, 18059–18072.
19. Verron, J.; Sengenes, P.; Lambin, J.; Noubel, J.; Steunou, N.; Guillot, A.; Picot, N.; Coutin-Faye, S.; Sharma, R.; Gairola, R.M.; et al. The SARAL/AltiKa Altimetry Satellite Mission. *Mar. Geod.* **2015**, *38*, 2–21.
20. Japan Aerospace Exploration Agency (JAXA). *ALOS Data Users Handbook, Revision C*; Japan Aerospace Exploration Agency: Chofu, Tokyo, Japan, 2008.
21. MDA. *RADARSAT-2 Product Description*; Report RN-SP-52-1238, Issue 1/13; MacDonald, Dettwiler and Associates Ltd.: Vancouver, BC, Canada, 2016.
22. Sentinel-1 Team. *Sentinel-1 User Handbook*; GMES-S1OP-EOPG-TN-13-0001, Issue Draft; European Space Agency: Paris, France, 2013.
23. Dierking W. Sea Ice Monitoring by Synthetic Aperture Radar. *Oceanography* **2013**, *26*, doi:10.5670/oceanog.2013.33.
24. Jackson, C.R.; Apel, J.R. *Synthetic Aperture Radar: Marine User's Manual*; US Department of Commerce, National Oceanic and Atmospheric Administration, National Environmental Satellite, Data, and Information Serve, Office of Research and Applications: Washington, DC, USA, 2004; pp. 81–115.

25. Tschudi, M.; Fowler, C.; Maslanik, J.; Stewart, J.S. *Polar Pathfinder Daily 25 km EASE-Grid Sea Ice Motion Vectors, Version 3*; Subset: Greenland Sea, Date Accessed: 16.11.2016; National Snow and Ice Data Center: Boulder, CO, USA, 2016.

26. Cavalieri, D.; Parkinson, C.; Gloersen, P.; Zwally, J.H. *Sea Ice Concentrations from Nimbus-7 SMMR and DMSP SSM/I-SSMIS Passive Microwave Data, Version 1*; Subset: Greenland Sea, Date Accessed: 16.11.2016; NASA National Snow and Ice Data Center Distributed Active Archive Center: Boulder, CO, USA, 1996.

27. Xu, R.; Wunsch, D.C. *Clustering*; IEEE Press Series on Computational Intelligence; IEEE Press: Hoboken, NJ, USA; Wiley: Piscataway, NJ, USA, 2009.

28. Hastie, T.; Tibshirani, R.; Friedman, J. *The Elements of Statistical Learning*; Springer: New York, NY, USA, 2009; Volume 2, pp. 337–387.

29. Kohonen, T. *Self-Organizing Maps*; Springer Series in Information Sciences, Springer: Berlin/Heidelberg, Germany, 2012.

30. Celebi, M. *Partitional Clustering Algorithms*; EBL-Schweitzer, Springer International Publishing: Cham, Switzerland, 2014.

31. Kaufman, L.; Rousseeuw, P.J. *Finding Groups in Data: An Introduction to Cluster Analysis*; Wiley Series in Probability and Mathematical Statistics; Wiley: New York, NY, USA, 1990.

32. Kvingedal, B. Sea-Ice Extent and Variability in the Nordic Seas, 1967–2002. In *The Nordic Seas: An Integrated Perspective*; American Geophysical Union: Washington, DC, USA, 2013; pp. 39–49.

33. Göttl, F.; Dettmering, D.; Müller, F.L.; Schwatke, C. Lake Level Estimation Based on CryoSat-2 SAR Altimetry and Multi-Looked Waveform Classification. *Remote Sens.* **2016**, *8*, 885.